From the most trusted name in the field....

The gold standard of research methods texts

This new edition of **The Practice of Social Research** offers much more than a thorough, accessible introduction to the field of research as practiced by social scientists. In this new Twelfth Edition, Earl Babbie emphasizes the research process by introducing the various observation modes currently in use and answering questions about research methods—such as how to conduct online surveys and analyze both qualitative and quantitative data. The new **Keeping Humanity in Focus** features, which profile exciting contemporary researchers, and plenty of opportunities for hands-on application offer a real-world view of the importance of social research.

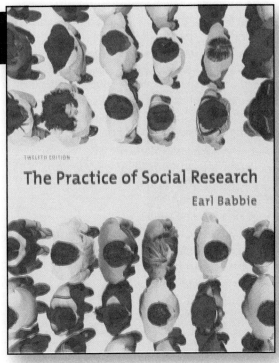

TWELFTH EDITION

The Practice of Social Research

Earl Babbie

Numerous examples (both real-world and hypothetical), abundant figures and tables, sample research tools, and the author's renowned sense of humor and compassion combine to make **The Practice of Social Research** the most accessible text on social research methods available.

With **The Practice of Social Research**, you're giving your students the tools they need to apply research concepts practically, as both researchers and consumers. *See inside!*

"It is the best methods text—nothing is close to this—over the years it has been perfected."

Craig Forsyth
University of Louisiana at LaFayette

Preview

New coverage and features engage students and convey the excitement of social research

Babbie has added new **Keeping Humanity in Focus** features that examine key qualitative researchers and their work. These profiles of exciting new researchers expand the book's coverage of qualitative methods for this edition and explore social research efforts that are able to undertake sophisticated analyses while keeping an immediate focus on the people involved. Some of the profiles in this edition include:

- Kathryn Edin and Maria Kefalas, *Promises I Can Keep: Why Poor Women Put Motherhood Before Marriage;*
- Elijah Anderson, *A Place on the Corner: A Study of Black Street Corner Men;*
- Rachel Sherman, *Class Acts: Service and Inequality in Luxury Hotels;*
- Kristen Schilt, *Just One of the Guys? How Transmen Make Gender Visible in the Workplace;*
- Sudhir Venkatesh, *Gang Leader for a Day: A Rogue Sociologist Takes to the Streets*

Keeping Humanity in Focus

As we have seen, a wide variety of research approaches can enhance our grasp of social dynamics. Much social research involves the analysis of masses of statistical data. As valuable as the examination of overall patterns can be, it can come at the risk of losing sight of the individual men and women those data represent. As such, some social research focuses specifically on the detailed particulars of real lives at the ground level of society. Throughout this book, I'll highlight some recent studies that reflect this latter approach to understanding social life, in an attempt to "keep humanity in focus" during our broader discussion of social science practice.

Statistics suggest that, in the United States, unwed mothers and their children, particularly those who are poor, will face a host of problems in the years to come. Both the child and the mother will likely struggle and suffer. The children are less likely to do well in school and in later life, and the mothers will probably have to struggle in low-paying jobs or live on welfare. The trend toward births out of wedlock has increased dramatically in recent decades, especially among the poor. As a reaction to these problems, the Bush administration launched a Healthy Marriage initiative in 2005 aimed at encouraging childbearing couples to marry. Voices for and against the program have been raised with vigor.

In their book *Promises I Can Keep*, Kathryn Edin and Maria Kefalas raise a question that might have been asked prior to the creation of a solution to the perceived problem: "Why do poor women bear children outside of wedlock?" The two social scientists spent five years speaking one-on-one with young women who had had children out of wedlock. Some of the things they learned dramatically contradicted various widespread images of unwed mothers. For instance, whereas many people have bemoaned the abandonment of marriage among the poor, the women interviewed tended to speak highly of the institution, indicating they hoped to be married one day. Further, many were only willing to settle down with someone trustworthy and stable—better to remain unmarried than to enter a marriage that will end in disaster.

At the same time, these young women felt strongly that their ultimate worth as women centered on their bearing children. Most felt it was preferable to be an unmarried mother than to be a childless woman, the real tragedy in their eyes.

This view of marriage may differ greatly from your own. As we have seen, assumptions about "what's real" are often contradicted by actual observations; this is crucial to understanding "what will work" to address social issues.

Promises I Can Keep: Why Poor Women Put Motherhood Before Marriage by Kathryn Edin and Maria Kefalas. © 2005 Regents of the University of California. Published by the University of California Press. Photograph © Michael Smyth.

Keeping Humanity in Focus

Transsexuals are individuals who choose to change their biological sex permanently though surgery and hormones. Clearly, such a radical change brings many adjustments and challenges that would make for interesting studies, but Kristen Schilt has taken an unusual tack.

While many kinds of research point to the disadvantaged status of women in the workplace, Schilt's research on transsexuals reveals the impact of gender on a personal level. In many of the cases, the subjects changed their sex while maintaining the same job in their employing organization. Following their sex change, female-to-male transsexuals were likely to enjoy pay raises and increased authority. In other studies, male-to-female transsexuals reported just the opposite experiences. Personal accounts such as these flesh out statistical studies that consistently show women earning less than men, even when they do the same work.

Source: Kristen Schilt, "Just One of the Guys? How Transmen Make Gender Visible in the Workplace," Gender and Society 20, no. 4 (2006): 465–90.

How to Do It: Framing a Hypothesis

As we have seen, the deductive method of research typically focuses on the testing of a hypothesis. Let's take a minute to look at how to create a hypothesis for testing.

Hypotheses state an expected causal relationship between two (or more) variables. Let's suppose you're interested in student political orientations, and your review of the literature and your own reasoning suggest to you that college major will play some part in determining students' political views. Already, we have two variables: *college major* and *political orientation*. Moreover, *political orientation* is the dependent variable—you believe it depends on something else, on the independent variable, which in this case is *college major*.

Now we need to specify the attributes comprising each of these variables. For simplicity's sake, let's assume political orientation includes only liberal or conservative. And to simplify the matter of major, let's suppose your research interests focus on the presumed differences between business students and those in the social sciences.

Even with these simplifications, you would need to specify more concretely how you would recognize a liberal or a conservative when you came across them in your study. This process of specification will be discussed at length in Chapter 5. For now, let's assume you will ask student-subjects whether they consider themselves liberals or conservatives, letting each student report on what the terms mean to them. (As we'll see later, this simple dichotomy is unlikely to work in practice, as some students would want to identify themselves as independents or something else.)

Identifying students' majors isn't as straightforward as you might think. For example, what disciplines compose the social sciences in your study? Also, must students be declared majors or simply be planning to major in one of the relevant fields?

Once these issues have been settled, you are ready to state your hypothesis. For example, it might be the following:

"Students majoring in the social sciences will be more likely to identify themselves as liberals than are those majoring in business."

In addition to this basic expectation, you may wish to specify "more likely" in terms of how *much* more likely. Chapter 16 will provide some options in this regard.

The empowering **How to Do It** boxes provide practical, step-by-step guidance to help students deal with what instructors have identified as especially elusive tasks. Some of the topics explored in the **How to Do It** boxes include

- Framing a Hypothesis
- Identifying the Unit of Analysis
- Conducting an Online Survey
- Establishing Rapport

New Coverage and Features

agreement reality
attributes
deduction
dependent variable
epistemology
idiographic
independent variable

induction
methodology
nomothetic
replication
theory
variables

PROPOSING SOCIAL RESEARCH: INTRODUCTION

This first chapter has given you an overview of some of the basic variations in social research, many of which can be useful in writing the introduction of your research proposal. For this assignment, you should first identify a topic or question you might like to explore in a research project. Perhaps you would like to explore some topic relating to race, gender, or social class. Perhaps there is some aspect of college life that you think needs study.

Once you have a research topic in mind, this chapter will offer some ideas on how the research might be organized. This is only a overview of the project and should take two to four paragraphs. It will work best if you can select a topic which you'll use in each of the chapters of the book, as you address different aspects of the research process.

Here are some examples of research questions to illustrate the kind of focus your project might take.

● Do women earn less money than men and, if so, why?
● What distinguishes juvenile gangs of different ethnic groups?
● Which academic departments at your college offer the broadest degree of liberal arts training?
● Is it true, as some suggest, that the United States was established as a "Christian nation"?
● Are American military actions in the Middle East reducing the threat of terrorist attacks or increasing those threats?
● What are the major functions of the American family and how have those been changing over time?
● Are official attempts to control illegal drug use succeeding or failing?
● Do undocumented immigrants overall represent a net economic cost or benefit to the United States?

Notice that you probably hear questions like these discussed frequently, both in your own interactions

and in the mass media. Most of those discussions are probably mostly based on opinions. Your opportunity in this course is to see how you might pursue such questions as a researcher, dealing with logic and facts in place of opinions.

REVIEW QUESTIONS AND EXERCISES

1. Review the common errors of human inquiry discussed in this chapter. Find a magazine or newspaper article, or perhaps a letter to the editor, that illustrates one of these errors. Discuss how a scientist would avoid it.

2. List five social variables and the attributes they comprise.

3. Go to one of the following websites, listed at http://www.academic.cengage.com/sociology/babbie, and find examples of both qualitative and quantitative data.
   a. UN High Commissioner for Refugees
   b. U.S. Centers for Disease Control and Prevention
   c. National Library of Australia

4. At InfoTrac College Edition, search for "post-modernism." Write a short report discussing the various fields or disciplines to which it has been applied. Give examples.

SPSS EXERCISES

See the booklet that accompanies your text for exercises using SPSS (Statistical Package for the Social Sciences). There are exercises offered for each chapter, and you'll also find a detailed primer on using SPSS.

Online Study Resources

If your book came with an access code card, visit http://www.academic.cengage.com/login to register. To purchase access, please visit http://www.ichapters.com.

1. Before you do your final review of the chapter, take the CengageNOW pretest to help identify the areas on which you should concentrate. You'll find information on this online tool, as well as instructions on how to access all of its great resources, in the front of the book.

The Ethics of Sampling

The key purpose of the sampling techniques discussed in this chapter is to allow researchers to make relatively few observations but gain an accurate picture of a much larger population. In the case of quantitative studies using probability sampling, the result should be a statistical profile, based on the sample, that closely mirrors the profile that would have been gained from observing the whole population. In addition to using legitimate sampling techniques, researchers should be careful to point out the possibility of errors: sampling error, flaws in the sampling frame, nonresponse error, or anything else that might make the results misleading.

Sometimes, more typically in qualitative studies, the purpose of sampling may be to tap into the breadth of variation within a population rather than to focus on the "average" or "typical" member of that population. While this is a legitimate and valuable approach, it poses the risk that readers may mistake the display of differences to reflect the distribution of characteristics in the population. In such a case, the researcher should make sure that the reader is not misled.

In addition, researchers studying a social group may make use of informants. Each of these techniques has its uses, but none of them ensures that the resulting sample will be representative of the population being sampled.

MAIN POINTS

Introduction

● Social researchers must select observations that will allow them to generalize to people and events not observed. Often this involves sampling a selection of people to observe.
● Understanding the logic of sampling is essential to doing social research.

A Brief History of Sampling

● Sometimes you can and should select probability samples using precise statistical techniques, but other times nonprobability techniques are more appropriate.

Nonprobability Sampling

● Nonprobability sampling techniques include relying on available subjects, purposive or judgmental sampling, snowball sampling, and quota sampling.

The Theory and Logic of Probability Sampling

● Probability-sampling methods provide an excellent way of selecting representative samples from large, known populations. These methods counter the problems of conscious and unconscious sampling bias by giving each element in the population a known (nonzero) probability of selection.
● Random selection is often a key element in probability sampling.
● The most carefully selected sample will never provide a perfect representation of the population from which it was selected. There will always be some degree of sampling error.
● By predicting the distribution of samples with respect to the target parameter, probability-sampling methods make it possible to estimate the amount of sampling error expected in a given sample.
● The expected error in a sample is expressed in terms of confidence levels and confidence intervals.

Populations and Sampling Frames

● A sampling frame is a list or quasi list of the members of a population. It is the resource used in the selection of a sample. A sample's representativeness depends directly on the extent to which a sampling frame contains all the members of the total population that the sample is intended to represent.

Types of Sampling Designs

● Several sampling designs are available to researchers.
● Simple random sampling is logically the most fundamental technique in probability sampling, but it is seldom used in practice.
● Systematic sampling involves the selection of every kth member from a sampling frame. This method is more practical than simple random sampling; with a few exceptions, it is functionally equivalent.

NEW!

Babbie demonstrates how to conduct a research proposal by including new **tips and questions to consider** in each chapter. This helps students apply specific chapter concepts to the research proposal process. Then, Chapter 17, *Reading and Writing Social Research*, pulls it all together.

NEW!

A **new section on research ethics** at the end of each chapter introduces students to the sensitivities that surround social research. As the concern for research ethics has increased among social researchers, Babbie has continually given this important topic more prominence over the years. In the Twelfth Edition, each chapter now ends with a new section that examines the ethical implications of the chapter topics.

PLUS discussion updates and new material in every chapter:

The "politicization of science" ■ The Zimbardo Prison Experiment ■ DHHS Confidentiality Certificate ■ Nomothetic Causal Analysis and Hypothesis-testing ■ Webometric Ranking of universities ■ Snowball sampling ■ The problems of bias that cell phones create for telephone survey sampling ■ Obama-Clinton primary as example for social desirability ■ The concept of emancipatory research ■ *Silent Racism* as an example of non-political, non-marketing focus groups ■ Major new section on the evaluation of qualitative research ■ New section on Qualitative Data Analysis programs ■ and more!

New Coverage and Features

Engaging features, topic coverage, and helpful pedagogy— along with Babbie's clear and accessible writing

Chapter-beginning overviews that explain chapter objectives and how the material fits into the overall scheme of social research, as well as **a chapter table of contents**, help students anticipate the material to come, and provide a roadmap for reviews and tests.

The **Proposing Social Research** section continually invites students to apply what they have learned in the chapter to the research proposal process.

Clear and provocative examples help students grasp difficult and/or abstract ideas. The Twelfth Edition boasts many new examples as well as some that have proven particularly valuable in earlier editions.

Babbie helps reinforce key ideas with a **running glossary** that appears at the foot of each page. These in-chapter definitions also make chapter reviews and exam preparation easier.

CHAPTER TWO

Paradigms, Theory, and Social Research

CHAPTER OVERVIEW

Social science inquiry is an interplay of theory and research, logic and observation, induction and deduction—and of the fundamental frames of reference known as paradigms.

PROPOSING SOCIAL RESEARCH: THEORY

As this chapter has indicated, social research can be pursued within numerous theoretical paradigms— each suggesting a somewhat different way to approach the research question. In this portion of your proposal, you should identify the paradigm(s) that will shape the design of your research.

We have also seen that paradigms provide frameworks within which causal theories ma[...] Perhaps your research project will expl[...] an existing theory. Or more ambitious[...] propose a theory or hypothesis for testi[...] section of the proposal in which to desc[...] of your project.

Not all research projects are formal[...] around the creation and/or testing of th[...] hypotheses. However, your research w[...] theoretical concepts, which should be a[...] this section of the proposal. As we'll see[...] Chapter 17, this portion of your propos[...] the literature on previous theory and r[...] shaped your own thinking and research[...]

In another example of ethnomethodology, Johen Heritage and David Greatbatch (1992) examined the role of applause in British political speeches: How did the speakers evoke applause, and what function did it serve (for example, to complete a topic)? Research within the ethnomethodological paradigm has often focused on communications.

There is no end to the opportunities you have for trying out the ethnomethodological paradigm. For instance, the next time you get on an elevator, don't face front, watching the floor numbers whip by (that's the norm, or expected behavior). Just stand quietly facing the rear. See how others react to this behavior. Just as important, notice how you feel about it. If you do this experiment a few times, you should begin to develop a feel for the ethnomethodological paradigm.*

We'll return to ethnomethodology in Chapter 10, when we discuss field research. For now, let's turn to a very different paradigm.

Structural Functionalism

Structural functionalism, sometimes also known as social systems theory, has grown out of a notion introduced by Comte and Spencer: A social entity, such as an organization or a whole society, can be viewed as an organism. Like other organisms, a social system is made up of parts, each of which contributes to the functioning of the whole.

By analogy, consider the human body. Each component—such as the heart, lungs, kidneys, skin, and brain—has a particular job to do. The body as a whole cannot survive unless each of these parts does its job, and none of the parts can survive except as a part of the whole body. Or consider an automobile. It is composed of the tires, the steering wheel, the gas tank, the spark plugs,

across town. None of the individual parts would be very useful to us by itself, however.

The view of society as a social system, then, looks for the "functions" served by its various components. Social scientists using the structural functional paradigm might note that the function of the police, for example, is to exercise social control—encouraging people to abide by the norms of society and bringing to justice those who do not. Notice, though, that the researchers could just as reasonably ask what functions criminals serve in society. Within the functionalist paradigm, we might say that criminals serve as job security for the police. In a related observation, Emile Durkheim (1858–1917) suggested that crimes and their punishment provide an opportunity to reaffirm society's values. By catching and punishing thieves, we reaffirm our collective respect for private property.

To get a sense of the structural functional paradigm, suppose you were interested in explaining how your college or university works. You might thumb through the institution's catalog and begin assembling a list of the administrators and support staff (such as the president, deans, registrar, campus security staff, maintenance personnel). Then you might figure out what each of them does and relate their roles and activities to the chief functions of your college or university, such as teaching or research. This way of looking at an institution of higher learning would clearly suggest a different line of inquiry than, say, a conflict paradigm, which might emphasize the clash of interests between people who have power in the institution and those who don't.

People often discuss "functions" in everyday conversation. Typically, however, the alleged functions are seldom tested empirically. Some people argue, for example, that welfare, intended to help

*I am grateful to my colleague, Bernard McGrane, for this experiment. Barney also has his students eat dinner with their hands, watch TV without turning it on, and engage in other strangely enlightening behavior (McGrane 1994).

structural functionalism A paradigm that divides social phenomena into parts, each of which serves a function for the operation of the whole.

Proven and Reliable Coverage

FIGURE 2-2
The Traditional Image of Science. The deductive model of scientific inquiry begins with a sometimes vague or general question, which is subjected to a process of specification, resulting in hypotheses that can be tested through empirical observations.

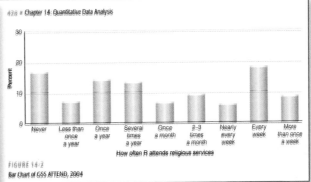

428 ■ Chapter 14: Quantitative Data Analysis

FIGURE 14-2
Bar Chart of GSS ATTEND, 2004

The Practice of Social Research has always been a text that helps visual learners grasp concepts through its graphics that instruct as well as illustrate. In this edition, the updated **graphics** translate mental pictures into words, helping students comprehend the concepts.

The book's **end-of-chapter features** include point-by-point chapter summaries, a list of key terms, review questions and exercises, **SPSS Exercises** and **Online Study Resources** to encourage students to study independently and beyond the text and the classroom.

Sociological Diagnostics
■ Sociological diagnostics is a quantitative analysis technique for determining the nature of social problems such as ethnic or gender discrimination.

Ethics and Quantitative Data Analysis
■ Unbiased analysis and reporting is as much an ethical concern in quantitative analysis as in qualitative analysis.
■ Subjects' privacy must be protected in quantitative data analysis and reporting.

KEY TERMS

The following terms are defined in context in the chapter and at the bottom of the page where the term is introduced, as well as in the comprehensive glossary at the back of the book.

average
bivariate analysis
codebook
contingency table
continuous variable
discrete variable
dispersion
frequency distribution

mean
median
mode
multivariate analysis
quantitative analysis
standard deviation
univariate analysis

PROPOSING SOCIAL RESEARCH: QUANTITATIVE DATA ANALYSIS

See the exercise for Chapter 16.

REVIEW QUESTIONS AND EXERCISES

1. How might the various majors at your college be classified into categories? Create a coding system that would allow you to categorize them according to some meaningful variable. Then create a different coding system, using a different variable.

2. How many ways could you be described in numerical terms? What are some of your intrinsically numerical attributes? Could you express some of your qualitative attributes in quantitative terms?

3. How would you construct and interpret a contingency table from the following information: 150 Democrats favor raising the minimum wage, and 50 oppose it; 100 Republicans favor raising the minimum wage, and 300 oppose it?

4. Using the hypothetical data in the following table, how would you construct and interpret tables showing the following?
   a. The bivariate relationship between age and attitude toward abortion
   b. The bivariate relationship between political orientation and attitude toward abortion
   c. The multivariate relationship linking age, political orientation, and attitude toward abortion

Age	Political Orientation	Attitude toward Abortion	Frequency
Young	Liberal	Favor	90
Young	Liberal	Oppose	10
Young	Conservative	Favor	60
Young	Conservative	Oppose	40
Old	Liberal	Favor	60
Old	Liberal	Oppose	40
Old	Conservative	Favor	20
Old	Conservative	Oppose	80

SPSS EXERCISES

See the booklet that accompanies your text for exercises using SPSS (Statistical Package for the Social Sciences). There are exercises offered for each chapter, and you'll also find a detailed primer on using SPSS.

Online Study Resources

If your book came with an access code card, visit www.cengage.com/login to register. To purchase access, please visit www.ichapters.com.

1. Before you do your final review of the chapter, take the CengageNOW pretest to help identify the areas on which you should concentrate. You'll find information on this online tool, as well as

Proven and Reliable Coverage

Exciting, interactive online resources take learning beyond the classroom

CengageNOW™ for Research Methods
... *with access to* InfoTrac College Edition *and the eBook*

1-Semester Instant Access Code
ISBN: ISBN-10: 0-495-59853-4 •
ISBN-13: 978-0-495-59853-4

This powerful, assignable online learning companion assesses each student's study needs. This combined resource includes **CengageNOW™** *(described on the inside front cover of the book)* an online teaching and learning resource that gives you more control in less time and delivers better outcomes. And, when you order the text packaged with **CengageNOW**, you give your students access to the **eBook** as well as **InfoTrac® College Edition**, an entire an online university library of more than 5,000 academic and popular magazines, newspapers, and journals.

Book Companion Website
—robust, chapter-by-chapter correlation
www.cengage.com/sociology/babbie

The helpful website includes a wonderful array of chapter-specific resources for instructors and students. For instructors, the site offers a password-protected *Instructor's Resource Manual*, Microsoft® PowerPoint® presentation slides, and more. Students will find a multitude of text-specific study aids including the following:

- Tutorial practice quizzes that can be e-mailed to the instructor
- Web links
- InfoTrac College Edition exercises
- Flashcards
- GSS data sets
- Guides to data analysis software, including SPSS, Nvivo, and Qualrus
- MicroCase® Online data exercises

WebTutor™ (on Blackboard or WebCT)
Contact your sales representative for packaging information.
Jumpstart your course with customizable, rich, text-specific content within your Course Management System. This web-based software for students and instructors takes a course beyond the classroom to an anywhere, anytime environment. Students gain access to the rich content from this book's companion websites.

So many great combinations ... You can order Babbie's text packaged with any of the student tools outlined on these pages. Just contact your Cengage Learning sales representative for details.

Online Resources

Additional resources for your students

SPSS® Integrated Student Version 16.0 CD-ROM

ISBN-10: 0-495-59827-5
ISBN-13: 978-0-495-59827-5

SPSS Student Version 16.0 for Windows® provides real-world software for students to conduct sociological data analysis. This student-friendly version is based on the professional version of one of the world's leading desktop statistical software packages.

Learning How to Use SPSS® with Exercises . . . *complimentary when packaged with the text!*

ISBN-10: 0-495-59846-1
ISBN-13: 978-0-495-59846-6

Coordinated with *The Practice of Social Research* and *SPSS* on CD-ROM, this handy guide helps students learn to navigate SPSS, including data entry; creating, saving, and retrieving files; producing and interpreting data summaries; and more. SPSS practice exercises are included.

GSS Data Disk for Sociology © 2008

ISBN-10: 0-495-09329-7
ISBN-13: 978-0-495-09329-9

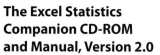

The **GSS Data Disk** provides valuable practice using data sets with SPSS Student Version or with any statistical software. They correspond with many of Babbie's in-text and Study Guide examples.

Guided Activities (Study Guide)

ISBN-10: 0-495-59847-X
ISBN-13: 978-0-495-59847-3

by Ted Wagenaar and Earl Babbie

The **Study Guide** contains chapter objectives, chapter summaries, key terms, multiple choice and open ended discussion questions for each chapter, along with four to six exercises that reinforce the text material with examples from everyday life. Also included are the answers to the matching and multiple-choice review questions, as well as a General Social Survey appendix.

The Excel Statistics Companion CD-ROM and Manual, Version 2.0

ISBN-10: 0-495-18695-3
ISBN-13: 978-0-495-18695-3

by Kenneth M. Rosenberg, State University of New York, Oswego

This collection of demonstrations, sampling experiments, and interactive exercises dovetails with traditional textbook coverage of descriptive and inferential statistics. The materials give students the hands-on experience they need to master abstract statistical concepts and computational procedures. Version 2.0 includes four new workbooks and several new worksheets within the workbooks.

Readings in Social Research, Third Edition

ISBN-10: 0-495-09337-8 • ISBN-13: 978-0-495-09337-4

by Diane Kholos Wysocki

The concepts and methodologies of social research come to life in this interesting collection of articles specifically designed to accompany *The Practice of Social Research*. The author includes an interdisciplinary range of readings from the fields of psychology, sociology, social work, criminal justice, and political science.

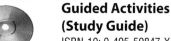

Online Resources

The most reliable name in research methods— now with a gold-standard support package to match

PowerLecture® Instructor's Resource CD-ROM with JoinIn® and ExamView®
ISBN-10: 0-495-59850-X • ISBN-13: 978-0-495-59850-3

PowerLecture

PowerLecture, a collection of book-specific lecture and class tools, is the fastest and easiest way to build powerful, customized media-rich lectures. **PowerLecture** includes chapter-specific PowerPoint® presentations, images, animations and video, instructor manuals, test banks, useful web links and more. The **JoinIn**™ Student Response System features instant classroom assessment and active learning, while **ExamView®** allows you to create, deliver, and customize tests and study guides of up to 250 questions in minutes.

Instructor's Manual with Test Bank
ISBN-10: 0-495-59848-8 • ISBN-13: 978-0-495-59848-0

by Andrew Martin and Lynette Martin of The Ohio State University
Streamline and maximize the effectiveness of your course preparation using such resources as brief chapter outlines, detailed chapter outlines, behavioral objectives, teaching suggestions and resources, **InfoTrac College Edition** exercises, Internet exercises, and possible study guide answers. This time-saving resource also includes a Test Bank that offers least 50 multiple-choice questions, 20 to 25 true/false questions, and five essay questions with answers and page references.

ABC Video: Research Methods, Volume 1
VHS ISBN-10: 0-495-80431-2 • ISBN-13: 978-0-495-80431-4
DVD ISBN-10: 0-495-80438-X • ISBN-13 978-0-495-80438-3

This series of videos comprised of footage from ABC broadcasts is specially selected and arranged to accompany your Research Methods course. The segments may be used in conjunction Babbie's text to help provide real-world examples to illustrate course concepts, or to instigate discussion. **ABC Videos** feature short, high interest clips from current news events as well as historic raw footage going back 40 years. Clips are drawn from such programs as *World News Tonight, Good Morning America, This Week, PrimeTime Live, 20/20*, and *Nightline*, as well as numerous *ABC News* specials and material from the *Associated Press Television News* and *British Movietone News* collections.

For more information on Wadsworth Sociology resources, visit
www.cengage.com/sociology

Resources for Instructors

The Practice
of Social Research

A Note from the Author

Writing is my joy, sociology my passion. I delight in putting words together in a way that makes people learn or laugh or both. Sociology shows up as a set of words, also. It represents our last, best hope for planet-training our race and finding ways for us to live together. I feel a special excitement at being present when sociology, at last, comes into focus as an idea whose time has come.

I grew up in small-town Vermont and New Hampshire. When I announced I wanted to be an auto-body mechanic, like my dad, my teacher told me I should go to college instead. When Malcolm X announced he wanted to be a lawyer, his teacher told him a colored boy should be something more like a carpenter. The difference in our experiences says something powerful about the idea of a level playing field. The inequalities among ethnic groups run deep.

I ventured into the outer world by way of Harvard, the USMC, U.C. Berkeley, and twelve years teaching at the University of Hawaii. I resigned from teaching in 1980 and wrote full-time for seven years, until the call of the classroom became too loud to ignore. For me, teaching is like playing jazz. Even if you perform the same number over and over, it never comes out the same twice and you don't know exactly what it'll sound like until you hear it. Teaching is like writing with your voice.

At last, I have matured enough to rediscover and appreciate my roots in Vermont each summer. Rather than a return to the past, it feels more like the next turn in a widening spiral. I can't wait to see what's around the next bend.

TWELFTH EDITION

The Practice of Social Research

Earl Babbie
Chapman University

WADSWORTH
CENGAGE Learning™

Australia • Brazil • Japan • Korea • Mexico • Singapore • Spain • United Kingdom • United States

WADSWORTH
CENGAGE Learning™

The Practice of Social Research, **Twelfth Edition**
Earl Babbie

Acquisitions Editor: Chris Caldeira

Development Editor: Dan Moneypenny

Assistant Editor: Rachael Krapf

Editorial Assistant: Rebecca Boorsma

Technology Project Manager: Lauren Keyes

Marketing Manager: Kim Russell

Marketing Assistant: Jillian Myers

Marketing Communications Manager: Martha Pfeiffer

Project Manager, Editorial Production: Cheri Palmer

Creative Director: Rob Hugel

Art Director: Caryl Gorska

Print Buyer: Linda Hsu

Permissions Editors: Bob Kauser and Leitha Etheridge-Sims

Production Service: Greg Hubit Bookworks

Text Designer: Carolyn Deacy

Photo Researcher: Pre-Press

Copy Editor: Molly D. Roth

Illustrator: Lotus Art

Cover Designer: RHDG/Brie Hettey

Cover, Part Opener, and Chapter Opener Image: Masterfile

Compositor: Newgen

For product information and technology assistance, contact us at **Cengage Learning Customer & Sales Support, 1-800-354-9706.**

For permission to use material from this text or product, submit all requests online at **www.cengage.com/permissions.** Further permissions questions can be e-mailed to **permissionrequest@cengage.com.**

Library of Congress Control Number: 2008933881

ISBN-13: 978-0-495-59841-1
ISBN-10: 0-495-59841-0

Wadsworth
10 Davis Drive
Belmont, CA 94002-3098
USA

Cengage Learning is a leading provider of customized learning solutions with office locations around the globe, including Singapore, the United Kingdom, Australia, Mexico, Brazil, and Japan. Locate your local office at **international.cengage.com/region.**

Cengage Learning products are represented in Canada by Nelson Education, Ltd.

For your course and learning solutions, visit **academic.cengage.com.** Purchase any of our products at your local college store or at our preferred online store **www.ichapters.com.**

Printed in the United States of America
1 2 3 4 5 6 7 12 11 10 09

Dedication

Suzanne Babbie

Contents in Brief

Contents in Detail

CHAPTER 6

Indexes, Scales, and Typologies 160

CHAPTER 7

The Logic of Sampling 187

Part 3 Modes of Observation: Quantitative and Qualitative 229

CHAPTER 8

Experiments 230

CHAPTER 9

Survey Research 253

Part 4   Analysis of Data: Quantitative and Qualitative 391

CHAPTER 13

Qualitative Data Analysis 393

CHAPTER 14

Quantitative Data Analysis 421

CHAPTER 15

The Elaboration Model 448

Preface

A "few" years ago (I hate to tell you how many), I began teaching my first course in social research methods. The course focused specifically on survey research methods, and I had only six students in the class. As the semester progressed, I became more relaxed as a teacher. Before long, my students and I began meeting in my office, where I could grab and lend books from my own library as their relevance occurred to me during class meetings.

One nagging problem I faced then was the lack of a good textbook on survey research. The available books fell into one of two groups. Some books presented the theoretical logic of research methods in such abstract terms that I didn't think students would be able to apply any of the general principles to the practical world of "doing" research. The other books were just the opposite. Often termed "cookbooks," they presented detailed, step-by-step instructions on how to conduct a survey. Unfortunately, this approach only prepared students to conduct surveys very much like the one described by the authors. Neither the abstract nor the "cookbook" approach seemed truly useful to students or their instructors.

One day I found myself jotting down the table of contents for my ideal research methods textbook. It was organized around three theoretical principles on which scientific research is based:

1. Understanding the theoretical principles on which scientific research is based.

2. Seeing how those principles are reflected in the established techniques for doing research.

3. Being prepared to make appropriate compromises whenever field conditions do not permit the routine application of established techniques.

The next day, unexpectedly, Wadsworth called and asked me to write a methods text!

Survey Research Methods was published in 1973. My editors and I immediately received some good news, some bad news, and some additional good news. The first good news was that all survey research instructors seemed to love the book, and it was being used in virtually every survey research course in the country. The bad news was that there weren't all that many survey research courses.

The final good news, however, was that many instructors who taught general social research courses—covering survey research alongside other research methods—were inclined to use our book and supplement it with other books dealing with field research, experiments, and so on. While adjusting to our specialized book, however, many instructors suggested that Wadsworth have "that same guy" write a more general social research text.

The preface of the first edition of *The Practice of Social Research* (1975) acknowledged the assistance of a dozen social research instructors from California to Florida. The book was a collaboration in a very real sense, even though only my name was on the cover and I was ultimately responsible for it.

The Practice of Social Research was an immediate success. Although it was initially written for sociology courses, subsequent editions have been increasingly used in fields such as psychology, public administration, urban studies, education, communications, social sciences, and political science—in some 30 different disciplines, I'm told. Moreover, it's being used by teachers and researchers in numerous countries around the world, and in 2000 a Beijing publisher released a two-volume Chinese edition.

I've laid out this lengthy history of the book for a couple of reasons. First, when I was a student, I suppose I thought of textbooks the same way that I thought about government buildings: They were just there. I never really thought about them as being written by human beings. I certainly never thought about textbooks as evolving: being updated, getting better, having errors corrected. As a student, I would have been horrified by the thought that any of my textbooks might contain mistakes!

Second, pointing out the evolution of the book sets the stage for a preview of the changes that have gone into this 12th edition. As with previous revisions, several factors have prompted changes. For example, because social research technology and practices are continually changing, the book must be updated to remain current and useful. In my own teaching, I frequently find improved ways to present standard materials. Colleagues also often share their ideas for ways to teach specific topics. Some of these appear as boxed inserts in the book. Both students and instructors often suggest that various topics be reorganized, expanded, clarified, shrunk, or—gasp—deleted.

New to the 12th Edition

In an earlier edition of this book, I said, "Revising a textbook such as this is a humbling experience. No matter how good it seems to be, there is no end of ideas about how it could be improved." That observation still holds true. When we asked instructors what could be improved, they once again thought of things, and I've considered all their suggestions, followed many of them, and chosen to "think some more" about others. I've also received numerous comments and suggestions from students who have been assigned the book; many of the changes come from them.

This edition of the book contains some new features, all of which were suggested by faculty reviewers and users.

Research Ethics: In 1973, *Survey Research Methods* contained an appendix on research ethics, which was one of the few things reviewers criticized: They were against it. Regarding the section "The Rights of Subjects," one reviewer angrily wrote, "What about the rights of science?" Over the years, concern for research ethics has steadily increased among social researchers, and the topic has become steadily more prominent in the various editions of this book. In the 12th edition, most chapters end with a section examining the ethical implications of the chapter's topics.

Research Proposal: Many instructors now use the research proposal as a vehicle for training students in social research, and several of the reviewers asked for more attention to that topic. As a consequence, each chapter concludes with a continuing exercise inviting students to apply what they've learned in the chapter to an evolving research proposal.

Keeping Humanity in Focus: Sometimes, social research requires us to delve deeply into the relationships among variables and/or take apart intricate social structures. This leads some researchers and research consumers to worry that we might lose sight of the human beings who lie at the core of our concerns. Some social research efforts, however, undertake sophisticated analyses while keeping an immediate focus on the people involved. A new series of boxes in this edition highlights some of those studies, as follows:

Chapter 1: Kathryn Edin and Maria Kefalas, *Promises I Can Keep: Why Poor Women Put Motherhood before Marriage* (Berkeley: University of California Press, 2005).

Chapter 5: Elijah Anderson, *A Place on the Corner: A Study of Black Street Corner Men* (Chicago: University of Chicago Press, 2004).

Chapter 10: Rachel Sherman, *Class Acts: Service and Inequality in Luxury Hotels* (Berkeley: University of California Press, 2005).

Chapter 14: Kristen Schilt, "Just One of the Guys? How Transmen Make Gender Visible in the Workplace," *Gender and Society* 20, no. 4 (2006): 465–90.

Chapter 17: Sudhir Venkatesh, *Gang Leader for a Day: A Rogue Sociologist Takes to the Streets* (New York: Penguin, 2008).

How to Do It: Another new series of boxes in the book provides practical, step-by-step guidance to assist students in dealing with what instructors have identified as especially elusive tasks. These are the boxes in the series, some of which adapted materials already existing in the book:

- Chapter 2: Framing a Hypothesis
- Chapter 4: Identifying the Unit of Analysis
- Chapter 7: Using a Table of Random Numbers
- Chapter 9: Conducting an Online Survey
- Chapter 10: Establishing Rapport
- Chapter 17: Using Google Scholar
- Chapter 17: Citing Bibliographic Sources

In addition to these identifiable features, I have continued to pursue my intention to demonstrate social research as an international, not just American, undertaking. Since researchers in different parts of the world sometimes face unique problems, the ways in which they deal with those problems often reveal new dimensions to the logic of social inquiry.

Here are some of the other changes in this edition, arranged by chapter:

Chapter 1, "Human Inquiry and Science"
- Provided new section on the purposes of social research
- Added materials on using both qualitative and quantitative approaches
- Clarified the original meaning of "the exception that proves the rule"
- Included glossary definitions of *epistemology* and *methodology*
- Coordinated definitions of *theory* in text and in glossary definition

- Clarified the top portion of Figure 1-4
- Cited evaluation research as a special kind of applied research, to be examined in Chapter 12

Chapter 2, "Paradigms, Theory, and Social Research"
- Added discussion of feminist standpoint theory
- Added discussion of critical realism
- Introduced the term *mesotheory*
- Dropped references to control variables as "test" variables
- Reviewed concepts of idiographic and nomothetic in "Elements of Social Theory"

Chapter 3, "The Ethics and Politics of Social Research"
- Included discussion of the "politicization of science"
- Added a comment on the positive results of the *Tearoom Trade* research
- Added discussion of Zimbardo's Stanford prison experiment
- Provided new discussion of DHHS "Certificate of Confidentiality"
- Mentioned Nuremberg trials
- Discussed Tuskegee syphilis experiments

Chapter 4, "Research Design"
- Provided example of using cohorts to infer process over time
- Included a new diagram to illustrate a cohort study
- Provided new section: "Nomothetic Causal Analysis and Hypothesis Testing"
- Changed Figure 4-5 in previous edition to 4-6 in current

Chapter 5, "Conceptualization, Operationalization, and Measurement"
- Gave new opening example of college satisfaction
- Discussed many names for *snow* to illustrate concepts and conceptualization
- Provided "fear of crime" example to illustrate dimensions

- Included a new box: "Measuring College Satisfaction"
- Discussed the Geertz concept of "thick description"
- Gave a new summary table reviewing direct and indirect observables and constructs
- Added a paragraph illustrating the importance of keeping the operationalization process open
- Clarified the definition of *validity*
- Added a paragraph distinguishing *sex* and *gender*
- Changed the diagram of the conceptualization process to a table with examples
- Discussed differences in qualitative and quantitative conceptualization and operationalization

Chapter 6, "Indexes, Scales, and Typologies"
- Discussed how indicators of a variable may be unrelated to one another
- Added discussion of "Webometrics Ranking" of universities to the box on *U.S. News* ranking
- Expanded the captions for Figures 6-2 and 6-3
- Provided new box on "Indexing the World" showing a variety of indexes rating various aspects of life

Chapter 7, "The Logic of Sampling"
- Mentioned that standard error can be calculated for means and other measures
- Discussed the problems of bias that cell phones create for telephone survey sampling
- Revised the glossary definition of *sampling error*
- Gave new example of snowball sampling: using social networks of nonheterosexual women in small-town U.K.
- Corrected the discussion of the finite population correction
- Worked in the quip: "Statistics means never having to say you are certain."
- Expanded discussion of cell-phone-only population
- Added a discussion of "theoretical sampling" in qualitative research projects
- Discussed recent address-based sampling (ABS)

- Added reference to registration-based sampling (RBS)
- Referenced discussion of random-digit dialing and problems of cell phones in Chapter 9

Chapter 8, "Experiments"
- Explained why experiments are discussed first
- Introduced the term *field experiment*
- Mentioned why we might use preexperimental methods

Chapter 9, "Survey Research"
- Updated and expanded section on online survey research
- Included box on using Survey Monkey
- Expanded description of random-digit dialing
- Expanded discussion of problems created by cell phones
- Referenced cognitive interviewing regarding pretesting questionnaires
- Reported that interviewer behavior changes over the course of a survey project
- Used Obama-Clinton primary as example for social desirability
- Noted suggestion that interviewers might be given more latitude in correcting respondent errors
- Discussed the use of PDAs for survey interviewing or self-administered surveys
- Expanded and updated the discussion of response rates
- Discussed the decline in interview survey response rates

Chapter 10, "Qualitative Field Research"
- Updated Lofland references with new, greatly revised version
- Gave example of using Internet for qualitative interviewing
- Discussed researcher responsibilities in participatory action research
- Discussed and defined *emancipatory research*
- Provided box on giving pencils and cameras to Peruvian Indians in rainforest

- Mentioned study of globalization impact among Irish young people that used drawings, poems, and songs
- Discussed study combining focus groups and in-depth interviews in Sweden
- Added pictures of some leading qualitative field researchers
- Updated discussion of focus groups
- Introduced the "ethnographic fallacy"
- Added example of Jeffrey Kidder's study of bike messengers in New York City
- Introduced Lofland's notion of "selective competence" in subjects and discussed it in terms of rapport
- Discussed how to continue interviewing difficult informants
- Referenced article on the difficulty of writing autoethnography
- Described kaupapa Maori research as an example of PAR
- Mentioned *Silent Racism* as an example of non-political, nonmarketing focus groups

Chapter 11, "Unobtrusive Research"
- Updated Megatrends studies
- Gave lengthy example of qualitative content analysis of rap music
- Cited the University of Michigan's archive of data sources
- Discussed and cited census website for mapping data
- Discussed and cited computer programs for qualitative and quantitative content analysis
- Discussed the uses of the Social Explorer for mapping data
- Updated Megatrends studies

Chapter 12, "Evaluation Research"
- Discussed the Sabido methodology for social change through mass media
- Gave example of "empowerment evaluation" in connection with Australian school program
- Updated examples of topics appropriate for evaluation research

- Used Peter Rossi et al. definition of evaluation research

Chapter 13, "Qualitative Data Analysis"
- Provided major new section on the evaluation of qualitative research
- Included new section on QDA programs
- Updated Lofland to newest edition

Chapter 14, "Quantitative Data Analysis"
- Clarified the base for percentaging tables
- Illustrated subgroup comparison with GSS example
- Introduced GapMinder as resource for displaying data
- Provided new discussion on how to read tables
- Updated and clarified discussion of Table 14-12
- Updated and clarified the discussion of Table 14-4
- Updated data on example of gender, age, and church attendance
- Updated data on example of income by education and gender
- Updated the data on example of age and attitude toward marijuana

Chapters 15, "The Elaboration Model"
- No significant changes made

Chapter 16, "Statistical Analyses"
- Added section on *t*-tests

Chapter 17, "Reading and Writing Social Research"
- Discussed Middlebury College history department banning of citing Wikipedia as a source
- Updated and expanded warnings of misconceptions and biases on the web
- Illustrated common bibliographic citation formats
- Introduced SourceWatch as a resource for judging websites

As always, I've updated materials throughout the book. As an instructor, I'm constantly searching for new and more effective ways of explaining social research to my own students; many of those new explanations take the form of diagrams. You'll find several new graphic illustrations in this edition. Once again, I've sought to replace older research examples (except for the classics) with more recent ones. I've also dropped some sections that I don't think do much for students anymore.

As with each new edition, I would appreciate any comments you have about how the book can be improved. Its evolution over the past 35 years has reflected countless comments from students and others.

Pedagogical Features

Although students and instructors alike have told me that the past editions of this book were effective tools for learning research methods, I have used this revision as an opportunity to review the book from a pedagogical standpoint, fine-tuning some elements, adding others. Here's the package we came up with in the 12th edition.

Chapter Overview Each chapter is preceded by a pithy focus paragraph that highlights the principal content of the chapter.

Chapter Introduction Each chapter opens with an introduction that lays out the main ideas in that chapter and, importantly, relates them to the content of other chapters in the book.

Clear and provocative examples Students often tell me that the examples—real and hypothetical—have helped them grasp difficult and/or abstract ideas, and this edition has many new examples as well as those that have proved particularly valuable in earlier editions.

Graphics From the first time I took a course in research methods, most of the key concepts have made sense to me in graphic form. Whereas my task here has been to translate those mental pictures into words, I've also included some graphics in the book. Advances in computer graphics have helped me communicate to the Wadsworth artists what I see in my head and would like to share with students. I'm delighted with the new graphics in this edition.

Boxed examples and discussions Students tell me they like the boxed materials that highlight particular ideas and studies, as well as varying the format of the book. Beginning in the tenth edition, I've been using boxes that focus on the ways the mass media use and misuse social research.

Running glossary Key terms are highlighted in the text, and definitions for each term are listed at the bottom of the page. This will help students learn the definitions of these terms and locate them in each chapter to review them in context.

Main Points At the end of each chapter, a concise list of main points provides both a brief chapter summary and a useful review. The main points let students know exactly what ideas they should focus on in each chapter.

Key Terms A list of key terms follows the main points. These lists reinforce the students' acquisition of necessary vocabulary. The new vocabulary in these lists is defined in context in the chapters. The terms are boldfaced in the text, defined in the running glossary that appears at the bottom of the page throughout the text, and included in the glossary at the back of the book.

Proposing Social Research This series of linked exercises invites students to apply what they've learned in the chapter by treating those topics in an evolving research proposal.

Review Questions and Exercises This review aid allows students to test their understanding of the chapter concepts and apply what they've learned.

SPSS Exercises and Online Study Resources This edition continues previous editions' movement into cyberspace. Students can use the annotated list of useful websites in this section, as well as other resources mentioned, to take their learning beyond the text and classroom.

Appendixes As in previous editions, a set of appendixes provides students with some research tools, such as a guide to the library, a table of random numbers, and so forth. There is an SPSS primer on the book's website along with primers for NVivo and Qualrus.

Clear and accessible writing This is perhaps the most important "pedagogical aid" of all. I know that all authors strive to write texts that are clear and accessible, and I take some pride in the fact that this "feature" of the book has been one of its most highly praised attributes through its 11 previous editions. It's the one thing students write most often about. For the 12th edition, the editors and I have taken special care to reexamine literally every line in the book, pruning, polishing, embellishing, and occasionally restructuring for a maximally "reader-friendly" text. Whether you're new to this book or intimately familiar with previous editions, I invite you to open any chapter and evaluate the writing for yourself.

Supplements

The Practice of Social Research, 12th edition, is accompanied by a wide array of supplements prepared for both the instructor and student to create the best learning environment inside as well as outside the classroom. All the continuing supplements for *The Practice of Social Research,* 12th edition, have been thoroughly revised and updated. I invite you to examine and take full advantage of the teaching and learning tools available to you.

For the Student

Guided Activities for The Practice of Social Research, *12th Edition* ISBN-10 049559847X

The student study guide and workbook Ted Wagenaar and I have prepared continues to be a mainstay of my own teaching. Students tell me they use it heavily as a review of the text, and I count the exercises as half their grade in the course.

In this edition, Ted and I have once again sorted through the exercises and added new ones

we've created in our own teaching or heard about from colleagues. These include matching, multiple-choice, and open-ended discussion questions for each chapter, along with four to six exercises that use examples from everyday life to reinforce the material learned in the text. Also included are the answers to the matching and multiple-choice review questions, as well as a General Social Survey appendix, plus chapter objectives, chapter summaries, and key terms.

SPSS Student Version CD-ROM 16.0 (Windows only)   ISBN-10 0495598275

Based on the professional version of one of the world's leading desktop statistical software packages, SPSS Student Version for Windows provides real-world software for students to do sociological data analysis, such as interpreting the GSS data sets found on the companion website.

SPSS Practice Workbook ISBN-10 0495598461

This handy guide is coordinated with the text and SPSS CD-ROM 16.0 to help students learn basic navigation in SPSS, including how to enter their own data; create, save, and retrieve files; produce and interpret data summaries; and much more. Also included are SPSS practice exercises correlated with each chapter. The guide comes free when bundled with the text.

GSS Data Disk   ISBN-10 0495093297

Over the years, the publisher and I have sought to provide up-to-date personal computer support for students and instructors. Because there are now many excellent programs for analyzing data, we've provided data to be used with them. With this edition, we've updated the data disk to include the 2006 GSS data.

Experiencing Social Research: An Introduction Using MicroCase, 2nd Edition ISBN-10 0534519113

This supplementary workbook and statistical package, written by David J. Ayers of Grove City

College, includes short discussions, quizzes, and computerized exercises in which students will learn and apply key methodological concepts and skills by analyzing, and in some cases collecting and building, simple data files for real sociological data. Designed to accompany *The Practice of Social Research*, the workbook and statistical package take a step-by-step approach to show students how to do real sociological research, using the same data and techniques used by professional researchers, to reinforce, build on, and complement course materials.

Readings in Social Research, 3rd Edition ISBN-10 0495093378

The concepts and methodologies of social research come to life in this interesting collection of articles specifically designed to accompany *The Practice of Social Research*. Diane Kholos Wysocki includes an interdisciplinary range of readings from the fields of psychology, sociology, social work, criminal justice, and political science. The articles focus on the important methods and concepts typically covered in the social research course and provide an illustrative advantage. Organized by key concepts, each of the reader's ten chapters begins with an introduction highlighting and explaining the research concept that each chapter's readings elucidate.

For the Instructor

Instructor's Resource Manual with Test Bank ISBN-10 0495598488

This supplement offers the instructor brief chapter outlines, behavioral objectives, ASA recommendations, detailed chapter outlines, teaching suggestions and resources, video suggestions, questions/activities to guide research projects, scenarios that apply social research to everyday life, InfoTrac® College Edition exercises, and Internet exercises. In addition, for each chapter of the text, the Test Bank has 50–70 multiple-choice questions, 20–25 true-false questions, and 5 essay questions with answers and page references. All questions are labeled as new, modified, or pickup so instructors know if the question is new to this edition of the Test Bank, picked up but modified from the previous edition of the Test Bank, or picked up straight from the previous edition.

PowerLecture with JoinIn™ and ExamView® CD-ROM ISBN-10 049559850X

This easy-to-use, one-stop digital library and presentation tool includes the following:

- Ready-to-use Microsoft® PowerPoint® lecture slides, making it easy for you to assemble, edit, publish, and present custom lectures for your course
- Video-based polling and quiz questions that can be used with the JoinIn™ on Turning-Point® personal response system
- ExamView® testing software, which provides all the test items from the text's printed test bank in electronic format, enabling you to create customized tests of up to 250 items that can be delivered in print or online
- The text's *Instructor's Resource Manual* in electronic format

ABC Video—Research Methods, Volume I VHS: ISBN-10 0495804312 DVD: ISBN-10 049580438X

This series of videos comprising footage from ABC broadcasts is specially selected and arranged to accompany your research methods course. The segments may be used in conjunction with Wadsworth's research methods texts to help provide a real-world example to illustrate course concepts or to instigate discussion. These ABC videos feature short, high-interest clips from current news events as well as historic raw footage going back 40 years. Clips are drawn from such programs as *World News Tonight, Good Morning America, This Week, PrimeTime Live, 20/20,* and *Nightline,* as well as numerous *ABC News* specials and material from the *Associated Press Television News* and *British Movietone News* collections.

Internet-Based Supplements

CengageNow™

This feature empowers students with the first assessment-centered student tutorial system for Social Research/Research Methods. Seamlessly tied to the new edition, this interactive web-based learning tool helps students gauge their unique study needs with a "pretest" for each chapter to assess their understanding of the material. They are then given a personalized study plan that offers interactive, visual, and audio resources to help them master the material. They can check their progress with an interactive posttest as well.

WebTutor™ on Blackboard® and WebCT®

This web-based software for students and instructors takes a course beyond the classroom to an anywhere, anytime environment. Students gain access to the rich content from this book's companion websites. Available for WebCT and Blackboard only.

InfoTrac® College Edition with InfoMarks™

Available as a free option with newly purchased texts, InfoTrac College Edition gives instructors and students four months of free access to an extensive online database of reliable, full-length articles (not just abstracts) from thousands of scholarly and popular publications going back as far as 22 years. Among the journals available are *American Journal of Sociology, Social Forces, Social Research,* and *Sociology.* InfoTrac College Edition now also comes with InfoMarks, a tool that allows you to save your search parameters, as well as save links to specific articles. (Available to North American college and university students only; journals are subject to change.)

Companion Website for The Practice of Social Research, 12th Edition

The book's companion website (**www.cengage .com/sociology/babbie**) includes chapter-specific resources for instructors and students. For instructors, the site offers a password-protected instructor's manual, Microsoft PowerPoint presentation slides, and more. For students, there is a multitude of text-specific study aids, including the following:

- Tutorial practice quizzing
- Web links
- InfoTrac College Edition key terms
- Flash cards
- Data-analysis primers
- Glossaries

Cengage InSite for Writing and Research™—with Turnitin® Originality Checker

InSite features a full suite of writing, peer review, online grading, and e-portfolio applications. It is an all-in-one tool that helps instructors manage the flow of papers electronically and allows students to submit papers and peer reviews online. Also included in the suite is Turnitin, an originality checker that offers a simple solution for instructors who want a strong deterrent against plagiarism, as well as encouragement for students to employ proper research techniques. Access is available for packaging with each copy of this book. For more information, visit http://insite.cengage.com.

Acknowledgments

It would be impossible to acknowledge adequately all the people who have influenced this book. My earlier methods text, *Survey Research Methods,* was dedicated to Samuel Stouffer, Paul Lazarsfeld, and Charles Glock. I again acknowledge my debt to them.

I also repeat my thanks to those colleagues acknowledged for their comments during the writing of the first, second, and third editions of this book. The present book still reflects their contributions. Many other colleagues helped me revise the book as well—including the amazing 110 instructors who took the time to respond to our electronic survey. Their feedback was invaluable. I also particularly want to thank the instructors who reviewed the manuscript of this edition and made helpful suggestions:

Emily Barman, Boston University

Dale Brooker, Saint Joseph's College of Maine

Dwight Haase, University of Dayton

Linda Lindsey, Maryville University

Elizabeth Monk-Turner, Old Dominion University

Sunil Thankam, The University of Texas at San Antonio

Also, I appreciate the insights and assistance of those who reviewed the previous edition:

Victor Agadjanian, Arizona State University

Melanie Arthur, Portland State University

Pat Christian, Canisius College

William T. Clute, University of Nebraska at Omaha

Marian A. O. Cohen, Framingham State College

Kimberly Dugan, Eastern Connecticut State University

Craig Forsyth, University of Louisiana at Lafayette

Herman Gibson, Henderson State University

Ellen Goldring, Peabody College, Vanderbilt

Susan Gore, University of Massachusetts at Boston

Sarah Hurley, Arkansas State University

Jana L. Jasinski, University of Central Florida

Robert Kleidman, Cleveland State University

Michael Kleiman, University of South Florida

Augustine Kposowa, University of California, Riverside

Marci B. Littlefield, Indiana State University

Patrick F. McManimon, Jr., William Patterson University

Jeanne Mekolichick, Radford University

Jared Schultz, Texas Tech University Health Sciences Center

Bruce H. Wade, Spelman College

Thomas C. Wilson, Florida Atlantic University

Gary Wyatt, Emporia State University

I would also like to thank survey participants who took the time to provide valuable information on several features of the book:

James T. Ault, III, Creighton University

Paul Calarco, SUNY at Albany

Roy Childs, University of the Pacific

Liz Depoy, University of Maine

Pat Fisher, University of Tennessee

Robert Gardner, Bowdoin College

Elizabeth Jones, California University of Pennsylvania

Barbara Keating, Minnesota State University, Mankato

J. David Martin, Midwestern State University

Patrick A. Moore, University of Great Falls

I also wish to thank Anne Baird, Morehouse College; Rae Banks, Syracuse University; Roland Chilton, University of Massachusetts, Amherst; M. Richard Cramer, University of North Carolina, Chapel Hill; Joseph Fletcher, University of Toronto; Shaul Gabbay, University of Illinois, Chicago; Marcia Ghidina, University of North Carolina, Asheville; Roland Hawkes, Southern Illinois University; Jeffrey Jacques, Florida A&M University; Daniel J. Klenow, North Dakota State University; Wanda Kosinski, Ramapo College, New Jersey; Manfred Kuechler, CUNY Hunter College; Cecilia Menjívar, Arizona State University; Joan Morris, University of Central Florida; Alisa Potter, Concordia College; Zhenchoa Qian, Arizona State University; Robert W. Reynolds, Weber State University; Laurie K. Scheuble, Doane College; Beth Anne Shelton, University of Texas, Arlington; Matthew Sloan, University of Wisconsin, Madison; Bernard Sorofman, University of Iowa; Ron Stewart; Randy Stoecker, University of Toledo; Theodore Wagenaar, Miami University, Ohio; Robert Wolf, Eastern Connecticut State University; and Jerome Wolfe, University of Miami.

Over the years, I've become more and more impressed by the important role played by editors in books like this. Although an author's name appears on the book's spine, much of its backbone derives from the strength of its editors. Since 1973 I've worked with many sociology editors at Wadsworth, which has involved the kinds of adjustments you might need to make in successive marriages. The quality of a book like this depends particularly on the wisdom, creativity, and perspiration of the acquisitions editor and the development editor, in this case Chris Caldeira and Dan Moneypenny, respectively. I am very grateful for their efforts on this book, and you should be, too.

There are also others at Wadsworth whose talents have had an impact on this book. I would like to acknowledge Meghan Pease for her inspired marketing efforts, making sure everyone on the planet is aware of the book; Lauren Keyes for her work on the website and other technology supplements; Erin Parkins for managing the development of all of the useful print supplements to round out the teaching package; and Cheri Palmer for shepherding the countless pieces and people required to turn a manuscript into a book.

I also wish to thank Greg Hubit for managing all the critical production processes with great skill.

Molly Roth is the standard by which copy editors should be judged, though that might set the standard too high. Molly and I have worked together on several books now, and she is simply the best. She successfully walks the thin line that separates a reluctance to say the author failed and a delight in saying it. I have never felt she let me get away with anything, nor have I felt anything but the highest support for my intention. Somehow, Molly can see what I'm trying to say and can often find ways of saying it more clearly and more powerfully.

Ted Wagenaar has contributed extensively to this book. Ted and I coauthor the accompanying student study guide, *Guided Activities for Practicing Social Research*, but that's only the tip of the iceberg. Ted is a cherished colleague, welcome critic, good friend, and altogether decent human being.

I have dedicated this book to my soul mate, best friend, and wife, Suzanne Babbie. The fact that she became my wife during the course of this revision makes it all the more fitting. I see in Suze those things I am most proud of in myself, except I see purer versions of those qualities in her. She ennobles what is possible in a human being, and I become a better person because of her example.

The Practice
of Social Research

*S*cience is a familiar word; everyone uses it. Yet, images of science differ greatly. For some, science is mathematics; for others, it's white coats and laboratories. It's often confused with technology or equated with tough high school or college courses.

Science is, of course, none of these things per se. It is difficult, however, to specify exactly what science is. Scientists themselves disagree on the proper definition. For the purposes of this book, we look at science as a method of inquiry—a way of learning and knowing things about the world around us. Contrasted with other ways of learning and knowing about the world, science has some special characteristics. It is a conscious, deliberate, and rigorous undertaking. Sometimes it uses statistical analyses, but often it does not. We'll examine these and other traits in this opening set of chapters.

Dr. Benjamin Spock, the renowned author and pediatrician, began his books on child care by assuring new parents that they already know more about child care than they think they do. I want to begin this book on a

An Introduction to Inquiry

similar note. Before you've read very far, you will realize that you already know a great deal about the practice of social research. In fact, you've been conducting research all your life. From that perspective, the purpose of this book is to help you sharpen skills you already have and perhaps to show you some tricks that may not have occurred to you.

Part 1 of this book lays the groundwork for the rest of the book by examining the fundamental characteristics and issues that make science different from other ways of knowing things. In Chapter 1, we'll begin with a look at native human inquiry, the sort of thing you've been doing all your life. In the course of that examination, we'll see some of the ways people go astray in trying to understand the world around them, and I'll summarize the primary characteristics of scientific inquiry that guard against those errors.

Chapter 2 deals with social theories and the links between theory and research. We'll look at some of the theoretical paradigms that shape the nature of inquiry and that largely determine what scientists look for and how they interpret what they see.

Whereas most of this book deals with the scientific concerns of social research, Chapter 3 introduces two other important concerns: the ethics and politics of research. Researchers are governed by a set of ethical constraints that reflect ideals and values aimed at helping, not harming, people. Social research is also shaped by the fact that it operates within the political codes and systems of the societies it seeks to study and understand. These two topics appear throughout the book as critical components of social research.

The overall purpose of Part 1 is to construct a backdrop against which to view the specifics of research design and execution. After completing Part 1, you'll be ready to look at some of the more concrete aspects of social research.

CHAPTER ONE

Human Inquiry and Science

All of us try to understand and predict the social world. Scientific inquiries—and social research in particular—are designed to avoid the pitfalls of ordinary human inquiry.

CengageNOW for Sociology

Use this online tool to help you make the grade on your next exam. After reading this chapter, go to "Online Study Resources" at the end of the chapter for instructions on how to benefit from *CengageNOW*.

Introduction

This book is about knowing things—not so much *what* we know as *how* we know it. Let's start by examining a few things you probably know already.

You know the world is round. You probably also know it's cold on the dark side of the moon, and you know people speak Chinese in China. You know that vitamin C can prevent colds and that unprotected sex can result in AIDS.

How do you know? Unless you've been to the dark side of the moon lately or done experimental research on the virtues of vitamin C, you know these things because somebody told them to you, and you believed what you were told. You may have read in *National Geographic* that people speak Chinese languages in China, and because that made sense to you, you didn't question it. Perhaps your physics or astronomy instructor told you it was cold on the dark side of the moon, or maybe you heard it on National Public Radio (NPR).

Some of the things you know seem absolutely obvious to you. If someone asked you how you know the world is round, you'd probably say, "Everybody knows that." There are a lot of things everybody knows. Of course, everyone used to "know" that the world was flat.

Most of what you and I know is a matter of agreement and belief. Little of it is based on personal experience and discovery. A big part of growing up in any society, in fact, is the process of learning to accept what everybody around us "knows" is so. If you don't know those same things, you can't really be a part of the group. If you were to question seriously whether the world is really round, you'd quickly find yourself set apart from other people. You might be sent to live in a hospital with other people who question things like that.

Although most of what we know is a matter of believing what we've been told, there is nothing wrong with us in that respect. It's simply the way human societies are structured, and it's a quite useful quality. The basis of knowledge is agreement. Because we can't learn all we need to know by means of personal experience and discovery alone, things are set up so we can simply believe what others tell us. We know some things through tradition and some things from "experts." I'm not saying you should never question this received knowledge; I'm just drawing your attention to the way you and society normally get along regarding what's so.

There are other ways of knowing things, however. In contrast to knowing things through agreement, we can know them through direct experience—through observation. If you dive into a glacial stream flowing through the Canadian Rockies, you don't need anyone to tell you it's cold. The first time you stepped on a thorn, you knew it hurt before anyone told you.

When our experience conflicts with what everyone else knows, though, there's a good chance we'll surrender our experience in favor of the agreement.

Let's take an example. Imagine you've come to a party at my house. It's a high-class affair, and the drinks and food are excellent. In particular, you're taken by one of the appetizers I bring around on a tray: a breaded, deep-fried appetizer that's especially zesty. You have a couple—they're so delicious! You have more. Soon you're subtly moving around the room to be wherever I am when I arrive with a tray of these nibblies.

Finally, you can't contain yourself any more. "What are they?" you ask. "How can I get the recipe?" And I let you in on the secret: "You've been eating breaded, deep-fried worms!" Your response is dramatic: Your stomach rebels, and you throw up all over the living-room rug. Argh! What a terrible thing to serve guests!

The point of the story is that both of your feelings about the appetizer were quite real. Your initial liking for them, based on your own direct experience, was certainly real. But so was your feeling of disgust when you found out that you'd been eating worms. It should be evident, however, that this feeling of disgust was strictly a product of the agreements you have with those around you that worms aren't fit to eat. That's an agreement you entered into the first time your parents found you sitting in a pile of dirt with half of a wriggling worm dangling from your lips. When they pried your mouth open and reached down your throat in search of the other half of the worm, you learned that worms are not acceptable food in our society.

Aside from these agreements, what's wrong with worms? They are probably high in protein and low in calories. Bite-sized and easily packaged, they are a distributor's dream. They are also a delicacy for some people who live in societies that lack our agreement that worms are disgusting. Some people might love the worms but be turned off by the deep-fried breading.

Here's another question to consider: "Are worms 'really' good or 'really' bad to eat?" And here's a more interesting question: "How could you know which was really so?" This book is about answering the second kind of question.

The rest of this chapter looks at how we know what is real. We'll begin by examining inquiry as a natural human activity, something we all have engaged in every day of our lives. We'll look at the source of everyday knowledge and at some kinds of errors we make in normal inquiry. We'll then examine what makes science—in particular, social science—different. After considering some of the underlying ideas of social research, we'll conclude with an initial consideration of issues in social research.

Looking for Reality

Reality is a tricky business. You probably already suspect that some of the things you "know" may not be true, but how can you really know what's real? People have grappled with this question for thousands of years.

---

epistemology The science of knowing; systems of knowledge.

methodology The science of finding out; procedures for scientific investigation.

Knowledge from Agreement Reality

One answer that has arisen out of that grappling is science, which offers an approach to both agreement reality and experiential reality. Scientists have certain criteria that must be met before they will accept the reality of something they have not personally experienced. In general, a scientific assertion must have both logical and empirical support: It must make sense, and it must not contradict actual observation. Why do earthbound scientists accept the assertion that the dark side of the moon is cold? First, it makes sense, because the moon's surface heat comes from the sun's rays, and the dark side of the moon is dark because it's always turned away from the sun. Second, scientific measurements made on the moon's dark side confirm this logical expectation. So, scientists accept the reality of things they don't personally experience— they accept an agreement reality—but they have special standards for doing so.

More to the point of this book, however, science offers a special approach to the discovery of reality through personal experience. In other words, it offers a special approach to the business of inquiry. **Epistemology** is the science of knowing; **methodology** (a subfield of epistemology) might be called the science of finding out. This book presents and examines social science methodology, or how social scientists find out about human social life.

Why do we need social science to discover the reality of social life? To find out, let's start by considering what happens in ordinary, nonscientific inquiry.

Ordinary Human Inquiry

Practically all people, and many other animals as well, exhibit a desire to predict their future circumstances. Humans seem predisposed to undertake this task by using causal and probabilistic reasoning. First, we generally recognize that future circumstances are somehow caused or conditioned by present ones. We learn that getting an education will affect how much money we earn later in life and that swimming beyond the reef may bring an unhappy encounter with a shark. Sharks, on the other hand—whether or not they reason the matter through—may learn that hanging around the reef often brings a happy encounter with unhappy swimmers.

Second, we also learn that such patterns of cause and effect are probabilistic. That is, the effects occur more often when the causes occur than when the causes are absent—but not always. Thus, students learn that studying hard produces good grades in most instances, but not every time. We recognize the danger of swimming beyond the reef, without believing that every such swim will be fatal. As we'll see throughout the book, science makes these concepts of causality and probability more explicit and provides techniques for dealing with them more rigorously than casual human inquiry does. It sharpens the skills we already have by making us more conscious, rigorous, and explicit in our inquiries.

In looking at ordinary human inquiry, we need to distinguish between prediction and understanding. Often, we can make predictions without understanding—perhaps you can predict rain when your trick knee aches. And often, even if we don't understand why, we're willing to act on the basis of a demonstrated predictive ability. A racetrack buff who discovers that the third-ranked horse in the third race of the day always seems to win will probably keep betting without knowing, or caring, why it works out that way. Of course, the drawback in predicting without understanding will become powerfully evident when one of the other horses wins and our buff loses a week's pay.

Whatever the primitive drives or instincts that motivate human beings and other animals, satisfying these drives depends heavily on the ability to predict future circumstances. For people, however, the attempt to predict is often placed in a context of knowledge and understanding. If you can understand why things are related to each other, why certain regular patterns occur, you can predict better than if you simply observe and remember those patterns. Thus, human inquiry aims at answering both "what" and "why" questions, and we pursue these goals by observing and figuring out.

As I suggested earlier in this chapter, our attempts to learn about the world are only partly linked to direct personal inquiry or experience. Another, much larger, part comes from the agreed-on knowledge that others give us, those things "everyone knows." This **agreement reality** both assists and hinders our attempts to find out for ourselves. To see how, consider two important sources of our secondhand knowledge—tradition and authority.

Tradition

Each of us inherits a culture made up, in part, of firmly accepted knowledge about the workings of the world and the values that guide our participation in it. We may learn from others that planting corn in the spring will garner the greatest assistance from the gods, that eating too much candy will decay our teeth, that the circumference of a circle is approximately twenty-two sevenths of its diameter, or that masturbation will blind us. Ideas about gender, race, religion, and different nations that you learned as you were growing up would fit in this category. We may test a few of these "truths" on our own, but we simply accept the great majority of them. These are the things that "everybody knows."

Tradition, in this sense of the term, offers some clear advantages to human inquiry. By accepting what everybody knows, we avoid the overwhelming task of starting from scratch in our search for regularities and understanding. Knowledge is cumulative, and an inherited body of information and understanding is the jumping-off point for the development of more knowledge. We often speak of "standing on the shoulders of giants," that is, on those of previous generations.

At the same time, tradition may hinder human inquiry. If we seek a fresh understanding of something everybody already understands and has always understood, we may be marked as fools for our efforts. More to the point, however, it rarely occurs to most of us to seek a different understanding of something we all "know" to be true.

---

agreement reality Those things we "know" as part of the culture we share with those around us.

Authority

Despite the power of tradition, new knowledge appears every day. Quite aside from our own personal inquiries, we benefit throughout our lives from new discoveries and understandings produced by others. Often, acceptance of these new acquisitions depends on the status of the discoverer. You're more likely to believe that the common cold can be transmitted through kissing, for example, when you hear it from an epidemiologist than when you hear it from your uncle Pete (unless, of course, he's also an epidemiologist).

Like tradition, authority can both assist and hinder human inquiry. We do well to trust the judgment of the person who has special training, expertise, and credentials in a given matter, especially in the face of controversy. At the same time, inquiry can be greatly hindered by the legitimate authorities who err within their own province. Biologists, after all, make their mistakes in the field of biology. Moreover, biological knowledge changes over time.

Inquiry is also hindered when we depend on the authority of experts speaking outside their realm of expertise. For example, consider the political or religious leader with no medical or biochemical expertise who declares that marijuana can fry your brain. The advertising industry plays heavily on this misuse of authority by, for example, having popular athletes discuss the nutritional value of breakfast cereals or having movie actors evaluate the performance of automobiles.

Both tradition and authority, then, act as double-edged swords in the search for knowledge about the world. Simply put, they provide us with a starting point for our own inquiry, but they can lead us to start at the wrong point and push us off in the wrong direction.

Errors in Inquiry, and Some Solutions

Besides the potential dangers of tradition and authority, other pitfalls often cause us to stumble and fall when we set out to learn for ourselves. Let's look at some of the common errors we make in our casual inquiries and at the ways science guards against those errors.

Inaccurate Observations

Quite frequently, we make mistakes in our observations. For example, what was your methodology instructor wearing on the first day of class? If you have to guess, it's because most of our daily observations are casual and semiconscious. That's why we often disagree about what really happened.

In contrast to casual human inquiry, scientific observation is a conscious activity. Just making observation more deliberate helps reduce error. If you had to guess what your instructor was wearing on the first day of class, you'd probably make a mistake. If you'd gone to the first class with a conscious plan to observe and record what your instructor was wearing, however, you'd be far more likely to be accurate. (You might also need a hobby.)

In many cases, both simple and complex measurement devices help guard against inaccurate observations. Moreover, they add a degree of precision well beyond the capacity of the unassisted human senses. Suppose, for example, that you'd taken color photographs of your instructor that day. (See earlier comment about needing a hobby.)

Overgeneralization

When we look for patterns among the specific things we observe around us, we often assume that a few similar events provide evidence of a general pattern. That is, we overgeneralize on the basis of limited observations. (Think back to our now-broke racetrack buff.)

Probably the tendency to overgeneralize peaks when the pressure to arrive at a general understanding is high. Yet it also occurs without such pressure. Whenever overgeneralization does occur, it can misdirect or impede inquiry.

Imagine you are a reporter covering an animal-rights demonstration. You have orders to turn in your story in just two hours, and you need to know why people are demonstrating. Rushing to the scene, you start interviewing them, asking for their reasons. The first three demonstrators you interview give you essentially the same reason, so you simply assume that the other 3,000 are also there for that reason. Unfortunately, when your story appears, your editor gets scores of letters from protesters who were there for an entirely different reason.

Scientists often guard against overgeneralization by committing themselves in advance to a sufficiently large and representative sample of observations. Another safeguard is provided by the replication of inquiry. Basically, **replication** means repeating a study and checking to see whether the same results are produced each time. Then, as a further test, the study may be repeated again under slightly varied conditions.

Selective Observation

One danger of overgeneralization is that it can lead to selective observation. Once we have concluded that a particular pattern exists and have developed a general understanding of why it exists, we tend to focus on future events and situations that fit the pattern, and we tend to ignore those that do not. Racial and ethnic prejudices depend heavily on selective observation for their persistence.

Sometimes a research design will specify in advance the number and kind of observations to be made as a basis for reaching a conclusion. If we wanted to learn whether women were more likely than men to support freedom to choose an abortion, we would commit ourselves to making a specified number of observations on that question in a research project. We might select a thousand carefully chosen people to be interviewed on the issue. Alternately, when making direct observations of an event, such as attending the animal-rights demonstration, we might make a special effort to find "deviant cases"—precisely those who do not fit into the general pattern. Concluding that one youth became delinquent largely because of a lack of positive adult role models draws attention to the part that role models play in keeping most youths on the straight and narrow.

In this recollection of growing up in rural Vermont, Lewis Hill (2000: 35) presents another example of selective observation:

Haying began right after the Fourth of July. The farmers in our neighborhood believed that anyone who started earlier was sure to suffer all the storms of late June in addition to those following the holiday which the old-timers said were caused by all the noise and smoke of gunpowder burning. My mother told me that my grandfather and other Civil War veterans claimed it always rained hard after a big battle. Things didn't always work out the way the older residents promised, of course, but everyone remembered only the times they did.

Illogical Reasoning

There are other ways in which we often deal with observations that contradict our understanding of the way things are in daily life. Surely one of the most remarkable creations of the human mind is "the exception that proves the rule." That idea doesn't make any sense at all. An exception can draw attention to a rule or to a supposed rule (in its original meaning, "prove" meant "test"), but in no system of logic can it validate the rule it contradicts. Even so, we often use this pithy saying to brush away contradictions with a simple stroke of illogic.

What statisticians have called the *gambler's fallacy* is another illustration of illogic in day-to-day reasoning. Often we assume that a consistent run of either good or bad luck foreshadows its opposite. An evening of bad luck at poker may kindle the belief that a winning hand is just around the corner. Many a poker player has stayed in a game much too long because of that mistaken belief. Conversely, an extended period of good weather may lead you to worry that rain will certainly ruin the weekend picnic.

Although all of us sometimes fall into embarrassingly illogical reasoning, scientists try to avoid this pitfall by using systems of logic consciously and explicitly. We'll examine the logic of science more deeply in Chapter 2. For now, simply note that

---

replication Repeating a research study to test and either confirm or question the findings of an earlier study.

logical reasoning is a conscious activity for scientists and that other scientists are always around to keep them honest.

Science, then, attempts to protect its inquiries from the common pitfalls of ordinary inquiry. Accurately observing and understanding reality is not an obvious or trivial matter. Indeed, it's more complicated than I've suggested.

What's Really Real?

Philosophers sometimes use the phrase *naive realism* to describe the way most of us operate in our daily lives. When you sit at a table to write, you probably don't spend a lot of time thinking about whether the table is made up of atoms, which in turn are mostly empty space. When you step into the street and see a city bus hurtling down on you, it's not the best time to reflect on methods for testing whether the bus really exists. We all live with a view that what's real is pretty obvious—and that view usually gets us through the day.

I don't want this book to interfere with your ability to deal with everyday life. I hope, however, that the preceding discussions have demonstrated that the nature of "reality" is perhaps more complex than we tend to assume in our everyday functioning. Here are three views on reality that will provide a philosophical backdrop for the discussions of science to follow. They are sometimes called *premodern, modern,* and *postmodern* views of reality (W. Anderson 1990).

The Premodern View

This view of reality has guided most of human history. Our early ancestors all assumed that they saw things as they really were. In fact, this assumption was so fundamental that they didn't even see it as an assumption. No cavemom said to her cavekid, "Our tribe makes an assumption that evil spirits reside in the Old Twisted Tree." No, she said, "STAY OUT OF THAT TREE OR YOU'LL TURN INTO A TOAD!"

As humans evolved and became aware of their diversity, they came to recognize that others did not always share their views of things. Thus, they may have discovered that another tribe didn't buy the wicked tree thing; in fact, the second tribe felt that the spirits in the tree were holy and beneficial. The discovery of this diversity led members of the first tribe to conclude that "some tribes I could name are pretty stupid." For them, the tree was still wicked, and they expected that some misguided people would soon be moving to Toad City.

The Modern View

What philosophers call the modern view accepts such diversity as legitimate, a philosophical "different strokes for different folks." As a modern thinker, you would say, "I regard the spirits in the tree as evil, but I know that others regard them as good. Neither of us is right or wrong. There are simply spirits in the tree. They are neither good nor evil, but different people have different ideas about them."

Adopting the modern view is easy for most of us. Some might regard a dandelion as a beautiful flower, whereas others see only an annoying weed. In the premodern view, a dandelion has to be either one or the other. If you think it is a weed, it is really a weed, though you may admit that some people have a warped sense of beauty. In the modern view, a dandelion is simply a dandelion. It is a plant with yellow petals and green leaves. The concepts "beautiful flower" and "annoying weed" are subjective points of view imposed on the plant by different people. Neither is a quality of the plant itself, just as "good" and "evil" were concepts imposed on the spirits in the tree in our example.

The Postmodern View

Increasingly, philosophers speak of a postmodern view of reality. In this view, the spirits don't exist. Neither does the dandelion. All that's "real" are the images we get through our points of view. Put differently, there's nothing "out there"; it's all "in here." As Gertrude Stein said of the city of Oakland, "There's no there, there."

No matter how bizarre the postmodern view may seem to you on first reflection, it has a certain ironic inevitability. Take a moment to notice the book you are reading; notice specifically what it looks like. Because you are reading these words, it probably looks something like Figure 1-1a.

FIGURE 1-1

A book. All of these are the same book, but it looks different when viewed from different locations, perspectives, or "points of view."

FIGURE 1-2

Wife's Point of View. There is no question in the wife's mind as to who is right and rational and who is out of control.

Does Figure 1-1a represent the way your book "really" looks? Or does it merely represent what the book looks like from your current point of view? Surely, Figures 1-1b, c, and d are equally valid representations. But these views of the book differ greatly from each other. Which is the "reality"?

As this example illustrates, there is no answer to the question "What does the book really look like?" All we can offer is the different ways it looks from different points of view. Thus, according to the postmodern view, there is no "book," only various images of it from different points of view. And all the different images are equally "true."

Now let's apply this logic to a social situation. Imagine a husband and wife arguing. When she looks over at her quarreling husband, Figure 1-2 is what the wife sees. Take a minute to imagine what you would feel and think if you were the woman in this drawing. How would you explain later to your best friend what had happened? What solutions to the conflict would seem appropriate if you were this woman?

Of course, what the woman's husband sees is another matter altogether, as shown in Figure 1-3. Take a minute to imagine experiencing the situation from his point of view. What thoughts and feelings would you have? How would you tell your

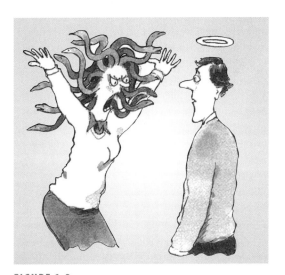

FIGURE 1-3

Husband's Point of View. The husband has a very different perception of the same set of events, of course.

best friend what had happened? What solutions would seem appropriate for resolving the conflict?

Now consider a third point of view. Suppose you were an outside observer, watching this interaction between a wife and husband. What would it look like to you now? Unfortunately, we can't easily portray the third point of view without knowing something about the personal feelings, beliefs, past experiences, and so forth that you would bring to your task as outside observer. (Though I call you an "outside" observer, you would be, of course, observing from inside your own mental system.)

To take an extreme example, if you were a confirmed male chauvinist, you'd probably see the fight pretty much the same way that the husband saw it. On the other hand, if you were committed to the view that men are generally unreasonable bums, you'd see things the way the wife saw them in the earlier picture.

Imagine that instead you see two unreasonable people quarreling irrationally with each other. Would you see them both as irresponsible jerks, equally responsible for the conflict? Or would you see them as two people facing a difficult human situation, each doing the best he or she can to resolve it? Imagine feeling compassion for them and noticing how each of them attempts to end the hostility, even though the gravity of the problem keeps them fighting.

Notice how different these several views are. Which is a "true" picture of what is happening between the wife and the husband? You win the prize if you notice that the personal viewpoint you bring to the observational task will again color your perception of what is happening.

The postmodern view represents a critical dilemma for scientists. Although their task is to observe and understand what is "really" happening, they are all human and, as such, bring along personal orientations that will color what they observe and how they explain it. There is ultimately no way people can totally step outside their humanness to see and understand the world as it "really" is—that is, independently of all human viewpoints.

Whereas the modern view acknowledges the inevitability of human subjectivity, the postmodern view suggests there is actually no "objective" reality to be observed in the first place. There are only our several subjective views.

You may want to ponder these three views of reality on your own for awhile. We'll return to them in Chapter 2 when we focus on specific scientific paradigms. Ultimately, two points will emerge. First, established scientific procedures sometimes allow us to deal effectively with this dilemma—that is, we can study people and help them through their difficulties without being able to view "reality" directly. Second, different philosophical stances suggest a powerful range of possibilities for structuring our research.

Let's turn now from general philosophical ideas to the foundations of social science approaches to understanding. A consideration of these underpinnings of social research will prepare the way for our exploration of specific research techniques.

The Foundations of Social Science

Science is sometimes characterized as logico-empirical. This ungainly term carries an important message: As we noted earlier, the two pillars of science are logic and observation. That is, a scientific understanding of the world must both make sense and correspond to what we observe. Both elements are essential to science and relate to the three major aspects of the enterprise of social science: theory, data collection, and data analysis.

To oversimplify just a bit, scientific **theory** deals with the logical aspect of science—providing systematic explanations—whereas data collection deals with the observational aspect. Data analysis looks for patterns in observations and, where appropriate, compares what is logically expected with what is actually observed. Although this book is primarily about data collection and data analysis—that is, how to conduct social research—the rest of Part 1 is devoted to the theoretical context of research. Parts 2 and 3 then focus on data collection, and Part 4 offers an introduction to the analysis of data.

Underlying the concepts presented in the rest of the book are some fundamental ideas that

---

theory A systematic explanation for the observations that relate to a particular aspect of life: juvenile delinquency, for example, or perhaps social stratification or political revolution.

distinguish social science—theory, data collection, and analysis—from other ways of looking at social phenomena. Let's consider these ideas.

Theory, Not Philosophy or Belief

Today, social theory has to do with what is, not with what should be. For many centuries, however, social theory did not distinguish between these two orientations. Social philosophers liberally mixed their observations of what happened around them, their speculations about why, and their ideas about how things ought to be. Although modern social researchers may do the same from time to time, as scientists they focus on how things actually are and why.

This means that scientific theory—and, more broadly, science itself—cannot settle debates about values. Science cannot determine whether capitalism is better or worse than socialism. What it can do is determine how these systems perform, but only in terms of some set of agreed-on criteria. For example, we could determine scientifically whether capitalism or socialism most supports human dignity and freedom only if we first agreed on some measurable definitions of dignity and freedom. Our conclusions would then be limited to the meanings specified in our definitions. They would have no general meaning beyond that.

By the same token, if we could agree that suicide rates, say, or giving to charity were good measures of the quality of a religion, then we could determine scientifically whether Buddhism or Christianity is the better religion. Again, our conclusion would be inextricably tied to our chosen criteria. As a practical matter, people seldom agree on precise criteria for determining issues of value, so science is seldom useful in settling such debates. In fact, questions like these are so much a matter of opinion and belief that scientific inquiry is often viewed as a threat to what is "already known."

We'll consider this issue in more detail in Chapter 12, when we look at evaluation research. As you'll see, researchers have become increasingly involved in studying social programs that reflect ideological points of view, such as affirmative action or welfare reform. One of the biggest problems they face is getting people to agree on criteria of success and failure. Yet such criteria are essential if social research is to tell us anything useful about matters of value. By analogy, a stopwatch cannot tell us if one sprinter is better than another unless we first agree that speed is the critical criterion.

Social science, then, can help us know only what is and why. We can use it to determine what ought to be, but only when people agree on the criteria for deciding what outcomes are better than others—an agreement that seldom occurs.

As I indicated earlier, even knowing "what is and why" is no simple task. Let's turn now to some of the fundamental ideas that underlie social science's efforts to describe and understand social reality.

Social Regularities

In large part, social research aims to find patterns of regularity in social life. Although all the sciences share that aim, it sometimes imposes a barrier for people when they first approach social science.

Certainly at first glance the subject matter of the physical sciences seems to be more governed by regularities than does that of the social sciences. A heavy object falls to earth every time we drop it, but a person may vote for a particular candidate in one election and against that same candidate in the next. Similarly, ice always melts when heated enough, but habitually honest people sometimes steal. Despite such examples, however, social affairs do exhibit a high degree of regularity that research can reveal and theory can explain.

To begin with, the tremendous number of formal norms in society create a considerable degree of regularity. For example, traffic laws in the United States induce the vast majority of people to drive on the right side of the street rather than the left. Registration requirements for voters lead to some predictable patterns in which classes of people vote in national elections. Labor laws create a high degree of uniformity in the minimum age of paid workers as well as the minimum amount they are paid. Such formal prescriptions regulate, or regularize, social behavior.

Aside from formal prescriptions, we can observe other social norms that create more regularities. Among registered voters, Republicans are more likely than Democrats to vote for Republican candidates. University professors tend to earn more money than unskilled laborers do. Men tend to earn more than women. And so on.

Three objections are sometimes raised in regard to such social regularities. First, some of the regularities may seem trivial. For example, Republicans vote for Republicans; everyone knows that. Second, contradictory cases may be cited, indicating that the "regularity" isn't totally regular. Some laborers make more money than some professors do. Third, it may be argued that, unlike the heavy objects that cannot decide not to fall when dropped, the people involved in the regularity could upset the whole thing if they wanted to.

Let's deal with each of these objections in turn.

The Charge of Triviality

During World War II, Samuel Stouffer, one of the greatest social science researchers, organized a research branch in the U.S. Army to conduct studies in support of the war effort (Stouffer et al. 1949–1950). Many of the studies focused on the morale among soldiers. Stouffer and his colleagues found there was a great deal of "common wisdom" regarding the bases of military morale. Much of their research was devoted to testing these "obvious" truths.

For example, people had long recognized that promotions affect morale in the military. When military personnel get promotions and the promotion system seems fair, morale rises. Moreover, it makes sense that people who are getting promoted will tend to think the system is fair, whereas those passed over will likely think the system is unfair. By extension, it seems sensible that soldiers in units with slow promotion rates will tend to think the system is unfair, and those in units with rapid promotions will think the system is fair. But was this the way they really felt?

Stouffer and his colleagues focused their studies on two units: the Military Police (MPs), which had the slowest promotions in the Army, and the Army Air Corps (forerunner of the U.S. Air Force), which had the fastest promotions. It stood to reason that MPs would say the promotion system was unfair, and the air corpsmen would say it was fair. The studies, however, showed just the opposite.

Notice the dilemma faced by a researcher in a situation such as this. On the one hand, the observations don't seem to make sense. On the other hand, an explanation that makes obvious good sense isn't supported by the facts.

A lesser person would have set the problem aside "for further study." Stouffer, however, looked for an explanation for his observations, and eventually he found it. Robert Merton and other sociologists at Columbia University had begun thinking and writing about something they called *reference group theory*. This theory says that people judge their lot in life less by objective conditions than by comparing themselves with others around them—their reference group. For example, if you lived among poor people, a salary of $50,000 a year would make you feel like a millionaire. But if you lived among people who earned $500,000 a year, that same $50,000 salary would make you feel impoverished.

Stouffer applied this line of reasoning to the soldiers he had studied. Even if a particular MP had not been promoted for a long time, it was unlikely that he knew some less-deserving person who had gotten promoted more quickly. Nobody got promoted in the MPs. Had he been in the Air Corps—even if he had gotten several promotions in rapid succession—he would probably have been able to point to someone less deserving who had gotten even faster promotions. An MP's reference group, then, was his fellow MPs, and the air corpsman compared himself with fellow corpsmen. Ultimately, then, Stouffer reached an understanding of soldiers' attitudes toward the promotion system that (1) made sense and (2) corresponded to the facts.

This story shows that documenting the obvious is a valuable function of any science, physical or social. Charles Darwin coined the phrase *fool's experiment* to describe much of his own research—research in which he tested things that everyone else "already knew." As Darwin understood, the obvious all too often turns out to be wrong; thus, apparent triviality is not a legitimate objection to any scientific endeavor.

What about Exceptions?

The objection that there are always exceptions to any social regularity does not mean that the regularity itself is unreal or unimportant. A particular woman may well earn more money than most men, but that provides small consolation to the majority of women, who earn less. The pattern still exists. Social regularities, in other words, are probabilistic patterns, and they are no less real simply because some cases don't fit the general pattern.

This point applies in physical science as well as social science. Subatomic physics, for example, is a science of probabilities. In genetics, the mating of a blue-eyed person with a brown-eyed person will probably result in a brown-eyed offspring. The birth of a blue-eyed child does not destroy the observed regularity, because the geneticist states only that the brown-eyed offspring is more likely and, further, that brown-eyed offspring will be born in a certain percentage of the cases. The social scientist makes a similar, probabilistic prediction—that women overall are likely to earn less than men. Once a pattern like this is observed, the social scientist has grounds for asking why it exists.

People Could Interfere

Finally, the objection that the conscious will of the actors could upset observed social regularities does not pose a serious challenge to social science. This is true even though a parallel situation does not appear to exist in the physical sciences. (Presumably, physical objects cannot violate the laws of physics, although the probabilistic nature of subatomic physics once led some observers to postulate that electrons had free will.) There is no denying that a religious, right-wing bigot could go to the polls and vote for an agnostic, left-wing African American if he wanted to upset political scientists studying the election. All voters in an election could suddenly switch to the underdog just to frustrate the pollsters. Similarly, workers could go to work early or stay home from work and thereby prevent the expected rush-hour traffic. But these things do not happen often enough to seriously threaten the observation of social regularities.

Social regularities, then, do exist, and social scientists can detect them and observe their effects.

When these regularities change over time, social scientists can observe and explain those changes.

Aggregates, Not Individuals

The regularities of social life that social scientists study generally reflect the collective behavior of many individuals. Although social scientists often study motivations that affect individuals, the individual as such is seldom the subject of social science. Instead, social scientists create theories about the nature of group, rather than individual, life. Similarly, the objects of their research are typically *aggregates,* or collections, rather than individuals.

Sometimes the collective regularities are amazing. Consider the birthrate, for example. People have babies for a wide variety of personal reasons. Some do it because their own parents want grandchildren. Some feel it's a way of completing their womanhood or manhood. Others want to hold their marriages together, enjoy the experience of raising children, perpetuate the family name, or achieve a kind of immortality. Still others have babies by accident.

If you have fathered or given birth to a baby, you could probably tell a much more detailed, idiosyncratic story. Why did you have the baby when you did, rather than a year earlier or later? Maybe you lost your job and had to delay a year before you could afford to have the baby. Maybe you only felt the urge to become a parent after someone close to you had a baby. Everyone who had a baby last year had his or her own reasons for doing so. Yet, despite this vast diversity, and despite the idiosyncrasy of each individual's reasons, the overall birthrate in a society—the number of live births per 1,000 population—is remarkably consistent from year to year. See Table 1-1 for recent birthrates for the United States.

If the U.S. birthrate were 15.9, 35.6, 7.8, 28.9, and 16.2 in five successive years, demographers would begin dropping like flies. As you can see, however, social life is far more orderly than that. Moreover, this regularity occurs without society-wide regulation. No one plans how many babies will be born or determines who will have them. You do not need a permit to have a baby; in fact,

TABLE 1-1

Birthrates, United States: 1980–2006*

1980	15.9	1994	15.0
1981	15.8	1995	14.6
1982	15.9	1996	14.4
1983	15.6	1997	14.2
1984	15.6	1998	14.3
1985	15.8	1999	14.2
1986	15.6	2000	14.4
1987	15.7	2001	14.1
1988	16.0	2002	13.9
1989	16.4	2003	14.1
1990	16.7	2004	14.0
1991	16.2	2005	14.0
1992	15.8	2006	14.3
1993	15.4		

*Live births per 1,000 population

Source: U.S. Bureau of the Census, *Statistical Abstract of the United States* (Washington, DC: U.S. Government Printing Office, 2008), Table 77, p. 63.

many babies are conceived unexpectedly, and some are borne unwillingly.

Social science theories, then, typically deal with aggregated, not individual, behavior. Their purpose is to explain why aggregate patterns of behavior are so regular even when the individuals participating in them may change over time. We could even say that social scientists don't seek to explain people at all. They try to understand the systems in which people operate, the systems that explain why people do what they do. The elements in such a system are not people but variables.

Concepts and Variables

Our most natural attempts at understanding usually take place at the level of the concrete and idiosyncratic. That's just the way we think.

---

variables Logical sets of attributes. The variable *gender* is made of up of the attributes *male* and *female*.

Imagine that someone says to you, "Women ought to get back into the kitchen where they belong." You're likely to hear that comment in terms of what you know about the speaker. If it's your old uncle Harry who is also strongly opposed to daylight saving time, zip codes, and personal computers, you're likely to think his latest pronouncement simply fits into his rather dated point of view about things in general. If, on the other hand, the statement is muttered by an incumbent politician trailing a female challenger in an electoral race, you'll probably explain his comment in a completely different way.

In both examples, you're trying to understand the behavior of a particular individual. Social research seeks insights into classes or types of individuals. Social researchers would want to find out about the kind of people who share that view of women's "proper" role. Do those people have other characteristics in common that may help explain their views?

Even when researchers focus their attention on a single case study—such as a community or a juvenile gang—their aim is to gain insights that would help people understand other communities and other juvenile gangs. Similarly, the attempt to fully understand one individual carries the broader purpose of understanding people or types of people in general.

When this venture into understanding and explanation ends, social researchers will be able to make sense out of more than one person. In understanding what makes a group of people hostile to women who are active outside the home, they gain insight into all the individuals who share that characteristic. This is possible because, in an important sense, they have not been studying antifeminists as much as they have been studying antifeminism. It might then turn out that Uncle Harry and the politician have more in common than first appeared.

Antifeminism is spoken of as a **variable** because it varies. Some people display the attitude more than others do. Social researchers are interested in understanding the system of variables that causes a particular attitude to be strong in one instance and weak in another.

The idea of a system composed of variables may seem rather strange, so let's look at an analogy. The subject of a physician's attention is the patient. If the patient is ill, the physician's purpose is to help the patient get well. By contrast, a medical researcher's subject matter is different—the variables that cause a disease, for example. The medical researcher may study the physician's patient, but for the researcher, that patient is relevant only as a carrier of the disease.

That is not to say that medical researchers don't care about real people. They certainly do. Their ultimate purpose in studying diseases is to protect people from them. But in their research, they are less interested in individual patients than they are in the patterns governing the appearance of the disease. In fact, when they can study a disease meaningfully without involving actual patients, they do so.

Social research, then, involves the study of variables and their relationships. Social theories are written in a language of variables, and people get involved only as the "carriers" of those variables.

Variables, in turn, have what social researchers call attributes (or categories or values). **Attributes** are characteristics or qualities that describe an object—in this case, a person. Examples include *female, Asian, alienated, conservative, dishonest, intelligent,* and *farmer.* Anything you might say to describe yourself or someone else involves an attribute.

Variables, on the other hand, are logical sets of attributes. Thus, for example, *male* and *female* are attributes, and *sex* or *gender* is the variable composed of those two attributes. The variable *occupation* is composed of attributes such as *farmer, professor,* and *truck driver. Social class* is a variable composed of a set of attributes such as *upper class, middle class,* and *lower class.* Sometimes it helps to think of attributes as the categories that make up a variable. (See Figure 1-4 for a schematic review of what social scientists mean by variables and attributes.)

The relationship between attributes and variables forms the heart of both description and explanation in science. For example, we might describe a college class in terms of the variable *gender* by reporting the observed frequencies of the attributes

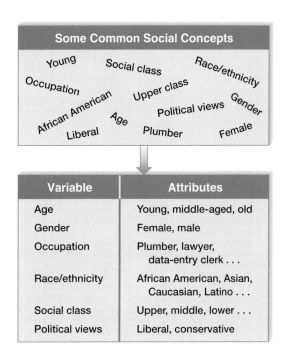

Some Common Social Concepts

Young   Social class   Race/ethnicity
Occupation   Upper class   Gender
African American   Age   Political views
Liberal   Plumber   Female

Variable	Attributes
Age	Young, middle-aged, old
Gender	Female, male
Occupation	Plumber, lawyer, data-entry clerk . . .
Race/ethnicity	African American, Asian, Caucasian, Latino . . .
Social class	Upper, middle, lower . . .
Political views	Liberal, conservative

FIGURE 1-4

Variables and Attributes. In social research and theory, both variables and attributes represent social concepts. Variables are sets of related attributes (categories, values).

male and *female:* "The class is 60 percent men and 40 percent women." An unemployment rate can be thought of as a description of the variable *employment status of a labor force* in terms of the attributes *employed* and *unemployed.* Even the report of *family income for a city* is a summary of attributes composing that variable: *$3,124; $10,980; $35,000;* and so forth.

Sometimes the meanings of the concepts that lie behind social science concepts are immediately clear. Other times they aren't. This point is discussed in "The Hardest Hit Was . . ."

The relationship between attributes and variables is more complicated in the case of explanation and gets to the heart of the variable language of scientific theory. Here's a simple example, involving two variables, *education* and *prejudice.* For the sake of simplicity, let's assume that the variable *education* has only two attributes: *educated* and *uneducated.*

attributes Characteristics of people or things.

The Hardest Hit Was . . .

In early 1982, a deadly storm ravaged the San Francisco Bay Area, leaving an aftermath of death, injury, and property damage. As the mass media sought to highlight the most tragic results of the storm, they sometimes focused on several people who were buried alive in a mud slide in Santa Cruz. Other times, they covered the plight of the 2,900 made homeless in Marin County.

Implicitly, everyone wanted to know where the worst damage was done, but the answer was not clear. Here are some data describing the results of the storm in two counties: Marin and Santa Cruz. Look over the comparisons and see if you can determine which county was "hardest hit."

Certainly, in terms of the loss of life, Santa Cruz was the "hardest hit" of the two counties. Yet more than seven times as many people were injured in Marin as in Santa Cruz; certainly, Marin County was "hardest hit" in that regard. Or consider the number of homes destroyed (worse in Santa Cruz) or damaged (worse in Marin): It matters which you focus on. The same dilemma holds true for the value of the damage done: Should we pay more attention to private damage or public damage?

So which county was "hardest hit"? Ultimately, the question as posed has no answer. Although you and I both have images in our minds about communities that are "devastated" or communities that are only "lightly touched," these images are not precise enough to permit rigorous measurements.

	Marin	Santa Cruz
Business destroyed	$1.50 million	$56.5 million
People killed	5	22
People injured	379	50
People displaced	370	400
Homes destroyed	28	135
Homes damaged	2,900	300
Businesses destroyed	25	10
Businesses damaged	800	35
Private damages	$65.1 million	$50.0 million
Public damages	$15.0 million	$56.5 million

The question can be answered only if we can specify what we mean by "hardest hit." If we measure it by death toll, then Santa Cruz was the hardest hit. If we choose to define the variable in terms of people injured and or displaced, then Marin suffered the bigger disaster. The simple fact is that we cannot answer the question without specifying exactly what we mean by the term *hardest hit*. This is a fundamental requirement that will arise again and again as we attempt to measure social science variables.

Data source: San Francisco Chronicle, January 13, 1982, p. 16.

Similarly, let's give the variable *prejudice* two attributes: *prejudiced* and *unprejudiced*.

Now let's suppose that 90 percent of the uneducated are prejudiced, and the other 10 percent are unprejudiced. And let's suppose that 30 percent of the educated people are prejudiced, and the other 70 percent are unprejudiced. This is illustrated in Figure 1-5a.

Figure 1-5a illustrates a relationship or association between the variables *education* and *prejudice*. This relationship can be seen in terms of the pairings of attributes on the two variables. There are two predominant pairings: (1) those who are educated and unprejudiced and (2) those who are uneducated and prejudiced. Here are two other useful ways of viewing that relationship.

First, let's suppose that we play a game in which we bet on your ability to guess whether a person is prejudiced or unprejudiced. I'll pick the people one at a time (not telling you which ones I've picked), and you have to guess whether each person is prejudiced. We'll do it for all 20 people in Figure 1-5a. Your best strategy in this case would be to guess prejudiced each time, because 12 out of the 20 are categorized that way. Thus, you'll get 12 right and 8 wrong, for a net success of 4.

Now let's suppose that when I pick a person from the figure, I tell you whether the person is educated or uneducated. Your best strategy now would be to guess prejudiced for each uneducated person and unprejudiced for each educated person. If you followed that strategy, you'd get 16 right and 4 wrong. Your improvement in guessing prejudice by knowing education is an illustration of what it means to say that the variables are related.

a. The uneducated are *more* prejudiced than the educated.

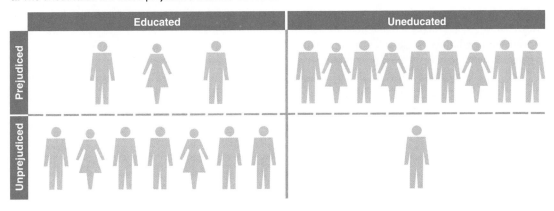

b. There is *no* apparent relationship between education and prejudice.

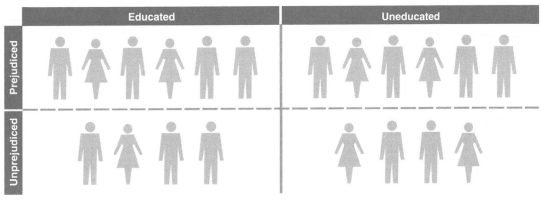

FIGURE 1-5

Relationship between Two Variables (Two Possibilities). Variables such as *education* and *prejudice* and their attributes *(educated/uneducated, prejudiced/unprejudiced)* are the foundation for the examination of causal relationships in social research.

Second, by contrast, let's consider how the 20 people would be distributed if education and prejudice were unrelated to each other (Figure 1-5b). Notice that half the people are educated, and half are uneducated. Also notice that 12 of the 20 (60 percent) are prejudiced. If 6 of the 10 people in each group were prejudiced, we would conclude that the two variables were unrelated to each other. Knowing a person's education would not be of any value to you in guessing whether that person was prejudiced.

We'll be looking at the nature of relationships between variables in some depth in Part 4. In particular, we'll explore some of the ways relationships can be discovered and interpreted in research analysis. For now, you need a general understanding of relationships in order to appreciate the logic of social science theories.

Theories describe the relationships we might logically expect between variables. Often, the expectation involves the idea of causation. That is, a person's attributes on one variable are expected to cause, predispose, or encourage a particular attribute on another variable. In the example just illustrated, we might theorize that a person's being educated or uneducated causes a lesser or greater likelihood of that person seeming prejudiced.

As I'll discuss in more detail later in the book, *education* and *prejudice* in this example would be regarded as an **independent variable** and a **dependent variable,** respectively. These two concepts are implicit in causal, or deterministic, models. In this example, we assume that the likelihood of being prejudiced is determined or caused by something. In other words, *prejudice* depends on something else, and so it is called the "dependent" variable. What the dependent variable depends on is an independent variable, in this case, *education.* For the purposes of this study, *education* is an "independent" variable because it is independent of *prejudice* (that is, people's level of education is not caused by whether or not they are prejudiced).

Of course, variations in levels of education can, in turn, be found to depend on something else. People whose parents have a lot of education, for example, are more likely to get a lot of education than are people whose parents have little education. In this relationship, the subject's education is the dependent variable, and the parents' education is the independent variable. We can say the independent variable is the cause, the dependent variable the effect.

In our discussion of Figure 1-5, we looked at the distribution of the 20 people in terms of the two variables. In constructing a social science theory, we would derive an expectation regarding the relationship between the two variables based on what we know about each. We know, for example, that education exposes people to a wide range of cultural variation and to diverse points of view—in short, it broadens their perspectives. Prejudice, on the other hand, represents a narrower perspective. Logically, then, we might expect education and prejudice to be somewhat incompatible. We might therefore arrive at an expectation that increasing education would reduce the occurrence of prejudice, an expectation that our observations would support.

Because Figure 1-5 has illustrated two possibilities—that education reduces the likelihood of prejudice or that it has no effect—you might be interested in knowing what is actually the case. There are, of course, many types of prejudice. For purposes of this illustration, let's consider prejudice against gays and lesbians. Over the years, the General Social Survey (GSS) has asked respondents whether homosexual relations between two adults is "always wrong, almost always wrong, sometimes wrong, or not wrong at all." In 2004, 56 percent of those interviewed said that homosexuality was always wrong. However, this response is strongly conditioned by respondents' education, as Table 1-2 indicates. (See the box for more about the GSS.)

Notice that the theory has to do with the two variables *education* and *prejudice,* not with people as such. People are the carriers of those two variables, so the relationship between the variables can only be seen when we observe people. Ultimately, however, the theory uses a language of variables. It describes the associations that we might logically expect to exist between particular attributes of different variables.

independent variable A variable with values that are not problematic in an analysis but are taken as simply given. An independent variable is presumed to cause or determine a dependent variable.

dependent variable A variable assumed to depend on or be caused by another (called the independent variable). If you find that income is partly a function of amount of formal education, income is being treated as a dependent variable.

TABLE 1-2
Education and Anti-Gay Prejudice

Level of Education	Percent saying Homosexuality is Always Wrong
Less than high school graduate	72%
High school graduate	62%
Junior college	56%
Bachelor's degree	44%
Graduate degree	30%

General Social Survey (GSS)

The National Opinion Research Center (NORC) at the University of Chicago conducts a periodic national survey of American public opinion for the purpose of making such data available for analysis by the social research community. Many data examples in this book come from that source. You can learn more about the GSS at the official website maintained by the University of Michigan; go to the link at http://www.norc.org/GSS+Website/

The Purposes of Social Research

Chapter 4 will examine the various purposes of social research in some detail, but a brief preview here will be useful. To begin, sometimes social research is a vehicle for mapping out a topic that may warrant further study later: looking into a new political or religious group, learning something about use of a new street drug, and so forth. The methods vary greatly and the conclusions are usually suggestive rather than definitive. Even so, such *exploratory* social research, if carefully done, can dispel some misconceptions and help focus future research.

Some social research is done for the purpose of *describing* the state of social affairs: What is the unemployment rate? What is the racial composition of a particular city? What percentage of the population plans to vote for a particular political candidate? Careful empirical description takes the place of speculation and impressions.

Often, social research has an *explanatory* purpose—providing reasons for phenomena in the form of causal relationships. Why do some cities have higher unemployment rates than others? Why are some people more prejudiced than others? Why are women likely to earn less than men for doing the same job? Although answers to such questions abound in ordinary, everyday discourse, some of those answers are simply wrong. Explanatory social research provides more trustworthy explanations.

Later in this chapter, we'll compare pure and applied research, but it's worth noting here that the purpose of some research is pretty much limited to understanding, while other research efforts are deliberately intended to bring about social change, creating a more workable and/or just society. Any kind of social science study, however, can change our view of society, in some cases challenging commonly accepted "truths" about certain groups of people (see "Keeping Humanity in Focus").

The Ethics of Human Inquiry

While most of this book is devoted to the logic and techniques of doing social research, you will soon discover another theme running throughout the discussion: the ethical dimension. You will learn that medical, social, and other studies of human beings have often used methods later condemned as unethical. In Chapter 3 and throughout the book, we are going to examine the various concerns that distinguish ethical from unethical research.

I suspect that such ethical concerns will make more sense to you as you learn more about the actual techniques of doing research, but I want to alert you to this important issue at the outset of our journey.

Some Dialectics of Social Research

There is no one way to do social research. (If there were, this would be a much shorter book.) In fact, much of the power and potential of social research lies in the many valid approaches it comprises.

Four broad and interrelated distinctions, however, underlie the variety of research approaches. Although one can see these distinctions

Keeping Humanity in Focus

As we have seen, a wide variety of research approaches can enhance our grasp of social dynamics. Much social research involves the analysis of masses of statistical data. As valuable as the examination of overall patterns can be, it can come at the risk of losing sight of the individual men and women those data represent. As such, some social research focuses specifically on the detailed particulars of real lives at the ground level of society. Throughout this book, I'll highlight some recent studies that reflect this latter approach to understanding social life, in an attempt to "keep humanity in focus" during our broader discussion of social science practice.

Statistics suggest that, in the United States, unwed mothers and their children, particularly those who are poor, will face a host of problems in the years to come. Both the child and the mother will likely struggle and suffer. The children are less likely to do well in school and in later life, and the mothers will probably have to struggle in low-paying jobs or live on welfare. The trend toward births out of wedlock has increased dramatically in recent decades, especially among the poor. As a reaction to these problems, the Bush administration launched a Healthy Marriage Initiative in 2005 aimed at encouraging childbearing couples to marry. Voices for and against the program have been raised with vigor.

In their book *Promises I Can Keep*, Kathryn Edin and Maria Kefalas raise a question that might have been asked prior to the creation of a solution to the perceived problem: "Why do poor women bear children outside of wedlock?" The two social scientists spent five years speaking one-on-one with young women who had had children out of wedlock. Some of the things they learned dramatically contradicted various widespread images of unwed mothers. For instance, whereas many people have bemoaned the abandonment of marriage among the poor, the

women interviewed tended to speak highly of the institution, indicating they hoped to be married one day. Further, many were only willing to settle down with someone trustworthy and stable—better to remain unmarried than to enter a marriage that will end in disaster.

At the same time, these young women felt strongly that their ultimate worth as women centered on their bearing children. Most felt it was preferable to be an unmarried mother than to be a childless woman, the real tragedy in their eyes.

This view of marriage may differ greatly from your own. As we have seen, assumptions about "what's real" are often contradicted by actual observations; this is crucial to understanding "what will work" to address social issues.

Promises I Can Keep: Why Poor Women Put Motherhood Before Marriage by Kathryn Edin and Maria Kefalas. © 2005 Regents of the University of California. Published by the University of California Press. Photograph © Michael Smyth

as competing choices, a good social researcher learns each of the orientations they represent. This is what I mean by the "dialectics" of social research: There is a fruitful tension between the complementary concepts I'm about to describe.

Idiographic and Nomothetic Explanation

All of us go through life explaining things. We do it every day. You explain why you did poorly or well on an exam, why your favorite team is winning or losing, why you may be having trouble getting good dates or a decent job. In our everyday explanations, we engage in two distinct forms of causal reasoning, though we do not ordinarily distinguish them.

Sometimes we attempt to explain a single situation in idiosyncratic detail. Thus, for example, you may have done poorly on an exam because (1) you forgot there was an exam that day, (2) it was in your worst subject, (3) a traffic jam made you late for class, (4) your roommate kept you up the night before the exam by playing loud music, (5) the police kept you until dawn demanding to know what you had done with your roommate's stereo—

and what you had done with your roommate, for that matter—and (6) a wild band of coyotes ate your textbook. Given all these circumstances, it's no wonder you did poorly.

This type of causal reasoning is called an **idiographic** explanation. *Idio-* in this context means unique, separate, peculiar, or distinct, as in the word *idiosyncrasy*. When we have completed an idiographic explanation, we feel that we fully understand the causes of what happened in this particular instance. At the same time, the scope of our explanation is limited to the single case at hand. Although parts of the idiographic explanation might apply to other situations, our intention is to explain one case fully.

Now consider a different kind of explanation. (1) Every time you study with a group, you do better on the exam than if you had studied alone. (2) Your favorite team does better at home than on the road. (3) Fraternity and sorority members get more dates than members of the biology club do. Notice that this type of explanation is more general, covering a wider range of experience or observation. It speaks implicitly of the relationship between variables: for example, (a) whether or not you study in a group and (b) how well you do on the exam. This type of explanation—labeled **nomothetic** —seeks to explain a class of situations or events rather than a single one. Moreover, it seeks to explain "economically," using only one or just a few explanatory factors. Finally, it settles for a partial rather than a full explanation.

In each of these examples, you might qualify your causal statements with such words or phrases as *on the whole, usually,* or *all else being equal.* Thus, you usually do better on exams when you've studied in a group, but not always. Similarly, your team has won some games on the road and lost some at home. And the attractive head of the biology club may get lots of good dates, while the homely members of sororities and fraternities spend a lot of Saturday nights alone working crossword puzzles. The existence of such exceptions is the price we pay for a broader range of overall explanation. As I noted earlier, patterns are real and important even when they are not perfect.

Both the idiographic and the nomothetic approaches to understanding can be useful in daily life. The nomothetic patterns you discover might offer a good guide for planning your study habits, for example, while the idiographic explanation might be more convincing to your parole officer. (See "Idiographic and Nomothetic Reasoning in Everyday Life.")

By the same token, both idiographic and nomothetic reasoning are powerful tools for social research. For example, A. Libin and J. Cohen-Mansfield (2000) contrast the way that the idiographic and nomothetic approaches are used in studying the elderly (gerontology). Some studies focus on the full experiences of individuals as they live their lives, whereas other studies look for statistical patterns describing the elderly in general. The authors conclude by suggesting ways to combine idiographic and nomothetic approaches in gerontology.

Social scientists, then, can access two distinct kinds of explanation. Just as physicists treat light sometimes as a particle and other times as a wave, so social scientists can search for broad relationships today and probe the narrowly particular tomorrow. Both are good science, both are rewarding, and both can be fun.

Inductive and Deductive Theory

Like idiographic and nomothetic forms of explanation, inductive and deductive thinking both play a role in our daily lives. They, too, represent an important variation within social research.

---

idiographic An approach to explanation in which we seek to exhaust the idiosyncratic causes of a particular condition or event. Imagine trying to list all the reasons why you chose to attend your particular college. Given all those reasons, it's difficult to imagine your making any other choice.

nomothetic An approach to explanation in which we seek to identify a few causal factors that generally impact a class of conditions or events. Imagine the two or three key factors that determine which colleges students choose—proximity, reputation, and so forth.

Idiographic and Nomothetic Reasoning in Everyday Life

The difference between idiographic and nomothetic explanations can be found in everyday life. Consider the following:

Idiographic: "He's like that because his father and mother kept giving him mixed signals. The fact that his family moved seven times by the time he was twelve years old didn't help. Moreover, his older brother is exactly the same and probably served as a role model."

Nomothetic: "Teenage boys are like that."

In the idiographic mode, we have a seemingly complete explanation for the behavior of the one boy in question. In the nomothetic mode, we have a simpler, more general, explanation, which wouldn't necessarily be true of all teenage boys but portrays a general pattern.

Be warned that neither explanation is necessarily true. When these models are used in social research, other elements of the inquiry—such as how subjects were chosen, how measurements were made, and so forth—strengthen the validity of conclusions drawn.

For example, there are two routes to the conclusion that you do better on exams if you study with others. On the one hand, you might find yourself puzzling, halfway through your college career, why you do so well on exams sometimes but poorly at other times. You might list all the exams you've taken, noting how well you did on each. Then you might try to recall any circumstances shared by all the good exams and by all the poor ones. Did you do better on multiple-choice exams or essay exams? Morning exams or afternoon exams? Exams in the natural sciences, the humanities, or the social sciences? Times when you studied alone or . . . SHAZAM! It occurs to you that you have almost always done best on exams when you studied with others. This mode of inquiry is known as induction.

Induction, or inductive reasoning, moves from the particular to the general, from a set of specific observations to the discovery of a pattern that represents some degree of order among all the given events. Notice, incidentally, that your

discovery doesn't necessarily tell you *why* the pattern exists—just that it does.

There is a second and very different way that you might arrive at the same conclusion about studying for exams. Imagine approaching your first set of exams in college. You wonder about the best ways to study—how much you should review the readings, how much you should focus on your class notes. You learn that some students prepare by rewriting their notes in an orderly fashion. Then you consider whether you should study at a measured pace or else pull an all-nighter just before the exam. Among these kinds of musings, you might ask whether you should get together with other students in the class or just study on your own. You could evaluate the pros and cons of both options.

Studying with others might not be as efficient, because a lot of time might be spent on things you already understand. On the other hand, you can understand something better when you've explained it to someone else. And other students might understand parts of the course that you haven't gotten yet. Several minds can reveal perspectives that might have escaped you. Also, your commitment to study with others makes it more likely that you'll study rather than watch the special *Brady Bunch* retrospective.

In this fashion, you might add up the pros and the cons and conclude, logically, that you'd benefit from studying with others. It seems reasonable to you, the way it seems reasonable that you'll do better if you study rather than not. Sometimes, we

---

induction The logical model in which general principles are developed from specific observations. Having noted that Jews and Catholics are more likely to vote Democratic than Protestants are, you might conclude that religious minorities in the United States are more affiliated with the Democratic party and then your task is to explain why. This would be an example of induction.

say things like this are true "in theory." To complete the process, we test whether they are true in practice. For a complete test, you might study alone for half your exams and study with others for the other exams. This procedure would test your logical reasoning.

This second mode of inquiry, known as **deduction** or deductive reasoning, moves from the general to the specific. It moves from (1) a pattern that might be logically or theoretically expected to (2) observations that test whether the expected pattern actually occurs. Notice that deduction begins with "why" and moves to "whether," whereas induction moves in the opposite direction.

As you'll see later in this book, these two very different approaches both serve as valid avenues for science. Each approach can stimulate the research process, prompting the researcher to take on specific questions and framing the manner in which they are addressed. Moreover, you'll see how induction and deduction work together to provide ever more powerful and complete understandings. Figure 1-6 shows how these two approaches interact in the practice of social research.

Notice, by the way, that the distinction between deductive and inductive reasoning is not necessarily linked to the distinction between nomothetic and idiographic modes of explanation. These four characterizations represent four possibilities, in everyday life as much as in social research.

For example, idiographically and deductively, you might prepare for a particular date by taking into account everything you know about the person you're dating, trying to anticipate logically how you can prepare—what type of clothing, behavior, hairstyle, oral hygiene, and so forth will likely produce a successful date. Or, idiographically and inductively, you might try to figure out what it was exactly that caused your date to call 911.

A nomothetic, deductive approach arises when you coach others on your "rules of dating," when you wisely explain why their dates will be impressed to hear them expound on the dangers of satanic messages concealed in rock and roll lyrics. When you later review your life and wonder why you didn't date more musicians, you might engage in nomothetic induction.

FIGURE 1-6

The Wheel of Science. The theory and research cycle can be compared to a relay race; although all participants do not necessarily start or stop at the same point, they share a common goal—to examine all levels of social life.

Source: Adapted from Walter Wallace, *The Logic of Science in Sociology* (New York: Aldine deGruyter, 1971). Copyright © 1971 by Walter L. Wallace. Used by permission.

We'll return to induction and deduction in Chapter 2. Let's turn now to a third broad distinction that generates rich variations in social research.

Qualitative and Quantitative Data

The distinction between quantitative and qualitative data in social research is essentially the distinction between numerical and nonnumerical data. When we say someone is intelligent, we've made a qualitative assertion. A corresponding assertion about someone less fortunately endowed would be that he or she is "unintelligent." When psychologists and others measure intelligence by IQ scores, they are attempting to quantify such qualitative assessments. For example, the psychologist might say that a person has an IQ of 120.

---

deduction The logical model in which specific expectations of hypotheses are developed on the basis of general principles. Starting from the general principle that all deans are meanies, you might anticipate that this one won't let you change courses. This anticipation would be the result of deduction.

Every observation is qualitative at the outset, whether it is our experience of someone's intelligence, the location of a pointer on a measuring scale, or a check mark entered in a questionnaire. None of these things is inherently numerical or quantitative, but converting them to a numerical form is sometimes useful. (Chapter 14 of this book will deal specifically with the quantification of data.)

Quantification often makes our observations more explicit. It also can make it easier to aggregate, compare, and summarize data. Further, it opens up the possibility of statistical analyses, ranging from simple averages to complex formulas and mathematical models.

Quantitative data, then, offer the advantages that numbers have over words as measures of some quality. On the other hand, they also carry the disadvantages that numbers have, including a potential loss in richness of meaning. For example, a social researcher might want to know whether college students aged 18–22 tend to date people older or younger than themselves. A quantitative answer to this question seems easily attained. The researcher asks a given number of college students how old each of their dates has been, calculates an average, and compares it with the age of the subject. Case closed.

Or is it? Although "age" here represents the number of years people have been alive, sometimes people use the term differently; perhaps for some "age" really means "maturity." You may date people who are younger than you but who act more maturely than others of their age and thus represent the same "age" as you. Or someone might see "age" as how young or old your dates look or maybe the degree of variation in their life experiences and worldliness. These latter meanings would be lost in the quantitative calculation of average age. Qualitative data, in short, can be richer in meaning than quantified data. This is implicit in the cliché, "He is older than his years." The poetic meaning of this expression would be lost in attempts to specify how much older.

On the other hand, qualitative data bring the disadvantages of purely verbal descriptions. For example, the richness of meaning I've mentioned is partly a function of ambiguity. If the expression "older than his years" meant something to you when you read it, that meaning came from your own experiences, from people you have known who might fit the description of being "older than their years" or perhaps the times you have heard others use that expression. Two things are certain: (1) You and I probably don't mean exactly the same thing, and (2) you don't know exactly what I mean, and vice versa.

I have a friend, Ray Zhang, who was responsible for communications at the 1989 freedom demonstrations in Tiananmen Square, Beijing. Following the Army clampdown, Ray fled south, was arrested, and was then released with orders to return to Beijing. Instead, he escaped from China and made his way to Paris. Eventually he came to the United States, where he resumed the graduate studies he had been forced to abandon in fleeing his homeland. I have seen him deal with the difficulties of getting enrolled in school without any transcripts from China, of studying in a foreign language, of meeting his financial needs—all on his own, thousands of miles from his family. Ray still speaks of one day returning to China to build a system of democracy.

Ray strikes me as someone "older than his years." You probably agree. The additional detail in my qualitative description, while it fleshes out the meaning of the phrase, still does not equip us to say how much older or even to compare two people in these terms without the risk of disagreeing as to which one is more "worldly."

It might be possible to quantify this concept, however. For example, we might establish a list of life experiences that would contribute to what we mean by worldliness, for example:

Getting married
Getting divorced
Having a parent die
Seeing a murder committed
Being arrested
Being exiled

Being fired from a job

Running away with the circus

We might quantify people's worldliness as the number of such experiences they've had: the more such experiences, the more worldly we'd say they were. If we thought of some experiences as more powerful than others, we could give those experiences more points. Once we had made our list and point system, scoring people and comparing their worldliness on a numerical scale would be straightforward. We would have no difficulty agreeing on who had more points than who.

To quantify a nonnumerical concept like worldliness, then, we need to be explicit about what the concept means. By focusing specifically on what we'll include in our measurement of the concept, however, we also exclude any other meanings. Inevitably, then, we face a trade-off: Any explicated, quantitative measure will be less rich in meaning than the corresponding qualitative description.

What a dilemma! Which approach should we choose? Which is better? Which is more appropriate to social research?

The good news is that we don't need to choose. In fact, we shouldn't. Both qualitative and quantitative methods are useful and legitimate in social research. Some research situations and topics are amenable to qualitative examination, others to quantification.

Although researchers may use both, these two approaches call for different skills and procedures. As a result, you may find that you feel more comfortable with—and become more adept in—one or the other. You will be a stronger researcher, however, to the extent that you can use both approaches effectively. Certainly, all researchers, whatever their personal inclinations, should recognize the legitimacy of both.

You may have noticed that the qualitative approach seems more aligned with idiographic explanations, while nomothetic explanations are more easily achieved through quantification. Although this is true, these relationships are not absolute. Moreover, both approaches present considerable "gray area." Recognizing the distinc-

tion between qualitative and quantitative research doesn't mean that you must identify your research activities with one to the exclusion of the other. A complete understanding of a topic often requires both techniques.

The contributions of these two approaches are widely recognized today. For example, when Stuart J. H. Biddle and his colleagues (2001) at the University of Wales set out to review the status of research in the field of sport and exercise psychology, they were careful to examine the uses of both quantitative and qualitative techniques, drawing attention to those they felt were underused.

The apparent conflict btween these two fundamental approaches has been neatly summarized by Paul Thompson (2004: 238–39):

> Only a few sociologists would openly deny the logic of combining the strengths of both quantitative and qualitative methods in social research. . . . In practice, however, despite such wider methodological aspirations in principle, social researchers have regrettably become increasingly divided into two camps, many of whose members know little of each other even if they are not explicitly hostile.

In reviewing the frequent disputes over the superiority of qualitative or quantitative methods, Anthony Onwuegbuzie and Nancy Leech (2005) suggest that the two approaches have more similarities than differences, and they urge that social research is strengthened by the use of both. My intention in this book is to focus on the complementarity of these two approaches rather than on any apparent competition between them.

Pure and Applied Research

From the beginning, social scientists have shown two distinct motivations: understanding and application. On the one hand, they are fascinated by the nature of human social life and are driven to explain it, to make sense out of apparent chaos. *Pure* research in all scientific fields is sometimes justified in terms of gaining "knowledge for knowledge's sake."

At the same time, perhaps inspired by their subject matter, social scientists are committed to having what they learn make a difference, to seeing their knowledge of society put into action. Sometimes they focus on making things better. When I study prejudice, for example, I'd like what I discover to result in a more tolerant society. This is no different from the AIDS researcher trying to defeat that disease. In Chapter 12, we'll focus on a special kind of applied research called evaluation research.

For some social scientists, professional activities are intimately interwoven with the intention of creating a more humane society. Today, there is no better role model than the Egyptian sociologist Saad Eddin Ibrahim. Having addressed a great many social issues, Ibrahim has focused most recently on the problems of modern Arab societies in general and Egypt in particular. After years of researching and writing on the edge of political tolerance, Ibrahim crossed the line in 2000.

> Following the publication of one of my articles on Arab presidents grooming their sons to succeed them in the North Korean tradition of the late Kim Il Sung, the old guard seemed to have gotten a green light to come after me. The day after the article appeared on Cairo newsstands—June 30, 2000—I was arrested.
>
> *(2003: 71)*

Ibrahim provides a good example of how social scientists deal with something like imprisonment, which is, after all, an all-too-common part of modern social life.

> In those initial 45 days, my human contacts in prison were limited to prison wardens and guards. I had little opportunity to do as much sociological research on the prison community as I would have liked. That would have to wait for the second and third rounds of my imprisonment which followed in 2001 and 2002.
>
> *(2003: 69)*

One of the charges brought against Ibrahim was Article 80D of the penal code, which prohib-

its "spreading false rumors and tarnishing Egypt's image abroad." A more serious charge was that he had accepted financial contributions from abroad without government permission, a violation of Military Order No. 4 of 1992. As Ibrahim was to learn, his research institute's acceptance of research grants—usually a valued achievement—was regarded as a federal crime in his case. As Ibrahim observes,

> Being an activist sociologist in a Third World country is tremendously challenging. While some elements of the work are gratifying, it is more often permeated with agony. One honestly never knows when one is breaking a law, violating a military order or simply stepping over an invisible red line.
>
> *(2003: 70)*

Eventually, because of his own efforts and the international uproar produced by his arrest and imprisonment, Ibrahim was given a new trial and was finally released from prison on his 64th birthday: December 3, 2002. (You can learn more about Ibrahim's experience at the link listed on this book's website: http://www.cengage.com/sociology/babbie.) Social researchers put their research into practice in many mundane ways as well. Experiments and surveys, for example, can be used in marketing products. In-depth interviewing techniques can be especially useful in social work encounters. Chapter 12 of this book deals with evaluation research, by which social scientists determine the effectiveness of social interventions.

Sometimes, seemingly mundane research efforts can powerfully affect people's lives. Imagine working alongside Crystal Eastman, an applied sociologist and settlement worker active in the Pittsburgh area in the early twentieth century:

> We got permission to use these [coroner's records] and made a record of every industrial fatality reported to the coroner during the twelve months from July 1906 to July 1907, taking down on a separate card for each case, the name and address of the man killed, his

age, occupation and conjugal condition, the name of his employer, the circumstances of the accident, the names of important witnesses, and the verdict. The plan was to learn from the evidence in the coroner's record, how each accident happened, and to learn from visiting family what happened after the accident, [for example,] how great a financial loss was suffered by the family of the workman killed, how much of this was made up by compensation received from the employer, and how the family was affected in its economic life by the accident. When we had done this with the fatalities, we followed the same course with the records of three months' industrial injuries which we secured from the hospitals.

(Eastman 1910: 789; quoted in Lengermann and Niebrugge-Brantley 2002: 13)

As a result of this and similar studies, U.S. workers now enjoy the protections of worker's compensation insurance.

As with each of the other dialectics just discussed, some social scientists are more inclined toward pure research, others toward application. Ultimately, both orientations are valid and vital elements in social research as a whole. In dealing with the basics of social research, whether pure or applied, one of the intentions of this book is to draw attention to the ways in which such research is used to make a difference.

These, then, are some of the foundations of social research. I hope this discussion has helped to show how social science is anything but routine or boring. At its best, it is a vibrant, exciting, and important activity. All we need is an open mind and a sense of adventure.

The Research Proposal

I conclude this chapter by introducing a feature that will run throughout the book: the preparation of a research proposal. Most organized research begins with a description of what is planned in the project: what questions it will raise and how it will answer them. Often, such proposals are created for the purpose of getting the resources needed to conduct the research envisioned.

One way to learn the topics of this course is to write a research proposal based on what you have learned. Even if you will not actually conduct a major research project, you can lay out a plan for doing so. Your instructor may use this as a course requirement, but even if that's not the case, you can use the "Proposing Social Research" exercise at the end of each chapter to test your mastery of the chapter.

There is a computer program, SAGrader, that is designed to assist you in writing exercises such as this one. It will accept a draft submission and critique it, pointing to elements that are missing, for example. You can learn more about SAGrader at the website listed at http://www.cengage.com/sociology/babbie.

There are many organizational structures for research proposals, and I've created a fairly typical one for you to use with this book. I've presented the proposal outline as follows, indicating which chapters in the book deal most directly with each topic.

Introduction (Chapter 1)

Review of the Literature (Chapters 2, 17; Appendix A)

Specifying the Problem/Question/Topic (Chapters 5, 6, 12)

Research Design (Chapter 4)

   Data-Collection Method (Chapters 4, 8, 9, 10, 11)

   Selection of Subjects (Chapter 7)

   Ethical Issues (Chapter 3)

Data Analysis (Chapters 13, 14, 15, 16)

Bibliography (Chapter 17; Appendix A)

I'll have more to say about each of these topics as we move through the book, beginning with this chapter's "Proposing Social Research" exercise. Chapter 4 will have an extended section on the research proposal, and Chapter 17 will give you an opportunity to pull together all the parts of the proposal into a coherent whole.

Introduction

- The subject of this book is how we find out about social reality.

Looking for Reality

- Inquiry is a natural human activity. Much of ordinary human inquiry seeks to explain events and predict future events.

- When we understand through direct experience, we make observations and seek patterns of regularities in what we observe.

- Much of what we know, we know by agreement rather than by experience. In particular, two important sources of agreed-on knowledge are tradition and authority. However, these useful sources of knowledge can also lead us astray.

- Science seeks to protect against the mistakes we make in day-to-day inquiry.

- Whereas we often observe inaccurately, researchers seek to avoid such errors by making observation a careful and deliberate activity.

- We sometimes jump to general conclusions on the basis of only a few observations, so scientists seek to avoid overgeneralization. They do this by committing themselves to a sufficient number of observations and by replicating studies.

- In everyday life we sometimes reason illogically. Researchers seek to avoid illogical reasoning by being as careful and deliberate in their reasoning as in their observations. Moreover, the public nature of science means that others are always there to challenge faulty reasoning.

- Three views of "reality" are the premodern, modern, and postmodern views. In the postmodern view, there is no "objective" reality independent of our subjective experiences. Different philosophical views suggest a range of possibilities for scientific research.

The Foundations of Social Science

- Social theory attempts to discuss and explain what is, not what should be. Theory should not be confused with philosophy or belief.

- Social science looks for regularities in social life.

- Social scientists are interested in explaining human aggregates, not individuals.

- Theories are written in the language of variables.

- A variable is a logical set of attributes. An attribute is a characteristic. *Gender,* for example, is a variable made up of the attributes *male* and *female.*

- In causal explanation, the presumed cause is the independent variable, and the affected variable is the dependent variable.

The Purposes of Social Research

- Three major purposes of social research are exploration, description, and explanation.

- Studies may aim to serve more than one of these purposes.

The Ethics of Human Inquiry

- It is important to recognize from start that ethical issues, particularly with reference to protecting subjects, may rule out certain research procedures and/or require certain elements in the research design.

Some Dialectics of Social Science

- Whereas idiographic explanations present specific cases fully, nomothetic explanations present a generalized understanding of many cases.

- Inductive theories reason from specific observations to general patterns. Deductive theories start from general statements and predict specific observations.

- Quantitative data are numerical; qualitative data are not. Both types of data are useful for different research purposes.

- Both pure and applied research are valid and vital parts of the social science enterprise.

The Research Proposal

- Research projects often begin with the preparation of a research proposal, describing the purpose and methods of the proposed study.

- In this book, each chapter will conclude with an exercise through which you can prepare part of a research proposal, thereby testing your mastery of the topics covered.

The following terms are defined in context in the chapter and at the bottom of the page where the term is introduced, as well as in the comprehensive glossary at the back of the book.

agreement reality

attributes

deduction

dependent variable

epistemology

idiographic

independent variable

induction

methodology

nomothetic

replication

theory

variables

PROPOSING SOCIAL RESEARCH: INTRODUCTION

This first chapter has given you an overview of some of the basic variations in social research, many of which can be useful in writing the introduction of your research proposal. For this assignment, you should first identify a topic or question you might like to explore in a research project. Perhaps you would like to explore some topic relating to race, gender, or social class. Perhaps there is some aspect of college life that you think needs study.

Once you have a research topic in mind, this chapter will offer some ideas on how the research might be organized. This is only a overview of the project and should take two to four paragraphs. It will work best if you can select a topic which you'll use in each of the chapters of the book, as you address different aspects of the research process.

Here are some examples of research questions to illustrate the kind of focus your project might take.

- Do women earn less money than men and, if so, why?
- What distinguishes juvenile gangs of different ethnic groups?
- Which academic departments at your college offer the broadest degree of liberal arts training?
- Is it true, as some suggest, that the United States was established as a "Christian nation"?
- Are American military actions in the Middle East reducing the threat of terrorist attacks or increasing those threats?
- What are the major functions of the American family and how have those been changing over time?
- Are official attempts to control illegal drug use succeeding or failing?
- Do undocumented immigrants overall represent a net economic cost or benefit to the United States?

Notice that you probably hear questions like these discussed frequently, both in your own interactions and in the mass media. Most of those discussions are probably mostly based in opinions. Your opportunity in this course is to see how you might pursue such questions as a researcher, dealing with logic and facts in place of opinions.

REVIEW QUESTIONS AND EXERCISES

1. Review the common errors of human inquiry discussed in this chapter. Find a magazine or newspaper article, or perhaps a letter to the editor, that illustrates one of these errors. Discuss how a scientist would avoid it.

2. List five social variables and the attributes they comprise.

3. Go to one of the following websites, listed at www.cengage.com/sociology/babbie, and find examples of both qualitative and quantitative data.

    a. UN High Commissioner for Refugees

    b. U.S. Centers for Disease Control and Prevention

    c. National Library of Australia

4. At InfoTrac College Edition, search for "postmodernism." Write a short report discussing the various fields or disciplines to which it has been applied. Give examples.

SPSS EXERCISES

See the booklet that accompanies your text for exercises using SPSS (Statistical Package for the Social Sciences). There are exercises offered for each chapter, and you'll also find a detailed primer on using SPSS.

Online Study Resources

If your book came with an access code card, visit www.cengage.com/login to register. To purchase access, please visit www.ichapters.com.

1. Before you do your final review of the chapter, take the *CengageNOW* pretest to help identify the areas on which you should concentrate. You'll find information on this online tool, as well as instructions on how to access all of its great resources, in the front of the book.

2. As you review, take advantage of the *CengageNOW* personalized study plan, based on your quiz results. Use this study plan with its interactive exercises and other resources to master the material.

3. When you're finished with your review, take the posttest to confirm that you're ready to move on to the next chapter.

WEBSITE FOR *THE PRACTICE OF SOCIAL RESEARCH* 12TH EDITION

Go to your book's website at www.cengage.com/ sociology/babbie for tools to aid you in studying for your exams. You'll find *Tutorial Quizzes* with feedback, *Internet Exercises, Flash Cards, Glossaries,* and *Essay Quizzes,* as well as *InfoTrac College Edition* search terms, suggestions for additional reading, *Web Links,* and primers for using data-analysis software such as SPSS.

CHAPTER TWO

Paradigms, Theory, and Social Research

CHAPTER OVERVIEW

Social science inquiry is an interplay of theory and research, logic and observation, induction and deduction—and of the fundamental frames of reference known as paradigms.

CengageNOW for Sociology

Use this online tool to help you make the grade on your next exam. After reading this chapter, go to "Online Study Resources" at the end of the chapter for instructions on how to benefit from *CengageNOW*.

Introduction

Certain restaurants in the United States are fond of conducting political polls among their diners whenever an election is in the offing. Some take these polls very seriously because of their uncanny history of predicting winners. Some movie theaters have achieved similar success by offering popcorn in bags picturing either donkeys or elephants. Years ago, granaries in the Midwest offered farmers a chance to indicate their political preferences through the bags of grain they selected.

Such idiosyncratic ways of determining trends, though interesting, all follow the same pattern over time: They work for a while, and then they fail. Moreover, we can't predict when or why they will fail.

These unusual polling techniques point to a significant shortcoming of "research findings" that are based only on the observation of patterns. Unless we can offer logical explanations for such patterns, the regularities we've observed may be mere flukes, chance occurrences. If you flip coins long enough, you'll get ten heads in a row. Scientists might adapt a street expression to describe this situation: "Patterns happen."

Logical explanations are what theories seek to provide. Theories function in three ways in research. First, they prevent our being taken in by flukes. If we can't explain why Ma's Diner has so successfully predicted elections, we run the risk of supporting a fluke. If we know why it has happened, we can anticipate whether or not it will work in the future.

Second, theories make sense of observed patterns in a way that can suggest other possibilities. If we understand the reasons why broken homes produce more juvenile delinquency than intact homes do—lack of supervision, for example—we can take effective action, such as after-school youth programs.

Finally, theories shape and direct research efforts, pointing toward likely discoveries through empirical observation. If you were looking for your lost keys on a dark street, you could whip your flashlight around randomly, hoping to chance upon the errant keys—or you could use your memory of where you had been and limit your search to more likely areas. Theories, by analogy, direct researchers' flashlights where they will most likely observe interesting patterns of social life.

This is not to say that all social science research is tightly intertwined with social theory. Sometimes social scientists undertake investigations simply to discover the state of affairs, such as an evaluation of whether an innovative social program is working or a poll to determine which candidate is winning a political race. Similarly, descriptive ethnographies, such as anthropological accounts of preliterate societies, produce valuable information and insights in and of themselves. However, even studies such as these often go beyond pure description to ask "why." Theory relates directly to "why" questions.

This chapter explores some specific ways theory and research work hand in hand during the adventure of inquiry into social life. We'll begin by looking at some fundamental frames of reference, called paradigms, that underlie social theories and inquiry. Whereas theories seek to explain, paradigms provide ways of looking. In and of themselves, paradigms don't explain anything; however, they provide logical frameworks within which theories are created. As you'll see in this chapter, theories and paradigms intertwine in the search for meaning in social life.

Some Social Science Paradigms

There is usually more than one way to make sense of things. In daily life, for example, liberals and conservatives often explain the same phenomenon—teenagers using guns at school, for example—quite differently. So might the parents and teenagers themselves. But underlying these different explanations, or theories, are

paradigms—the fundamental models or frames of reference we use to organize our observations and reasoning.

Paradigms are often difficult to recognize as such, because they are so implicit, assumed, taken for granted. They seem more like "the way things are" than like one possible point of view among many. Here's an illustration of what I mean.

Where do you stand on the issue of human rights? Do you feel that individual human beings are sacred? Are they "endowed by their creator with certain inalienable rights," as asserted by the U.S. Declaration of Independence? Are there some things that no government should do to its citizens?

Let's get more concrete. In wartime, civilians are sometimes used as human shields to protect military targets. Sometimes they are impressed into slave labor or even used as mobile blood banks for military hospitals. How about organized programs of rape and murder in support of "ethnic cleansing"?

Those of us who are horrified and incensed by such practices probably find it difficult to see our individualistic paradigm as only one possible point of view among many. However, many cultures in today's world regard the Western (and particularly U.S.) commitment to the sanctity of the individual as bizarre. Historically, it has decidedly been a minority viewpoint.

Although many Asian countries, for example, now subscribe to some "rights" that belong to individuals, those are balanced against the "rights" of families, organizations, and the society at large. Criticized for violating human rights, Asian leaders often point to high crime rates and social disorganization in Western societies as the cost of what they see as our radical "cult of the individual."

I won't try to change your point of view on individual human dignity, nor have I given up my own. It's useful, however, to recognize that our views and feelings in this matter result from the paradigm we have been socialized into. The sanctity of the individual is not an objective fact of nature; it is a point of view, a paradigm. All of us operate within many such paradigms.

A traditional Western view holds that the world you experience has an objective reality separate from your experience of it. As we saw in Chapter 1, however, the postmodern paradigm suggests that only the experience is real: The book in your hands right now is not real; only your experience of it is. Whether you think the book really exists or not reflects the paradigm you operate within.

When we recognize that we are operating within a paradigm, two benefits accrue. First, we can better understand the seemingly bizarre views and actions of others who are operating from a different paradigm. Second, at times we can profit from stepping outside our paradigm. Suddenly we can see new ways of seeing and explaining things. We can't do that as long as we mistake our paradigm for reality.

Paradigms play a fundamental role in science, just as they do in daily life. Thomas Kuhn (1970) draws attention to the role of paradigms in the history of the natural sciences. Major scientific paradigms have included such fundamental viewpoints as Copernicus's conception of the earth moving around the sun (instead of the reverse), Darwin's theory of evolution, Newtonian mechanics, and Einstein's relativity. Which scientific theories "make sense" depends on which paradigm scientists are maintaining.

Although we sometimes think of science as developing gradually over time, marked by important discoveries and inventions, Kuhn says that scientific paradigms typically become entrenched, resisting substantial change. Thus, theories and research alike tend to follow a given fundamental direction. Eventually, however, as the shortcomings of a particular paradigm became obvious, a new one emerges and supplants the old. The seemingly natural view that the rest of the universe revolves around the earth, for example, compelled astronomers to devise ever more elaborate ways to account for the motions of heavenly bodies that they actually observed. Eventually this paradigm was supplanted by the view that the earth and other planets revolve around the sun. This was

---

paradigm A model or frame of reference through which to observe and understand.

nothing less than a revolutionary change in perspective, which fundamentally altered the direction of theory and research. Kuhn's classic book on this subject is entitled, appropriately enough, *The Structure of Scientific Revolutions*.

Social scientists have developed several paradigms for understanding social behavior. The fate of supplanted paradigms in the social sciences, however, has differed from what Kuhn observed in the natural sciences. Natural scientists generally believe that the succession from one paradigm to another represents progress from a false view to a true one. For example, no modern astronomer believes that the sun revolves around the earth.

In the social sciences, on the other hand, theoretical paradigms may gain or lose popularity, but they are seldom discarded altogether. The paradigms of the social sciences offer a variety of views, each of which offers insights the others lack and ignores aspects of social life that the others reveal.

Ultimately, paradigms are neither true nor false; as ways of looking, they are only more or less useful. Each of the paradigms we are about to examine offers a different way of looking at human social life. Each makes its own assumptions about the nature of social reality. As we'll see, each can open up new understandings, suggest different kinds of theories, and inspire different kinds of research.

Macrotheory and Microtheory

Let's begin with a difference concerning focus, a difference that stretches across many of the paradigms we'll discuss. Some social theorists focus their attention on society at large, or at least on large portions of it. Topics of study for such

macrotheories include the struggle between economic classes in a society, international relations, or the interrelations among major institutions in society, such as government, religion, and family. **Macrotheory** deals with large, aggregate entities of society or even whole societies. (Note that some researchers prefer to limit the macrolevel to whole societies, using the term *mesotheory* for an intermediate level between macro and micro: studying organizations, communities, and perhaps social categories such as gender.)

Some scholars have taken a more intimate view of social life. **Microtheory** deals with issues of social life at the level of individuals and small groups. Dating behavior, jury deliberations, and student–faculty interactions are apt subjects for a microtheoretical perspective. Such studies often come close to the realm of psychology, but whereas psychologists typically focus on what goes on inside humans, social scientists study what goes on between them.

The basic distinction between macro- and microtheory cuts across the other paradigms we'll examine. Some of them, such as symbolic interactionism and ethnomethodology, are often limited to the microlevel. Others, such as the conflict paradigm, can be pursued at either the micro- or the macrolevel.

Early Positivism

When the French philosopher Auguste Comte (1798–1857) coined the term *sociologie* in 1822, he launched an intellectual adventure that continues to unfold today. Most importantly, Comte identified society as a phenomenon that can be studied scientifically. (Initially, he wanted to label his enterprise *social physics*, but that term was taken over by another scholar.)

Prior to Comte's time, society simply was. To the extent that people recognized different kinds of societies or changes in society over time, religious paradigms generally predominated in explanations of such differences. People often saw the state of social affairs as a reflection of God's will. Alternatively, people were challenged to create a "City of God" on earth to replace sin and godlessness.

---

macrotheory A theory aimed at understanding the "big picture" of institutions, whole societies, and the interactions among societies. Karl Marx's examination of the class struggle is an example of macrotheory.

microtheory A theory aimed at understanding social life at the intimate level of individuals and their interactions. Examining how the play behavior of girls differs from that of boys would be an example of microtheory.

Comte separated his inquiry from religion. He felt that religious belief could be replaced with scientific study and objectivity. His "positive philosophy" postulated three stages of history. A theological stage predominated throughout the world until about 1300 C.E. During the next five hundred years, a metaphysical stage replaced God with philosophical ideas such as "nature" and "natural law."

Comte felt he was launching the third stage of history, in which science would replace religion and metaphysics by basing knowledge on observations through the five senses rather than on belief or logic alone. Comte felt that society could be observed and then explained logically and rationally and that sociology could be as scientific as biology or physics.

In a sense, all social research descends from Comte. His view that society could be studied scientifically formed the foundation for subsequent development of the social sciences. In his optimism for the future, he coined the term **positivism** to describe this scientific approach, in contrast to what he regarded as negative elements in the Enlightenment. As we'll see later in this discussion, positivism has been seriously challenged only in recent decades.

Social Darwinism

Comte's major work on his positivist philosophy was published between 1830 and 1842. One year after the publication of the first volume in that series, a young British naturalist set sail on HMS *Beagle,* beginning a cruise that would profoundly affect the way we think of ourselves and our place in the world.

In 1858, when Charles Darwin published *On the Origin of Species,* he set forth the idea of evolution through natural selection. Simply put, the theory states that as a species coped with its environment, those individuals most suited to success would be the most likely to survive long enough to reproduce. Those less well suited would perish. Over time the traits of the survivor would come to dominate the species. As later Darwinians put it,

species evolved into different forms through the "survival of the fittest."

As scholars began to study society analytically, it was perhaps inevitable that they would apply Darwin's ideas to changes in the structure of human affairs. The journey from simple hunting-and-gathering tribes to large, industrial civilizations was easily seen as the evolution of progressively "fitter" forms of society.

Among others, Herbert Spencer (1820–1903) concluded that society was getting better and better. Indeed, his native England had profited greatly from the development of industrial capitalism, and Spencer favored a system of free competition, which he felt would insure continued progress and improvement. Spencer may even have coined the phrase "the survival of the fittest." He certainly believed that this principle was a primary force shaping the nature of society. Social Darwinism or social evolution was a popular view in Spencer's time, although it was not universally accepted.

This excerpt from a social science methods textbook published in 1950 illustrates the long-term popularity of the notion that things are getting better and better.

> The use of atomic energy as an explosive offers most interesting prospects in the civil as in the military field. Atomic explosives may be used for transforming the landscape. They may be used for blasting great holes and trenches in the earth, which can be transformed into lakes and canals. In this way, it may become possible to produce lakes in the midst of deserts, and thus convert some of the worst places in the world into oases and fertile countries. It may also be possible to make the Arctic regions comfortable by providing immense and constant sources of heat. The North Pole might be converted into a holiday resort.
>
> *(Gee 1950: 339–40)*

positivism Introduced by August Comte, this philosophical system is grounded on the rational proof/disproof of scientific assertions; assumes a knowable, objective reality.

Quite aside from the widespread disenchantment with nuclear power, contemporary concerns over global warming and the threat of rising sea levels illustrate a growing consciousness that "progress" is often a two-edged sword. Clearly, most of us operate today from a different paradigm.

Conflict Paradigm

One of Spencer's contemporaries took a sharply different view of the evolution of capitalism. Karl Marx (1818–1883) suggested that social behavior could best be seen as a process of conflict: the attempt to dominate others and to avoid being dominated. Marx focused primarily on the struggle among economic classes. Specifically, he examined the way capitalism produced the oppression of workers by the owners of industry. Marx's interest in this topic did not end with analytical study; he was also ideologically committed to restructuring economic relations to end the oppression he observed.

The contrast between the views set forth by Spencer and Marx indicates the influence of paradigms on research. These fundamental viewpoints shape the kinds of observations we are likely to make, the sorts of facts we seek to discover, and the conclusions we draw from those facts. Paradigms also help determine which concepts we see as relevant and important. Whereas economic classes were essential to Marx's analysis, for example, Spencer was more interested in the relationship between individuals and society—particularly the amount of freedom individuals had to surrender for society to function.

The **conflict paradigm** proved to be fruitful outside the realm of purely economic analyses. Georg Simmel (1858–1918) was especially interested in small-scale conflict, in contrast to the class struggle that interested Marx. Simmel noted, for example, that conflicts among members of a tightly knit group tended to be more intense than those among people who did not share feelings of belonging and intimacy.

In a more recent application of the conflict paradigm, when Michel Chossudovsky's (1997) analysis of the International Monetary Fund (IMF) and World Bank suggested that these two international organizations were increasing global poverty rather than eradicating it, he directed his attention to the competing interests involved in the process. In theory, the chief interest being served should be the poor people of the world or perhaps the impoverished, Third World nations. The researcher's inquiry, however, identified many other interested parties who benefited: the commercial lending institutions who made loans in conjunction with the IMF and World Bank, as well as multinational corporations seeking cheap labor and markets for their goods, for example. Chossudovsky concluded that the interests of the banks and corporations tended to take precedence over those of the poor people. Moreover, he found that many policies were weakening national economies in the Third World, as well as undermining democratic governments.

Although the conflict paradigm often focuses on class, gender, and ethnic struggles, we could appropriately apply it whenever different groups have competing interests. For example, we could fruitfully apply it to understanding relations among different departments in an organization, fraternity and sorority rush weeks, or student–faculty–administrative relations, to name just a few.

Symbolic Interactionism

In his overall focus, Georg Simmel differed from both Spencer and Marx. Whereas they were chiefly concerned with macrotheoretical issues—large institutions and whole societies in their evolution through the course of history—Simmel was more interested in how individuals interacted with one another. In other words, his thinking and research took a "micro" turn, thus calling attention to aspects of social reality that are invisible in Marx's or Spencer's theory. For example, he began by examining dyads (groups of two people) and triads (groups of three). Similarly, he wrote about "the web of group affiliations."

---

conflict paradigm A paradigm that views human behavior as attempts to dominate others or avoid being dominated by others.

Simmel was one of the first European sociologists to influence the development of U.S. sociology. His focus on the nature of interactions particularly influenced George Herbert Mead (1863–1931), Charles Horton Cooley (1864–1929), and others who took up the cause and developed it into a powerful paradigm for research.

Cooley, for example, introduced the idea of the "primary group," those intimate associates with whom we share a sense of belonging, such as our family and friends. Cooley also wrote of the "looking-glass self" we form by looking into the reactions of people around us. If everyone treats us as beautiful, for example, we conclude that we are. Notice how fundamentally the concepts and theoretical focus inspired by this paradigm differ from the society-level concerns of Spencer and Marx.

Mead emphasized the importance of our human ability to "take the role of the other," imagining how others feel and how they might behave in certain circumstances. As we gain an idea of how people in general see things, we develop a sense of what Mead called the "generalized other."

Mead also showed a special interest in the role of communications in human affairs. Most interactions, he felt, revolved around the process of individuals reaching common understanding through the use of language and other such systems, hence the term **symbolic interactionism.**

This paradigm can lend insights into the nature of interactions in ordinary social life, but it can also help us understand unusual forms of interaction, as in the following case. Robert Emerson, Kerry Ferris, and Carol Gardner (1998) set out to understand the nature of "stalking." Through interviews with numerous stalking victims, they came to identify different motivations among stalkers, stages in the development of a stalking scenario, how people can recognize if they are being stalked, and what they can do about it.

Here's one way you might apply the symbolic interactionism paradigm to an examination of your own life. The next time you meet someone new, pay attention to how you get to know each other. To begin, what assumptions do you make about the other person based merely on appearances, how

he or she talks, and the circumstances under which you've met. ("What's someone like you doing in a place like this?") Then watch how your knowledge of each other unfolds through the process of interaction. Notice also any attempts you make to manage the image you are creating in the other person's mind.

Ethnomethodology

Whereas some social science paradigms emphasize the impact of social structure on human behavior—that is, the effect of norms, values, control agents, and so forth—other paradigms do not. Harold Garfinkel, a contemporary sociologist, claims that people are continually creating social structure through their actions and interactions—that they are, in fact, creating their realities. Thus, when you and your instructor meet to discuss your term paper, even though there are myriad expectations about how you both should act, your conversation will differ somewhat from any of those that have occurred before, and how you each act will somewhat modify your expectations in the future. That is, discussing your term paper will impact the interactions each of you have with other professors and students in the future.

Given the tentativeness of reality in this view, Garfinkel suggests that people are continuously trying to make sense of the life they experience. In a sense, he suggests that everyone is acting like a social scientist, hence the term *ethnomethodology,* or "methodology of the people."

How would you go about learning about people's expectations and how they make sense out of their world? One technique ethnomethodologists use is to break the rules, to violate people's expectations. Thus, if you try to talk to me about your term paper but I keep talking about football, this might reveal the expectations you had for my behavior. We might also see how you make sense out of my

---

symbolic interactionism A paradigm that views human behavior as the creation of meaning through social interactions, with those meanings conditioning subsequent interactions.

behavior. ("Maybe he's using football as an analogy for understanding social systems theory.")

In another example of ethnomethodology, Johen Heritage and David Greatbatch (1992) examined the role of applause in British political speeches: How did the speakers evoke applause, and what function did it serve (for example, to complete a topic)? Research within the ethnomethodological paradigm has often focused on communications.

There is no end to the opportunities you have for trying out the ethnomethodological paradigm. For instance, the next time you get on an elevator, don't face front, watching the floor numbers whip by (that's the norm, or expected behavior). Just stand quietly facing the rear. See how others react to this behavior. Just as important, notice how you feel about it. If you do this experiment a few times, you should begin to develop a feel for the ethnomethodological paradigm.*

We'll return to ethnomethodology in Chapter 10, when we discuss field research. For now, let's turn to a very different paradigm.

Structural Functionalism

Structural functionalism, sometimes also known as social systems theory, has grown out of a notion introduced by Comte and Spencer: A social entity, such as an organization or a whole society, can be viewed as an organism. Like other organisms, a social system is made up of parts, each of which contributes to the functioning of the whole.

By analogy, consider the human body. Each component—such as the heart, lungs, kidneys, skin, and brain—has a particular job to do. The body as a whole cannot survive unless each of these parts does its job, and none of the parts can survive except as a part of the whole body. Or consider an automobile. It is composed of the tires, the steering wheel, the gas tank, the spark plugs,

and so forth. Each of the parts serves a function for the whole; taken together, that system can get us across town. None of the individual parts would be very useful to us by itself, however.

The view of society as a social system, then, looks for the "functions" served by its various components. Social scientists using the structural functional paradigm might note that the function of the police, for example, is to exercise social control—encouraging people to abide by the norms of society and bringing to justice those who do not. Notice, though, that the researchers could just as reasonably ask what functions criminals serve in society. Within the functionalist paradigm, we might say that criminals serve as job security for the police. In a related observation, Emile Durkheim (1858–1917) suggested that crimes and their punishment provide an opportunity to reaffirm society's values. By catching and punishing thieves, we reaffirm our collective respect for private property.

To get a sense of the structural functional paradigm, suppose you were interested in explaining how your college or university works. You might thumb through the institution's catalog and begin assembling a list of the administrators and support staff (such as the president, deans, registrar, campus security staff, maintenance personnel). Then you might figure out what each of them does and relate their roles and activities to the chief functions of your college or university, such as teaching or research. This way of looking at an institution of higher learning would clearly suggest a different line of inquiry than, say, a conflict paradigm, which might emphasize the clash of interests between people who have power in the institution and those who don't.

People often discuss "functions" in everyday conversation. Typically, however, the alleged functions are seldom tested empirically. Some people argue, for example, that welfare, intended to help

---

structural functionalism A paradigm that divides social phenomena into parts, each of which serves a function for the operation of the whole.

*I am grateful to my colleague, Bernard McGrane, for this experiment. Barney also has his students eat dinner with their hands, watch TV without turning it on, and engage in other strangely enlightening behavior (McGrane 1994).

the poor, actually harms them in a variety of ways. It is sometimes alleged that welfare creates a deviant, violent subculture in society, at odds with the mainstream. From this viewpoint, welfare programs actually result in increased crime rates.

Lance Hannon and James Defronzo (1998) decided to test this last assertion. Working with data drawn from 406 urban counties in the United States, they examined the relationship between welfare payments and crime rates. Contrary to the beliefs of some, their data indicated that higher welfare payments were associated with lower crime rates. In other words, welfare programs have the function of decreasing rather than increasing lawlessness.

In applying the functionalist paradigm to everyday life, people sometimes make the mistake of thinking that "functionality," stability, and integration are necessarily good, or that the functionalist paradigm makes that assumption. However, when social researchers look for the functions served by poverty, racial discrimination, or the oppression of women, they are not justifying them. Just the opposite: They seek to understand the functions such things play in the larger society, as a way of understanding why they persist and how they could be eliminated.

Feminist Paradigms

When Ralph Linton concluded his anthropological classic, *The Study of Man* (1937: 490), speaking of "a store of knowledge that promises to give man a better life than any he has known," no one complained that he had left out women. Linton was using the linguistic conventions of his time; he implicitly included women in all his references to men. Or did he?

When feminists first began questioning the use of masculine pronouns and nouns whenever gender was ambiguous, their concerns were often viewed as petty, even silly. At most, many felt the issue was one of women having their feelings hurt, their egos bruised. But be honest: When you read Linton's words, what did you picture? An amorphous, genderless human being, a hermaphrodite at once male and female, or a male persona?

In a similar way, researchers looking at the social world from a **feminist paradigm** have called attention to aspects of social life that other paradigms do not reveal. In part, feminist theory and research have focused on gender differences and how they relate to the rest of social organization. These lines of inquiry have drawn attention to the oppression of women in many societies, which in turn has shed light on oppression generally.

Feminist paradigms not only reveal the treatment of women or the experience of oppression but often point to limitations in how other aspects of social life are examined and understood. Thus, feminist perspectives are often related to a concern for the environment, for example. As Greta Gard suggests,

> The way in which women and nature have been conceptualized historically in Western intellectual tradition has resulted in devaluing whatever is associated with women, emotion, animals, nature, and the body, while simultaneously elevating in value those things associated with men, reason, humans, culture, and the mind. One task of ecofeminism has been to expose these dualisms and the ways in which feminizing nature and naturalizing or animalizing women has served as justification for the domination of women, animals and the earth.
>
> *(1993: 5; quoted in Rynbrandt and Deegan 2002: 60)*

Feminist paradigms have also challenged the prevailing notions concerning consensus in society. Most descriptions of the predominant beliefs, values, and norms of a society are written by people representing only portions of society. In the United States, for example, such analyses have typically been written by middle-class white men—not surprisingly, they have written about the beliefs, values, and norms they themselves share. Though George Herbert Mead spoke of the

feminist paradigms Paradigms that (1) view and understand society through the experiences of women and/or (2) examine the generally deprived status of women in society.

"generalized other" that each of us becomes aware of and can "take the role of," feminist paradigms question whether such a generalized other even exists.

Further, whereas Mead used the example of learning to play baseball to illustrate how we learn about the generalized other, Janet Lever's research suggests that understanding the experience of boys may tell us little about girls.

> Girls' play and games are very different. They are mostly spontaneous, imaginative, and free of structure or rules. Turn-taking activities like jumprope may be played without setting explicit goals. Girls have far less experience with interpersonal competition. The style of their competition is indirect, rather than face to face, individual rather than team affiliated. Leadership roles are either missing or randomly filled.
>
> *(Lever 1986: 86)*

Social researchers' growing recognition of the general intellectual differences between men and women led the psychologist Mary Field Belenky and her colleagues to speak of *Women's Ways of Knowing* (1986). In-depth interviews with 45 women led the researchers to distinguish five perspectives on knowing that should challenge the view of inquiry as obvious and straightforward:

Silence: Some women, especially early in life, feel themselves isolated from the world of knowledge, their lives largely determined by external authorities.

Received knowledge: From this perspective, women feel themselves capable of taking in and holding knowledge originating with external authorities.

Subjective knowledge: This perspective opens up the possibility of personal, subjective knowledge, including intuition.

Procedural knowledge: Some women feel they have fully learned the ways of gaining knowledge through objective procedures.

Constructed knowledge: The authors describe this perspective as "a position in which women view all knowledge as contextual, experience themselves as creators of knowledge, and value

both subjective and objective strategies for knowing" (Belenky et al. 1986: 15).

"Constructed knowledge" is particularly interesting in the context of paradigms. The positivistic paradigm of Comte would have a place neither for "subjective knowledge" nor for the idea that truth might vary according to its context. The ethnomethodological paradigm, on the other hand, would accommodate these ideas.

Feminist standpoint theory is a term often used in reference to the fact that women have knowledge about their status and experience that is not available to men. Introduced by Nancy Hartsock (1983), this viewpoint has evolved over time. For example, scholars have come to recognize that there is no single female experience, that different kinds of women (varying by wealth, ethnicity, or age, for example) have very different experiences of life in society, all the while sharing some things in common because of their gender. This sensitivity to variations in the female experience is also a main element in what is referred to as *third-wave feminism,* which began in the 1990s.

To try out feminist paradigms, you might want to explore whether discrimination against women exists at your college or university. Are the top administrative positions held equally by men and women? How about secretarial and clerical positions? Are men's and women's sports supported equally? Read through the official history of your school; is it a history that includes men and women equally? (If you attend an all-male or all-female school, of course, some of these questions won't apply.)

As we just saw, feminist paradigms reflect both a concern for the unequal treatment of women but also an epistemological recognition that men and women overall perceive and understand society differently. Social theories created solely by men, which has been the norm, run the risk of an unrecognized bias. A similar case can be made for theories created almost exclusively by white people.

Critical Race Theory

The roots of **critical race theory** are generally associated with the civil rights movement of the mid-1950s and race-related legislation of the 1960s. By the mid-1970s, with fears that the strides toward equality were beginning to bog down, civil rights activists and social scientists began the codification of a paradigm based on race awareness and a commitment to racial justice.

This was not the first time sociologists paid attention to the status of nonwhites in U.S. society. Perhaps the best known African American sociologist in the history of the discipline was W. E. B. DuBois, who published *The Souls of Black Folk* in 1903. Among other things, DuBois pointed out that African Americans lived their lives through a "dual consciousness": as Americans and as black people. By contrast, white Americans seldom reflect on being white. If you are American, white is simply assumed. If you are not white, you are seen and feel like the exception. So imagine the difference between an African American sociologist and a white sociologist creating a theory of social identity. Their theories of identity would likely differ in some fundamental ways, even if they were not limiting their analyses to their own race.

Much of the contemporary scholarship in critical race theory has to do with the role of race in politics and government, often undertaken by legal scholars as well as social scientists. Thus, for example, Derrick Bell (1980) critiqued the Supreme Court's landmark *Brown vs. Board of Education* decision, which struck down the "separate but equal" system of school segregation. He suggested that the Court was motivated by the economic and political interests of the white majority, not by educational equality for African American students. In his analysis, he introduced the concept of **interest convergence,** suggesting that laws will only be changed to benefit African Americans if and when those changes are seen to further the interests of whites. Richard Delgado (2002) provides an excellent overview of how Bell's reasoning has been pursued by subsequent critical race theory scholars.

As a general rule, whenever you find the word *critical* in the name of a paradigm or theory, it will likely refer to a nontraditional view, one that may be at odds with the prevailing paradigms of an academic discipline and also at odds with the mainstream structure of society.

Rational Objectivity Reconsidered

We began this discussion of paradigms with Comte's assertion that society can be studied rationally and objectively. Since his time, the growth of science and technology, together with the relative decline of superstition, have put rationality more and more at the center of social life. As fundamental as rationality is to most of us, however, some contemporary scholars have raised questions about it.

For example, positivistic social scientists have sometimes erred in assuming that humans always act rationally. I'm sure your own experience offers ample evidence to the contrary. Yet many modern economic models fundamentally assume that people will make rational choices in the economic sector: They will choose the highest-paying job, pay the lowest price, and so forth. This assumption ignores the power of tradition, loyalty, image, and other factors that compete with reason and calculation in determining human behavior.

A more sophisticated positivism would assert that we can rationally understand and predict even nonrational behavior. An example is the famous Asch experiment (Asch 1958). In this experiment, a group of subjects is presented with a set of lines on a screen and asked to identify the two lines that are equal in length.

Imagine yourself a subject in such an experiment. You are sitting in the front row of a classroom in a group of six subjects. A set of lines is projected on the wall in front of you (see Figure 2-1). The experimenter asks each of you, one at a time, to identify the line to the right (A, B, or C) that matches the length of line X. The correct answer (B) is pretty obvious to you. To your surprise, however, you find that all the other subjects agree on a different answer!

The experimenter announces that all but one of the group has gotten the correct answer. Because

---

critical race theory A paradigm grounded in race awareness and an intention to achieve racial justice.

interest convergence The thesis that majority group members will only support the interests of minorities when those actions also support the interests of the majority group.

FIGURE 2-1

The Asch Experiment. Subjects in the Asch experiment have a seemingly easy task: to determine whether A, B, or C is the same length as X. But there's more here than meets the eye.

you are the only one who chose B, this amounts to saying that you've gotten it wrong. Then a new set of lines is presented, and you have the same experience. What seems to be the obviously correct answer is said by everyone else to be wrong.

As it turns out, of course, you are the only real subject in this experiment—all the others are working with the experimenter. The purpose of the experiment is to see whether you will be swayed by public pressure to go along with the incorrect answer. In his initial experiments, all of which involved young men, Asch found that a little over one-third of his subjects did just that.

Choosing an obviously wrong answer in a simple experiment is an example of nonrational behavior. But as Asch went on to show, experimenters can examine the circumstances that lead more or fewer subjects to go along with the incorrect answer. For example, in subsequent studies, Asch varied the size of one group and the number of "dissenters" who chose the "wrong" (that is, the correct) answer. Thus, it is possible to study nonrational behavior rationally and scientifically.

More radically, we can question whether social life abides by rational principles at all. In the physical sciences, developments such as chaos theory, fuzzy logic, and complexity have suggested that we may need to rethink fundamentally the orderliness of events in the physical world. Certainly the social world might be no tidier than the world of physics.

The contemporary challenge to positivism, however, goes beyond the question of whether people behave rationally. In part, the criticism of positivism challenges the idea that scientists can be as objective as the positivistic ideal assumes. Most scientists would agree that personal feelings can and do influence the problems scientists choose to study, what they choose to observe, and the conclusions they draw from their observations.

There is an even more radical critique of the ideal of objectivity. As we glimpsed in the discussions of feminism and ethnomethodology, some contemporary researchers suggest that subjectivity might actually be preferable in some situations. Let's take a moment to return to the dialectic of subjectivity and objectivity.

To begin, all our experiences are inescapably subjective. There is no way out. We can see only through our own eyes, and anything peculiar to our eyes will shape what we see. We can hear things only the way our particular ears and brain transmit and interpret sound waves. You and I, to some extent, hear and see different realities. And both of us experience quite different physical "realities" than, say, do bats. In what to us is total darkness, a bat "sees" things such as flying insects by emitting a sound we humans can't hear. The reflection of the bat's sound creates a "sound picture" precise enough for the bat to home in on the moving insect and snatch it up in its teeth. In a similar vein, scientists on the planet Xandu might develop theories of the physical world based on a sensory apparatus that we humans can't even imagine. Maybe they see X-rays or hear colors.

Despite the inescapable subjectivity of our experience, we humans seem to be wired to seek an agreement on what is really real, what is objectively so. Objectivity is a conceptual attempt to get beyond our individual views. It is ultimately a matter of communication, as you and I attempt to find a common ground in our subjective experiences. Whenever we succeed in our search, we say we are dealing with objective reality. This is the agreement reality discussed in Chapter 1.

To this point, perhaps the most significant studies in the history of social science were conducted in the 1930s by a Turkish American social psychologist, Muzafer Sherif (1935), who slyly said he wanted to study "auto-kinetic effects." To do this, he put small groups in totally darkened rooms, save for a single point of light in the center of the wall in front of the participants. Sherif explained that the light would soon begin to move about, and the subjects were to determine how far it was moving—a difficult task with nothing else visible as a gauge of length or distance.

Amazingly, each of the groups agreed on the distance the point of light moved about. Oddly, however, the different groups of subjects arrived at quite different conclusions. Strangest of all—as you may have guessed—the point of light had remained stationary. If you stare at a fixed point of light long enough it will seem to move about (Sherif's "auto-kinetic effect"). Notice, however, that each of the groups agreed on a specific delusion. The movement of the light was real to them, but it was a reality created out of nothing: a socially constructed reality.

Whereas our subjectivity is individual, then, our search for objectivity is social. This is true in all aspects of life, not just in science. While you and I prefer different foods, we must agree to some extent on what is fit to eat and what is not, or else there could be no restaurants or grocery stores. The same argument could be made regarding every other form of consumption. Without agreement reality, there could be no movies or television, no sports.

Social scientists as well have found benefits in the concept of a socially agreed-on objective reality. As people seek to impose order on their experience of life, they find it useful to pursue this goal as a collective venture. What are the causes and cures of prejudice? Working together, social researchers have uncovered some answers that hold up to intersubjective scrutiny. Whatever your subjective experience of things, for example, you can discover for yourself that as education increases, prejudice generally tends to decrease. Because each of us can discover this independently, we say that it is objectively true.

From the seventeenth century through the middle of the twentieth, however, the belief in an objective reality that was independent of individual perceptions predominated in science. For the most part, it was not simply held as a useful paradigm but held as The Truth. The term *positivism* has generally represented the belief in a logically ordered, objective reality that we can come to know better and better through science. This is the view challenged today by postmodernists and others.

Some say that the ideal of objectivity conceals as much as it reveals. As we saw earlier, in years past much of what was regarded as objectivity in Western social science was actually an agreement primarily among white, middle-class European men. Equally real experiences common to women, to ethnic minorities, to non-Western cultures, or to the poor were not necessarily represented in that reality.

Thus, early anthropologists are now criticized for often making modern, Westernized "sense" out of the beliefs and practices of nonliterate tribes around the world, sometimes by portraying their subjects as superstitious savages. We often call orally transmitted beliefs about the distant past "creation myth," whereas we speak of our own beliefs as "history." Increasingly today, there is a demand to find the native logic by which various peoples make sense out of life and to understand it on its own terms.

Ultimately, we'll never be able to distinguish completely between an objective reality and our subjective experience. We can't know whether our concepts correspond to an objective reality or are simply useful in allowing us to predict and control our environment. So desperate is our need to know what is really real, however, that both positivists and postmodernists are sometimes drawn into the belief that their own view is real and true. There is a dual irony in this. On the one hand, the positivist's belief that science precisely mirrors the objective world must ultimately be based on faith; it cannot be proved by "objective" science, because that's precisely what's at issue. And the postmodernists, who say nothing is objectively so and everything is ultimately subjective, do at least feel that that is really the way things are.

Postmodernism is often portrayed as a denial of the possibility of social science. Because this book has already expressed sympathy for some postmodern views and concerns, a word of explanation may be in order. This textbook makes no assumption about the existence or absence of an objective reality. At the same time, human beings demonstrate an extensive and robust ability to establish agreements as to what's "real." This appears in regard to rocks and trees, as well as ghosts and gods, and even more elusive ideas such as loyalty and treason. Whether something like "prejudice" really exists, research into its nature can take place, because enough people agree that prejudice does exist, and researchers can use agreed-on techniques of inquiry to study it.

Another social science paradigm, **critical realism,** suggests that we define "reality" as that which can be seen to have an effect. Since prejudice clearly has an observable effect in our lives, it must be judged "real" in terms of this point of view. This paradigm fits interestingly with a statement attributed to an early U.S. sociologist, W. I. Thomas: "If men define situations as real, they are real in their consequences."

This book will not require or even encourage you to choose among positivism, postmodernism, or any of the other paradigms discussed in this chapter. In fact, I invite you to look for value in any and all as you seek to understand the world that may or may not exist around you.

Similarly, as social researchers, we are not forced to align ourselves entirely with either positivism or postmodernism. Instead, we can treat them as two distinct arrows in our quiver. Each approach compensates for the weaknesses of the other by suggesting complementary perspectives that can produce useful lines of inquiry.

For example, the renowned British physicist Stephen Hawking has elegantly described the appealing simplicity of the positivistic model but tempers his remarks with a recognition of the way science is practiced.

---

postmodernism A paradigm that questions the assumptions of positivism and theories describing an "objective" reality.

critical realism A paradigm that holds things are real insofar as they produce effects.

According to this way of thinking, a scientific theory is a mathematical model that describes and codifies the observations we make. A good theory will describe a large range of phenomena on the basis of a few simple postulates and will make definite predictions that can be tested. If the predictions agree with the observations, the theory survives that test, though it can never be proved to be correct. On the other hand, if the observations disagree with the predictions, one has to discard or modify the theory. (At least, that is what is supposed to happen. In practice, people often question the accuracy of the observations and the reliability and moral character of those making the observations.)

(2001: 31)

In summary, a rich variety of theoretical paradigms can be brought to bear on the study of social life. With each of these fundamental frames of reference, useful theories can be constructed. We turn now to some of the issues involved in theory construction, which are of interest and use to all social researchers, from positivists to postmodernists—and all those in between.

Elements of Social Theory

As we have seen, paradigms are general frameworks or viewpoints: literally "points from which to view." They provide ways of looking at life and are grounded in sets of assumptions about the nature of reality.

Theories, by contrast, are systematic sets of interrelated statements intended to explain some aspect of social life. Thus, theories flesh out and specify paradigms. Whereas a paradigm offers a way of looking, a theory aims at explaining what we see. Recall from Chapter 1 that social scientists engage in both idiographic and nomothetic explanations. Idiographic explanations seek to explain a limited phenomenon as completely as possible—explaining why a particular woman voted as she did, for example—while nomothetic explanations attempt to explain a broad range of phenomena at least partially: identifying a few factors that account for much voting behavior in general.

Let's look a little more deliberately now at some of the elements of a theory. As I mentioned in Chapter 1, science is based on observation. In social research, *observation* typically refers to seeing, hearing, and (less commonly) touching. A corresponding idea is *fact*. Although for philosophers "fact" is as complex a notion as "reality," social scientists generally use it to refer to some phenomenon that has been observed. It is a fact, for example, that Barack Obama defeated John McCain in the 2008 presidential election.

Scientists aspire to organize many facts under "rules" called *laws*. Abraham Kaplan (1964: 91) defines *laws* as universal generalizations about classes of facts. The law of gravity is a classic example: Bodies are attracted to each other in proportion to their masses and in inverse proportion to the distance separating them.

Laws must be truly universal, however, not merely accidental patterns found among a specific set of facts. It is a fact, Kaplan points out (1964: 92), that in each of the U.S. presidential elections from 1920 to 1960, the major candidate with the longest name won. That is not a law, however, as shown by the next three elections. The earlier pattern was a coincidence.

Sometimes called principles, laws are important statements about what is so. We speak of them as being "discovered," granting, of course, that our paradigms affect what we choose to look for and what we see. Laws in and of themselves do not explain anything. They just summarize the way things are. Explanation is a function of theory, as we'll see shortly.

There are no social science laws that claim the universal certainty of those of the natural sciences. Social scientists debate among themselves whether such laws will ever be discovered. Perhaps social life essentially does not abide by invariant laws. This does not mean that social life is so chaotic as to defy prediction and explanation. As we saw in Chapter 1, social behavior falls into patterns, and those patterns quite often make perfect sense, although we may have to look below the surface to find the logic.

As I just indicated, laws should not be confused with theories. Whereas a law is an observed regularity, a *theory* is a systematic explanation for observations that relate to a particular aspect of life. For example, someone might offer a theory of juvenile delinquency, prejudice, or political revolution.

Theories explain observations by means of concepts. Jonathan Turner (1989: 5) calls concepts the "basic building blocks of theory." *Concepts* are abstract elements representing classes of phenomena within the field of study. The concepts relevant to a theory of juvenile delinquency, for example, include "juvenile" and "delinquency," for starters. A "peer group"—the people you hang around with and identify with—is another relevant concept. "Social class" and "ethnicity" are undoubtedly relevant concepts in a theory of juvenile delinquency. "School performance" might also be relevant.

A *variable* is a special kind of concept. Some of the concepts just mentioned refer to things, and others refer to sets of things. As we saw in Chapter 1, each variable comprises a set of attributes; thus, *delinquency*, in the simplest case, is made up of *delinquent* and *not delinquent*. A theory of delinquency would aim at explaining why some juveniles are delinquent and others are not.

Axioms or *postulates* are fundamental assertions, taken to be true, on which a theory is grounded. In a theory of juvenile delinquency, we might begin with axioms such as "Everyone desires material comforts" and "The ability to obtain material comforts legally is greater for the wealthy than for the poor." From these we might proceed to *propositions:* specific conclusions, derived from the axiomatic groundwork, about the relationships among concepts. From our beginning axioms about juvenile delinquency, for example, we might reasonably formulate the proposition that poor youths are more likely to break the law to gain material comforts than are rich youths.

This proposition, incidentally, accords with Robert Merton's classic attempt to account for deviance in society. Merton (1957: 139–57) spoke of the agreed-on means and ends of a society. In Merton's model, nondeviants are those who share the societal agreement as to desired ends (such as a new car) and the means prescribed for achieving them (such as to buy it). One type of deviant—Merton called this type the "innovator"—agrees on the desired end but does not have access to the prescribed means for achieving it. Innovators find another method, such as crime, of getting the desired end.

From propositions, in turn, we can derive hypotheses. A **hypothesis** is a specified testable expectation about empirical reality that follows from a more general proposition. Thus, a researcher might formulate the hypothesis, "Poor youths have higher delinquency rates than do rich youths." Research is designed to test hypotheses. In other words, research will support (or fail to support) a theory only indirectly—by testing specific hypotheses that are derived from theories and propositions.

Let's look more clearly at how theory and research come together.

Two Logical Systems Revisited

In Chapter 1, I introduced deductive and inductive reasoning, with a promise that we would return to them later. It's later.

The Traditional Model of Science

Most of us have a somewhat idealized picture of "the scientific method," a view gained from science instruction ever since elementary school, especially in the physical sciences. Although this traditional model of science tells only a part of the story, it's helpful to understand its logic.

---

hypothesis A specified testable expectation about empirical reality that follows from a more general proposition; more generally, an expectation about the nature of things derived from a theory. It is a statement of something that ought to be observed in the real world if the theory is correct.

operationalization One step beyond conceptualization. Operationalization is the process of developing operational definitions, or specifying the exact operations involved in measuring a variable.

There are three main elements in the traditional model of science: theory, operationalization, and observation. At this point we're already well acquainted with the idea of theory.

Theory

According to the traditional model of science, scientists begin with a thing, from which they derive testable hypotheses. So, for example, as social scientists we might have a theory about the causes of juvenile delinquency. Let's assume that we have arrived at the hypothesis that delinquency is inversely related to social class. That is, as social class goes up, delinquency goes down.

Operationalization

To test any hypothesis, we must specify the meanings of all the variables involved in it, in observational terms. In the present case, the variables are *social class* and *delinquency*. To give these terms specific meaning, we might define delinquency as "being arrested for a crime," "being convicted of a crime," or some other plausible phrase, whereas social class might be specified in terms of family income, for the purposes of this particular study.

Once we have defined our variables, we need to specify how we'll measure them. (Recall from Chapter 1 that science, in the classical ideal, depends on measurable observations.) **Operationalization** literally means specifying the exact operations involved in measuring a variable. There are many ways we can attempt to test our hypothesis, each of which allows for different ways of measuring our variables.

For simplicity, let's assume we're planning to conduct a survey of high school students. We might operationalize delinquency in the form of the question "Have you ever stolen anything?" Those who answer "yes" will be classified as delinquents in our study; those who say "no" will be classified as nondelinquents. Similarly, we might operationalize social class by asking respondents, "What was your family's income last year?" and providing them with a set of family income categories: under $10,000; $10,000–$24,999; $25,000–$49,999; and $50,000 and above.

At this point someone might object that delinquency can mean something more than or different from having stolen something at one time or another, or that social class isn't necessarily the same as family income. Some parents might think body piercing is a sign of delinquency even if their children don't steal, and to some social class might include an element of prestige or community standing as well as how much money a family has. For the researcher testing a hypothesis, however, the meaning of variables is exactly and only what the operational definition specifies.

In this respect, scientists are very much like Humpty Dumpty in Lewis Carroll's *Through the Looking Glass.* "When *I* use a word," Humpty Dumpty tells Alice, "it means just what I choose it to mean—neither more nor less."

"The question is," Alice replies, "whether you *can* make words mean so many different things." To which Humpty Dumpty responds, "The question is, which is to be master—that's all."

Scientists have to be "masters" of their **operational definitions** for the sake of precision in observation, measurement, and communication. Otherwise, we would never know whether a study that contradicted ours did so only because it used a different set of procedures to measure one of the variables and thus changed the meaning of the hypothesis being tested. Of course, this also means that to evaluate a study's conclusions about juvenile delinquency and social class, or any other variables, we need to know how those variables were operationalized.

The way we have operationalized the variables in our imaginary study could be open to other problems, however. Perhaps some respondents will lie about having stolen anything, in which cases we'll misclassify them as nondelinquent. Some respondents will not know their family incomes and will give mistaken answers; others may be embarrassed and lie. We'll consider issues like these in detail in Part 2.

Our operationalized hypothesis now is that the highest incidence of delinquents will be found among respondents who select the lowest family income category (under $10,000); a lower percentage of delinquents will be found in the $10,000–

$24,999 category; still fewer delinquents will be found in the $25,000–$49,999 category; and the lowest percentage of delinquents will be found in the $50,000-and-above category. Now we're ready for the final step in the traditional model of science—observation. Having developed theoretical clarity and specific expectations, and having created a strategy for looking, all that remains is to look at the way things actually are.

Observation

The final step in the traditional model of science involves actual observation, looking at the world and making measurements of what is seen.

Let's suppose our survey produced the following data:

	Percent delinquent
Under $10,000	20
$10,000–$24,999	15
$25,000–$49,999	10
$50,000 and above	5

Observations producing such data would confirm our hypothesis. But suppose our findings were as follows:

	Percent delinquent
Under $10,000	15
$10,000–$24,999	15
$25,000–$49,999	15
$50,000 and above	15

These findings would disconfirm our hypothesis regarding family income and delinquency. *Disconfirmability*—the possibility that observations may not support our expectations—is an essential quality of any hypothesis. In other words, if

---

operational definition The concrete and specific definition of something in terms of the operations by which observations are to be categorized. The operational definition of "earning an A in this course" might be "correctly answering at least 90 percent of the final exam questions."

there is no chance that our hypothesis will be disconfirmed, it hasn't said anything meaningful.

For example, the hypothesis that juvenile delinquents commit more crimes than do non-delinquents cannot possibly be disconfirmed, because criminal behavior is intrinsic to the idea of delinquency. Even if we recognize that some young people commit crimes without being caught and labeled as delinquents, they couldn't threaten our hypothesis, because our actual observations would lead us to conclude they were law-abiding nondelinquents.

Figure 2-2 provides a schematic diagram of the traditional model of scientific inquiry. In it we see the researcher beginning with an interest in a phenomenon (such as juvenile delinquency). Next comes the development of a theoretical under-standing, in this case that a single concept (such as social class) might explain others. The theoretical considerations result in an expectation about what should be observed if the theory is correct. The notation $X = f(Y)$ is a conventional way of saying that X (for example, delinquency) is a function of (depends on) Y (for example, social class). At that level, however, X and Y still have rather general meanings that could give rise to quite different observations and measurements. Operationalization specifies the procedures that will be used to mea-sure the variables. The lowercase x in Figure 2-2, for example, is a precisely measurable indicator of capital X. This operationalization process results in the formation of a testable hypothesis: for example, self-reported theft is a function of family income. Observations aimed at finding out whether this statement accurately describes reality are part of what is typically called *hypothesis testing*. (See "Hints for Stating Hypotheses" for more on the process of formulating hypotheses.)

Deductive and Inductive Reasoning: A Case Illustration

As you probably recognized, the traditional model of science just described is a nice example of deductive reasoning: From a general theoretical understanding, the researcher derives (deduces) an expectation and finally a testable hypothesis. This

THEORETICAL UNDERSTANDING

Y causes *X*

HYPOTHESIS

$X = f(Y)$ — Theoretical expectation

— Operationalization

$x = f(y)$ — Testable hypothesis

$x \stackrel{?}{=} f(y)$ — Observation (hypothesis testing)

FIGURE 2-2

The Traditional Image of Science. The deductive model of scientific inquiry begins with a sometimes vague or general question, which is subjected to a process of specification, resulting in hypotheses that can be tested through empirical observations.

picture is tidy, but in reality science uses inductive reasoning as well. Let's consider a real research ex-ample as a vehicle for comparing the deductive and inductive linkages between theory and research. Years ago, Charles Glock, Benjamin Ringer, and I (1967) set out to discover what caused differing levels of church involvement among U.S. Episco-palians. Several theoretical or quasi-theoretical positions suggested possible answers. I'll focus on only one here: what we came to call the "Comfort Hypothesis."

In part, we took our lead from the Christian injunction to care for "the halt, the lame, and the blind" and those who are "weary and heavy laden." At the same time, ironically, we noted the Marxist assertion that religion is an "opiate for the masses." Given both, it made sense to expect the following, which was our hypothesis: "Parishioners whose life situations most deprive them of satisfac-tion and fulfillment in the secular society turn to

Hints for Stating Hypotheses

Riley E. Dunlap

Department of Sociology,
Oklahoma State University

A hypothesis is the basic statement that is tested in research. Typically a hypothesis states a relationship between two variables. (Although it is possible to use more than two variables, you should stick to two for now.) Because a hypothesis makes a prediction about the relationship between the two variables, it must be testable so you can determine if the prediction is right or wrong when you examine the results obtained in your study. A hypothesis must be stated in an unambiguous manner to be clearly testable. What follows are suggestions for developing testable hypotheses.

Assume you have an interest in trying to predict some phenomenon such as "attitudes toward women's liberation," and that you can measure such attitudes on a continuum ranging from "opposed to women's liberation" to "neutral" to "supportive of women's liberation." Also assume that, lacking a theory, you'll rely on "hunches" to come up with variables that might be related to attitudes toward women's liberation.

In a sense, you can think of hypothesis construction as a case of filling in the blank: "_____ is related to attitudes toward women's liberation." Your job is to think of a variable that might plausibly be related to such attitudes, and then to word a hypothesis that states a relationship between the two variables (the one that fills in the "blank" and "attitudes toward women's liberation"). You need to do so in a precise manner so that you can determine clearly whether the hypothesis is supported or not when you examine the results (in this case, most likely the results of a survey).

The key is to word the hypothesis carefully so that the prediction it makes is quite clear to you as well as others. If you use age, note that saying "Age is related to attitudes toward women's liberation" does not say precisely how you think the two are related (in fact, the only way this hypothesis could be falsified is if you fail to find a statistically significant relationship of any type between age and attitudes toward women's liberation). In this case a couple of steps are necessary. You have two options:

1. "Age is related to attitudes toward women's liberation, with younger adults being more supportive than older adults." (Or, you could state the opposite, if you believed older people are likely to be more supportive.)

2. "Age is negatively related to support for women's liberation." Note here that I specify "support" for women's liberation (SWL) and then predict a negative relationship—that is, as age goes up, I predict that SWL will go down.

In this hypothesis, note that both of the variables (*age*, the independent variable or likely "cause," and *SWL*, the dependent variable or likely "effect") range from low to high. This feature of the two variables is what allows you to use "negatively" (or "positively") to describe the relationship.

Notice what happens if you hypothesize a relationship between gender and SWL. Because *gender* is a nominal variable (as you'll learn in Chapter 5) it does not range from low to high—people are either male or female (the two attributes of the variable *gender*). Consequently, you must be careful in stating the hypothesis unambiguously:

1. "Gender is positively (or negatively) related to SWL" is not an adequate hypothesis, because it doesn't specify how you expect gender to be related to SWL—that is, whether you think men or women will be more supportive of women's liberation.

2. It's tempting to say something like "Women are positively related to SWL," but this really doesn't work, because *female* is only an attribute, not a full variable (*gender* is the variable).

3. "Gender is related to SWL, with women being more supportive than men" would be my recommendation. Or, you could say, "with men being less supportive than women," which makes the identical prediction. (Of course, you could also make the opposite prediction, that men are more supportive than women are, if you wished.)

4. Equally legitimate would be "Women are more likely to support women's liberation than are men." (Note the need for the second "are," or you could be construed as hypothesizing that women support women's liberation more than they support men—not quite the same idea.)

The above examples hypothesized relationships between a "characteristic" (age or gender) and an "orientation" (attitudes toward women's liberation). Because the causal order is pretty clear (obviously age and gender come before attitudes, and are less alterable), we could state the hypotheses as I've done, and everyone would assume that we were stating causal hypotheses.

Finally, you may run across references to the **null hypothesis,** especially in statistics. Such a hypothesis predicts no relationship (technically, no statistically significant relationship) between the two variables, and it is always implicit in testing hypotheses. Basically, if you have hypothesized a positive (or negative) relationship, you are hoping that the results will allow you to reject the null hypothesis and verify your hypothesized relationship.

the church for comfort and substitute rewards" (Glock, Ringer, and Babbie 1967: 107–8).

Having framed this general hypothesis, we set about testing it. Were those deprived of satisfaction in the secular society in fact more religious than those who received more satisfaction from the secular society? To answer this, we needed to distinguish who was deprived. The questionnaire, which was constructed for the purpose of testing the Comfort Hypothesis, included items that seemed to offer indicators of whether parishioners were relatively deprived or gratified in secular society.

To start, we reasoned that men enjoy more status than women do in our generally male-dominated society. Though hardly novel, this conclusion laid the groundwork for testing the Comfort Hypothesis. If we were correct in our hypothesis, women should appear more religious than men. Once the survey data had been collected and analyzed, our expectation about gender and religion was clearly confirmed. On three separate measures of religious involvement—ritual (such as church attendance), organizational (such as belonging to church organizations), and intellectual (such as reading church publications)—women were more religious than men. On our overall measure, women scored 50 percent higher than men.

In another test of the Comfort Hypothesis, we reasoned that in a youth-oriented society, old people would be more deprived of secular gratification than the young would. Once again, the data confirmed our expectation. The oldest parishioners were more religious than the middle-aged, who were more religious than young adults.

Social class—measured by education and income—afforded another test of the Comfort Hypothesis. Once again, the test succeeded. Those with low social status were more involved in the church than those with high social status were.

null hypothesis In connection with hypothesis testing and tests of statistical significance, that hypothesis that suggests there is no relationship among the variables under study. You may conclude that the variables are related after having statistically rejected the null hypothesis.

The hypothesis was even confirmed in a test that went against everyone's commonsense expectations. Despite church posters showing worshipful young families and bearing the slogan "The Family That Prays Together Stays Together," the Comfort Hypothesis suggested that parishioners who were married and had children—the clear American ideal at that time—would enjoy secular gratification in that regard. As a consequence, they should be less religious than those who lacked one or both family components. Thus, we hypothesized that parishioners who were both single and childless should be the most religious; those with either spouse or child should be somewhat less religious; and those married with children—representing the ideal pictured on all those posters—should be the least religious of all. That's exactly what we found.

Finally, the Comfort Hypothesis suggested that the various kinds of secular deprivation should be cumulative: Those with all the characteristics associated with deprivation should be the most religious; those with none should be the least. When we combined the four individual measures of deprivation into a composite measure, the theoretical expectation was exactly confirmed. Comparing the two extremes, we found that single, childless, elderly, lower-class female parishioners scored more than three times as high on the measure of church involvement than did young, married, upper-class fathers. Thus was the Comfort Hypothesis confirmed.

I like this research example because it so clearly illustrates the logic of the deductive model. Beginning with general, theoretical expectations about the impact of social deprivation on church involvement, one could derive concrete hypotheses linking specific measurable variables, such as *age* and *church attendance*. The actual empirical data could then be analyzed to determine whether empirical reality supported the deductive expectations.

I say this example shows how it was possible to do it that way, but, alas, I've been fibbing. To tell the truth, although we began with an interest in discovering what caused variations in church involvement among Episcopalians, we didn't actually begin with a Comfort Hypothesis, or any other hypothesis for that matter. (In the interest of further honesty, Glock and Ringer initiated the study, and I joined it years after the data had been collected.) A questionnaire was designed to collect information that might shed some light on why some parishioners participated in the church more than others, but the construction of the questionnaire was not guided by any precise, deductive theory. Once the data were collected, the task of explaining differences in religiosity began with an analysis of variables that have a wide impact on people's lives, including *gender, age, social class,* and *family status.* Each of these four variables was found to relate strongly to church involvement, in the ways already described. Indeed, they had a cumulative effect, also already described. Rather than being good news, however, this presented a dilemma.

Glock recalls discussing his findings with colleagues over lunch at the Columbia faculty club. Once he had displayed the tables illustrating the impact of each individual variable as well as their powerful composite effect, a colleague asked, "What does it all mean, Charlie?" Glock was at a loss. Why *were* those variables so strongly related to church involvement?

That question launched a process of reasoning about what the several variables had in common, aside from their impact on religiosity. Eventually he saw that each of the four variables also reflected *differential status in the secular society.* He then had the thought that perhaps the issue of comfort was involved. Thus, the inductive process had moved from concrete observations to a general theoretical explanation.

It seems easier to lay out the steps involved in deductive than inductive research. Deductive research begins with a theory, from which we may derive hypotheses—which are then tested through observations. Inductive research begins with observations and proceeds with a search for patterns in what we have observed. In a quantitative study, we can search for correlations or relationships between variables (discussed further in Chapter 16). Thus, once a relationship has been discovered between gender and religiosity, our attention turns to figuring out logical reasons why that is so.

Most qualitative research is oriented toward the inductive rather than the deductive approach. However, qualitative research does not, by definition, allow us to use statistical tools to find correlations that point toward patterns in need of explanation (see Chapter 14). Although there are computer programs designed for recording and analyzing qualitative data, the qualitative inductive analyst needs a strong reserve of insight and reflection to tease important patterns out of a body of observations.

A Graphic Contrast

As the preceding case illustration shows, theory and research can usefully be done both inductively and deductively. Figure 2-3 shows a graphic comparison of the two approaches as applied to an inquiry into study habits and performance on exams. In both cases, we are interested in the relationship between the number of hours spent studying for an exam and the grade earned on that exam. Using the deductive method, we would begin by examining the matter logically. Doing well on an exam reflects a student's ability to recall and manipulate information. Both of these abilities should be increased by exposure to the information before the exam. In this fashion, we would arrive at a hypothesis suggesting a positive relationship between the number of hours spent studying and the grade earned on the exam. We say "positive" because we expect grades to increase as the hours of studying increase. If increased hours produced decreased grades, that would be called a negative, or inverse, relationship. The hypothesis is represented by the line in part 1(a) of Figure 2-3.

Our next step would be to make observations relevant to testing our hypothesis. The shaded area in part 1(b) of the figure represents perhaps hundreds of observations of different students, specifically, how many hours they studied and what grades they received. Finally, in part 1(c), we compare the hypothesis and the observations. Because observations in the real world seldom if ever match our expectations perfectly, we must decide whether the match is close enough to consider

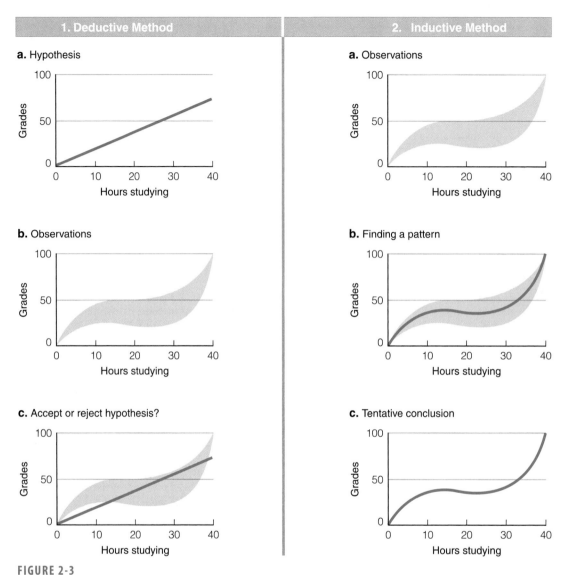

FIGURE 2-3

Deductive and Inductive Methods. Both deduction and induction are legitimate and valuable approaches to understanding. Deduction begins with an expected pattern that is tested against observations, whereas induction begins with observations and seeks to find a pattern within them.

the hypothesis confirmed. Put differently, can we conclude that the hypothesis describes the general pattern that exists, granting some variations in real life? Sometimes, answering this question necessitates methods of statistical analysis, which will be discussed in Part 4.

Now suppose we used the inductive method to address the same research question. In this case, we

would begin with a set of observations, as in part 2(a) of Figure 2-3. Curious about the relationship between hours spent studying and grades earned, we might simply arrange to collect relevant data. Then we'd look for a pattern that best represented or summarized our observations. In part 2(b) of the figure, the pattern is shown as a curved line running through the center of our observations.

The pattern found among the points in this case suggests that with 1 to 15 hours of studying, each additional hour generally produces a higher grade on the exam. With 15 to about 25 hours, however, more study seems to lower the grade slightly. Studying more than 25 hours, on the other hand, results in a return to the initial pattern: More hours produce higher grades. Using the inductive method, then, we end up with a tentative conclusion about the pattern of the relationship between the two variables. The conclusion is tentative because the observations we have made cannot be taken as a test of the pattern—those observations are the source of the pattern we've created.

As I discussed in Chapter 1, in actual practice, theory and research interact through a never-ending alternation of deduction and induction. A good example is the classic work of Emile Durkheim on suicide ([1897] 1951). When Durkheim pored over table after table of official statistics on suicide rates in different areas, he was struck by the fact that Protestant countries consistently had higher suicide rates than Catholic ones did. Why should that be the case? His initial observations led him to create inductively a theory of religion, social integration, anomie, and suicide. His theoretical explanations in turn led deductively to further hypotheses and further observations.

In summary, the scientific norm of logical reasoning provides a two-way bridge between theory and research. Scientific inquiry in practice typically involves alternating between deduction and induction. Both methods involve an interplay of logic and observation. And both are routes to the construction of social theories.

Although both inductive and deductive methods are valid in scientific inquiry, individuals may feel more comfortable with one approach than the other. Consider this exchange in Sir Arthur Conan Doyle's story "A Scandal in Bohemia," as Sherlock Holmes answers Dr. Watson's inquiry (Doyle [1891] 1892: 13):

> "What do you imagine that it means?"
> "I have no data yet. It is a capital mistake to theorise before one has data. Insensibly one begins to twist facts to suit theories, instead of theories to suit facts."

Some social scientists would more or less agree with this inductive position (see especially the dis-cussion of grounded theory in Chapter 10), whereas others would take a more deductive stance. Most, however, concede the legitimacy of both approaches.

With this understanding of the deductive and inductive links between theory and research in hand, let's now delve a little more deeply into how theories are constructed using either of these two different approaches.

Deductive Theory Construction

To see what's involved in deductive theory construction and hypothesis testing, imagine that you're going to construct a deductive theory. How would you go about it?

Getting Started

The first step in deductive theory construction is to pick a topic that interests you. The topic can be very broad, such as "What is the structure of society?" or it can be narrower, as in "Why do people support or oppose the idea of a woman's right to an abortion?" Whatever the topic, it should be something you're interested in understanding and explaining.

Once you've picked your topic, the next step is to undertake an inventory of what's already known or thought about it. In part, this means writing down your own observations and ideas. Beyond that, it means learning what other scholars have said about it. You can talk to other people, and you'll want to read the scholarly literature on the topic. Appendix A provides guidelines for using the library—you'll likely spend a lot of time there.

Your preliminary research will probably uncover consistent patterns discovered by prior scholars. For example, religious and political variables will stand out as important determinants of attitudes about abortion. Findings such as these will be very useful to you in creating your own theory.

In this process, don't overlook the value of introspection. Whenever we can look at our own personal processes—including reactions, fears, and prejudices—we may gain important insights into human behavior in general. I don't mean to say that everyone thinks like you or me, but introspection can provide a useful source of insights that can inform our inquiries.

Constructing Your Theory

Now that you've reviewed previous work on the topic, you're ready to begin constructing your theory. Although theory construction is not a lock-step affair, the process generally involves something like the following steps.

1. Specify the topic.
2. Specify the range of phenomena your theory addresses. Will your theory apply to all of human social life, will it apply only to U.S. citizens, only to young people, or what?
3. Identify and specify your major concepts and variables.
4. Find out what is known (propositions) about the relationships among those variables.
5. Reason logically from those propositions to the specific topic you're examining.

We've already discussed items (1) through (3), so let's focus now on (4) and (5). As you identify the relevant concepts and discover what's already been learned about them, you can begin to create a propositional structure that explains the topic under study.

Let's look now at an example of how these building blocks fit together in deductive theory construction and empirical research.

An Example of Deductive Theory: Distributive Justice

A topic of interest to scholars is the concept of distributive justice, people's perceptions of whether they are being treated fairly by life, whether they are getting "their share." Guillermina Jasso describes the theory of distributive justice more formally, as follows:

The theory provides a mathematical description of the process whereby individuals, reflecting on their holdings of the goods they value (such as beauty, intelligence, or wealth), compare themselves to others, experiencing a fundamental instantaneous magnitude of the justice evaluation (J), which captures their sense of being fairly or unfairly treated in the distributions of natural and social goods.

(Jasso 1988: 11)

Notice that Jasso has assigned a symbolic representation for her key variable: J will stand for distributive justice. She does this to support her intention of stating her theory in mathematical formulas. Though theories are often expressed mathematically, we'll not delve too deeply into that practice here.

Jasso indicates that there are three kinds of postulates in her theory. "The first makes explicit the fundamental axiom which represents the substantive point of departure for the theory." She elaborates as follows: "The theory begins with the received Axiom of Comparison, which formalizes the long-held view that a wide class of phenomena, including happiness, self-esteem, and the sense of distributive justice, may be understood as the product of a comparison process" (Jasso 1988: 11).

Thus, your sense of whether you're receiving a "fair" share of the good things of life comes from comparing yourself with others. If this seems obvious to you, that's not a shortcoming of the axiom. Remember, axioms are the taken-for-granted beginnings of theory.

Jasso continues to do the groundwork for her theory. First, she indicates that our sense of distributive justice is a function of "Actual Holdings (A)" and "Comparison Holdings (C)" of some good. Let's consider money, for example. My sense of justice in this regard is a function of how much I actually have, compared with how much others have. By specifying the two components of the comparison, Jasso can use them as variables in her theory.

Next, Jasso offers a "measurement rule" that further specifies how the two variables, A and C, will be conceptualized. This step is needed because some of the goods to be examined are concrete and commonly measured (such as money), whereas others are less tangible (such as respect). The former kind, she says, will be measured conventionally, whereas the latter will be measured "by the individual's relative rank . . . within a specially selected comparison group." The theory will provide a formula for making that measurement (Jasso 1988: 13).

Jasso continues in this fashion to introduce additional elements, weaving them into mathematical formulas to be used in deriving predictions about the workings of distributive justice in a variety of social settings. Here is just a sampling of where her theorizing takes her (1988: 14–15).

- Other things [being] the same, a person will prefer to steal from a fellow group member rather than from an outsider.

- The preference to steal from a fellow group member is more pronounced in poor groups than in rich groups.

- In the case of theft, informants arise only in cross-group theft, in which case they are members of the thief's group.

- Persons who arrive a week late at summer camp or for freshman year of college are more likely to become friends of persons who play games of chance than of persons who play games of skill.

- A society becomes more vulnerable to deficit spending as its wealth increases.

- Societies in which population growth is welcomed must be societies in which the set of valued goods includes at least one quantity-good, such as wealth.

Jasso's theory leads to many other propositions, but this sampling should provide a good sense of where deductive theorizing can take you. To get a feeling for how she reasons her way to these propositions, let's look briefly at the logic involved in two of the propositions that relate to theft within and outside one's group.

- Other things [being] the same, a person will prefer to steal from a fellow group member rather than from an outsider.

Beginning with the assumption that thieves want to maximize their relative wealth, ask yourself whether that goal would be best served by stealing from those you compare yourself with or from outsiders. In each case, stealing will increase your

Actual Holdings, but what about your Comparison Holdings?

A moment's thought should suggest that stealing from people in your comparison group will lower their holdings, further increasing your relative wealth. To simplify, imagine there are only two people in your comparison group: you and I. Suppose we each have $100. If you steal $50 from someone outside our group, you will have increased your relative wealth by 50 percent compared with me: $150 versus $100. But if you steal $50 from me, you will have increased your relative wealth 200 percent: $150 to my $50. Your goal is best served by stealing from within the comparison group.

- In the case of theft, informants arise only in cross-group theft, in which case they are members of the thief's group.

Can you see why it would make sense for informants (1) to arise only in the case of cross-group theft and (2) to come from the thief's comparison group? This proposition again depends on the fundamental assumption that everyone wants to increase his or her relative standing. Suppose you and I are in the same comparison group, but this time the group contains additional people. If you steal from someone else within our comparison group, my relative standing in the group does not change. Although your wealth has increased, the average wealth in the group remains the same (because someone else's wealth has decreased by the same amount). So my relative standing remains the same. I have no incentive to inform on you.

If you steal from someone outside our comparison group, however, your nefarious income increases the total wealth in our group. Now my own wealth relative to that total is diminished. Because my relative wealth has suffered, I'm more likely to inform on you in order to bring an end to your stealing. Hence, informants arise only in cross-group theft.

This last deduction also begins to explain why these informants come from the thief's own comparison group. We've just seen how your theft decreased my relative standing. How about members of the other group (other than the individual you stole from)? Each of them actually profits from

How to Do It: Framing a Hypothesis

As we have seen, the deductive method of research typically focuses on the testing of a hypothesis. Let's take a minute to look at how to create a hypothesis for testing.

Hypotheses state an expected causal relationship between two (or more) variables. Let's suppose you're interested in student political orientations, and your review of the literature and your own reasoning suggest to you that college major will play some part in determining students' political views. Already, we have two variables: *college major* and *political orientation*. Moreover, *political orientation* is the dependent variable—you believe it depends on something else, on the independent variable, which in this case is *college major*.

Now we need to specify the attributes comprising each of these variables. For simplicity's sake, let's assume political orientation includes only liberal or conservative. And to simplify the matter of major, let's suppose your research interests focus on the presumed differences between business students and those in the social sciences.

Even with these simplifications, you would need to specify more concretely how you would recognize a liberal or a conservative when you came across them in your study. This process of specification will be discussed at length in Chapter 5. For now, let's assume you will ask student-subjects whether they consider themselves liberals or conservatives, letting each student report on what the terms mean to them. (As we'll see later, this simple dichotomy is unlikely to work in practice, as some students would want to identify themselves as independents or something else.)

Identifying students' majors isn't as straightforward as you might think. For example, what disciplines compose the social sciences in your study? Also, must students be declared majors or simply be planning to major in one of the relevant fields?

Once these issues have been settled, you are ready to state your hypothesis. For example, it might be the following:

"Students majoring in the social sciences will be more likely to identify themselves as liberals than are those majoring in business."

In addition to this basic expectation, you may wish to specify "more likely" in terms of how *much* more likely. Chapter 16 will provide some options in this regard.

---

the theft, because you have reduced the total with which they compare themselves. Hence, they have no reason to inform on you. Thus, the theory of distributive justice predicts that informants arise from the thief's own comparison group.

This brief peek into Jasso's derivations should give you some sense of the enterprise of deductive theory. Of course, the theory guarantees none of the given predictions. The role of research is to test each of them to determine whether what makes sense (logic) actually occurs in practice (observation).

See "How to Do It: Framing a Hypothesis" for a look at creating hypotheses for deductive purposes.

Inductive Theory Construction

As we have seen, quite often social scientists begin constructing a theory through the inductive method by first observing aspects of social life and then seeking to discover patterns that may point to relatively universal principles. Barney Glaser and Anselm Strauss (1967) coined the term *grounded theory* in reference to this method.

Field research—the direct observation of events in progress—is frequently used to develop theories through observation. In a long and rich tradition, anthropologists have used this method to good advantage.

Among modern social scientists, no one has been more adept at seeing the patterns of human behavior through observation than Erving Goffman:

> A game such as chess generates a habitable universe for those who can follow it, a plane of being, a cast of characters with a seemingly un-limited number of different situations and acts through which to realize their natures and des-tinies. Yet much of this is reducible to a small set of interdependent rules and practices. If the meaningfulness of everyday activity is similarly dependent on a closed, finite set of rules, then

explication of them would give one a powerful means of analyzing social life.

(1974: 5)

In a variety of research efforts, Goffman uncovered the rules of such diverse behaviors as living in a mental institution (1961) and managing the "spoiled identity" of being disfigured (1963). In each case, Goffman observed the phenomenon in depth and teased out the rules governing behavior. Goffman's research provides an excellent example of qualitative field research as a source of grounded theory.

Our earlier discussion of the Comfort Hypothesis and church involvement shows that qualitative field research is not the only method of observation appropriate to the development of inductive theory. Here's another detailed example to illustrate further the construction of inductive theory using quantitative methods.

An Example of Inductive Theory: Why Do People Smoke Marijuana?

During the 1960s and 1970s, marijuana use on U.S. college campuses was a subject of considerable discussion in the popular press. Some people were troubled by marijuana's popularity; others welcomed it. What interests us here is why some students smoked marijuana and others didn't. A survey of students at the University of Hawaii by David Takeuchi (1974) provided the data to answer that question.

At the time of the study, a huge number of explanations were being offered for drug use. People who opposed drug use, for example, often suggested that marijuana smokers were academic failures trying to avoid the rigors of college life. Those in favor of marijuana, on the other hand, often spoke of the search for new values: Marijuana smokers, they said, were people who had seen through the hypocrisy of middle-class values.

Takeuchi's analysis of the data gathered from University of Hawaii students, however, did not support any of the explanations being offered.

Those who reported smoking marijuana had essentially the same academic records as those who didn't smoke it, and both groups were equally involved in traditional "school spirit" activities. Both groups seemed to feel equally well integrated into campus life.

There were other differences between the groups, however:

1. Women were less likely than men to smoke marijuana.

2. Asian students (a large proportion of the student body) were less likely to smoke marijuana than non-Asians were.

3. Students living at home were less likely to smoke marijuana than those living in apartments were.

As in the case of religiosity, the three variables independently affected the likelihood of a student's smoking marijuana. About 10 percent of the Asian women living at home had smoked marijuana, in contrast to about 80 percent of the non-Asian men living in apartments. And, as in the religiosity study, the researchers discovered a powerful pattern of drug use before they had an explanation for that pattern.

In this instance, the explanation took a peculiar turn. Instead of explaining why some students smoked marijuana, the researchers explained why some didn't. Assuming that all students had some motivation for trying drugs, the researchers suggested that students differed in the degree of "social constraints" preventing them from following through on that motivation.

U.S. society is, on the whole, more permissive with men than with women when it comes to deviant behavior. Consider, for example, a group of men getting drunk and boisterous. We tend to dismiss such behavior with references to "camaraderie" and "having a good time," whereas a group of women behaving similarly would probably be regarded with disapproval. We have an idiom, "Boys will be boys," but no comparable idiom for girls. The researchers reasoned, therefore, that women would have more to lose by smoking marijuana than men would. In other words, being female provided a constraint against smoking marijuana.

Students living at home had obvious constraints against smoking marijuana, compared with students living on their own. Quite aside from differences in opportunity, those living at home were seen as being more dependent on their parents—hence more vulnerable to additional punishment for breaking the law.

Finally, the Asian subculture in Hawaii has traditionally placed a higher premium on obedience to the law than other subcultures have, so Asian students would have more to lose if they were caught violating the law by smoking marijuana.

Overall, then, a "social constraints" theory was offered as the explanation for observed differences in the likelihood of smoking marijuana. The more constraints a student had, the less likely he or she would be to smoke marijuana. It bears repeating that the researchers had no thoughts about such a theory when their research began. The theory came from an examination of the data.

The Links between Theory and Research

Throughout this chapter, we have seen various aspects of the links between theory and research in social science inquiry. In the deductive model, research is used to test theories. In the inductive model, theories are developed from the analysis of research data. This final section looks more closely into the ways theory and research are related in actual social science inquiry.

Whereas we have discussed two idealized logical models for linking theory and research, social science inquiries have developed a great many variations on these themes. Sometimes theoretical issues are introduced merely as a background for empirical analyses. Other studies cite selected empirical data to bolster theoretical arguments. In neither case do theory and research really interact for the purpose of developing new explanations. Some studies make no use of theory at all, aiming specifically, for example, at an ethnographic description of a particular social situation, such as

an anthropological account of food and dress in a particular society.

As you read social research reports, however, you'll often find that the authors are conscious of the implications of their research for social theories and vice versa. Here are a few examples to illustrate this point.

When W. Lawrence Neuman (1998) set out to examine the problem of monopolies (the "trust problem") in U.S. history, he saw the relevance of theories about how social movements transform society ("state transformation"). He became convinced, however, that existing theories were inadequate for the task before him:

> State transformation theory links social movements to state policy formation processes by focussing on the role of cultural meaning in organized political struggles. Despite a resemblance among concepts and concerns, constructionist ideas found in the social problems, social movements, and symbolic politics literatures have not been incorporated into the theory. In this paper, I draw on these three literatures to enhance state transformation theory.
>
> *(Neuman 1998: 315)*

Having thus modified state transformation theory, Neuman had a theoretical tool that could guide his inquiry and analysis into the political maneuverings related to monopolies beginning in the 1880s and continuing until World War I. Thus, theory served as a resource for research and at the same time was modified by it.

In a somewhat similar study, Alemseghed Kebede and J. David Knottnerus (1998) set out to investigate the rise of Rastafarianism in the Caribbean. However, they felt that recent theories on social movements had become too positivistic in focusing on the mobilization of resources. Resource mobilization theory, they felt, downplays

> the motivation, perceptions, and behavior of movement participants . . . and concentrates instead on the whys and hows of mobilization. Typically theoretical and research problems include: How do emerging movement organizations seek to mobilize and routinize the flow

of resources and how does the existing political apparatus affect the organization of resources?

(1998: 500)

To study Rastafarianism more appropriately, the researchers felt the need to include several concepts from contemporary social psychology. In particular, they sought models to use in dealing with problems of meaning and collective thought.

Frederika Schmitt and Patricia Martin (1999) were particularly interested in discovering what made for successful rape crisis centers and how they dealt with the organizational and political environments within which they operated. The researchers found theoretical constructs appropriate to their inquiry:

> This case study of unobtrusive mobilizing by Southern California Rape Crisis Center uses archival, observational, and interview data to explore how a feminist organization worked to change police, schools, prosecutor, and some state and national organizations from 1974 to 1994. Mansbridge's concept of street theory and Katzenstein's concepts of unobtrusive mobilization and discursive politics guide the analysis.
>
> *(1999: 364)*

In summary, there is no simple recipe for conducting social science research. It is far more open-ended than the traditional view of science suggests. Ultimately, science depends on two categories of activity: logic and observation. As you'll see throughout this book, they can be fit together in many patterns.

Research Ethics and Theory

In Chapter 1, I introduced the subject of research ethics and said we would return to that topic throughout the book. At this point, what ethical issues do you suppose theory engenders?

In this chapter, we have seen how the paradigms and theories that guide research inevitably impact what is observed and how it is interpreted.

Choosing a particular paradigm or theory does not guarantee a particular research conclusion, but it will affect what you look for and what you ignore. Whether you choose a functionalist or a conflict paradigm to organize your research on police–community relations will made a big difference.

This is a difficult issue to resolve in practice. Choosing a theoretical orientation for the purpose of encouraging a particular conclusion would be regarded as unethical as a general matter, but when research is linked to an intention to bring about social change, the researcher will likely choose a theoretical orientation appropriate to that intention. Let's say you're concerned about the treatment of homeless people by the police in your community. You might organize your research in terms of interactionist or conflict paradigms and theories that would reveal any instances of mistreatment that may occur.

Two factors counter the potential problem of bias from theoretical orientation. First, as we'll see in the remainder of the book, social science research techniques—the various methods of observation and analysis—place a damper on our simply seeing what we expect. Even if you expect to find the police mistreating the homeless and use theories and methods that will reveal such mistreatment, you will not observe what isn't there if you apply those theories and methods appropriately.

Second, the collective nature of social research offers further protection. As indicated in Chapter 1, peer review in which researchers evaluate each other's efforts will point to instances of shoddy and/or biased research. Moreover, with several researchers studying the same phenomenon, perhaps using different paradigms, theories, and methods, the risk of biased research findings is further reduced.

MAIN POINTS

Introduction

- Theories function in three ways in research: (1) helping to avoid flukes, (2) making sense of observed patterns, and (3) shaping and directing research efforts.

Some Social Science Paradigms

- Social scientists use a variety of paradigms to organize how they understand and inquire into social life.

- A distinction between types of theories that cuts across various paradigms is macrotheory (theories about large-scale features of society) versus microtheory (theories about smaller units or features of society).

- The positivistic paradigm assumes that we can scientifically discover the rules governing social life.

- The Social Darwinist paradigm sees a progressive evolution in social life.

- The conflict paradigm focuses on the attempt of individuals and groups to dominate others and to avoid being dominated.

- The symbolic interaction paradigm examines how shared meanings and social patterns develop in the course of social interactions.

- Ethnomethodology focuses on the ways people make sense out of social life in the process of living it, as though each were a researcher engaged in an inquiry.

- The structural functionalist (or social systems) paradigm seeks to discover what functions the many elements of society perform for the whole system.

- Feminist paradigms, in addition to drawing attention to the oppression of women in most societies, highlight how previous images of social reality have often come from and reinforced the experiences of men.

- Like feminist paradigms, critical race theory both examines the disadvantaged position of a social group (African Americans) and offers a different vantage point from which to view and understand society.

- Some contemporary theorists and researchers have challenged the long-standing belief in an objective reality that abides by rational rules. They point out that it is possible to agree on an "intersubjective" reality, a view that characterizes postmodernism.

Elements of Social Theory

- The elements of social theory include observations, facts, and laws (which relate to the reality being observed), as well as concepts, variables, axioms or postulates, propositions, and hypotheses (which are logical building blocks of the theory itself).

Two Logical Systems Revisited

- In the traditional image of science, scientists proceed from theory to operationalization to observation. But this image does not accurately depict how scientific research is actually done.

- Social scientific theory and research are linked through the two logical methods of deduction (the derivation of expectations and hypotheses from theories) and induction (the development of generalizations from specific observations).

- In practice, science is a process involving an alternation of deduction and induction.

Deductive Theory Construction

- Guillermina Jasso's theory of distributive justice illustrates how formal reasoning can lead to a variety of theoretical expectations that can be tested by observation.

Inductive Theory Construction

- David Takeuchi's study of factors influencing marijuana smoking among University of Hawaii students illustrates how collecting observations can lead to generalizations and an explanatory theory.

The Links between Theory and Research

- In practice, there are many possible links between theory and research and many ways of going about social inquiry.

Research Ethics and Theory

- Researchers should not use paradigm and theory selection as a means of achieving desired research results.

- The collective nature of social research offers protection against biased research findings.

KEY TERMS

The following terms are defined in context in the chapter and at the bottom of the page where the term is introduced, as well as in the comprehensive glossary at the back of the book.

conflict paradigm	interest convergence
critical race theory	macrotheory
critical realism	microtheory
feminist paradigms	null hypothesis
hypothesis	operational definition

operationalization

paradigm

positivism

postmodernism

structural functionalism

symbolic interactionism

PROPOSING SOCIAL RESEARCH: THEORY

As this chapter has indicated, social research can be pursued within numerous theoretical paradigms—each suggesting a somewhat different way to approach the research question. In this portion of your proposal, you should identify the paradigm(s) that will shape the design of your research.

We have also seen that paradigms provide frameworks within which causal theories may be developed. Perhaps your research project will explore or test an existing theory. Or more ambitiously, you may propose a theory or hypothesis for testing. This is the section of the proposal in which to describe this aspect of your project.

Not all research projects are formally organized around the creation and/or testing of theories and hypotheses. However, your research will involve theoretical concepts, which should be described in this section of the proposal. As we'll see more fully in Chapter 17, this portion of your proposal will reflect the literature on previous theory and research that has shaped your own thinking and research plans.

REVIEW QUESTIONS AND EXERCISES

1. Consider the possible relationship between education and prejudice that was mentioned in Chapter 1. Describe how you might examine that relationship through (a) deductive and (b) inductive methods.

2. Review the relationships between theory and research discussed in this chapter. Select a research article from an academic journal and classify the relationship between theory and research you find there.

3. Using one of the many search engines (such as Google, Excite, HotBot, Ask Jeeves, LookSmart, Lycos, Netscape, WebCrawler, or Yahoo), find information on the web concerning at least three of the following paradigms: functionalism, symbolic interactionism, conflict theory, ethnomethodology, feminist paradigms, critical race paradigms, rational choice paradigm. Give the web locations and report on the theorists discussed in connection with the discussions you found.

4. Using InfoTrac College Edition (Article A67051613) or the library, locate Judith A. Howard (2000), "Social Psychology of Identities," *Annual Review of Sociology* 26:367–93. What paradigm does she find most useful for the study of social identities? Explain why she feels that it is the appropriate paradigm. Do you agree? Why or why not?

SPSS EXERCISES

See the booklet that accompanies your text for exercises using SPSS (Statistical Package for the Social Sciences). There are exercises offered for each chapter, and you'll also find a detailed primer on using SPSS.

Online Study Resources

If your book came with an access code card, visit www.cengage.com/login to register. To purchase access, please visit www.ichapters.com.

1. Before you do your final review of the chapter, take the *CengageNOW* pretest to help identify the areas on which you should concentrate. You'll find information on this online tool, as well as instructions on how to access all of its great resources, in the front of the book.

2. As you review, take advantage of the *CengageNOW* personalized study plan, based on your quiz results. Use this study plan with its interactive exercises and other resources to master the material.

3. When you're finished with your review, take the posttest to confirm that you're ready to move on to the next chapter.

WEBSITE FOR *THE PRACTICE OF SOCIAL RESEARCH* 12TH EDITION

Go to your book's website at www.cengage.com/sociology/babbie for tools to aid you in studying for your exams. You'll find *Tutorial Quizzes* with feedback, *Internet Exercises, Flash Cards, Glossaries,* and *Essay Quizzes,* as well as *InfoTrac College Edition* search terms, suggestions for additional reading, *Web Links,* and primers for using data-analysis software such as SPSS.

CHAPTER THREE

The Ethics and Politics of Social Research

CHAPTER OVERVIEW

Social research takes place in a social context. Researchers must therefore take into account many ethical and political considerations alongside scientific ones in designing and executing their research. Often, however, clear-cut answers to thorny ethical and political issues are hard to come by.

CengageNOW for Sociology

Use this online tool to help you make the grade on your next exam. After reading this chapter, go to "Online Study Resources" at the end of the chapter for instructions on how to benefit from *CengageNOW*.

Introduction

My purpose in this book is to present a realistic and useful introduction to doing social research. For this introduction to be fully realistic, it must include four main constraints on research projects: scientific, administrative, ethical, and political.

Most of the book focuses on scientific and administrative constraints. We'll see that the logic of science suggests certain research procedures, but we'll also see that some scientifically "perfect" study designs are not administratively feasible, because they would be too expensive or take too long to execute. Throughout the book, therefore, we'll deal with workable compromises.

Before we get to the scientific and administrative constraints on research, it's useful to explore the two other important considerations in doing research in the real world—ethics and politics—which this chapter covers. Just as certain procedures are too impractical to use, others are either ethically prohibitive or politically difficult or impossible. Here's a story to illustrate what I mean.

Several years ago, I was invited to sit in on a planning session to design a study of legal education in California. The joint project was to be conducted by a university research center and the state bar association. The purpose of the project was to improve legal education by learning which aspects of the law school experience were related to success on the bar exam. Essentially, the plan was to prepare a questionnaire that would get detailed information about the law school experiences of individuals. People would be required to answer the questionnaire when they took the bar exam. By analyzing how people with different kinds of law school experiences did on the bar exam, we could find out what sorts of things worked and what didn't. The findings of the research could be made available to law schools, and ultimately legal education could be improved.

The exciting thing about collaborating with the bar association was that all the normally irritating logistical hassles would be handled. There would be no problem getting permission to administer questionnaires in conjunction with the exam, for example, and the problem of nonresponse could be eliminated altogether.

I left the meeting excited about the prospects for the study. When I told a colleague about it, I glowed about the absolute handling of the nonresponse problem. Her immediate comment turned everything around completely. "That's unethical. There's no law requiring the questionnaire, and participation in research has to be voluntary." The study wasn't done.

In retelling this story, I can easily see that requiring participation would have been inappropriate. You may have seen this even before I told you about my colleague's comment. I still feel a little embarrassed over the matter, but I have a specific purpose in telling this story about myself.

All of us consider ourselves ethical—not perfect perhaps, but as ethical as anyone else and perhaps more so than most. The problem in social research, as probably in life, is that ethical considerations are not always apparent to us. As a result, we often plunge into things without seeing ethical issues that may be apparent to others and may even be obvious to us when pointed out. When I reported back to the others in the planning group, for example, no one disagreed with the inappropriateness of requiring participation. Everyone was a bit embarrassed about not having seen it.

Any of us can immediately see that a study requiring small children to be tortured is unethical. I know you'd speak out immediately if I suggested that we interview people about their sex lives and then publish what they said in the local newspaper. But, as ethical as you are, you'll totally miss the ethical issues in some other situations—we all do.

The first half of this chapter deals with the ethics of social research. In part, it presents some of the broadly agreed-on norms describing what's ethical in research and what's not. More important than simply knowing the guidelines, however, is becoming sensitized to the ethical component in research so that you'll look for it whenever you plan a study. Even when the ethical aspects of a situation are debatable, you should know that there's something to argue about. It's worth noting in this context that many professions operate under ethical constraints and that these constraints differ from one profession to another. Thus, priests, physicians, lawyers, reporters, and television producers operate under different ethical constraints. In this chapter, we'll look only at the ethical principles that govern social research.

Political considerations in research are also subtle, ambiguous, and arguable. Notice that the law school example involves politics as well as ethics. Although social researchers have an ethical norm that participation in research should be voluntary, this norm clearly grows out of U.S. political norms protecting civil liberties. In some nations, the proposed study would have been considered quite ethical.

In the second half of this chapter, we'll look at social research projects that were crushed or nearly crushed by political considerations. As with ethical concerns, there is often no "correct" take on a given situation. People of goodwill disagree. I won't try to give you a party line about what is and is not politically acceptable. As with ethics, the point is to become sensitive to the political dimension of social research.

Ethical Issues in Social Research

In most dictionaries and in common usage, ethics is typically associated with morality, and both words concern matters of right and wrong. But what is right and what wrong? What is the source of the distinction? For individuals the sources vary. They may be religions, political ideologies, or the pragmatic observation of what seems to work and what doesn't.

Webster's New World Dictionary is typical among dictionaries in defining *ethical* as "conforming to the standards of conduct of a given profession or group." Although this definition may frustrate those in search of moral absolutes, what we regard as morality and ethics in day-to-day life is a matter of agreement among members of a group. And, not

surprisingly, different groups have agreed on different codes of conduct. Part of living successfully in a particular society is knowing what that society considers ethical and unethical. The same holds true for the social research community.

Anyone involved in social science research, then, needs to be aware of the general agreements shared by researchers about what is proper and improper in the conduct of scientific inquiry. This section summarizes some of the most important ethical agreements that prevail in social research.

Voluntary Participation

Often, though not always, social research represents an intrusion into people's lives. The interviewer's knock on the door or the arrival of a questionnaire in the mail signals the beginning of an activity that the respondent has not requested and that may require significant time and energy. Participation in a social experiment disrupts the subject's regular activities.

Social research, moreover, often requires that people reveal personal information about themselves—information that may be unknown to their friends and associates. And social research often requires that such information be revealed to strangers. Other professionals, such as physicians and lawyers, also ask for such information. Their requests may be justified, however, by their aims: They need the information in order to serve the personal interests of the respondent. Social researchers can seldom make this claim. Like medical scientists, they can only argue that the research effort may ultimately help all humanity.

A major tenet of medical research ethics is that experimental participation must be voluntary. The same norm applies to social research. No one should be forced to participate. This norm is far easier to accept in theory than to apply in practice, however.

Again, medical research provides a useful parallel. Many experimental drugs used to be tested on prisoners. In the most rigorously ethical cases, the prisoners were told the nature and the possible dangers of the experiment, they were told that participation was completely voluntary, and they were further instructed that they could expect no special rewards—such as early parole—for participation. Even under these conditions, it was often clear that volunteers were motivated by the belief that they would personally benefit from their cooperation.

When the instructor in an introductory sociology class asks students to fill out a questionnaire that he or she hopes to analyze and publish, students should always be told that participation in the survey is completely voluntary. Even so, most students will fear that nonparticipation will somehow affect their grade. The instructor should therefore be sensitive to such implications and make special provisions to eliminate them. For example, the instructor could insure anonymity by leaving the room while the questionnaires are being completed. Or, students could be asked to return the questionnaires by mail or to drop them in a box near the door just before the next course meeting.

This norm of voluntary participation, though, goes directly against several scientific concerns. In the most general terms, the scientific goal of generalizability is threatened if experimental subjects or survey respondents are all the kind of people who willingly participate in such things. Because this orientation probably reflects more general personality traits, the results of the research might not be generalizable to all people. Most clearly, in the case of a descriptive survey, a researcher cannot generalize the sample survey findings to an entire population unless a substantial majority of the scientifically selected sample actually participates—the willing respondents and the somewhat unwilling.

As you'll see in Chapter 10, field research has its own ethical dilemmas in this regard. Very often the researcher cannot even reveal that a study is being done, for fear that that revelation might significantly affect the social processes being studied. Clearly, the subjects of study in such cases are not given the opportunity to volunteer or refuse to participate.

Though the norm of voluntary participation is important, it is often impossible to follow. In cases where researchers feel ultimately justified in violating it, their observing the other ethical norms of scientific research, such as bringing no harm to the people under study, becomes all the more important.

No Harm to the Participants

The need for norms against harming research subjects has stemmed in part from horrendous actions by medical researchers. Perhaps at the top of list stand the medical experiments on prisoners of war by Nazi researchers in World War II. The subsequent war-crimes trials at Nuremberg added the phrase *crimes against humanity* to the language of research and political ethics.

Less well-known were the Tuskegee syphilis experiments conducted by the U.S. Public Health Service between 1932 and 1972. The study followed the fate of nearly 400 impoverished, rural African American men suffering from syphilis. Even after penicillin had been accepted as an effective treatment for syphilis, the subjects were denied treatment—even kept from seeking treatment in the community—because the researchers wanted to observe the full progression of the disease. At times, diagnostic procedures such as spinal taps were falsely presented to subjects as cures for syphilis.

When the details of the Tuskegee syphilis experiments became widely known, the U.S. government took action, including a formal apology by President Bill Clinton and a program of financial reparations to the families of the subjects. (You can learn more about this sad history in medical research at the link on this book's website: http://www.cengage.com/sociology/babbie.)

Human research should never injure the people being studied, regardless of whether they volunteer for the study. In social research practice, this often concerns being careful not to reveal information that would embarrass subjects or endanger their home lives, friendships, jobs, and so forth. We'll discuss this aspect of the norm more fully in a moment.

Because subjects can be harmed psychologically in the course of a social research study, the researcher must look for the subtlest dangers and guard against them. Quite often, research subjects are asked to reveal deviant behavior, attitudes they feel are unpopular, or personal characteristics that may seem demeaning, such as low income, the receipt of welfare payments, and the like. Revealing such information usually makes subjects feel at least uncomfortable.

Social research projects may also force participants to face aspects of themselves that they don't normally consider. This can happen even when the information is not revealed directly to the researcher. In retrospect, a certain past behavior may appear unjust or immoral. The project, then, can cause continuing personal agony for the subject. If the study concerns codes of ethical conduct, for example, the subject may begin questioning his or her own morality, and that personal concern may last long after the research has been completed and reported. For instance, probing questions can injure a fragile self-esteem.

In 1971 the psychologist Philip Zimbardo created his famous simulation of prison life, widely known as the Stanford prison experiment, to study the dynamics of prisoner–guard interactions. Zimbardo employed Stanford students as subjects and randomly assigned them to roles as prisoners or guards. As you may be aware, the simulation became quickly and increasingly real for all the participants, including Zimbardo, who served as prison superintendent. It became evident that many of the student-prisoners were suffering psychological damage as a consequence of their mock incarceration, and some of the student-guards were soon exhibiting degrees of sadism that would later challenge their own self-images.

As these developments became apparent to Zimbardo, he terminated the experiment. He then created a debriefing program in which all the participants were counseled so as to avoid any lasting damage from the experience. (See the link on this book's website, http://www.cengage.com/sociology/babbie, for a link to Zimbardo's discussion of the experiment.)

As you can see, just about any research you might conduct runs the risk of injuring other people in some way. It isn't possible to insure against all these possible injuries, but some study designs make such injuries more likely than others do. If a particular research procedure seems likely to produce unpleasant effects for subjects—asking survey respondents to report deviant behavior, for example—the researcher should have the firmest of scientific grounds for doing it. If your research design is essential and also likely to be unpleasant for subjects, you'll find yourself in an ethical netherworld and may go through some personal agonizing. Although agonizing has little value in itself, it may be a healthy sign that you've become sensitive to the problem.

Increasingly, the ethical norms of voluntary participation and no harm to participants have become formalized in the concept of **informed consent.** This norm means that subjects must base their voluntary participation in research projects on a full understanding of the possible risks involved. In a medical experiment, for example, prospective subjects are presented with a discussion of the experiment and all the possible risks to themselves. They are required to sign a statement indicating that they are aware of the risks and that they choose to participate anyway. Although the value of such a procedure is obvious when subjects will be injected with drugs designed to produce physical effects, for example, it's hardly appropriate when a participant observer rushes to a scene of urban rioting to study deviant behavior. Whereas the researcher in this latter case must still bring no harm to those observed, gaining informed consent is not the means to achieving that end.

Although the fact often goes unrecognized, another possible source of harm to subjects lies in the analysis and reporting of data. Every now and then, research subjects read the books published about the studies they participated in. Reasonably sophisticated subjects can locate themselves in the various indexes and tables. Having done so, they may find themselves characterized—though not identified by name—as bigoted, unpatriotic, irreligious, and so forth. At the very least, such characterizations are likely to trouble them and

---

informed consent A norm in which subjects base their voluntary participation in research projects on a full understanding of the possible risks involved.

threaten their self-images. Yet the whole purpose of the research project may be to explain why some people are prejudiced and others are not.

In one survey of churchwomen (Babbie 1967), ministers in a sample of churches were asked to distribute questionnaires to a specified sample of members, collect them, and return them to the research office. One of these ministers read through the questionnaires from his sample before returning them, and then he delivered a hellfire and brimstone sermon to his congregation, saying that many of them were atheists and were going to hell. Even though he could not identify the people who gave particular responses, many respondents certainly endured personal harm from his tirade.

Like voluntary participation, avoiding harm to people is easy in theory but often difficult in practice. Sensitivity to the issue and experience with its applications, however, should improve the researcher's tact in delicate areas of research.

In recent years, social researchers have been gaining support for abiding by this norm. Federal and other funding agencies typically require an independent evaluation of the treatment of human subjects for research proposals, and most universities now have human-subject committees to serve this evaluative function. Although sometimes troublesome and inappropriately applied, such requirements not only guard against unethical research but also can reveal ethical issues overlooked by even the most scrupulous researchers.

Anonymity and Confidentiality

The clearest concern in the protection of the subjects' interests and well-being is the protection of their identity, especially in survey research. If revealing their survey responses would injure them in any way, adherence to this norm becomes all the more important. Two techniques—anonymity and confidentiality—assist researchers in this regard, although people often confuse the two.

Anonymity

A research project guarantees **anonymity** when the researcher—not just the people who read about the research—cannot identify a given

response with a given respondent. This implies that a typical interview-survey respondent can never be considered anonymous, because an interviewer collects the information from an identifiable respondent. An example of anonymity is a mail survey in which no identification numbers are put on the questionnaires before their return to the research office.

As we'll see in Chapter 9 (on survey research), assuring anonymity makes keeping track of who has or hasn't returned the questionnaires difficult. Despite this problem, paying the necessary price is advisable in certain situations. For example, in one study of drug use among university students, I decided that I specifically did not want to know the identity of respondents. I felt that honestly assuring anonymity would increase the likelihood and accuracy of responses. Also, I did not want to be in the position of being asked by authorities for the names of drug offenders. In the few instances in which respondents volunteered their names, such information was immediately obliterated from the questionnaires.

Confidentiality

A research project guarantees **confidentiality** when the researcher can identify a given person's responses but essentially promises not to do so publicly. In an interview survey, for example, the researcher could make public the income reported by a given respondent, but the respondent is assured that this will not be done.

Whenever a research project is confidential rather than anonymous, it is the researcher's responsibility to make that fact clear to the

---

anonymity Anonymity is achieved in a research project when neither the researchers nor the readers of the findings can identify a given response with a given respondent.

confidentiality A research project guarantees confidentiality when the researcher can identify a given person's responses but promises not to do so publicly.

respondent. Moreover, researchers should never use the term *anonymous* to mean *confidential.*

With few exceptions (such as surveys of public figures who agree to have their responses published), the information respondents give must at least be kept confidential. This is not always an easy norm to follow, because for example the courts have not recognized social research data as the kind of "privileged communication" priests and attorneys have.

This unprotected guarantee of confidentiality produced a near disaster in 1991. Two years earlier, the Exxon *Valdez* supertanker had run aground near the port of Valdez in Alaska, spilling ten million gallons of oil into the bay. The economic and environmental damage was widely reported.

The media paid less attention to the psychological and sociological damage suffered by residents of the area. There were anecdotal reports of increased alcoholism, family violence, and other secondary consequences of the disruptions caused by the oil spill. Eventually, 22 communities in Prince William Sound and the Gulf of Alaska sued Exxon for the economic, social, and psychological damages suffered by their residents.

To determine the amount of damage done, the communities commissioned a San Diego research firm to undertake a household survey asking residents very personal questions about increased problems in their families. The sample of residents were asked to reveal painful and embarrassing information, under the guarantee of absolute confidentiality. Ultimately, the results of the survey confirmed that a variety of personal and family problems had increased substantially following the oil spill.

When Exxon learned that survey data would be presented to document the suffering, they took an unusual step: They asked the court to subpoena the survey questionnaires. The court granted the request and ordered the researchers to turn over the questionnaires—with all identifying information. It appeared that Exxon's intention was to call survey respondents to the stand and cross-examine them regarding answers they had given to interviewers under the guarantee of confidentiality.

Moreover, many of the respondents were Native Americans, whose cultural norms made such public revelations all the more painful.

Happily, the Exxon *Valdez* case was settled before the court decided whether it would force survey respondents to testify in open court. Unhappily, the potential for disaster remains. (For more information on this ecological disaster, see Picou, Gill, and Cohen (1999).)

The seriousness of this issue is not limited to established research firms. Rik Scarce was a graduate student at Washington State University when he undertook participant observation among animal-rights activists. In 1990 he published a book based on his research: *Ecowarriors: Understanding the Radical Environmental Movement.* In 1993, Scarce was called before a grand jury and asked to identify the activists he had studied. In keeping with the norm of confidentiality, the young researcher refused to answer the grand jury's questions and spent 159 days in the Spokane County jail. He reports,

> Although I answered many of the prosecutor's questions, on 32 occasions I refused to answer, saying, "Your question calls for information that I have only by virtue of a confidential disclosure given to me in the course of my research activities. I cannot answer the question without actually breaching a confidential communication. Consequently, I decline to answer the question under my ethical obligations as a member of the American Sociological Association and pursuant to any privilege that may extend to journalists, researchers, and writers under the First Amendment."
>
> *(Scarce 1999: 982)*

At the time of his grand jury appearance and his incarceration, Scarce felt that the American Sociological Association (ASA) code of ethics strongly supported his ethical stand, and the ASA filed a friend of the court brief on his behalf. In 1997, the ASA revised its code and, while still upholding the norm of confidentiality, warned researchers to inform themselves regarding laws and rules that may limit their ability to promise confidentiality to research subjects.

You can use several techniques to guard against such dangers and ensure better performance on the guarantee of confidentiality. To begin, interviewers and others with access to respondent identifications should be trained in their ethical responsibilities. Beyond training, the most fundamental technique is to remove identifying information as soon as it's no longer necessary. In a survey, for example, all names and addresses should be removed from questionnaires and replaced by identification numbers. An identification file should be created that links numbers to names to permit the later correction of missing or contradictory information, but this file should not be available except for legitimate purposes.

Similarly, in an interview survey you may need to identify respondents initially so that you can recontact them to verify that the interview was conducted and perhaps to get information that was missing in the original interview. As soon as you've verified an interview and assured yourself that you don't need any further information from the respondent, however, you can safely remove all identifying information from the interview booklet. Often, interview booklets are printed so that the first page contains all the identifiers—it can be torn off once the respondent's identification is no longer needed. J. Steven Picou (1996a, 1996b) points out that even removing identifiers from data files does not always sufficiently protect respondent confidentiality, a lesson he learned during nearly a year in federal court. A careful examination of all the responses of a particular respondent sometimes allows others to deduce that person's identity. Imagine, for example, that someone said he or she was a former employee of a particular company. Knowing the person's gender, age, ethnicity, and other characteristics could enable the company to identify that person.

Even if you intend to remove all identifying information, suppose you have not yet done so. What do you do when the police or a judge orders you to provide the responses given by your research subjects?

In 2002, the U.S. Department of Health and Human Services announced a program to issue a "Certificate of Confidentiality" to protect the confidentiality of research subject data against forced disclosure by the police and other authorities. Not all research projects qualify for such protection, but it can provide an important support for research ethics in many cases.

> Under section 301(d) of the Public Health Service Act (42 U.S.C. 241(d)) the Secretary of Health and Human Services may authorize persons engaged in biomedical, behavioral, clinical, or other research to protect the privacy of individuals who are the subjects of that research. This authority has been delegated to the National Institutes of Health (NIH).
> Persons authorized by the NIH to protect the privacy of research subjects may not be compelled in any Federal, State, or local civil, criminal, administrative, legislative, or other proceedings to identify them by name or other identifying characteristic.
>
> *(U.S. Department of Health and Human Services 2002)*
> *Source: http://grants.nih.gov/grants/policy/COC/*
> *background.htm?print=yes⊋*

In all the aspects of research ethics discussed in this chapter, professional researchers avoid settling for mere rote compliance with established ethical rules. Rather, they continually ask what actions would be most appropriate in protecting the interests of those being studied. Here's the way Penny Becker (1998: 452) addressed the issue of confidentiality in connection with a qualitative research project studying religious life in a community:

> Following the lead of several recent studies, I identify the real name of the community, Oak Park, rather than reducing the complexity of the community's history to a few underlying dimensions or creating an "insider/outsider" dynamic where some small group of fellow researchers knows the community's real name and the rest of the world is kept in the dark. . . . In all cases individual identities are disguised, except for Jack Finney, the Lutheran pastor, who gave permission to be identified. "City Baptist" is a pseudonym used at the request of the church's leadership. The leaders of Good Shepherd Lutheran Church (GSLC) gave permission to use the church's real name.

Deception

We've seen that the handling of subjects' identities is an important ethical consideration. Handling your own identity as a researcher can also be tricky. Sometimes it's useful and even necessary to identify yourself as a researcher to those you want to study. You'd have to be an experienced con artist to get people to participate in a laboratory experiment or complete a lengthy questionnaire without letting on that you were conducting research.

Even when you must conceal your research identity, you need to consider the following. Because deceiving people is unethical, deception within social research needs to be justified by compelling scientific or administrative concerns. Even then, the justification will be arguable.

Sometimes researchers admit that they're doing research but fudge about why they're doing it or for whom. Suppose you've been asked by a public welfare agency to conduct a study of living standards among aid recipients. Even if the agency is looking for ways of improving conditions, the recipient-subjects are likely to fear a witch hunt for "cheaters." They might be tempted, therefore, to give answers that make them seem more destitute than they really are. Unless they provide truthful answers, however, the study will not produce accurate data that will contribute to an improvement of living conditions. What do you do?

One solution would be to tell subjects that you're conducting the study as part of a university research program—concealing your affiliation with the welfare agency. Although doing that improves the scientific quality of the study, it raises serious ethical questions.

Lying about research purposes is common in laboratory experiments. Although it's difficult to conceal that you're conducting research, it's usually simple—and sometimes appropriate—to conceal your purpose. Many experiments in social psychology, for example, test the extent to which subjects will abandon the evidence of their own observations in favor of the views expressed by others. Recall Figure 2-1 (p. 42), which shows the stimulus from the classic Asch experiment—frequently replicated by psychology classes—in which subjects are shown three lines of differing lengths (A, B, and C) and asked to compare them with a fourth line (X). Subjects are then asked, "Which of the first three lines is the same length as the fourth?"

You'd probably find it a fairly simple task to identify "B" as the correct answer. Your job would be complicated, however, by the fact that several other "subjects" sitting beside you all agree that A is the same length as X! In reality, of course, the others in the experiment are all confederates of the researcher, told to agree on the wrong answer. As we saw in Chapter 2, the purpose of the experiment is to see whether you'd give up your own judgment in favor of the group agreement. I think you can see that conformity is a useful phenomenon to study and understand, and it couldn't be studied experimentally without deceiving the subjects. We'll examine a similar situation in the discussion of a famous experiment by Stanley Milgram later in this chapter. The question is, how do we get around the ethical issue that deception is necessary for an experiment to work?

One appropriate solution researchers have found is to debrief subjects following an experiment. **Debriefing** entails interviews to discover any problems generated by the research experience so that those problems can be corrected. Even though subjects can't be told the true purpose of the study prior to their participation in it, there's usually no reason they can't know afterward. Telling them the truth afterward may make up for having to lie to them at the outset. This must be done with care, however, making sure the subjects aren't left with bad feelings or doubts about themselves based on their performance in the experiment. If this seems complicated, it's simply the price we pay for using other people's lives as the subject matter for our research.

As a social researcher, then, you have many ethical obligations to the subjects in your studies. "Ethical Issues in Research on Human Sexuality"

---

debriefing Interviewing subjects to learn about their experience of participation in the project. Especially important if there's a possibility that they have been damaged by that participation.

Ethical Issues in Research on Human Sexuality

Kathleen McKinney

Department of Sociology
Illinois State University

When studying any form of human behavior, ethical concerns are paramount. This statement may be even truer for studies of human sexuality because of the topic's highly personal, salient, and perhaps threatening nature. Concern has been expressed by the public and by legislators about human sexuality research. Three commonly discussed ethical criteria have been related specifically to research in the area of human sexuality.

Informed Consent This criterion emphasizes the importance of both accurately informing your subject or respondent as to the nature of the research and obtaining his or her verbal or written consent to participate. Coercion is not to be used to force participation, and subjects may terminate their involvement in the research at any time. There are many possible violations of this standard. Misrepresentation or deception may be used when describing an embarrassing or personal topic of study, because the researchers fear high rates of refusal or false data. Covert research, such as some observational studies, also violate the informed consent standard since subjects are unaware that they are being studied. Informed consent may create special problems with certain populations. For example, studies of the sexuality of children are limited by the concern that children may be cognitively and emotionally unable to give informed consent. Although there can be problems such as those discussed, most research is clearly voluntary, with informed consent from those participating.

Right to Privacy Given the highly personal nature of sexuality and society's tremendous concern with social control of sexuality, the right to privacy is a very important ethical concern for research in this area. Individuals may risk losing their jobs, having family difficulties, or being ostracized by peers if certain facets of their sexual lives are revealed. This is especially true for individuals involved in sexual behavior categorized as deviant (such as transvestism). Violations of right to privacy occur when researchers identify members of certain groups they have studied, release or share an individual's data or responses, or covertly observe sexual behavior. In most cases, right to privacy is easily maintained by the researchers. In survey research, self-administered questionnaires can be anonymous and interviews can be kept confidential. In case and observational studies, the identity of the person or group studied can be disguised in any publications. In most research methods, analysis and reporting of data should be at the group or aggregate level.

Protection from Harm Harm may include emotional or psychological distress, as well as physical harm. Potential for harm varies by research method; it is more likely in experimental studies where the researcher manipulates or does something to the subject than in observational or survey research. Emotional distress, however, is a possibility in all studies of human sexuality. Respondents may be asked questions that elicit anxiety, dredge up unpleasant memories, or cause them to evaluate themselves critically. Researchers can reduce the potential for such distress during a study by using anonymous, self-administered questionnaires or well-trained interviewers and by wording sensitive questions carefully.

All three of these ethical criteria are quite subjective. Violations are sometimes justified by arguing that risks to subjects are outweighed by benefits to society. The issue here, of course, is who makes that critical decision. Usually, such decisions are made by the researcher and often a screening committee that deals with ethical concerns. Most creative researchers have been able to follow all three ethical guidelines and still do important research.

illustrates some of the ethical questions involved in a specific research area.

Analysis and Reporting

In addition to their ethical obligations to subjects, researchers have ethical obligations to their colleagues in the scientific community. These obligations concern the analysis of data and the way the results are reported.

In any rigorous study, the researcher should be more familiar than anyone else with the study's technical limitations and failures. Researchers have an obligation to make such shortcomings known to their readers—even if admitting qualifications and mistakes makes them feel foolish.

Negative findings, for example, should be reported if they are at all related to the analysis. There is an unfortunate myth in scientific reporting that only positive discoveries are worth reporting (journal editors are sometimes guilty of believing this as well). In science, however, it's often as important to know that two variables are not related as to know that they are.

Similarly, researchers must avoid the temptation to save face by describing their findings as the product of a carefully preplanned analytic strategy when that is not the case. Many findings arrive unexpectedly—even though they may seem obvious in retrospect. So an interesting relationship was uncovered by accident—so what? Embroidering such situations with descriptions of fictitious hypotheses is dishonest. It also does a disservice to less-experienced researchers by leading them into thinking that all scientific inquiry is rigorously preplanned and organized.

In general, science progresses through honesty and openness; ego defenses and deception retard it. Researchers can best serve their peers—and scientific discovery as a whole—by telling the truth about all the pitfalls and problems they've experienced in a particular line of inquiry. Perhaps they'll save others from the same problems.

Finally, there is a sense in which simple carelessness or sloppiness can be considered an ethical problem. If the research project uses up limited resources and/or imposes on subjects with no benefit produced by the research, many in the research community would consider that an ethical violation. This is not to say that all research must produce positive results, but it should be conducted in a manner that promotes that possibility.

Institutional Review Boards

The issue of research ethics in studies involving humans is now also governed by federal law. Any agency (such as a university or a hospital) wishing to receive federal research support must establish an Institutional Review Board (IRB), a panel of faculty (and possibly others) who review all research proposals involving human subjects so that they can guarantee that the subjects' rights and interests will be protected. Although the law applies specifically to federally funded research, many universities apply the same standards and procedures

to all research, including that funded by nonfederal sources and even research done at no cost, such as student projects.

The chief responsibility of an IRB is to ensure that the risks faced by human participants in research are minimal. In some cases, the IRB may ask the researcher to revise the study design; in others, the IRB may refuse to approve a study. Where some minimal risks are deemed unavoidable, researchers are required to prepare an "informed consent" form that describes those risks clearly. Subjects may participate in the study only after they have read the statement and signed it as an indication that they know the risks and voluntarily accept them.

Much of the impetus for establishing IRBs had to do with medical experimentation on humans, and many social research study designs are generally regarded as exempt from IRB review. An example is an anonymous survey sent to a large sample of respondents. The guideline to be followed by IRBs, as contained in the Federal Exemption Categories (45 CFR 46.101 [b]), exempts a variety of research situations:

(1) Research conducted in established or commonly accepted educational settings, involving normal educational practices, such as (i) research on regular and special education instructional strategies, or (ii) research on the effectiveness of or the comparison among instructional techniques, curricula, or classroom management methods.

(2) Research involving the use of educational tests (cognitive, diagnostic, aptitude, achievement), survey procedures, interview procedures or observation of public behavior, unless:

(i) information obtained is recorded in such a manner that human subjects can be identified, directly or through identifiers linked to the subjects; and (ii) any disclosure of the human subjects' responses outside the research could reasonably place the subjects at risk of criminal or civil liability or be damaging to the subjects' financial standing, employability, or reputation.

(3) Research involving the use of educational tests (cognitive, diagnostic, aptitude,

achievement), survey procedures, interview procedures, or observation of public behavior that is not exempt under paragraph (b)(2) of this section, if:

(i) the human subjects are elected or appointed public officials or candidates for public office; or (ii) Federal statute(s) require(s) without exception that the confidentiality of the personally identifiable information will be maintained throughout the research and thereafter.

(4) Research involving the collection or study of existing data, documents, records, pathological specimens, or diagnostic specimens, if these sources are publicly available or if the information is recorded by the investigator in such a manner that subjects cannot be identified, directly or through identifiers linked to the subjects.

(5) Research and demonstration projects which are conducted by or subject to the approval of Department or Agency heads, and which are designed to study, evaluate, or otherwise examine:

(i) Public benefit or service programs; (ii) procedures for obtaining benefits or services under those programs; (iii) possible changes in or alternatives to those programs or procedures; or (iv) possible changes in methods or levels of payment for benefits or services under those programs.

(6) Taste and food quality evaluation and consumer acceptance studies, (i) if wholesome foods without additives are consumed or (ii) if a food is consumed that contains a food ingredient at or below the level and for a use found to be safe, or agricultural chemical or environmental contaminant at or below the level found to be safe, by the Food and Drug Administration or approved by the Environmental Protection Agency or the Food Safety and Inspection Service of the U.S. Department of Agriculture.

Paragraph (2) of the excerpt exempts much of the social research described in this book. Nonetheless, universities sometimes apply the law's provisions inappropriately. As chair of a university IRB, for example, I was once asked to review the letter of informed consent that was to be sent to medical insurance companies, requesting their agreement to participate in a survey that would ask which medical treatments were covered under their programs. Clearly the humans involved were not at risk in the sense anticipated by the law. In a case like that, the appropriate technique for gaining informed consent is to mail the questionnaire. If a company returns it, they've consented. If they don't, they haven't.

Other IRBs have suggested that researchers need to obtain permission before observing participants in public gatherings and events, before conducting surveys on the most mundane matters, and so forth. Christopher Shea (2000) has chronicled several such questionable applications of the law while supporting the ethical logic that originally prompted the law.

Don't think that these critiques of IRBs minimize the importance of protecting human subjects. Indeed, some universities exceed the federal requirements in reasonable and responsible ways: requiring IRB review of nonfederally funded projects, for example.

Research ethics is an ever-evolving subject, because new research techniques often require revisiting old concerns. Thus, for example, the increased use of public databases for secondary research has caused some IRBs to worry whether they need to reexamine such projects as the General Social Survey every time a researcher proposes to use those data. (Most have decided this is unnecessary; see Skedsvold 2002 for a discussion of issues relating to public databases.)

Similarly, the prospects for research of and through the Internet has raised ethical concerns. In November 1999, the American Association for the Advancement of Science sponsored a workshop on this subject. The overall conclusion of the report produced by the workshop summarizes some of the primary concerns already examined in this chapter:

The current ethical and legal framework for protecting human subjects rests on the principles of autonomy, beneficence, and

SURVEY PRACTICES THAT AAPOR CONDEMNS

AAPOR joins the Research Industry Coalition and the National Council on Public Polls in condemning certain misleading practices sometimes performed in the name of research. In no case are the following practices deemed legitimate or acceptable elements of professionally conducted research:

1. **Requiring a monetary payment or soliciting monetary contributions from members of the public as part of a research process.**
   This set of practices amounts to fund raising under the guise of research. It takes unfair advantage of the cooperative attitude that a majority of the public manifests when asked to take part in a legitimate information gathering process. In some cases, unwary members of the public are enticed to contribute money as a condition of gaining some future "benefit" from their participation.

2. **Offering products or services for sale, or using participant contacts as a means of generating sales leads.**
   A common practice is to gain entry or acceptance in order to make a sales pitch by initially defining the contact as being made for "research" purposes. This trades on the prestige of science, and it exploits the willingness of the public to reveal information about themselves in the public interest. In some cases, questions establish respondents' susceptibility to sales pressure or their interest in some product or service. Follow-up contacts are then made to those so identified, all under the guise of "research."

3. **Revealing the identity of individual respondents to a survey or participants in a research process without their permission.**
   It is normal research practice to pledge anonymity or confidentiality to the public in order to secure their cooperation and frankness in responding to questions. Revealing the identity of individuals, for whatever purpose, is a violation of that pledge unless a respondent's prior informed consent has been obtained.

4. **Representing the results of a 900-number or other type of self-selected "poll" as if they were the outcome of legitimate research.**
   900-number and other types of write-in, call-in, and interactive polls have become increasingly common. These "polls" report the opinions of only those people who called in, and not those of the general public. AAPOR believes that any publicizing or promotion of such activities not only damages legitimate market and survey research, but can be very misleading when used to influence public policy or simply to disseminate information about the general public.

5. **Conducting a so-called "push poll," a telemarketing technique in which telephone calls are used to canvass potential voters, feeding them false or misleading "information" about a candidate under the pretense of taking a poll to see how this "information" affects voter preferences.**
   So-called "Push polls" are not polls at all. They are a form of political telemarketing whose intent is not to measure public opinion but to manipulate — "push" — voters away from one candidate and toward the opposing candidate. Such polls defame selected candidates by spreading false or misleading information about them. The intent is to disseminate campaign propaganda under the guise of conducting a legitimate public opinion poll. Read AAPOR's recent statement on "push" polls.

As members of AAPOR, a professional organization which relies on public cooperation to gather information that is useful in formulating public policy as well as in understanding the public's preferences for products and services, we condemn these practices in the strongest terms.

FIGURE 3-1

Excerpt from the Code of Conduct of the American Association for Public Opinion Research

Source: American Association for Public Opinion Research, By-Laws (2005). The entire code of conduct can be found at the link on this book's website: http://www.cengage.com/sociology/babbie.

justice. The first principle, autonomy, requires that subjects be treated with respect as autonomous agents and affirms that those persons with diminished autonomy are entitled to special protection. In practice, this principle is reflected in the process of informed consent, in which the risks and benefits of the research are disclosed to the subject. The second principle, beneficence, involves maximizing possible benefits and good for the subject, while minimizing the amount of possible harm and risks resulting from the research. Since the fruits of knowledge can come at a cost to those participating in research, the last principle, justice, seeks a fair distribution of the burdens and benefits associated with research, so that certain individuals or groups do not bear disproportionate risks while others reap the benefits.

(Frankel and Siang 1999: 2–3)

Professional Codes of Ethics

Ethical issues in social research are both important and ambiguous. For this reason, most of the professional associations of social researchers have created and published formal codes of conduct describing what is considered acceptable and unacceptable professional behavior. As one example, Figure 3-1 presents a portion of the code of conduct of the American Association for Public Opinion Research (AAOPR), an interdisciplinary research association in the social sciences. Most professional associations have such codes of ethics. See, for example, the American Sociological Association, the American Psychological Association, the American Political Science Association, and so forth. You can find many of these on the associations' websites. In addition, the Association of Internet Researchers (AoIR) has a code of ethics accessible online. The excerpt presented details several pseudoresearch practices that are denounced by AAPOR and other professional researchers.

Two Ethical Controversies

As you may already have guessed, the adoption and publication of professional codes of conduct have not totally resolved the issue of research eth-

ics. Social researchers still disagree on some general principles, and those who agree in principle often debate specifics.

This section briefly describes two research projects that have provoked ethical controversy and discussion. The first project studied homosexual behavior in public restrooms, and the second examined obedience in a laboratory setting.

Trouble in the Tearoom

As a graduate student, Laud Humphreys became interested in the study of homosexual behavior. He developed a special interest in the casual and fleeting same-sex acts engaged in by some male nonhomosexuals. In particular, his research interest focused on homosexual acts between strangers meeting in the public restrooms in parks, called "tearooms" among homosexuals. The result was the publication in 1970 of *Tearoom Trade*.

What particularly interested Humphreys about the tearoom activity was that the participants seemed otherwise to live conventional lives as "family men" and accepted members of the community. They did nothing else that might qualify them as homosexuals. Thus, it was important to them that they remain anonymous in their tearoom visits. How would you study something like that?

Humphreys decided to take advantage of the social structure of the situation. Typically, the tearoom encounter involved three people: the two men actually engaging in the sexual act and a lookout, called the "watchqueen." Humphreys began showing up at public restrooms, offering to serve as watchqueen whenever it seemed appropriate. Because the watchqueen's payoff was the chance to watch the action, Humphreys was able to conduct field observations as he would in a study of political rallies or jaywalking behavior at intersections.

To round out his understanding of the tearoom trade, Humphreys needed to know something more about the people who participated. Because the men probably would not have been thrilled about being interviewed, Humphreys developed a different solution. Whenever possible, he noted the license numbers of participants' cars and tracked down their names and addresses through the police. Humphreys then visited the men at their homes, disguising himself enough to avoid recognition, and announced that he was conducting a survey. In that fashion, he collected the personal information he couldn't get in the restrooms.

As you can imagine, Humphreys' research provoked considerable controversy both inside and outside the social science community. Some critics charged Humphreys with a gross invasion of privacy in the name of science. What men did in public restrooms was their own business. Others were mostly concerned about the deceit involved—Humphreys had lied to the participants by leading them to believe he was only a voyeur-participant. Even people who felt that the tearoom participants were fair game for observation because they used a public facility protested the follow-up survey. They felt it was unethical for Humphreys to trace the participants to their homes and to interview them under false pretenses.

Still others justified Humphreys' research. The topic, they said, was worth study. It couldn't be studied any other way, and they regarded the deceit as essentially harmless, noting that Humphreys was careful not to harm his subjects by disclosing their tearoom activities. One result of Humphrey's research was to challenge some of the common stereotypes about the participants in anonymous sexual encounters in public places, showing them to be basically conventional in other aspects of their lives.

The tearoom trade controversy has never been resolved. It's still debated, and it probably always will be, because it stirs emotions and involves ethical issues people disagree about. What do you think? Was Humphreys ethical in doing what he did? Are there parts of the research that you believe were acceptable and other parts that were not? (For more on the political and ethical context of the "tearoom" research, see the discussion by Joan Sieber at the link on this book's website: http://www.cengage.com/sociology/babbie.)

Observing Human Obedience

The second illustration differs from the first in many ways. Whereas Humphreys' study involved participant observation, this study took place in the laboratory. Humphreys' study was sociological, this one psychological. And whereas Humphreys examined behavior considered by many to be deviant, the researcher in this study examined obedience and conformity.

One of the most unsettling clichés to come out of World War II was the German soldier's common excuse for atrocities: "I was only following orders." From the point of view that gave rise to this comment, any behavior—no matter how reprehensible—could be justified if someone else could be assigned responsibility for it. If a superior officer ordered a soldier to kill a baby, the fact of the order supposedly exempted the soldier from personal responsibility for the action.

Although the military tribunals that tried the war crime cases did not accept this excuse, social researchers and others have recognized the extent to which this point of view pervades social life. People often seem willing to do things they know would be considered wrong, if they can claim that some higher authority ordered them to do it. Such was the pattern of justification in the 1968 My Lai tragedy of Vietnam, when U.S. soldiers killed more than 300 unarmed civilians—some of them young children—simply because their village, My Lai, was believed to be a Viet Cong stronghold. This sort of justification appears less dramatically in day-to-day civilian life. Few would disagree that this reliance on authority exists, yet Stanley Milgram's study (1963, 1965) of the topic provoked considerable controversy.

To observe people's willingness to harm others when following orders, Milgram brought 40 adult men from many different walks of life into a laboratory setting designed to create the phenomenon under study. If you had been a subject in the experiment, you would have had something like the following experience.

You've been informed that you and another subject are about to participate in a learning experiment. Through a draw of lots, you're assigned the job of "teacher" and your fellow subject the job of "pupil." The pupil is led into another room and strapped into a chair; an electrode is attached to his wrist. As the teacher, you're seated in front of an impressive electric control panel covered with dials, gauges, and switches. You notice that each switch has a label giving a different number of volts, ranging from 15 to 315. The switches have other labels, too, some with the ominous phrases "Extreme-Intensity Shock," "Danger—Severe Shock," and "XXX."

The experiment runs like this. You read a list of word pairs to the learner and then test his ability to match them up. Because you can't see him, a light on your control panel indicates his answer. Whenever the learner makes a mistake, you're instructed by the experimenter to throw one of the switches—beginning with the mildest—and administer a shock to your pupil. Through an open door between the two rooms, you hear your pupil's response to the shock. Then you read another list of word pairs and test him again.

As the experiment progresses, you administer ever more intense shocks, until your pupil screams for mercy and begs for the experiment to end. You're instructed to administer the next shock anyway. After a while, your pupil begins kicking the wall between the two rooms and continues to scream. The implacable experimenter tells you to give the next shock. Finally, you read a list and ask for the pupil's answer—but there is no reply whatever, only silence from the other room. The experimenter informs you that no answer is considered an error and instructs you to administer the next higher shock. This continues up to the "XXX" shock at the end of the series.

What do you suppose you really would have done when the pupil first began screaming? When he began kicking on the wall? Or when he became totally silent and gave no indication of life? You'd refuse to continue giving shocks, right? And surely the same would be true of most people.

So we might think—but Milgram found otherwise. Of the first 40 adult men Milgram tested, nobody refused to continue administering the shocks until they heard the pupil begin kicking the wall between the two rooms. Of the 40, 5 did so then. Two-thirds of the subjects, 26 of the 40, continued doing as they were told through the entire series—up to and including the administration of the highest shock.

As you've probably guessed, the shocks were phony, and the "pupil" was a confederate of the experimenter. Only the "teacher" was a real subject in the experiment. As a subject, you wouldn't actually have been hurting another person, but you would have been led to think you were. The experiment was designed to test your willingness to follow orders to the point of presumably killing someone.

Milgram's experiments have been criticized both methodologically and ethically. On the ethical side, critics have particularly cited the effects of the experiment on the subjects. Many seemed to have experienced personally about as much pain as they thought they were administering to someone else. They pleaded with the experimenter to let them stop giving the shocks. They became extremely upset and nervous. Some had uncontrollable seizures.

How do you feel about this research? Do you think the topic was important enough to justify such measures? Would debriefing the subjects be sufficient to ameliorate any possible harm? Can you think of other ways the researcher might have examined obedience?

There is a wealth of discussion regarding the Milgram experiments on the web. Search for "Milgram experiments," "human obedience experiments," or "Stanley Milgram."

The Politics of Social Research

As I indicated earlier, both ethics and politics hinge on ideological points of view. What is unacceptable from one point of view will be acceptable from another. Although political and ethical issues are often closely intertwined, I want to distinguish between them in two ways.

First, the ethics of social research deals mostly with the methods employed; political issues tend to center on the substance and use of research. Thus, for example, some critics raise ethical objections to the Milgram experiments, saying that the methods harm the subjects. A political objection would be that obedience is not a suitable topic for study, either because (1) we should not tinker with people's willingness to follow orders from higher authority or (2), from the opposite political point of view, because the results of the research could be used to make people *more* obedient.

The second distinction between the ethical and political aspects of social research is that there are no formal codes of accepted political conduct. Although some ethical norms have political aspects—for example, specific guidelines for not harming subjects clearly relate to Western ideas about the protection of civil liberties—no one has developed a set of political norms that all social researchers accept.

The only partial exception to the lack of political norms is the generally accepted view that a researcher's personal political orientation should not interfere with or unduly influence his or her scientific research. It would be considered improper for a researcher to use shoddy techniques or to distort or lie about his or her research as a way of furthering the researcher's political views. As you can imagine, however, studies are often enough attacked for allegedly violating this norm.

Objectivity and Ideology

In Chapter 1, I suggested that social research can never be totally objective, because researchers are human and therefore necessarily subjective. As a collective enterprise, science achieves the equivalent of objectivity through intersubjectivity. That is, different scientists, having different subjective views, can and should arrive at the same results when they employ accepted research techniques. Essentially, this will happen to the extent that each can set personal values and views aside for the duration of the research.

The classic statement on objectivity and neutrality in social science is Max Weber's lecture "Science as a Vocation" ([1925] 1946). In this talk, Weber coined the phrase *value-free sociology* and urged that sociology, like other sciences, needed to be unencumbered by personal values if it were to make a special contribution to society. Liberals and conservatives alike could recognize the "facts" of social science, regardless of how those facts accorded with their personal politics.

Most social researchers have agreed with this abstract ideal, but not all. Marxist and neo-Marxist scholars, for example, have argued that social science and social action cannot and should not be separated. Explanations of the status quo in society, they contend, shade subtly into defenses of that same status quo. Simple explanations of the social functions of, say, discrimination can easily become justifications for its continuance. By the same token, merely studying society and its ills without a commitment to making society more humane has been called irresponsible.

In Chapter 10, we'll examine *participatory action research*, which is explicitly committed to using social research for purposes designed and valued by the subjects of the research. Thus, for example, researchers committed to improving the working conditions for workers at a factory would ask the workers to define the outcomes they would like to see and to have a hand in conducting social research relevant to achieving the desired ends. The role of the researchers is to ensure that the workers have access to professional research methods.

Quite aside from abstract disagreements about whether social science can or *should* be value-free, many have argued about whether particular research undertakings *are* value-free or whether they represent an intrusion of the researcher's own political values. Typically, researchers have denied such intrusion, and their denials have then been challenged. Let's look at some examples of the controversies this issue has produced.

Social Research and Race

Nowhere have social research and politics been more controversially intertwined than in the area of racial relations. Social researchers studied the topic for a long time, and the products of the social research have often found their way into practical politics. A few brief references should illustrate the point.

In 1896, when the U.S. Supreme Court established the principle of "separate but equal" as a means of reconciling the Fourteenth Amendment's guarantee of equality to African Americans with the norms of segregation, it neither asked for nor cited social research. Nonetheless, it is widely believed that the Court was influenced by the writings of William Graham Sumner, a leading social scientist of his era. Sumner was noted for his view that the mores and folkways of a society were relatively impervious to legislation and social planning. His view has often been paraphrased as "stateways do not make folkways." Thus, the Court ruled that it could not accept the assumption that "social prejudices may be overcome by legislation" and denied the wisdom of "laws which conflict with the general sentiment of the community" (Blaunstein and Zangrando 1970: 308). As many a politician has said, "You can't legislate morality."

When the doctrine of "separate but equal" was overturned in 1954 (*Brown v. Board of Education*), the new Supreme Court decision was based in part on the conclusion that segregation had a detrimental effect on African American children. In drawing that conclusion, the Court cited several sociological and psychological research reports (Blaunstein and Zangrando 1970).

For the most part, social researchers in this century have supported the cause of African American equality in the United States, and their convictions often have been the impetus for their research. Moreover, they've hoped that their research will lead to social change. There is no doubt, for example, that Gunnar Myrdal's classic two-volume study (1944) of race relations in the United States had a significant impact on the topic of his research. Myrdal amassed a great deal of data to show that the position of African Americans directly contradicted U.S. values of social and political equality. Further, Myrdal did not attempt to hide his own point of view in the matter. (You can pursue Myrdal's landmark research further online by searching for "Gunnar Myrdal" or "An American Dilemma.")

Many social researchers have become directly involved in the civil rights movement, some more radically than others. Given the broad support for ideals of equality, research conclusions supporting the cause of equality draw little or no criticism. To recognize how solid the general social science position is in this matter, we need only examine a few research projects that have produced conclusions disagreeing with the predominant ideological position.

Most social researchers have—overtly, at least—supported the end of school segregation. Thus, an immediate and heated controversy arose

in 1966 when James Coleman, a respected sociologist, published the results of a major national study of race and education. Contrary to general agreement, Coleman found little difference in academic performance between African American students attending integrated schools and those attending segregated ones. Indeed, such obvious things as libraries, laboratory facilities, and high expenditures per student made little difference. Instead, Coleman reported that family and neighborhood factors had the most influence on academic achievement.

Coleman's findings were not well received by many of the social researchers who had been active in the civil rights movement. Some scholars criticized Coleman's work on methodological grounds, but many others objected hotly on the grounds that the findings would have segregationist political consequences. The controversy that raged around the Coleman report was reminiscent of that provoked a year earlier by Daniel Moynihan (1965) in his critical analysis of the African American family in the United States. Whereas some felt Moynihan was blaming the victims, others objected to his tracing those problems to the legacy of slavery.

Another example of political controversy surrounding social research in connection with race concerns IQ scores. In 1969, Arthur Jensen, a Harvard psychologist, was asked to prepare an article for the *Harvard Educational Review* examining the data on racial differences in IQ test results (Jensen 1969). In the article, Jensen concluded that genetic differences between African Americans and whites accounted for the lower average IQ scores of African Americans. Jensen became so identified with that position that he appeared on college campuses across the country discussing it.

Jensen's research has been attacked on numerous methodological bases. Critics charged that much of the data on which Jensen's conclusion was based were inadequate and sloppy—there are many IQ tests, some worse than others. Similarly, it was argued that Jensen had not taken social-environmental factors sufficiently into account. Other social researchers raised still other methodological objections.

Beyond the scientific critique, however, many condemned Jensen as a racist. Hostile crowds booed him, drowning out his public presentations. Ironically, Jensen's reception by several university audiences did not differ significantly from the reception received by abolitionists over a century before, when the prevailing opinion favored leaving the institution of slavery intact.

Many social researchers limited their objections to the Moynihan, Coleman, and Jensen research to scientific, methodological grounds. The political firestorms ignited by these studies, however, point out how ideology often shows up in matters of social research. Although the abstract model of science is divorced from ideology, the practice of science is not.

To examine a more recent version of the controversy surrounding race and achievement, search the web for differing points of view concerning "The Bell Curve"—sparked by a book with that title by Richard J. Herrnstein and Charles Murray.

The controversies relating to research and race are hardly over at present, as we saw in the Chapter 2 discussion of critical race theory.

The Politics of Sexual Research

As I indicated earlier, the Laud Humphreys study of tearoom trade raised ethical issues that researchers still discuss and debate. At the same time, it seems clear that much of the furor raised by the research was related to the subject matter itself. As I have written elsewhere,

> Laud Humphreys didn't just study S-E-X but observed and discussed *homo*sexuality. And it wasn't even the caring-and-committed-relationships-between-two-people-who-just-happen-to-be-of-the-same-sex homosexuality but tawdry encounters between strangers in public toilets. Only adding the sacrifice of Christian babies could have made this more inflammatory for the great majority of Americans in 1970.
>
> *(Babbie 2004: 12)*

Whereas Humphreys' research topic proved unusually provocative for many, much tamer sexuality research has also engendered outcries of public horror. During the 1940s and 1950s, the biologist Alfred Kinsey and his colleagues published landmark studies of sexual practices of American men (1948) and women (1953). Kinsey's extensive interviewing allowed him to report on frequency of sexual activity, premarital and extramarital sex, homosexual behavior, and so forth. His studies produced public outrage and efforts to close his research institute at Indiana University.

Although today most people no longer get worked up about the Kinsey reports, Americans tend to remain touchy about research on sex. In 1987, the National Institutes of Health (NIH), charged with finding ways to combat the AIDS epidemic, found they needed hard data on contemporary sexual practices if they were to design effective anti-AIDS programs. Their request for research proposals resulted in a sophisticated study design by Edward O. Laumann and colleagues. The proposed study focused on the different patterns of sexual activity characterizing different periods of life, and it received rave reviews from the NIH and their consultants.

Enter Senator Jesse Helms (R-North Carolina) and Congressman William Dannemeyer (R-California). In 1989, having learned of the Laumann study, Helms and Dannemeyer began a campaign to block the study and shift the same amount of money to a teen celibacy program. Anne Fausto-Sterling, a biologist, sought to understand the opposition to the Laumann study.

> The surveys, Helms argued, are not really intended "to stop the spread of AIDS. The real purpose is to compile supposedly scientific facts to support the left-wing liberal argument that homosexuality is a normal, acceptable life-style. . . . As long as I am able to stand on the floor of the U.S. Senate," he added, "I am never going to yield to that sort of thing, because it is not just another life-style; it is sodomy."
>
> *(Fausto-Sterling 1992)*

Helms won a 66–34 vote in favor of his amendment in the U.S. Senate. Although the House of Representatives rejected the amendment, and it was dropped in conference committee, government funding for the study was put on hold. Laumann and his colleagues then turned to the private sector and obtained funding, albeit for a smaller study, from private foundations. Their research results were published in 1994 as *The Social Organization of Sexuality.*

Politics and the Census

There is probably a political dimension to every attempt to study human social behavior. Consider the matter of the U.S. decennial census, mandated by the Constitution. The original purpose was to discover the population sizes of the various states to determine their proper representation in the House of Representatives. Whereas each state gets two senators, large states get more representatives than small ones do. So what could be simpler? Just count the number of people in each state.

From the beginning, there was nothing simple about counting heads in a dispersed, national population like the United States. Even the definition of a "person" was anything but straightforward. A slave, for example, counted as only three-fifths of a person for purposes of the census. This decreased the representation of the slaveholding Southern states, though counting slaves as whole people might have raised the dangerously radiscal idea that they should be allowed to vote.

Further, the logistical problems of counting people who reside in suburban tract houses, urban apartments, college dorms, military barracks, farms, cabins in the woods, and illegal housing units, as well as counting those who have no place to live, has always presented a daunting task. It's the sort of challenge social researchers tackle with relish. However, the difficulty of finding the hard-to-reach and the techniques created for doing so cannot escape the political net.

Kenneth Prewitt, who directed the Census Bureau from 1998 to 2001, describes some of the political aspects of counting heads:

> Between 1910 and 1920, there was a massive wartime population movement from the rural, Southern states to industrial Northern cities. In 1920, for the first time in American history, the census included more city dwellers than rural residents. An urban America was something new and disturbing, especially to those who held to the Jeffersonian belief that independent farmers best protected democracy. Among those of this persuasion were rural, conservative congressmen in the South and West. They saw that reapportionment would shift power to factory-based unions and politically radical immigrants concentrated in Northeastern cities. Conservatives in Congress blocked reapportionment, complaining among other things that because January 1 was then census day, transient agricultural workers were "incorrectly" counted in cities rather than on the farms to which they would return in time for spring planting. (Census day was later shifted to April 1, where it has remained.) The arguments dragged out for a decade, and Congress was not reapportioned until after the next census.
>
> *(Prewitt 2003)*

In more recent years, concern for undercounting the urban poor has become a political issue. The big cities, which have the most to lose from the undercounting, typically vote Democratic rather than Republican, so you can probably guess which party supports efforts to improve the counting and which party is less enthusiastic. By the same token, when social scientists have argued in favor of replacing the attempt at a total enumeration of the population with modern survey sampling methods (see Chapter 7), they have enjoyed more support from Democrats, who would stand to gain from such a methodological shift, than from Republicans, who would stand to lose. Rather than suggesting Democrats support science more than Republicans do, this situation offers another example of how the political context in which we live and conduct social research often affects that research.

Politics with a Little "p"

Social research is often confounded by political ideologies, but the "politics" of social research runs far deeper still. Social research in relation to contested social issues simply cannot remain antiseptically objective—particularly when differing ideologies are pitted against each other in a field of social science data.

The same is true when research is invoked in disputes between people with conflicting interests. For instance, social researchers who have served as "expert witnesses" in court would probably agree that the scientific ideal of a "search for truth" seems hopelessly naive in a trial or lawsuit. Although expert witnesses technically do not represent either side in court, they are, nonetheless, engaged by only one side to appear, and their testimony tends to support the side of the party who pays for their time. This doesn't necessarily mean that these witnesses will lie on behalf of their patrons, but the contenders in a lawsuit are understandably more likely to pay for expert testimony that supports their case than for testimony that attacks it.

Thus, as an expert witness, you appear in court only because your presumably scientific and honest judgment happens to coincide with the interests of the party paying you to testify. Once you arrive in court and swear to tell the truth, the whole truth, and nothing but the truth, however, you find yourself in a world foreign to the ideals of objective contemplation. Suddenly, the norms are those of winning and losing. As an expert witness, of course, all you have to lose is your respectability (and perhaps the chance to earn fees as an expert witness in the future). Still, such stakes are high enough to create discomfort for most social researchers.

I recall one case in federal court when I was testifying on behalf of some civil service workers who had had their cost-of-living allowance (COLA) cut on the basis of research I thought was rather shoddy. I was engaged to conduct more "scientific" research that would demonstrate the injustice worked against the civil servants (Babbie 1982: 232–43).

I took the stand, feeling pretty much like a respected professor and textbook author. In short order, however, I found I had moved from the academy to the hockey rink. Tests of statistical significance and sampling error were suddenly less relevant than a slap shot. At one point, an attorney

from Washington lured me into casually agreeing that I was familiar with a certain professional journal. Unfortunately, the journal did not exist. I was mortified and suddenly found myself shifting domains. Without really thinking about it, I now was less committed to being a friendly Mr. Chips and more aligned with ninja-professor. I would not be fully satisfied until I, in turn, could mortify the attorney, which I succeeded in doing.

Even though the civil servants got their cost-of-living allowance back, I have to admit I was also concerned with how I looked in front of the courtroom assemblage. I tell you this anecdote to illustrate the personal "politics" of human interactions involving presumably scientific and objective research. We need to realize that as human beings social researchers are going to act like human beings, and we must take this into account in assessing their findings. This recognition does not invalidate their research or provide an excuse for rejecting findings we happen to dislike, but it does need to be taken into account.

Politics in Perspective

Although the ethical and the political dimensions of research are in principle distinct, they do intersect. Whenever politicians or the public feel that social research is violating ethical or moral standards, they'll be quick to respond with remedies of their own. Moreover, the standards they defend may not be those of the research community. Even when researchers support the goals of measures directed at the way research is done, the means specified by regulations or legislation can hamstring research.

Legislators show special concern for research on children. Although the social research norms discussed in this chapter would guard against bringing any physical or emotional harm to children, some of the restrictive legislation introduced from time to time borders on the actions of one particular western city, which shall remain nameless. In response to concerns that a public school teacher had been playing New Age music in class and encouraging students to meditate, the city council passed legislation stating that no teacher could do anything that would "affect the minds of students"!

In recent years, the "politicization of science" has become a particularly hot topic, with charges flung from both sides of the political spectrum. On the one hand, renewed objections to the teaching of evolution have coupled with demands for the teaching of Intelligent Design (replacing Creationism). In many of these regards, science is seen as a threat to religiously-based views, and scientists are sometimes accused of an antireligious agenda.

On the other hand, a statement by the Union of Concerned Scientists (2005), cosigned by thousands of scientists, illustrates the concern that the concentration of political power in the hands of one party can threaten the independent functioning of scientific research:

> The United States has an impressive history of investing in scientific research and respecting the independence of scientists. As a result, we have enjoyed sustained economic progress and public health, as well as unequaled leadership within the global scientific community. Recent actions by political appointees, however, threaten to undermine this legacy by preventing the best available science from informing policy decisions that have serious consequences for our health, safety, and environment.
>
> Across a broad range of issues—from childhood lead poisoning and mercury emissions to climate change, reproductive health, and nuclear weapons—political appointees have distorted and censored scientific findings that contradict established policies. In some cases, they have manipulated the underlying science to align results with predetermined political decisions.

I hope you take away four main lessons from this discussion. First, science is not untouched by politics. The intrusion of politics and related ideologies is not unique to social research; the natural sciences have experienced and continue to experience similar intrusions. But social science is particularly linked to social life. Social researchers study things that matter to people—things that people have firm, personal feelings about and that affect their lives. Moreover, researchers are human beings, and their feelings often surface in their professional lives. To think otherwise would be naive.

Second, science manages to proceed in the midst of political controversy and hostility. Even when researchers get angry and call each other names, or when the research community comes under attack from the outside, scientific inquiry persists. Studies are done, reports are published, and new things are learned. In short, ideological disputes do not bring science to a halt, but they do make it more challenging—and exciting.

Third, an awareness of ideological considerations enriches the study and practice of social research methods. Many of the established characteristics of science, such as intersubjectivity, function to cancel out or hold in check our human shortcomings, especially those we are unaware of. Otherwise, we might look into the world and never see anything but a reflection of our personal biases and beliefs.

Finally, whereas researchers should not let their own values interfere with the quality and honesty of their research, this does not mean that researchers cannot or should not participate in public debates and express both their scientific expertise and personal values. You can do scientifically excellent research on racial prejudice, all the while being opposed to prejudice and saying so. Some would argue that social scientists, because of their scientific expertise in the workings of society, have an obligation to speak out, rather than leaving that role to politicians, journalists, and talk-show hosts. Herbert Gans (2002) writes of the need for "public sociologists":

> A public sociologist is a public intellectual who applies sociological ideas and findings to social (defined broadly) issues about which sociology (also defined broadly) has something to say. Public intellectuals comment on whatever issues show up on the public agenda; public sociologists do so only on issues to which they can apply their sociological insights and findings.

MAIN POINTS

Introduction

- In addition to technical, scientific considerations, social research projects are likely to be shaped by administrative, ethical, and political considerations.

Ethical Issues in Social Research

- What is ethical and unethical in research is ultimately a matter of what a community of people agree is right and wrong.
- Researchers agree that participation in research should normally be voluntary. This norm, however, can conflict with the scientific need for generalizability.
- Researchers agree that research should not harm those who participate in it, unless they give their informed consent, thereby willingly and knowingly accepting the risks of harm.
- Whereas anonymity refers to the situation in which even the researcher cannot identify specific information with the individuals it describes, confidentiality refers to the situation in which the researcher promises to keep information about subjects private. The most straightforward way to ensure confidentiality is to destroy identifying information as soon as it's no longer needed.
- Many research designs involve a greater or lesser degree of deception of subjects. Because deceiving people violates common standards of ethical behavior, deception in research requires a strong justification—and even then the justification may be challenged.
- Social researchers have ethical obligations to the community of researchers as well as to subjects. These obligations include reporting results fully and accurately as well as disclosing errors, limitations, and other shortcomings in the research.
- Professional associations in several disciplines publish codes of ethics to guide researchers. These codes are necessary and helpful, but they do not resolve all ethical questions.

Two Ethical Controversies

- Laud Humphreys' study of "tearoom" encounters and Stanley Milgram's study of obedience raise ethical issues that are debated to this day.

The Politics of Social Research

- Social research inevitably has a political and ideological dimension. Although science is neutral on political matters, scientists are not. Moreover, much social research inevitably involves the political beliefs of people outside the research community.
- Although most researchers agree that political orientation should not unduly influence research, in practice separating politics and ideology from the conduct of research can be quite difficult. Some researchers maintain that research can and should be an instrument of social action and change. More subtly, a shared ideology can affect the way other researchers receive one's research.
- Even though the norms of science cannot force individual researchers to give up their personal values, the intersubjective character of science provides a guard against scientific findings being the product of bias only.

KEY TERMS

The following terms are defined in context in the chapter and at the bottom of the page where the term is introduced, as well as in the comprehensive glossary at the back of the book.

anonymity	debriefing
confidentiality	informed consent

PROPOSING SOCIAL RESEARCH: ETHICAL ISSUES

If you are actually proposing a research project, you may be required to submit your proposal to your campus Institutional Review Board (IRB). In that case, you will need to inform yourself as to the forms and procedures involved locally. The key concern here is the protection of research subjects: avoiding harm, safeguarding their privacy, and the other such topics discussed in this chapter.

In this section of the proposal, you will discuss the ethical risks involved in your study and the steps you will take to avoid them. Perhaps you will prepare forms to insure that subjects are aware of and give *informed consent* to the risks attendant on their participation. The terms "anonymous" and/or "confidential" are likely to appear in your discussion.

REVIEW QUESTIONS AND EXERCISES

1. Consider the following real and hypothetical research situations. What is the ethical component in each example? How do you feel about it? Do you think the procedures described are ultimately acceptable or unacceptable? You might find

discussing some of these situations with class-mates useful.

a. A psychology instructor asks students in an introductory psychology class to complete questionnaires that the instructor will analyze and use in preparing a journal article for publication.

b. After a field study of deviant behavior during a riot, law enforcement officials demand that the researcher identify those people who were observed looting. Rather than risk arrest as an accomplice after the fact, the researcher complies.

c. After completing the final draft of a book reporting a research project, the researcher-author discovers that 25 of the 2,000 survey interviews were falsified by interviewers. To protect the bulk of the research, the author leaves out this information and publishes the book.

d. Researchers obtain a list of right-wing radicals they wish to study. They contact the radicals with the explanation that each has been selected "at random" from among the general population to take a sampling of "public opinion."

e. A college instructor who wants to test the effect of unfair berating administers an hour exam to both sections of a specific course. The overall performance of the two sections is essentially the same. The grades of one section are artificially lowered, however, and the instructor berates the students for performing so badly. The instructor then administers the same final exam to both sections and discovers that the performance of the unfairly berated section is worse. The hypothesis is confirmed, and the research report is published.

f. In a study of sexual behavior, the investigator wants to overcome subjects' reluctance to report what they might regard as shameful behavior. To get past their reluctance, subjects are asked, "Everyone masturbates now and then; about how much do you masturbate?"

g. A researcher studying dorm life on campus discovers that 60 percent of the residents regularly violate restrictions on alcohol consumption. Publication of this finding would probably create a furor in the campus community. Because no extensive analysis of

alcohol use is planned, the researcher decides to keep this finding quiet.

h. To test the extent to which people may try to save face by expressing attitudes on matters they are wholly uninformed about, the researcher asks for their attitudes regarding a fictitious issue.

i. A research questionnaire is circulated among students as part of their university registration packet. Although students are not told they must complete the questionnaire, the hope is that they will believe they must—thus ensuring a higher completion rate.

j. A researcher pretends to join a radical political group in order to study it and is successfully accepted as a member of the inner planning circle. What should the researcher do if the group makes plans for the following?

  (1) A peaceful, though illegal, demonstration

  (2) The bombing of a public building during a time it is sure to be unoccupied

  (3) The assassination of a public official

2. Review the discussion of the Milgram experiment on obedience. How would you design a study to accomplish the same purpose while avoiding the ethical criticisms leveled at Milgram? Would your design be equally valid? Would it have the same effect?

3. Suppose a researcher who is personally in favor of small families—as a response to the problem of overpopulation—wants to conduct a survey to determine why some people want many children and others don't. What personal-involvement problems would the researcher face, and how could she or he avoid them? What ethical issues should the researcher take into account in designing the survey?

4. Using InfoTrac College Edition, search for "informed content" and then narrow your search to "research." Skim the resulting articles and begin to identify groups of people for whom informed consent may be problematic—people who may not be able to give it. Suggest some ways in which the problem might be overcome.

SPSS EXERCISES

See the booklet that accompanies your text for exercises using SPSS (Statistical Package for the Social Sciences). There are exercises offered for each chapter, and you'll also find a detailed primer on using SPSS.

Online Study Resources

If your book came with an access code card, visit www.cengage.com/login to register. To purchase access, please visit www.ichapters.com.

1. Before you do your final review of the chapter, take the *CengageNOW* pretest to help identify the areas on which you should concentrate. You'll find information on this online tool, as well as instructions on how to access all of its great resources, in the front of the book.

2. As you review, take advantage of the *CengageNOW* personalized study plan, based on your quiz results. Use this study plan with its interactive exercises and other resources to master the material.

3. When you're finished with your review, take the posttest to confirm that you're ready to move on to the next chapter.

WEBSITE FOR *THE PRACTICE OF SOCIAL RESEARCH* 12TH EDITION

Go to your book's website at www.cengage.com/sociology/babbie for tools to aid you in studying for your exams. You'll find *Tutorial Quizzes* with feedback, *Internet Exercises, Flash Cards, Glossaries,* and *Essay Quizzes,* as well as *InfoTrac College Edition* search terms, suggestions for additional reading, *Web Links,* and primers for using data-analysis software such as SPSS.

Posing problems properly is often more difficult than answering them. Indeed, a properly phrased question often seems to answer itself. You may have discovered the answer to a question just in the process of making the question clear to someone else.

Part 2 deals with what should be observed; that is, Part 2 considers the posing of proper scientific questions, the structuring of inquiry. Part 3 will describe some of the specific methods of social science observation.

Chapter 4 addresses the beginnings of research. It examines some of the purposes of inquiry, units of analysis, and the reasons scientists get involved in research projects.

Chapter 5 deals with the specification of what it is you want to measure—the processes of conceptualization and operationalization. It looks at some of the terms that you and I use quite casually in everyday life—*prejudice, liberalism, happiness,* and so forth—and shows how essential it is to clarify what we really mean

The Structuring of Inquiry: Quantitative and Qualitative

by such terms when we do research. This process of clarification is called conceptualization.

Once we clarify what we mean by certain terms, we can then measure the referents of those terms. The process of devising steps or operations for measuring what we want to study is called operationalization. Chapter 5 deals with the topic of operationalization in general, paying special attention to the framing of questions for interviews and questionnaires.

To complete the introduction to measurement, Chapter 6 breaks with the chronological discussion of how research is conducted. In this chapter, we'll examine techniques for measuring variables in quantitative research through the combination of several indicators: indexes, scales, and typologies. As an example, we might ask survey respondents five different questions about their attitudes toward gender equality and then combine the answers to all five questions into a composite measure of gender-based egalitarianism.

Although such composite measures are constructed during the analysis of data (see Part 4), the raw materials for them must be provided for in the design and execution of data collection.

Finally, we'll look at how social researchers select people or things for observation. Chapter 7, on sampling, addresses the fundamental scientific issue of generalizability. As you'll see, we can select a few people or things for observation and then apply what we observe to a much larger group. For example, by surveying 2,000 U.S. citizens about whom they favor for president of the United States, we can accurately predict how tens of millions will vote. In this chapter, we'll examine techniques that increase the generalizability of what we observe.

What you learn in Part 2 will bring you to the verge of making controlled social science observations. Part 3 will then show you how to take that next step.

CHAPTER FOUR

Research Design

CHAPTER OVERVIEW

Here you'll see the wide variety of research designs available to social researchers as well as how to design a study—that is, specifying exactly who or what is to be studied when, how, and for what purpose.

CengageNOW for Sociology

Use this online tool to help you make the grade on your next exam. After reading this chapter, go to "Online Study Resources" at the end of the chapter for instructions on how to benefit from *CengageNOW*.

Introduction

Science is an enterprise dedicated to "finding out." No matter what you want to find out, though, there will likely be a great many ways of doing it. That's true in life generally. Suppose, for example, that you want to find out whether a particular automobile—say, the new Burpo-Blasto—would be a good car for you. You could, of course, buy one and find out that way. Or you could talk to a lot of B-B owners or to people who considered buying one but didn't. You might check the classified ads to see if there are a lot of B-Bs being sold cheap. You could read a consumer magazine evaluation of Burpo-Blastos. A similar situation occurs in scientific inquiry.

Ultimately, scientific inquiry comes down to making observations and interpreting what you've observed, the subjects of Parts 3 and 4 of this book. Before you can observe and analyze, however, you need a plan. You need to determine what you're going to observe and analyze: why and how. That's what research design is all about.

Although the details vary according to what you wish to study, you face two major tasks in any research design. First, you must specify as clearly as possible what you want to find out. Second, you must determine the best way to do it. Interestingly, if you can handle the first consideration fully, you'll probably handle the second in the same process. As mathematicians say, a properly framed question contains the answer.

Let's say you're interested in conducting social research on terrorism. When Jeffrey Ross (2004) addressed this issue, he found the existing studies used a variety of qualitative and quantitative approaches. Qualitative researchers, for example, generated original data through

Autobiographies

Incident Reports and Accounts

Hostages' Experiences with Terrorists

Firsthand Accounts of Implementing Policies

Ross goes on to discuss some of the secondary materials used by qualitative researchers: "bi-ographies of terrorists, case studies of terrorist organizations, case studies on types of terrorism, case studies on particular terrorist incidents, and case studies of terrorism in selected regions and countries" (2004: 27). Quantitative researchers, on the other hand, have addressed terrorism in a variety of ways, including analyses of media coverage, statistical modeling of terrorist events, and the use of various databases relevant to the topic. As you'll see in this chapter, any research topic can be approached from many different directions. Each of the topics we'll examine is relevant to both qualitative and quantitative studies, though some topics may be more relevant to one than to the other approach.

This chapter provides a general introduction to research design, whereas the other chapters in Part 2 elaborate on specific aspects of it. In practice, all aspects of research design are interrelated. As you read through Part 2, the interrelationships among parts will become clearer.

We'll start by briefly examining the main purposes of social research. Then, we'll consider units of analysis—the what or whom you want to study. Next we'll consider ways of handling time in social research, or how to study a moving target that changes over time.

With these ideas in hand, we'll turn to how to design a research project. This overview of the research process serves two purposes: Besides describing how you might go about designing a study, it provides a map of the remainder of this book.

Next, we'll look at the elements of research proposals. Often, you'll need to detail your intentions before you actually conduct your research; this might be required in order to obtain funding for a major project or perhaps to get your instructor's approval for a class project. You'll see that the research proposal provides an excellent opportunity for you to consider all aspects of your research in advance. Also, this section should help you with the end-of-chapter exercise concerning the research proposal, if you are doing that. Finally, the last section of this chapter focuses on the ethical dimension of research design.

Three Purposes of Research

Social research can serve many purposes. Three of the most common and useful purposes are exploration, description, and explanation. Although a given study can have more than one of these purposes—and most do—examining them separately is useful because each has different implications for other aspects of research design.

Exploration

Much of social research is conducted to explore a topic, that is, to start to familiarize a researcher with that topic. This approach typically occurs when a researcher examines a new interest or when the subject of study itself is relatively new.

As an example, let's suppose that widespread taxpayer dissatisfaction with the government erupts into a taxpayers' revolt. People begin refusing to pay their taxes, and they organize themselves around that issue. You might like to learn more about the movement: How widespread is it? What levels and degrees of support are there within the community? How is the movement organized? What kinds of people are active in it? An exploratory study could help you find at least approximate answers to some of these questions. You might check figures with tax-collecting officials, collect and study the literature of the movement, attend meetings, and interview leaders.

Exploratory studies are also appropriate for more persistent phenomena. Suppose you're unhappy with your college's graduation requirements and want to help change them. You might study the history of such requirements at the college and meet with college officials to learn the reasons for the current standards. You could talk to several students to get a rough idea of their sentiments on the subject. Though this last activity would not necessarily yield an accurate picture of student opinion, it could suggest what the results of a more extensive study might be.

Sometimes exploratory research is pursued through the use of focus groups, or guided small-group discussions. This technique is frequently used in market research; we'll examine it further in Chapter 10.

Exploratory studies are most typically done for three purposes: (1) to satisfy the researcher's curiosity and desire for better understanding, (2) to test the feasibility of undertaking a more extensive study, and (3) to develop the methods to be employed in any subsequent study.

A while back, for example, I became aware of the growing popularity of something called "channeling," in which a person known as a channel or medium enters a trance state and begins speaking with a voice that claims it originates outside the channel. Some of the voices say they come from a spirit world of the dead, some say they are from other planets, and still others say they exist in dimensions of reality difficult to explain in ordinary human terms.

The channeled voices, often referred to as entities, sometimes use the metaphor of radio or television for the phenomenon they represent. "When you watch the news," one told me in the course of an interview, "you don't believe Dan Rather is really inside the television set. The same is true of me. I use this medium's body the way Dan Rather uses your television set."

The idea of channeling interested me from several perspectives, not the least of which was the methodological question of how to study scientifically something that violates so much of what we take for granted, including scientific staples such as space, time, causation, and individuality.

Lacking any rigorous theory or precise expectations, I merely set out to learn more. Using some of the techniques of qualitative field research discussed in Chapter 10, I began amassing information and forming categories for making sense of what I observed. I read books and articles about the phenomenon and talked to people who had attended channeling sessions. I then attended channeling sessions myself, observing those who attended as well as the channel and entity. Next, I conducted personal interviews with numerous channels and entities.

In most interviews, I began by asking the human channels questions about how they first began channeling, what it was like, and why they continued, as well as standard biographical questions. The channel would then go into a trance, whereby the interview continued with the entity speaking. "Who are you?" I might ask. "Where do you come from?" "Why are you doing this?" "How can I tell if you are real or a fake?" Although I went into these interview sessions with several questions prepared in advance, each of the interviews followed whatever course seemed appropriate in light of the answers given.

This example of exploration illustrates where social research often begins. Whereas researchers working from deductive theories have the key variables laid out in advance, one of my first tasks was to identify some of the possibly relevant variables. For example, I noted a channel's gender, age, education, religious background, regional origins, and previous participation in things metaphysical. I chose most of these variables because they commonly affect behavior.

I also noted differences in the circumstances of channeling sessions. Some channels said they must go into deep trances, some use light trances, and others remain conscious. Most sit down while channeling, but others stand and walk about. Some channels operate under pretty ordinary conditions; others seem to require props such as dim lights, incense, and chanting. Many of these differences became apparent to me only in the course of my initial observations.

Regarding the entities, I have been interested in classifying where they say they come from. Over the course of my interviews, I've developed a set of questions about specific aspects of "reality," attempting to classify the answers they give. Similarly, I ask each to speak about future events.

Over the course of this research, my examination of specific topics has become increasingly focused as I've identified variables that seem worth pursuing: gender, education, and religion, for example. Note, however, that I began with a reasonably blank slate.

Exploratory studies are quite valuable in social science research. They're essential whenever a researcher is breaking new ground, and they almost always yield new insights into a topic for research. Exploratory studies are also a source of grounded theory, as discussed in Chapter 2.

The chief shortcoming of exploratory studies is that they seldom provide satisfactory answers to research questions, though they can hint at the answers and can suggest which research methods could provide definitive ones. The reason exploratory studies are seldom definitive in themselves has to do with representativeness; that is, the people you study in your exploratory research may not be typical of the larger population that interests you. Once you understand representativeness, you'll be able to know whether a given exploratory study actually answered its research problem or only pointed the way toward an answer. (Representativeness is discussed at length in Chapter 7.)

Description

A major purpose of many social science studies is to describe situations and events. The researcher observes and then describes what was observed. Because scientific observation is careful and deliberate, however, scientific descriptions are typically more accurate and precise than casual ones are.

The U.S. Census is an excellent example of descriptive social research. The goal of the census is to describe accurately and precisely a wide variety of characteristics of the U.S. population, as well as the populations of smaller areas such as states and counties. Other examples of descriptive studies are the computation of age-gender profiles of populations done by demographers, the computation of crime rates for different cities, and a product-marketing survey that describes the people who use, or would use, a particular product. A researcher who carefully chronicles the events that take place on a labor union picket line has, or at least serves, a descriptive purpose. A researcher who computes and reports the number of times individual legislators voted for or against organized labor also fulfills a descriptive purpose.

Many qualitative studies aim primarily at description. An anthropological ethnography, for example, may try to detail the particular culture of some preliterate society. At the same time, such studies are seldom limited to a merely descriptive purpose. Researchers usually go on to examine why the observed patterns exist and what they imply.

Explanation

The third general purpose of social science research is to explain things. Descriptive studies answer questions of what, where, when, and how; explanatory questions, of why. So when William Sanders (1994) set about describing the varieties of gang violence, he also wanted to reconstruct the process that brought about violent episodes among the gangs of different ethnic groups.

Reporting the voting intentions of an electorate is descriptive, but reporting why some people plan to vote for Candidate A and others for Candidate B is explanatory. Identifying variables that explain why some cities have higher crime rates than others involves explanation. A researcher who sets out to know why an antiabortion demonstration ended in a violent confrontation with police, as opposed to simply describing what happened, has an explanatory purpose.

Let's look at a specific case. What factors do you suppose might shape people's attitudes toward the legalization of marijuana? To answer this, you might first consider whether men and women differ in their opinions. An explanatory analysis of the 2002 General Social Survey (GSS) data indicates that 38 percent of men and 30 percent of women said marijuana should be legalized.

What about political orientation? The GSS data show that 55 percent of liberals said marijuana should be legalized, compared with 29 percent of moderates and 27 percent of conservatives. Further, 41 percent of Democrats, compared with 34 percent of Independents and 28 percent of Republicans, supported legalization.

Given these statistics, you might begin to develop an explanation for attitudes toward marijuana legalization. Further study of gender and political orientation might then lead to a deeper explanation of these attitudes.

The Logic of Nomothetic Explanation

The preceding examination of what factors might cause attitudes about legalizing marijuana illustrates nomothetic explanation, as discussed in Chapter 1. Recall that in this model, we try to find a few factors (independent variables) that can account for many of the variations in a given phenomenon. This explanatory model stands in contrast to the idiographic model, in which we seek a complete, in-depth understanding of a single case.

In our example, an idiographic approach would suggest all the reasons that one person was opposed to legalization—involving what her parents, teachers, and clergy told her about it; any bad experiences experimenting with it; and so forth. When we understand something idiographically, we feel we really understand it. When we know all the reasons why someone opposed legalizing marijuana, we couldn't imagine that person having any other attitude.

In contrast, a nomothetic approach might suggest that overall political orientations account for much of the difference of opinion about legalizing marijuana. Because this model is inherently probabilistic, it is more open than the idiographic model to misunderstanding and misinterpretation. Let's examine what social researchers mean when they say one variable (nomothetically) causes another. Then, we'll look at what they don't mean.

Criteria for Nomothetic Causality

There are three main criteria for nomothetic causal relationships in social research: (1) the variables must be correlated, (2) the cause takes place before the effect, and (3) the variables are nonspurious.

Correlation

Unless some actual relationship—or **correlation**—is found between two variables, we can't say that a causal relationship exists. Our analysis of GSS data suggested that political orientation was a cause of attitudes about legalizing marijuana. Had the same percentage of liberals and conservatives supported legalization, we could hardly say that political orientations caused the attitude. Though this criterion is obvious, it emphasizes the need to base social research assertions on actual observations rather than assumptions.

Time Order

Next, we can't say a causal relationship exists unless the cause precedes the effect in time. Notice that it makes more sense to say that most children's religious affiliations are caused by those of their parents than to say that parents' affiliations are caused by those of their children—even though it would be possible for you to change your religion and for your parents to follow suit. Remember, nomothetic explanation deals with "most cases" but not all.

In our marijuana example, it would make sense to say that gender causes, to some extent, attitudes toward legalization, whereas it would make no sense to say that opinions about marijuana determine a person's gender. Notice, however, that the time order connecting political orientations and attitudes about legalization is less clear, though we sometimes reason that general orientations cause specific opinions. And sometimes our analyses involve two or more independent variables that were established at the same time: looking at the effects of gender and race on voting behavior, for example. As we'll see in the next chapter, the issue of time order can be a complex matter.

Nonspuriousness

The third requirement for a causal relationship is that the effect cannot be explained in terms of some third variable. For example, there is a correlation between ice-cream sales and deaths due to drowning: the more ice cream sold, the more drownings, and vice versa. There is, however, no direct link between ice cream and drowning.

The third variable at work here is *season* or *temperature*. Most drowning deaths occur during summer—the peak period for ice-cream sales.

Here are a couple of other examples of **spurious relationships,** or ones that aren't genuine. There is a negative relationship between the number of mules and the number of Ph.D.'s in towns and cities: the more mules, the fewer Ph.D.'s and vice versa. Perhaps you can think of another variable that would explain this apparent relationship. The answer is rural versus urban settings. There are more mules (and fewer Ph.D.'s) in rural areas, whereas the opposite is true in cities.

Or, consider the positive correlation between shoe size and math ability among schoolchildren. Here, the third variable that explains the puzzling relationship is *age*. Older children have bigger feet and more highly developed math skills, on average, than younger children do. See Figure 4-1 for an illustration of this spurious relationship. Notice that observed associations go in both directions. That is, as one variable occurs or changes, so does the other.

The list goes on. Areas with many storks have high birthrates. Those with few storks have low birthrates. Do storks really deliver babies? Birthrates are higher in the country than in the city; more storks live in the country than the city. The third variable here is *urban/rural areas.*

Finally, the more fire trucks that put out a fire, the more damage to the structure. Can you guess what the third variable is? In this case, it's the *size of the fire.*

Thus, when social researchers say there is a causal relationship between, say, *education* and *racial tolerance*, they mean (1) there is a statistical

---

correlation An empirical relationship between two variables such that (1) changes in one are associated with changes in the other or (2) particular attributes of one variable are associated with particular attributes of the other. Correlation in and of itself does not constitute a causal relationship between the two variables, but it is one criterion of causality.

spurious relationship A coincidental statistical correlation between two variables, shown to be caused by some third variable.

Observed Correlation

Positive (direct) correlation

Bigger shoe size is associated with greater math skill, and vice versa.

Spurious causal relationships

Neither shoe size nor math skill is a cause of the other.

Actual causal relationships

The underlying variable of age causes both bigger shoe size and greater math skill, thus explaining the observed correlation.

FIGURE 4-1

An Example of a Spurious Causal Relationship. Finding an empirical correlation between two variables does not necessarily establish a causal relationship. Sometimes the observed correlation is the incidental result of other causal relationships, involving other variables.

correlation between the two variables, (2) a person's educational level occurred before their current level of tolerance or prejudice, and (3) there is no third variable that can explain away the observed correlation as spurious.

Nomothetic Causal Analysis and Hypothesis-Testing

The nomothetic model of causal analysis lends itself to hypothesis-testing (see Chapter 1), though hypotheses are not required in nomothetical research. To test a hypothesis, you would carefully specify the variables you think are causally related, as well as specifying the manner in which you will measure them. (These steps will be discussed in detail in the following chapter under the terms *conceptualization* and *operationalization*.)

In addition to hypothesizing that two variables will be correlated with each other, you may specify the strength of the relationship you expect within the study design you are using. Often this specification will take the form of a level of *statistical significance:* the chance you are willing to take that a given relationship might have been caused by chance in the selection of subjects for study. (This will be discussed further in Chapter 7, on sampling.)

Finally, you may specify the tests for spuriousness that any observed relationship must survive. Not only will you hypothesize, for example, that increased education will reduce levels of prejudice, but you will specify further that the hypothesized relationship will not be the product of, say, political orientations.

False Criteria for Nomothetic Causality

Because notions of cause and effect are well entrenched in everyday language and logic, it's important to specify some of the things social researchers do *not* mean when they speak of causal relationships. When they say that one variable causes another, they do not necessarily mean to suggest complete causation, to account for exceptional cases, or to claim that the causation exists in a majority of cases.

Complete Causation

Whereas an idiographic explanation of causation is relatively complete, a nomothetic explanation is probabilistic and usually incomplete. As we've seen, social researchers may say that political orientations cause attitudes toward legalizing marijuana even though not all liberals approve nor all conservatives disapprove. Thus, we say that political orientation is one of the causes of the attitude, but not the only one.

Exceptional Cases

In nomothetic explanations, exceptions do not disprove a causal relationship. For example, it is consistently found that women are more religious than men in the United States. Thus, gender may be a cause of religiosity, even if your uncle is a religious zealot or you know a woman who is an avowed atheist. Those exceptional cases do not disprove the overall, causal pattern.

Majority of Cases

Causal relationships can be true even if they don't apply in a majority of cases. For example, we say that children who are not supervised after school are more likely to become delinquent than those who are supervised are; hence, lack of supervision is a cause of delinquency. This causal relationship holds true even if only a small percentage of those not supervised become delinquent. As long as they are *more likely* than those who are supervised to be delinquent, we say there is a causal relationship.

The social science view of causation may vary from what you are accustomed to, because people commonly use the term *cause* to mean something that completely causes another thing. The somewhat different standard used by social researchers can be seen more clearly in terms of necessary and sufficient causes.

Necessary and Sufficient Causes

A *necessary cause* represents a condition that *must* be present for the effect to follow. For example, it is necessary for you to take college courses in order to get a degree. Take away the courses, and the degree never follows. However, simply taking the courses is not a sufficient cause of getting a degree. You need to take the right ones and pass them. Similarly, being female is a necessary condition of becoming pregnant, but it is not a sufficient cause. Otherwise, all women would get pregnant. Figure 4-2 illustrates this relationship between the variables of *gender* and *pregnancy* as a matrix showing the possible outcomes of combining these variables.

A *sufficient cause,* on the other hand, represents a condition that, if it is present, guarantees the effect in question. This is not to say that a sufficient cause is the *only* possible cause of a particular effect. For example, skipping an exam in this course would be a sufficient cause for failing it, though students could fail it other ways as well. Thus, a cause can be sufficient, but not necessary. Figure 4-3 illustrates the relationship between taking or not taking the exam and either passing or failing it.

The discovery of a cause that is both necessary *and* sufficient is, of course, the most satisfying outcome in research. If juvenile delinquency were the effect under examination, it would be nice to discover a single condition that (1) must be present for delinquency to develop and (2) always results in delinquency. In such a case, you would surely feel that you knew precisely what caused juvenile delinquency.

Unfortunately, we never discover single causes that are absolutely necessary and absolutely sufficient when analyzing the nomothetic relationships among variables. It is not uncommon, however, to find causal factors that are either 100 percent necessary (you must be female to become pregnant) or 100 percent sufficient (skipping an exam will inevitably cause you to fail it).

In the idiographic analysis of single cases, you may reach a depth of explanation from which it is reasonable to assume that things could not

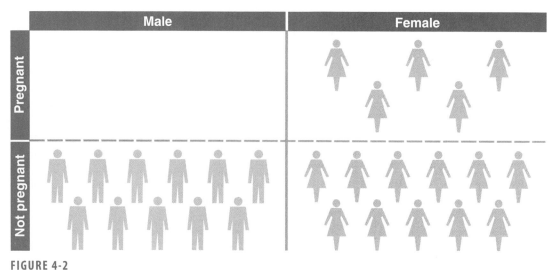

FIGURE 4-2

Necessary Cause. Being female is a necessary cause of pregnancy; that is, you can't get pregnant unless you are female.

FIGURE 4-3

Sufficient Cause. Not taking the exam is a sufficient cause of failing it, even though there are other ways of failing (such as answering randomly).

have turned out differently, suggesting you have determined the *sufficient* causes for a particular result. (Anyone with all the same details of your genetic inheritance, upbringing, and subsequent

experiences would have ended up going to college.) At the same time, there could always be other causal paths to the same result. Thus, the idiographic causes are sufficient but not necessary.

Units of Analysis

In social research, there is virtually no limit to what or whom can be studied, or the **units of analysis.** This topic is relevant to all forms of social research, although its implications are clearest in the case of nomothetic, quantitative studies.

The idea for units of analysis may seem slippery at first, because research—especially nomothetic research—often studies large collections of people or things, or aggregates. It's important to distinguish between the unit of analysis and the aggregates that we generalize about. For instance, a researcher may study a class of people, such as Democrats, college undergraduates, African American women under 30, or some other collection. But if the researcher is interested in exploring, describing, or explaining how different groups of individuals behave *as individuals,* the unit of analysis is the individual, not the group. This is true even though the researcher uses the information about individuals

units of analysis The what or whom being studied. In social science research, the most typical units of analysis are individual people.

to generalize about aggregates of individuals, as in saying that more Democrats than Republicans favor legalizing marijuana. Think of it this way: Having an attitude about marijuana is something that can only be an attribute of an individual, not a group; that is, there is no one group "mind" that can have an attitude. So even when we generalize about Democrats, we're generalizing about an attribute they possess as individuals.

In contrast, we may sometimes want to study groups, considered as individual "actors" or entities that have attributes as groups. For instance, we might want to compare the characteristics of different types of street gangs. In that case our unit of analysis would be gangs (not members of gangs), and we might proceed to make generalizations about different types of gangs. For example, we might conclude that male gangs are more violent than female gangs. Each gang (unit of analysis) would be described in terms of two variables: (1) What sex are the members? and (2) How violent are its activities? So we might study 52 gangs, reporting that 40 were male and 12 were female, and so forth. The "gang" would be the unit of analysis, even though some of the characteristics were drawn from the components (members) of the gangs.

Social researchers tend to choose individual people as their units of analysis. You may note the characteristics of individual people—gender, age, region of birth, attitudes, and so forth. You can then combine these descriptions to provide a composite picture of the group the individuals represent, whether a street-corner gang or a whole society.

For example, you may note the age and gender of each student enrolled in Political Science 110 and then characterize the group of students as being 53 percent men and 47 percent women and as having a mean age of 18.6 years. Although the final description would be of the class as a whole, the description is based on characteristics that members of the class have as individuals.

The same distinction between units of analysis and aggregates occurs in explanatory studies. Suppose you wished to discover whether students with good study habits received better grades in

Political Science 110 than students with poor study habits did. You would operationalize the variable *study habits* and measure this variable, perhaps in terms of hours of study per week. You might then aggregate students with good study habits and those with poor study habits and see which group received the best grades in the course. The purpose of the study would be to explain why some groups of students do better in the course than others do, but the unit of analysis is still individual students.

Units of analysis in a study are usually also the units of observation. Thus, to study success in a political science course, we would observe individual students. Sometimes, however, we "observe" our units of analysis indirectly. For example, suppose we want to find out whether disagreements about the death penalty tend to cause divorce. In this case, we might "observe" individual husbands and wives by asking them about their attitudes about capital punishment, in order to distinguish couples who agree and disagree on this issue. In this case, our units of observation are individual wives and husbands, but our units of analysis (the things we want to study) are couples.

Units of analysis, then, are those things we examine in order to create summary descriptions of all such units and to explain differences among them. In most research projects, the unit of analysis will probably be clear to you. When the unit of analysis is not clear, however, it's essential to determine what it is; otherwise, you cannot determine what observations are to be made about whom or what.

Some studies try to describe or explain more than one unit of analysis. In these cases, the researcher must anticipate what conclusions she or he wishes to draw with regard to which units of analysis. For example, we may want to discover what kinds of college students (individuals) are most successful in their careers; we may also want to learn what kinds of colleges (organizations) produce the most-successful graduates.

Here's an example that illustrates the complexity of units of analysis. Murder is a fairly personal matter: One individual kills another individual. However, when Charis Kubrin and Ronald Weitzer (2003: 157) ask, "Why do these neighborhoods generate high homicide rates?" the unit of analysis in that phrase is *neighborhood*. You can probably imagine some kinds of neighborhoods (e.g., poor, urban) that would have high homicide rates and some (e.g., wealthy, suburban) that would have low rates. In this particular conversation, the unit of analysis (neighborhood) would be categorized in terms of variables such as *economic level, locale,* and *homicide rate*.

In their analysis, however, Kubrin and Weitzer were also interested in different types of homicide: in particular, those that occurred in retaliation for some earlier event, such as an assault or insult. Can you identify the unit of analysis common to all of the following excerpts?

1. The sample of killings . . .
2. The coding instrument includes over 80 items related to the homicide.
3. Of the 2,161 homicides that occurred from 1985 [to] 1995 . . .
4. Of those with an identified motive, 19.5 percent (n = 337) are retaliatory.

(Kubrin and Weitzer 2003: 163)

In each of these excerpts, the unit of analysis is *homicide* (also called killing or murder). Sometimes you can identify the unit of analysis in the description of the sampling methods, as in the first excerpt. A discussion of classification methods might also identify the unit of analysis, as in the second excerpt (80 ways to code the homicides). Often, numerical summaries point the way: 2,161 homicides; 19.5 percent (of the homicides). With a little practice you'll be able to identify the units of analysis in most social research reports, even when more than one is used in a given analysis.

To explore this topic in more depth, let's consider several common units of analysis in social research.

Individuals

As mentioned, individual human beings are perhaps the most typical units of analysis for social research. Social researchers tend to describe and explain social groups and interactions by aggregating and manipulating the descriptions of individuals.

Any type of individual may be the unit of analysis for social research. This point is more important than it may seem at first. The norm of generalized understanding in social research should suggest that scientific findings are most valuable when they apply to all kinds of people. In practice, however, social researchers seldom study all kinds of people. At the very least, their studies are typically limited to the people living in a single country, though some comparative studies stretch across national boundaries. Often, though, studies are quite circumscribed.

Examples of classes of individuals that might be chosen for study include students, gays and lesbians, auto workers, voters, single parents, and faculty members. Note that each of these terms implies some population of individuals. Descriptive studies with individuals as their units of analysis typically aim to describe the population that comprises those individuals, whereas explanatory studies aim to discover the social dynamics operating within that population.

As the units of analysis, individuals may be characterized in terms of their membership in social groupings. Thus, an individual may be described as belonging to a rich family or to a poor one, or a person may be described as having a college-educated mother or not. We might examine in a research project whether people with college-educated mothers are more likely to attend college than are those with non-college-educated mothers or whether high school graduates in rich families are more likely than those in poor families to attend college. In each case, the unit of analysis—the "thing" whose characteristics we are seeking to describe or explain—is the individual. We then aggregate these individuals and make generalizations about the population they belong to.

Groups

Social groups can also be units of analysis in social research. That is, we may be interested in characteristics that belong to one group, considered as a single entity. If you were to study the members of a criminal gang to learn about criminals, the individual (criminal) would be the unit of analysis; but if you studied all the gangs in a city to learn the differences, say, between big gangs and small ones, between "uptown" and "downtown" gangs, and so forth, you would be interested in gangs rather than their individual members. In this case, the unit of analysis would be the gang, a social group.

Here's another example. Suppose you were interested in the question of access to computers in different segments of society. You might describe families in terms of total annual income and according to whether or not they had computers. You could then aggregate families and describe the mean income of families and the percentage with computers. You would then be in a position to determine whether families with higher incomes were more likely to have computers than were those with lower incomes. In this case, the unit of analysis would be families.

As with other units of analysis, we can derive the characteristics of social groups from those of their individual members. Thus, we might describe a family in terms of the age, race, or education of its head. In a descriptive study, we might find the percentage of all families that have a college-educated head of family. In an explanatory study, we might determine whether such families have, on average, more or fewer children than do families headed by people who have not graduated from college. In each of these examples, the family is the unit of analysis. In contrast, had we asked whether college-educated individuals have more or fewer children than do their less-educated counterparts, then the individual would have been the unit of analysis.

Other units of analysis at the group level could be friendship cliques, married couples, census blocks, cities, or geographic regions. As with individuals, each of these terms implies some population. *Street gangs* implies some population that includes all street gangs, perhaps in a given city. You might then describe this population by generalizing from your findings about individual gangs. For instance, you might describe the geographic distribution of gangs throughout a city. In an explanatory

study of street gangs, you might discover whether large gangs are more likely than small ones to engage in intergang warfare. Thus, you would arrive at conclusions about the population of gangs by using individual groups as your unit of analysis.

Organizations

Formal social organizations may also be the units of analysis in social research. For example, a researcher might study corporations, by which he or she implies a population of all corporations. Individual corporations might be characterized in terms of their number of employees, net annual profits, gross assets, number of defense contracts, percentage of employees from racial or ethnic minority groups, and so forth. We might determine whether large corporations hire a larger or smaller percentage of minority group employees than small corporations do. Other examples of formal social organizations suitable as units of analysis include church congregations, colleges, army divisions, academic departments, and supermarkets.

Figure 4-4 provides a graphic illustration of some different units of analysis and the statements that might be made about them.

Social Interactions

Sometimes social interactions are the relevant units of analysis. Instead of individual humans, you can study what goes on between them: telephone calls, kisses, dancing, arguments, fistfights, e-mail exchanges, chat-room discussions, and so forth. As you saw in Chapter 2, social interaction is the basis for one of the primary theoretical paradigms in the social sciences, and the number of units of analysis that social interactions provide is nearly infinite.

Even though individuals are usually the actors in social interactions, there is a difference between (1) comparing the kinds of people who subscribe to different Internet service providers (individuals being the units of analysis) and (2) comparing the length of chat-room interactions on those same ISPs (interactions being the units of analysis).

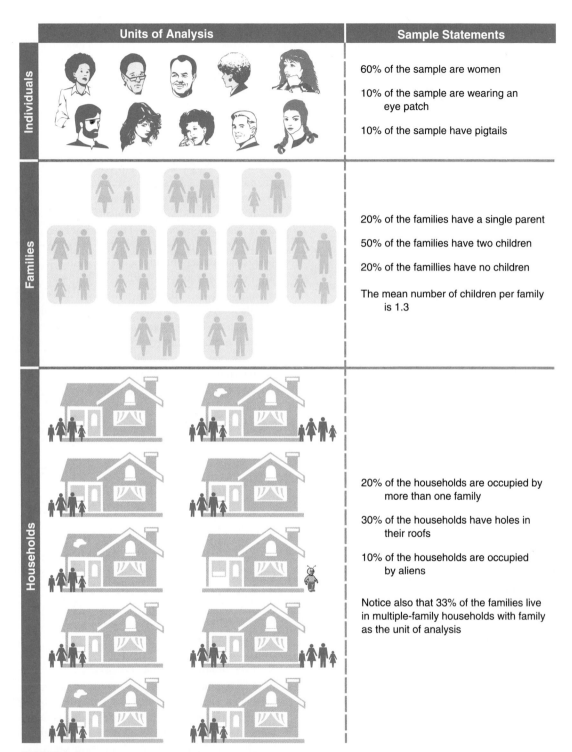

Units of Analysis	Sample Statements
Individuals	60% of the sample are women 10% of the sample are wearing an eye patch 10% of the sample have pigtails
Families	20% of the families have a single parent 50% of the families have two children 20% of the famillies have no children The mean number of children per family is 1.3
Households	20% of the households are occupied by more than one family 30% of the households have holes in their roofs 10% of the households are occupied by aliens Notice also that 33% of the families live in multiple-family households with family as the unit of analysis

FIGURE 4-4

Illustrations of Units of Analysis. Units of analysis in social research can be individuals, groups, or even nonhuman entities.

Social Artifacts

Another unit of analysis is the **social artifact,** or any product of social beings or their behavior. One class of artifacts includes concrete objects such as books, poems, paintings, automobiles, buildings, songs, pottery, jokes, student excuses for missing exams, and scientific discoveries.

For example, Lenore Weitzman and her associates (1972) were interested in learning how gender roles are taught. They chose children's picture books as their unit of analysis. Specifically, they examined books that had received the Caldecott Medal. Their results were as follows:

> We found that females were underrepresented in the titles, central roles, pictures, and stories of every sample of books we examined. Most children's books are about boys, men, male animals, and deal exclusively with male adventures. Most pictures show men singly or in groups. Even when women can be found in the books, they often play insignificant roles, remaining both inconspicuous and nameless.
>
> *(Weitzman et al. 1972: 1128)*

In a more recent study, Roger Clark, Rachel Lennon, and Leana Morris (1993) concluded that male and female characters at that time were portrayed less stereotypically than before, observing a clear progress toward portraying men and women in nontraditional roles. However, they did not find total equality between the sexes.

As this example suggests, just as people or social groups imply populations, each social object implies a set of all objects of the same class: all books, all novels, all biographies, all introductory sociology textbooks, all cookbooks, all press conferences. In a study using books as the units of analysis, an individual book might be characterized by its size, weight, length, price, content, number of pictures, number sold, or description of the author. Then the population of all books or of a particular kind of book could be analyzed for the purpose of description or explanation: What kinds of books sell best and why, for example.

Similarly, a social researcher could analyze whether paintings by Russian, Chinese, or U.S. artists showed the greatest degree of working-class consciousness, taking paintings as the units of analysis and describing each, in part, by the nationality of its creator. Or you might examine a newspaper's editorials regarding a local university, for the purpose of describing, or perhaps explaining, changes in the newspaper's editorial position on the university over time. In this example, individual editorials would be the units of analysis. See "How to Do It: Identifying the Unit of Analysis" for more.

Units of Analysis in Review

The examples in this section should suggest the nearly infinite variety of possible units of analysis in social research. Although individual human beings are typical objects of study, many research questions can be answered more appropriately through the examination of other units of analysis. Indeed, social researchers can study just about anything that bears on social life.

Moreover, the types of units of analysis named in this section do not begin to exhaust the possibilities. Morris Rosenberg (1968: 234–48), for example, speaks of individual, group, organizational, institutional, spatial, cultural, and societal units of analysis. John Lofland and his associates (2006: 122–32) speak of practices, episodes, encounters, roles and social types, social and personal relationships, groups and cliques, organizations, settlements and habitats, subcultures, and lifestyles as suitable units of study. The important thing here is to grasp the logic of units of analysis. Once you do, the possibilities for fruitful research are limited only by your imagination.

Categorizing possible units of analysis might make the concept seem more complicated than it needs to be. What you call a given unit of analysis—a group, a formal organization, or a social artifact—is irrelevant. The key is to be clear about what your unit of analysis is. When you embark on a research project, you must decide whether you're studying marriages or marriage partners, crimes or criminals, corporations or corporate executives.

---

social artifact Any product of social beings or their behavior. Can be a unit of analysis.

How to Do It: Identifying the Unit of Analysis

The unit of analysis is an important element in research design and later in data analysis. However, students sometimes find identifying it elusive. The easiest way to identify the unit of analysis is to examine a statement regarding the variables under study.

Consider the following: "The average household income was $40,000." *Income* is the variable of interest, but who or what *has* income? Households, in this instance. We would arrive at the given statement by examining the incomes of several households. To calculate the mean (average) income, we would add up all the household incomes and divide by the number of households. Household is the unit of analysis. It is the unit being analyzed in terms of the variable, *income.*

Consider another statement: "Italian movies show more nudity than do American movies." The variable here is the extent to which

nudity is shown, but who or what *shows* nudity? Movies. Movies are the units of analysis.

Finally, how about this statement: "Twenty-four percent of the families have more than one adult earning $30,000 or more." To be sure, adults are earning the income, but the statement is about whether *families* have such adults. To make this statement, we would study several families. For each, we would ask whether they had more than one adult earning in excess of $30,000; each family would be scored as "yes" or "no" in that respect. Finally, we would calculate the percentage of families scored as "yes." The family, therefore, is the unit of analysis.

Otherwise, you run the risk of drawing invalid conclusions because your assertions about one unit of analysis are actually based on the examination of another. We'll see an example of this issue in the next section as we look at the ecological fallacy.

Faulty Reasoning about Units of Analysis: The Ecological Fallacy and Reductionism

At this point, it's appropriate to introduce two types of faulty reasoning that you should be aware of: the ecological fallacy and reductionism. Each represents a potential pitfall regarding units of analysis, and either can occur in doing research and drawing conclusions from the results.

The Ecological Fallacy

In this context, "ecological" refers to groups or sets or systems: something larger than individuals. The **ecological fallacy** is the assumption that

something learned about an ecological unit says something about the individuals making up that unit. Let's consider a hypothetical illustration of this fallacy.

Suppose we're interested in learning something about the nature of electoral support received by a female political candidate in a recent citywide election. Let's assume we have the vote tally for each precinct so we can tell which precincts gave her the greatest support and which the least. Assume also that we have census data describing some characteristics of these precincts. Our analysis of such data might show that precincts with relatively young voters gave the female candidate a greater proportion of their votes than precincts with older voters did. We might be tempted to conclude from these findings that younger voters are more likely to vote for female candidates than older voters are—in other words, that age affects support for the woman. In reaching such a conclusion, we run the risk of committing the ecological fallacy, because it may have been the older voters in those "young" precincts who voted for the woman. Our problem is that we have examined *precincts* as our units of analysis but wish to draw conclusions about *voters.*

The same problem would arise if we discovered that crime rates were higher in cities having large African American populations than in those

---

ecological fallacy Erroneously drawing conclusions about individuals solely from the observation of groups.

with few African Americans. We would not know if the crimes were actually committed by African Americans. Or if we found suicide rates higher in Protestant countries than in Catholic ones, we still could not know for sure that more Protestants than Catholics committed suicide.

In spite of these hazards, social researchers often have little choice but to address a particular research question through an ecological analysis. Perhaps the most appropriate data are simply not available. For example, the precinct vote tallies and the precinct characteristics mentioned in our initial example may be easy to obtain, but we may not have the resources to conduct a postelection survey of individual voters. In such cases, we may reach a tentative conclusion, recognizing and noting the risk of an ecological fallacy.

Although you should be careful not to commit the ecological fallacy, don't let these warnings lead you into committing what we might call the individualistic fallacy. Some people who approach social research for the first time have trouble reconciling general patterns of attitudes and actions with individual exceptions. But generalizations and probabilistic statements are not invalidated by individual exceptions. Your knowing a rich Democrat, for example, doesn't deny the fact that most rich people vote Republican—as a general pattern. Similarly, if you know someone who has gotten rich without any formal education, that doesn't deny the general pattern of higher education relating to higher income.

The ecological fallacy deals with something else altogether—confusing units of analysis in such a way that we draw conclusions about individuals solely from the observation of groups. Although the patterns observed between variables at the level of groups may be genuine, the danger lies in reasoning from the observed attributes of groups to the attributes of the individuals who made up those groups, even though we have not actually observed individuals.

Reductionism

A second type of faulty reasoning related to units of analysis is reductionism. **Reductionism** involves attempts to explain a particular phenomenon in terms of limited and/or lower-order concepts. The reductionist explanation is not altogether wrong; it is simply too limited. Thus, you might attempt to predict this year's winners and losers in the National Basketball Association by focusing on the abilities of the individual players on each team. This is certainly not stupid or irrelevant, but the success or failure of *teams* involves more than just the individuals in them; it involves coaching, teamwork, strategies, finances, facilities, fan loyalty, and so forth. To understand why some teams do better than others, you would make *team* the unit of analysis, and the *quality of players* would be one variable you would probably want to use in describing and classifying the teams.

Further, different academic disciplines approach the same phenomenon quite differently. Sociologists tend to consider sociological variables (such as *values, norms,* and *roles*), economists ponder economic variables (such as *supply and demand* and *marginal value*), and psychologists examine psychological variables (such as *personality types* and *traumas*). Explaining all or most human behavior in terms of economic factors is called economic reductionism, explaining it in terms of psychological factors is called psychological reductionism, and so forth. Notice how this issue relates to the discussion of theoretical paradigms in Chapter 2.

For many social scientists, the field of **sociobiology** is a prime example of reductionism, suggesting that all social phenomena can be explained in terms of biological factors. Thus, for example, Edward O. Wilson (1975) sought to explain altruistic behavior in human beings in terms of genetic makeup. In his neo-Darwinian view, Wilson suggests that humans have evolved in such a way that individuals sometimes need to sacrifice themselves for the benefit of the whole species. Some people

---

reductionism A fault of some researchers: a strict limitation (reduction) of the kinds of concepts to be considered relevant to the phenomenon under study.

sociobiology A paradigm based in the view that social behavior can be explained solely in terms of genetic characteristics and behavior.

might explain such sacrifice in terms of ideals or warm feelings between humans. However, genes are the essential unit in Wilson's paradigm, producing his famous dictum that human beings are "only DNA's way of making more DNA."

Reductionism of any type tends to suggest that particular units of analysis or variables are more relevant than others. Suppose we ask what caused the American Revolution. Was it a shared commitment to the value of individual liberty? The economic plight of the colonies in relation to Britain? The megalomania of the founders? As soon as we inquire about *the* single cause, we run the risk of reductionism. If we were to regard shared values as the cause of the American Revolution, our unit of analysis would be the individual colonist. An economist, though, might choose the 13 colonies as units of analysis and examine the economic organizations and conditions of each. A psychologist might choose individual leaders as the units of analysis for purposes of examining their personalities. Of course, there's nothing wrong in choosing these units of analysis as part of an explanation of the American Revolution, but I think you can see how each alone would not produce a complete answer.

Like the ecological fallacy, reductionism can occur when we use inappropriate units of analysis. The appropriate unit of analysis for a given research question, however, is not always clear. Social researchers, especially across disciplinary boundaries, often debate this issue.

The Time Dimension

So far in this chapter, we've regarded research design as a process for deciding what aspects we'll observe, of whom, and for what purpose. Now we must consider a set of time-related options that cuts across each of these earlier considerations. We can choose to make observations more or less at one time or over a long period.

Time plays many roles in the design and execution of research, quite aside from the time it takes to do research. Earlier we noted that the time sequence of events and situations is critical to determining causation (a point we'll return to in Part 4). Time also affects the generalizability of research findings. Do the descriptions and explanations resulting from a particular study accurately represent the situation of ten years ago, ten years from now, or only the present? Researchers have two principal options available to deal with the issue of time in the design of their research: cross-sectional studies and longitudinal studies.

Cross-Sectional Studies

A **cross-sectional study** involves observations of a sample, or cross section, of a population or phenomenon that are made at one point in time. Exploratory and descriptive studies are often cross-sectional. A single U.S. Census, for instance, is a study aimed at describing the U.S. population at a given time.

Many explanatory studies are also cross-sectional. A researcher conducting a large-scale national survey to examine the sources of racial and religious prejudice would, in all likelihood, be dealing with a single time frame—taking a snapshot, so to speak, of the sources of prejudice at a particular point in history.

Explanatory cross-sectional studies have an inherent problem. Although their conclusions are based on observations made at only one time, typically they aim at understanding causal processes that occur over time. This problem is somewhat akin to that of determining the speed of a moving object on the basis of a high-speed, still photograph that freezes the movement of the object.

Yanjie Bian, for example, conducted a survey of workers in Tianjin, China, for the purpose of studying stratification in contemporary, urban Chinese society. In undertaking the survey in 1988, however, he was conscious of the important changes brought about by a series of national campaigns, such as the Great Proletarian Cultural Revolution, dating from the Chinese Revolution in 1949 (which

---

cross-sectional study A study based on observations representing a single point in time.

brought the Chinese Communists into power) and continuing into the present.

> These campaigns altered political atmospheres and affected people's work and nonwork activities. Because of these campaigns, it is difficult to draw conclusions from a cross-sectional social survey, such as the one presented in this book, about general patterns of Chinese workplaces and their effects on workers. Such conclusions may be limited to one period of time and are subject to further tests based on data collected at other times.
>
> *(1994: 19)*

The problem of generalizations about social life from a "snapshot" is one this book repeatedly addresses. One solution is suggested by Bian's final comment—about data collected "at other times": Social research often involves revisiting phenomena and building on the results of earlier research.

Longitudinal Studies

In contrast to cross-sectional studies, a **longitudinal study** is designed to permit observations of the same phenomenon over an extended period. For example, a researcher can participate in and observe the activities of a UFO cult from its inception to its demise. Other longitudinal studies use records or artifacts to study changes over time. In analyses of newspaper editorials or Supreme Court decisions over time, for example, the studies are longitudinal whether the researcher's actual observations and analyses were made at one time or over the course of the actual events under study.

Many field research projects, involving direct observation and perhaps in-depth interviews, are naturally longitudinal. Thus, for example, when Ramona Asher and Gary Fine (1991) studied the life experiences of the wives of alcoholic men, they were in a position to examine the evolution of troubled marital relationships over time, sometimes even including the reactions of the subjects to the research itself.

In the classic study *When Prophecy Fails* (1956), Leon Festinger, Henry Reicker, and Stanley

Schachter were specifically interested in learning what happened to a flying saucer cult when their predictions of an alien encounter failed to come true. Would the cult members close down the group, or would they become all the more committed to their beliefs? A longitudinal study was required to provide an answer. (They redoubled their efforts to get new members.)

Longitudinal studies can be more difficult for quantitative studies such as large-scale surveys. Nonetheless, they are often the best way to study changes over time. There are three special types of longitudinal studies that you should know about: trend studies, cohort studies, and panel studies.

Trend Studies

A **trend study** is a type of longitudinal study that examines changes within a population over time. A simple example is a comparison of U.S. Censuses over a period of decades, showing shifts in the makeup of the national population. A similar use of archival data was made by Michael Carpini and Scott Keeter (1991), who wanted to know whether contemporary U.S. citizens were better or more poorly informed about politics than citizens of an earlier generation were. To find out, they compared the results of several Gallup Polls conducted during the 1940s and 1950s with a 1989 survey that asked several of the same questions tapping political knowledge.

Overall, the analysis suggested that contemporary citizens were slightly better informed than earlier generations were. In 1989, 74 percent of the sample could name the vice president of the United States, compared with 67 percent in 1952. Substantially higher percentages of people in 1989

---

longitudinal study A study design involving the collection of data at different points in time.

trend study A type of longitudinal study in which a given characteristic of some population is monitored over time. An example would be the series of Gallup Polls showing the electorate's preferences for political candidates over the course of a campaign, even though different samples were interviewed at each point.

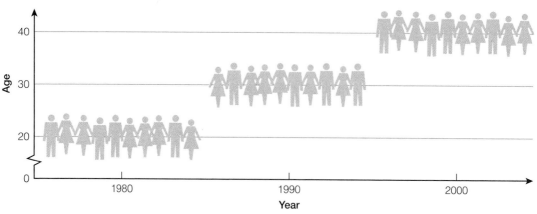

FIGURE 4-5

A Cohort Study Design. Each of the three groups shown here is a sample representing people who were born in 1960.

than in 1947 could explain presidential vetoes and congressional overrides of vetoes. On the other hand, more of the 1947 sample could identify their U.S. representative (38 percent) than the 1989 sample (29 percent) could.

An in-depth analysis, however, indicates that the slight increase in political knowledge resulted from the fact that the people in the 1989 sample were more highly educated than those from earlier samples were. When educational levels were taken into account, the researchers concluded that political knowledge has actually declined within specific educational groups.

Cohort Studies

In a **cohort study,** a researcher examines specific subpopulations, or *cohorts*, as they change over time. Typically, a cohort is an age group, such as people born during the 1950s, but it can also be some other time grouping, such as people born during the Vietnam War, people who got

married in 1994, and so forth. An example of a cohort study would be a series of national surveys, conducted perhaps every 20 years, to study the attitudes of the cohort born during World War II toward U.S. involvement in global affairs. A sample of people 15–20 years old might be surveyed in 1960, another sample of those 35–40 years old in 1980, and another sample of those 55–60 years old in 2000. Although the specific set of people studied in each survey would differ, each sample would represent the cohort born between 1940 and 1945.

Figure 4-5 offers a graphic illustration of a cohort design. In the example, three studies are being compared: one was conducted in 1980, another in 1990, and the third in 2000. Those who were 20 years old in the 1980 study are compared with those who were 30 in the 1990 study and those who were 40 in the 2000 study. Although the subjects being described in each of the three groups are different, each set of subjects represents the same cohort: those who were born in 1960.

James Davis (1992) turned to a cohort analysis in an attempt to understand shifting political orientations during the 1970s and 1980s in the United States. Overall, he found a liberal trend on issues such as race, gender, religion, politics, crime, and free speech. But did this trend represent people in general getting a bit more liberal, or did it merely reflect liberal younger generations replacing the conservative older ones?

cohort study A study in which some specific subpopulation, or cohort, is studied over time, although data may be collected from different members in each set of observations. For example, a study of the occupational history of the class of 1970 in which questionnaires were sent every five years would be a cohort study.

TABLE 4-1
Age and Political Liberalism

Survey dates	1972 to 1974	1977 to 1980	1982 to 1984	1987 to 1989
Age of cohort	20–24	25–29	30–34	35–39
Percent who would let the Communist speak	72%	68%	73%	73%

To answer this question, Davis examined national surveys (from the General Social Survey, of which he is a founder) conducted in four time periods, five years apart. In each survey, he grouped the respondents into age groups, also five years apart. This strategy allowed him to compare different age groups at any given point in time as well as to follow the political development of each age group over time.

One of the questions he examined was whether a person who admitted to being a Communist should be allowed to speak in the respondents' communities. Consistently, the younger respondents in each time period were more willing to let the Communist speak than the older ones were. Among those aged 20–40 in the first set of the survey, for example, 72 percent took this liberal position, contrasted with 27 percent among respondents 80 and older. What Davis found when he examined the youngest cohort over time is shown in Table 4-1. This pattern of a slight, conservative shift in the 1970s, followed by a liberal rebound in the 1980s, typifies the several cohorts Davis analyzed (J. Davis 1992: 269).

In another study, Eric Plutzer and Michael Berkman (2005) used a cohort design to completely reverse a prior conclusion regarding aging and support for education. Logically, as people grow well beyond the child-rearing years, we might expect them to reduce their commitment to educational funding. Moreover, cross-sectional data support that expectation. The researchers present several data sets showing those over 65 voicing less support for educational funding than those under 65 did.

Such simplistic analyses, however, leave out an important variable: increasing support for educational funding in U.S. society over time in general. The researchers add to this the concept of "generational replacement," meaning that the older respondents in a survey grew up during a time when there was less support for education in general, whereas the younger respondents grew up during a time of greater overall support.

A cohort analysis allowed the researchers to determine what happened to the attitudes of specific cohorts over time. Here, for example, are the percentages of Americans born during the 1940s who felt educational spending was too low, when members of that cohort were interviewed over time (Plutzer and Berkman 2005: 76):

Year Interviewed	Percent Who Say Educational Funding Is Too Low
1970s	58
1980s	66
1990s	74
2000s	79

As these data indicate, those who were born during the 1940s have steadily increased their support for educational funding as they have passed through and beyond the child-rearing years.

Panel Studies

Though similar to trend and cohort studies, a **panel study** examines the same set of people each time. For example, we could interview the same

---

panel study A type of longitudinal study, in which data are collected from the same set of people (the sample or panel) at several points in time.

sample of voters every month during an election campaign, asking for whom they intended to vote. Though such a study would allow us to analyze overall trends in voter preferences for different candidates, it would also show the precise patterns of persistence and change in intentions. For example, a trend study that showed that Candidates A and B each had exactly half of the voters on September 1 and on October 1 as well could indicate that none of the electorate had changed voting plans, that all of the voters had changed their intentions, or something in-between. A panel study would eliminate this confusion by showing what kinds of voters switched from A to B and what kinds switched from B to A, as well as other facts.

Joseph Veroff, Shirley Hatchett, and Elizabeth Douvan (1992) wanted to learn about marital adjustment among newlyweds, specifically regarding differences between white and African American couples. To get subjects for study, they selected a sample of couples who applied for marriage licenses in Wayne County, Michigan, April through June 1986.

Concerned about the possible impact their research might have on the couples' marital adjustment, the researchers divided their sample in half at random: an *experimental* group and a *control* group (concepts we'll explore further in Chapter 8). Couples in the former group were intensively interviewed over a four-year period, whereas the latter group was contacted only briefly each year.

By studying the same couples over time, the researchers could follow the specific problems that arose and the way the couples dealt with them. As a by-product of their research, they found that those studied the most intensely seemed to achieve a somewhat better marital adjustment. The researchers felt that the interviews could have forced couples to discuss matters they might have otherwise buried.

Comparing the Three Types of Longitudinal Studies

To reinforce the distinctions among trend, cohort, and panel studies, let's contrast the three study designs in terms of the same variable: *religious*

affiliation. A trend study might look at shifts in U.S. religious affiliations over time, as the Gallup Poll does on a regular basis. A cohort study might follow shifts in religious affiliations among "the Depression generation," specifically, say, people who were 20 to 30 years old in 1932. We could study a sample of people 30–40 years old in 1942, a new sample of people aged 40–50 in 1952, and so forth throughout their lifespan. A panel study could start with a sample of the whole population or of some special subset and study those specific individuals over time. Notice that only the panel study would give a full picture of the shifts among the various categories of affiliations, including "none." Cohort and trend studies would uncover only net changes.

Longitudinal studies have an obvious advantage over cross-sectional ones in providing information describing processes over time. But this advantage often comes at a heavy cost in both time and money, especially in a large-scale survey. Observations may have to be made at the time events are occurring, and the method of observation may require many research workers.

Panel studies, which offer the most comprehensive data on changes over time, face a special problem: panel attrition. Some of the respondents studied in the first wave of the survey might not participate in later waves. (This is comparable to the problem of experimental mortality discussed in Chapter 8.) The danger is that those who drop out of the study may be atypical, thereby distorting the results of the study. Thus, when Carol Aneshensel and her colleagues conducted a panel study of adolescent girls (comparing Latinas and non-Latinas), they looked for and found differences in characteristics of survey dropouts among Latinas born in the United States and those born in Mexico. These differences needed to be taken into account to avoid misleading conclusions about differences between Latinas and non-Latinas (Aneshensel et al. 1989).

Approximating Longitudinal Studies

Longitudinal studies do not always provide a feasible or practical means of studying processes that take place over time. Fortunately, researchers often can draw approximate conclusions about such processes even when only cross-sectional data are available. Here are some ways to do that.

Sometimes cross-sectional data imply processes over time on the basis of simple logic. For example, in the study of student drug use conducted at the University of Hawaii (Chapter 2), students were asked to report whether they had ever tried each of several illegal drugs. The study found that some students had tried both marijuana and LSD, some had tried only one, and others had tried neither. Because these data were collected at one time, and because some students presumably would experiment with drugs later on, it would appear that such a study could not tell whether students were more likely to try marijuana or LSD first.

A closer examination of the data showed, however, that although some students reported having tried marijuana but not LSD, there were no students in the study who had tried only LSD. From this finding it was inferred—as common sense suggested—that marijuana use preceded LSD use. If the process of drug experimentation occurred in the opposite time order, then a study at a given time should have found some students who had tried LSD but not marijuana, and it should have found no students who had tried only marijuana.

Researchers can also make logical inferences whenever the time order of variables is clear. If we discovered in a cross-sectional study of college students that those educated in private high schools received better college grades than those educated in public high schools did, we would conclude that the type of high school attended affected college grades, not the other way around. Thus, even though we made our observations at only one time, we would feel justified in drawing conclusions about processes taking place across time.

Very often, age differences discovered in a cross-sectional study form the basis for inferring processes across time. Suppose you're interested in the pattern of worsening health over the course of the typical life cycle. You might study the results of annual checkups in a large hospital. You could group health records according to the ages of those examined and rate each age group in terms of several health conditions—sight, hearing, blood

pressure, and so forth. By reading across the age-group ratings for each health condition, you would have something approximating the health history of individuals. Thus, you might conclude that the average person develops vision problems before hearing problems. You would need to be cautious in this assumption, however, because the differences might reflect societywide trends. Perhaps improved hearing examinations instituted in the schools had affected only the young people in your study.

Asking people to recall their pasts is another common way of approximating observations over time. Researchers use that method when they ask people where they were born or when they graduated from high school or whom they voted for in 1988. Qualitative researchers often conduct in-depth "life history" interviews. For example, C. Lynn Carr (1998) used this technique in a study of "tomboyism." Her respondents, aged 25–40, were asked to reconstruct aspects of their lives from childhood on, including experiences of identifying themselves as tomboys.

The danger in this technique is evident. Sometimes people have faulty memories; sometimes they lie. When people are asked in postelection polls whom they voted for, the results inevitably show more people voting for the winner than actually did so on election day. As part of a series of in-depth interviews, such a report can be validated in the context of other reported details; however, results based on a single question in a survey must be regarded with caution.

Cohorts can also be used to infer processes over time from cross-sectional data. For example, when Prem Saxena and his colleagues (2004) wanted to examine whether wartime conditions would affect the age at which people married, he used cross-sectional data from a survey of Lebanese women. During the Lebanese Civil War from 1975 to 1990, many young men migrated to other countries. By noting the year in which the survey respondents first married, he could determine that the average age-at-first-marriage increased with the onset of the war.

This discussion of the ways that time figures into social research suggests several questions you should confront in your own research projects. In designing any study, be sure to look at both the explicit and implicit assumptions you're making about time. Are you interested in describing some process that occurs over time, or are you simply going to describe what exists now? If you want to describe a process occurring over time, will you be able to make observations at different points in the process, or will you have to approximate such observations by drawing logical inferences from what you can observe now? If you opt for a longitudinal design, which method best serves your research purposes?

Examples of Research Strategies

As the preceding discussions have implied, social research follows many paths. The following short excerpts further illustrate this point. As you read each excerpt, note both the content of each study and the method used to study the chosen topic. Does the study seem to be exploring, describing, or explaining (or some combination of these)? What are the sources of data in each study? Can you identify the unit of analysis? Is the dimension of time relevant? If so, how will it be handled?

- This case study of unobtrusive mobilizing by Southern California Rape Crisis Center uses archival, observational, and interview data to explore how a feminist organization worked to change police, schools, prosecutors, and some state and national organizations from 1974 to 1994. (Schmitt and Martin 1999: 364)

- Using life history narratives, the present study investigates processes of agency and consciousness among 14 women who identified themselves as tomboys. (Carr 1998: 528)

- By drawing on interviews with activists in the former Estonian Soviet Socialist Republic, we specify the conditions by which accommodative and oppositional subcultures exist and are successfully transformed into social movements. (Johnston and Snow 1998: 473)

- This paper presents the results of an ethnographic study of an AIDS service organization located in a small city. It is based on a combi-

nation of participant observation, interviews with participants, and review of organizational records. (Kilburn 1998: 89)

- Using interviews obtained during fieldwork in Palestine in 1992, 1993, and 1994, and employing historical and archival records, I argue that Palestinian feminist discourses were shaped and influenced by the sociopolitical context in which Palestinian women acted and with which they interacted. (Abdulhadi 1998: 649)

- This article reports on women's experiences of breastfeeding in public as revealed through in-depth interviews with 51 women. (Stearns 1999: 308)

- Using interview and observational field data, I demonstrate how a system of temporary employment in a participative workplace both exploited and shaped entry-level workers' aspirations and occupational goals. (V. Smith 1998: 411)

- I collected data [on White Separatist rhetoric] from several media of public discourse, including periodicals, books, pamphlets, transcripts from radio and television talk shows, and newspaper and magazine accounts. (Berbrier 1998: 435)

- In the analysis that follows, racial and gender inequality in employment and retirement will be analyzed, using a national sample of persons who began receiving Social Security Old Age benefits in 1980–81. (Hogan and Perrucci 1998: 528)

- Drawing from interviews with female crack dealers, this paper explores the techniques they use to avoid arrest. (Jacobs and Miller 1998: 550)

How to Design a Research Project

You've now seen some of the options available to social researchers in designing projects. I know there are a lot of components, and the relationships among them may not be totally clear, so here's a way of pulling them together. Let's assume you were to undertake research. Where would you start? Then, where would you go?

Although research design occurs at the beginning of a research project, it involves all the steps of the subsequent project. This discussion, then, provides both guidance on how to start a research project and an overview of the topics that follow in later chapters of this book.

Figure 4-6 presents a schematic view of the traditional image of research design. I present this view reluctantly, because it may suggest more of a step-by-step order to research than actual practice bears out. Nonetheless, this idealized overview of the process provides a context for the specific details of particular components of social research. Essentially, it is another and more detailed picture of the scientific process presented in Chapter 2.

At the top of the diagram are interests, ideas, and theories, the possible beginning points for a line of research. The letters (A, B, X, Y, and so forth) represent variables or concepts such as prejudice or alienation. Thus, you might have a general interest in finding out what causes some people to be more prejudiced than others, or you might want to know some of the consequences of alienation. Alternatively, your inquiry might begin with a specific idea about the way things are. For example, you might have the idea that working on an assembly line causes alienation. The question marks in the diagram indicate that you aren't sure things are the way you suspect they are—that's why you're doing the research. Notice that a theory is represented as a set of complex relationships among several variables.

The double arrows between "interest," "idea," and "theory" suggest that a movement back and forth across these several possible beginnings often takes place. An initial interest may lead to the formulation of an idea, which may be fit into a larger theory, and the theory may produce new ideas and create new interests.

Any or all of these three may suggest the need for empirical research. The purpose of such research can be to explore an interest, test a specific idea, or validate a complex theory. Whatever the

purpose, the researcher needs to make a variety of decisions, as indicated in the remainder of the diagram.

To make this discussion more concrete, let's take a specific research example. Suppose you're concerned with the issue of abortion and have a special interest in learning why some college students support abortion rights and others oppose them. Going a step further, let's say you've formed the impression that students in the humanities and social sciences seem generally more inclined to support the idea of abortion rights than those in the natural sciences do. (That kind of thinking often leads people to design and conduct social research.)

So, where do you start? You have an idea you want to pursue, one that involves abortion attitudes and choice of college major. In terms of the options we've discussed in this chapter, you probably have both descriptive and explanatory interests, but you might decide you only want to explore the issue. You might wonder what sorts of attitudes students with different majors have about abortion (exploratory), what percentage of the student body supports a woman's right to an abortion (descriptive), or what causes some to support it and others to oppose it (explanation). The units of analysis in this case would be individuals: college students. But we're jumping the gun. As you can see, even before we've "started," we've started. The reciprocal processes described in Figure 4-6 begin even before you've made a commitment to a project. Let's look more formally at the various steps, then, keeping this reciprocal motion in mind.

Getting Started

At the outset of your project, your aim would probably be exploratory. At this point, you might choose among several possible activities in pursuing your interest in student attitudes about abortion rights. To begin with, you might want to read something about the issue. If you have a hunch that attitudes are somehow related to college major, you might find out what other researchers may have written about that. Appendix A of this book will help you make use of your college library. In addition, you

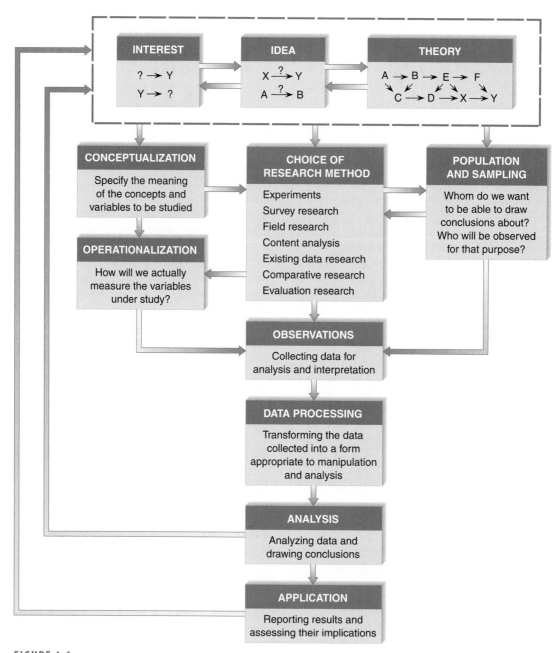

FIGURE 4-6

Traditional Image of Research Design.

would probably talk to some people who support abortion rights and some who don't. You might attend meetings of abortion-related groups. All these activities could help prepare you to handle the various decisions of research design we're about to examine.

Before designing your study, you must define the purpose of your project. What kind of study will you undertake—exploratory, descriptive, explanatory? Do you plan to write a research paper to satisfy a course or thesis requirement? Is your purpose to gain information that will support you in arguing for or against abortion rights? Do you want to write an article for the campus newspaper or an academic journal? In reviewing the previous research literature regarding abortion rights, you should note the design decisions other researchers have made, always asking whether the same decisions would satisfy your purpose.

Usually, your purpose for undertaking research can be expressed as a report. A good first step in designing your project is to outline such a report (see Chapter 17 for help on this). Although your final report may not look much like your initial image of it, this exercise will help you figure out which research designs are most appropriate. During this step, clearly describe the kinds of statements you want to make when the research is complete. Here are some examples of such statements: "Students frequently mentioned abortion rights in the context of discussing social issues that concerned them personally." "X percent of State U. students favor a woman's right to choose an abortion." "Engineers are (more/less) likely than sociologists to favor abortion rights."

Conceptualization

Once you have a well-defined purpose and a clear description of the kinds of outcomes you want to achieve, you can proceed to the next step in the design of your study—conceptualization. We often talk pretty casually about social science concepts such as prejudice, alienation, religiosity, and liberalism, but it's necessary to clarify what we mean by these concepts, in order to draw meaningful conclusions about them. Chapter 5 examines this

process of conceptualization in depth. For now, let's see what it might involve in the case of our hypothetical example.

If you're going to study how college students feel about abortion and why, the first thing you'll have to specify is what you mean by "the right to an abortion." Because support for abortion probably varies according to the circumstances, you'll want to pay attention to the different conditions under which people might approve or disapprove of abortion: for example, when the woman's life is in danger, in the case of rape or incest, or simply as a matter of personal choice.

Similarly, you'll need to specify exact meanings for all the other concepts you plan to study. If you want to study the relationship of opinion about abortion to college major, you'll have to decide whether you want to consider only officially declared majors or to include students' intentions as well. What will you do with those who have no major?

In surveys and experiments, you need to specify such concepts in advance. In less tightly structured research, such as open-ended interviews, an important part of the research may involve the discovery of different dimensions, aspects, or nuances of concepts. In such cases, the research itself may uncover and report aspects of social life that were not evident at the outset of the project.

Choice of Research Method

As we'll discuss in Part 3, each research method has its strengths and weaknesses, and certain concepts are more appropriately studied through some methods than through others. In our study of attitudes toward abortion rights, a survey might be the most appropriate method: either interviewing students or asking them to fill out a questionnaire. Surveys are particularly well suited to the study of public opinion. This is not to say that you couldn't make good use of the other methods presented in Part 3. For example, you might use the method of content analysis to examine letters to the editor and analyze the different images of abortion that letter writers have. Field research would provide an avenue to understanding how people interact with one another regarding the issue of abortion, how they discuss it, and how they change their minds. Other research methods introduced in Part 3 could also be used in studying this topic. Usually, the best study design uses more than one research method, taking advantage of their different strengths. If you look back at the brief examples of actual studies at the end of the preceding section, you'll see several instances where the researchers used many methods in a single study.

Operationalization

Once you've specified the concepts to be studied and chosen a research method, the next step is operationalization, or deciding on your measurement techniques (discussed further in Chapters 5 and 6). The meaning of variables in a study is determined in part by how they are measured. Part of the task here is deciding how the desired data will be collected: direct observation, review of official documents, a questionnaire, or some other technique.

If you decided to use a survey to study attitudes toward abortion rights, part of operationalization is determining the wording of questionnaire items. For example, you might operationalize your main variable by asking respondents whether they would approve of a woman's right to have an abortion under each of the conditions you've conceptualized: in the case of rape or incest, if her life were threatened by the pregnancy, and so forth. You'd design the questionnaire so that it asked respondents to express approval or disapproval for each situation. Similarly, you would specify exactly how respondents would indicate their college major, as well as what choices to provide those who have not declared a major.

Population and Sampling

In addition to refining concepts and measurements, you must decide whom or what to study. The *population* for a study is that group (usually of people) about whom we want to draw conclusions. We're almost never able to study all the members of the population that interests us, however, and we can never make every possible observation of them.

In every case, then, we select a *sample* from among the data that might be collected and studied. The sampling of information, of course, occurs in everyday life and often produces biased observations. (Recall the discussion of "selective observation" in Chapter 1.) Social researchers are more deliberate in their sampling of what will be observed.

Chapter 7 describes methods for selecting samples that adequately reflect the whole population that interests us. Notice in Figure 4-6 that decisions about population and sampling are related to decisions about the research method to be used. Whereas probability sampling techniques would be relevant to a large-scale survey or a content analysis, a field researcher might need to select only those informants who will yield a balanced picture of the situation under study, and an experimenter might assign subjects to experimental and control groups in a manner that creates comparability.

In your hypothetical study of abortion attitudes, the relevant population would be the student population of your college. As you'll discover in Chapter 7, however, selecting a sample will require you to get more specific than that. Will you include part-time as well as full-time students? Only degree candidates or everyone? International students as well as U.S. citizens? Undergraduates, graduate students, or both? There are many such questions—each of which must be answered in terms of your research purpose. If your purpose is to predict how students would vote in a local referendum on abortion, you might want to limit your population to those eligible and likely to vote.

Observations

Having decided what to study among whom by what method, you're now ready to make observations—to collect empirical data. The chapters of Part 3, which describe the various research methods, give the different observation techniques appropriate to each.

To conduct a survey on abortion, you might want to print questionnaires and mail them to a sample selected from the student body. Alternatively, you could arrange to have a team of interviewers conduct the survey over the telephone. The relative advantages and disadvantages of these and other possibilities are discussed in Chapter 9.

Data Processing

Depending on the research method chosen, you'll have amassed a volume of observations in a form that probably isn't immediately interpretable. If you've spent a month observing a street-corner gang firsthand, you'll now have enough field notes to fill a book. In a historical study of ethnic diversity at your school, you may have amassed volumes of official documents, interviews with administrators and others, and so forth. Chapters 13 and 14 describe some of the ways social science data are processed or transformed for qualitative or quantitative analysis.

In the case of a survey, the "raw" observations are typically in the form of questionnaires with boxes checked, answers written in spaces, and the like. The data-processing phase of a survey typically involves the classification (coding) of written-in answers and the transfer of all information to a computer.

Analysis

Once the collected data are in a suitable form, you're ready to interpret them for the purpose of drawing conclusions that reflect the interests, ideas, and theories that initiated the inquiry. Chapters 13 and 14 describe a few of the many options available to you in analyzing data. In Figure 4-6, notice that the results of your analyses feed back into your initial interests, ideas, and theories. Often this feedback represents the beginning of another cycle of inquiry.

In the survey of student attitudes about abortion rights, the analysis phase would pursue both descriptive and explanatory aims. You might begin by calculating the percentages of students who favored or opposed each of the several different versions of abortion rights. Taken together, these several percentages would provide a good picture of student opinion on the issue.

Moving beyond simple description, you might describe the opinions of subsets of the student body, such as different college majors. Provided that your design called for trapping other information about respondents, you could also look at men versus women; freshmen, sophomores, juniors, seniors, and graduate students; or other categories that you've included. The description of subgroups could then lead you into an explanatory analysis.

Application

The final stage of the research process involves the uses made of the research you've conducted and the conclusions you've reached. To start, you'll probably want to communicate your findings so that others will know what you've learned. It may be appropriate to prepare—and even publish—a written report. Perhaps you'll make oral presentations, such as papers delivered to professional and scientific meetings. Other students would also be interested in hearing what you've learned about them.

You may want to go beyond simply reporting what you've learned to discussing the implications of your findings. Do they say anything about actions that might be taken in support of policy goals? Both the proponents and the opponents of abortion rights would be interested.

Finally, be sure to consider what your research suggests in regard to further research on your subject. What mistakes should be corrected in future studies? What avenues—opened up slightly in your study—should be pursued further?

Research Design in Review

As this overview shows, research design involves a set of decisions regarding what topic is to be studied among what population with what research methods for what purpose. Although you'll want to consider many ways of studying a subject—and use your imagination as well as your knowledge of a variety of methods—research design is the process of focusing your perspective for the purposes of a particular study.

If you're doing a research project for one of your courses, many aspects of research design may be specified for you in advance, including the method (such as an experiment) or the topic (as in a course on a particular subject, such as prejudice). The following summary assumes that you're free to choose both your topic and your research strategy.

In designing a research project, you'll find it useful to begin by assessing three things: your interests, your abilities, and the available resources. Each of these considerations will suggest a large number of possible studies.

Simulate the beginning of a somewhat conventional research project: Ask yourself what you're interested in understanding. Surely you have several questions about social behavior and attitudes. Why are some people politically liberal and others politically conservative? Why are some people more religious than others? Why do people join militia groups? Do colleges and universities still discriminate against minority faculty members? Why would a woman stay in an abusive relationship? Spend some time thinking about the kinds of questions that interest and concern you.

Once you have a few questions you'd be interested in answering for yourself, think about the kind of information needed to answer them. What research units of analysis would provide the most relevant information: college students, corporations, voters, cities, or corporations? This question will probably be inseparable in your thoughts from the question of research topics. Then ask which aspects of the units of analysis would provide the information you need in order to answer your research question.

Once you have some ideas about the kind of information relevant to your purpose, ask yourself how you might go about getting that information. Are the relevant data likely to be already available somewhere (say, in a government publication), or would you have to collect them yourself? If you think you would have to collect them, how would you go about doing it? Would you need to survey a large number of people, or interview a few people in depth? Could you learn what you need to know by attending meetings of certain groups? Could you glean the data you need from books in the library?

As you answer these questions, you'll find yourself well into the process of research design. Keep in mind your own research abilities and the resources available to you. There's little point in designing a perfect study that you can't actually carry out. You may want to try a research method you haven't used before so you can learn from it, but be careful not to put yourself at too great a disadvantage.

Once you have a general idea of what you want to study and how, carefully review previous research in journals and books to see how other researchers have addressed the topic and what they have learned about it. Your review of the literature may lead you to revise your research design: Perhaps you'll decide to use a previous researcher's method or even replicate an earlier study. A standard procedure in the physical sciences, the independent replication of research projects is just as important in the social sciences, although social researchers tend to overlook that. Or, you might want to go beyond replication and study some aspect of the topic that you feel previous researchers have overlooked.

Here's another approach you might take. Suppose a topic has been studied previously using field research methods. Can you design an experiment that would test the findings those earlier researchers produced? Or, can you think of existing statistics that could be used to test their conclusions? Did a mass survey yield results that you'd like to explore in greater detail through on-the-spot observations and in-depth interviews? The use of several different research methods to test the same finding is sometimes called *triangulation,* and you should always keep it in mind as a valuable research strategy. Because each research method has particular strengths and weaknesses, there is always a danger that research findings will reflect, at least in part, the method of inquiry. In the best of all worlds, your own research design should bring more than one research method to bear on the topic.

The Research Proposal

Quite often, in the design of a research project, you'll have to lay out the details of your plan for someone else's review and/or approval. In the case of a course project, for example, your instructor might very well want to see a "proposal" before you set off to work. Later in your career, if you wanted to undertake a major project, you might need to obtain funding from a foundation or government agency, who would most definitely want a detailed proposal that describes how you would spend their money. You might respond to a Request for Proposals (RFP), which both public and private agencies often circulate in search of someone to do research for them.

This chapter continues with a brief discussion of how you might prepare a research proposal. This will give you one more overview of the whole research process that the rest of this book details.

Elements of a Research Proposal

Although some funding agencies (or your instructor, for that matter) may have specific requirements for the elements or structure of a research proposal, here are some basic elements you should include.

Problem or Objective

What exactly do you want to study? Why is it worth studying? Does the proposed study have practical significance? Does it contribute to the construction of social theories?

Literature Review

What have others said about this topic? What theories address it and what do they say? What previous research exists? Are there consistent findings, or do past studies disagree? Are there flaws in the body of existing research that you think you can remedy?

Chapter 17 has a lengthier discussion of this topic. You'll find that reading social science research reports requires special skills. If you need to undertake a review of the literature at this point in your course, you may want to skip ahead to Chapter 17. It will familiarize you with the different types of research literature, how to find what you want, and how to read it. There is a special discussion of how to use online resources and how to avoid being misled by information on the Internet.

In part, your review of the literature will be shaped by the data-collection method(s) you intend to use in your study. Reviewing the designs of previous studies using that same technique can give you a head start in planning your own study. At the same time, you should focus your search on your research topic, regardless of the methods other researchers have used. So, if you're planning field research on, say, interracial marriages, you might gain some useful insights from the findings of surveys on the topic; further, past field research on interracial marriages could be invaluable in your designing a survey on the topic.

Because the literature review will appear early in your research proposal, you should write it with an eye to introducing the reader to the topic you will address, laying out in a logical manner what has already been learned on the topic by past researchers, then leading up to the holes or loose ends in our knowledge of the topic, which you propose to remedy. Or, a little differently, your review of the literature may point to inconsistencies or disagreements to be found among the existing research findings. In that case, your proposed research will aim to resolve the ambiguities that plague us. I don't know about you, but I'm already excited about the research you're proposing to undertake.

Subjects for Study

Whom or what will you study in order to collect data? Identify the subjects in general, theoretical terms; then, in specific, more concrete terms, identify who is available for study and how you'll reach them. Will it be appropriate to select a sample? If so, how will you do that? If there is any possibility that your research will affect those you study, how will you insure that the research does not harm them?

Beyond these general questions, the specific research method you'll use will further specify the matter. If you're planning to undertake an experiment, a survey, or field research, for example, the techniques for subject selection will vary quite a bit. Happily, Chapter 7 of this book discusses sampling techniques for both qualitative and quantitative studies.

Measurement

What are the key variables in your study? How will you define and measure them? Do your definitions and measurement methods duplicate or differ from those of previous research on this topic? If you have already developed your measurement device (a questionnaire, for example) or will be using something previously developed by others, it might be appropriate to include a copy in an appendix to your proposal.

Data-Collection Methods

How will you actually collect the data for your study? Will you conduct an experiment or a survey? Will you undertake field research or will you focus on the reanalysis of statistics already created by others? Perhaps you'll use more than one method.

Analysis

Indicate the kind of analysis you plan to conduct. Spell out the purpose and logic of your analysis. Are you interested in precise description? Do you intend to explain why things are the way they are? Do you plan to account for variations in some quality: for example, why some students are more liberal than others? What possible explanatory variables will your analysis consider, and how will you know if you've explained variations adequately?

Schedule

It's often appropriate to provide a schedule for the various stages of research. Even if you don't do this for the proposal, do it for yourself. Unless you have a timeline for accomplishing the several stages of research and keeping track of how you're doing, you may end up in trouble.

Budget

When you ask someone to cover the costs of your research, you need to provide a budget that specifies where the money will go. Large, expensive projects include budgetary categories such as personnel, equipment, supplies, telephones, and postage. Even for a project you'll pay for yourself, it's a good idea to spend some time anticipating expenses: office supplies, photocopying, CD-ROMs, telephone calls, transportation, and so on.

As you can see, if you're interested in conducting a social research project, it's a good idea to prepare a research proposal for your own purposes, even if you aren't required to do so by your instructor or a funding agency. If you're going to invest your time and energy in such a project, you should do what you can to insure a return on that investment.

Now that you've had a broad overview of social research, you can move on to the remaining chapters in this book and learn exactly how to design and execute each specific step. If you've found a research topic that really interests you, you'll want to keep it in mind as you see how you might go about studying it. As always, however, you should keep the ethical dimension of research design in mind as you explore your options.

The Ethics of Research Design

Designing a research project must include serious consideration of the ethical dimension of social science research. To begin, if your study requires the participation of human subjects, you must determine whether the likely benefits of the research will justify the time and effort you'll ask of them.

Then you'll want to design the study in concurrence with the ethical concerns discussed in Chapter 3. For example, you'll want to insure that the subjects' privacy and well-being is protected. As I've indicated earlier, it may be appropriate for your research design to be reviewed by an Institutional Review Board.

MAIN POINTS

Introduction

- Any research design requires researchers to specify as clearly as possible what they want to find out and then determine the best way to do it.

Three Purposes of Research

- The principal purposes of social research include exploration, description, and explanation. Research studies often combine more than one purpose.
- Exploration is the attempt to develop an initial, rough understanding of some phenomenon.
- Description is the precise measurement and reporting of the characteristics of some population or phenomenon under study.
- Explanation is the discovery and reporting of relationships among different aspects of the phenomenon under study. Whereas descriptive studies answer the question "What's so?" explanatory ones tend to answer the question "Why?"

The Logic of Nomothetic Explanation

- Both idiographic and nomothetic models of explanation rest on the idea of causation. The idiographic model aims at a complete understanding of a particular phenomenon, using all relevant causal factors. The nomothetic model aims at a general understanding—not necessarily complete—of a class of phenomena, using a small number of relevant causal factors.
- There are three basic criteria for establishing causation in nomothetic analyses: (1) The variables must be empirically associated, or correlated, (2) the causal variable must occur earlier in time than the variable it is said to affect, and (3) the observed effect cannot be explained as the effect of a different variable.

Necessary and Sufficient Causes

- Mere association, or correlation, does not in itself establish causation. A spurious causal relationship is an association that in reality is caused by one or more other variables.

Units of Analysis

- Units of analysis are the people or things whose characteristics social researchers observe, describe, and explain. Typically, the unit of analysis in social research is the individual person, but it may also be a social group, a formal organization, a social interaction, a social artifact, or some other phenomenon such as a lifestyle.
- The ecological fallacy involves taking conclusions drawn solely from the analysis of groups (e.g., corporations) and applying them to individuals (e.g., the employees of corporations).
- Reductionism is the attempt to understand a complex phenomenon in terms of a narrow set of concepts, such as attempting to explain the American Revolution solely in terms of economics (or political idealism or psychology).

The Time Dimension

- Research into processes that occur over time presents social challenges that can be addressed through cross-sectional studies or longitudinal studies.
- Cross-sectional studies are based on observations made at one time. Although this characteristic limits such studies, researchers can sometimes use them to make inferences about processes that occur over time.
- In longitudinal studies, observations are made at many times. Such observations may be made of samples drawn from general populations (trend studies), samples drawn from more specific subpopulations (cohort studies), or the same sample of people each time (panel studies).

How to Design a Research Project

- Research design starts with an initial interest, idea, or theoretical expectation and proceeds through a series of interrelated steps to narrow the focus of the study so that concepts, methods, and procedures are well defined. A good research plan accounts for all these steps in advance.
- At the outset, a researcher specifies the meaning of the concepts or variables to be studied (conceptualization), chooses a research method or methods (e.g., experiments versus surveys), and specifies the population to be studied and, if applicable, how it will be sampled.
- To operationalize the concepts to be studied, the researcher states precisely how variables in the study will be measured. Research then proceeds through observation, data processing, analysis, and application, such as reporting the results and assessing their implications.

The Research Proposal

- A research proposal provides a preview of why a study will be undertaken and how it will be conducted. A research proposal is often required to get permission or necessary resources. Even when not required, a proposal is a useful device for planning.

The Ethics of Research Design

- Your research design should indicate how your study will abide by the ethical requirements of social research.
- It may be appropriate for your research proposal to be reviewed by an Institutional Review Board.

KEY TERMS

The following terms are defined in context in the chapter and at the bottom of the page where the term is introduced, as well as in the comprehensive glossary at the back of the book.

cohort study	reductionism
correlation	social artifact
cross-sectional study	sociobiology
ecological fallacy	spurious relationship
longitudinal study	trend study
panel study	units of analysis

PROPOSING SOCIAL RESEARCH: DESIGN

This chapter has laid out many different ways social research can be structured. In designing your research project, you will need to specify which among these you will use. Is your purpose that of exploring a topic, providing a detailed description, or explaining the social differences and processes you may observe? If you are planning a causal analysis, you should say something about how you will organize and pursue that goal.

Further, will your project collect data at one point in time or compare data across time? What data collection technique(s) will you employ? You will revisit these and similar questions as you delve into your project.

REVIEW QUESTIONS AND EXERCISES

1. One example in this chapter suggested that political orientations cause attitudes toward legalizing marijuana. Can you make an argument that the time order is just the opposite of what was assumed?

2. Here are some examples of real research topics. For each one, can you name the unit of analysis? (The answers are at the end of this chapter.)

   a. Women watch TV more than men because they are likely to work fewer hours outside the home than men. . . . Black people watch an average of approximately three-quarters of an hour more television per day than white people. (Hughes 1980: 290)

   b. Of the 130 incorporated U.S. cities with more than 100,000 inhabitants in 1960, 126 had at least two short-term nonproprietary general hospitals accredited by the American Hospital Association. (Turk 1980: 317)

   c. The early TM [transcendental meditation] organizations were small and informal. The Los Angeles group, begun in June 1959, met at a member's house where, incidentally, Maharishi was living. (Johnston 1980: 337)

   d. However, it appears that the nursing staffs exercise strong influence over . . . a decision to change the nursing care system. . . . Conversely, among those decisions dominated by the administration and the medical staffs . . . (Comstock 1980: 77)

   e. Though 667,000 out of 2 million farmers in the United States are women, women historically have not been viewed as farmers, but rather, as the farmer's wife. (Votaw 1979: 8)

   f. The analysis of community opposition to group homes for the mentally handicapped . . . indicates that deteriorating neighborhoods are most likely to organize in opposition, but that upper-middle class neighborhoods are most likely to enjoy private access to local officials. (Graham and Hogan 1990: 513)

   g. Some analysts during the 1960s predicted that the rise of economic ambition and political militancy among blacks would foster discontent with the "otherworldly" black mainline churches. (Ellison and Sherkat 1990: 551)

   h. This analysis explores whether propositions and empirical findings of contemporary theories of organizations directly apply to both private product producing organizations (PPOs) and public human service organizations (PSOs). (Shiflett and Zey 1990: 569)

   i. This paper examines variations in job title structures across work roles. Analyzing 3,173 job titles in the California civil service system in 1985, we investigate how and why lines of work vary in the proliferation of job categories that differentiate ranks, functions, or particular organizational locations. (Strang and Baron 1990: 479)

3. Review the logic of spuriousness. Can you think up an example where an observed relationship between two variables could actually be explained away by a third variable?

4. Using InfoTrac College Edition or printed journals in the library, locate a research project involving a panel study. Describe the nature of the study design and its primary findings.

SPSS EXERCISES

See the booklet that accompanies your text for exercises using SPSS (Statistical Package for the Social Sciences). There are exercises offered for each chapter, and you'll also find a detailed primer on using SPSS.

Online Study Resources

If your book came with an access code card, visit www.cengage.com/login to register. To purchase access, please visit www.ichapters.com.

1. Before you do your final review of the chapter, take the *CengageNOW* pretest to help identify the areas on which you should concentrate. You'll find information on this online tool, as well as instructions on how to access all of its great resources, in the front of the book.

2. As you review, take advantage of the *CengageNOW* personalized study plan, based on your quiz results. Use this study plan with its interactive exercises and other resources to master the material.

3. When you're finished with your review, take the posttest to confirm that you're ready to move on to the next chapter.

WEBSITE FOR *THE PRACTICE OF SOCIAL RESEARCH* 12TH EDITION

Go to your book's website at www.cengage.com/sociology/babbie for tools to aid you in studying for your exams. You'll find *Tutorial Quizzes* with feedback, *Internet Exercises, Flash Cards, Glossaries,* and *Essay Quizzes,* as well as *InfoTrac College Edition* search terms, suggestions for additional reading, *Web Links,* and primers for using data-analysis software such as SPSS.

ANSWERS TO UNITS OF ANALYSIS QUIZ, EXERCISE 2

a. Men and women, black and white people (individuals)

b. Incorporated U.S. cities (groups)

c. Transcendental meditation organizations (groups)

d. Nursing staffs (groups)

e. Farmers (individuals)

f. Neighborhoods (groups)

g. Blacks (individuals)

h. Service and production organizations (formal organizations)

i. Job titles (artifacts)

CHAPTER FIVE

Conceptualization, Operationalization, and Measurement

CHAPTER OVERVIEW

The interrelated steps of conceptualization, operationalization, and measurement allow researchers to turn a general idea for a research topic into useful and valid measurements in the real world. An essential part of this process involves transforming the relatively vague terms of ordinary language into precise objects of study with well-defined and measurable meanings.

CengageNOW for Sociology

Use this online tool to help you make the grade on your next exam. After reading this chapter, go to "Online Study Resources" at the end of the chapter for instructions on how to benefit from *CengageNOW*.

Introduction

This chapter and the next deal with how researchers move from a general idea about what they want to study to effective and well-defined measurements in the real world. This chapter discusses the interrelated processes of conceptualization, operationalization, and measurement. Chapter 6 builds on this foundation to discuss types of measurements that are more complex.

Consider a notion such as "satisfaction with college." I'm sure you know some people who are very satisfied, some who are very dissatisfied, and many who are between those extremes. Moreover, you can probably place yourself somewhere along that satisfaction spectrum. While this probably makes sense to you as a general matter, how would you go about measuring how different students were, so you could place them along that spectrum?

There are some comments students make in conversations (such as "This place sucks") that would tip you off as to where they stood. Or, in a more active effort, you can probably think of questions you might ask students to learn about their satisfaction (such as "How satisfied are you with . . . ?"). Perhaps there are certain behaviors (class attendance, use of campus facilities, setting the dean's office on fire) that would suggest different levels of satisfaction. As you think about ways of measuring satisfaction with college, you are engaging in the subject matter of this chapter.

We begin by confronting the hidden concern people sometimes have about whether it's truly possible to measure the stuff of life: love, hate, prejudice, religiosity, radicalism, alienation. The answer is yes, but it will take a few pages to see how. Once we establish that researchers can measure anything that exists, we'll turn to the steps involved in doing just that.

Measuring Anything That Exists

Earlier in this book, I said that one of the two pillars of science is observation. Because this word can suggest a casual, passive activity, scientists often use the term *measurement* instead, meaning careful, deliberate observations of the real world for the purpose of describing objects and events in terms of the attributes composing a variable.

You may have some reservations about the ability of science to measure the really important aspects of human social existence. If you've read research reports dealing with something like liberalism or religion or prejudice, you may have been dissatisfied with the way the researchers measured whatever they were studying. You may have felt that they were too superficial, that they missed the aspects that really matter most. Maybe they measured religiosity as the number of times a person went to religious services, or maybe they measured liberalism by how people voted in a single election. Your dissatisfaction would surely have increased if you had found yourself being misclassified by the measurement system.

Your feeling of dissatisfaction reflects an important fact about social research: Most of the variables we want to study don't actually exist in the way that rocks exist. Indeed, they are made up. Moreover, they seldom have a single, unambiguous meaning.

To see what I mean, suppose we want to study *political party affiliation.* To measure this variable, we might consult the list of registered voters to note whether the people we were studying were registered as Democrats or Republicans and take that as a measure of their party affiliation. But we could also simply ask someone what party they identify with and take their response as our measure. Notice that these two different measurement possibilities reflect somewhat different definitions of *political party affiliation*. They might even produce different results: Someone may have registered as a Democrat years ago but gravitated more and more toward a Republican philosophy over time. Or someone who is registered with neither political party may, when asked, say she is affiliated with the one she feels the most kinship with.

Similar points apply to *religious affiliation*. Sometimes this variable refers to official membership in a particular church, temple, mosque, and so forth; other times it simply means whatever religion, if any, you identify yourself with. Perhaps to you it means something else, such as attendance at religious services.

The truth is that neither *party affiliation* nor *religious affiliation* has any real meaning, if by "real" we mean corresponding to some objective aspect of reality. These variables do not exist in nature. They are merely terms we've made up and assigned specific meanings to for some purpose, such as doing social research.

But, you might object, *political affiliation* and *religious affiliation*—and a host of other things social researchers are interested in, such as prejudice or compassion—have some reality. After all, researchers make statements about them, such as "In Happytown, 55 percent of the adults affiliate with the Republican Party, and 45 percent of them are Episcopalians. Overall, people in Happytown are low in prejudice and high in compassion." Even ordinary people, not just social researchers, have been known to make statements like that. If these things don't exist in reality, what is it that we're measuring and talking about?

What indeed? Let's take a closer look by considering a variable of interest to many social researchers (and many other people as well)—*prejudice*.

Conceptions, Concepts, and Reality

As you and I wandered down the road of life, we observed a lot of things and knew they were real through our observations, and we heard reports from other people that seemed real. For example:

- We personally heard people say nasty things about minority groups.
- We heard people say that women were inferior to men.
- We read about African Americans being lynched.
- We read that women and minorities earned less for the same work.
- We learned about "ethnic cleansing" and wars in which one ethnic group tried to eradicate another.

With additional experience, we noticed something more. People who participated in lynching were also quite likely to call African Americans ugly names. A lot of them, moreover, seemed to want women to "stay in their place." Eventually it dawned on us that these several tendencies often appeared together in the same people and also had something in common. At some point, someone had a bright idea: "Let's use the word *prejudiced* as a shorthand notation for people like that. We can use the term even if they don't do all those things—as long as they're pretty much like that."

Being basically agreeable and interested in efficiency, we went along with the system. That's where "prejudice" came from. We never observed it. We just agreed to use it as a shortcut, a name that represents a collection of apparently related phenomena that we've each observed in the course of life. In short, we made it up.

Here's another clue that prejudice isn't something that exists apart from our rough agreement to use the term in a certain way. Each of us develops our own mental image of what the set of real phenomena we've observed represents in general and what these phenomena have in common. When I say the word *prejudice*, it evokes a mental image in your mind, just as it evokes one in mine. It's as though file drawers in our minds contained thousands of sheets of paper, with each sheet of paper labeled in the upper right-hand corner. A sheet of paper in each of our minds has the term *prejudice* on it. On your sheet are all the things you've been told about prejudice and everything you've

Keeping Humanity in Focus

In the early 1970s, Elijah Anderson spent three years observing life in a black, working-class neighborhood in South Chicago, focusing on Jelly's, a combination bar and liquor store. While some people still believe that impoverished neighborhoods in the inner city are socially chaotic and disorganized, Anderson's study and others like it have clearly demonstrated a definite social structure there that guides the behavior of its participants. Much of his interest centered on systems of social status and how the 55 or so regulars at Jelly's worked those systems to establish themselves among their peers.

In the second edition of this classic study of urban life, Elijah Anderson returned to Jelly's and the surrounding neighborhood. There he found several changes, largely due to the outsourcing of manufacturing jobs overseas that has brought economic and mental depression to many of the residents. These changes, in turn, had also altered the nature of social organization.

For a research methods student, the book offers many insights into the process of establishing rapport with people being observed in their natural surroundings. Further, he offers excellent examples of how concepts are established in qualitative research.

You can read excerpts of the book online and can hear Anderson discuss the book in an interview with BBC's Laurie Taylor at the links offered on this book's website: http://www.cengage.com/sociology/babbie.

A Place on the Corner: A Study of Black Street Corner Men by Elijah Anderson. © 2004 University of Chicago Press

---

observed that seems to be an example of it. My sheet has what I've been told about it plus all the things I've observed that seem examples of it—and mine isn't the same as yours.

The technical term for those mental images, those sheets of paper in our mental file drawers, is *conception*. That is, I have a conception of prejudice, and so do you. We can't communicate these mental images directly, so we use the terms written in the upper right-hand corner of our own mental sheets of paper as a way of communicating about our conceptions and the things we observe that are related to those conceptions. These terms make it possible for us to communicate and eventually agree on what we specifically mean by those terms. In social research, the process of coming to an agreement about what terms mean is **conceptualization,** and the result is called a *concept*. See "Keeping Humanity in Focus" for a glimpse at a project that reveals a lot about conceptualization.

I'm sure you've heard some reference to the many words Eskimos have for *snow*, as an example of how environment can shape language. Here's an exercise you might enjoy when you're ready to take a break from reading. Search the web for "Eskimo words for snow." You may be surprised by what you find. You're likely to discover wide disagreement on the number of, say, Inuit, words—ranging from 1 to 400. Several sources, moreover, will suggest that if the Inuit have several words for *snow*, so does English. Cecil Adams, for example, lists "snow, slush, sleet, hail, powder, hard pack, blizzard, flurries, flake, dusting, crust, avalanche,

---

conceptualization The mental process whereby fuzzy and imprecise notions (concepts) are made more specific and precise. So you want to study prejudice. What do you mean by "prejudice"? Are there different kinds of prejudice? What are they?

drift, frost, and iceberg" (Straight Dope 2001). This illustrates the ambiguities in the field with regard to the concepts and words that we use in everyday communications and that also serve as the grounding for social research.

Let's take another example of a conception. Suppose that I'm going to meet someone named Pat, whom you already know. I ask you what Pat is like. Now suppose that you've seen Pat help lost children find their parents and put a tiny bird back in its nest. Pat got you to take turkeys to poor families on Thanksgiving and to visit a children's hospital on Christmas. You've seen Pat weep through a movie about a mother overcoming adversities to save and protect her child. As you search through your mental files, you may find all or most of those phenomena recorded on a single sheet labeled "compassionate." You look over the other entries on the page, and you find they seem to provide an accurate description of Pat. So you say, "Pat is compassionate."

Now I leaf through my own mental file drawer until I find a sheet marked "compassionate." I then look over the things written on my sheet, and I say, "Oh, that's nice." I now feel I know what Pat is like, but my expectations reflect the entries on my file sheet, not yours. Later, when I meet Pat, I happen to find that my own experiences correspond to the entries I have on my "compassionate" file sheet, and I say that you sure were right.

But suppose my observations of Pat contradict the things I have on my file sheet. I tell you that I don't think Pat is very compassionate, and we begin to compare notes.

You say, "I once saw Pat weep through a movie about a mother overcoming adversity to save and protect her child." I look at my "compassionate sheet" and can't find anything like that. Looking elsewhere in my file, I locate that sort of phenomenon on a sheet labeled "sentimental." I retort, "That's not compassion. That's just sentimentality."

To further strengthen my case, I tell you that I saw Pat refuse to give money to an organization dedicated to saving whales from extinction. "That represents a lack of compassion," I argue. You search through your files and find saving the whales on two sheets—"environmental activism"

and "cross-species dating"—and you say so. Eventually, we set about comparing the entries we have on our respective sheets labeled "compassionate." We then discover that many of our mental images corresponding to that term differ.

In the big picture, language and communication work only to the extent that you and I have considerable overlap in the kinds of entries we have on our corresponding mental file sheets. The similarities we have on those sheets represent the agreements existing in our society. As we grow up, we're told approximately the same thing when we're first introduced to a particular term, though our nationality, gender, race, ethnicity, region, language, or other cultural factors may shade our understanding of concepts.

Dictionaries formalize the agreements our society has about such terms. Each of us, then, shapes his or her mental images to correspond with such agreements. But because all of us have different experiences and observations, no two people end up with exactly the same set of entries on any sheet in their file systems. If we want to measure "prejudice" or "compassion," we must first stipulate what, exactly, counts as prejudice or compassion for our purposes.

Returning to the assertion made at the outset of this chapter, we can measure anything that's real. We can measure, for example, whether Pat actually puts the little bird back in its nest, visits the hospital on Christmas, weeps at the movie, or refuses to contribute to saving the whales. All of those behaviors exist, so we can measure them. But is Pat really compassionate? We can't answer that question; we can't measure compassion in any objective sense, because compassion doesn't exist in the way that those things I just described exist. Compassion exists only in the form of the agreements we have about how to use the term in communicating about things that are real.

Concepts as Constructs

If you recall the discussions of postmodernism in Chapter 1, you'll recognize that some people would object to the degree of "reality" I've allowed in the preceding comments. Did Pat "really" visit the

TABLE 5-1
What Social Scientists Measure

	Examples
Direct observables	Physical characteristics (sex, height, skin color) of a person being observed and/or interviewed
Indirect observables	Characteristics of a person as indicated by answers given in a self-administered questionnaire
Constructs	Level of alienation, as measured by a scale that is created by combining several direct and/or indirect observables

hospital on Christmas? Does the hospital "really" exist? Does Christmas? Though we aren't going to be radically postmodern in this chapter, I think you'll recognize the importance of an intellectually tough view of what's real and what's not. (When the intellectual going gets tough, the tough become social scientists.)

In this context, Abraham Kaplan (1964) distinguishes three classes of things that scientists measure. The first class is *direct observables:* those things we can observe rather simply and directly, like the color of an apple or the check mark on a questionnaire. The second class, *indirect observables,* require "relatively more subtle, complex, or indirect observations" (1964: 55). We note a person's check mark beside "female" in a questionnaire and have indirectly observed that person's gender. History books or minutes of corporate board meetings provide indirect observations of past social actions. Finally, the third class of observables consists of *constructs*— theoretical creations that are based on observations but that cannot be observed directly or indirectly. A good example is intelligence quotient, or IQ. It is constructed mathematically from observations of the answers given to a large number of questions on an IQ test. No one can directly or indirectly observe IQ. It is no more a "real" characteristic of people than is compassion or prejudice. See Table 5-1 for more examples of what social scientists measure.

Kaplan (1964: 49) defines *concept* as a "family of conceptions." A concept is, as Kaplan notes, a construct, something we create. Concepts such as compassion and prejudice are constructs created from your conception of them, my conception of

them, and the conceptions of all those who have ever used these terms. They cannot be observed directly or indirectly, because they don't exist. We made them up.

To summarize, *concepts* are constructs derived by mutual agreement from mental images (conceptions). Our *conceptions* summarize collections of seemingly related observations and experiences. Although the observations and experiences are real, at least subjectively, conceptions, and the concepts derived from them, are only mental creations. The terms associated with concepts are merely devices created for the purposes of filing and communication. A term such as *prejudice* is, objectively speaking, only a collection of letters. It has no intrinsic reality beyond that. Is has only the meaning we agree to give it. See "A Concept in Search of a Label" for one example of such an agreement.

Usually, however, we fall into the trap of believing that terms for constructs do have intrinsic meaning, that they name real entities in the world. That danger seems to grow stronger when we begin to take terms seriously and attempt to use them precisely. Further, the danger is all the greater in the presence of experts who appear to know more than we do about what the terms really mean: It's easy to yield to authority in such a situation.

Once we assume that terms like *prejudice* and *compassion* have real meanings, we begin the tortured task of discovering what those real meanings are and what constitutes a genuine measurement of them. Regarding constructs as real is called *reification.* The reification of concepts in day-to-day life is quite common. In science, we want to be quite clear about what it is we are actually

A Concept in Search of a Label

In the late 1950s, in the heat of the Cold War between the United States and the Communist bloc, American foreign policy came under criticism for being sometimes arrogant and thoughtless regarding the cultures and concerns of other countries—especially the recipients of U.S. aid. The image of arrogant and thoughtless Americans abroad became an entrenched and vivid image that required a label. From the late 1950s to this day, a rude and insensitive American abroad has been known as an "ugly American."

This term was taken from a 1958 political novel of the same name, by William J. Lederer and Eugene Burdick. Ironically, however, the meaning of the term was completely perverted. In the book, Homer Atkins is the hero—a retired engineer, volunteering to help villagers in a country strikingly similar to Vietnam. Rather than pushing the villagers around and superimposing his will, he is a thoughtful and considerate listener. When the local people discuss problems and possible solutions, he looks for ways his engineering training and lifetime of experience can be of

use. If they need to get river water up a hill to irrigate fields, he uses his knowledge of hydrology to show them how *they* can do it. In retrospect, he was an early model for participatory action research, which we'll examine in Chapter 10 of this book.

So, why was Atkins called "the ugly American"? The authors used the term as a somewhat ironic reference to the main character's homeliness—that's it. The public's misappropriation of the term points to the strong societal need to name a concept that was very real for many people.

Each of the concepts studied by social researchers began as a mental image, ultimately requiring a label to allow communication about it, just as the concept of the boorish, bossy American found "ugly American" as its label.

Source: William Lederer and Eugene Burdick, *The Ugly American* (New York: Norton, 1958).

---

measuring, but this aim brings a pitfall with it. Settling on the "best" way of measuring a variable in a particular study may imply that we've discovered the "real" meaning of the concept involved. In fact, concepts have no real, true, or objective meanings—only those we agree are best for a particular purpose.

Does this discussion imply that compassion, prejudice, and similar constructs can't be measured? Interestingly, the answer is no. (And a good thing, too, or a lot of us social researcher types would be out of work.) I've said that we can measure anything that's real. Constructs aren't real in the way that trees are real, but they do have another important virtue: They are useful. That is, they help us organize, communicate about, and understand things that are real. They help us make predictions about real things. Some of those predictions even turn out to be true. Constructs can work this way because, although not real or observable in themselves, they have a definite relationship to things that are real and observable. The bridge from direct and indirect observables to useful constructs is the process called conceptualization.

Conceptualization

As we've seen, day-to-day communication usually occurs through a system of vague and general agreements about the use of terms. Although you and I do not agree completely about the use of the term *compassionate*, I'm probably safe in assuming that Pat won't pull the wings off flies. A wide range of misunderstandings and conflict—from the interpersonal to the international—is the price we pay for our imprecision, but somehow we muddle through. Science, however, aims at more than muddling; it cannot operate in a context of such imprecision.

The process through which we specify what we mean when we use particular terms in research is called *conceptualization*. Suppose we want to find out, for example, whether women are more compassionate than men. I suspect many people assume this is the case, but it might be interesting to find out if it's really so. We can't meaningfully study the question, let alone agree on the answer, without some working agreements about the meaning of compassion. They are "working" agreements in the sense that they allow us to work on the question. We don't need to agree or even pretend to agree that a particular specification is ultimately the best one.

Conceptualization, then, produces a specific, agreed-on meaning for a concept for the purposes of research. This process of specifying exact meaning involves describing the indicators we'll be using to measure our concept and the different aspects of the concept, called dimensions.

Indicators and Dimensions

Conceptualization gives definite meaning to a concept by specifying one or more indicators of what we have in mind. An **indicator** is a sign of the presence or absence of the concept we're studying. Here's an example.

We might agree that visiting children's hospitals during Christmas and Hanukkah is an indicator of compassion. Putting little birds back in their nests might be agreed on as another indicator, and so forth. If the unit of analysis for our study is the individual person, we can then observe the presence or absence of each indicator for each person under study. Going beyond that, we can add up the number of indicators of compassion observed for each individual. We might agree on ten specific indicators, for example, and find six present in our study of Pat, three for John, nine for Mary, and so forth.

Returning to our question about whether men or women are more compassionate, we might calculate that the women we studied displayed an average of 6.5 indicators of compassion, the men an average of 3.2. On the basis of our quantitative analysis of group difference, we might therefore conclude that women are, on the whole, more compassionate than men.

Usually, though, it's not that simple. Imagine you're interested in understanding a small fundamentalist religious cult, particularly their harsh views on various groups: gays, nonbelievers, feminists, and others. In fact, they suggest that anyone who refuses to join their group and abide by its teachings will "burn in hell." In the context of your interest in compassion, they don't seem to have

much. And yet, the group's literature often speaks of their compassion for others. You want to explore this seeming paradox.

To pursue this research interest, you might arrange to interact with cult members, getting to know them and learning more about their views. You could tell them you were a social researcher interested in learning about their group, or perhaps you would just express an interest in learning more, without saying why.

In the course of your conversations with group members and perhaps attendance of religious services, you would put yourself in situations where you could come to understand what the cult members mean by compassion. You might learn, for example, that members of the group were so deeply concerned about sinners burning in hell that they were willing to be aggressive, even violent, to make people change their sinful ways. Within their own paradigm, then, cult members would see beating up gays, prostitutes, and abortion doctors as acts of compassion.

Social researchers focus their attention on the meanings that the people under study give to words and actions. Doing so can often clarify the behaviors observed: At least now you understand how the cult can see violent acts as compassionate. On the other hand, paying attention to what words and actions mean to the people under study almost always complicates the concepts researchers are interested in. (We'll return to this issue when we discuss the validity of measures, toward the end of this chapter.)

Whenever we take our concepts seriously and set about specifying what we mean by them, we discover disagreements and inconsistencies. Not only do you and I disagree, but each of us is likely to find a good deal of muddiness within our own mental images. If you take a moment to look at what you mean by compassion, you'll probably find that your image contains several kinds of

---

indicator An observation that we choose to consider as a reflection of a variable we wish to study. Thus, for example, attending religious services might be considered an indicator of *religiosity*.

compassion. That is, the entries on your mental file sheet can be combined into groups and subgroups, say, compassion toward friends, co-religionists, humans, and birds. You may also find several different strategies for making combinations. For example, you might group the entries into feelings and actions.

The technical term for such groupings is **dimension:** a specifiable aspect of a concept. For instance, we might speak of the "feeling dimension" of compassion and the "action dimension" of compassion. In a different grouping scheme, we might distinguish "compassion for humans" from "compassion for animals." Or we might see compassion as helping people have what we want for them versus what they want for themselves. Still differently, we might distinguish compassion as forgiveness from compassion as pity.

Thus, we could subdivide compassion into several clearly defined dimensions. A complete conceptualization involves both specifying dimensions and identifying the various indicators for each.

When Jonathan Jackson (2005: 301) set out to measure "fear of crime," he considered seven different dimensions:

- The frequency of worry about becoming a victim of three personal crimes and two property crimes in the immediate neighbourhood . . .

- Estimates of likelihood of falling victim to each crime locally

- Perceptions of control over the possibility of becoming a victim of each crime locally

- Perceptions of the seriousness of the consequences of each crime

- Beliefs about the incidence of each crime locally

- Perceptions of the extent of social physical incivilities in the neighbourhood

- Perceptions of community cohesion, including informal social control and trust/social capital

---

dimension A specifiable aspect of a concept. "Religiosity," for example, might be specified in terms of a belief dimension, a ritual dimension, a devotional dimension, a knowledge dimension, and so forth.

Sometimes conceptualization aimed at identifying different dimensions of a variable leads to a different kind of distinction. We may conclude that we've been using the same word for meaningfully distinguishable concepts. In the following example, the researchers find (1) that "violence" is not a sufficient description of "genocide" and (2) that the concept "genocide" itself comprises several distinct phenomena. Let's look at the process they went through to come to this conclusion.

When Daniel Chirot and Jennifer Edwards attempted to define the concept of "genocide," they found existing assumptions were not precise enough for their purposes:

> The United Nations originally defined it as an attempt to destroy "in whole or in part, a national, ethnic, racial, or religious group." If genocide is distinct from other types of violence, it requires its own unique explanation.
>
> *(2003: 14)*

Notice the final comment in this excerpt, as it provides an important insight into why researchers are so careful in specifying the concepts they study. If genocide, such as the Holocaust, were simply another example of violence, like assaults and homicides, then what we know about violence in general might explain genocide. If it differs from other forms of violence, then we may need a different explanation for it. So, the researchers began by suggesting that "genocide" was a concept distinct from "violence" for their purposes.

Then, as Chirot and Edwards examined historical instances of genocide, they began concluding that the motivations for launching genocidal mayhem differed sufficiently to represent four distinct phenomena that were all called "genocide" (2003: 15–18).

1. *Convenience:* Sometimes the attempt to eradicate a group of people serves a function for the eradicators, such as Julius Caesar's attempt to eradicate tribes defeated in battle, fearing they would be difficult to rule. Or when gold was discovered on Cherokee land in the Southeastern United States in the early nineteenth century, the Cherokee were forcibly relocated

to Oklahoma in an event known as the "Trail of Tears," which ultimately killed as many as half of those forced to leave.

2. *Revenge:* When the Chinese of Nanking bravely resisted the Japanese invaders in the early years of World War II, the conquerors felt they had been insulted by those they regarded as inferior beings. Tens of thousands were slaughtered in the "Rape of Nanking" in 1937–1938.

3. *Fear:* The ethnic cleansing that recently occurred in the former Yugoslavia was at least partly motivated by economic competition and worries that the growing Albanian population of Kosovo was gaining political strength through numbers. Similarly, the Hutu attempt to eradicate the Tutsis of Rwanda grew out of a fear that returning Tutsi refugees would seize control of the country. Often intergroup fears such as these grow out of long histories of atrocities, often inflicted in both directions.

4. *Purification:* The Nazi Holocaust, probably the most publicized case of genocide, was intended as a purification of the "Aryan race." While Jews were the main target, gypsies, homosexuals, and many other groups were also included. Other examples include the Indonesian witch-hunt against communists in 1965–1966 and the attempt to eradicate all non-Khmer Cambodians under Pol Pot in the 1970s.

No single theory of genocide could explain these various forms of mayhem. Indeed, this act of conceptualization suggests four distinct phenomena, each needing a different set of explanations.

Specifying the different dimensions of a concept often paves the way for a more sophisticated understanding of what we're studying. We might observe, for example, that women are more compassionate in terms of feelings, and men more so in terms of actions—or vice versa. Whichever turned out to be the case, we would not be able to say whether men or women are really more compassionate. Our research would have shown that there is no single answer to the question. That alone represents an advance in our understanding of reality. To get a better feel for concepts, variables, and indicators, go to the General Social Survey codebook and explore some of the ways the researchers have measured various concepts (see the link on this book's website: http://www.cengage.com/sociology/babbie).

The Interchangeability of Indicators

There is another way that the notion of indicators can help us in our attempts to understand reality by means of "unreal" constructs. Suppose, for the moment, that you and I have compiled a list of 100 indicators of compassion and its various dimensions. Suppose further that we disagree widely on which indicators give the clearest evidence of compassion or its absence. If we pretty much agree on some indicators, we could focus our attention on those, and we would probably agree on the answer they provided. We would then be able to say that some people are more compassionate than others in some dimension. But suppose we don't really agree on any of the possible indicators. Surprisingly, we can still reach an agreement on whether men or women are the more compassionate. How we do that has to do with the interchangeability of indicators.

The logic works like this. If we disagree totally on the value of the indicators, one solution would be to study all of them. Suppose that women turn out to be more compassionate than men on all 100 indicators—on all the indicators you favor and on all of mine. Then we would be able to agree that women are more compassionate than men, even though we still disagree on exactly what compassion means in general.

The interchangeability of indicators means that if several different indicators all represent, to some degree, the same concept, then all of them will behave the same way that the concept would behave if it were real and could be observed. Thus, given a basic agreement about what "compassion" is, if women are generally more compassionate than men, we should be able to observe that difference by using any reasonable measure of compassion. If, on the other hand, women are more compassionate than men on some indicators but not on others, we should see if the two sets of indicators represent different dimensions of compassion.

You have now seen the fundamental logic of conceptualization and measurement. The discussions that follow are mainly refinements and extensions of what you've just read. Before turning to a technical elaboration of measurement, however, we need to fill out the picture of conceptualization by looking at some of the ways social researchers provide standards, consistency, and commonality for the meanings of terms.

Real, Nominal, and Operational Definitions

As we have seen, the design and execution of social research requires us to clear away the confusion over concepts and reality. To this end, logicians and scientists have found it useful to distinguish three kinds of definitions: real, nominal, and operational.

The first of these reflects the reification of terms. As Carl Hempel cautions,

> A "real" definition, according to traditional logic, is not a stipulation determining the meaning of some expression but a statement of the "essential nature" or the "essential attributes" of some entity. The notion of essential nature, however, is so vague as to render this characterization useless for the purposes of rigorous inquiry.
>
> (1952: 6)

In other words, trying to specify the "real" meaning of concepts only leads to a quagmire: It mistakes a construct for a real entity.

The **specification** of concepts in scientific inquiry depends instead on nominal and operational definitions. A nominal definition is one that is simply assigned to a term without any claim that the definition represents a "real" entity. Nominal definitions are arbitrary—I could define compas-

---

specification The process through which concepts are made more specific.

sion as "plucking feathers off helpless birds" if I wanted to—but they can be more or less useful. For most purposes, especially communication, that last definition of compassion would be pretty useless. Most nominal definitions represent some consensus, or convention, about how a particular term is to be used.

An operational definition, as you may remember from Chapter 4, specifies precisely how a concept will be measured—that is, the operations we'll perform. An operational definition is nominal rather than real, but it has the advantage of achieving maximum clarity about what a concept means in the context of a given study. In the midst of disagreement and confusion over what a term "really" means, we can specify a working definition for the purposes of an inquiry. Wishing to examine socioeconomic status (SES) in a study, for example, we may simply specify that we are going to treat SES as a combination of income and educational attainment. In this decision, we rule out other possible aspects of SES: occupational status, money in the bank, property, lineage, lifestyle, and so forth. Our findings will then be interesting to the extent that our definition of SES is useful for our purpose.

Creating Conceptual Order

The clarification of concepts is a continuing process in social research. Catherine Marshall and Gretchen Rossman (1995: 18) speak of a "conceptual funnel" through which a researcher's interest becomes increasingly focused. Thus, a general interest in social activism could narrow to "individuals who are committed to empowerment and social change" and further focus on discovering "what experiences shaped the development of fully committed social activists." This focusing process is inescapably linked to the language we use.

In some forms of qualitative research, the clarification of concepts is a key element in the collection of data. Suppose you were conducting interviews and observations in a radical political group devoted to combating oppression in U.S. society. Imagine how the meaning of oppression would shift as you delved more and more deeply into the members' experiences and worldviews.

For example, you might start out thinking of oppression in physical and perhaps economic terms. The more you learned about the group, however, the more you might appreciate the possibility of psychological oppression.

The same point applies even to contexts where meanings might seem more fixed. In the analysis of textual materials, for example, social researchers sometimes speak of the "hermeneutic circle," a cyclical process of ever-deeper understanding.

> The understanding of a text takes place through a process in which the meaning of the separate parts is determined by the global meaning of the text as it is anticipated. The closer determination of the meaning of the separate parts may eventually change the originally anticipated meaning of the totality, which again influences the meaning of the separate parts, and so on.
>
> *(Kvale 1996: 47)*

Consider the concept "prejudice." Suppose you needed to write a definition of the term. You might start out thinking about racial/ethnic prejudice. At some point you would realize you should probably allow for gender prejudice, religious prejudice, antigay prejudice, and the like in your definition. Examining each of these specific types of prejudice would affect your overall understanding of the general concept. As your general understanding changed, however, you would likely see each of the individual forms somewhat differently.

The continual refinement of concepts occurs in all social research methods. Often you will find yourself refining the meaning of important concepts even as you write up your final report.

Although conceptualization is a continuing process, it is vital to address it specifically at the beginning of any study design, especially rigorously structured research designs such as surveys and experiments. In a survey, for example, operationalization results in a commitment to a specific set of questionnaire items that will represent the concepts under study. Without that commitment, the study could not proceed.

Even in less-structured research methods, however, it's important to begin with an initial set of anticipated meanings that can be refined during data collection and interpretation. No one seriously believes we can observe life with no preconceptions; for this reason, scientific observers must be conscious of and explicit about these conceptual starting points.

Let's explore initial conceptualization the way it applies to structured inquiries such as surveys and experiments. Though specifying nominal definitions focuses our observational strategy, it does not allow us to observe. As a next step we must specify exactly what we are going to observe, how we will do it, and what interpretations we are going to place on various possible observations. All these further specifications make up the operational definition of the concept.

In the example of socioeconomic status, we might decide to ask survey respondents two questions, corresponding to the decision to measure SES in terms of income and educational attainment:

1. What was your total family income during the past 12 months?
2. What is the highest level of school you completed?

To organize our data, we'd probably want to specify a system for categorizing the answers people give us. For income, we might use categories such as "under $5,000," "$5,000 to $10,000," and so on. Educational attainment might be similarly grouped in categories: less than high school, high school, college, graduate degree. Finally, we would specify the way a person's responses to these two questions would be combined in creating a measure of SES.

In this way we would create a working and workable definition of SES. Although others might disagree with our conceptualization and operationalization, the definition would have one essential scientific virtue: It would be absolutely specific and unambiguous. Even if someone disagreed with our definition, that person would have a good idea how to interpret our research results, because what we meant by SES—reflected in our analyses and conclusions—would be precise and clear.

Table 5-2 shows the progression of measurement steps from our vague sense of what a term

TABLE 5-2

Progression of Measurement

Measurement Step	Example: Social Class
Conceptualization	What are the different meanings and dimensions of the concept "social class"?
Nominal definition	For our study, we will define "social class" as representing economic differences: specifically, income.
Operational definition	We will measure economic differences via responses to the survey question "What was your annual income, before taxes, last year?"
Measurements in the real world	The interviewer will ask, "What was your annual income, before taxes, last year?"

means to specific measurements in a fully structured scientific study.

An Example of Conceptualization: The Concept of Anomie

To bring this discussion of conceptualization in research together, let's look briefly at the history of a specific social science concept. Researchers studying urban riots are often interested in the part played by feelings of powerlessness. Social scientists sometimes use the word *anomie* in this context. This term was first introduced into social science by Emile Durkheim, the great French sociologist, in his classic 1897 study, *Suicide.*

Using only government publications on suicide rates in different regions and countries, Durkheim produced a work of analytic genius. To determine the effects of religion on suicide, he compared the suicide rates of predominantly Protestant countries with those of predominantly Catholic ones, Protestant regions of Catholic countries with Catholic regions of Protestant countries, and so forth. To determine the possible effects of the weather, he compared suicide rates in northern and southern countries and regions, and he examined the different suicide rates across the months and seasons of the year. Thus, he could draw conclusions about a supremely individualistic and personal act without having any data about the individuals engaging in it.

At a more general level, Durkheim suggested that suicide also reflects the extent to which a so-

ciety's agreements are clear and stable. Noting that times of social upheaval and change often present individuals with grave uncertainties about what is expected of them, Durkheim suggested that such uncertainties cause confusion, anxiety, and even self-destruction. To describe this societal condition of normlessness, Durkheim chose the term *anomie.* Durkheim did not make this word up. Used in both German and French, it literally meant "without law." The English term *anomy* had been used for at least three centuries before Durkheim to mean disregard for divine law. However, Durkheim created the social science concept of anomie.

In the years that have followed the publication of *Suicide,* social scientists have found anomie a useful concept, and many have expanded on Durkheim's use. Robert Merton, in a classic article entitled "Social Structure and Anomie" (1938), concluded that anomie results from a disparity between the goals and means prescribed by a society. Monetary success, for example, is a widely shared goal in our society, yet not all individuals have the resources to achieve it through acceptable means. An emphasis on the goal itself, Merton suggested, produces normlessness, because those denied the traditional avenues to wealth go about getting it through illegitimate means. Merton's discussion, then, could be considered a further conceptualization of the concept of anomie.

Although Durkheim originally used the concept of anomie as a characteristic of societies, as did Merton after him, other social scientists have used it to describe individuals. To clarify this distinction, some scholars have chosen to use *anomie* in reference to its original, societal meaning and to use the term *anomia* in reference to the individual characteristic. In a given society, then, some individuals experience anomia, and others do not. Elwin Powell, writing 20 years after Merton, provided the following conceptualization of anomia (though using the term *anomie*) as a characteristic of individuals:

> When the ends of action become contradictory, inaccessible or insignificant, a condition of anomie arises. Characterized by a general loss of orientation and accompanied by feelings of "emptiness" and apathy, anomie can be simply conceived as meaninglessness.
>
> *(1958: 132)*

Powell went on to suggest there were two distinct kinds of anomia and to examine how the two rose out of different occupational experiences to result at times in suicide. In his study, however, Powell did not measure anomia per se; he studied the relationship between suicide and occupation, making inferences about the two kinds of anomia. Thus, the study did not provide an operational definition of anomia, only a further conceptualization.

Although many researchers have offered operational definitions of anomia, one name stands out over all. Two years before Powell's article appeared, Leo Srole (1956) published a set of questionnaire items that he said provided a good measure of anomia as experienced by individuals. It consists of five statements that subjects were asked to agree or disagree with:

1. In spite of what some people say, the lot of the average man is getting worse.
2. It's hardly fair to bring children into the world with the way things look for the future.
3. Nowadays a person has to live pretty much for today and let tomorrow take care of itself.
4. These days a person doesn't really know who he can count on.
5. There's little use writing to public officials because they aren't really interested in the problems of the average man.

(1956: 713)

In the half-century following its publication, the Srole scale has become a research staple for social scientists. You'll likely find this particular operationalization of anomia used in many of the research projects reported in academic journals. Srole touches on this in the accompanying box, "The Origins of Anomia," which he prepared for this book before his death.

This abbreviated history of anomie and anomia as social science concepts illustrates several points. First, it's a good example of the process through which general concepts become operationalized measurements. This is not to say that the issue of how to operationalize anomie/anomia has been resolved once and for all. Scholars will surely continue to reconceptualize and reoperationalize these concepts for years to come, continually seeking more-useful measures.

The Srole scale illustrates another important point. Letting conceptualization and operationalization be open-ended does not necessarily produce anarchy and chaos, as you might expect. Order often emerges. For one thing, although we could define anomia any way we chose—in terms of, say, shoe size—we're likely to define it in ways not too different from other people's mental images. If you were to use a really offbeat definition, people would probably ignore you.

A second source of order is that, as researchers discover the utility of a particular conceptualization and operationalization of a concept, they're likely to adopt it, which leads to standardized definitions of concepts. Besides the Srole scale, examples include IQ tests and a host of demographic and economic measures developed by the U.S. Census Bureau. Using such established measures has two advantages: They have been extensively pretested and debugged, and studies using the same scales can be compared. If you and I do separate studies of two different groups and use the Srole scale, we can compare our two groups on the basis of anomia.

Social scientists, then, can measure anything that's real; through conceptualization and operationalization, they can even do a pretty good job of measuring things that aren't. Granting that such concepts as socioeconomic status, prejudice, compassion, and anomia aren't ultimately real, social

The Origins of Anomia

Leo Srole

My career-long fixation on anomie began with reading Durkheim's *Le Suicide* as a Harvard undergraduate. Later, as a graduate student at Chicago, I studied under two Durkheimian anthropologists: William Lloyd Warner and Alfred Radcliffe-Brown. Radcliffe-Brown had carried on a lively correspondence with Durkheim, making me a collateral "descendant" of the great French sociologist.

For me, the early impact of Durkheim's work on suicide was mixed but permanent. On the one hand, I had serious reservations about his strenuous, ingenious, and often awkward efforts to force the crude, bureaucratic records on suicide rates to fit with his unidirectional sociological determinism. On the other hand, I was moved by Durkheim's unswerving preoccupation with the moral force of the interpersonal ties that bind us to our time, place, and past, and also his insights about the lethal consequences that can follow from shrinkage and decay in those ties.

My interest in anomie received an eyewitness jolt at the finale of World War II, when I served with the United Nations Relief and Rehabilitation Administration, helping to rebuild a war-torn Europe. At the Nazi concentration camp of Dachau, I saw firsthand the depths of dehumanization that macro-social forces, such as those that engaged Durkheim, could produce in individuals like Hitler, Eichmann, and the others serving their dictates at all levels in the Nazi death factories.

Returning from my UNRRA post, I felt most urgently that the time was long overdue to come to an understanding of the dynamics underlying disintegrated social bonds. We needed to work expeditiously, deemphasizing proliferation of macro-level theory in favor of a direct exploratory encounter with individuals, using newly developed state-of-the-art survey research methodology. Such research, I also felt, should focus on a broader spectrum of behavioral pathologies than suicide.

My initial investigations were a diverse effort. In 1950, for example, I was able to interview a sample of 401 bus riders in Springfield, Massachusetts. Four years later, the Midtown Manhattan Mental Health Study provided a much larger population reach. These and other field projects gave me scope to expand and refine my measurements of that quality in individuals which reflected the macro-social quality Durkheim had called anomie.

While I began by using Durkheim's term in my own work, I soon decided that it was necessary to limit the use of that concept to its macro-social meaning and to sharply segregate it from its individual manifestations. For the latter purpose, the cognate but hitherto obsolete Greek term, *anomia*, readily suggested itself.

I first published the anomia construct in a 1956 article in the *American Sociological Review*, describing ways of operationalizing it, and presenting the results of its initial field application research. By 1982, the Science Citation Index and Social Science Citation Index had listed some 400 publications in political science, psychology, social work, and sociology journals here and abroad that had cited use of that article's instruments or findings, warranting the American Institute for Scientific Information to designate it a "citation classic."

Source: From "Social Integration and Certain Corollaries," Center for Geriatrics and Gerontology, Columbia University, Leo Srole, *American Sociological Review* 21, 1956, 709–716. Copyright © 1956. Reprinted by permission.

scientists can create order in handling them. It is an order based on utility, however, not on ultimate truth.

Definitions in Descriptive and Explanatory Studies

As you'll recall from Chapter 4, two general purposes of research are description and explanation. The distinction between them has important implications for definition and measurement. If it seems that description is simpler than explanation, you may be surprised to learn that definitions are more problematic for descriptive research than for explanatory research. Before we turn to other aspects of measurement, you'll need a basic understanding of why this is so (we'll discuss this point more fully in Part 4).

It's easy to see the importance of clear and precise definitions for descriptive research. If we want to describe and report the unemployment rate in a city, our definition of being unemployed is obviously critical. That definition will depend on our definition of another term: the labor force. If it seems patently absurd to regard a three-year-old child as being unemployed, it is because such a child is not considered a member of the labor force. Thus, we might follow the U.S. Census Bureau's convention and exclude all people under 14 years of age from the labor force.

This convention alone, however, would not give us a satisfactory definition, because it would count as unemployed such people as high school students, the retired, the disabled, and homemakers. We might follow the census convention further by defining the labor force as "all persons 14 years of age and over who are employed, looking for work, or waiting to be called back to a job from which they have been laid off or furloughed." If a student, homemaker, or retired person is not looking for work, such a person would not be included in the labor force. Unemployed people, then, would be those members of the labor force, as defined, who are not employed.

But what does "looking for work" mean? Must a person register with the state employment service or go from door to door asking for employment? Or would it be sufficient to want a job or be open to an offer of employment? Conventionally, "looking for work" is defined operationally as saying yes in response to an interviewer's asking "Have you been looking for a job during the past seven days?" (Seven days is the period most often specified, but for some research purposes it might make more sense to shorten or lengthen it.)

As you can see, the conclusion of a descriptive study about the unemployment rate depends directly on how each issue of definition is resolved. Increasing the period during which people are counted as looking for work would add more unemployed people to the labor force as defined, thereby increasing the reported unemployment rate. If we follow another convention and speak of the civilian labor force and the civilian unemployment rate, we're excluding military personnel; that, too, increases the reported unemployment rate, because military personnel would be employed—by definition. Thus, the descriptive statement that the unemployment rate in a city is 3 percent, or 9 percent, or whatever it might be, depends directly on the operational definitions used.

This example is relatively clear because there are several accepted conventions relating to the labor force and unemployment. Now, consider how difficult it would be to get agreement about the definitions you would need in order to say, "Forty-five percent of the students at this institution are politically conservative." Like the unemployment rate, this percentage would depend directly on the definition of what is being measured—in this case, political conservatism. A different definition might result in the conclusion "Five percent of the student body are politically conservative."

Ironically, definitions are less problematic in the case of explanatory research. Let's suppose we're interested in explaining political conservatism. Why are some people conservative and others not? More specifically, let's suppose we're interested in whether conservatism increases with age. What if you and I have 25 different operational definitions of *conservative,* and we can't agree on which definition is best? As we saw in the discussion of indicators, this is not necessarily an insurmountable obstacle to our research. Suppose we found old people to be more conservative than young people in terms of all 25 definitions. Clearly, the exact definition wouldn't matter much. We would conclude that old people are generally more conservative than young people—even though we couldn't agree about exactly what *conservative* means.

In practice, explanatory research seldom results in findings quite as unambiguous as this example suggests; nonetheless, the general pattern is quite common in actual research. There are consistent patterns of relationships in human social life that result in consistent research findings. However, such consistency does not appear in a descriptive situation. Changing definitions almost inevitably results in different descriptive conclusions. "The Importance of Variable Names" explores this issue in connection with the variable *citizen participation.*

Operationalization Choices

In discussing conceptualization, I frequently have referred to operationalization, for the two are intimately linked. To recap: Conceptualization is the refinement and specification of abstract concepts, and operationalization is the development of

The Importance of Variable Names

Patricia Fisher

Graduate School of Planning, University of Tennessee

Operationalization is one of those things that's easier said than done. It is quite simple to explain to someone the purpose and importance of operational definitions for variables, and even to describe how operationalization typically takes place. However, until you've tried to operationalize a rather complex variable, you may not appreciate some of the subtle difficulties involved. Of considerable importance to the operationalization effort is the particular name that you have chosen for a variable. Let's consider an example from the field of Urban Planning.

A variable of interest to planners is *citizen participation*. Planners are convinced that participation in the planning process by citizens is important to the success of plan implementation. Citizen participation is an aid to planners' understanding of the real and perceived needs of a community, and such involvement by citizens tends to enhance their cooperation with and support for planning efforts. Although many different conceptual definitions might be offered by different planners, there would be little misunderstanding over what is meant by citizen participation. The name of the variable seems adequate.

However, if we ask different planners to provide very simple operational measures for citizen participation, we are likely to find a variety among their responses that does generate confusion. One planner might keep a tally of attendance by private citizens at city commission and other local government meetings; another might maintain a record of the different topics addressed by private citizens at similar meetings; while a third might record the number of local government meeting attendees, letters and phone calls received by the mayor and other public officials, and meetings held by special interest groups during a particular time period. As skilled researchers, we can readily see that each planner would be measuring (in a very simplistic fashion) a different dimension of citizen participation: extent of citizen participation, issues prompting citizen participation, and form of citizen participation. Therefore, the original naming of our variable, *citizen participation*, which was quite satisfactory from a conceptual point of view, proved inadequate for purposes of operationalization.

The precise and exact naming of variables is important in research. It is both essential to and a result of good operationalization. Variable names quite often evolve from an iterative process of forming a conceptual definition, then an operational definition, then renaming the concept to better match what can or will be measured. This looping process continues (our example illustrates only one iteration), resulting in a gradual refinement of the variable name and its measurement until a reasonable fit is obtained. Sometimes the concept of the variable that you end up with is a bit different from the original one that you started with, but at least you are measuring what you are talking about, if only because you are talking about what you are measuring!

Source: From Patricia Fisher, *The Importance of Variable Names,* Copyright © Patricia Fisher. Reprinted by permission.

specific research procedures (operations) that will result in empirical observations representing those concepts in the real world.

As with the methods of data collection, social researchers have a variety of choices when operationalizing a concept. Although the several choices are intimately interconnected, I've separated them for the sake of discussion. Realize, though, that operationalization does not proceed through a systematic checklist.

Range of Variation

In operationalizing any concept, researchers must be clear about the range of variation that interests them. The question is, to what extent are they willing to combine attributes in fairly gross categories?

Let's suppose you want to measure people's incomes in a study by collecting the information from either records or interviews. The highest annual incomes people receive run into the millions of dollars, but not many people get that much. Unless you're studying the very rich, it probably won't add much to your study to keep track of extremely high categories. Depending on whom you study, you'll probably want to establish a highest income category with a much lower floor—maybe $100,000 or more. Although this decision will lead you to throw together people who earn a trillion dollars a year with paupers earning a mere $100,000, they'll survive it, and that mixing probably won't hurt your research any, either. The same decision faces you at the other end of the income spectrum. In studies of the general U.S. population, a bottom category of $5,000 or less usually works fine.

In studies of attitudes and orientations, the question of range of variation has another dimension. Unless you're careful, you may end up measuring only half an attitude without really meaning to. Here's an example of what I mean.

Suppose you're interested in people's attitudes toward expanding the use of nuclear power generators. You'd anticipate that some people consider nuclear power the greatest thing since the wheel, whereas other people have absolutely no interest in it. Given that anticipation, it would seem to make sense to ask people how much they favor expanding the use of nuclear energy and to give them answer categories ranging from "Favor it very much" to "Don't favor it at all."

This operationalization, however, conceals half the attitudinal spectrum regarding nuclear energy. Many people have feelings that go beyond simply not favoring it: They are, with greater or lesser degrees of intensity, actively opposed to it. In this instance, there is considerable variation on the left side of zero. Some oppose it a little, some quite a bit, and others a great deal. To measure the full range of variation, then, you'd want to operationalize attitudes toward nuclear energy with a range from favoring it very much, through no feelings one way or the other, to opposing it very much.

This consideration applies to many of the variables social scientists study. Virtually any public issue involves both support and opposition, each in varying degrees. In measuring religiosity, people are not just more or less religious; some are positively antireligious. Political orientations range from very liberal to very conservative, and depending on the people you're studying, you may want to allow for radicals on one or both ends.

The point is not that you must measure the full range of variation in every case. You should, however, consider whether you need to, given your particular research purpose. If the difference between not religious and antireligious isn't relevant to your research, forget it. Someone has defined pragmatism as "any difference that makes no difference is no difference." Be pragmatic.

Finally, decisions on the range of variation should be governed by the expected distribution of attributes among the subjects of the study. In a study of college professors' attitudes toward the value of higher education, you could probably stop at no value and not worry about those who might consider higher education dangerous to students' health. (If you were studying students, however . . .)

Variations between the Extremes

Degree of precision is a second consideration in operationalizing variables. What it boils down to is how fine you will make distinctions among the various possible attributes composing a given variable. Does it matter for your purposes whether a person is 17 or 18 years old, or could you conduct your inquiry by throwing them together in a group labeled 10 to 19 years old? Don't answer too quickly. If you wanted to study rates of voter registration and participation, you'd definitely want to know whether the people you studied were old enough to vote. In general, if you're going to measure age, you must look at the purpose and procedures of your study and decide whether fine or gross differences in age are important to you. In a survey, you'll need to make these decisions in order to design an appropriate questionnaire. In the case of in-depth interviews, these decisions will condition the extent to which you probe for details.

The same thing applies to other variables. If you measure *political affiliation,* will it matter to your inquiry whether a person is a conservative Democrat rather than a liberal Democrat, or will it be sufficient to know the party? In measuring *religious affiliation,* is it enough to know that a person is Protestant, or do you need to know the denomination? Do you simply need to know whether or not a person is married, or will it make a difference to know if he or she has never married or is separated, widowed, or divorced?

There is, of course, no general answer to such questions. The answers come out of the purpose of a given study, or why we are making a particular measurement. I can give you a useful guideline, though. Whenever you're not sure how much detail to pursue in a measurement, get too much rather than too little. When a subject in an in-depth interview volunteers that she is 37 years old, record "37" in your notes, not "in her thirties." When you're analyzing the data, you can always combine precise attributes into more general categories, but you can never separate any variations you lumped together during observation and measurement.

A Note on Dimensions

We've already discussed dimensions as a characteristic of concepts. When researchers get down to the business of creating operational measures of variables, they often discover—or worse, never notice—that they're not exactly clear about which dimensions of a variable they're really interested in. Here's an example.

Let's suppose you're studying people's attitudes toward government, and you want to include an examination of how people feel about corruption. Here are just a few of the dimensions you might examine:

- Do people think there is corruption in government?
- How much corruption do they think there is?
- How certain are they in their judgment of how much corruption there is?
- How do they feel about corruption in government as a problem in society?
- What do they think causes it?
- Do they think it's inevitable?
- What do they feel should be done about it?
- What are they willing to do personally to eliminate corruption in government?
- How certain are they that they would be willing to do what they say they would do?

The list could go on and on—how people feel about corruption in government has many dimensions. It's essential to be clear about which ones are important in our inquiry; otherwise, you may measure how people feel about corruption when you really wanted to know how much they think there is, or vice versa.

Once you've determined how you're going to collect your data (for example, survey, field research) and have decided on the relevant range of variation, the degree of precision needed between the extremes of variation, and the specific

dimensions of the variables that interest you, you may have another choice: a mathematical-logical one. That is, you may need to decide what level of measurement to use. To discuss this point, we need to take another look at attributes and their relationship to variables.

Defining Variables and Attributes

An attribute, you'll recall, is a characteristic or quality of something. *Female* is an example. So is *old* or *student*. Variables, on the other hand, are logical sets of attributes. Thus, *gender* is a variable composed of the attributes *female* and *male*. What could be simpler?

Actually, some would insist that *sex* is the proper name of the variable composed of the physical attributes *female* and *male*, while *gender* is a social-identity and behavioral variable composed of the attributes *feminine* and *masculine*. In most social science research, biological differences are less important than how people treat those differences in terms of their own behavior as well as their expectations and treatment of others. Despite this distinction, the two terms are commonly used interchangeably, both in everyday language and by social scientists. As long as the terms are defined for the purposes of research, there should be little confusion.

In any case, the conceptualization and operationalization processes can be seen as the specification of variables and the attributes composing them. Thus, in the context of a study of unemployment, *employment status* is a variable having the attributes *employed* and *unemployed;* the list of attributes could also be expanded to include the other possibilities discussed earlier, such as *homemaker.*

Every variable must have two important qualities. First, the attributes composing it should be exhaustive. For the variable to have any utility in research, we must be able to classify every observation in terms of one of the attributes composing the variable. We'll run into trouble if we conceptualize the variable *political party affiliation* in terms of the attributes *Republican* and *Democrat,* because some of the people we set out to study will identify with the Green Party, the Reform Party, or some other organization, and some (often a large percentage) will tell us they have no party affiliation. We could make the list of attributes exhaustive by adding *other* and *no affiliation*. Whatever we do, we must be able to classify every observation.

At the same time, attributes composing a variable must be mutually exclusive. Every observation must be able to be classified in terms of one and only one attribute. For example, we need to define *employed* and *unemployed* in such a way that nobody can be both at the same time. That means being able to classify the person who is working at a job but is also looking for work. (We might run across a fully employed mud wrestler who is looking for the glamour and excitement of being a social researcher.) In this case, we might define the attributes so that *employed* takes precedence over *unemployed*, and anyone working at a job is employed regardless of whether he or she is looking for something better.

Levels of Measurement

Attributes operationalized as mutually exclusive and exhaustive may be related in other ways as well. For example, the attributes composing variables may represent different levels of measurement. In this section, we'll examine four levels of measurement: nominal, ordinal, interval, and ratio.

Nominal Measures

Variables whose attributes have only the characteristics of exhaustiveness and mutual exclusiveness are *nominal measures*. Examples include *gender, religious affiliation, political party affiliation, birthplace, college major,* and *hair color.* Although the attributes composing each of these variables—as *male* and *female* compose the variable *gender*—are distinct from one another (and exhaust the possibilities of gender among people), they have no additional structures. **Nominal measures** merely offer names or labels for characteristics.

Imagine a group of people characterized in terms of one such variable and physically grouped by the applicable attributes. For example, say

we've asked a large gathering of people to stand together in groups according to the states in which they were born: all those born in Vermont in one group, those born in California in another, and so forth. The variable is *place of birth;* the attributes are *born in California, born in Vermont,* and so on. All the people standing in a given group have at least one thing in common and differ from the people in all other groups in that same regard. Where the individual groups form, how close they are to one another, or how the groups are arranged in the room is irrelevant. All that matters is that all the members of a given group share the same state of birth and that each group has a different shared state of birth. All we can say about two people in terms of a nominal variable is that they are either the same or different.

Ordinal Measures

Variables with attributes we can logically rank-order are *ordinal measures.* The different attributes of ordinal variables represent relatively more or less of the variable. Variables of this type are *social class, conservatism, alienation, prejudice, intellectual sophistication,* and the like. In addition to saying whether two people are the same or different in terms of an ordinal variable, you can also say one is "more" than the other—that is, more conservative, more religious, older, and so forth.

In the physical sciences, *hardness* is the most frequently cited example of an ordinal measure. We may say that one material (for example, diamond) is harder than another (say, glass) if the former can scratch the latter and not vice versa. By attempting to scratch various materials with other materials, we might eventually be able to arrange several materials in a row, ranging from

nominal measure A variable whose attributes have only the characteristics of exhaustiveness and mutual exclusiveness. In other words, a level of measurement describing a variable that has attributes that are merely different, as distinguished from ordinal, interval, or ratio measures. *Gender* is an example of a nominal measure.

the softest to the hardest. We could never say how hard a given material was in absolute terms; we could only say how hard in relative terms—which materials it is harder than and which softer than.

Let's pursue the earlier example of grouping the people at a social gathering. This time imagine that we ask all the people who have graduated from college to stand in one group, all those with only a high school diploma to stand in another group, and all those who have not graduated from high school to stand in a third group. This manner of grouping people satisfies the requirements for exhaustiveness and mutual exclusiveness discussed earlier. In addition, however, we might logically arrange the three groups in terms of the relative amount of formal education (the shared attribute) each had. We might arrange the three groups in a row, ranging from most to least formal education. This arrangement would provide a physical representation of an **ordinal measure.** If we knew which groups two individuals were in, we could determine that one had more, less, or the same formal education as the other.

Notice in this example that it is irrelevant how close or far apart the educational groups are from one another. The college and high school groups might be 5 feet apart, and the less-than-high-school group 500 feet farther down the line. These actual distances don't have any meaning. The high school group, however, should be between the less-than-high-school group and the college group, or else the rank order will be incorrect.

Interval Measures

For the attributes composing some variables, the actual distance separating those attributes does have meaning. Such variables are **interval measures.** For these, the logical distance between attributes can be expressed in meaningful standard intervals.

For example, in the Fahrenheit temperature scale, the difference, or distance, between 80 degrees and 90 degrees is the same as that between 40 degrees and 50 degrees. However, 80 degrees Fahrenheit is not twice as hot as 40 degrees, because the zero point in the Fahrenheit scale is arbitrary; zero degrees does not really mean lack of heat. Similarly, minus 30 degrees on this scale doesn't represent 30 degrees less than no heat. (This is true for the Celsius scale as well. In contrast, the Kelvin scale is based on an absolute zero, which does mean a complete lack of heat.)

About the only interval measures commonly used in social science research are constructed measures such as standardized intelligence tests that have been more or less accepted. The interval separating IQ scores of 100 and 110 may be regarded as the same as the interval separating scores of 110 and 120 by virtue of the distribution of observed scores obtained by many thousands of people who have taken the tests over the years. But it would be incorrect to infer that someone with an IQ of 150 is 50 percent more intelligent than someone with an IQ of 100. (A person who received a score of 0 on a standard IQ test could not be regarded, strictly speaking, as having no intelligence, although we might feel he or she was unsuited to be a college professor or even a college student. But perhaps a dean . . . ?)

When comparing two people in terms of an interval variable, we can say they are different from each other (nominal), and that one is more than the other (ordinal). In addition, we can say "how much" more.

Ratio Measures

Most of the social science variables meeting the minimum requirements for interval measures also meet the requirements for ratio measures. In **ratio measures,** the attributes composing a variable, besides having all the structural characteristics mentioned previously, are based on a true zero point.

---

ordinal measure A level of measurement describing a variable with attributes we can rank-order along some dimension. An example is *socioeconomic status* as composed of the attributes *high, medium, low.*

interval measure A level of measurement describing a variable whose attributes are rank-ordered and have equal distances between adjacent attributes. The Fahrenheit temperature scale is an example of this, because the distance between 17 and 18 is the same as that between 89 and 90.

The Kelvin temperature scale is one such measure. Examples from social science research include *age, length of residence in a given place, number of organizations belonged to, number of times attending religious services during a particular period of time, number of times married,* and *number of Arab friends.*

Returning to the illustration of methodological party games, we might ask a gathering of people to group themselves by age. All the one-year-olds would stand (or sit or lie) together, the two-year-olds together, the three-year-olds, and so forth. The fact that members of a single group share the same age and that each different group has a different shared age satisfies the minimum requirements for a nominal measure. Arranging the several groups in a line from youngest to oldest meets the additional requirements of an ordinal measure and lets us determine if one person is older than, younger than, or the same age as another. If we space the groups equally far apart, we satisfy the additional requirements of an interval measure and can say how much older one person is than another. Finally, because one of the attributes included in age represents a true zero (babies carried by women about to give birth), the phalanx of hapless party goers also meets the requirements of a ratio measure, permitting us to say that one person is twice as old as another. (Remember this in case you're asked about it in a workbook assignment.) Another example of a ratio measure is *income,* which extends from an absolute zero to approximately infinity, if you happen to be the founder of Microsoft.

Comparing two people in terms of a ratio variable, then, allows us to conclude (1) whether they are different (or the same), (2) whether one is more than the other, (3) how much they differ, and (4) what the ratio of one to another is. Figure 5-1 summarizes this discussion by presenting a graphic illustration of the four levels of measurement.

Implications of Levels of Measurement

Because it's unlikely that you'll undertake the physical grouping of people just described (try it once, and you won't be invited to many parties),

I should draw your attention to some of the practical implications of the differences that have been distinguished. These implications appear primarily in the analysis of data (discussed in Part 4), but you need to anticipate such implications when you're structuring any research project.

Certain quantitative analysis techniques require variables that meet certain minimum levels of measurement. To the extent that the variables to be examined in a research project are limited to a particular level of measurement—say, ordinal—you should plan your analytic techniques accordingly. More precisely, you should anticipate drawing research conclusions appropriate to the levels of measurement used in your variables. For example, you might reasonably plan to determine and report the mean age of a population under study (add up all the individual ages and divide by the number of people), but you should not plan to report the mean religious affiliation, because that is a nominal variable, and the mean requires ratio-level data. (You could report the modal—the most common—religious affiliation.)

At the same time, you can treat some variables as representing different levels of measurement. Ratio measures are the highest level, descending through interval and ordinal to nominal, the lowest level of measurement. A variable representing a higher level of measurement—say, ratio—can also be treated as representing a lower level of measurement—say, ordinal. Recall, for example, that age is a ratio measure. If you wished to examine only the relationship between age and some ordinal-level variable—say, self-perceived religiosity: high, medium, and low—you might choose to treat age as an ordinal-level variable as well. You might characterize the subjects of your study as being young, middle-aged, and old, specifying what age range composed each of these groupings. Finally,

---

ratio measure A level of measurement describing a variable with attributes that have all the qualities of nominal, ordinal, and interval measures and in addition are based on a "true zero" point. *Age* is an example of a ratio measure.

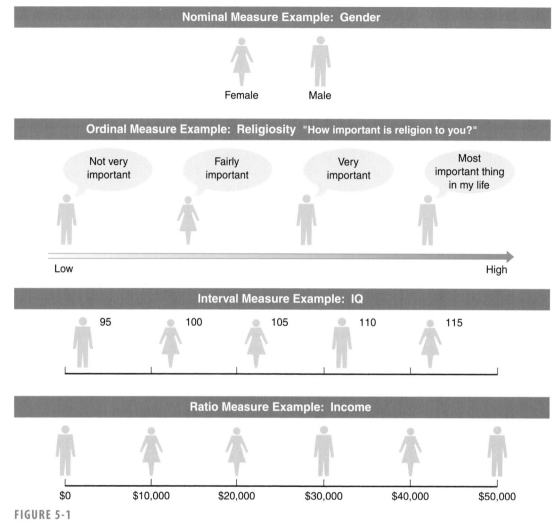

FIGURE 5-1

Levels of Measurement. Often you can choose among different levels of measurement—nominal, ordinal, interval, or ratio—carrying progressively more amounts of information.

age might be used as a nominal-level variable for certain research purposes. People might be grouped as being born during the Depression or not. Another nominal measurement, based on birth date rather than just age, would be the grouping of people by astrological signs.

The level of measurement you'll seek, then, is determined by the analytic uses you've planned for a given variable, keeping in mind that some variables are inherently limited to a certain level. If a variable is to be used in a variety of ways, requiring different levels of measurement, the study should be designed to achieve the highest level required. For example, if the subjects in a study are asked their exact ages, they can later be organized into ordinal or nominal groupings.

Again, you need not necessarily measure variables at their highest level of measurement. If you're sure to have no need for ages of people at higher than the ordinal level of measurement, you may simply ask people to indicate their age range, such as 20 to 29, 30 to 39, and so forth. In a study of the wealth of corporations, rather than seek more precise information, you may use Dun & Bradstreet ratings to rank corporations. Whenever your research purposes are not altogether clear, however, seek the highest level of measurement possible. As we've discussed, although ratio measures can later be reduced to ordinal ones, you cannot convert an ordinal measure to a ratio one. More generally, you cannot convert a lower-level measure to a higher-level one. That is a one-way street worth remembering.

Typically a research project will tap variables at different levels of measurement. For example, William Bielby and Denise Bielby (1999) set out to examine the world of film and television, using a nomothetic, longitudinal approach (take a moment to remind yourself what that means). In what they referred to as the "culture industry," the authors found that *reputation* (an ordinal variable) is the best predictor of screenwriters' future productivity. More interestingly, they found that screenwriters who were represented by "core" (or elite) agencies were not only far more likely to find jobs (a nominal variable), but also jobs that paid more (a ratio variable). In other words, the researchers found that agencies' reputations (ordinal) was a key independent variable for predicting a screenwriter's career success. The researchers also found that being older (ratio), female (nominal), an ethnic minority (nominal), and having more years of experience (ratio) were disadvantageous for a writer's career. On the other hand, higher earnings from previous years (measured in ordinal categories) led to more success in the future. In Bielby and Bielby's terms, "success breeds success" (1999: 80).

Single or Multiple Indicators

With so many alternatives for operationalizing social science variables, you may find yourself worrying about making the right choices. To counter this feeling, let me add a momentary dash of certainty and stability.

Many social research variables have fairly obvious, straightforward measures. No matter how you

cut it, gender usually turns out to be a matter of male or female: a nominal-level variable that can be measured by a single observation—either by looking (well, not always) or by asking a question (usually). In a study involving the size of families, you'll want to think about adopted and foster children, as well as blended families, but it's usually pretty easy to find out how many children a family has. For most research purposes, the resident population of a country is the resident population of that country—you can look it up in an almanac and know the answer. A great many variables, then, have obvious single indicators. If you can get one piece of information, you have what you need.

Sometimes, however, there is no single indicator that will give you the measure of a variable you really want. As discussed earlier in this chapter, many concepts are subject to varying interpretations—each with several possible indicators. In these cases, you'll want to make several observations for a given variable. You can then combine the several pieces of information you've collected, creating a composite measurement of the variable in question. Chapter 6 is devoted to ways of doing that, so here let's just discuss one simple illustration.

Consider the concept "college performance." All of us have noticed that some students perform well in college courses and others don't. In studying these differences, we might ask what characteristics and experiences are related to high levels of performance (many researchers have done just that). How should we measure overall performance? Each grade in any single course is a potential indicator of college performance, but it also may not typify the student's general performance. The solution to this problem is so firmly established that it is, of course, obvious: the grade point average (GPA). We assign numerical scores to each letter grade, total the points earned by a given student, and divide by the number of courses taken, thus obtaining a composite measure. (If the courses vary in number of credits, we adjust the point values accordingly.) Creating such composite measures in social research is often appropriate.

Some Illustrations of Operationalization Choices

To bring together all the operationalization choices available to the social researcher and to show the potential in those possibilities, let's look at some of the distinct ways you might address various research problems. The alternative ways of operationalizing the variables in each case should demonstrate the opportunities that social research can present to our ingenuity and imaginations. To simplify matters, I have not attempted to describe all the research conditions that would make one alternative superior to the others, though in a given situation they would not all be equally appropriate.

Here are specific research questions, then, and some of the ways you could address them. We'll begin with an example discussed earlier in the chapter. It has the added advantage that one of the variables is straightforward to operationalize.

1. Are women more compassionate than men?

    a. Select a group of subjects for study, with equal numbers of men and women. Present them with hypothetical situations that involve someone's being in trouble. Ask them what they would do if they were confronted with that situation. What would they do, for example, if they came across a small child who was lost and crying for his or her parents? Consider any answer that involves helping or comforting the child as an indicator of compassion. See whether men or women are more likely to indicate they would be compassionate.

    b. Set up an experiment in which you pay a small child to pretend that he or she is lost. Put the child to work on a busy sidewalk and observe whether men or women are more likely to offer assistance. Also be sure to count the total number of men and women who walk by, because there may be more of one than the other. If that's the case, simply calculate the percentage of men and the percentage of women who help.

    c. Select a sample of people and do a survey in which you ask them what organizations they belong to. Calculate whether women or men are more likely to belong to those that seem to reflect compassionate feelings. To account for the case in which one group belongs to more organizations than the other does, do this: For each person you study, calculate the percentage of his or her organizational memberships that reflect compassion. See if men or women have a higher average percentage.

2. Are sociology students or accounting students better informed about world affairs?

    a. Prepare a short quiz on world affairs and arrange to administer it to the students in a sociology class and in an accounting class at a comparable level. If you want to compare sociology and accounting majors, be sure to ask students what they are majoring in.

    b. Get the instructor of a course in world affairs to give you the average grades of sociology and accounting students in the course.

    c. Take a petition to sociology and accounting classes that urges that "the United Nations headquarters be moved to New York City." Keep a count of how many in each class sign the petition and how many inform you that the UN headquarters is already located in New York City.

3. Do people consider New York or California the better place to live?

    a. Consulting the *Statistical Abstract of the United States* or a similar publication, check the migration rates into and out of each state. See if you can find the numbers moving directly from New York to California and vice versa.

    b. The national polling companies—Gallup, Harris, Roper, and so forth—often ask people what they consider the best state to live in. Look up some recent results in the library or through your local newspaper.

    c. Compare suicide rates in the two states.

4. Who are the most popular instructors on your campus, those in the social sciences, the natural sciences, or the humanities?

Measuring College Satisfaction

Early in this chapter, we considered "college satisfaction" as an example of a concept people often talk about casually. To study such a concept, however, we need to engage in the processes of conceptualization and operationalization. I'll sketch out the process briefly, then you might try your hand at expanding on my comments.

What are some of the dimensions of college satisfaction? Here are a few to get you started, but feel free to add your own:

Academic quality: faculty, courses, majors
Physical facilities: classrooms, dorms, cafeteria, grounds
Athletics and extracurricular activities
Costs and availability of financial aid
Sociability of students, faculty, staff
Security, crime on campus

How would you measure each of these dimensions? One method would be to ask a sample of students, "How would you rate your level of satisfaction with each of the following?" giving them a list of items similar to those listed here and providing a set of categories for them to use (such as very satisfied, satisfied, dissatisfied, very dissatisfied).

But suppose you didn't have the time and/or money to conduct a survey and were interested in comparing overall levels of satisfaction at several schools. What data about schools (the unit of analysis) might give you the answer you were interested in? Retention rates might be one general indicator. Can you think of others?

a. If your school has a provision for student evaluation of instructors, review some recent results and compute the average rating of each of the three groups.

b. Begin visiting the introductory courses given in each group of disciplines and measure the attendance rate of each class.

c. In December, select a group of faculty in each of the three divisions and ask them to keep a record of the numbers of holiday greeting cards and presents they receive from admiring students. See who wins.

The point of these examples is not necessarily to suggest respectable research projects but to illustrate the many ways variables can be operationalized.

"Measuring College Satisfaction" briefly overviews the preceding steps in terms of a concept mentioned at the outset of this chapter.

Operationalization Goes On and On

Although I've discussed conceptualization and operationalization as activities that precede data collection and analysis—for example, you must design questionnaire items before you send out a questionnaire—these two processes continue throughout any research project, even if the data have been collected in a structured mass survey. As we've seen, in less-structured methods such as field research, the identification and specification of relevant concepts is inseparable from the ongoing process of observation.

Imagine, for example, that you're doing a qualitative, observational study of members of a new religious cult, and, in part, you want to identify those members who are more religious and those who are less religious. You may begin with a focus on certain kinds of ritual behavior, only to eventually discover that the members of the group place a higher premium on religious experience or steadfast beliefs.

The open-endedness of conceptualization and operationalization is perhaps more obvious in qualitative than in quantitative research, since changes can be made at any point during data collection and analysis. In quantitative methods such as survey research or experiments, you will be required to commit yourself to particular measurement structures. Once a questionnaire has been printed and administered, for example, altering it would be impractical if not impossible, even when the unfolding of the research might suggest changes. Even in the case of a survey questionnaire, however, you may have some flexibility in how you measure variables during the analysis phase, as we'll see in the following chapter.

As I mentioned, however, the qualitative researcher has a greater flexibility in this regard. Things you notice during in-depth interviews, for example, may suggest a different set of questions than you initially planned, allowing you to pursue unanticipated avenues. Then later, as you review and organize your notes for analysis, you may again see unanticipated patterns and redirect your analysis.

Regardless of whether you are using qualitative or quantitative methods, you should always be open to reexamining your concepts and definitions. The ultimate purpose of social research is to clarify the nature of social life. The validity and utility of what you learn in this regard doesn't depend on when you first figured out how to look at things any more than it matters whether you got the idea from a learned textbook, a dream, or your brother-in-law.

Criteria of Measurement Quality

This chapter has come some distance. It began with the bald assertion that social scientists can measure anything that exists. Then we discovered that most of the things we might want to measure and study don't really exist. Next we learned that it's possible to measure them anyway. Now we'll discuss of some of the yardsticks against which we judge our relative success or failure in measuring things—even things that don't exist.

reliability That quality of measurement method that suggests that the same data would have been collected each time in repeated observations of the same phenomenon. In the context of a survey, we would expect that the question "Did you attend religious services last week?" would have higher reliability than the question "About how many times have you attended religious services in your life?" This is not to be confused with validity.

Precision and Accuracy

To begin, measurements can be made with varying degrees of precision. As we saw in the discussion of operationalization, precision concerns the fineness of distinctions made between the attributes that compose a variable. The description of a woman as "43 years old" is more precise than "in her forties." Saying a street-corner gang was formed "in the summer of 1996" is more precise than saying "during the 1990s."

As a general rule, precise measurements are superior to imprecise ones, as common sense dictates. There are no conditions under which imprecise measurements are intrinsically superior to precise ones. Even so, exact precision is not always necessary or desirable. If knowing that a woman is in her forties satisfies your research requirements, then any additional effort invested in learning her precise age is wasted. The operationalization of concepts, then, must be guided partly by an understanding of the degree of precision required. If your needs are not clear, be more precise rather than less.

Don't confuse precision with accuracy, however. Describing someone as "born in New England" is less precise than "born in Stowe, Vermont"—but suppose the person in question was actually born in Boston. The less-precise description, in this instance, is more accurate, a better reflection of the real world.

Precision and accuracy are obviously important qualities in research measurement, and they probably need no further explanation. When social scientists construct and evaluate measurements, however, they pay special attention to two technical considerations: reliability and validity.

Reliability

In the abstract, **reliability** is a matter of whether a particular technique, applied repeatedly to the same object, yields the same result each time. Let's say you want to know how much I weigh. (No, I don't know why.) As one technique, say you ask two different people to estimate my weight. If the first person estimates 150 pounds and the other estimates 300, we have to conclude that the technique of having people estimate my weight isn't very reliable.

Suppose, as an alternative, that you use a bathroom scale as your measurement technique. I step on the scale twice, and you note the same result each time. The scale has presumably reported the same weight for me both times, indicating that the scale provides a more reliable technique for measuring a person's weight than asking people to estimate it does.

Reliability, however, does not ensure accuracy any more than precision does. Suppose I've set my bathroom scale to shave five pounds off my weight just to make me feel better. Although you would (reliably) report the same weight for me each time, you would always be wrong. This new element, called *bias,* is discussed in Chapter 8. For now, just be warned that reliability does not ensure accuracy.

Let's suppose we're interested in studying morale among factory workers in two different kinds of factories. In one set of factories, workers have specialized jobs, reflecting an extreme division of labor. Each worker contributes a tiny part to the overall process performed on a long assembly line. In the other set of factories, each worker performs many tasks, and small teams of workers complete the whole process.

How should we measure morale? Following one strategy, we could observe the workers in each factory, noticing such things as whether they joke with one another, whether they smile and laugh a lot, and so forth. We could ask them how they like their work and even ask them whether they think they would prefer their current arrangement or the other one being studied. By comparing what we observed in the different factories, we might reach a conclusion about which assembly process produces the higher morale. Notice that I've just described a qualitative measurement procedure.

Now let's look at some reliability problems inherent in this method. First, how you and I are feeling when we do the observing will likely color what we see. We may misinterpret what we see. We may see workers kidding each other but think they're having an argument. We may catch them on an off day. If we were to observe the same group of workers several days in a row, we might arrive at different evaluations on each day. Further, even if several observers evaluated the same behavior, they might arrive at different conclusions about the workers' morale.

Here's another strategy for assessing morale, a quantitative approach. Suppose we check the company records to see how many grievances have been filed with the union during some fixed period. Presumably this would be an indicator of morale: the more grievances, the lower the morale. This measurement strategy would appear to be more reliable: Counting up the grievances over and over, we should keep arriving at the same number.

If you find yourself thinking that the number of grievances doesn't necessarily measure morale, you're worrying about validity, not reliability. We'll discuss validity in a moment. The point for now is that the last method is more like my bathroom scale—it gives consistent results.

In social research, reliability problems crop up in many forms. Reliability is a concern every time a single observer is the source of data, because we have no certain guard against the impact of that observer's subjectivity. We can't tell for sure how much of what's reported originated in the situation observed and how much in the observer.

Subjectivity is not only a problem with single observers, however. Survey researchers have known for a long time that different interviewers, because of their own attitudes and demeanors, get different answers from respondents. Or, if we were to conduct a study of newspapers' editorial positions on some public issue, we might create a team of coders to take on the job of reading hundreds of editorials and classifying them in terms of their position on the issue. Unfortunately, different coders will code the same editorial differently. Or we might want to classify a few hundred specific occupations in terms of some standard coding scheme, say a set of categories created by the Department of Labor or by the Census Bureau. You and I would not place all those occupations in the same categories.

Each of these examples illustrates problems of reliability. Similar problems arise whenever we ask people to give us information about themselves.

Sometimes we ask questions that people don't know the answers to: How many times have you been to religious services? Sometimes we ask people about things they consider totally irrelevant: Are you satisfied with China's current relationship with Albania? In such cases, people will answer differently at different times because they're making up answers as they go. Sometimes we explore issues so complicated that a person who had a clear opinion in the matter might arrive at a different interpretation of the question when asked a second time.

So how do you create reliable measures? If your research design calls for asking people for information, you can be careful to ask only about things the respondents are likely to know the answer to. Ask about things relevant to them, and be clear in what you're asking. Of course, these techniques don't solve every possible reliability problem. Fortunately, social researchers have developed several techniques for cross-checking the reliability of the measures they devise.

Test-Retest Method

Sometimes it's appropriate to make the same measurement more than once, a technique called the *test-retest method*. If you don't expect the sought-after information to change, then you should expect the same response both times. If answers vary, the measurement method may, to the extent of that variation, be unreliable. Here's an illustration.

In their research on Health Hazard Appraisal (HHA), a part of preventive medicine, Jeffrey Sacks, W. Mark Krushat, and Jeffrey Newman (1980) wanted to determine the risks associated with various background and lifestyle factors, making it possible for physicians to counsel their patients appropriately. By knowing patients' life situations, physicians could advise them on their potential for survival and on how to improve it. This purpose, of course, depended heavily on the accuracy of the information gathered about each subject in the study.

To test the reliability of their information, Sacks and his colleagues had all 207 subjects complete a baseline questionnaire that asked about their

characteristics and behavior. Three months later, a follow-up questionnaire asked the same subjects for the same information, and the results of the two surveys were compared. Overall, only 15 percent of the subjects reported the same information in both studies.

Sacks and his colleagues report the following:

> Almost 10 percent of subjects reported a different height at follow-up examination. Parental age was changed by over one in three subjects. One parent reportedly aged 20 chronologic years in three months. One in five ex-smokers and ex-drinkers have apparent difficulty in reliably recalling their previous consumption pattern.
>
> *(1980: 730)*

Some subjects erased all trace of previously reported heart murmur, diabetes, emphysema, arrest record, and thoughts of suicide. One subject's mother, deceased in the first questionnaire, was apparently alive and well in time for the second. One subject had one ovary missing in the first study but present in the second. In another case, an ovary present in the first study was missing in the second study—and had been for ten years! One subject was reportedly 55 years old in the first study and 50 years old three months later. (You have to wonder whether the physician-counselors could ever have nearly the impact on their patients that their patients' memories did.) Thus, test-retest revealed that this data-collection method was not especially reliable.

Split-Half Method

As a general rule, it's always good to make more than one measurement of any subtle or complex social concept, such as prejudice, alienation, or social class. This procedure lays the groundwork for another check on reliability. Let's say you've created a questionnaire that contains ten items you believe measure prejudice against women. Using the split-half technique, you would randomly assign those ten items to two sets of five. Each set should provide a good measure of prejudice against women, and the two sets should classify respondents the same way. If the two sets of items classify people differently, you most likely have a problem of reliability in your measure of the variable.

Using Established Measures

Another way to help ensure reliability in getting information from people is to use measures that have proved their reliability in previous research. If you want to measure anomia, for example, you might want to follow Srole's lead.

The heavy use of measures, though, does not guarantee their reliability. For example, the Scholastic Assessment Tests (SATs) and the Minnesota Multiphasic Personality Inventory (MMPI) have been accepted as established standards in their respective domains for decades. In recent years, though, they've needed fundamental overhauling to reflect changes in society, eliminating outdated topics and gender bias in wording.

Reliability of Research Workers

As we've seen, it's also possible for measurement unreliability to be generated by research workers: interviewers and coders, for example. There are several ways to check on reliability in such cases. To guard against interviewer unreliability in surveys, for example, a supervisor will call a subsample of the respondents on the telephone and verify selected pieces of information.

Replication works in other situations also. If you're worried that newspaper editorials or occupations may not be classified reliably, you could have each independently coded by several coders. Those cases that are classified inconsistently can then be evaluated more carefully and resolved.

Finally, clarity, specificity, training, and practice can prevent a great deal of unreliability and grief. If you and I spent some time reaching a clear agreement on how to evaluate editorial positions on an issue—discussing various positions and reading through several together—we could probably do a good job of classifying them in the same way independently.

The reliability of measurements is a fundamental issue in social research, and we'll return to it more than once in the chapters ahead. For

now, however, let's recall that even total reliability doesn't ensure that our measures actually measure what we think they measure. Now let's plunge into the question of validity.

Validity

In conventional usage, **validity** refers to the extent to which an empirical measure adequately reflects the real meaning of the concept under consideration. A measure of social class should measure social class, not political orientations. A measure of political orientations should measure political orientations, not sexual permissiveness. Validity means that we are actually measuring what we say we are measuring.

Whoops! I've already committed us to the view that concepts don't have real meanings. How can we ever say whether a particular measure adequately reflects the concept's meaning, then? Ultimately, of course, we can't. At the same time, as we've already seen, all of social life, including social research, operates on agreements about the terms we use and the concepts they represent. There are several criteria of success in making measurements that are appropriate to these agreed-on meanings of concepts.

First, there's something called **face validity.** Particular empirical measures may or may not jibe with our common agreements and our individual

---

validity A term describing a measure that accurately reflects the concept it is intended to measure. For example, your IQ would seem a more valid measure of your intelligence than the number of hours you spend in the library would. Though the ultimate validity of a measure can never be proved, we may agree to its relative validity on the basis of face validity, criterion-related validity, construct validity, content validity, internal validation, and external validation (see Chapter 6). This must not be confused with reliability.

face validity That quality of an indicator that makes it seem a reasonable measure of some variable. That the frequency of attendance at religious services is some indication of a person's religiosity seems to make sense without a lot of explanation. It has face validity.

mental images concerning a particular concept. For example, you and I might quarrel about whether counting the number of grievances filed with the union will adequately measure morale. Still, we'd surely agree that the number of grievances has *something* to do with morale. That is, the measure is valid "on its face," whether or not it's adequate. If I were to suggest that we measure morale by finding out how many books the workers took out of the library during their off-duty hours, you'd undoubtedly raise a more serious objection: That measure wouldn't have much face validity.

Second, I've already pointed to many of the more formally established agreements that define some concepts. The Census Bureau, for example, has created operational definitions of such concepts as family, household, and employment status that seem to have a workable validity in most studies using these concepts.

Three additional types of validity also specify particular ways of testing the validity of measures. The first, **criterion-related validity,** sometimes called *predictive validity,* is based on some external criterion. For example, the validity of College Board exams is shown in their ability to predict students' success in college. The validity of a written driver's test is determined, in this sense, by the relationship between the scores people get on the test and their subsequent driving records. In these examples, college success and driving ability are the criteria.

To test your understanding of criterion-related validity, see whether you can think of behaviors that might be used to validate each of the following attitudes:

Is very religious

Supports equality of men and women

Supports far-right militia groups

Is concerned about the environment

Some possible validators would be, respectively, attends religious services, votes for women candidates, belongs to the NRA, and belongs to the Sierra Club.

Sometimes it's difficult to find behavioral criteria that can be taken to validate measures as directly as in such examples. In those instances, however, we can often approximate such criteria by applying a different test. We can consider how the variable in question ought, theoretically, to relate to other variables. **Construct validity** is based on the logical relationships among variables.

Suppose, for example, that you want to study the sources and consequences of marital satisfaction. As part of your research, you develop a measure of marital satisfaction, and you want to assess its validity.

In addition to developing your measure, you'll have developed certain theoretical expectations about the way the variable *marital satisfaction* relates to other variables. For example, you might reasonably conclude that satisfied husbands and wives will be less likely than dissatisfied ones to cheat on their spouses. If your measure relates to marital fidelity in the expected fashion, that constitutes evidence of your measure's construct validity. If satisfied marriage partners are as likely to cheat on their spouses as the dissatisfied ones are, however, that would challenge the validity of your measure.

Tests of construct validity, then, can offer a weight of evidence that your measure either does or doesn't tap the quality you want it to measure, without providing definitive proof. Although I have suggested that tests of construct validity are less compelling than those of criterion validity, there is room for disagreement about which kind of test a particular comparison variable (*driving record, marital fidelity*) represents in a given situation. It's less important to distinguish the two types of validity tests than to understand the logic of validation that they have in common: If we've succeeded in measuring some variable, then our measures should relate in some logical way to other measures.

criterion-related validity The degree to which a measure relates to some external criterion. For example, the validity of College Board tests is shown in their ability to predict the college success of students. Also called predictive validity.

construct validity The degree to which a measure relates to other variables as expected within a system of theoretical relationships.

Reliable but not valid

Valid but not reliable

Valid *and* reliable

FIGURE 5-2

An Analogy to Validity and Reliability. A good measurement technique should be both valid (measuring what it is intended to measure) and reliable (yielding a given measurement dependably).

Finally, **content validity** refers to how much a measure covers the range of meanings included within a concept. For example, a test of mathematical ability cannot be limited to addition but also needs to cover subtraction, multiplication, division, and so forth. Or, if we're measuring prejudice, do our measurements reflect all types of prejudice, including prejudice against racial and ethnic groups, religious minorities, women, the elderly, and so on?

Figure 5-2 presents a graphic portrayal of the difference between validity and reliability. If you think of measurement as analogous to repeatedly shooting at the bull's-eye on a target, you'll see that reliability looks like a "tight pattern," regardless of where the shots hit, because reliability is a function of consistency. Validity, on the other hand, is a function of shots being arranged around the bull's-eye. The failure of reliability in the figure is randomly distributed around the target; the failure of validity is systematically off the mark. Notice that neither an unreliable nor an invalid measure is likely to be very useful.

Who Decides What's Valid?

Our discussion of validity began with a reminder that we depend on agreements to determine what's real, and we've just seen some of the ways social scientists can agree among themselves that they have made valid measurements. There is yet another way of looking at validity.

Social researchers sometimes criticize themselves and one another for implicitly assuming they are somewhat superior to those they study. For example, researchers often seek to uncover motivations that the social actors themselves are unaware of. You think you bought that new Burpo-Blasto because of its high performance and good looks, but *we* know you're really trying to achieve a higher social status.

This implicit sense of superiority would fit comfortably with a totally positivistic approach (the biologist feels superior to the frog on the lab table), but it clashes with the more humanistic and typically qualitative approach taken by many social scientists. We'll explore this issue more deeply in Chapter 10. In seeking to understand the way ordinary people make sense of their worlds, ethnomethodologists have urged all social scientists to pay more respect to the natural social processes of conceptualization and shared meaning. At the very least, behavior that may seem irrational from the scientist's paradigm may make logical sense when viewed through the actor's paradigm.

Clifford Geertz (1973) applies the term *thick description* in reference to the goal of understanding, as deeply as possible, the meanings that elements

content validity The degree to which a measure covers the range of meanings included within a concept.

of a culture have for those who live within that culture. He recognizes that the outside observer will never grasp those meanings fully, however, and warns, "Cultural analysis is intrinsically incomplete." He then elaborates:

> There are a number of ways to escape this—turning culture into folklore and collecting it, turning it into traits and counting it, turning it into institutions and classifying it, turning it into structures and toying with it. But they are escapes. The fact is that to commit oneself to a semiotic concept of culture and an interpretive approach to the study of it is to commit oneself to a view of ethnographic assertion as, to borrow W. B. Gallie's by now famous phrase, "essentially contestable." Anthropology, or at least interpretive anthropology, is a science whose progress is marked less by a perfection of consensus than by a refinement of debate. What gets better is the precision with which we vex each other.
>
> *(1973: 29)*

Ultimately, social researchers should look both to their colleagues and to their subjects as sources of agreement on the most useful meanings and measurements of the concepts they study. Sometimes one source will be more useful, sometimes the other. But neither one should be dismissed.

Tension between Reliability and Validity

Clearly, we want our measures to be both reliable and valid. However, a tension often arises between the criteria of reliability and validity, forcing a trade-off between the two.

Recall the example of measuring morale in different factories. The strategy of immersing yourself in the day-to-day routine of the assembly line, observing what goes on, and talking to the workers would seem to provide a more valid measure of morale than counting grievances would. It just seems obvious that we'd get a clearer sense of whether the morale was high or low using this first method.

As I pointed out earlier, however, the counting strategy would be more reliable. This situation reflects a more general strain in research measurement. Most of the really interesting concepts we want to study have many subtle nuances, so specifying precisely what we mean by them is hard. Researchers sometimes speak of such concepts as having a "richness of meaning." Although scores of books and articles have been written on the topic of anomie/anomia, for example, they still haven't exhausted its meaning.

Very often, then, specifying reliable operational definitions and measurements seems to rob concepts of their richness of meaning. Positive morale is much more than a lack of grievances filed with the union; anomia is much more than what is measured by the five items created by Leo Srole. Yet, the more variation and richness we allow for a concept, the more opportunity there is for disagreement on how it applies to a particular situation, thus reducing reliability.

To some extent, this dilemma explains the persistence of two quite different approaches to social research: quantitative, nomothetic, structured techniques such as surveys and experiments on the one hand, and qualitative, idiographic methods such as field research and historical studies on the other. In the simplest generalization, the former methods tend to be more reliable, the latter more valid.

By being forewarned, you'll be effectively forearmed against this persistent and inevitable dilemma. If there is no clear agreement on how to measure a concept, measure it several different ways. If the concept has several dimensions, measure them all. Above all, know that the concept does not have any meaning other than what you and I give it. The only justification for giving any concept a particular meaning is utility. Measure concepts in ways that help us understand the world around us.

The Ethics of Measurement

Measurement decisions can sometimes be judged by ethical standards. We have seen that most of the concepts of interest to social researchers are open to varied meanings. Suppose, for example, that you are interested in sampling public opinion on the abortion issue in the United States. Notice the difference it would make if you conceptualized one side of the debate as "pro-choice" or as "pro-abortion." If your personal bias made you want to minimize support for having an abortion, you might be tempted to frame the concept and the measurements based on it in terms of people being "pro-abortion," thereby eliminating all those who were not especially fond of abortion per se but felt a woman should have the right to make that choice for herself. To pursue this strategy, however, would violate accepted research ethics.

Consider the choices available to you in conceptualizing attitudes toward the U.S. invasion of Iraq in 2003. Imagine the different levels of support you would "discover" if you framed the position as an unprovoked invasion of a sovereign nation, as a retaliation for the September 11, 2001, attack on the World Trade Towers (many Americans still believe Saddam Hussein masterminded that attack), as a defensive act against a perceived threat, as part of a global war on terrorism, or in any of the other way this event has been portrayed. There is no one, correct way to conceptualize this issue, but it would be unethical to seek to slant the results through a biased definition of the issue.

MAIN POINTS

Introduction

- The interrelated processes of conceptualization, operationalization, and measurement allow researchers to move from a general idea about what they want to study to effective and well-defined measurements in the real world.

Measuring Anything That Exists

- Concepts are mental images we use as summary devices for bringing together observations and experiences that seem to have something in common. We use terms or labels to reference these concepts.
- Concepts are constructs; they represent the agreed-on meanings we assign to terms. Our

concepts don't exist in the real world, so they can't be measured directly, but we can measure the things that our concepts summarize.

Conceptualization

- Conceptualization is the process of specifying observations and measurements that give concepts definite meaning for the purposes of a research study.
- Conceptualization includes specifying the indicators of a concept and describing its dimensions. Operational definitions specify how variables relevant to a concept will be measured.

Definitions in Descriptive and Explanatory Studies

- Precise definitions are even more important in descriptive than in explanatory studies. The degree of precision needed varies with the type and purpose of a study.

Operationalization Choices

- Operationalization is an extension of conceptualization that specifies the exact procedures that will be used to measure the attributes of variables.
- Operationalization involves a series of interrelated choices: specifying the range of variation that is appropriate for the purposes of a study, determining how precisely to measure variables, accounting for relevant dimensions of variables, clearly defining the attributes of variables and their relationships, and deciding on an appropriate level of measurement.
- Researchers must choose from four levels of measurement, which capture increasing amounts of information: nominal, ordinal, interval, and ratio. The most appropriate level depends on the purpose of the measurement.
- A given variable can sometimes be measured at different levels. When in doubt, researchers should use the highest level of measurement appropriate to that variable so they can capture the greatest amount of information.
- Operationalization begins in the design phase of a study and continues through all phases of the research project, including the analysis of data.

Criteria of Measurement Quality

- Criteria of the quality of measures include precision, accuracy, reliability, and validity.

- Whereas reliability means getting consistent results from the same measure, validity refers to getting results that accurately reflect the concept being measured.

- Researchers can test or improve the reliability of measures through the test-retest method, the split-half method, the use of established measures, and the examination of work performed by research workers.

- The yardsticks for assessing a measure's validity include face validity, criterion-related validity, construct validity, and content validity.

- Creating specific, reliable measures often seems to diminish the richness of meaning our general concepts have. This problem is inevitable. The best solution is to use several different measures, tapping the different aspects of a concept.

The Ethics of Measurement

- Conceptualization and measurement must never be guided by bias or preferences for particular research outcomes.

KEY TERMS

The following terms are defined in context in the chapter and at the bottom of the page where the term is introduced, as well as in the comprehensive glossary at the back of the book.

conceptualization	interval measure
construct validity	nominal measure
content validity	ordinal measure
criterion-related validity	ratio measure
dimension	reliability
face validity	specification
indicator	validity

PROPOSING SOCIAL RESEARCH: MEASUREMENT

This chapter has taken us deeper into the matter of measurement. In previous exercises, you've identified the concepts and variables you want to address in your research project. Now you'll need to get more specific in terms of conceptualization and operationalization. You should conclude this portion of the proposal with a description of how, precisely, you will make distinctions regarding your variables. If you want to compare liberals and conservatives, for example, how exactly will you identify subjects' political orientations?

The ease or difficulty of this exercise may vary with the type of data collection you're planning. It will probably be easier to accomplish in the case of quantitative studies, such as surveys, where you can report the questionnaire items you'll use for measurements. In qualitative research, however, you'll have more opportunities to modify the ways variables are measured as the study unfolds, taking advantage of insights gained "in the trenches." Even so, you'll still need to begin with some clear ideas about how you'll begin your measurements.

Criteria such as precision, accuracy, validity, and reliability matter greatly in all kinds of social research projects.

REVIEW QUESTIONS AND EXERCISES

1. Pick a social science concept such as liberalism or alienation, then specify that concept so that it could be studied in a research project. Be sure to specify the indicators you'll use as well as the dimensions you wish to include in and exclude from your conceptualization.

2. What level of measurement—nominal, ordinal, interval, or ratio—describes each of the following variables?

   a. Race (white, African American, Asian, and so on)

   b. Order of finish in a race (first, second, third, and so on)

   c. Number of children in families

   d. Populations of nations

   e. Attitudes toward nuclear energy (strongly approve, approve, disapprove, strongly disapprove)

   f. Region of birth (Northeast, Midwest, and so on)

   g. Political orientation (very liberal, somewhat liberal, somewhat conservative, very conservative)

3. To conceptualize the variable *prejudice,* use your favorite web browser to search for this term. After reviewing several of the websites resulting from your search, make a list of some different forms of prejudice that might be studied in an omnibus project dealing with that topic.

4. In a good dictionary, look up *truth* and *true,* then copy out the definitions. Note the key terms used in those definitions (such as *reality*), look up the definitions of those terms, and copy out these

definitions as well. Continue this process until no new terms appear. Comment on what you've learned from this exercise. Did you discover "truth"?

SPSS EXERCISES

See the booklet that accompanies your text for exercises using SPSS (Statistical Package for the Social Sciences). There are exercises offered for each chapter, and you'll also find a detailed primer on using SPSS.

Online Study Resources

If your book came with an access code card, visit www.cengage.com/login to register. To purchase access, please visit www.ichapters.com.

1. Before you do your final review of the chapter, take the *CengageNOW* pretest to help identify the areas on which you should concentrate. You'll find information on this online tool, as well as instructions on how to access all of its great resources, in the front of the book.

2. As you review, take advantage of the *CengageNOW* personalized study plan, based on your quiz results. Use this study plan with its interactive exercises and other resources to master the material.

3. When you're finished with your review, take the posttest to confirm that you're ready to move on to the next chapter.

WEBSITE FOR *THE PRACTICE OF SOCIAL RESEARCH* 12TH EDITION

Go to your book's website at www.cengage.com/sociology/babbie for tools to aid you in studying for your exams. You'll find *Tutorial Quizzes* with feedback, *Internet Exercises, Flash Cards, Glossaries,* and *Essay Quizzes,* as well as *InfoTrac College Edition* search terms, suggestions for additional reading, *Web Links,* and primers for using data-analysis software such as SPSS.

CHAPTER SIX

Indexes, Scales, and Typologies

CHAPTER OVERVIEW

Researchers often need to employ multiple indicators to measure a variable adequately and validly. Indexes, scales, and typologies are useful composite measures made up of several indicators of variables.

CengageNOW for Sociology

Use this online tool to help you make the grade on your next exam. After reading this chapter, go to "Online Study Resources" at the end of the chapter for instructions on how to benefit from *CengageNOW*.

Introduction

As we saw in Chapter 5, many social science concepts have complex and varied meanings. Making measurements that capture such concepts can be a challenge. Recall our discussion of content validity, which concerns whether we have captured all the different dimensions of a concept.

To achieve broad coverage of the various dimensions of a concept, we usually need to make multiple observations pertaining to that concept. Thus, for example, Bruce Berg (1989: 21) advises in-depth interviewers to prepare essential questions, which are "geared toward eliciting specific, desired information." In addition, the researcher should prepare extra questions: "questions roughly equivalent to certain essential ones, but worded slightly differently."

Multiple indicators are used with quantitative data as well. Suppose you're designing a survey. Although you can sometimes construct a single questionnaire item that captures the variable of interest—"Gender: ☐ Male ☐ Female" is a simple example—other variables are less straightforward and may require you to use several questionnaire items to measure them adequately.

Quantitative data analysts have developed specific techniques for combining indicators into a single measure. This chapter discusses the construction of two types of composite measures of variables—indexes and scales. Although these measures can be used in any form of social research, they are most common in survey research and other quantitative methods. A short section at the end of this chapter considers typologies, which are relevant to both qualitative and quantitative research.

Composite measures are frequently used in quantitative research, for several reasons. First, social scientists often wish to study variables that have no clear and unambiguous single indicators. Single indicators do suffice for some variables, such as age. We can determine a survey respondent's age by simply asking, "How old are you?" Similarly, we can determine a newspaper's circulation by merely looking at the figure the newspaper reports. In the case of complex concepts, however, researchers can seldom develop single indicators before they actually do the research. This is especially true with regard to attitudes and orientations. Rarely can a survey researcher, for example, devise single questionnaire items that adequately tap respondents' degrees of prejudice, religiosity, political orientation, alienation, and the like. More likely, the researcher will devise several items, each of which provides some indication of the variables. Taken individually, each of these items is likely to prove invalid or unreliable for many respondents. A composite measure, however, can overcome this problem.

Second, researchers may wish to employ a rather refined ordinal measure of a particular variable (*alienation,* say), arranging cases in several ordinal categories from very low to very high, for example. A single data item might not have enough categories to provide the desired range of variation. However, an index or scale formed from several items can provide the needed range.

Finally, indexes and scales are efficient devices for data analysis. If considering a single data item gives us only a rough indication of a given variable, considering several data items can give us a more comprehensive and more accurate indication. For example, a single newspaper editorial may give us some indication of the political orientations of that newspaper. Examining several editorials would probably give us a better assessment, but the manipulation of several data items simultaneously could be very complicated. Indexes and scales (especially scales) are efficient data-reduction devices: They allow us to summarize several indicators in a single numerical score, while sometimes nearly maintaining the specific details of all the individual indicators.

Indexes versus Scales

The terms **index** and **scale** are typically used imprecisely and interchangeably in social research literature. The two types of measures do have some characteristics in common, but in this book we'll distinguish between the two. However, you should be warned of a growing tendency in the literature to use the term *scale* to refer to both indexes and scales, as they are distinguished here.

First, let's consider what they have in common. Both scales and indexes are ordinal measures of variables. Both rank-order the units of analysis in terms of specific variables such as religiosity, alienation, socioeconomic status, prejudice, or intellectual sophistication. A person's score on either a scale or an index of religiosity, for example, gives an indication of his or her relative religiosity vis-à-vis other people.

Further, both scales and indexes are composite measures of variables—that is, measurements based on more than one data item. Thus, a survey respondent's score on an index or scale of religiosity is determined by the responses given to several questionnaire items, each of which provides some indication of religiosity. Similarly, a person's IQ score is based on answers to a large number of test questions. The political orientation of a newspaper might be represented by an index or scale score reflecting the newspaper's editorial policy on various political issues.

Despite these shared characteristics, it's useful to distinguish between indexes and scales. In this book, we'll distinguish them by the way scores are assigned in each. We construct an index simply by accumulating scores assigned to individual attributes. We might measure prejudice, for example, by adding up the number of prejudiced statements each respondent agreed with. We construct a scale, however, by assigning scores to patterns of responses, recognizing that some items reflect a relatively weak degree of the variable while others

reflect something stronger. For example, agreeing that "Women are different from men" is, at best, weak evidence of sexism compared with agreeing that "Women should not be allowed to vote." A scale takes advantage of differences in intensity among the attributes of the same variable to identify distinct patterns of response.

Let's consider this simple example of sexism a bit further. Imagine asking people to agree or disagree with the two statements just presented. Some might agree with both, some might disagree with both. But suppose I told you someone agreed with one and disagreed with the other: Could you guess which statement they agreed with and which they did not? I'd guess the person in question agreed that women were different but disagreed that they should be prohibited from voting. On the other hand, I doubt that anyone would want to prohibit women from voting, while asserting that there is no difference between men and women. That would make no sense.

Now consider this. The two responses we wanted from each person would technically yield four possible response patterns: agree/agree, agree/disagree, disagree/agree, and disagree/disagree. We've just seen, however, that only three of the four patterns make any sense or are likely to occur. Where indexes score people based on their responses, scales score people on the basis of response patterns: We determine what the logical response patterns are and score people in terms of the pattern their responses most closely resemble.

Figure 6-1 provides a graphic illustration of the difference between indexes and scales. Let's assume we want to develop a measure of political activism, distinguishing those people who are very active in political affairs, those who don't participate much at all, and those who are somewhere in between.

The first part of Figure 6-1 illustrates the logic of indexes. The figure shows six different political actions. Although you and I might disagree on some specifics, I think we could agree that the six actions represent roughly the same degree of political activism.

Using these six items, we could construct an index of political activism by giving each person 1 point for each of the actions he or she has taken.

index A type of composite measure that summarizes and rank-orders several specific observations and represents some more general dimension.

scale A type of composite measure composed of several items that have a logical or empirical structure among them. Examples of scales include Bogardus social distance, Guttman, Likert, and Thurstone scales.

Index-Construction Logic

Here are several types of political actions people may have taken. By and large, the different actions represent similar *degrees* of political activism.

To create an *index* of overall political activism, we might give people 1 point for each of the actions they've taken.

Wrote a letter to a public official	Signed a political petition	Gave money to a political cause

Gave money to a political candidate	Wrote a political letter to the editor	Persuaded someone to change her or his voting plans

Scale-Construction Logic

Here are some political actions that represent very different degrees of activism: e.g., running for office represents a higher degree of activism than simply voting does. It seems likely, moreover, that anyone who has taken one of the more demanding actions would have taken all the easier ones as well.

To construct a *scale* of political activism, we might score people according to which of the following "ideal" patterns comes closest to describing them.

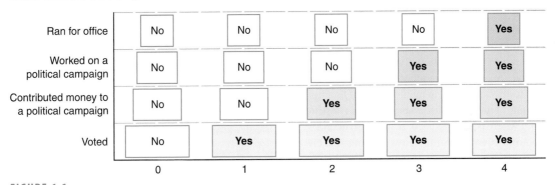

	0	1	2	3	4
Ran for office	No	No	No	No	**Yes**
Worked on a political campaign	No	No	No	**Yes**	**Yes**
Contributed money to a political campaign	No	No	**Yes**	**Yes**	**Yes**
Voted	No	**Yes**	**Yes**	**Yes**	**Yes**

FIGURE 6-1

Indexes versus Scales. Both indexes and scales seek to measure variables such as *political activism*. Whereas indexes count the number of indicators of the variable, scales take account of the differing intensities of those indicators.

If you wrote to a public official and signed a petition, you'd get a total of 2 points. If I gave money to a candidate and persuaded someone to change her or his vote, I'd get the same score as you. Using this approach, we'd conclude that you and I had the same degree of political activism, even though we had taken different actions.

The second part of Figure 6-1 describes the logic of scale construction. In this case, the actions clearly represent different degrees of political activism, ranging from simply voting to running for office. Moreover, it seems safe to assume a pattern of actions in this case. For example, all those who

contributed money probably also voted. Those who worked on a campaign probably also gave some money and voted. This suggests that most people will fall into only one of five idealized action patterns, represented by the illustrations at the bottom of the figure. The discussion of scales, later in this chapter, describes ways of identifying people with the type they most closely represent.

As you might surmise, scales are generally superior to indexes, because scales take into consideration the intensity with which different items reflect the variable being measured. Also, as the example in Figure 6-1 shows, scale scores convey more information than index scores do. Again, be aware that the term *scale* is commonly misused to refer to measures that are only indexes. Merely calling a measure a scale instead of an index doesn't make it better.

There are two other misconceptions about scaling that you should know about. First, whether the combination of several data items results in a scale almost always depends on the particular sample of observations under study. Certain items may form a scale within one sample but not within another. For this reason, do not assume that a given set of items is a scale simply because it has turned out that way in an earlier study.

Second, the use of specific scaling techniques— such as Guttman scaling, to be discussed—does not ensure the creation of a scale. Rather, such techniques let us determine whether or not a set of items constitutes a scale.

An examination of actual social science research reports will show that researchers use indexes much more frequently than they do scales. Ironically, however, the methodological literature contains little if any discussion of index construction, whereas discussions of scale construction abound. There appear to be two reasons for this disparity. First, indexes are more frequently used because scales are often difficult or impossible to construct from the data at hand. Second, methods of index construction seem so obvious and straightforward that they aren't discussed much.

Constructing indexes is not a simple undertaking, however. The general failure to develop index-construction techniques has resulted in many bad indexes in social research. With this in mind, I've devoted over half of this chapter to the methods of index construction. With a solid understanding of the logic of this activity, you'll be better equipped to try constructing both indexes and scales.

Index Construction

Let's look now at four main steps in the construction of an index: selecting possible items, examining their empirical relationships, scoring the index,

and validating it. We'll conclude this discussion by examining the construction of an index that provided interesting findings about the status of women in different countries.

Item Selection

The first step in creating an index is selecting items for a composite index, which is created to measure some variable.

Face Validity

The first criterion for selecting items to be included in an index is face validity (or logical validity). If you want to measure political conservatism, for example, each of your items should appear on its face to indicate conservatism (or its opposite, liberalism). Political party affiliation would be one such item. Another would be an item asking people to approve or disapprove of the views of a well-known conservative public figure. In constructing an index of religiosity, you might consider items such as attendance at religious services, acceptance of certain religious beliefs, and frequency of prayer; each of these appears to offer some indication of religiosity.

Unidimensionality

The methodological literature on conceptualization and measurement stresses the need for unidimensionality in scale and index construction. That is, a composite measure should represent only one dimension of a concept. Thus, items reflecting religious fundamentalism should not be included in a measure of political conservatism, even though the two variables might be empirically related to each other.

General or Specific

Although measures should tap the same dimension, the general dimension you're attempting to measure may have many nuances. In the example of religiosity, the indicators mentioned previously—ritual participation, belief, and so on—represent different types of religiosity. If you want to focus on ritual participation in religion, you should choose items specifically indicating this type of religiosity: attendance at religious services and other rituals such as confession, bar mitzvah, bowing toward Mecca, and the like. If you want to measure religiosity in a more general way, you should include a balanced set of items, representing each of the different types of religiosity. Ultimately, the nature of the items you include will determine how specifically or generally the variable is measured.

Variance

In selecting items for an index, you must also be concerned with the amount of variance they provide. If an item is intended to indicate political conservatism, for example, you should note what proportion of respondents would be identified as conservatives by that item. If a given item identified no one as a conservative or everyone as a conservative—for example, if nobody indicated approval of a radical-right political figure—that item would not be very useful in the construction of an index.

To guarantee variance, you have two options. First, you may select several items the responses to which divide people about equally in terms of the variable, for example, about half conservative and half liberal. Although no single response would justify the characterization of a person as very conservative, a person who responded as a conservative on all items might be so characterized.

The second option is to select items differing in variance. One item might identify about half of the subjects as conservative, while another might identify few of the respondents as conservatives. Note that this second option is necessary for scaling, and it is reasonable for index construction as well.

Examination of Empirical Relationships

The second step in index construction is to examine the empirical relationships among the items being considered for inclusion. (See Chapter 14 for more.) An empirical relationship is established when respondents' answers to one question—in a questionnaire, for example—help us predict how they'll answer other questions. If two items are empirically related to each other, we can reasonably argue that each reflects the same variable, and we may include them both in the same index. There are two types of possible relationships among items: bivariate and multivariate.

Bivariate Relationships

A *bivariate relationship* is, simply put, a relationship between two variables. Suppose we want to measure respondents' support for U.S. participation in the United Nations. One indicator of different levels of support might be the question "Do you feel the U.S. financial support of the UN is ☐ Too high ☐ About right ☐ Too low?"

A second indicator of support for the United Nations might be the question "Should the United States contribute military personnel to UN peacekeeping actions? ☐ Strongly approve ☐ Mostly approve ☐ Mostly disapprove ☐ Strongly disapprove."

Both of these questions, on their face, seem to reflect different degrees of support for the United Nations. Nonetheless, some people might feel the United States should give more money but not provide troops. Others might favor sending troops but cutting back on financial support.

If the two items both reflect degrees of the same thing, however, we should expect responses to the two items to correspond with each other. Specifically, those who approve of military support should be more likely to favor financial support than those who disapprove of military support would. Conversely, those who favor financial support should be more likely to favor military support than those disapproving of financial support would. If these expectations are met, we say there is a bivariate relationship between the two items.

Here's another example. Suppose we want to determine the degree to which respondents feel women have the right to an abortion. We might ask (1) "Do you feel a woman should have the right to an abortion when her pregnancy was the result of rape?" and (2) "Do you feel a woman should have the right to an abortion if continuing her pregnancy would seriously threaten her life?"

"Cause" and "Effect" Indicators

Kenneth Bollen

Department of Sociology, University of North Carolina, Chapel Hill

While it often makes sense to expect indicators of the same variable to be positively related to one another, as discussed in the text, this is not always the case.

Indicators should be related to one another if they are essentially "effects" of a variable. For example, to measure self-esteem, we might ask a person to indicate whether he or she agrees or disagrees with the statements (1) "I am a good person" and (2) "I am happy with myself." A person with high self-esteem should agree with both statements while one with low self-esteem would probably disagree with both. Since each indicator depends on or "reflects" self-esteem, we expect them to be positively correlated. More generally, indicators that depend on the same variable should be associated with one another if they are valid measures.

But, this is not the case when the indicators are the "cause" rather than the "effect" of a variable. In this situation the indicators may correlate positively, negatively, or not at all. For example, we could use gender and race as indicators of the variable *exposure to discrimination*. Being nonwhite or female increases the likelihood of experiencing discrimina-tion, so both are good indicators of the variable. But we would not expect the race and gender of individuals to be strongly associated.

Or, we may measure social interaction with three indicators: time spent with friends, time spent with family, and time spent with coworkers. Though each indicator is valid, they need not be positively correlated. Time spent with friends, for instance, may be inversely related to time spent with family. Here, the three indicators "cause" the degree of social interaction.

As a final example, exposure to stress may be measured by whether a person recently experienced divorce, death of a spouse, or loss of a job. Though any of these events may indicate stress, they need not correlate with one another.

In short, we expect an association between indicators that depend on or "reflect" a variable, that is, if they are the "effects" of the variable. But if the variable depends on the indicators—if the indicators are the "causes"—those indicators may be either positively or negatively corre-lated, or even unrelated. Therefore, we should decide whether indicators are causes or effects of a variable before using their intercorrelations to assess their validity.

Granted, some respondents might agree with item (1) and disagree with item (2); others will do just the reverse. However, if both items tap into some general opinion people have about the issue of abortion, then the responses to these two items should be related to each other. Those who support the right to an abortion in the case of rape should be more likely to support it if the woman's life is threatened than those who disapproved of abortion in the case of rape would. This would be another example of a bivariate relationship.

You should examine all the possible bivari-ate relationships among the several items being considered for inclusion in an index, in order to determine the relative strengths of relationships among the several pairs of items. Percentage tables, correlation coefficients (see Chapter 16), or both may be used for this purpose. How we evaluate the strength of the relationships, however, can be rather subtle. "'Cause' and 'Effect' Indicators" examines some of these subtleties.

Be wary of items that are not related to one another empirically: It's unlikely that they measure the same variable. You should probably drop any item that is not related to several other items.

At the same time, a very strong relationship between two items presents a different problem. If two items are perfectly related to each other, then only one needs to be included in the index; because it completely conveys the indications pro-vided by the other, nothing more would be added by including the other item. (This problem will become even clearer in the next section.)

Here's an example to illustrate the testing of bivariate relationships in index construction. I once conducted a survey of medical school faculty members to find out about the consequences of a "scientific perspective" on the quality of patient care provided by physicians. The primary intent was to determine whether scientifically inclined doctors treated patients more impersonally than other doctors did.

The survey questionnaire offered several possible indicators of respondents' scientific perspectives. Of those, three items appeared to provide especially clear indications of whether the doctors were scientifically oriented:

1. As a medical school faculty member, in what capacity do you feel you can make your greatest *teaching* contribution: as a practicing physician or as a medical researcher?

2. As you continue to advance your own medical knowledge, would you say your ultimate medical interests lie primarily in the direction of total patient management or the understanding of basic mechanisms? [The purpose of this item was to distinguish those who were mostly interested in overall patient care from those mostly interested in biological processes.]

3. In the field of therapeutic research, are you generally more interested in articles reporting evaluations of the effectiveness of various treatments or articles exploring the basic rationale underlying the treatments? [Similarly, I wanted to distinguish those more interested in articles dealing with patient care from those more interested in biological processes.]

(Babbie 1970: 27–31)

For each of these items, we might conclude that those respondents who chose the second answer are more scientifically oriented than respondents who chose the first answer. Though this comparative conclusion is reasonable, we should not be misled into thinking that respondents who chose the second answer to a given item are scientists in any absolute sense. They are simply more scientifically oriented than those who chose the first answer to the item.

To see this point more clearly, let's examine the distribution of responses to each item. From the first item—greatest teaching contribution— only about one-third of the respondents appeared scientifically oriented. That is, approximately one-third said they could make their greatest teaching

contribution as medical researchers. In response to the second item—ultimate medical interests— approximately two-thirds chose the scientific answer, saying they were more interested in learning about basic mechanisms than learning about total patient management. In response to the third item—reading preferences—about 80 percent chose the scientific answer.

These three questionnaire items can't tell us how many "scientists" there are in the sample, for none of them is related to a set of criteria for what constitutes being a scientist in any absolute sense. Using the items for this purpose would present us with the problem of three quite different estimates of how many scientists there were in the sample.

However, these items do provide us with three independent indicators of respondents' relative inclinations toward science in medicine. Each item separates respondents into the more scientific and the less scientific. But each grouping of more or less scientific respondents will have a somewhat different membership from the others. Respondents who seem scientific in terms of one item will not seem scientific in terms of another. Nevertheless, to the extent that each item measures the same general dimension, we should find some correspondence among the several groupings. Respondents who appear scientific in terms of one item should be more likely to appear scientific in their response to another item than would those who appeared nonscientific in their response to the first. In other words, we should find an association or correlation between the responses given to two items.

Figure 6-2 shows the associations among the responses to the three items. Three bivariate tables are presented, showing the distribution of responses for each possible pairing of items. An examination of the three bivariate relationships presented in the figure supports the suggestion that the three items all measure the same variable: *scientific orientation*. To see why this is so, let's begin by looking at the first bivariate relationship in the table. The table shows that faculty who responded that "researcher" was the role in which they could make their greatest teaching contribution were more likely to identify their ultimate medical interests as "basic mechanisms" (87 percent) than were

a.

		Greatest Teaching Contribution	
		Physician	Researcher
Ultimate Medical Interest	Total patient management	49%	13%
	Basic mechanisms	51%	87%
		100% (268)	100% (159)

b.

		Reading Preferences	
		Effectiveness	Rationale
Ultimate Medical Interest	Total patient management	68%	30%
	Basic mechanisms	32%	70%
		100% (78)	100% (349)

c.

		Reading Preferences	
		Effectiveness	Rationale
Greatest Teaching Contribution	Physician	85%	64%
	Researcher	15%	36%
		100% (78)	100% (349)

FIGURE 6-2

Bivariate Relationships among Scientific Orientation Items. If several indicators are measures of the same variable, then they should be empirically correlated with one another, as you can observe in this case. Those who choose the scientific orientation on one item are more likely to choose the scientific orientation on other items.

those who answered "physician" (51 percent). The fact that the "physicians" are about evenly split in their ultimate medical interests is irrelevant for our purposes. It is only relevant that they are less scientific in their medical interests than the

"researchers." The strength of this relationship may be summarized as a 36 percentage point difference.

The same general conclusion applies to the other bivariate relationships. The strength of the relationship between reading preferences and ultimate medical interests may be summarized as a 38 percentage point difference, and the strength of the relationship between reading preferences and greatest teaching contribution as a 21 percentage point difference. In summary, then, each single item produces a different grouping of "scientific" and "nonscientific" respondents. However, the responses given to each of the items correspond, to a greater or lesser degree, to the responses given to each of the other items.

Initially, the three items were selected on the basis of face validity—each appeared to give some indication of faculty members' orientations to science. By examining the bivariate relationship between the pairs of items, we have found support for the expectation that they all measure basically the same thing. However, that support does not sufficiently justify including the items in a composite index. Before combining them in a single index, we need to examine the multivariate relationships among the several variables.

Multivariate Relationships among Items

Figure 6-3 categorizes the sample respondents into four groups according to (1) their greatest teaching contribution and (2) their reading preferences. The numbers in parentheses indicate the number of respondents in each group. Thus, 66 of the faculty members who said they could best teach as physicians also said they preferred articles dealing with the effectiveness of treatments. For each of the four groups, the figure presents the percentage of those who say they are ultimately more interested in basic mechanisms. So, for example, of the 66 faculty mentioned, 27 percent are primarily interested in basic mechanisms.

The arrangement of the four groups is based on a previously drawn conclusion regarding scientific orientations. The group in the upper left corner of the table is presumably the least scientifically oriented, based on greatest teaching contribution

Percent Interested in Basic Mechanisms

		Greatest Teaching Contribution	
		Physician	Researcher
Reading Preferences	Effectiveness of treatments	27% (66)	58% (12)
	Rationale behind treatments	58% (219)	89% (130)

FIGURE 6-3

Trivariate Relationships among Scientific Orientation Items. Indicators of the same variable should be correlated in a multivariate analysis as well as in bivariate analyses. Those who choose the scientific responses on greatest teaching contribution and reading preferences are the most likely to choose the scientific response on the third item.

Percent Interested in Basic Mechanisms

		Greatest Teaching Contribution	
		Physician	Researcher
Reading Preferences	Effectiveness of treatments	51% (66)	87% (12)
	Rationale behind treatments	51% (219)	87% (130)

FIGURE 6-4

Hypothetical Trivariate Relationship among Scientific Orientation Items. This hypothetical relationship suggests that not all three indicators would contribute effectively to a composite index.

and reading preferences. The group in the lower right corner is presumably the most scientifically oriented in terms of those items.

Recall that expressing a primary interest in basic mechanisms was also taken as an indication of scientific orientation. As we should expect, then, those in the lower right corner are the most likely to give this response (89 percent), and those in the upper left corner are the least likely (27 percent). The respondents who gave mixed responses in terms of teaching contributions and reading preferences have an intermediate rank in their concern for basic mechanisms (58 percent in both cases).

This table tells us many things. First, we may note that the original relationships between pairs of items are not significantly affected by the presence of a third item. Recall, for example, that the relationship between teaching contribution and ultimate medical interest was summarized as a 36 percentage point difference. Looking at Figure 6-3, we see that among only those respondents who are most interested in articles dealing with the effectiveness of treatments, the relationship between teaching contribution and ultimate medical interest is 31 percentage points (58 percent minus 27 percent: first row). The same is true among those most interested in articles dealing with the rationale for treatments (89 percent minus 58 percent: second row). The

original relationship between teaching contribution and ultimate medical interest is essentially the same as in Figure 6-2, even among those respondents judged as scientific or nonscientific in terms of reading preferences.

We can draw the same conclusion from the columns in Figure 6-3. Recall that the original relationship between reading preferences and ultimate medical interest was summarized as a 38 percentage point difference. Looking only at the "physicians" in Figure 6-3, we see that the relationship between the other two items is now 31 percentage points. The same relationship is found among the "researchers" in the second column.

The importance of these observations becomes clearer when we consider what might have happened. In Figure 6-4, hypothetical data tell a much different story than the actual data in Figure 6-3 do. As you can see, Figure 6-4 shows that the original relationship between teaching role and ultimate medical interest persists, even when reading preferences are introduced into the picture. In each row of the table, the "researchers" are more likely to express an interest in basic mechanisms than the "physicians" are. Looking down the columns, however, we note that there is no relationship between reading preferences and ultimate medical interest. If we know whether a respondent feels he or she can best teach as a physician or as a researcher, knowing the respondent's reading preference adds nothing to our evaluation of his or her scientific orientation. If something like Figure 6-4 resulted from the actual data, we would conclude that reading preference should not be included in the same index as teaching role, because it contributed nothing to the composite index.

This example used only three questionnaire items. If more were being considered, then more-complex multivariate tables would be in order, constructed of four, five, or more variables. The purpose of this step in index construction, again, is to discover the simultaneous interaction of the items in order to determine which should be included in the same index. These kinds of data analyses are easily accomplished using programs such as SPSS and MicroCase. They are usually referred to as cross-tabulations.

Index Scoring

When you've chosen the best items for your index, you next assign scores for particular responses, thereby creating a single composite measure out of the several items. There are two basic decisions to be made in this step.

First, you must decide the desirable range of the index scores. A primary advantage of an index over a single item is the range of gradations it offers in the measurement of a variable. As noted earlier, political conservatism might be measured from "very conservative" to "not at all conservative" or "very liberal." How far to the extremes, then, should the index extend?

In this decision, the question of variance enters once more. Almost always, as the possible extremes of an index are extended, fewer cases are to be found at each end. The researcher who wishes to measure political conservatism to its greatest extreme (somewhere to the right of Attila the Hun, as the saying goes) may find there is almost no one in that category. At some point, additional gradations do not add meaning to the results.

The first decision, then, concerns the conflicting desire for (1) a range of measurement in the index and (2) an adequate number of cases at each point in the index. You'll be forced to reach some

kind of compromise between these conflicting desires.

The second decision concerns the actual assignment of scores for each particular response. Basically you must decide whether to give items in the index equal weight or different weights. Although there are no firm rules, I suggest—and practice tends to support this method—that items be weighted equally unless there are compelling reasons for differential weighting. That is, the burden of proof should be on differential weighting; equal weighting should be the norm.

Of course, this decision must be related to the earlier issue regarding the balance of items chosen. If the index is to represent the composite of slightly different aspects of a given variable, then you should give each aspect the same weight. In some instances, however, you may feel that two items reflect essentially the same aspect, and the third reflects a different aspect. If you want to have both aspects equally represented by the index, you might give the different item a weight equal to the combination of the two similar ones. For instance, you could assign a maximum score of 2 to the different item and a maximum score of 1 to each of the similar ones.

Although the rationale for scoring responses should take such concerns as these into account, typically researchers experiment with different scoring methods, examining the relative weights given to different aspects but at the same time worrying about the range and distribution of cases provided. Ultimately, the scoring method chosen will represent a compromise among these several demands. Of course, as in most research activities, such a decision is open to revision on the basis of later examinations. Validation of the index, to be discussed shortly, may lead the researcher to recycle his or her efforts by constructing a completely different index.

In the example taken from the medical school faculty survey, I decided to weight the items equally, since I'd chosen them, in part, because they represent slightly different aspects of the overall variable *scientific orientation*. On each of the items, the respondents were given a score of 1 for choosing the "scientific" response to the item and a score of 0 for choosing the "nonscientific" response. Each respondent, then, could receive a score of 0, 1, 2, or 3. This scoring method provided what I considered a useful range of variation—four index categories—and also provided enough cases for analysis in each category.

Here's a similar example of index scoring, from a study of work satisfaction. One of the key variables was *job-related depression,* measured by an index composed of the following four items, which asked workers how they felt when thinking about themselves and their jobs:

- "I feel downhearted and blue."
- "I get tired for no reason."
- "I find myself restless and can't keep still."
- "I am more irritable than usual."

The researchers, Amy Wharton and James Baron, report, "Each of these items was coded: 4 = often, 3 = sometimes, 2 = rarely, 1 = never." They go on to explain how they measured another variable, job-related self-esteem:

> Job-related self-esteem was based on four items asking respondents how they saw themselves in their work: happy/sad; successful/not successful; important/not important; doing their best/not doing their best. Each item ranged from 1 to 7, where 1 indicates a self-perception of not being happy, successful, important, or doing one's best.
>
> *(1987: 578)*

As you look through the social research literature, you'll find numerous similar examples of cumulative indexes being used to measure variables. Sometimes the indexing procedures are controversial, as evidenced in "What Is the Best College in the United States?"

Although it is often appropriate to examine the relationships among indicators of a variable being measured by an index or scale, you should realize that the indicators are sometimes independent of one another. For example, Stacy De Coster notes that the indicators of family stress may be independent of one another, though they contribute to the same variable.

> Family Stress is a scale of stressful events within the family. The experience of any one of these events—parent job loss, parent separation, parent illness—is independent of the other events. Indeed, prior research on events utilized in stress scales has demonstrated that the events in these scales typically are independent of one another and reliabilities on the scales low.
>
> *(2005: 176)*

If the indicators of a variable are logically related to one another, on the other hand, it is important to use that relationship as a criterion for determining which are the better indicators.

Handling Missing Data

Regardless of your data-collection method, you'll frequently face the problem of missing data. In a content analysis of the political orientations of newspapers, for example, you may discover that a particular newspaper has never taken an editorial position on one of the issues being studied. In an experimental design involving several retests of subjects over time, some subjects may be unable to participate in some of the sessions. In virtually every survey, some respondents fail to answer some questions (or choose a "don't know" response). Although missing data present problems at all stages of analysis, they're especially troublesome in index construction. There are, however, several methods of dealing with these problems.

First, if there are relatively few cases with missing data, you may decide to exclude them from the construction of the index and the analysis. (I did this in the medical school faculty example.) The primary concerns in this instance are whether the numbers available for analysis will remain sufficient and whether the exclusion will result in an unrepresentative sample whenever the index, excluding some of the respondents, is used in the analysis. The latter possibility can be examined through a comparison—on other relevant

What Is the Best College in the United States?

Each year the newsmagazine *U.S. News and World Report* issues a special report ranking the nation's colleges and universities. Their rankings reflect an index, created from several items: educational expenditures per student, graduation rates, selectivity (percentage accepted of those applying), average SAT scores of first-year students, and similar indicators of quality.

Typically, Harvard is ranked the number one school in the nation, followed by Yale and Princeton. However, the 1999 "America's Best Colleges" issue shocked educators, prospective college students, and their parents. The California Institute of Technology had leaped from ninth place in 1998 to first place a year later. While Harvard, Yale, and Princeton still did well, they had been supplanted. What had happened at Caltech to produce such a remarkable surge in quality?

The answer was to be found at *U.S. News and World Report*, not at Caltech. The newsmagazine changed the structure of the ranking index in 1999, which made a big difference in how schools fared.

Bruce Gottlieb (1999) gives this example of how the altered scoring made a difference.

So, how did Caltech come out on top? Well, one variable in a school's ranking has long been educational expenditures per student, and Caltech has traditionally been tops in this category. But until this year, *U.S. News* considered only a school's ranking in this category—first, second, etc.—rather than how much it spent relative to other schools. It didn't matter whether Caltech beat Harvard by $1 or by $100,000. Two other schools that rose in their rankings this year were MIT (from fourth to third) and Johns Hopkins (from 14th to seventh). All three have high per-student expenditures and all three are especially strong in the hard sciences. Universities are allowed to count their research budgets in their per-student expenditures, though students get no direct benefit from costly research their professors are doing outside of class.

In its "best colleges" issue two years ago, *U.S. News* made precisely this point, saying it considered only the rank ordering of per-student expenditures, rather than the actual amounts, on the grounds that expenditures at institutions with large research programs and medical schools are substantially higher than those at the rest of the schools in the category. In other words, just two years ago, the magazine felt it unfair to give Caltech, MIT, and Johns Hopkins credit for having lots of fancy laboratories that don't actually improve undergraduate education.

Gottlieb reviewed each of the changes in the index and then asked how 1998's ninth-ranked Caltech would have done had the revised indexing formula been in place a year earlier. His conclusion: Caltech would have been first in 1998 as well. In other words, the apparent improvement was solely a function of how the index was scored.

Clearly, composite measures such as scales and indexes are valuable tools for understanding society. However, it's important that we know how those measures are constructed and what that construction implies.

For a very different ranking of colleges and universities, you might be interested in the "Webometrics Ranking," which can be found at the link on this book's website: http://www.cengage.com/sociology/babbie. This link details the items included in the index, as well as how they are combined to produce an overall ranking of the world's institutions of higher education. As of January 2008, MIT was the top-ranked American university, but you'll have to examine the methodological description to know what that means.

So, what's really the best college in the United States? It depends on how you define "best." There is no "really best," only the various social constructions we can create.

Sources: "America's Best Colleges," *U.S. News and World Report*, August 30, 1999; Bruce Gottlieb, "Cooking the School Books: How U.S. News Cheats in Picking Its 'Best American Colleges,'" *Slate*, August 31, 1999 (http://slate.msn.com/default .aspx?id534027).

variables—of those who would be included in and excluded from the index.

Second, you may sometimes have grounds for treating missing data as one of the available responses. For example, if a questionnaire has asked respondents to indicate their participation in various activities by checking "yes" or "no" for each, many respondents may have checked some

of the activities "yes" and left the remainder blank. In such a case, you might decide that a failure to answer meant "no," and score missing data in this case as though the respondents had checked the "no" space.

Third, a careful analysis of missing data may yield an interpretation of their meaning. In constructing a measure of political conservatism, for example, you may discover that respondents who failed to answer a given question were generally as conservative on other items as those who gave the conservative answer were. In another example, a recent study measuring religious beliefs found that people who answered "don't know" about a given belief were almost identical to the "disbelievers" in their answers about other beliefs. (*Note:* You should take these examples not as empirical guides in your own studies but only as suggestions of general ways to analyze your own data.) Whenever the analysis of missing data yields such interpretations, then, you may decide to score such cases accordingly.

There are many other ways of handling the problem of missing data. If an item has several possible values, you might assign the middle value to cases with missing data; for example, you could assign a 2 if the values are 0, 1, 2, 3, and 4. For a continuous variable such as age, you could similarly assign the mean to cases with missing data (more on this in Chapter 14). Or, missing data can be supplied by assigning values at random. All of these are conservative solutions because they weaken the "purity" of your index and reduce the likelihood that it will relate to other variables in ways you may have hypothesized.

If you're creating an index out of a large number of items, you can sometimes handle missing data by using proportions based on what is observed. Suppose your index is composed of six indicators, and you only have four observations for a particular subject. If the subject has earned 4 points out of a possible 4, you might assign an index score of 6; if the subject has 2 points (half the possible score on four items), you could assign a score of 3 (half the possible score on six observations).

The choice of a particular method to be used depends so much on the research situation that I can't reasonably suggest a single "best" method or rank the several I've described. Excluding all cases with missing data can bias the representativeness of the findings, but including such cases by assigning scores to missing data can influence the nature of the findings. The safest and best method is to construct the index using more than one of these methods and see whether you reach the same

conclusions using each of the indexes. Understanding your data is the final goal of analysis anyway.

Index Validation

Up to this point, we've discussed all the steps in the selection and scoring of items that result in an index purporting to measure some variable. If each of the preceding steps is carried out carefully, the likelihood of the index actually measuring the variable is enhanced. To demonstrate success, however, we must show that the index is valid. Following the basic logic of validation, we assume that the index provides a measure of some variable; that is, the scores on the index arrange cases in a rank order in terms of that variable. An index of political conservatism rank-orders people in terms of their relative conservatism. If the index does that successfully, then people scored as relatively conservative on the index should appear relatively conservative in all other indications of political orientation, such as their responses to other questionnaire items. There are several methods of validating an index.

Item Analysis

The first step in index validation is an internal validation called **item analysis.** In item analysis, you examine the extent to which the index is related to (or predicts responses to) the individual items it comprises. Here's an illustration of this step.

In the index of scientific orientations among medical school faculty, index scores ranged from 0 (most interested in patient care) to 3 (most interested in research). Now let's consider one of the items in the index: whether respondents wanted to advance their own knowledge more with regard to total patient management or more in the area of basic mechanisms. The latter were treated as being more scientifically oriented than the former. The following empty table shows how we would

---

item analysis An assessment of whether each of the items included in a composite measure makes an independent contribution or merely duplicates the contribution of other items in the measure.

examine the relationship between the index and the individual item.

	Index of Scientific Orientations			
	0	1	2	3
Percent who said they were more interested in basic mechanisms		??	??	??	??

If you take a minute to reflect on the table, you may see that we already know the numbers that go in two of the cells. To get a score of 3 on the index, respondents had to say "basic mechanisms" in response to this question and give the "scientific" answers to the other two items as well. Thus, 100 percent of the 3's on the index said "basic mechanisms." By the same token, all the 0's had to answer this item with "total patient management." Thus, 0 percent of those respondents said "basic mechanisms." Here's how the table looks with the information we already know.

	Index of Scientific Orientations			
	0	1	2	3
Percent who said they were more interested in basic mechanisms	0	??	??	100

If the individual item is a good reflection of the overall index, we should expect the 1's and 2's to fill in a progression between 0 percent and 100 percent. More of the 2's should choose "basic mechanisms" than 1's. This result is not guaranteed by the way the index was constructed, however; it is an empirical question—one we answer in an item analysis. Here's how this particular item analysis turned out.

	Index of Scientific Orientations			
	0	1	2	3
Percent who said they were more interested in basic mechanisms	0	16	91	100

As you can see, in accord with our assumption that the 2's are more scientifically oriented than

the 1's, we find that a higher percentage of the 2's (91 percent) say "basic mechanisms" than the 1's (16 percent).

An item analysis of the other two components of the index yields similar results, as shown here.

	Index of Scientific Orientations			
	0	1	2	3
Percent who said they could teach best as medical researchers	0	4	14	100
Percent who said they preferred reading about rationales	0	80	97	100

Each of the items, then, seems an appropriate component in the index. Each seems to reflect the same quality that the index as a whole measures.

In a complex index containing many items, this step provides a convenient test of the independent contribution of each item to the index. If a given item is found to be poorly related to the index, it may be assumed that other items in the index cancel out the contribution of that item, and it should be excluded from the index. If the item in question contributes nothing to the index's power, it should be excluded.

Although item analysis is an important first test of an index's validity, it is not a sufficient test. If the index adequately measures a given variable, it should successfully predict other indications of that variable. To test this, we must turn to items not included in the index.

External Validation

People scored as politically conservative on an index should appear conservative by other measures as well, such as their responses to other items in a questionnaire. Of course, we're talking about relative conservatism, because we can't define conservatism in any absolute way. However, those respondents scored as the most conservative on the index should score as the most conservative in answering other questions. Those scored as the least conservative on the index should score as the least conservative on other items. Indeed, the ranking of groups of respondents on the index should predict the ranking of those groups in answering other questions dealing with political orientations.

In our example of the scientific orientation index, several questions in the questionnaire offered the possibility of such **external validation.** Table 6-1 presents some of these items, which provide several lessons regarding index validation. First, we note that the index strongly predicts the responses to the validating items in the sense that the rank order of scientific responses among the four groups is the same as the rank order provided by the index itself. That is, the percentages reflect greater scientific orientation as you read across the rows of the table. At the same time, each item gives a different description of scientific orientation overall. For example, the last validating item indicates that the great majority of all faculty were engaged in research during the preceding year. If this were the only indicator of scientific orientation, we would conclude that nearly all faculty were scientific. Nevertheless, those scored as more scientific on the index are more likely to have engaged in research than were those scored as relatively less scientific. The third validating item provides a different descriptive picture: Only a minority of the faculty overall say they would prefer duties limited exclusively to research. Nevertheless, the relative percentages giving this answer correspond to the scores assigned on the index.

Bad Index versus Bad Validators

Nearly every index constructor at some time must face the apparent failure of external items to validate the index. If the internal item analysis shows inconsistent relationships between the items included in the index and the index itself, something is wrong with the index. But if the index fails to predict strongly the external validation items, the conclusion to be drawn is more ambiguous. In this situation we must choose between two possibilities: (1) the index does not adequately measure the variable in question, or (2) the validation items do not adequately measure the variable and thereby do not provide a sufficient test of the index.

Having worked long and conscientiously on the construction of an index, you'll likely find the

TABLE 6-1

Validation of Scientific Orientation Index

| | Index of Scientific Orientation | | | |
| | Low | | | High |
	0	1	2	3
Percent interested in attending scientific lectures at the medical school	34	42	46	65
Percent who say faculty members should have experience as medical researchers	43	60	65	89
Percent who would prefer faculty duties involving research activities only	0	8	32	66
Percent who engaged in research during the preceding academic year	61	76	94	99

second conclusion compelling. Typically, you'll feel you have included the best indicators of the variable in the index; the validating items are, therefore, second-rate indicators. Nevertheless, you should recognize that the index is purportedly a very powerful measure of the variable; thus, it should be somewhat related to any item that taps the variable even poorly.

When external validation fails, you should reexamine the index before deciding that the validating items are insufficient. One way to do this is to examine the relationships between the validating items and the individual items included in the index. If you discover that some of the index items relate to the validators and others do not, you'll have improved your understanding of the index as it was initially constituted.

There's no cookbook solution to this problem; it is an agony serious researchers must learn to

external validation The process of testing the validity of a measure, such as an index or scale, by examining its relationship to other, presumed indicators of the same variable. If the index really measures prejudice, for example, it should correlate with other indicators of prejudice.

survive. Ultimately, the wisdom of your decision to accept an index will be determined by the usefulness of that index in your later analyses. Perhaps you'll initially decide that the index is a good one and that the validators are defective, but you'll later find that the variable in question (as measured by the index) is not related to other variables in the ways you expected. You may then have to compose a new index.

The Status of Women: An Illustration of Index Construction

For the most part, our discussion of index construction has focused on the specific context of survey research, but other types of research also lend themselves to this kind of composite measure. For example, when the United Nations (1995) set out to examine the status of women in the world, they chose to create two indexes, reflecting two different dimensions.

The Gender-related Development Index (GDI) compared women to men in terms of three indicators: life expectancy, education, and income. These indicators are commonly used in monitoring the status of women in the world. The Scandinavian countries of Norway, Sweden, Finland, and Denmark ranked highest on this measure.

The second index, the Gender Empowerment Measure (GEM), aimed more at power issues and comprised three different indicators:

- The proportion of parliamentary seats held by women
- The proportion of administrative, managerial, professional, and technical positions held by women
- A measure of access to jobs and wages

Once again, the Scandinavian countries ranked high but were joined by Canada, New Zealand, the Netherlands, the United States, and Austria. Having two different measures of gender equality rather than one allowed the researchers to make more-sophisticated distinctions. For example, in several countries, most notably Greece, France, and Japan, women fared relatively well on the GDI but quite

poorly on the GEM. Thus, while women were doing fairly well in terms of income, education, and life expectancy, they were still denied access to power. And whereas the GDI scores were higher in the wealthier nations than in the poorer ones, GEM scores showed that women's empowerment depended less on national wealth, with many poor, developing countries outpacing some rich, industrial ones in regard to such empowerment.

By examining several different dimensions of the variables involved in their study, the UN researchers also uncovered an aspect of women's earnings that generally goes unnoticed. Population Communications International (1996: 1) summarizes the finding nicely:

> Every year, women make an invisible contribution of eleven trillion U.S. dollars to the global economy, the UNDP [United Nations Development Programme] report says, counting both unpaid work and the underpayment of women's work at prevailing market prices. This "underevaluation" of women's work not only undermines their purchasing power, says the 1995 HDR [Human Development Report], but also reduces their already low social status and affects their ability to own property and use credit. Mahbub ul Haq, the principal author of the report, says that "if women's work were accurately reflected in national statistics, it would shatter the myth that men are the main breadwinner of the world." The UNDP report finds that women work longer hours than men in almost every country, including both paid and unpaid duties. In developing countries, women do approximately 53% of all work and spend two-thirds of their work time on unremunerated activities. In industrialized countries, women do an average of 51% of the total work, and—like their counterparts in the developing world—perform about two-thirds of their total labor without pay. Men in industrialized countries are compensated for two-thirds of their work.

"Indexing the World" gives some other examples of indexes that have been created to monitor the state of the world.

"Indexing the World"

If you browse the web in search of indexes, you'll be handsomely rewarded. Here are just a few examples of the ways in which people have used the logic of social indexes to monitor the state of the world. Go to http://www.cengage.com/sociology/babbie for links to each of the following examples.

The well-being of nations is commonly measured in economic terms, such as the Gross Domestic Product per capita, average income, or stock market averages. In 1972, however, the mountainous kingdom of Bhutan drew global attention by proposing an index of "Gross National Happiness," augmenting economic factors with measures of physical and mental health, freedom, environment, marital stability, and other indicators of noneconomic well-being. The World Data Base of Happiness expands this general idea to 24 countries.

Columbia University's Environmental Sustainability Index is one of several measures that seek to monitor the environmental impact of the nations of the planet.

The well-being of America's young people is the focus of the Child and Youth Well-Being Index, housed at Duke University.

Money Magazine has indexed the 100 best places to live in America, using factors such as economics, housing, schools, health, crime, weather, and public facilities.

The Heritage Foundation offers the Index of Economic Freedom for those planning business ventures around the world.

For Christians who believe in prophecies of the end of times, the Rapture Index uses 45 indicators—including inflation, famine, floods, liberalism, and Satanism—and offers a gauge of how close or far away the end is.

Can you find other, similar indexes online?

As you can see, indexes can be constructed from many different kinds of data for a variety of purposes. Now we'll turn our attention from the construction of indexes to an examination of scaling techniques.

Scale Construction

Good indexes provide an ordinal ranking of cases on a given variable. All indexes are based on this kind of assumption: A senator who voted for seven conservative bills is considered to be more conservative than one who voted for only four of them. What an index may fail to take into account, however, is that not all indicators of a variable are equally important or equally strong. The first senator might have voted in favor of seven mildly conservative bills, whereas the second senator might have voted in favor of four extremely conservative bills. (The second senator might have considered the other seven bills too liberal and voted against them.)

Scales offer more assurance of ordinality by tapping the intensity structures among the indicators. The several items going into a composite measure may have different intensities in terms of the variable. Many methods of scaling are available. We'll look at four scaling procedures to illustrate the variety of techniques available, along with a technique called the semantic differential. Although these examples focus on questionnaires, the logic of scaling, like that of indexing, applies to other research methods as well.

Bogardus Social Distance Scale

Let's suppose you're interested in the extent to which U.S. citizens are willing to associate with, say, sex offenders. You might ask the following questions:

1. Are you willing to permit sex offenders to live in your country?

2. Are you willing to permit sex offenders to live in your community?

3. Are you willing to permit sex offenders to live in your neighborhood?

4. Would you be willing to let a sex offender live next door to you?

5. Would you let your child marry a sex offender?

These questions increase in terms of the closeness of contact with sex offenders. Beginning with the original concern to measure willingness to associate with sex offenders, you have thus developed several questions indicating differing degrees of intensity on this variable. The kinds of items presented constitute a **Bogardus social distance scale** (created by Emory Bogardus). This scale is a measurement technique for determining the willingness of people to participate in social relations—of varying degrees of closeness—with other kinds of people.

The clear differences of intensity suggest a structure among the items. Presumably if a person is willing to accept a given kind of association, he or she would be willing to accept all those preceding it in the list—those with lesser intensities. For example, the person who is willing to permit sex offenders to live in the neighborhood will surely accept them in the community and the nation but may or may not be willing to accept them as next-door neighbors or relatives. This, then, is the logical structure of intensity inherent among the items.

Empirically, one would expect to find the largest number of people accepting co-citizenship and the fewest accepting intermarriage. In this sense, we speak of "easy items" (for example, residence in the United States) and "hard items" (for example, intermarriage). More people agree to the easy items than to the hard ones. With some inevitable exceptions, logic demands that once a person has refused a relationship presented in the scale, he or she will also refuse all the harder ones that follow it.

Bogardus social distance scale A measurement technique for determining the willingness of people to participate in social relations—of varying degrees of closeness—with other kinds of people. It is an especially efficient technique in that one can summarize several discrete answers without losing any of the original details of the data.

The Bogardus social distance scale illustrates the important economy of scaling as a data-reduction device. By knowing how many relationships with sex offenders a given respondent will accept, we know which relationships were accepted. Thus, a single number can accurately summarize five or six data items without a loss of information.

Motoko Lee, Stephen Sapp, and Melvin Ray (1996) noticed an implicit element in the Bogardus social distance scale: It looks at social distance from the point of view of the majority group in a society. These researchers decided to turn the tables and create a "reverse social distance" scale: looking at social distance from the perspective of the minority group. Here's how they framed their questions (1996: 19):

> Considering typical Caucasian Americans you have known, not any specific person nor the worst or the best, circle Y or N to express your opinion.
>
> Y N 5. Do they mind your being a citizen in this country?
>
> Y N 4. Do they mind your living in the same neighborhood?
>
> Y N 3. Would they mind your living next to them?
>
> Y N 2. Would they mind your becoming a close friend to them?
>
> Y N 1. Would they mind your becoming their kin by marriage?

As with the original scale, the researchers found that knowing the number of items minority respondents agreed with also told the researchers which ones were agreed with, 98.9 percent of the time in this case.

Thurstone Scales

Often, the inherent structure of the Bogardus social distance scale is not appropriate to the variable being measured. Indeed, such a logical structure among several indicators is seldom apparent. A **Thurstone scale** (created by Louis Thurstone) is an attempt to develop a format for generating groups of indicators of a variable that have at least an empirical structure among them. A group of judges is given perhaps a hundred items that are thought to be indicators of a given variable. Each judge is then asked to estimate how strong an indicator of a variable each item is—by assigning scores of perhaps 1 to 13. If the variable were prejudice, for example, the judges would be asked to assign the score of 1 to the very weakest indicators of prejudice, the score of 13 to the strongest indicators, and intermediate scores to those felt to be somewhere in between.

Once the judges have completed this task, the researcher examines the scores assigned to each item by all the judges, then determines which items produced the greatest agreement among the judges. Those items on which the judges disagreed broadly would be rejected as ambiguous. Among those items producing general agreement in scoring, one or more would be selected to represent each scale score from 1 to 13.

The items selected in this manner might then be included in a survey questionnaire. Respondents who appeared prejudiced on those items representing a strength of 5 would then be expected to appear prejudiced on those having lesser strengths, and if some of those respondents did not appear prejudiced on the items with a strength of 6, it would be expected that they would also not appear prejudiced on those with greater strengths.

If the Thurstone scale items were adequately developed and scored, the economy and effectiveness of data reduction inherent in the Bogardus social distance scale would appear. A single score might be assigned to each respondent (the strength of the hardest item accepted), and that score would adequately represent the responses to several questionnaire items. And as is true of the Bogardus scale, a respondent who scored 6 might be regarded as more prejudiced than one who scored 5 or less.

Thurstone scaling is not often used in research today, primarily because of the tremendous expenditure of energy and time required to have 10 to 15 judges score the items. Because the quality of their judgments would depend on their experience with the variable under consideration, they might need to be professional researchers. Moreover, the meanings conveyed by the several items indicating a given variable tend to change over time. Thus, an item having a given weight at one time might have quite a different weight later on. For a Thurstone scale to be effective, it would have to be updated periodically.

Likert Scaling

You may sometimes hear people refer to a questionnaire item containing response categories such as "strongly agree," "agree," "disagree," and "strongly disagree" as a Likert scale. This is technically a misnomer, although Rensis Likert (pronounced "LICK-ert") did create this commonly used question format. Likert also created a technique for combining the items into a scale, but while Likert's scaling technique is rarely used, his answer format is one of the most frequently used in survey research.

The particular value of this format is the unambiguous ordinality of response categories. If respondents were permitted to volunteer or select such answers as "sort of agree," "pretty much agree," "really agree," and so forth, you would find it impossible to judge the relative strength of agreement intended by the various respondents. The Likert format solves this problem.

Though seldom used, Likert's scaling method is fairly easy to understand, based on the relative intensity of different items. As a simple example, suppose we wish to measure prejudice against women. To do this, we create a set of 20 statements, each of which reflects that prejudice. One of the items might be "Women can't drive as well as men." Another might be "Women shouldn't be allowed to vote." Likert's scaling technique would demonstrate the difference in intensity between these items as well as pegging the intensity of the other 18 statements.

Thurstone scale A type of composite measure, constructed in accord with the weights assigned by "judges" to various indicators of some variables.

Let's suppose we ask a sample of people to agree or disagree with each of the 20 statements. Simply giving one point for each of the indicators of prejudice against women would yield the possibility of index scores ranging from 0 to 20. A true **Likert scale** goes one step beyond that and calculates the average index score for those agreeing with each of the individual statements. Let's say that all those who agreed that women are poorer drivers than men had an average index score of 1.5 (out of a possible 20). Those who agreed that women should be denied the right to vote might have an average index score of, say, 19.5—indicating the greater degree of prejudice reflected in that response.

As a result of this item analysis, respondents could be rescored to form a scale: 1.5 points for agreeing that women are poorer drivers, 19.5 points for saying women shouldn't vote, and points for other responses reflecting how those items related to the initial, simple index. If those who disagreed with the statement "I might vote for a woman for president" had an average index score of 15, then the scale would give 15 points to people disagreeing with that statement.

As I've said earlier, Likert scaling is seldom used today. The item format devised by Likert, however, is one of the most commonly used formats in contemporary questionnaire design. Typically, it is now used in the creation of simple indexes. With, say, five response categories (including "no opinion" or something similar), scores of 0 to 4 or 1 to 5 might be assigned, taking the direction of the items into account (for example, assign a score of 5 to "strongly agree" for positive items and to "strongly disagree" for negative items). Each respondent would then be assigned an overall score representing the summation of the scores he or she received for responses to the individual items.

Semantic Differential

Like the Likert format, the **semantic differential** asks questionnaire respondents to choose between two opposite positions by using qualifiers to bridge the distance between the two opposites. Here's how it works.

Suppose you're evaluating the effectiveness of a new music-appreciation lecture on subjects' appreciation of music. As a part of your study, you want to play some musical selections and have the subjects report their feelings about them. A good way to tap those feelings would be to use a semantic differential format.

To begin, you must determine the dimensions along which subjects should judge each selection. Then you need to find two opposite terms, representing the polar extremes along each dimension. Let's suppose one dimension that interests you is simply whether subjects enjoyed the piece or not. Two opposite terms in this case could be "enjoyable" and "unenjoyable." Similarly, you might want to know whether they regarded the individual selections as "complex" or "simple," "harmonic" or "discordant," and so forth.

Once you have determined the relevant dimensions and have found terms to represent the extremes of each, you might prepare a rating sheet each subject would complete for each piece of music. Figure 6-5 shows what it might look like.

On each line of the rating sheet, the subject would indicate how he or she felt about the piece of music: whether it was enjoyable or unenjoyable, for example, and whether it was "somewhat" that way or "very much" so. To avoid creating a biased pattern of responses to such items, it's a good idea

---

Likert scale A type of composite measure developed by Rensis Likert, in an attempt to improve the levels of measurement in social research through the use of standardized response categories in survey questionnaires, to determine the relative intensity of different items. Likert items are those using such response categories as strongly agree, agree, disagree, and strongly disagree. Such items may be used in the construction of true Likert scales as well as other types of composite measures.

semantic differential A questionnaire format in which the respondent is asked to rate something in terms of two, opposite adjectives (e.g., rate textbooks as "boring" or "exciting"), using qualifiers such as "very," "somewhat," "neither," "somewhat," and "very" to bridge the distance between the two opposites.

	Very Much	Somewhat	Neither	Somewhat	Very Much	
Enjoyable	☐	☐	☐	☐	☐	**Unenjoyable**
Simple	☐	☐	☐	☐	☐	**Complex**
Discordant	☐	☐	☐	☐	☐	**Harmonic**
Traditional	☐	☐	☐	☐	☐	**Modern**

FIGURE 6-5

Semantic Differential: Feelings about Musical Selections. The semantic differential asks respondents to describe something or someone in terms of opposing adjectives.

to vary the placement of terms that are likely to be related to each other. Notice, for example, that "discordant" and "traditional" are on the left side of the sheet, with "harmonic" and "modern" on the right. Most likely, those selections scored as "discordant" would also be scored as "modern" rather than "traditional."

Both the Likert and semantic differential formats have a greater rigor and structure than other question formats do. As I indicated earlier, these formats produce data suitable to both indexing and scaling.

Guttman Scaling

Researchers today often use the scale developed by Louis Guttman. Like Bogardus, Thurstone, and Likert scaling, Guttman scaling is based on the fact that some items under consideration may prove to be more-extreme indicators of the variable than others. Here's an example to illustrate this pattern.

In the earlier example of measuring scientific orientation among medical school faculty members, you'll recall that a simple index was constructed. As it happens, however, the three items included in the index essentially form a **Guttman scale.**

The construction of a Guttman scale begins with some of the same steps that initiate index construction. You begin by examining the face validity of items available for analysis. Then, you examine the bivariate and perhaps multivariate relations among those items. In scale construction, however, you also look for relatively "hard" and "easy" indicators of the variable being examined.

Earlier, when we talked about attitudes regarding a woman's right to have an abortion, we discussed several conditions that can affect people's opinions: whether the woman is married, whether her health is endangered, and so forth. These differing conditions provide an excellent illustration of Guttman scaling.

Here are the percentages of the people in the 2000 GSS sample who supported a woman's right to an abortion, under three different conditions:

Woman's health is seriously endangered	89%
Pregnant as a result of rape	81%
Woman is not married	39%

The different percentages supporting abortion under the three conditions suggest something about the different levels of support that each item indicates. For example, if someone supported abortion when the mother's life is seriously endangered, that's not a very strong indicator of general support for abortion, because almost everyone agreed with that. Supporting abortion for unmarried women seems a much stronger indicator of support for abortion in general—fewer than half the sample took that position.

Guttman scaling is based on the idea that anyone who gives a strong indicator of some variable will also give the weaker indicators. In this case, we would assume that anyone who supported

---

Guttman scale A type of composite measure used to summarize several discrete observations and to represent some more-general variable.

TABLE 6-2
Scaling Support for Choice of Abortion

	Women's Health	Result of Rape	Woman Unmarried	Number of Cases
	+	+	+	677
	+	+	−	607
Scale types	+	−	−	165
	−	−	−	147
			Total =	1,596
	−	+	−	42
	+	−	+	5
Mixed types	−	−	+	2
	−	+	+	4
			Total =	53

abortion for unmarried women would also support it in the case of rape or of the woman's health being threatened. Table 6-2 tests this assumption by presenting the number of respondents who gave each of the possible response patterns.

The first four response patterns in the table compose what we would call the *scale types:* those patterns that form a scalar structure. Following those respondents who supported abortion under all three conditions (line 1), we see (line 2) that those with only two pro-choice responses have chosen the two easier ones; those with only one such response (line 3) chose the easiest of the three (the woman's health being endangered). And finally, there are some respondents who opposed abortion in all three circumstances (line 4).

The second part of the table presents those response patterns that violate the scalar structure of the items. The most radical departures from the scalar structure are the last two response patterns: those who accepted only the hardest item and those who rejected only the easiest one.

The final column in the table indicates the number of survey respondents who gave each of the response patterns. The great majority (1,596, or 99 percent) fit into one of the scale types. The presence of mixed types, however, indicates that the items do not form a perfect Guttman scale.

(It would be extremely rare for such data to form a Guttman scale perfectly.)

Recall at this point that one of the chief functions of scaling is efficient data reduction. Scales provide a technique for presenting data in a summary form while maintaining as much of the original information as possible. When the scientific orientation items were formed into an index in our earlier discussion, respondents were given one point for each scientific response they gave. If these same three items were scored as a Guttman scale, some respondents would be assigned scale scores that would permit the most accurate reproduction of their original responses to all three items.

In the present example of attitudes regarding abortion, respondents fitting into the scale types would receive the same scores as would be assigned in the construction of an index. Persons selecting all three pro-choice responses (+ + +) would still be scored 3, those who selected pro-choice responses to the two easier items and were opposed on the hardest item (+ + −) would be scored 2, and so on. For each of the four scale types we could predict accurately all the actual responses given by all the respondents based on their scores.

The mixed types in the table present a problem, however. The first mixed type (− + −) was scored 1 on the index to indicate only one pro-choice response. But, if 1 were assigned as a scale score, we would predict that the 42 respondents in this group had chosen only the easiest item (approving abortion when the woman's life was endangered), and we would be making two errors for each such respondent: thinking their response pattern was (+ − −) instead of (− + −). Scale scores are assigned, therefore, with the aim of minimizing the errors that would be made in reconstructing the original responses.

Table 6-3 illustrates the index and scale scores that would be assigned to each of the response patterns in our example. Note that one error is made for each respondent in the mixed types. This is the minimum we can hope for in a mixed-type pattern. In the first mixed type, for example, we would erroneously predict a pro-choice response to the easiest item for each of the 42 respondents in this group, making a total of 42 errors.

TABLE 6-3

Index and Scale Scores

	Response Pattern	Number of Cases	Index Scores	Scale Scores	Total Scale Errors
Scale Types	+ + +	677	3	3	0
	+ + +	607	2	2	0
	+ − −	165	1	1	0
	− − −	147	0	0	0
Mixed Types	− + −	42	1	2	42
	+ − +	5	2	3	5
	− − +	2	1	0	2
	− + +	4	2	3	4

Total Scale errors = 53

$$\text{Coefficient of reproducibility} = 1 - \frac{\text{number of errors}}{\text{number of guesses}}$$

This table presents one common method for scoring mixed types, but you should be advised that other methods are also used.

The extent to which a set of empirical responses form a Guttman scale is determined by the accuracy with which the original responses can be reconstructed from the scale scores. For each of the 1,649 respondents in this example, we'll predict three questionnaire responses, for a total of 4,947 predictions. Table 6-3 indicates that we'll make 53 errors using the scale scores assigned. The percentage of correct predictions is called the *coefficient of reproducibility:* the percentage of original responses that could be reproduced by knowing the scale scores used to summarize them. In the present example, the coefficient of reproducibility is 4,894/4,947, or 98.9 percent.

Except for the case of perfect (100 percent) reproducibility, there is no way of saying that a set of items does or does not form a Guttman scale in any absolute sense. Virtually all sets of such items approximate a scale. As a general guideline, however, coefficients of 90 or 95 percent are the commonly used standards. If the observed reproducibility exceeds the level you've set, you'll probably decide to score and use the items as a scale.

The decision concerning criteria in this regard is, of course, arbitrary. Moreover, a high degree of reproducibility does not insure that the scale con-

structed in fact measures the concept under consideration. What it does is increase confidence that all the component items measure *the same thing.* Also, you should realize that a high coefficient of reproducibility is most likely when few items are involved.

One concluding remark with regard to Guttman scaling: It's based on the structure observed among the actual data under examination. This is an important point that is often misunderstood. It does not make sense to say that a set of questionnaire items (perhaps developed and used by a previous researcher) constitutes a Guttman scale. Rather, we can say only that they form a scale within a given body of data being analyzed. Scalability, then, is a sample-dependent, empirical matter. Although a set of items may form a Guttman scale among one sample of survey respondents, for example, there is no guarantee that this set will form such a scale among another sample. In this sense, then, a set of questionnaire items in and of itself never forms a scale, but a set of empirical observations may.

This concludes our discussion of indexing and scaling. Like indexes, scales are composite measures of a variable, typically broadening the meaning of the variable beyond what might be captured by a single indicator. Both scales and indexes seek to measure variables at the ordinal level of measurement. Unlike indexes, however, scales take advantage of any intensity structure that may be present among the individual indicators. To the extent that such an intensity structure is found and the data from the people or other units of analysis comply with the logic of that intensity structure, we can have confidence that we have created an ordinal measure.

Typologies

Indexes and scales, then, are constructed to provide ordinal measures of given variables. We attempt to assign index or scale scores to cases in such a way as to indicate a rising degree of prejudice, religiosity, conservatism, and so forth. In such cases, we're dealing with single dimensions.

TABLE 6-4

A Political Typology of Newspapers

		Foreign Policy	
		Conservative	Liberal
Domestic Policy	Conservative	A	B
	Liberal	C	D

Often, however, the researcher wishes to summarize the intersection of two or more variables, thereby creating a set of categories or types—a nominal variable—called a **typology.** You may, for example, wish to examine the political orientations of newspapers separately in terms of domestic issues and foreign policy. The fourfold presentation in Table 6-4 describes such a typology.

Newspapers in cell A of the table are conservative on both foreign policy and domestic policy; those in cell D are liberal on both. Those in cells B and C are conservative on one and liberal on the other.

As another example, Rodney Coates (2006) created a typology of "racial hegemony" from two dimensions:

1. Political Ideology
    a. Democratic
    b. Nondemocratic
2. Military and Industrial Sophistication
    a. Low
    b. High

He then used the typology to examine modern examples of colonial rule, with specific reference to race relations. The specific cases he examined allowed him to illustrate and refine the typology. He points out that such a device represents Max Weber's "ideal type": "As stipulated by Weber, ideal types represent a type of abstraction from reality. These abstractions, constructed from the logical

---

typology The classification (typically nominal) of observations in terms of their attributes on two or more variables. The classification of newspapers as liberal-urban, liberal-rural, conservative-urban, or conservative-rural would be an example.

extraction of elements derived from specific examples, provide a theoretical model by which and from which we may examine reality" (2006: 87).

Frequently, you arrive at a typology in the course of an attempt to construct an index or scale. The items that you felt represented a single variable appear to represent two. We might have been attempting to construct a single index of political orientations for newspapers but discovered—empirically—that foreign and domestic politics had to be kept separate.

In any event, you should be warned against a difficulty inherent in typological analysis. Whenever the typology is used as the independent variable, there will probably be no problem. In the preceding example, you might compute the percentages of newspapers in each cell that normally endorse Democratic candidates; you could then easily examine the effects of both foreign and domestic policies on political endorsements.

It's extremely difficult, however, to analyze a typology as a dependent variable. If you want to discover *why* newspapers fall into the different cells of typology, you're in trouble. That becomes apparent when we consider the ways you might construct and read your tables. Assume, for example, that you want to examine the effects of community size on political policies. With a single dimension, you could easily determine the percentages of rural and urban newspapers that were scored conservative and liberal on your index or scale.

With a typology, however, you would have to present the distribution of the urban newspapers in your sample among types A, B, C, and D. Then you would repeat the procedure for the rural ones in the sample and compare the two distributions. Let's suppose that 80 percent of the rural newspapers are scored as type A (conservative on both dimensions), compared with 30 percent of the urban ones. Moreover, suppose that only 5 percent of the rural newspapers are scored as type B (conservative only on domestic issues), compared with 40 percent of the urban ones. It would be incorrect to conclude from an examination of type B that urban newspapers are more conservative on domestic issues than rural ones are, because 85 percent of the rural newspapers, compared with 70 percent of the urban ones, have this characteristic. The relative sparsity of rural newspapers in type B is due to their concentration in type A. It should be apparent that an interpretation of such data would be very difficult for anything other than description.

In reality, you'd probably examine two such dimensions separately, especially if the dependent variable has more categories of responses than the given example does.

Don't think that typologies should always be avoided in social research; often they provide the most appropriate device for understanding the data. To examine the pro-life orientation in depth, for example, you might create a typology involving both abortion and capital punishment. Libertarianism could be seen in terms of both economic and social permissiveness. You've now been warned, however, against the special difficulties involved in using typologies as dependent variables.

MAIN POINTS

Introduction

- Single indicators of variables seldom (1) capture all the dimensions of a concept, (2) have sufficiently clear validity to warrant their use, or (3) permit the desired range of variation to allow ordinal rankings. Composite measures, such as scales and indexes, solve these problems by including several indicators of a variable in one summary measure.

Indexes versus Scales

- Although both indexes and scales are intended as ordinal measures of variables, scales typically satisfy this intention better than indexes do.

- Whereas indexes are based on the simple cumulation of indicators of a variable, scales take advantage of any logical or empirical intensity structures that exist among a variable's indicators.

Index Construction

- The principal steps in constructing an index include selecting possible items, examining their empirical relationships, scoring the index, and validating it.

- Criteria of item selection include face validity, unidimensionality, the degree of specificity with which a dimension is to be measured, and the amount of variance provided by the items.

- If different items are indeed indicators of the same variable, then they should be related empirically to one another. In constructing an index, the researcher needs to examine bivariate and multivariate relationships among the items.

- Index scoring involves deciding the desirable range of scores and determining whether items will have equal or different weights.

- There are various techniques that allow items to be used in an index in spite of missing data.

- Item analysis is a type of internal validation, based on the relationship between individual items in the composite measure and the measure itself. External validation refers to the relationships between the composite measure and other indicators of the variable—indicators not included in the measure.

Scale Construction

- Four types of scaling techniques are represented by the Bogardus social distance scale, a device for measuring the varying degrees to which a person would be willing to associate with a given class of people; Thurstone scaling, a technique that uses judges to determine the intensities of different indicators; Likert scaling, a measurement technique based on the use of standardized response categories; and Guttman scaling, a method of discovering and using the empirical intensity structure among several indicators of a given variable. Guttman scaling is probably the most popular scaling technique in social research today.

- The semantic differential is a question format that asks respondents to make ratings that lie between two extremes, such as "very positive" and "very negative."

Typologies

- A typology is a nominal composite measure often used in social research. Typologies may be used effectively as independent variables, but interpretation is difficult when they are used as dependent variables.

KEY TERMS

The following terms are defined in context in the chapter and at the bottom of the page where the term is introduced, as well as in the comprehensive glossary at the back of the book.

Bogardus social distance scale

external validation

Guttman scale

index

item analysis

Likert scale scale

semantic differential

Thurstone scale

typology

PROPOSING SOCIAL RESEARCH: COMPOSITE MEASURES

This chapter has extended the issue of measurement to include those in which variables are measured by more than one indicator. What you have learned here may extend the discussion of measurement in your proposal. As in the case of operationalization, you may find this easier to formulate in the case of quantitative studies, but the logic of multiple indicators may be applied to all research methods.

If your study will involve the use of composite measures, you should identify the type(s), the indicators to be used in their construction, and the methods you'll use to create and validate them. If the study you are planning in this series of exercises will not include composite measures, you can test your understanding of the chapter by exploring ways in which they could be used, even if you need to temporarily vary the data-collection method and/or variables you have in mind.

REVIEW QUESTIONS AND EXERCISES

1. In your own words, describe the difference between an index and a scale.

2. Suppose you wanted to create an index for rating the quality of colleges and universities. Name three data items that might be included in such an index.

3. Make up three questionnaire items that measure attitudes toward nuclear power and that would probably form a Guttman scale.

4. Construct a typology of pro-life attitudes as discussed in the chapter.

5. Economists often use indexes to measure economic variables, such as the cost of living. Go to the Bureau of Labor Statistics link on this book's website and find the Consumer Price Index survey. What are some of the dimensions of living costs included in this measure?

SPSS EXERCISES

See the booklet that accompanies your text for exercises using SPSS (Statistical Package for the Social Sciences). There are exercises offered for each chapter, and you'll also find a detailed primer on using SPSS.

Online Study Resources

If your book came with an access code card, visit www.cengage.com/login to register. To purchase access, please visit www.ichapters.com.

1. Before you do your final review of the chapter, take the *CengageNOW* pretest to help identify the areas on which you should concentrate. You'll find information on this online tool, as well as instructions on how to access all of its great resources, in the front of the book.

2. As you review, take advantage of the *CengageNOW* personalized study plan, based on your quiz results. Use this study plan with its interactive exercises and other resources to master the material.

3. When you're finished with your review, take the posttest to confirm that you're ready to move on to the next chapter.

WEBSITE FOR *THE PRACTICE OF SOCIAL RESEARCH* 12TH EDITION

Go to your book's website at www.cengage.com/ sociology/babbie for tools to aid you in studying for your exams. You'll find *Tutorial Quizzes* with feedback, *Internet Exercises, Flash Cards, Glossaries,* and *Essay Quizzes,* as well as *InfoTrac College Edition* search terms, suggestions for additional reading, *Web Links,* and primers for using data-analysis software such as SPSS.

CHAPTER SEVEN

The Logic of Sampling

CHAPTER OVERVIEW

Now you'll see how social scientists can select a few people for study—and discover things that apply to hundreds of millions of people not studied.

CengageNOW for Sociology

Use this online tool to help you make the grade on your next exam. After reading this chapter, go to "Online Study Resources" at the end of the chapter for instructions on how to benefit from *CengageNOW*.

Introduction

One of the most visible uses of survey sampling lies in the political polling that is subsequently tested by election results. Whereas some people doubt the accuracy of sample surveys, others complain that political polls take all the suspense out of campaigns by foretelling the result. In recent presidential elections, however, the polls have not removed the suspense.

Going into the 2004 presidential elections, pollsters generally agreed that the election was "too close to call," a repeat of their experience four years earlier. The Roper Center has compiled a list of polls conducted throughout the campaign; Table 7-1 reports those conducted during the few days preceding the election. Despite some variations, the overall picture they present is amazingly consistent and was played out in the election results.

Now, how many interviews do you suppose it took each of these pollsters to come within a couple of percentage points in estimating the behavior of more than 115 million voters? Often fewer than 2,000! In this chapter, we're going to find out how social researchers can pull off such wizardry.

For another powerful illustration of the potency of sampling, look at this graphic portrayal of President George W. Bush's approval ratings prior to and following the September 11, 2001, terrorist attack on the U.S. (see Figure 7-1). The data reported by several different polling agencies describe the same pattern.

Political polling, like other forms of social research, rests on observations. But neither pollsters nor other social researchers can observe everything that might be relevant to their interests. A critical part of social research, then, is deciding what to observe and what not. If you want to study voters, for example, which voters should you study?

The process of selecting observations is called *sampling*. Although sampling can mean any procedure for selecting units of observation—for example, interviewing every tenth passerby on a busy street—the key to generalizing from a sample to

TABLE 7-1

Election Eve Polls Reporting Percentage of Population Voting for U.S. Presidential Candidates, 2004

Poll	Date Begun	Bush	Kerry
Fox/OpinDynamics	Oct 28	50	50
TIPP	Oct 28	53	47
CBS/NYT	Oct 28	52	48
ARG	Oct 28	50	50
ABC	Oct 28	51	49
Fox/OpinDynamics	Oct 29	49	51
Gallup/CNN/USA	Oct 29	49	51
NBC/WSJ	Oct 29	51	49
TIPP	Oct 29	51	49
Harris	Oct 29	52	48
Democracy Corps	Oct 29	49	51
Harris	Oct 29	51	49
CBS	Oct 29	51	49
Fox/OpinDynamics	Oct 30	49	52
TIPP	Oct 30	51	49
Marist	Oct 31	50	50
GWU Battleground 2004	Oct 31	52	48
Actual vote	**Nov 2**	**52**	**48**

Source: Poll data adapted from the Roper Center, *Election 2004* (http://www.ropercenter.uconn.edu/elect_2004/pres_trial_heats.html), accessed November 16, 2004. I've apportioned the undecided and other votes according to the percentages saying they were voting for Bush or Kerry.

a larger population is probability sampling, which involves the important idea of random selection.

Much of this chapter is devoted to the logic and skills of probability sampling. This topic is more rigorous and precise than some of the other topics in this book. Whereas social research as a whole is both art and science, sampling leans toward science. Although this subject is somewhat technical, the basic logic of sampling is not difficult to understand. In fact, the logical neatness of this topic can make it easier to comprehend than, say, conceptualization.

Although probability sampling is central to social research today, we'll take some time to

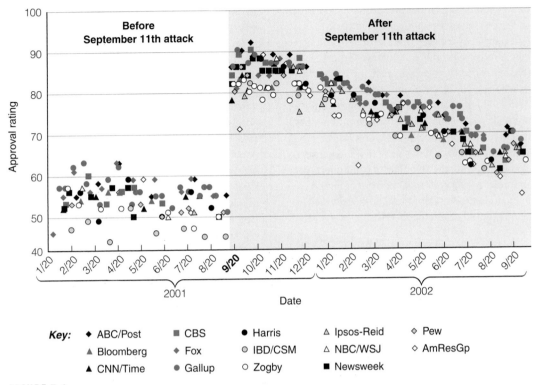

FIGURE 7-1

Bush Approval: Raw Poll Data. This graph demonstrates how independent polls produce the same picture of reality. This also shows the impact of a national crisis on the president's popularity: in this case, the September 11 terrorist attack and President George W. Bush's popularity.

Source: Copyright © 2001, 2002 by drlimerick.com. (http://www.pollkatz.homestead.com/files/MyHTML2.gif). All rights reserved.

examine a variety of nonprobability methods as well. These methods have their own logic and can provide useful samples for social inquiry.

Before we discuss the two major types of sampling, I'll introduce you to some basic ideas by way of a brief history of sampling. As you'll see, the pollsters who correctly predicted the election cliff-hanger of 2000 did so in part because researchers had learned to avoid some pitfalls that earlier pollsters had fallen into.

A Brief History of Sampling

Sampling in social research has developed hand in hand with political polling. This is the case, no doubt, because political polling is one of the few

opportunities social researchers have to discover the accuracy of their estimates. On election day, they find out how well or how poorly they did.

President Alf Landon

President Alf Landon? Who's he? Did you sleep through an entire presidency in your U.S. history class? No—but Alf Landon would have been president if a famous poll conducted by the *Literary Digest* had proved to be accurate. The *Literary Digest* was a popular newsmagazine published between 1890 and 1938. In 1920, *Digest* editors mailed postcards to people in six states, asking them whom they were planning to vote for in the presidential campaign between Warren Harding and James Cox. Names were selected for the poll from telephone directories and automobile registration lists. Based on the postcards sent back, the *Digest* correctly predicted that Harding would be elected. In the elections that followed, the *Literary Digest* expanded the size of its poll and made correct predictions in 1924, 1928, and 1932.

In 1936, the *Digest* conducted its most ambitious poll: Ten million ballots were sent to people listed in telephone directories and on lists of automobile owners. Over two million people responded, giving the Republican contender, Alf Landon, a stunning 57 to 43 percent landslide over the incumbent, President Franklin Roosevelt. The editors modestly cautioned,

> We make no claim to infallibility. We did not coin the phrase "uncanny accuracy" which has been so freely applied to our Polls. We know only too well the limitations of every straw vote, however enormous the sample gathered, however scientific the method. It would be a miracle if every State of the forty-eight behaved on Election Day exactly as forecast by the Poll.
>
> *(Literary Digest 1936a: 6)*

Two weeks later, the *Digest* editors knew the limitations of straw polls even better: The voters gave Roosevelt a second term in office by the largest landslide in history, with 61 percent of the vote. Landon won only 8 electoral votes to Roosevelt's 523.

The editors were puzzled by their unfortunate turn of luck. A part of the problem surely lay in the 22 percent return rate garnered by the poll. The editors asked,

> Why did only one in five voters in Chicago to whom the Digest sent ballots take the trouble to reply? And why was there a preponderance of Republicans in the one-fifth that did reply? . . . We were getting better cooperation in what we have always regarded as a public service from Republicans than we were getting from Democrats. Do Republicans live nearer to mailboxes? Do Democrats generally disapprove of straw polls?
>
> *(Literary Digest 1936b: 7)*

Actually, there was a better explanation—what is technically called the *sampling frame* used by the *Digest*. In this case the sampling frame consisted of telephone subscribers and automobile owners. In the context of 1936, this design selected a disproportionately wealthy sample of the voting population, especially coming on the tail end of the worst economic depression in the nation's history. The sample effectively excluded poor people, and the poor voted predominantly for Roosevelt's New Deal recovery program. The *Digest's* poll may or may not have correctly represented the voting intentions of telephone subscribers and automobile owners. Unfortunately for the editors, it decidedly did not represent the voting intentions of the population as a whole.

President Thomas E. Dewey

The 1936 election also saw the emergence of a young pollster whose name would become synonymous with public opinion. In contrast to the *Literary Digest*, George Gallup correctly predicted that Roosevelt would beat Landon. Gallup's success in 1936 hinged on his use of something called quota sampling, which we'll look at more closely later in the chapter. For now, it's enough to know that quota sampling is based on a knowledge of the characteristics of the population being sampled: what proportion are men, what proportion are women, what proportions are of various incomes, ages, and so on. Quota sampling selects people to match a set of these characteristics: the right number of poor, white, rural men; the right number of rich, African American, urban women; and so on. The quotas are based on those variables most relevant to the study. In the case of Gallup's poll, the sample selection was based on levels of income; the selection procedure ensured the right proportion of respondents at each income level.

Gallup and his American Institute of Public Opinion used quota sampling to good effect in 1936, 1940, and 1944—correctly picking the presidential winner each of those years. Then, in 1948, Gallup and most political pollsters suffered the embarrassment of picking Governor Thomas Dewey of New York over the incumbent, President

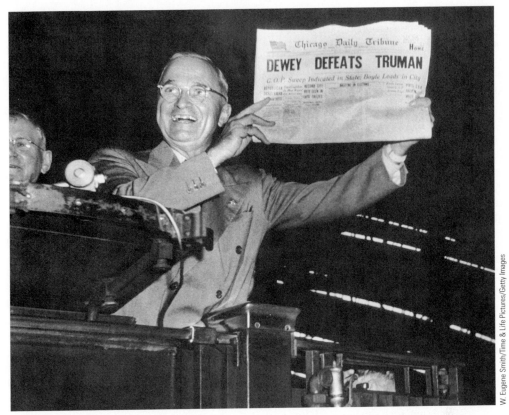

Based on early political polls that showed Dewey leading Truman, the *Chicago Daily Tribune* sought to scoop the competition with this unfortunate headline.

Harry Truman. The pollsters' embarrassing miscue continued right up to election night. A famous photograph shows a jubilant Truman—whose followers' battle cry was "Give 'em hell, Harry!"—holding aloft a newspaper with the banner headline "Dewey Defeats Truman."

Several factors accounted for the pollsters' failure in 1948. First, most pollsters stopped polling in early October despite a steady trend toward Truman during the campaign. In addition, many voters were undecided throughout the campaign, and these went disproportionately for Truman when they stepped into the voting booth.

More important, Gallup's failure rested on the unrepresentativeness of his samples. Quota sampling—which had been effective in earlier years—was Gallup's undoing in 1948. This technique requires that the researcher know something about

the total population (of voters in this instance). For national political polls, such information came primarily from census data. By 1948, however, World War II had produced a massive movement from the country to cities, radically changing the character of the U.S. population from what the 1940 census showed, and Gallup relied on 1940 census data. City dwellers, moreover, tended to vote Democratic; hence, the overrepresentation of rural voters in his poll had the effect of underestimating the number of Democratic votes.

Two Types of Sampling Methods

By 1948, some academic researchers had already been experimenting with a form of sampling based on probability theory. This technique involves the selection of a "random sample" from a list containing the names of everyone in the population being sampled. By and large, the probability-sampling methods used in 1948 were far more accurate than quota-sampling techniques.

Today, probability sampling remains the primary method of selecting large, representative samples for social research, including national political polls. At the same time, probability sampling can be impossible or inappropriate in many research situations. Accordingly, before turning to the logic and techniques of probability sampling, we'll first take a look at techniques for nonprobability sampling and how they're used in social research.

Nonprobability Sampling

Social research is often conducted in situations that do not permit the kinds of probability samples used in large-scale social surveys. Suppose you wanted to study homelessness: There is no list of all homeless individuals, nor are you likely to create such a list. Moreover, as you'll see, there are times when probability sampling wouldn't be appropriate even if it were possible. Many such situations call for **nonprobability sampling.**

In this section, we'll examine four types of nonprobability sampling: reliance on available subjects, purposive (judgmental) sampling, snowball sampling, and quota sampling. We'll conclude with a brief discussion of techniques for obtaining information about social groups through the use of informants.

Reliance on Available Subjects

Relying on available subjects, such as stopping people at a street corner or some other location, is sometimes called "convenience" or "haphazard"

---

nonprobability sampling Any technique in which samples are selected in some way not suggested by probability theory. Examples include reliance on available subjects as well as purposive (judgmental), quota, and snowball sampling.

---

sampling. This is a common method for journalists in their "person-on-the-street" interviews, but it is an extremely risky sampling method for social research. Clearly, this method does not permit any control over the representativeness of a sample. It's justified only if the researcher wants to study the characteristics of people passing the sampling point at specified times or if less-risky sampling methods are not feasible. Even when this method is justified on grounds of feasibility, researchers must exercise great caution in generalizing from their data. Also, they should alert readers to the risks associated with this method.

University researchers frequently conduct surveys among the students enrolled in large lecture classes. The ease and frugality of such a method explains its popularity, but it seldom produces data of any general value. It may be useful for pretesting a questionnaire, but such a sampling method should not be used for a study purportedly describing students as a whole.

Consider this report on the sampling design in an examination of knowledge and opinions about nutrition and cancer among medical students and family physicians:

> The fourth-year medical students of the University of Minnesota Medical School in Minneapolis comprised the student population in this study. The physician population consisted of all physicians attending a "Family Practice Review and Update" course sponsored by the University of Minnesota Department of Continuing Medical Education.
>
> *(Cooper-Stephenson and Theologides 1981: 472)*

After all is said and done, what will the results of this study represent? They do not provide a meaningful comparison of medical students and family physicians in the United States or even in Minnesota. Who were the physicians who attended the course? We can guess that they were probably more concerned about their continuing education than other physicians were, but we can't say for sure. Although such studies can provide useful insights, we must take care not to overgeneralize from them.

Purposive or Judgmental Sampling

Sometimes it's appropriate to select a sample on the basis of knowledge of a population, its elements, and the purpose of the study. This type of sampling is called **purposive** or **judgmental sampling.** In the initial design of a questionnaire, for example, you might wish to select the widest variety of respondents to test the broad applicability of questions. Although the study findings would not represent any meaningful population, the test run might effectively uncover any peculiar defects in your questionnaire. This situation would be considered a pretest, however, rather than a final study.

In some instances, you may wish to study a small subset of a larger population in which many members of the subset are easily identified, but the enumeration of them all would be nearly impossible. For example, you might want to study the leadership of a student protest movement; many of the leaders are easily visible, but it would not be feasible to define and sample all the leaders. In studying all or a sample of the most visible leaders, you may collect data sufficient for your purposes.

Or let's say you want to compare left-wing and right-wing students. Because you may not be able to enumerate and sample from all such students, you might decide to sample the memberships of left- and right-leaning groups, such as the Green Party and the Young Americans for Freedom. Although such a sample design would not provide a good description of either left-wing or right-wing students as a whole, it might suffice for general comparative purposes.

Field researchers are often particularly interested in studying deviant cases—cases that don't fit into fairly regular patterns of attitudes and behaviors—in order to improve their understanding of the more-regular pattern. For example, you might gain important insights into the nature of school spirit, as exhibited at a pep rally, by interviewing people who did not appear to be caught up in the emotions of the crowd or by interviewing students who did not attend the rally at all. Selecting deviant cases for study is another example of purposive study.

In qualitative research projects, the sampling of subjects may evolve as the structure of the situation being studied becomes clearer and certain types of subjects seem more central to understanding than others do. Let's say you're conducting an interview study among the members of a radical political group on campus. You may initially focus on friendship networks as a vehicle for the spread of group membership and participation. In the course of your analysis of the earlier interviews, you may find several references to interactions with faculty members in one of the social science departments. As a consequence, you may expand your sample to include faculty in that department and other students that they interact with. This is called "theoretical sampling," since the evolving theoretical understanding of the subject directs the sampling in certain directions.

Snowball Sampling

Another nonprobability sampling technique, which some consider to be a form of accidental sampling, is called **snowball sampling.** This procedure is appropriate when the members of a special population are difficult to locate, such as homeless individuals, migrant workers, or undocumented immigrants. In snowball sampling, the researcher collects data on the few members of the target population he or she can locate, then asks those individuals to provide the information needed to locate other members of that population whom they happen to know. "Snowball" refers to the process of accumulation as each located subject suggests other subjects. Because this procedure also results in samples with questionable representativeness, it's used primarily for exploratory purposes.

---

purposive (judgmental) sampling A type of nonprobability sampling in which the units to be observed are selected on the basis of the researcher's judgment about which ones will be the most useful or representative.

snowball sampling A nonprobability sampling method, often employed in field research, whereby each person interviewed may be asked to suggest additional people for interviewing.

Suppose you wish to learn a community organization's pattern of recruitment over time. You might begin by interviewing fairly recent recruits, asking them who introduced them to the group. You might then interview the people named, asking them who introduced *them* to the group. You might then interview those people named, asking, in part, who introduced *them*. Or, in studying a loosely structured political group, you might ask one of the participants who he or she believes to be the most influential members of the group. You might interview those people and, in the course of the interviews, ask who *they* believe to be the most influential. In each of these examples, your sample would "snowball" as each of your interviewees suggested other people to interview.

Examples of this technique in social science research abound. Karen Farquharson (2005) provides a detailed discussion of how she used snowball sampling to discover a network of tobacco policy makers in Australia: both those at the core of the network and those on the periphery. Kath Browne (2005) used snowballing through social networks to develop a sample of nonheterosexual women in a small town in the United Kingdom. She reports that her own membership in such networks greatly facilitated this type of sampling, and that potential subjects in the study were more likely to trust her than to trust heterosexual researchers.

Quota Sampling

Quota sampling is the method that helped George Gallup avoid disaster in 1936—and set up the disaster of 1948. Like probability sampling, quota sampling addresses the issue of representativeness, although the two methods approach the issue quite differently.

Quota sampling begins with a matrix, or table, describing the characteristics of the target population. Depending on your research purposes, you may need to know what proportion of the population is male and what proportion female as well as what proportions of each gender fall into various age categories, educational levels, ethnic groups, and so forth. In establishing a national quota sample, you might need to know what proportion of the national population is urban, eastern, male, under 25, white, working class, and the like, and all the possible combinations of these attributes.

Once you've created such a matrix and assigned a relative proportion to each cell in the matrix, you proceed to collect data from people having all the characteristics of a given cell. You then assign to all the people in a given cell a weight appropriate to their portion of the total population. When all the sample elements are so weighted, the overall data should provide a reasonable representation of the total population.

Although quota sampling resembles probability sampling, it has several inherent problems. First, the *quota frame* (the proportions that different cells represent) must be accurate, and it's often difficult to get up-to-date information for this purpose. The Gallup failure to predict Truman as the presidential victor in 1948 was due partly to this problem. Second, the selection of sample elements within a given cell may be biased even though its proportion of the population is accurately estimated. Instructed to interview five people who meet a given, complex set of characteristics, an interviewer may still avoid people living at the top of seven-story walk-ups, having particularly run-down homes, or owning vicious dogs.

In recent years, attempts have been made to combine probability- and quota-sampling methods, but the effectiveness of this effort remains to be seen. At present, you would be advised to treat quota sampling warily if your purpose is statistical description.

At the same time, the logic of quota sampling can sometimes be applied usefully to a field research project. In the study of a formal group, for example, you might wish to interview both leaders and nonleaders. In studying a student political organization, you might want to interview radical, moderate, and conservative members of that group. You may be able to achieve sufficient representativeness

---

quota sampling A type of nonprobability sampling in which units are selected into a sample on the basis of prespecified characteristics, so that the total sample will have the same distribution of characteristics assumed to exist in the population being studied.

in such cases by using quota sampling to ensure that you interview both men and women, both younger and older people, and so forth.

Selecting Informants

When field research involves the researcher's attempt to understand some social setting—a juvenile gang or local neighborhood, for example—much of that understanding will come from a collaboration with some members of the group being studied. Whereas social researchers speak of *respondents* as people who provide information about themselves, allowing the researcher to construct a composite picture of the group those respondents represent, an **informant** is a member of the group who can talk directly about the group per se.

Especially important to anthropologists, informants are important to other social researchers as well. If you wanted to learn about informal social networks in a local public-housing project, for example, you would do well to locate individuals who could understand what you were looking for and help you find it.

When Jeffrey Johnson (1990) set out to study a salmon-fishing community in North Carolina, he used several criteria to evaluate potential informants. Did their positions allow them to interact regularly with other members of the camp, for example, or were they isolated? (He found that the carpenter had a wider range of interactions than the boat captain did.) Was their information about the camp pretty much limited to their specific jobs, or did it cover many aspects of the operation? These and other criteria helped determine how useful the potential informants might be.

Usually, you'll want to select informants somewhat typical of the groups you're studying. Otherwise, their observations and opinions may be misleading. Interviewing only physicians will not give you a well-rounded view of how a community medical clinic is working, for example. Along the same lines, an anthropologist who interviews only men in a society where women are sheltered from outsiders will get a biased view. Similarly, although informants fluent in English are convenient for English-speaking researchers from the United States, they do not typify the members of many societies nor even many subgroups within English-speaking countries.

Simply because they're the ones willing to work with outside investigators, informants will almost always be somewhat "marginal" or atypical within their group. Sometimes this is obvious. Other times, however, you'll learn about their marginality only in the course of your research.

In Jeffrey Johnson's study, the county agent identified one fisherman who seemed squarely in the mainstream of the community. Moreover, he was cooperative and helpful to Johnson's research. The more Johnson worked with the fisherman, however, the more he found the man to be a marginal member of the fishing community.

> First, he was a Yankee in a southern town. Second, he had a pension from the Navy [so he was not seen as a "serious fisherman" by others in the community]. . . . Third, he was a major Republican activist in a mostly Democratic village. Finally, he kept his boat in an isolated anchorage, far from the community harbor.
>
> *(1990: 56)*

Informants' marginality may not only bias the view you get, but their marginal status may also limit their access (and hence yours) to the different sectors of the community you wish to study.

These comments should give you some sense of the concerns involved in nonprobability sampling, typically used in qualitative research projects. I conclude with the following injunction:

> Your overall goal is to collect the *richest possible data.* By rich data, we mean a wide and diverse range of information collected over a relatively prolonged period of time in a persistent and systematic manner. Ideally, such data enable you to grasp the meanings associated with the actions of those you are studying and to understand the contexts in which those actions are embedded.
>
> *(Lofland et al. 2006: 15)*

informant Someone who is well versed in the social phenomenon that you wish to study and who is willing to tell you what he or she knows about it. Not to be confused with a *respondent*.

In other words, nonprobability sampling does have its uses, particularly in qualitative research projects. But researchers must take care to acknowledge the limitations of nonprobability sampling, especially regarding accurate and precise representations of populations. This point will become clearer as we discuss the logic and techniques of probability sampling.

The Theory and Logic of Probability Sampling

However appropriate to some research purposes, nonprobability sampling methods cannot guarantee that the sample we observed is representative of the whole population. When researchers want precise, statistical descriptions of large populations— for example, the percentage of the population who are unemployed, plan to vote for Candidate X, or feel a rape victim should have the right to an abortion—they turn to **probability sampling.** All large-scale surveys use probability-sampling methods.

Although the application of probability sampling involves some sophisticated use of statistics, the basic logic of probability sampling is not difficult to understand. If all members of a population were identical in all respects—all demographic characteristics, attitudes, experiences, behaviors, and so on—there would be no need for careful sampling procedures. In this extreme case of perfect homogeneity, in fact, any single case would suffice as a sample to study characteristics of the whole population.

In fact, of course, the human beings who compose any real population are quite heterogeneous, varying in many ways. Figure 7-2 offers a simplified illustration of a heterogeneous popula-

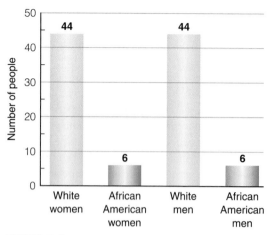

FIGURE 7-2

A Population of 100 Folks. Typically, sampling aims to reflect the characteristics and dynamics of large populations. For the purpose of some simple illustrations, let's assume our total population only has 100 members.

tion: The 100 members of this small population differ by gender and race. We'll use this hypothetical micropopulation to illustrate various aspects of probability sampling.

The fundamental idea behind probability sampling is this: To provide useful descriptions of the total population, a sample of individuals from a population must contain essentially the same variations that exist in the population. This isn't as simple as it might seem, however. Let's take a minute to look at some of the ways researchers might go astray. Then, we'll see how probability sampling provides an efficient method for selecting a sample that should adequately reflect variations that exist in the population.

Conscious and Unconscious Sampling Bias

At first glance, it may look as though sampling is pretty straightforward. To select a sample of 100 university students, you might simply interview the first 100 students you find walking around campus. This kind of sampling method is often used by untrained researchers, but it runs a high risk of introducing biases into the samples.

---

probability sampling The general term for samples selected in accord with probability theory, typically involving some random-selection mechanism. Specific types of probability sampling include EPSEM, PPS, simple random sampling, and systematic sampling.

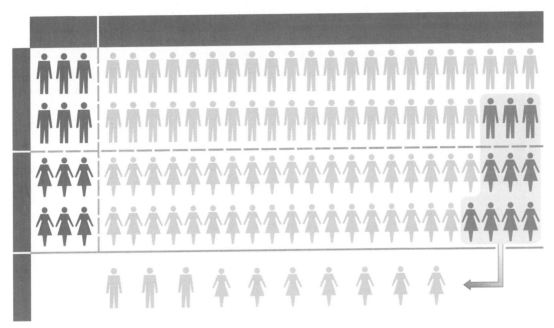

FIGURE 7-3

A Sample of Convenience: Easy, but Not Representative. Simply selecting and observing those people who are most readily at hand is the simplest method, perhaps, but it's unlikely to provide a sample that accurately reflects the total population.

In connection with sampling, *bias* simply means that those selected are not typical nor representative of the larger populations they have been chosen from. This kind of bias does not have to be intentional. In fact, it is virtually inevitable when you pick people by the seat of your pants.

Figure 7-3 illustrates what can happen when researchers simply select people who are convenient for study. Although women are only 50 percent of our micropopulation, the people closest to the researcher (in the lower right corner) happen to be 70 percent women, and although the population is 12 percent African American, none was selected into the sample.

Beyond the risks inherent in simply studying people who are convenient, other problems can arise. To begin with, the researcher's personal leanings may affect the sample to the point where it does not truly represent the student population. Suppose you're a little intimidated by students who look particularly "cool," feeling they might ridicule your research effort. You might consciously or unconsciously avoid interviewing such people. Or,

you might feel that the attitudes of "super-straight-looking" students would be irrelevant to your research purposes and so avoid interviewing them.

Even if you sought to interview a "balanced" group of students, you wouldn't know the exact proportions of different types of students making up such a balance, and you wouldn't always be able to identify the different types just by watching them walk by.

Even if you made a conscientious effort to interview, say, every tenth student entering the university library, you could not be sure of a representative sample, because different types of students visit the library with different frequencies. Your sample would overrepresent students who visit the library more often than others do.

Similarly, the "public opinion" call-in polls—in which radio stations or newspapers ask people to call specified telephone numbers to register their opinions—cannot be trusted to represent general populations. At the very least, not everyone in the population will even be aware of the poll. This problem also invalidates polls by magazines and newspapers that publish coupons for readers to complete and mail in. Even among those who are aware of such polls, not all will express an opinion, especially if doing so will cost them a stamp, an envelope, or a telephone charge. Similar considerations apply to polls taken over the Internet.

Ironically, the failure of such polls to represent all opinions equally was inadvertently acknowledged by Phillip Perinelli (1986), a staff manager of AT&T Communications' DIAL-IT 900 Service, which offers a call-in poll facility to organizations. Perinelli attempted to counter criticisms by saying, "The 50-cent charge assures that only interested parties respond and helps assure also that no individual 'stuffs' the ballot box." We cannot determine general public opinion while considering "only interested parties." This excludes those who don't care 50-cents' worth, as well as those who recognize that such polls are not valid. Both types of people may have opinions and may even vote on election day. Perinelli's assertion that the 50-cent charge will prevent ballot stuffing actually means that only those who can afford it will engage in ballot stuffing.

The possibilities for inadvertent sampling bias are endless and not always obvious. Fortunately, many techniques can help us avoid bias.

Representativeness and Probability of Selection

Although the term **representativeness** has no precise, scientific meaning, it carries a common-sense meaning that makes it useful here. For our purpose, a sample is representative of the population from which it is selected if the aggregate characteristics of the sample closely approximate those same aggregate characteristics in the population. If, for example, the population contains 50 percent women, then a sample must contain "close to" 50 percent women to be representative. Later, we'll discuss "how close" in detail.

Note that samples need not be representative in all respects; representativeness is limited to those characteristics that are relevant to the substantive interests of the study. However, you may not know in advance which characteristics are relevant.

A basic principle of probability sampling is that a sample will be representative of the population from which it is selected if all members of the population have an equal chance of being selected in the sample. (We'll see shortly that the size of the sample selected also affects the degree of representativeness.) Samples that have this quality are often labeled **EPSEM** samples (EPSEM stands for "equal probability of selection method"). Later, we'll discuss variations of this principle, which forms the basis of probability sampling.

Moving beyond this basic principle, we must realize that samples—even carefully selected EPSEM samples—seldom if ever perfectly represent the populations from which they are drawn. Nevertheless, probability sampling offers two special advantages.

First, probability samples, although never perfectly representative, are typically more representative than other types of samples, because the biases previously discussed are avoided. In practice, a probability sample is more likely than a nonprobability sample to be representative of the population from which it is drawn.

Second, and more important, probability theory permits us to estimate the accuracy or representativeness of the sample. Conceivably, an uninformed researcher might, through wholly haphazard means, select a sample that nearly perfectly represents the larger population. The odds are against doing so, however, and we would be unable to estimate the likelihood that he or she has achieved representativeness. The probability sampler, on the other hand, can provide an accurate

---

representativeness That quality of a sample of having the same distribution of characteristics as the population from which it was selected. By implication, descriptions and explanations derived from an analysis of the sample may be assumed to represent similar ones in the population. Representativeness is enhanced by probability sampling and provides for generalizability and the use of inferential statistics.

EPSEM (equal probability of selection method) A sample design in which each member of a population has the same chance of being selected into the sample.

estimate of success or failure. We'll shortly see exactly how this estimate can be achieved.

I've said that probability sampling ensures that samples are representative of the population we wish to study. As we'll see in a moment, probability sampling rests on the use of a random-selection procedure. To develop this idea, though, we need to give more-precise meaning to two important terms: *element* and *population.**

An **element** is that unit about which information is collected and that provides the basis of analysis. Typically, in survey research, elements are people or certain types of people. However, other kinds of units can constitute the elements for social research: Families, social clubs, or corporations might be the elements of a study. In a given study, elements are often the same as units of analysis, though the former are used in sample selection and the latter in data analysis.

Up to now we've used the term **population** to mean the group or collection that we're interested in generalizing about. More formally, a *population* is the theoretically specified aggregation of study elements. Whereas the vague term *Americans* might be the target for a study, the delineation of the population would include the definition of the element *Americans* (for example, citizenship, residence) and the time referent for the study (Americans as of when?). Translating the abstract "adult New Yorkers" into a workable population would require a specification of the age defining *adult* and the boundaries of New York. Specifying the term *college student* would include a consideration of full- and part-time students, degree candidates and nondegree candidates, undergraduate and graduate students, and so forth.

A **study population** is that aggregation of elements from which the sample is actually selected. As a practical matter, researchers are seldom in a position to guarantee that every element meeting the theoretical definitions laid down actually

has a chance of being selected in the sample. Even where lists of elements exist for sampling purposes, the lists are usually somewhat incomplete. Some students are always inadvertently omitted from student rosters. Some telephone subscribers request that their names and numbers be unlisted.

Often, researchers decide to limit their study populations more severely than indicated in the preceding examples. National polling firms may limit their national samples to the 48 adjacent states, omitting Alaska and Hawaii for practical reasons. A researcher wishing to sample psychology professors may limit the study population to those in psychology departments, omitting those in other departments. Whenever the population under examination is altered in such fashions, you must make the revisions clear to your readers.

Random Selection

With these definitions in hand, we can define the ultimate purpose of sampling: to select a set of elements from a population in such a way that descriptions of those elements accurately portray the total population from which the elements are selected. Probability sampling enhances the likelihood of accomplishing this aim and also provides methods for estimating the degree of probable success.

Random selection is the key to this process. In **random selection,** each element has an equal chance of selection independent of any other event in the selection process. Flipping a coin is the most frequently cited example: Provided that the coin is perfect (that is, not biased in terms of coming up

---

*I would like to acknowledge a debt to Leslie Kish and his excellent textbook *Survey Sampling*. Although I've modified some of the conventions used by Kish, his presentation is easily the most important source of this discussion.

---

element That unit of which a population is composed and which is selected in a sample. Distinguished from *units of analysis,* which are used in data analysis.

population The theoretically specified aggregation of the elements in a study.

study population That aggregation of elements from which a sample is actually selected.

random selection A sampling method in which each element has an equal chance of selection independent of any other event in the selection process.

heads or tails), the "selection" of a head or a tail is independent of previous selections of heads or tails. No matter how many heads turn up in a row, the chance that the next flip will produce "heads" is exactly 50–50. Rolling a perfect set of dice is another example.

Such images of random selection, although useful, seldom apply directly to sampling methods in social research. More typically, social researchers use tables of random numbers or computer programs that provide a random selection of sampling units. A **sampling unit** is that element or set of elements considered for selection in some stage of sampling. In Chapter 9, on survey research, we'll see how computers are used to select random telephone numbers for interviewing, a technique called *random-digit dialing.*

The reasons for using random-selection methods are twofold. First, this procedure serves as a check on conscious or unconscious bias on the part of the researcher. The researcher who selects cases on an intuitive basis might very well select cases that would support his or her research expectations or hypotheses. Random selection erases this danger. More importantly, random selection offers access to the body of probability theory, which provides the basis for estimating the characteristics of the population as well as estimating the accuracy of samples. Let's now examine probability theory in greater detail.

Probability Theory, Sampling Distributions, and Estimates of Sampling Error

Probability theory is a branch of mathematics that provides the tools researchers need to devise sampling techniques that produce representative samples and to analyze the results of their sampling

statistically. More formally, probability theory provides the basis for estimating the parameters of a population. A **parameter** is the summary description of a given variable in a population. The mean income of all families in a city is a parameter; so is the age distribution of the city's population. When researchers generalize from a sample, they're using sample observations to estimate population parameters. Probability theory enables them to both make these estimates and arrive at a judgment of how likely the estimates will accurately represent the actual parameters in the population. For example, probability theory allows pollsters to infer from a sample of 2,000 voters how a population of 100 million voters is likely to vote—and to specify exactly what the probable margin of error of the estimates is.

Probability theory accomplishes these seemingly magical feats by way of the concept of sampling distributions. A single sample selected from a population will give an estimate of the population parameter. Other samples would give the same or slightly different estimates. Probability theory tells us about the distribution of estimates that would be produced by a large number of such samples. To see how this works, we'll look at two examples of sampling distributions, beginning with a simple example in which our population consists of just ten cases, then moving on to a case of percentages that allows a clear illustration of probable margin of error.

The Sampling Distribution of Ten Cases

Suppose there are ten people in a group, and each has a certain amount of money in his or her pocket. To simplify, let's assume that one person has no money, another has one dollar, another has two dollars, and so forth up to the person with nine dollars. Figure 7-4 presents the population of ten people.*

Our task is to determine the average amount of money one person has: specifically, the mean number of dollars. If you simply add up the money

---

sampling unit That element or set of elements considered for selection in some stage of sampling.

parameter The summary description of a given variable in a population.

---

*I want to thank Hanan Selvin for suggesting this method of introducing probability sampling.

FIGURE 7-4

A Population of 10 People with $0–$9. Let's simplify matters even more now by imagining a population of only 10 people with differing amounts of money in their pockets—ranging from $0 to $9.

shown in Figure 7-4, you'll find that the total is $45, so the mean is $4.50. Our purpose in the rest of this exercise is to estimate that mean without actually observing all ten individuals. We'll do that by selecting random samples from the population and using the means of those samples to estimate the mean of the whole population.

To start, suppose we were to select—at random—a sample of only one person from the ten. Our ten possible samples thus consist of the ten cases shown in Figure 7-4.

The ten dots shown on the graph in Figure 7-5 represent these ten samples. Because we're taking samples of only one, they also represent the "means" we would get as estimates of the population. The distribution of the dots on the graph is called the *sampling distribution*. Obviously, it wouldn't be a very good idea to select a sample of only one, because the chances are great that we'll miss the true mean of $4.50 by quite a bit.

Now suppose we take a sample of two. As shown in Figure 7-6, increasing the sample size improves our estimations. There are now 45 pos-

sible samples: [$0 $1], [$0 $2], . . . [$7 $8], [$8 $9]. Moreover, some of those samples produce the same means. For example, [$0 $6], [$1 $5], and [$2 $4] all produce means of $3. In Figure 7-6, the three dots shown above the $3 mean represent those three samples.

Moreover, the 45 samples are not evenly distributed, as they were when the sample size was only one. Rather, they're somewhat clustered around the true value of $4.50. Only two possible samples deviate by as much as $4 from the true value ([$0 $1] and [$8 $9]), whereas five of the samples would give the true estimate of $4.50; another eight samples miss the mark by only 50 cents (plus or minus).

Now suppose we select even larger samples. What do you think that will do to our estimates of the mean? Figure 7-7 presents the sampling distributions of samples of 3, 4, 5, and 6.

The progression of sampling distributions is clear. Every increase in sample size improves the distribution of estimates of the mean. The limiting case in this procedure, of course, is to select a

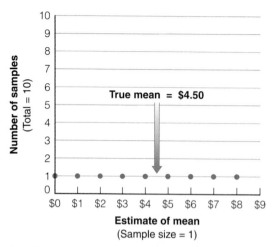

FIGURE 7-5

The Sampling Distribution of Samples of 1. In this simple example, the mean amount of money these people have is $4.50 ($45/10). If we picked 10 different samples of 1 person each, our "estimates" of the mean would range all across the board.

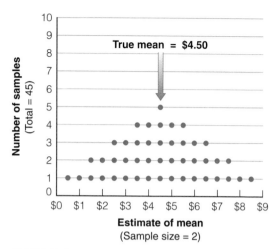

FIGURE 7-6

The Sampling Distribution of Samples of 2. By merely increasing our sample size to 2, we get possible samples that provide somewhat better estimates of the mean. We couldn't get either $0 or $9, and the estimates are beginning to cluster around the true value of the mean: $4.50.

sample of ten. There would be only one possible sample (everyone) and it would give us the true mean of $4.50. As we'll see shortly, this principle applies to actual sampling of meaningful populations. The larger the sample selected, the more accurate it is as an estimation of the population from which it was drawn.

Sampling Distribution and Estimates of Sampling Error

Let's turn now to a more realistic sampling situation involving a much larger population and see how the notion of sampling distribution applies. Assume that we wish to study the student population of State University (SU) to determine the percentage of students who approve or disapprove of a student-conduct code proposed by the administration. The study population will be the aggregation of, say, 20,000 students contained in a student roster: the sampling frame. The elements will be the individual students at SU. We'll select a random sample of, say, 100 students for the

purposes of estimating the entire student body. The variable under consideration will be *attitudes toward the code*, a binomial variable: *approve* and *disapprove*. (The logic of probability sampling applies to the examination of other types of variables, such as mean income, but the computations are somewhat more complicated. Consequently, this introduction focuses on binomials.)

The horizontal axis of Figure 7-8 presents all possible values of this parameter in the population—from 0 percent to 100 percent approval. The midpoint of the axis—50 percent—represents half the students approving of the code and the other half disapproving.

To choose our sample, we give each student on the student roster a number and select 100 random numbers from a table of random numbers. Then we interview the 100 students whose numbers have been selected and ask for their attitudes toward the student code: whether they approve or disapprove. Suppose this operation gives us 48 students who approve of the code and 52 who disapprove. This summary description of a variable in a sample is called a **statistic.** We present this statistic by placing a dot on the *x* axis at the point representing 48 percent.

statistic The summary description of a variable in a sample, used to estimate a population parameter.

a. Samples of 3

b. Samples of 4

c. Samples of 5

d. Samples of 6

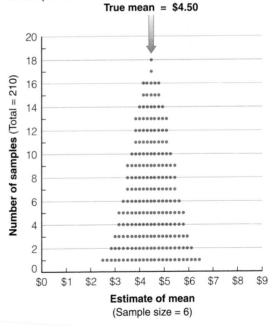

FIGURE 7-7

The Sampling Distributions of Samples of 3, 4, 5, and 6. As we increase the sample size, the possible samples cluster ever more tightly around the true value of the mean. The chance of extremely inaccurate estimates is reduced at the two ends of the distribution, and the percentage of the samples near the true value keeps increasing.

Percent of students approving of the student code

FIGURE 7-8

Range of Possible Sample Study Results. Shifting to a more realistic example, let's assume that we want to sample student attitudes concerning a proposed conduct code. Let's assume that 50 percent of the whole student body approves and 50 percent disapproves—though the researcher doesn't know that.

Now let's suppose we select another sample of 100 students in exactly the same fashion and measure their approval or disapproval of the student code. Perhaps 51 students in the second sample approve of the code. We place another dot in the appropriate place on the x axis. Repeating this process once more, we may discover that 52 students in the third sample approve of the code.

Figure 7-9 presents the three different sample statistics representing the percentages of students in each of the three random samples who approved of the student code. The basic rule of random sampling is that such samples drawn from a population give estimates of the parameter that exists in the total population. Each of the random samples, then, gives us an estimate of the percentage of students in the total student body who approve of the student code. Unhappily, however, we have selected three samples and now have three separate estimates.

To retrieve ourselves from this problem, let's draw more and more samples of 100 students each, question each of the samples concerning their approval or disapproval of the code, and plot the new sample statistics on our summary graph. In drawing many such samples, we discover that some of the new samples provide duplicate estimates, as in the illustration of ten cases. Figure 7-10 shows

---

sampling error The degree of error to be expected by virtue of studying a sample instead of everyone. For probability sampling, the maximum error depends on three factors: the sample size, the diversity of the population, and the confidence level.

Percent of students approving of the student code

FIGURE 7-9

Results Produced by Three Hypothetical Studies. Assuming a large student body, let's suppose that we selected three different samples, each of substantial size. We would not necessarily expect those samples to perfectly reflect attitudes in the whole student body, but they should come reasonably close.

the sampling distribution of, say, hundreds of samples. This is often referred to as a *normal curve.*

Note that by increasing the number of samples selected and interviewed, we've also increased the range of estimates provided by the sampling operation. In one sense we've increased our dilemma in attempting to guess the parameter in the population. Probability theory, however, provides certain important rules regarding the sampling distribution presented in Figure 7-10.

First, if many independent random samples are selected from a population, the sample statistics provided by those samples will be distributed around the population parameter in a known way. Thus, although Figure 7-10 shows a wide range of estimates, more of them are in the vicinity of 50 percent than elsewhere in the graph. Probability theory tells us, then, that the true value is in the vicinity of 50 percent.

Second, probability theory gives us a formula for estimating how closely the sample statistics are clustered around the true value. To put it another way, probability theory enables us to estimate the **sampling error**—the degree of error to be expected for a given sample design. This formula contains three factors: the parameter, the sample size, and the standard error (a measure of sampling error):

$$s = \sqrt{\frac{P \times Q}{n}}$$

FIGURE 7-10

The Sampling Distribution. If we were to select a large number of good samples, we would expect them to cluster around the true value (50 percent), but given enough such samples, a few would fall far from the mark.

The symbols P and Q in the formula equal the population parameters for the binomial: If 60 percent of the student body approve of the code and 40 percent disapprove, P and Q are 60 percent and 40 percent, respectively, or 0.6 and 0.4. Note that $Q = 1 - P$ and $P = 1 - Q$. The symbol n equals the number of cases in each sample, and s is the standard error.

Let's assume that the population parameter in the student example is 50 percent approving of the code and 50 percent disapproving. Recall that we've been selecting samples of 100 cases each. When these numbers are put into the formula, we find that the standard error equals 0.05, or 5 percent.

In probability theory, the standard error is a valuable piece of information because it indicates the extent to which the sample estimates will be distributed around the population parameter. (If you're familiar with the standard deviation in statistics, you may recognize that the standard error, in this case, is the standard deviation of the sampling distribution.) Specifically, probability theory indicates that certain proportions of the sample estimates will fall within specified increments—each equal to one standard error—from the population parameter. Approximately 34 percent (0.3413) of the sample estimates will fall within one standard error increment above the population parameter, and another 34 percent will fall within one stan-

dard error below the parameter. In our example, the standard error increment is 5 percent, so we know that 34 percent of our samples will give estimates of student approval between 50 percent (the parameter) and 55 percent (one standard error above); another 34 percent of the samples will give estimates between 50 percent and 45 percent (one standard error below the parameter). Taken together, then, we know that roughly two-thirds (68 percent) of the samples will give estimates within 5 percent of the parameter.

Moreover, probability theory dictates that roughly 95 percent of the samples will fall within plus or minus two standard errors of the true value, and 99.9 percent of the samples will fall within plus or minus three standard errors. In our present example, then, we know that only one sample out of a thousand would give an estimate lower than 35 percent approval or higher than 65 percent.

The proportion of samples falling within one, two, or three standard errors of the parameter is constant for any random sampling procedure such as the one just described, providing that a large number of samples are selected. The size of the standard error in any given case, however, is a function of the population parameter and the sample size. If we return to the formula for a moment, we note that the standard error will increase as a function of an increase in the quantity P times Q. Note further that this quantity reaches its maximum in the situation of an even split in the population. If $P = 0.5$, $PQ = 0.25$; if $P = 0.6$, $PQ = 0.24$; if $P = 0.8$, $PQ = 0.16$; if $P = 0.99$, $PQ = 0.0099$. By extension, if P is either 0.0 or 1.0 (either 0 percent or 100 percent approve of the student code), the standard error will be 0. If everyone in the population has the same attitude (no variation), then every sample will give exactly that estimate.

The standard error is also a function of the sample size—an inverse function. As the sample size increases, the standard error decreases. As the sample size increases, the several samples will be clustered nearer to the true value. Another general guideline is evident in the formula: Because of the square root formula, the standard error is reduced by half if the sample size is quadrupled. In our present example, samples of 100 produce a standard error of 5 percent; to reduce the standard error to 2.5 percent, we must increase the sample size to 400.

All of this information is provided by established probability theory in reference to the selection of large numbers of random samples. (If you've taken a statistics course, you may know this as the Central Tendency Theorem.) If the population parameter is known and many random samples are selected, we can predict how many of the sample estimates will fall within specified intervals from the parameter.

Recognize that this discussion illustrates only the logic of probability sampling; it does not describe the way research is actually conducted. Usually, we don't know the parameter: The very reason we conduct a sample survey is to estimate that value. Moreover, we don't actually select large numbers of samples: We select only one sample. Nevertheless, the preceding discussion of probabil-

ity theory provides the basis for inferences about the typical social research situation. Knowing what it would be like to select thousands of samples allows us to make assumptions about the one sample we do select and study.

Confidence Levels and Confidence Intervals

Whereas probability theory specifies that 68 percent of that fictitious large number of samples would produce estimates falling within one standard error of the parameter, we can turn the logic around and infer that any single random sample estimate has a 68 percent chance of falling within that range. This observation leads us to the two key components of sampling error estimates: **confidence level** and **confidence interval.** We express the accuracy of our sample statistics in terms of a level of confidence that the statistics fall within a specified interval from the parameter. For example, we may say we are 95 percent confident that our sample statistics (for example, 50 percent favor the new student code) are within plus or minus 5 percentage points of the population parameter. As the confidence interval is expanded for a given statistic, our confidence increases. For example, we may say that we are 99.9 percent confident that our statistic falls within three standard errors of the true value. (Now perhaps you can appreciate the humorous quip of unknown origin: Statistics means never having to say you are certain.)

Although we may be confident (at some level) of being within a certain range of the parameter, we've already noted that we seldom know what the parameter is. To resolve this problem, we substitute our sample estimate for the parameter in the formula; that is, lacking the true value, we substitute the best available guess.

The result of these inferences and estimations is that we can estimate a population parameter and also the expected degree of error on the basis of one sample drawn from a population. Beginning with the question "What percentage of the student body approves of the student code?" you could select a random sample of 100 students and

---

confidence level The estimated probability that a population parameter lies within a given confidence interval. Thus, we might be 95 percent confident that between 35 and 45 percent of all voters favor Candidate A.

confidence interval The range of values within which a population parameter is estimated to lie.

interview them. You might then report that your best estimate is that 50 percent of the student body approves of the code and that you are 95 percent confident that between 40 and 60 percent (plus or minus two standard errors) approve. The range from 40 to 60 percent is the confidence interval. (At the 68 percent confidence level, the confidence interval would be 45–55 percent.)

The logic of confidence levels and confidence intervals also provides the basis for determining the appropriate sample size for a study. Once you've decided on the degree of sampling error you can tolerate, you'll be able to calculate the number of cases needed in your sample. Thus, for example, if you want to be 95 percent confident that your study findings are accurate within plus or minus 5 percentage points of the population parameters, you should select a sample of at least 400. (Appendix F is a convenient guide in this regard.)

This, then, is the basic logic of probability sampling. Random selection permits the researcher to link findings from a sample to the body of probability theory so as to estimate the accuracy of those findings. All statements of accuracy in sampling must specify both a confidence level and a confidence interval. The researcher must report that he or she is x percent confident that the population parameter is between two specific values. In this example, I've demonstrated the logic of sampling error using a variable analyzed in percentages. A different statistical procedure would be required to calculate the standard error for a mean, for example, but the overall logic is the same.

Notice that nowhere in this discussion of sample size and accuracy of estimates did we consider the size of the population being studied. This is because the population size is almost always irrelevant. A sample of 2,000 respondents drawn properly to represent Vermont voters will be no more accurate than a sample of 2,000 drawn properly to represent all voters in the United States, even though the Vermont sample would be a substantially larger proportion of that small state's voters than would the same number chosen to represent the nation's voters. The reason for this counter-

intuitive fact is that the equations for calculating sampling error all assume that the populations being sampled are infinitely large, so every sample would equal 0 percent of the whole.

Of course, this is not literally true in practice. However, a sample of 2,000 represents only 0.68 percent of the Vermonters who voted for president in the 2000 election, and a sample of 2,000 U.S. voters represents a mere 0.002 percent of the national electorate. Both of these proportions are sufficiently small as to approach the situation with infinitely large populations.

Unless a sample represents, say, 5 percent or more of the population it's drawn from, that proportion is irrelevant. In those rare cases of large proportions being selected, a "finite population correction" can be calculated to adjust the confidence intervals. The following formula calculates the proportion to be multiplied against the calculated error.

$$\text{finite population correction} = \sqrt{\frac{N - n}{N - 1}}$$

In the formula, N is the population size and n is the size of the sample. Notice that in the extreme case where you studied the whole population (hence $N = n$), the formula would yield zero as the finite population correction. Multiplying zero times the sampling error calculated by the earlier formula would give a final sampling error of zero, which would, of course, be precisely the case since you wouldn't have sampled at all.

Two cautions are in order before we conclude this discussion of the basic logic of probability sampling. First, the survey uses of probability theory as discussed here are technically not wholly justified. The theory of sampling distribution makes assumptions that almost never apply in survey conditions. The exact proportion of samples contained within specified increments of standard errors, for example, mathematically assumes an infinitely large population, an infinite number of samples, and sampling with replacement—that is, every sampling unit selected is "thrown back into the pot" and could be selected again. Second, our discussion has greatly oversimplified the inferential jump from the distribution of several samples to the probable characteristics of one sample.

I offer these cautions to provide perspective on the uses of probability theory in sampling. Social researchers often appear to overestimate the precision of estimates produced by the use of probability theory. As I'll mention elsewhere in this chapter and throughout the book, variations in sampling techniques and nonsampling factors may further reduce the legitimacy of such estimates. For example, those selected in a sample who fail or refuse to participate detract further from the representativeness of the sample.

Nevertheless, the calculations discussed in this section can be extremely valuable to you in understanding and evaluating your data. Although the calculations do not provide as precise estimates as some researchers might assume, they can be quite valid for practical purposes. They are unquestionably more valid than less rigorously derived estimates based on less-rigorous sampling methods. Most important, being familiar with the basic logic underlying the calculations can help you react sensibly both to your own data and to those reported by others.

Populations and Sampling Frames

The preceding section introduced the theoretical model for social research sampling. Although as students, research consumers, and researchers we need to understand that theory, it's no less important to appreciate the less-than-perfect conditions that exist in the field. In this section we'll look

---

sampling frame That list or quasi list of units composing a population from which a sample is selected. If the sample is to be representative of the population, it is essential that the sampling frame include all (or nearly all) members of the population.

---

at one aspect of field conditions that requires a compromise with idealized theoretical conditions and assumptions: the congruence of or disparity between populations of sampling frames.

Simply put, a **sampling frame** is the list or quasi list of elements from which a probability sample is selected. If a sample of students is selected from a student roster, the roster is the sampling frame. If the primary sampling unit for a complex population sample is the census block, the list of census blocks composes the sampling frame—in the form of a printed booklet or, better, some digital format permitting computer manipulation. Here are some reports of sampling frames appearing in research journals. In each example I've italicized the actual sampling frames.

> The data for this research were obtained from a random sample of *parents of children in the third grade in public and parochial schools in Yakima County, Washington.*
>
> *(Petersen and Maynard 1981: 92)*

> The sample at Time 1 consisted of 160 names drawn randomly from the *telephone directory of Lubbock, Texas.*
>
> *(Tan 1980: 242)*

> The data reported in this paper . . . were gathered from a probability sample of *adults aged 18 and over residing in households in the 48 contiguous United States.* Personal interviews with 1,914 respondents were conducted by the Survey Research Center of the University of Michigan during the fall of 1975.
>
> *(Jackman and Senter 1980: 345)*

Properly drawn samples provide information appropriate for describing the population of elements composing the sampling frame—nothing more. I emphasize this point in view of the all-too-common tendency for researchers to select samples from a given sampling frame and then make assertions about a population similar to, but not identical to, the population defined by the sampling frame.

For example, take a look at this report, which discusses the drugs most frequently prescribed by U.S. physicians:

Information on prescription drug sales is not easy to obtain. But Rinaldo V. DeNuzzo, a professor of pharmacy at the Albany College of Pharmacy, Union University, Albany, NY, has been tracking prescription drug sales for 25 years by polling nearby drugstores. He publishes the results in an industry trade magazine, *MM&M*.

DeNuzzo's latest survey, covering 1980, is based on reports from 66 pharmacies in 48 communities in New York and New Jersey. Unless there is something peculiar about that part of the country, his findings can be taken as representative of what happens across the country.

(Moskowitz 1981: 33)

What is striking in the excerpt is the casual comment about whether there is anything peculiar about New York and New Jersey. There is. The lifestyle in these two states hardly typifies the other 48. We cannot assume that residents in these large, urbanized, eastern seaboard states necessarily have the same drug-use patterns that residents of Mississippi, Nebraska, or Vermont do.

Does the survey even represent prescription patterns in New York and New Jersey? To determine that, we would have to know something about the way the 48 communities and the 66 pharmacies were selected. We should be wary in this regard, in view of the reference to "polling nearby drugstores." As we'll see, there are several methods for selecting samples that ensure representativeness, and unless they're used, we shouldn't generalize from the study findings.

A sampling frame, then, must be consonant with the population we wish to study. In the simplest sample design, the sampling frame is a list of the elements composing the study population. In practice, though, existing sampling frames often define the study population rather than the other way around. That is, we often begin with a population in mind for our study; then we search for possible sampling frames. Having examined and evaluated the frames available for our use, we decide which frame presents a study population most appropriate to our needs.

Studies of organizations are often the simplest from a sampling standpoint because organizations typically have membership lists. In such cases, the list of members constitutes an excellent sampling frame. If a random sample is selected from a membership list, the data collected from that sample may be taken as representative of all members—if all members are included in the list.

Populations that can be sampled from good organizational lists include elementary school, high school, and university students and faculty; church members; factory workers; fraternity or sorority members; members of social, service, or political clubs; and members of professional associations.

The preceding comments apply primarily to local organizations. Often, statewide or national organizations do not have a single membership list. There is, for example, no single list of Episcopalian church members. However, a slightly more complex sample design could take advantage of local church membership lists by first sampling churches and then subsampling the membership lists of those churches selected. (More about that later.)

Other lists of individuals may be especially relevant to the research needs of a particular study. Government agencies maintain lists of registered voters, for example, and some political pollsters use registration-based sampling (RBS), using those lists. In some cases, there may be delays in keeping such files up-to-date, and a person who is registered to vote may not actually do so in the election of interest.

Other lists that may be available contain the names of automobile owners, welfare recipients, taxpayers, business permit holders, licensed professionals, and so forth. Although it may be difficult to gain access to some of these lists, they provide excellent sampling frames for specialized research purposes.

Of course, the sampling elements in a study need not be individuals. Social researchers might use lists of universities, businesses, cities, academic journals, newspapers, unions, political clubs, professional associations, and so forth.

Telephone directories are frequently used for "quick-and-dirty" public opinion polls. They're easy and inexpensive to use—no doubt the reason for their popularity. And, if you want to make assertions about telephone subscribers, the directory is a fairly good sampling frame. (Realize, of course, that a given directory will not include new subscribers or those who have requested unlisted numbers. Sampling is further complicated by the directories' inclusion of nonresidential listings.) Unfortunately, telephone directories are all too often used as a listing of a city's population or of its voters. Of the many defects in this reasoning, the chief one involves a bias, as we have seen. Poor people are less likely to have telephones; rich people may have more than one line. A telephone directory sample, therefore, is likely to have a middle- or upper-class bias.

The class bias inherent in telephone directory samples is often hidden. Preelection polls conducted in this fashion are sometimes quite accurate, perhaps because of the class bias evident in voting itself: Poor people are less likely to vote. Frequently, then, these two biases nearly coincide, so that the results of a telephone poll may come very close to the final election outcome. Unhappily, you never know for sure until after the election. And sometimes, as in the case of the 1936 *Literary Digest* poll, you may discover that the voters have not acted according to the expected class biases. The ultimate disadvantage of this method, then, is the researcher's inability to estimate the degree of error to be expected in the sample findings.

In Chapter 9 we'll return to the matter of sampling telephones, in connection with survey research. We'll examine random-digit dialing, which was developed to resolve some of the problems just discussed, and we'll see that the growth in popularity of cell phones has further complicated matters.

Street directories and tax maps are sometimes used for easy samples of households, but they may present incompleteness and bias. For example, in strictly zoned urban regions, illegal housing units are unlikely to appear on official records. As a result, such units could not be selected, and sample findings could not be representative of those units, which are often poorer and more crowded than the average.

The preceding comments apply to the United States but not to all countries. In Japan, for example, the government maintains quite accurate population registration lists. Moreover, citizens are required by law to keep their information up-to-date, such as changes in residence or births and deaths in the household. As a consequence, you can select simple random samples of the population more easily in Japan than in the United States. Such a registration list in the United States would conflict directly with this country's norms regarding individual privacy.

In recent years, American researchers have begun experimenting with address files maintained by the U.S. Postal Service, such as the Special Delivery Sequence File. As problems have increasingly arisen with regard to the sampling of telephone numbers (discussed further in Chapter 9), address-based sampling (ABS) for use in mail surveys has been improving (Link et al. 2008).

Review of Populations and Sampling Frames

Because social research literature gives surprisingly little attention to the issues of populations and sampling frames, I've devoted special attention to them. Here is a summary of the main guidelines to remember:

1. Findings based on a sample can be taken as representing only the aggregation of elements that compose the sampling frame.

2. Often, sampling frames do not truly include all the elements their names might imply. Omissions are almost inevitable. Thus, a first concern of the researcher must be to assess the extent of the omissions and to correct them if possible. (Of course, the researcher may feel that he or she can safely ignore a small number of omissions that cannot easily be corrected.)

3. To be generalized even to the population composing the sampling frame, all elements must have equal representation in the frame. Typically, each element should appear only once. Elements that appear more than once will have a greater probability of selection, and the sample will, overall, overrepresent those elements.

Other, more practical matters relating to populations and sampling frames will be treated elsewhere in this book. For example, the form of the sampling frame—such as a list in a publication, a 3-by-5 card file, CD-ROM, or magnetic tape—can affect how easy it is to use. And ease of use may often take priority over scientific considerations: An "easier" list may be chosen over a "harder" one, even though the latter is more appropriate to the target population. We should not take a dogmatic position in this regard, but every researcher should carefully weigh the relative advantages and disadvantages of such alternatives.

Types of Sampling Designs

Up to this point, we've focused on simple random sampling. Indeed, the body of statistics typically used by social researchers assumes such a sample. As you'll see shortly, however, you have several options in choosing your sampling method, and you'll seldom if ever choose simple random sampling. There are two reasons for this. First, with all but the simplest sampling frame, simple random sampling is not feasible. Second, and probably surprisingly, simple random sampling may not be the most accurate method available. Let's turn now to a discussion of simple random sampling and the other options available.

Simple Random Sampling

As noted, **simple random sampling (SRS)** is the basic sampling method assumed in the statistical computations of social research. Because the mathematics of random sampling are especially complex, we'll detour around them in favor of describing the ways of employing this method in the field.

Once a sampling frame has been properly established, to use simple random sampling the researcher assigns a single number to each element in the list, not skipping any number in the process. A table of random numbers (Appendix C) is then used to select elements for the sample. See "How to Do It: Using a Table of Random Numbers" for more.

If your sampling frame is in a machine-readable form, such as CD-ROM or magnetic tape, a computer can automatically select a simple random sample. (In effect, the computer program numbers the elements in the sampling frame, generates its own series of random numbers, and prints out the list of elements selected.)

Figure 7-11 offers a graphic illustration of simple random sampling. Note that the members of our hypothetical micropopulation have been numbered from 1 to 100. Moving to Appendix C, we decide to use the last two digits of the first column and to begin with the third number from the top. This yields person number 30 as the first one selected into the sample. Number 67 is next, and so forth. (Person 100 would have been selected if "00" had come up in the list.)

Systematic Sampling

Simple random sampling is seldom used in practice. As you'll see, it's not usually the most efficient method, and it can be laborious if done manually. Typically, simple random sampling requires a list of elements. When such a list is available, researchers usually employ systematic sampling instead.

In **systematic sampling,** every kth element in the total list is chosen (systematically) for inclusion in the sample. If the list contained 10,000 elements and you wanted a sample of 1,000, you would select every tenth element for your sample. To en-

---

simple random sampling (SRS) A type of probability sampling in which the units composing a population are assigned numbers. A set of random numbers is then generated, and the units having those numbers are included in the sample.

systematic sampling A type of probability sampling in which every kth unit in a list is selected for inclusion in the sample—for example, every 25th student in the college directory of students. You compute k by dividing the size of the population by the desired sample size; k is called the sampling interval. Within certain constraints, systematic sampling is a functional equivalent of simple random sampling and usually easier to do. Typically, the first unit is selected at random.

How to Do It: Using a Table of Random Numbers

In social research, it's often appropriate to select a set of random numbers from a table such as the one in Appendix C. Here's how to do that.

Suppose you want to select a simple random sample of 100 people (or other units) out of a population totaling 980.

1. To begin, number the members of the population: in this case, from 1 to 980. Now the task is to select 100 random numbers. Once you've done that, your sample will consist of the people having the numbers you've selected. (*Note:* It's not essential to actually number them, as long as you're sure of the total. If you have them in a list, for example, you can always count through the list after you've selected the numbers.)

2. The next step is to determine the number of digits you'll need in the random numbers you select. In our example, there are 980 members of the population, so you'll need three-digit numbers to give everyone a chance of selection. (If there were 11,825 members of the population, you'd need to select five-digit numbers.) Thus, we want to select 100 random numbers in the range from 001 to 980.

3. Now turn to the first page of Appendix C. Notice there are several rows and columns of five-digit numbers, with the columns continuing from the first page to the second. The table represents a series of random numbers in the range from 00001 to 99999. To use the table for your hypothetical sample, you have to answer these questions:

   a. How will you create three-digit numbers out of five-digit numbers?

   b. What pattern will you follow in moving through the table to select your numbers?

   c. Where will you start?

Each of these questions has several satisfactory answers. The key is to create a plan and follow it. Here's an example.

4. To create three-digit numbers from five-digit numbers, let's agree to select five-digit numbers from the table but consider only the left-most three digits in each case. If we picked the first number on the first page—10480—we'd consider only the 104. (We could

agree to take the digits farthest to the right, 480, or the middle three digits, 048, and any of these plans would work.) They key is to make a plan and stick with it. For convenience, let's use the left-most three digits.

5. We can also choose to progress through the tables any way we want: down the columns, up them, across to the right or to the left, or diagonally. Again, any of these plans will work just fine as long as we stick to it. For convenience, let's agree to move down the columns. When we get to the bottom of one column, we'll go to the top of the next.

6. Now, where do we start? You can close your eyes and stick a pencil into the table and start wherever the pencil point lands. (I know it doesn't sound scientific, but it works.) Or, if you're afraid you'll hurt the book or miss it altogether, close your eyes and make up a column number and a row number. ("I'll pick the number in the fifth row of column 2.") Start with that number.

7. Let's suppose we decide to start with the fifth number in column 2. If you look on the first page of Appendix C, you'll see that the starting number is 39975. We've selected 399 as our first random number, and we have 99 more to go. Moving down the second column, we select 069, 729, 919, 143, 368, 695, 409, 939, and so forth. At the bottom of column 2 (on the second page of this table), we select number 017 and continue to the top of column 3: 015, 255, and so on.

8. See how easy it is? But trouble lies ahead. When we reach column 5, we're speeding along, selecting 816, 309, 763, 078, 061, 277, 988 …Wait a minute! There are only 980 students in the senior class. How can we pick number 988? The solution is simple: Ignore it. Any time you come across a number that lies outside your range, skip it and continue on your way: 188, 174, and so forth. The same solution applies if the same number comes up more than once. If you select 399 again, for example, just ignore it the second time.

9. That's it. You keep up the procedure until you've selected 100 random numbers. Returning to your list, your sample consists of person number 399, person number 69, person number 729, and so forth.

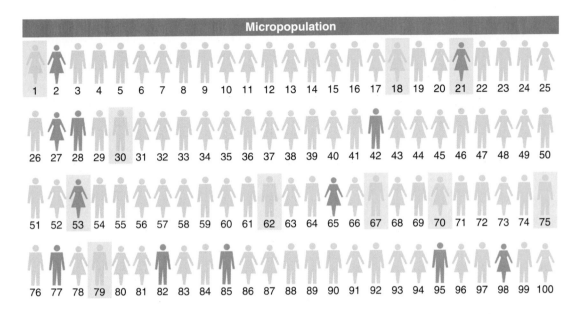

Micropopulation

Appendix C: Table of Random Numbers

10480	15011	01536
22368	46573	25595
24130	48360	22527
42167	93093	06243
37570	39975	81837
77921	06907	11008
99562	72905	56420
96301	91977	05463
89579	14342	63661
85475	36857	53342
28918	69578	88231
63553	40961	48235
09429	93969	52636

The Sample

30 67 70 21 62

1 79 75 18 53

FIGURE 7-11

A Simple Random Sample. Having numbered everyone in the population, we can use a table of random numbers to select a representative sample from the overall population. Anyone whose number is chosen from the table is in the sample.

sure against any possible human bias in using this method, you should select the first element at random. Thus, in the preceding example, you would begin by selecting a random number between one and ten. The element having that number is included in the sample, plus every tenth element following it. This method is technically referred to as a *systematic sample with a random start*. Two terms are frequently used in connection with systematic

sampling. The **sampling interval** is the standard distance between elements selected in the sample: ten in the preceding sample. The **sampling ratio**

sampling interval The standard distance between elements selected from a population for a sample.

sampling ratio The proportion of elements in the population that are selected to be in a sample.

is the proportion of elements in the population that are selected: 1/10 in the example.

$$\text{sampling interval} = \frac{\text{population size}}{\text{sample size}}$$

$$\text{sampling ratio} = \frac{\text{sample size}}{\text{population size}}$$

In practice, systematic sampling is virtually identical to simple random sampling. If the list of elements is indeed randomized before sampling, one might argue that a systematic sample drawn from that list is in fact a simple random sample. By now, debates over the relative merits of simple random sampling and systematic sampling have been resolved largely in favor of the latter, simpler method. Empirically, the results are virtually identical. And, as you'll see in a later section, systematic sampling, in some instances, is slightly more accurate than simple random sampling.

There is one danger involved in systematic sampling. The arrangement of elements in the list can make systematic sampling unwise, especially an arrangement usually called *periodicity*. If the list of elements is arranged in a cyclical pattern that coincides with the sampling interval, a grossly biased sample might be drawn. Here are two examples that illustrate this danger.

In a classic study of soldiers during World War II, the researchers selected a systematic sample from unit rosters. Every tenth soldier on the roster was selected for the study. The rosters, however, were arranged in a table of organizations: sergeants first, then corporals and privates, squad by squad. Each squad had ten members. As a result, every tenth person on the roster was a squad sergeant. The systematic sample selected contained only sergeants. It could, of course, have been the case that no sergeants were selected for the same reason.

stratification The grouping of the units composing a population into homogeneous groups (or strata) before sampling. This procedure, which may be used in conjunction with simple random, systematic, or cluster sampling, improves the representativeness of a sample, at least in terms of the stratification variables.

As another example, suppose we select a sample of apartments in an apartment building. If the sample is drawn from a list of apartments arranged in numerical order (for example, 101, 102, 103, 104, 201, 202, and so on), there is a danger of the sampling interval coinciding with the number of apartments on a floor or some multiple thereof. Then the samples might include only northwest-corner apartments or only apartments near the elevator. If these types of apartments have some other particular characteristic in common (for example, higher rent), the sample will be biased. The same danger would appear in a systematic sample of houses in a subdivision arranged with the same number of houses on a block.

In considering a systematic sample from a list, then, you should carefully examine the nature of that list. If the elements are arranged in any particular order, you should figure out whether that order will bias the sample to be selected, then you should take steps to counteract any possible bias (for example, take a simple random sample from cyclical portions).

Usually, however, systematic sampling is superior to simple random sampling, in convenience if nothing else. Problems in the ordering of elements in the sampling frame can usually be remedied quite easily.

Stratified Sampling

So far we've discussed two methods of sample selection from a list: random and systematic. **Stratification** is not an alternative to these methods; rather, it represents a possible modification of their use.

Simple random sampling and systematic sampling both ensure a degree of representativeness and permit an estimate of the error present. Stratified sampling is a method for obtaining a greater degree of representativeness by decreasing the probable sampling error. To understand this method, we must return briefly to the basic theory of sampling distribution.

Recall that sampling error is reduced by two factors in the sample design. First, a large sample produces a smaller sampling error than a small sample does. Second, a homogeneous population produces samples with smaller sampling errors than a heterogeneous population does. If 99 percent of the population agrees with a certain statement, it's extremely unlikely that any probability sample will greatly misrepresent the extent of agreement. If the population is split 50–50 on the statement, then the sampling error will be much greater.

Stratified sampling is based on this second factor in sampling theory. Rather than selecting a sample from the total population at large, the researcher ensures that appropriate numbers of elements are drawn from homogeneous subsets of that population. To get a stratified sample of university students, for example, you would first organize your population by college class and then draw appropriate numbers of freshmen, sophomores, juniors, and seniors. In a nonstratified sample, representation by class would be subjected to the same sampling error as other variables would. In a sample stratified by class, the sampling error on this variable is reduced to zero.

More-complex stratification methods are also possible. In addition to stratifying by class, you might also stratify by gender, by GPA, and so forth. In this fashion you might be able to ensure that your sample would contain the proper numbers of male sophomores with a 3.5 average, of female sophomores with a 4.0 average, and so forth.

The ultimate function of stratification, then, is to organize the population into homogeneous subsets (with heterogeneity between subsets) and to select the appropriate number of elements from each. To the extent that the subsets are homogeneous on the stratification variables, they may be homogeneous on other variables as well. Because age is related to college class, a sample stratified by class will be more representative in terms of age as well, compared with an unstratified sample. Because occupational aspirations still seem to be related to gender, a sample stratified by gender will be more representative in terms of occupational aspirations.

The choice of stratification variables typically depends on what variables are available. Gender can often be determined in a list of names. Uni-

versity lists are typically arranged by class. Lists of faculty members may indicate their departmental affiliation. Government agency files may be arranged by geographic region. Voter registration lists are arranged according to precinct.

In selecting stratification variables from among those available, however, you should be concerned primarily with those that are presumably related to variables you want to represent accurately. Because gender is related to many variables and is often available for stratification, it's often used. Education is related to many variables, but it's often not available for stratification. Geographic location within a city, state, or nation is related to many things. Within a city, stratification by geographic location usually increases representativeness in social class, ethnic group, and so forth. Within a nation, it increases representativeness in a broad range of attitudes as well as in social class and ethnicity.

When you're working with a simple list of all elements in the population, two methods of stratification predominate. In one method, you sort the population elements into discrete groups based on whatever stratification variables are being used. On the basis of the relative proportion of the population represented by a given group, you select— randomly or systematically—several elements from that group constituting the same proportion of your desired sample size. For example, if sophomore men with a 4.0 average compose 1 percent of the student population and you desire a sample of 1,000 students, you would select 10 sophomore men with a 4.0 average.

The other method is to group students as described and then put those groups together in a continuous list, beginning with all freshmen men with a 4.0 average and ending with all senior women with a 1.0 or below. You would then select a systematic sample, with a random start, from the entire list. Given the arrangement of the list, a systematic sample would select proper numbers (within an error range of 1 or 2) from each subgroup. (*Note:* A simple random sample drawn from such a composite list would cancel out the stratification.)

Figure 7-12 offers a graphic illustration of stratified, systematic sampling. As you can see, we

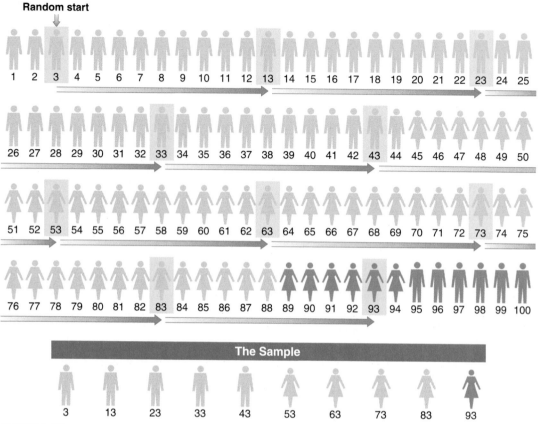

FIGURE 7-12

A Stratified, Systematic Sample with a Random Start. A stratified, systematic sample involves two stages. First the members of the population are gathered into homogeneous strata; this simple example merely uses gender and race as stratification variables, but more could be used. Then every *k*th (in this case, every 10th) person in the stratified arrangement is selected into the sample.

lined up our micropopulation according to gender and race. Then, beginning with a random start of "3," we've taken every tenth person thereafter: 3, 13, 23, . . . , 93.

Stratified sampling ensures the proper representation of the stratification variables; this, in turn, enhances the representation of other variables related to them. Taken as a whole, then, a stratified sample is more likely than a simple random sample to be more representative on several variables. Although the simple random sample is still regarded as somewhat sacred, it should now be clear that you can often do better.

Implicit Stratification in Systematic Sampling

I mentioned that systematic sampling can, under certain conditions, be more accurate than simple random sampling. This is the case whenever the arrangement of the list creates an implicit stratification. As already noted, if a list of university students is arranged by class, then a systematic sample provides a stratification by class where a simple random sample would not.

In a study of students at the University of Hawaii, after stratification by school class, the students were arranged by their student identification numbers. These numbers, however, were their social security numbers. The first three digits of the social security number indicate the state in which the number was issued. As a result, within a class, students were arranged by the state in which they were issued a social security number, providing a rough stratification by geographic origin.

An ordered list of elements, therefore, may be more useful to you than an unordered, randomized list. I've stressed this point in view of the unfortunate belief that lists should be randomized before systematic sampling. Only if the arrangement presents the problems discussed earlier should the list be rearranged.

Illustration: Sampling University Students

Let's put these principles into practice by looking at an actual sampling design used to select a sample of university students. The purpose of the study was to survey, with a mail-out questionnaire, a representative cross section of students attending the main campus of the University of Hawaii. The following sections describe the steps and decisions involved in selecting that sample.

Study Population and Sampling Frame

The obvious sampling frame available for use in this sample selection was the computerized file maintained by the university administration. The file contained students' names, local and permanent addresses, and social security numbers, as well as a variety of other information such as field of study, class, age, and gender.

The computer database, however, contained entries on all people who could, by any conceivable definition, be called students, many of whom seemed inappropriate for the purposes of the study. As a result, researchers needed to define the study population in a somewhat more restricted fashion. The final definition included those 15,225 day-program degree candidates who were registered for the fall semester on the Manoa campus of the

university, including all colleges and departments, both undergraduate and graduate students, and both U.S. and foreign students. The computer program used for sampling then limited consideration to students fitting this definition.

Stratification

The sampling program also permitted stratification of students before sample selection. The researchers decided that stratification by college class would be sufficient, although the students might have been further stratified within class, if desired, by gender, college, major, and so forth.

Sample Selection

Once the students had been arranged by class, a systematic sample was selected across the entire rearranged list. The sample size for the study was initially set at 1,100. To achieve this sample, the sampling program was set for a 1/14 sampling ratio. The program generated a random number between 1 and 14; the student having that number and every 14th student thereafter was selected in the sample.

Once the sample had been selected, the computer was instructed to print each student's name and mailing address on self-adhesive mailing labels. These labels were then simply transferred to envelopes for mailing the questionnaires.

Sample Modification

This initial design of the sample had to be modified. Before the mailing of questionnaires, the researchers discovered that, because of unexpected expenses in the production of the questionnaires, they couldn't cover the costs of mailing to all 1,100 students. As a result, one-third of the mailing labels were systematically selected (with a random start) for exclusion from the sample. The final sample for the study was thereby reduced to 733 students.

I mention this modification in order to illustrate the frequent need to alter a study plan in midstream. Because the excluded students were systematically omitted from the initial systematic sample, the remaining 733 students could still be taken as reasonably representing the study population. The reduction in sample size did, of course, increase the range of sampling error.

Multistage Cluster Sampling

The preceding sections have dealt with reasonably simple procedures for sampling from lists of elements. Such a situation is ideal. Unfortunately, however, much interesting social research requires the selection of samples from populations that cannot easily be listed for sampling purposes: the population of a city, state, or nation; all university students in the United States; and so forth. In such cases, the sample design must be much more complex. Such a design typically involves the initial sampling of groups of elements—*clusters*—followed by the selection of elements within each of the selected clusters.

Cluster sampling may be used when it's either impossible or impractical to compile an exhaustive list of the elements composing the target population, such as all church members in the United States. Often, however, the population elements are already grouped into subpopulations, and a list of those subpopulations either exists or can be created practically. For example, church members in the United States belong to discrete churches, which are either listed or could be. Following a cluster sample format, then, researchers could sample the list of churches in some manner (for example, a stratified, systematic sample). Next, they would obtain lists of members from each of the selected churches. Each of the lists would then be sampled, to provide samples of church members for study.

Another typical situation concerns sampling among population areas such as a city. Although there is no single list of a city's population, citizens reside on discrete city blocks or census blocks. Researchers can, therefore, select a sample of blocks initially, create a list of people living on each of the selected blocks, and take a subsample of the people on each block.

In a more complex design, researchers might sample blocks, list the households on each selected block, sample the households, list the people residing in each household, and, finally, sample the people within each selected household. This multistage sample design leads ultimately to a selection of a sample of individuals but does not require the initial listing of all individuals in the city's population.

Multistage cluster sampling, then, involves the repetition of two basic steps: listing and sampling. The list of primary sampling units (churches, blocks) is compiled and, perhaps, stratified for sampling. Then a sample of those units is selected. The selected primary sampling units are then listed and perhaps stratified. The list of secondary sampling units is then sampled, and so forth.

The listing of households on even the selected blocks is, of course, a labor-intensive and costly activity—one of the elements making face-to-face, household surveys quite expensive. Vincent Iannacchione, Jennifer Staab, and David Redden (2003) report some initial success using postal mailing lists for this purpose. Although the lists are not perfect, they may be close enough to warrant the significant savings in cost.

Multistage cluster sampling makes possible those studies that would otherwise be impossible. Specific research circumstances often call for special designs, as "Sampling Iran" demonstrates.

cluster sampling A multistage sampling in which natural groups (clusters) are sampled initially, with the members of each selected group being subsampled afterward. For example, you might select a sample of U.S. colleges and universities from a directory, get lists of the students at all the selected schools, then draw samples of students from each.

Multistage Designs and Sampling Error

Although cluster sampling is highly efficient, the price of that efficiency is a less-accurate sample. A simple random sample drawn from a population list is subject to a single sampling error, but a two-stage cluster sample is subject to two sampling

Sampling Iran

Whereas most of the examples given in this textbook are taken from its country of origin, the United States, the basic methods of sampling would apply in other national settings as well. At the same time, researchers may need to make modifications appropriate to local conditions. In selecting a national sample of Iran, for example, Hamid Abdollahyan and Taghi Azadarmaki (2000: 21) from the University of Tehran began by stratifying the nation on the basis of cultural differences, dividing the country into nine cultural zones as follows:

1. Tehran
2. Central region including Isfahan, Arak, Qum, Yazd and Kerman
3. The southern provinces including Hormozgan, Khuzistan, Bushehr and Fars
4. The marginal western region including Lorestan, Charmahal and Bakhtiari, Kogiluyeh and Eelam
5. The western provinces including western and eastern Azarbaijan, Zanjan, Ghazvin and Ardebil

6. The eastern provinces including Khorasan and Semnan
7. The northern provinces including Gilan, Mazandran and Golestan
8. Systan
9. Kurdistan

Within each of these cultural areas, the researchers selected samples of census blocks and, on each selected block, a sample of households. Their sample design made provisions for getting the proper numbers of men and women as respondents within households and provisions for replacing those households where no one was at home.

Source: Hamid Abdollahyan and Taghi Azadarmaki, *Sampling Design in a Survey Research: The Sampling Practice in Iran,* paper presented to the meetings of the American Sociological Association, August 12–16, 2000, Washington, DC.

---

errors. First, the initial sample of clusters will represent the population of clusters only within a range of sampling error. Second, the sample of elements selected within a given cluster will represent all the elements in that cluster only within a range of sampling error. Thus, for example, a researcher runs a certain risk of selecting a sample of disproportionately wealthy city blocks, plus a sample of disproportionately wealthy households within those blocks. The best solution to this problem lies in the number of clusters selected initially and the number of elements within each cluster.

Typically, researchers are restricted to a total sample size; for example, you may be limited to conducting 2,000 interviews in a city. Given this broad limitation, however, you have several options in designing your cluster sample. At the extremes you could choose one cluster and select 2,000 elements within that cluster, or you could select 2,000 clusters with one element selected within each. Of course, neither approach is advisable, but a broad range of choices lies between them. Fortunately, the logic of sampling distributions provides a general guideline for this task.

Recall that sampling error is reduced by two factors: an increase in the sample size and increased homogeneity of the elements being sampled. These factors operate at each level of a multistage sample design. A sample of clusters will best represent all clusters if a large number are selected and if all clusters are very much alike. A sample of elements will best represent all elements in a given cluster if a large number are selected from the cluster and if all the elements in the cluster are very much alike.

With a given total sample size, however, if the number of clusters is increased, the number of elements within a cluster must be decreased. In this respect, the representativeness of the clusters is increased at the expense of more poorly representing the elements composing each cluster, or vice versa. Fortunately, homogeneity can be used to ease this dilemma.

Typically, the elements composing a given natural cluster within a population are more homogeneous than all elements composing the total population are. The members of a given church are more alike than all church members are; the residents of a given city block are more alike than the residents of a whole city are. As a result, relatively few elements may be needed to represent a given natural cluster adequately, although a larger number of clusters may be needed to represent adequately the diversity found among the clusters. This fact is most clearly seen in the extreme case of very different clusters composed of identical elements within each. In such a situation, a large number of clusters would adequately represent all its members. Although this extreme situation never exists in reality, it's closer to the truth in most cases than its opposite: identical clusters composed of grossly divergent elements.

The general guideline for cluster design, then, is to maximize the number of clusters selected while decreasing the number of elements within each cluster. However, this scientific guideline must be balanced against an administrative constraint. The efficiency of cluster sampling is based on the ability to minimize the listing of population elements. By initially selecting clusters, you need only list the elements composing the selected clusters, not all elements in the entire population. Increasing the number of clusters, however, goes directly against this efficiency factor. A small number of clusters may be listed more quickly and more cheaply than a large number. (Remember that all the elements in a selected cluster must be listed even if only a few are to be chosen in the sample.)

The final sample design will reflect these two constraints. In effect, you'll probably select as many clusters as you can afford. Lest this issue be left too open-ended at this point, here's one general guideline. Population researchers conventionally aim at the selection of 5 households per census block. If a total of 2,000 households are to be interviewed, you would aim at 400 blocks with 5 household interviews on each. Figure 7-13 presents a graphic overview of this process.

Before we turn to other, more detailed procedures available to cluster sampling, let me reiterate that this method almost inevitably involves a loss of accuracy. The manner in which this appears, however, is somewhat complex. First, as noted earlier, a multistage sample design is subject to a sampling error at each stage. Because the sample size is nec-

essarily smaller at each stage than the total sample size, the sampling error at each stage will be greater than would be the case for a single-stage random sample of elements. Second, sampling error is estimated on the basis of observed variance among the sample elements. When those elements are drawn from among relatively homogeneous clusters, the estimated sampling error will be too optimistic and must be corrected in the light of the cluster sample design.

Stratification in Multistage Cluster Sampling

Thus far, we've looked at cluster sampling as though a simple random sample were selected at each stage of the design. In fact, stratification techniques can be used to refine and improve the sample being selected.

The basic options here are essentially the same as those in single-stage sampling from a list. In selecting a national sample of churches, for example, you might initially stratify your list of churches by denomination, geographic region, size, rural or urban location, and perhaps by some measure of social class.

Once the primary sampling units (churches, blocks) have been grouped according to the relevant, available stratification variables, either simple random or systematic-sampling techniques can be used to select the sample. You might select a specified number of units from each group, or stratum, or you might arrange the stratified clusters in a continuous list and systematically sample that list.

To the extent that clusters are combined into homogeneous strata, the sampling error at this stage will be reduced. The primary goal of stratification, as before, is homogeneity.

There's no reason why stratification couldn't take place at each level of sampling. The elements listed within a selected cluster might be stratified before the next stage of sampling. Typically, however, this is not done. (Recall the assumption of relative homogeneity within clusters.)

Stage One: Identify blocks and select a sample.
(Selected blocks are shaded.)

Stage Two: Go to each selected block and list all households in order.
(Example of one listed block.)

1. 491 Rosemary Ave.
2. 487 Rosemary Ave.
3. 473 Rosemary Ave.
4. 455 Rosemary Ave.
5. 437 Rosemary Ave.
6. 423 Rosemary Ave.
7. 411 Rosemary Ave.
8. 403 Rosemary Ave.
9. 1101 4th St.
10. 1123 4th St.
11. 1137 4th St.
12. 1157 4th St.
13. 1169 4th St.
14. 1187 4th St.
15. 402 Thyme Ave.

Stage Three: For each list, select sample of households.
(In this example, every sixth household has been selected starting with #5, which was selected at random.)

16. 408 Thyme Ave.
17. 424 Thyme Ave.
18. 446 Thyme Ave.
19. 458 Thyme Ave.
20. 480 Thyme Ave.
21. 498 Thyme Ave.
22. 1186 5th St.
23. 1174 5th St.
24. 1160 5th St.
25. 1140 5th St.
26. 1122 5th St.
27. 1118 5th St.
28. 1116 5th St.
29. 1104 5th St.
30. 1102 5th St.

FIGURE 7-13

Multistage Cluster Sampling. In multistage cluster sampling, we begin by selecting a sample of the clusters (in this case, city blocks). Then, we make a list of the elements (households, in this case) and select a sample of elements from each of the selected clusters.

Probability Proportionate to Size (PPS) Sampling

This section introduces you to a more sophisticated form of cluster sampling, one that is used in many large-scale survey-sampling projects. In the preced-ing discussion, I talked about selecting a random or systematic sample of clusters and then a random or systematic sample of elements within each cluster selected. Notice that this produces an overall sampling scheme in which every element in the whole population has the same probability of selection.

Let's say we're selecting households within a city. If there are 1,000 city blocks and we initially select a sample of 100, that means that each block has a 100/1,000 or 0.1 chance of being selected. If we next select 1 household in 10 from those residing on the selected blocks, each household has a 0.1 chance of selection within its block. To calculate the overall probability of a household being selected, we simply multiply the probabilities at the individual steps in sampling. That is, each household has a 1/10 chance of its block being selected and a 1/10 chance of that specific household being selected if the block is one of those chosen. Each household, in this case, has a $1/10 \times 1/10 = 1/100$ chance of selection overall. Because each household would have the same chance of selection, the sample so selected should be representative of all households in the city.

There are dangers in this procedure, however. In particular, the variation in the size of blocks (measured in numbers of households) presents a problem. Let's suppose that half the city's population resides in 10 densely packed blocks filled with high-rise apartment buildings, and suppose that the rest of the population lives in single-family dwellings spread out over the remaining 900 blocks. When we first select our sample of 1/10 of the blocks, it's quite possible that we'll miss all of the 10 densely packed high-rise blocks. No matter what happens in the second stage of sampling, our final sample of households will be grossly unrepresentative of the city, comprising only single-family dwellings.

Whenever the clusters sampled are of greatly differing sizes, it's appropriate to use a modified sampling design called **PPS** (probability proportionate to size). This design guards against the problem I've just described and still produces a final sample in which each element has the same chance of selection.

As the name suggests, each cluster is given a chance of selection proportionate to its size. Thus, a city block with 200 households has twice the chance of selection as one with only 100 households. Within each cluster, however, a fixed number of elements is selected, say, 5 households per block. Notice how this procedure results in each household having the same probability of selection overall.

Let's look at households of two different city blocks. Block A has 100 households; Block B has only 10. In PPS sampling, we would give Block A ten times as good a chance of being selected as Block B. So if, in the overall sample design, Block A has a 1/20 chance of being selected, that means Block B would only have a 1/200 chance. Notice that this means that all the households on Block A would have a 1/20 chance of having their block selected; Block B households have only a 1/200 chance.

If Block A is selected and we're taking 5 households from each selected block, then the households on Block A have a 5/100 chance of being selected into the block's sample. Because we can multiply probabilities in a case like this, we see that every household on Block A has an overall chance of selection equal to $1/20 \times 5/100 = 5/2000 = 1/400$.

If Block B happens to be selected, on the other hand, its households stand a much better chance of being among the 5 chosen there: 5/10. When this is combined with their relatively poorer chance of having their block selected in the first place, however, they end up with the same chance of selection as those on Block A: $1/200 \times 5/10 = 5/2000 = 1/400$.

Further refinements to this design make it a very efficient and effective method for selecting large cluster samples. For now, however, it's enough to understand the basic logic involved.

Disproportionate Sampling and Weighting

Ultimately, a probability sample is representative of a population if all elements in the population have an equal chance of selection in that sample. Thus,

PPS (probability proportionate to size) This refers to a type of multistage cluster sample in which clusters are selected, not with equal probabilities (see *EPSEM*) but with probabilities proportionate to their sizes—as measured by the number of units to be subsampled.

in each of the preceding discussions, we've noted that the various sampling procedures result in an equal chance of selection—even though the ultimate selection probability is the product of several partial probabilities.

More generally, however, a probability sample is one in which each population element has a known nonzero probability of selection—even though different elements may have different probabilities. If controlled probability sampling procedures have been used, any such sample may be representative of the population from which it is drawn if each sample element is assigned a weight equal to the inverse of its probability of selection. Thus, where all sample elements have had the same chance of selection, each is given the same weight: 1. This is called a *self-weighting* sample.

Sometimes it's appropriate to give some cases more weight than others, a process called **weighting.** Disproportionate sampling and weighting come into play in two basic ways. First, you may sample subpopulations disproportionately to ensure sufficient numbers of cases from each for analysis. For example, a given city may have a suburban area containing one-fourth of its total population. Yet you might be especially interested in a detailed analysis of households in that area and may feel that one-fourth of this total sample size would be too few. As a result, you might decide to select the same number of households from the suburban area as from the remainder of the city. Households in the suburban area, then, are given a disproportionately better chance of selection than those located elsewhere in the city are.

As long as you analyze the two area samples separately or comparatively, you need not worry about the differential sampling. If you want to combine the two samples to create a composite picture of the entire city, however, you must take the disproportionate sampling into account. If n is the number of households selected from each area, then the households in the suburban area had a chance of selection equal to n divided by one-fourth of the total city population. Because the total city population and the sample size are the same for both areas, the suburban-area households should be given a weight of $\frac{1}{4}$ n, and the remaining households should be given a weight of $\frac{3}{4}$ n. This weighting procedure could be simplified by merely giving a weight of 3 to each of the households selected outside the suburban area.

Here's an example of the problems that can be created when disproportionate sampling is not accompanied by a weighting scheme. When the *Harvard Business Review* decided to survey its subscribers on the issue of sexual harassment at work, it seemed appropriate to oversample women because female subscribers were vastly outnumbered by male subscribers. Here's how G. C. Collins and Timothy Blodgett explained the matter:

> We also skewed the sample another way: to ensure a representative response from women, we mailed a questionnaire to virtually every female subscriber, for a male/female ratio of 68% to 32%. This bias resulted in a response of 52% male and 44% female (and 4% who gave no indication of gender)—compared to HBR's U.S. subscriber proportion of 93% male and 7% female.
>
> *(1981: 78)*

Notice a couple of things in this excerpt. First, it would be nice to know a little more about what "virtually every female" means. Evidently, the authors of the study didn't send questionnaires to all female subscribers, but there's no indication of who was omitted and why. Second, they didn't use the term *representative* with its normal social science usage. What they mean, of course, is that they wanted to get a substantial or "large enough" response from women, and oversampling is a perfectly acceptable way of accomplishing that.

By sampling more women than a straightforward probability sample would have produced, the authors were able to "select" enough women (812) to compare with the men (960). Thus, when

weighting Assigning different weights to cases that were selected into a sample with different probabilities of selection. In the simplest scenario, each case is given a weight equal to the inverse of its probability of selection. When all cases have the same chance of selection, no weighting is necessary.

they report, for example, that 32 percent of the women and 66 percent of the men agree that "the amount of sexual harassment at work is greatly exaggerated," we know that the female response is based on a substantial number of cases. That's good. There are problems, however.

To begin with, subscriber surveys are always problematic. In this case, the best the researchers can hope to talk about is "what subscribers to *Harvard Business Review* think." In a loose way, it might make sense to think of that population as representing the more sophisticated portion of corporate management. Unfortunately, the overall response rate was 25 percent. Although that's quite good for subscriber surveys, it's a low response rate in terms of generalizing from probability samples.

Beyond that, however, the disproportionate sample design creates another problem. When the authors state that 73 percent of respondents favor company policies against harassment (Collins and Blodgett 1981: 78), that figure is undoubtedly too high, because the sample contains a disproportionately high percentage of women—who are more likely than men to favor such policies. And, when the researchers report that top managers are more likely to feel that claims of sexual harassment are exaggerated than are middle- and lower-level managers (1981: 81), that finding is also suspect. As the researchers report, women are disproportionately represented in lower management. That alone might account for the apparent differences among levels of management. In short, the failure to take account of the oversampling of women confounds all survey results that don't separate the findings by gender. The solution to this problem would have been to weight the responses by gender, as described earlier in this section.

In the 2000 and 2004 election campaign polling, survey weighting became a controversial topic, as some polling agencies weighted their results on the basis of party affiliation and other variables, whereas others did not. Weighting in this instance involved assumptions regarding the differential participation of Republicans and Democrats in opinion polls and on election day—plus a determination of how many Republicans and Democrats there were. This is likely to be a topic of debate among pollsters and politicians in the years to come. Alan Reifman has created a website devoted to a discussion of this topic (go to the link on this book's website: http://www.cengage.com/sociology/babbie).

Probability Sampling in Review

Much of this chapter has been devoted to the key sampling method used in controlled survey research: probability sampling. In each of the variations examined, we've seen that elements are chosen for study from a population on a basis of random selection with known nonzero probabilities.

Depending on the field situation, probability sampling can be either very simple or extremely difficult, time-consuming, and expensive. Whatever the situation, however, it remains the most effective method for the selection of study elements. There are two reasons for this.

First, probability sampling avoids researchers' conscious or unconscious biases in element selection. If all elements in the population have an equal (or unequal and subsequently weighted) chance of selection, there is an excellent chance that the sample so selected will closely represent the population of all elements.

Second, probability sampling permits estimates of sampling error. Although no probability sample will be perfectly representative in all respects, controlled selection methods permit the researcher to estimate the degree of expected error.

In this lengthy chapter, we've taken on a basic issue in much social research: selecting observations that will tell us something more general than the specifics we've actually observed. This issue confronts field researchers, who face more action and more actors than they can observe and record fully, as well as political pollsters who want to predict an election but can't interview all voters. As we proceed through the book, we'll see in greater detail how social researchers have found ways to deal with this issue.

The Ethics of Sampling

The key purpose of the sampling techniques discussed in this chapter is to allow researchers to make relatively few observations but gain an accurate picture of a much larger population. In the case of quantitative studies using probability sampling, the result should be a statistical profile, based on the sample, that closely mirrors the profile that would have been gained from observing the whole population. In addition to using legitimate sampling techniques, researchers should be careful to point out the possibility of errors: sampling error, flaws in the sampling frame, nonresponse error, or anything else that might make the results misleading.

Sometimes, more typically in qualitative studies, the purpose of sampling may be to tap into the breadth of variation within a population rather than to focus on the "average" or "typical" member of that population. While this is a legitimate and valuable approach, it poses the risk that readers may mistake the display of differences to reflect the distribution of characteristics in the population. In such a case, the researcher should make sure that the reader is not misled.

MAIN POINTS

Introduction

- Social researchers must select observations that will allow them to generalize to people and events not observed. Often this involves sampling a selection of people to observe.
- Understanding the logic of sampling is essential to doing social research.

A Brief History of Sampling

- Sometimes you can and should select probability samples using precise statistical techniques, but other times nonprobability techniques are more appropriate.

Nonprobability Sampling

- Nonprobability sampling techniques include relying on available subjects, purposive or judgmental sampling, snowball sampling, and quota sampling.

In addition, researchers studying a social group may make use of informants. Each of these techniques has its uses, but none of them ensures that the resulting sample will be representative of the population being sampled.

The Theory and Logic of Probability Sampling

- Probability-sampling methods provide an excellent way of selecting representative samples from large, known populations. These methods counter the problems of conscious and unconscious sampling bias by giving each element in the population a known (nonzero) probability of selection.
- Random selection is often a key element in probability sampling.
- The most carefully selected sample will never provide a perfect representation of the population from which it was selected. There will always be some degree of sampling error.
- By predicting the distribution of samples with respect to the target parameter, probability-sampling methods make it possible to estimate the amount of sampling error expected in a given sample.
- The expected error in a sample is expressed in terms of confidence levels and confidence intervals.

Populations and Sampling Frames

- A sampling frame is a list or quasi list of the members of a population. It is the resource used in the selection of a sample. A sample's representativeness depends directly on the extent to which a sampling frame contains all the members of the total population that the sample is intended to represent.

Types of Sampling Designs

- Several sampling designs are available to researchers.
- Simple random sampling is logically the most fundamental technique in probability sampling, but it is seldom used in practice.
- Systematic sampling involves the selection of every kth member from a sampling frame. This method is more practical than simple random sampling; with a few exceptions, it is functionally equivalent.

- Stratification, the process of grouping the members of a population into relatively homogeneous strata before sampling, improves the representativeness of a sample by reducing the degree of sampling error.

Multistage Cluster Sampling

- Multistage cluster sampling is a relatively complex sampling technique that frequently is used when a list of all the members of a population does not exist. Typically, researchers must balance the number of clusters and the size of each cluster to achieve a given sample size. Stratification can be used to reduce the sampling error involved in multistage cluster sampling.

- Probability proportionate to size (PPS) is a special, efficient method for multistage cluster sampling.

- If the members of a population have unequal probabilities of selection into the sample, researchers must assign weights to the different observations made, in order to provide a representative picture of the total population. The weight assigned to a particular sample member should be the inverse of its probability of selection.

Probability Sampling in Review

- Probability sampling remains the most effective method for the selection of study elements for two reasons: it avoids researcher bias in element selection and it permits estimates of sampling error.

The Ethics of Sampling

- Because probability sampling always carries a risk of error, the researcher must inform readers of any errors that might make results misleading.

- Sometimes, nonprobability sampling methods are used to obtain the breadth of variations in a population. In this case, the researcher must ensure that readers do not confuse variations with what's typical in the population.

KEY TERMS

The following terms are defined in context in the chapter and at the bottom of the page where the term is introduced, as well as in the comprehensive glossary at the back of the book.

cluster sampling
confidence interval

representativeness
sampling error

confidence level
element
EPSEM
informant
nonprobability sampling
parameter
population
PPS
probability sampling
purposive (judgmental) sampling
quota sampling
random selection

sampling frame
sampling interval
sampling ratio
sampling unit
simple random sampling
snowball sampling
statistic
stratification
study population
systematic sampling
weighting

PROPOSING SOCIAL RESEARCH: SAMPLING

In this portion of the proposal, you'll describe how you'll select from among all the possible observations you might make. Depending on the data collection method you plan to employ, either probability or nonprobability sampling may be more appropriate to your study. Similarly, this aspect of your proposal may involve the sampling of subjects or informants, or it could involve the sampling of corporations, cities, books, and so forth.

Your proposal, then, must specify what units you'll be sampling among, the data you'll use (such as a sampling frame) for purposes of your sample selection, and the actual sampling methods you'll use.

REVIEW QUESTIONS AND EXERCISES

1. Review the discussion of the 1948 Gallup Poll that predicted that Thomas Dewey would defeat Harry Truman for president. What are some ways Gallup could have modified his quota sample design to avoid the error?

2. Using Appendix C of this book, select a simple random sample of 10 numbers in the range of 1 to 9,876. What is each step in the process?

3. What are the steps involved in selecting a multistage cluster sample of students taking first-year English in U.S. colleges and universities?

4. In Chapter 9 we'll discuss surveys conducted on the Internet. Can you anticipate possible problems concerning sampling frames, representativeness, and the like? Do you see any solutions?

5. Using InfoTrac College Edition, locate studies using (1) a quota sample, (2) a multistage cluster sample, and (3) a systematic sample. Write a brief description of each study.

SPSS EXERCISES

See the booklet that accompanies your text for exercises using SPSS (Statistical Package for the Social Sciences). There are exercises offered for each chapter, and you'll also find a detailed primer on using SPSS.

Online Study Resources

If your book came with an access code card, visit www.cengage.com/login to register. To purchase access, please visit www.ichapters.com.

1. Before you do your final review of the chapter, take the *CengageNOW* pretest to help identify the areas on which you should concentrate. You'll find information on this online tool, as well as instructions on how to access all of its great resources, in the front of the book.

2. As you review, take advantage of the *CengageNOW* personalized study plan, based on your quiz results. Use this study plan with its interactive exercises and other resources to master the material.

3. When you're finished with your review, take the posttest to confirm that you're ready to move on to the next chapter.

WEBSITE FOR *THE PRACTICE OF SOCIAL RESEARCH* 12TH EDITION

Go to your book's website at www.cengage.com/sociology/babbie for tools to aid you in studying for your exams. You'll find *Tutorial Quizzes* with feedback, *Internet Exercises, Flash Cards, Glossaries,* and *Essay Quizzes,* as well as *InfoTrac College Edition* search terms, suggestions for additional reading, *Web Links,* and primers for using data-analysis software such as SPSS.

Having explored the structuring of inquiry in depth, we're now ready to dive into the various observational techniques available to social scientists.

Experiments are usually thought of in connection with the physical sciences. In Chapter 8 we'll see how social scientists use experiments. This is the most rigorously controllable of the methods we'll examine. Understanding experiments is also a useful way to enhance your understanding of the general logic of social science research.

Chapter 9 will describe survey research, one of the most popular methods in social science. This type of research involves collecting data by asking people questions—either in self-administered questionnaires or through interviews, which, in turn, can be conducted face-to-face or over the telephone.

Chapter 10, on qualitative field research, examines perhaps the most natural form of data collection used by social scientists: the direct observation of social phenomena in natural settings. As you'll see, some researchers go beyond mere observation to participate in what they're studying, because they want a more intimate view and a fuller understanding of it.

Chapter 11 discusses three forms of unobtrusive data collection that take advantage of some of the data

PART 3

Modes of Observation: Quantitative and Qualitative

available all around us. For example, content analysis is a method of collecting social data through carefully specifying and counting social artifacts such as books, songs, speeches, and paintings. Without making any personal contact with people, you can use this method to examine a wide variety of social phenomena. The analysis of existing statistics offers another way of studying people without having to talk to them. Governments and a variety of private organizations regularly compile great masses of data, which you often can use with little or no modification to answer properly posed questions. Finally, historical documents are a valuable resource for social science analysis.

Chapter 12, on evaluation research, looks at a rapidly growing subfield in social science, involving the application of experimental and quasi-experimental models to the testing of social interventions in real life. You might use evaluation research, for example, to test the effectiveness of a drug rehabilitation program or the efficiency of a new school cafeteria. In the same chapter, we'll look briefly at social indicators as a way of assessing broader social processes.

Before we turn to the actual descriptions of these research methods, two points should be made. First, you'll probably discover that you've been using these scientific methods casually in your daily life for as long as you can remember. You use some form of field research every day. You employ a crude form of content analysis every time you judge an author's motivation from her or his writings. You engage in at least casual experiments frequently. Part 3 will show you how to improve your use of these methods so as to avoid certain pitfalls.

Second, none of the data-collection methods described in these chapters is appropriate to all research topics and situations. I give you some ideas, early in each chapter, regarding when a given method might be appropriate. Still, I could never anticipate all the research topics that may one day interest you. As a general guideline, you should always use a variety of techniques in the study of any topic. Because each method has its weaknesses, the use of several methods can help fill any gaps; if the different, independent approaches to the topic all yield the same conclusion, you've achieved a form of replication.

CHAPTER EIGHT

Experiments

CHAPTER OVERVIEW

An experiment is a mode of observation that enables researchers to probe causal relationships. Many experiments in social research are conducted under the controlled conditions of a laboratory, but experimenters can also take advantage of natural occurrences to study the effects of events in the social world.

CengageNOW for Sociology

Use this online tool to help you make the grade on your next exam. After reading this chapter, go to "Online Study Resources" at the end of the chapter for instructions on how to benefit from *CengageNOW*.

Introduction

This chapter addresses the controlled experiment: a research method associated more with the natural than the social sciences. We begin Part 3 with this method because the logic and basic techniques of the controlled experiment provide a useful backdrop for understanding other techniques more commonly used in social science, especially for explanatory purposes. We'll also see in this chapter some of the inventive ways social scientists have conducted experiments.

At base, experiments involve (1) taking action and (2) observing the consequences of that action. Social researchers typically select a group of subjects, do something to them, and observe the effect of what was done.

It's worth noting at the outset that we often use experiments in nonscientific inquiry. In preparing a stew, for example, we add salt, taste, add more salt, and taste again. In defusing a bomb, we clip the red wire, observe whether the bomb explodes, clip another, and . . .

We also experiment copiously in our attempts to develop generalized understandings about the world we live in. All skills are learned through experimentation: eating, walking, talking, riding a bicycle, swimming, and so forth. Through experimentation, students discover how much studying is required for academic success. Through experimentation, professors learn how much preparation is required for successful lectures. This chapter discusses how social researchers use experiments to develop generalized understandings. We'll see that, like other methods available to the social researcher, experimenting has its special strengths and weaknesses.

Topics Appropriate for Experiments

Experiments are more appropriate for some topics and research purposes than others. Experiments are especially well suited to research projects in-volving relatively limited and well-defined concepts and propositions. In terms of the traditional image of science, discussed earlier in this book, the experimental model is especially appropriate for hypothesis testing. Because experiments focus on determining causation, they're also better suited to explanatory than to descriptive purposes.

Let's assume, for example, that we want to discover ways of reducing prejudice against African Americans. We hypothesize that learning about the contribution of African Americans to U.S. history will reduce prejudice, and we decide to test this hypothesis experimentally. To begin, we might test a group of experimental subjects to determine their levels of prejudice against African Americans. Next, we might show them a documentary film depicting the many important ways African Americans have contributed to the scientific, literary, political, and social development of the nation. Finally, we would measure our subjects' levels of prejudice against African Americans to determine whether the film has actually reduced prejudice.

Experimentation has also been successful in the study of small-group interaction. Thus, we might bring together a small group of experimental subjects and assign them a task, such as making recommendations for popularizing car pools. We observe, then, how the group organizes itself and deals with the problem. Over the course of several such experiments, we might systematically vary the nature of the task or the rewards for handling the task successfully. By observing differences in the way groups organize themselves and operate under these varying conditions, we can learn a great deal about the nature of small-group interaction and the factors that influence it. For example, attorneys sometimes present evidence in different ways to different mock juries, to see which method is the most effective.

We typically think of experiments as being conducted in laboratories. Indeed, most of the examples in this chapter involve such a setting. This need not be the case, however. Increasingly, social researchers are using the World Wide Web as a vehicle for conducting experiments. Further, sometimes we can construct what are called natural experiments: "experiments" that occur in the regular course of social events. The latter portion of this chapter deals with such research.

The Classical Experiment

In both the natural and the social sciences, the most conventional type of experiment involves three major pairs of components: (1) independent and dependent variables, (2) pretesting and posttesting, and (3) experimental and control groups. This section looks at each of these components and the way they're put together in the execution of the experiment.

Independent and Dependent Variables

Essentially, an experiment examines the effect of an independent variable on a dependent variable. Typically, the independent variable takes the form of an experimental stimulus, which is either present or absent. That is, the stimulus is a dichotomous variable, having two attributes, present or not present. In this typical model, the experimenter compares what happens when the stimulus is present to what happens when it is not.

In the example concerning prejudice against African Americans, *prejudice* is the dependent variable and *exposure to African American history* is the independent variable. The researcher's hypothesis suggests that prejudice depends, in part, on a lack of knowledge of African American history. The purpose of the experiment is to test the validity of this hypothesis by presenting some subjects with an appropriate stimulus, such as a documentary film. In other terms, the independent variable is

---

pretesting The measurement of a dependent variable among subjects.

posttesting The remeasurement of a dependent variable among subjects after they've been exposed to an independent variable.

---

the cause and the dependent variable is the effect. Thus, we might say that watching the film caused a change in prejudice or that reduced prejudice was an effect of watching the film.

The independent and dependent variables appropriate for experimentation are nearly limitless. Moreover, a given variable might serve as an independent variable in one experiment and as a dependent variable in another. For example, *prejudice* is the dependent variable in our example, but it might be the independent variable in an experiment examining the effect of prejudice on voting behavior.

To be used in an experiment, both independent and dependent variables must be operationally defined. Such operational definitions might involve a variety of observation methods. Responses to a questionnaire, for example, might be the basis for defining prejudice. Speaking to or ignoring African Americans, or agreeing or disagreeing with them, might be elements in the operational definition of interaction with African Americans in a small-group setting.

Conventionally, in the experimental model, dependent and independent variables must be operationally defined before the experiment begins. However, as you'll see in connection with survey research and other methods, it's sometimes appropriate to make a wide variety of observations during data collection and then determine the most useful operational definitions of variables during later analyses. Ultimately, however, experimentation, like other quantitative methods, requires specific and standardized measurements and observations.

Pretesting and Posttesting

In the simplest experimental design, subjects are measured in terms of a dependent variable (**pretesting**), exposed to a stimulus representing an independent variable, and then remeasured in terms of the dependent variable (**posttesting**). Any differences between the first and last measurements of the dependent variable are then attributed to the independent variable.

In the example of prejudice and exposure to African American history, we'd begin by pretesting the extent of prejudice among our experimental subjects. Using a questionnaire asking about attitudes toward African Americans, for example, we could measure both the extent of prejudice exhibited by each individual subject and the average prejudice level of the whole group. After exposing the subjects to the African American history film, we could administer the same questionnaire again. Responses given in this posttest would permit us to measure the later extent of prejudice for each subject and the average prejudice level of the group as a whole. If we discovered a lower level of prejudice during the second administration of the questionnaire, we might conclude that the film had indeed reduced prejudice.

In the experimental examination of attitudes such as prejudice, we face a special practical problem relating to validity. As you may already have imagined, the subjects might respond differently to the questionnaires the second time even if their attitudes remain unchanged. During the first administration of the questionnaire, the subjects might be unaware of its purpose. By the second measurement, they might have figured out that the researchers were interested in measuring their prejudice. Because no one wishes to seem prejudiced, the subjects might "clean up" their answers the second time around. Thus, the film would seem to have reduced prejudice although, in fact, it had not.

This is an example of a more general problem that plagues many forms of social research: The very act of studying something may change it. The techniques for dealing with this problem in the context of experimentation will be discussed in various places throughout the chapter. The first technique involves the use of control groups.

Experimental and Control Groups

Laboratory experiments seldom, if ever, involve only the observation of an **experimental group** to which a stimulus has been administered. In addition, the researchers also observe a **control group,** which does not receive the experimental stimulus.

In the example of prejudice and African American history, we might examine two groups of subjects. To begin, we give each group a questionnaire designed to measure their prejudice against African Americans. Then we show the film to only the experimental group. Finally, we administer a posttest of prejudice to both groups. Figure 8-1 illustrates this basic experimental design.

Using a control group allows the researcher to detect any effects of the experiment itself. If the posttest shows that the overall level of prejudice exhibited by the control group has dropped as much as that of the experimental group, then the apparent reduction in prejudice must be a function of the experiment or of some external factor rather than a function of the film. If, on the other hand, prejudice is reduced only in the experimental group, this reduction would seem to be a consequence of exposure to the film, because that's the only difference between the two groups. Alternatively, if prejudice is reduced in both groups but to a greater degree in the experimental group than in the control group, that, too, would be grounds for assuming that the film reduced prejudice.

The need for control groups in social research became clear in connection with a series of studies of employee satisfaction conducted by F. J. Roethlisberger and W. J. Dickson (1939) in the late 1920s and early 1930s. These two researchers were interested in discovering what changes in working conditions would improve employee satisfaction and productivity. To pursue this objective, they studied working conditions in the telephone "bank wiring room" of the Western Electric Works in the Chicago suburb of Hawthorne, Illinois.

---

experimental group In experimentation, a group of subjects to whom an experimental stimulus is administered.

control group In experimentation, a group of subjects to whom no experimental stimulus is administered and who should resemble the experimental group in all other respects. The comparison of the control group and the experimental group at the end of the experiment points to the effect of the experimental stimulus.

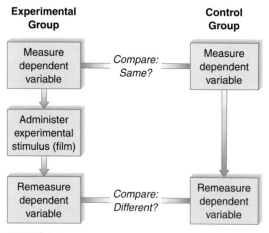

FIGURE 8-1

Diagram of Basic Experimental Design. The fundamental purpose of an experiment is to isolate the possible effect of an independent variable (called the *stimulus* in experiments) on a dependent variable. Members of the experimental group(s) are exposed to the stimulus, while those in the control group(s) are not.

To the researchers' great satisfaction, they discovered that improving the working conditions increased satisfaction and productivity consistently. As the workroom was brightened up through better lighting, for example, productivity went up. When lighting was further improved, productivity went up again.

To further substantiate their scientific conclusion, the researchers then *dimmed* the lights. Whoops—productivity improved again!

At this point it became evident that the wiring-room workers were responding more to the attention given them by the researchers than to improved working conditions. As a result of this phenomenon, often called the *Hawthorne effect,* social researchers have become more sensitive to and cautious about the possible effects of experiments themselves. In the wiring-room study, the use of a proper control group—one that was studied intensively without any other changes in the working conditions—would have pointed to the presence of this effect.

The need for control groups in experimentation has been nowhere more evident than in medical research. Time and again, patients who participate in medical experiments have appeared to improve, but it has been unclear how much of the improvement has come from the experimental treatment and how much from the experiment. In testing the effects of new drugs, then, medical researchers frequently administer a *placebo*—a "drug" with no relevant effect, such as sugar pills—to a control group. Thus, the control-group patients believe that they, like the experimental group, are receiving an experimental drug. Often, they improve. If the new drug is effective, however, those receiving the actual drug will improve more than those receiving the placebo.

In social science experiments, control groups guard against not only the effects of the experiments themselves but also the effects of any events outside the laboratory during the experiments. In the example of the study of prejudice, suppose that a popular African American leader is assassinated in the middle of, say, a weeklong experiment. Such an event may very well horrify the experimental subjects, requiring them to examine their own attitudes toward African Americans, with the result of reduced prejudice. Because such an effect should happen about equally for members of the control and experimental groups, a greater reduction of prejudice among the experimental group would, again, point to the impact of the experimental stimulus: the documentary film.

Sometimes an experimental design requires more than one experimental or control group. In the case of the documentary film, for example, we might also want to examine the impact of reading a book about African American history. In that case, we might have one group see the film and read the book, another group only see the movie, still another group only read the book, and the control group do neither. With this kind of design, we could determine the impact of each stimulus separately, as well as their combined effect.

The Double-Blind Experiment

Like patients who improve when they merely think they're receiving a new drug, sometimes experimenters tend to prejudge results. In medical research, the experimenters may be more likely to "observe" improvements among patients receiving the experimental drug than among those receiving the placebo. (This would be most likely, perhaps, for the researcher who developed the drug.) A **double-blind experiment** eliminates this possibility, because in this design neither the subjects nor the experimenters know which is the experimental group and which is the control. In the medical case, those researchers who were responsible for administering the drug and for noting improvements would not be told which subjects were receiving the drug and which the placebo. Conversely, the researcher who knew which subjects were in which group would not administer the experiment.

In social science experiments, as in medical experiments, the danger of experimenter bias is further reduced to the extent that the operational definitions of the dependent variables are clear and precise. Thus, medical researchers would be less likely to unconsciously bias their reading of a patient's temperature than they would be to bias their assessment of how lethargic the patient was. For the same reason, the small-group researcher would be less likely to misperceive which subject spoke, or to whom he or she spoke, than whether the subject's comments sounded cooperative or competitive, a more subjective judgment that's difficult to define in precise behavioral terms.

As I've indicated several times, seldom can we devise operational definitions and measurements that are wholly precise and unambiguous. This is another reason why it can be appropriate to employ a double-blind design in social research experiments.

Selecting Subjects

In Chapter 7 we discussed the logic of sampling, which involves selecting a sample that is representative of some population. Similar considerations apply to experiments. Because most social researchers work in colleges and universities, it seems likely that research laboratory experiments would be conducted with college undergraduates

as subjects. Typically, the experimenter asks students enrolled in his or her classes to participate in experiments or advertises for subjects in a college newspaper. Subjects may or may not be paid for participating in such experiments (recall also from Chapter 3 the ethical issues involved in asking students to participate in such studies).

In relation to the norm of generalizability in science, this tendency clearly represents a potential defect in social research. Simply put, college undergraduates are not typical of the public at large. There is a danger, therefore, that we may learn much about the attitudes and actions of college undergraduates but not about social attitudes and actions in general.

However, this potential defect is less significant in explanatory research than in descriptive research. True, having noted the level of prejudice among a group of college undergraduates in our pretesting, we would have little confidence that the same level existed among the public at large. On the other hand, if we found that a documentary film reduced whatever level of prejudice existed among those undergraduates, we would have more confidence—without being certain—that it would have a comparable effect in the community at large. Social processes and patterns of causal relationships appear to be more generalizable and more stable than specific characteristics such as an individual's level of prejudice.

Aside from the question of generalizability, the cardinal rule of subject selection in experimentation concerns the comparability of experimental and control groups. Ideally, the control group represents what the experimental group would be like if it had *not* been exposed to the experimental stimulus. The logic of experiments requires, therefore, that experimental and control groups be as similar as possible. There are several ways to accomplish this.

---

double-blind experiment An experimental design in which neither the subjects nor the experimenters know which is the experimental group and which is the control.

Probability Sampling

The discussions of the logic and techniques of probability sampling in Chapter 7 provide one method for selecting two groups of people that are similar to each other. Beginning with a sampling frame composed of all the people in the population under study, the researcher might select two probability samples. If these samples each resemble the total population from which they're selected, they'll also resemble each other.

Recall also, however, that the degree of resemblance (representativeness) achieved by probability sampling is largely a function of the sample size. As a general guideline, probability samples of less than 100 are not likely to be terribly representative, and social science experiments seldom involve that many subjects in either experimental or control groups. As a result, then, probability sampling is seldom used in experiments to select subjects from a larger population. Researchers do, however, use the logic of random selection when they assign subjects to groups.

Randomization

Having recruited, by whatever means, a total group of subjects, the experimenter may randomly assign those subjects to either the experimental or the control group. The researcher might accomplish such **randomization** by numbering all of the subjects serially and selecting numbers by means of a random number table. Alternatively, the experimenter might assign the odd-numbered subjects to the experimental group and the even-numbered subjects to the control group.

randomization A technique for assigning experimental subjects to experimental and control groups randomly.

matching In connection with experiments, the procedure whereby pairs of subjects are matched on the basis of their similarities on one or more variables, and one member of the pair is assigned to the experimental group and the other to the control group.

Let's return again to the basic concept of probability sampling. If we recruit 40 subjects altogether, in response to a newspaper advertisement, for example, there's no reason to believe that the 40 subjects represent the entire population from which they've been drawn. Nor can we assume that the 20 subjects randomly assigned to the experimental group represent that larger population. We can have greater confidence, however, that the 20 subjects randomly assigned to the experimental group will be reasonably similar to the 20 assigned to the control group.

Following the logic of our earlier discussions of sampling, we can see our 40 subjects as a population from which we select two probability samples—each consisting of half the population. Because each sample reflects the characteristics of the total population, the two samples will mirror each other.

As we saw in Chapter 7, our assumption of similarity in the two groups depends in part on the number of subjects involved. In the extreme case, if we recruited only two subjects and assigned, by the flip of a coin, one as the experimental subject and one as the control, there would be no reason to assume that the two subjects are similar to each other. With larger numbers of subjects, however, randomization makes good sense.

Matching

Another way to achieve comparability between the experimental and control groups is through **matching.** This process is similar to the quota-sampling methods discussed in Chapter 7. If 12 of our subjects are young white men, we might assign 6 of them at random to the experimental group and the other 6 to the control group. If 14 are middle-aged African American women, we might assign 7 to each group. We repeat this process for every relevant grouping of subjects.

The overall matching process could be most efficiently achieved through the creation of a quota matrix constructed of all the most relevant characteristics. Figure 8-2 provides a simplified illustration of such a matrix. In this example, the experimenter

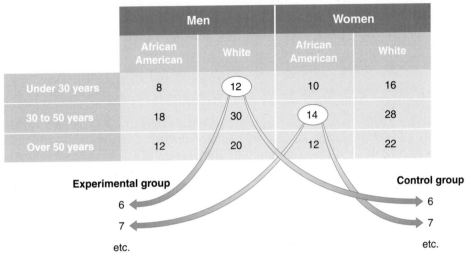

FIGURE 8-2

Quota Matrix Illustration. Sometimes the experimental and control groups are created by finding pairs of matching subjects and assigning one to the experimental group and the other to the control group.

has decided that the relevant characteristics are race, age, and gender. Ideally, the quota matrix is constructed to result in an even number of subjects in each cell of the matrix. Then, half the subjects in each cell go into the experimental group and half into the control group.

Alternatively, we might recruit more subjects than our experimental design requires. We might then examine many characteristics of the large initial group of subjects. Whenever we discover a pair of quite similar subjects, we might assign one at random to the experimental group and the other to the control group. Potential subjects who are unlike anyone else in the initial group might be left out of the experiment altogether.

Whatever method we employ, the desired result is the same. The overall average description of the experimental group should be the same as that of the control group. For example, on average both groups should have about the same ages, the same gender composition, the same racial composition, and so forth. This test of comparability should be used whether the two groups are created through probability sampling or through randomization.

Thus far I've referred to the "relevant" variables without saying clearly what those variables

are. Of course, these variables cannot be specified in any definite way, any more than I could specify in Chapter 7 which variables should be used in stratified sampling. Which variables are relevant ultimately depends on the nature and purpose of an experiment. As a general rule, however, the control and experimental groups should be comparable in terms of those variables that are most likely to be related to the dependent variable under study. In a study of prejudice, for example, the two groups should be alike in terms of education, ethnicity, and age, among other characteristics. In some cases, moreover, we may delay assigning subjects to experimental and control groups until we have initially measured the dependent variable. Thus, for example, we might administer a questionnaire measuring subjects' prejudice and then match the experimental and control groups on this variable to assure ourselves that the two groups exhibit the same overall level of prejudice.

Matching or Randomization?

When assigning subjects to the experimental and control groups, you should be aware of two arguments in favor of randomization over matching.

First, you may not be in a position to know in advance which variables will be relevant for the matching process. Second, most of the statistics used to analyze the results of experiments assume randomization. Failure to design your experiment that way, then, makes your later use of those statistics less meaningful.

On the other hand, randomization only makes sense if you have a fairly large pool of subjects, so that the laws of probability sampling apply. With only a few subjects, matching would be a better procedure.

Sometimes researchers can combine matching and randomization. When conducting an experiment on the educational enrichment of young adolescents, for example, J. Milton Yinger and his colleagues (1977) needed to assign a large number of students, aged 13 and 14, to several different experimental and control groups to ensure the comparability of students composing each of the groups. They achieved this goal by the following method.

Beginning with a pool of subjects, the researchers first created strata of students nearly identical to one another in terms of some 15 variables. From each of the strata, students were randomly assigned to the different experimental and control groups. In this fashion, the researchers actually improved on conventional randomization. Essentially, they had used a stratified-sampling procedure (Chapter 7), except that they had employed far more stratification variables than are typically used in, say, survey sampling.

Thus far I've described the classical experiment—the experimental design that best represents the logic of causal analysis in the laboratory. In practice, however, social researchers use a great variety of experimental designs. Let's look at some now.

Variations on Experimental Design

Donald Campbell and Julian Stanley (1963), in a classic book on research design, describe some 16 different experimental and quasi-experimental designs. This section describes some of these variations to better show the potential for experimentation in social research.

Preexperimental Research Designs

To begin, Campbell and Stanley discuss three "preexperimental" designs, not to recommend them but because they're frequently used in less-than-professional research. These designs are called preexperimental to indicate that they do not meet the scientific standards of experimental designs, and sometimes they may be used because the conditions for full-fledged experiments are impossible to meet. In the first such design—the *one-shot case study*—the researcher measures a single group of subjects on a dependent variable following the administration of some experimental stimulus. Suppose, for example, that we show the African American history film, mentioned earlier, to a group of people and then administer a questionnaire that seems to measure prejudice against African Americans. Suppose further that the answers given to the questionnaire seem to represent a low level of prejudice. We might be tempted to conclude that the film reduced prejudice. Lacking a pretest, however, we can't be sure. Perhaps the questionnaire doesn't really represent a sensitive measure of prejudice, or perhaps the group we're studying was low in prejudice to begin with. In either case, the film might have made no difference, though our experimental results might have misled us into thinking it did.

The second preexperimental design discussed by Campbell and Stanley adds a pretest for the experimental group but lacks a control group. This design—which the authors call the *one-group pretest-posttest design*—suffers from the possibility that some factor other than the independent variable might cause a change between the pretest and posttest results, such as the assassination of a respected African American leader. Thus, although we can see that prejudice has been reduced, we can't be sure that the film is what caused that reduction.

To round out the possibilities for preexperimental designs, Campbell and Stanley point out that

One-Shot Case Study

A man who exercises
is observed to be in
trim shape.

Time 1   *Time 2*   *Time 3*

Comparison

Some intuitive
standard of
what constitutes
a trim shape

One-Group Pretest-Posttest Design

An overweight man who
exercises is later observed
to be in trim shape.

Time 1   *Time 2*   *Time 3*

Comparison

Static-Group Comparison

A man who exercises is
observed to be in trim
shape while one who
doesn't is observed to
be overweight.

Comparison

Time 1   *Time 2*   *Time 3*

FIGURE 8-3

Three Preexperimental Research Designs. These preexperimental designs anticipate the logic of true experiments but leave themselves open to errors of interpretation. Can you see the errors that might be made in each of these designs? The various risks are solved by the addition of control groups, pretesting, and posttesting.

some research is based on experimental and control groups but has no pretests. They call this design the *static-group comparison*. For example, we might show the African American history film to one group and not to another and then measure prejudice in both groups. If the experimental group had less prejudice at the conclusion of the experiment, we might assume the film was responsible. But

unless we had randomized our subjects, we would have no way of knowing that the two groups had the same degree of prejudice initially; perhaps the experimental group started out with less.

Figure 8-3 graphically illustrates these three preexperimental research designs by using a different research question: Does exercise cause weight reduction? To make the several designs clearer, the figure shows individuals rather than groups, but the same logic pertains to group comparisons. Let's review the three preexperimental designs in this new example.

The one-shot case study represents a common form of logical reasoning in everyday life. Asked whether exercise causes weight reduction, we may bring to mind an example that would seem to support the proposition: someone who exercises and is thin. There are problems with this reasoning, however. Perhaps the person was thin long before beginning to exercise. Or perhaps he became thin for some other reason, like eating less or getting sick. The observations shown in the diagram do not guard against these other possibilities. Moreover, the observation that the man in the diagram is in trim shape depends on our intuitive idea of what constitutes trim and overweight body shapes. All told, this is very weak evidence for testing the relationship between exercise and weight loss.

The one-group pretest-posttest design offers somewhat better evidence that exercise produces weight loss. Specifically, we've ruled out the possibility that the man was thin before beginning to exercise. However, we still have no assurance that his exercising is what caused him to lose weight.

Finally, the static-group comparison eliminates the problem of our questionable definition of what constitutes trim or overweight body shapes. In this case, we can compare the shapes of the man who exercises and the one who does not. This design, however, reopens the possibility that the man who exercises was thin to begin with.

Validity Issues in Experimental Research

At this point I want to present in a more systematic way the factors that affect the validity of experimental research. First we'll look at what Campbell

and Stanley call the sources of internal invalidity, reviewed and expanded in a follow-up book by Thomas Cook and Donald Campbell (1979). Then we'll consider the problem of generalizing experimental results to the "real" world, referred to as external invalidity. Having examined these, we'll be in a position to appreciate the advantages of some of the more sophisticated experimental and quasi-experimental designs social science researchers sometimes use.

Sources of Internal Invalidity

The problem of **internal invalidity** refers to the possibility that the conclusions drawn from experimental results may not accurately reflect what has gone on in the experiment itself. The threat of internal invalidity is present whenever anything other than the experimental stimulus can affect the dependent variable.

Campbell and Stanley (1963: 5–6) and Cook and Campbell (1979: 51–55) point to several sources of internal invalidity. Here are 12:

1. *History.* During the course of the experiment, historical events may occur that will confound the experimental results. The assassination of an African American leader during the course of an experiment on reducing anti–African American prejudice is one example; the arrest of an African American leader for some heinous crime, which might increase prejudice, is another.

2. *Maturation.* People are continually growing and changing, and such changes can affect the results of the experiment. In a long-term experiment, the fact that the subjects grow older (and wiser?) may have an effect. In shorter experiments, they may grow tired, sleepy, bored, or hungry, or change in other ways that affect their behavior in the experiment.

3. *Testing.* As we've seen, often the process of testing and retesting influences people's behavior, thereby confounding the experimental results. Suppose we administer a questionnaire to a group as a way of measuring their prejudice. Then we administer an experimental stimulus and remeasure their prejudice. By the time we conduct the

---

internal invalidity Refers to the possibility that the conclusions drawn from experimental results may not accurately reflect what went on in the experiment itself.

posttest, the subjects will probably have become more sensitive to the issue of prejudice and will be more thoughtful in their answers. In fact, they may have figured out that we're trying to find out how prejudiced they are, and, because few people like to appear prejudiced, they may give answers that they think we want or that will make them look good.

4. *Instrumentation.* The process of measurement in pretesting and posttesting brings in some of the issues of conceptualization and operationalization discussed earlier in the book. If we use different measures of the dependent variable in the pretest and posttest (say, different questionnaires about prejudice), how can we be sure they're comparable to each other? Perhaps prejudice will seem to decrease simply because the pretest measure was more sensitive than the posttest measure. Or if the measurements are being made by the experimenters, their standards or their abilities may change over the course of the experiment.

5. *Statistical regression.* Sometimes it's appropriate to conduct experiments on subjects who start out with extreme scores on the dependent variable. If you were testing a new method for teaching math to hard-core failures in math, you'd want to conduct your experiment on people who previously had done extremely poorly in math. But consider for a minute what's likely to happen to the math achievement of such people over time without any experimental interference. They're starting out so low that they can only stay at the bottom or improve: They can't get worse. Even without any experimental stimulus, then, the group as a whole is likely to show some improvement over time. Referring to a *regression to the mean,* statisticians often point out that extremely tall people as a group are likely to have children shorter than themselves, and extremely short people as a group are likely to have children taller than themselves. There is a danger, then, that changes occurring by virtue of subjects' starting out in extreme positions will be attributed erroneously to the effects of the experimental stimulus.

6. *Selection biases.* We discussed selection bias earlier when we examined different ways of selecting subjects for experiments and assigning them to experimental and control groups. Comparisons don't have any meaning unless the groups are comparable at the start of an experiment.

7. *Experimental mortality.* Although some social experiments could, I suppose, kill subjects, *experimental mortality* refers to a more general and less-extreme problem. Often, experimental subjects will drop out of the experiment before it's completed, and this can affect statistical comparisons and conclusions. In the classical experiment involving an experimental and a control group, each with a pretest and posttest, suppose that the bigots in the experimental group are so offended by the African American history film that they tell the experimenter to forget it, and they leave. Those subjects sticking around for the posttest will have been less prejudiced to start with, so the group results will reflect a substantial "decrease" in prejudice.

8. *Causal time order.* Though rare in social research, ambiguity about the time order of the experimental stimulus and the dependent variable can arise. Whenever this occurs, the research conclusion that the stimulus caused the dependent variable can be challenged with the explanation that the "dependent" variable actually caused changes in the stimulus.

9. *Diffusion or imitation of treatments.* When experimental and control-group subjects can communicate with each other, experimental subjects may pass on some elements of the experimental stimulus to the control group. For example, suppose there's a lapse of time between our showing of the African American history film and the posttest administration of the questionnaire. Members of the experimental group might tell control-group subjects about the film. In that case, the control group becomes affected by the stimulus and is not a real control. Sometimes we speak of the control group as having been "contaminated."

10. *Compensation.* As you'll see in Chapter 12, in experiments in real-life situations—such as a special educational program—subjects in the control group are often deprived of something considered to be of value. In such cases, there may be pressures to offer some form of compensation. For example, hospital staff might feel sorry for

control-group patients and give them extra "tender loving care." In such a situation, the control group is no longer a genuine control group.

11. *Compensatory rivalry.* In real-life experiments, the subjects deprived of the experimental stimulus may try to compensate for the missing stimulus by working harder. Suppose an experimental math program is the experimental stimulus; the control group may work harder than before on their math in an attempt to beat the "special" experimental subjects.

12. *Demoralization.* On the other hand, feelings of deprivation within the control group may result in their giving up. In educational experiments, demoralized control-group subjects may stop studying, act up, or get angry.

These, then, are some of the sources of internal invalidity in experiments. Aware of these, experimenters have devised designs aimed at handling them. The classical experiment, if coupled with proper subject selection and assignment, addresses each of these problems. Let's look again at that study design, presented graphically in Figure 8-4.

If we use the experimental design shown in Figure 8-4, we should expect two findings. For the experimental group, the level of prejudice measured in their posttest should be less than was found in their pretest. In addition, when the two posttests are compared, less prejudice should be found in the experimental group than in the control group.

This design also guards against the problem of history in that anything occurring outside the experiment that might affect the experimental group should also affect the control group. Consequently, there should still be a difference in the two posttest results. The same comparison guards against problems of maturation as long as the subjects have been randomly assigned to the two groups. Testing and instrumentation can't be problems, because both the experimental and control groups are subject to the same tests and experimenter effects. If the subjects have been assigned to the two groups randomly, statistical regression should affect both equally, even if people with extreme scores on prejudice are being studied. Selection bias is ruled out by the random assignment of subjects. Experimental mortality is more complicated to handle, but the data provided in this study design offer several ways to deal with it. Slight modifications to the design—administering a placebo (such as a film having nothing to do with African Americans) to the control group, for example—can make the problem even easier to manage.

The remaining five problems of internal invalidity are avoided through the careful administration of a controlled experimental design. The experimental design we've been discussing facilitates the clear specification of independent and dependent variables. Experimental and control subjects can be kept separate, reducing the possibility of diffusion or imitation of treatments. Administrative controls can avoid compensations given to the control group, and compensatory rivalry can be watched for and taken into account in evaluating the results of the experiment, as can the problem of demoralization.

Sources of External Invalidity

Internal invalidity accounts for only some of the complications faced by experimenters. In addition, there is the problem that Campbell and Stanley call **external invalidity,** which relates to the generalizability of experimental findings to the "real" world. Even if the results of an experiment provide an accurate gauge of what happened during that experiment, do they really tell us anything about life in the wilds of society?

Campbell and Stanley describe four forms of this problem; I'll present one as an illustration. The generalizability of experimental findings is jeopardized, as the authors point out, if there's an interaction between the testing situation and the experimental stimulus (1963: 18). Here's an example of what they mean.

Staying with the study of prejudice and the African American history film, let's suppose that our

---

external invalidity Refers to the possibility that conclusions drawn from experimental results may not be generalizable to the "real" world.

FIGURE 8-4

The Classical Experiment: Using an African American History Film to Reduce Prejudice. This diagram illustrates the basic structure of the classical experiment as a vehicle for testing the impact of a film on prejudice. Notice how the control group, the pretesting, and the posttesting function.

experimental group—in the classical experiment—has less prejudice in its posttest than in its pretest and that its posttest shows less prejudice than that of the control group. We can be confident that the film actually reduced prejudice among our experimental subjects. But would it have the same effect if the film were shown in theaters or on television? We can't be sure, because the film might be effective only when people have been sensitized to the issue of prejudice, as the subjects may have been in taking the pretest. This is an example of interaction between the testing and the stimulus. The classical experimental design cannot control for that possibility. Fortunately, experimenters have devised other designs that can.

The *Solomon four-group design* (D. Campbell and Stanley 1963: 24–25) addresses the problem of testing interaction with the stimulus. As the name suggests, it involves four groups of subjects, assigned randomly from a pool. Figure 8-5 presents this design graphically.

Notice that Groups 1 and 2 in Figure 8-5 compose the classical experiment, with Group 2

being the control group. Group 3 is administered the experimental stimulus without a pretest, and Group 4 is only posttested. This experimental design permits four meaningful comparisons, which are described in the figure. If the African American history film really reduces prejudice—unaccounted for by the problem of internal validity and unaccounted for by an interaction between the testing and the stimulus—we should expect four findings:

1. In Group 1, posttest prejudice should be less than pretest prejudice.

2. In Group 2, prejudice should be the same in the pretest and the posttest.

3. The Group 1 posttest should show less prejudice than the Group 2 posttest does.

4. The Group 3 posttest should show less prejudice than the Group 4 posttest does.

Notice that finding (4) rules out any interaction between the testing and the stimulus. And remember that these comparisons are meaningful

Expected Findings

1 In Group 1, posttest prejudice should be less than pretest prejudice.

2 In Group 2, prejudice should be the same in the pretest and the posttest.

3 The Group 1 posttest should show less prejudice than the Group 2 posttest does.

4 The Group 3 posttest should show less prejudice than the Group 4 posttest does.

FIGURE 8-5

The Solomon Four-Group Design. The classical experiment runs the risk that pretesting will have an effect on subjects, so the Solomon Four Group Design adds experimental and control groups that skip the pretest.

only if subjects have been assigned randomly to the different groups, thereby providing groups of equal prejudice initially, even though their preexperimental prejudice is measured only in Groups 1 and 2.

There is a side benefit to this research design, as the authors point out. Not only does the Solomon four-group design rule out interactions between testing and the stimulus, it also provides data for comparisons that will reveal how much of this interaction has occurred in a classical experiment. This knowledge allows a researcher to review and evaluate the value of any prior research that used the simpler design.

The last experimental design I'll mention here is what Campbell and Stanley (1963: 25–26) call the *posttest-only control group design;* it consists of the second half—Groups 3 and 4—of the Solomon

design. As the authors argue persuasively, with proper randomization, only Groups 3 and 4 are needed for a true experiment that controls for the problems of internal invalidity as well as for the interaction between testing and stimulus. With randomized assignment to experimental and control groups (which distinguishes this design from the static-group comparison discussed earlier), the subjects will be initially comparable on the dependent variable—comparable enough to satisfy the conventional statistical tests used to evaluate the results—so it's not necessary to measure them. Indeed, Campbell and Stanley suggest that the only justification for pretesting in this situation is tradition. Experimenters have simply grown accustomed to pretesting and feel more secure with research designs that include it. Be clear, however, that this point applies only to experiments in which subjects have been assigned to experimental and control groups randomly, because that's what justifies the assumption that the groups are equivalent without having been measured to find out.

This discussion has introduced the intricacies of experimental design, its problems, and some solutions. There are, of course, a great many other experimental designs in use. Some involve more than one stimulus and combinations of stimuli. Others involve several tests of the dependent variable over time and the administration of the stimulus at different times for different groups. If you're interested in pursuing this topic, you might look at the Campbell and Stanley book.

An Illustration of Experimentation

Experiments have been used to study a wide variety of topics in the social sciences. Some experiments have been conducted within laboratory situations; others occur out in the "real world" and are referred to as *field experiments.* The following discussion provides a glimpse of both. We'll begin with an example of a field experiment.

In George Bernard Shaw's well-loved play *Pygmalion*—the basis of the long-running Broadway musical *My Fair Lady*—Eliza Doolittle speaks of the powers others have in determining our social identity. Here's how she distinguishes the way she's treated by her tutor, Professor Higgins, and by Higgins's friend, Colonel Pickering:

> You see, really and truly, apart from the things anyone can pick up (the dressing and the proper way of speaking, and so on), the difference between a lady and a flower girl is not how she behaves, but how she's treated. I shall always be a flower girl to Professor Higgins, because he always treats me as a flower girl, and always will, but I know I can be a lady to you, because you always treat me as a lady, and always will.
>
> *(Act V)*

The sentiment Eliza expresses here is basic social science, addressed more formally by sociologists such as Charles Horton Cooley (the "looking-glass self") and George Herbert Mead ("the generalized other"). The basic point is that who we think we are—our self-concept—and how we behave are largely a function of how others see and treat us. Related to this, the way others perceive us is largely conditioned by expectations they have in advance. If they've been told we're stupid, for example, they're likely to see us that way—and we may come to see ourselves that way and actually act stupidly. "Labeling theory" addresses the phenomenon of people acting in accord with the ways that others perceive and label them. These theories have served as the premise for numerous movies, such as the 1983 film *Trading Places,* in which Eddie Murphy and Dan Ackroyd play a derelict converted into a stockbroker and vice versa.

The tendency to see in others what we've been led to expect takes its name from Shaw's play. Called the *Pygmalion effect,* it's nicely suited to controlled experiments. In one of the best-known experimental investigations of the Pygmalion effect, Robert Rosenthal and Lenore Jacobson (1968) administered what they called the "Harvard Test of Inflected Acquisition" to students in a West Coast school. Subsequently, they met with the students'

teachers to present the results of the test. In particular, Rosenthal and Jacobson identified certain students as very likely to exhibit a sudden spurt in academic abilities during the coming year, based on the results of the test.

When IQ test scores were compared later, the researchers' predictions proved accurate. The students identified as "spurters" far exceeded their classmates during the following year, suggesting that the predictive test was a powerful one. In fact, the test was a hoax! The researchers had made their predictions randomly among both good and poor students. What they told the teachers did not really reflect students' test scores at all. The progress made by the "spurters" was simply a result of the teachers expecting the improvement and paying more attention to those students, encouraging them, and rewarding them for achievements. (Notice the similarity between this situation and the Hawthorne effect discussed earlier in this chapter.)

The Rosenthal-Jacobson study attracted a great deal of popular as well as scientific attention. Subsequent experiments have focused on specific aspects of what has become known as the *attribution process,* or the *expectations communication model.* This research, largely conducted by psychologists, parallels research primarily by sociologists, which takes a slightly different focus and is often gathered under the label *expectations-states theory.* Psychological studies focus on situations in which the expectations of a dominant individual affect the performance of subordinates—as in the case of a teacher and students, or a boss and employees. The sociological research has tended to focus more on the role of expectations among equals in small, task-oriented groups. In a jury, for example, how do jurors initially evaluate each other, and how do those initial assessments affect their later interactions? (You can learn more about this phenomenon, including attempts to find practical applications, by searching the web for "Pygmalion Effect.")

Here's an example of an experiment conducted to examine the way our perceptions of our abilities and the abilities of others affect our willingness to accept the other person's ideas. Martha Foschi, G. Keith Warriner, and Stephen Hart (1985) were particularly interested in the role "standards" play in that respect:

> In general terms, by "standards" we mean how well or how poorly a person has to perform in order for an ability to be attributed or denied him/her. In our view, standards are a key variable affecting how evaluations are processed and what expectations result. For example, depending on the standards used, the same level of success may be interpreted as a major accomplishment or dismissed as unimportant.
>
> *(1985: 108–9)*

To begin examining the role of standards, the researchers designed an experiment involving four experimental groups and a control. Subjects were told that the experiment involved something called "pattern recognition ability," defined as an innate ability some people had and others didn't. The researchers said subjects would be working in pairs on pattern recognition problems.

In fact, of course, there's no such thing as pattern recognition ability. The object of the experiment was to determine how information about this supposed ability affected subjects' subsequent behavior.

The first stage of the experiment was to "test" each subject's pattern recognition abilities. If you had been a subject in the experiment, you would have been shown a geometric pattern for 8 seconds, followed by two more patterns, each of which was similar to but not the same as the first one. Your task would be to choose which of the subsequent set had a pattern closest to the first one you saw. You would be asked to do this 20 times, and a computer would print out your "score." Half the subjects would be told that they had gotten 14 correct; the other half would be told that they had gotten only 6 correct—regardless of which patterns they matched with which. Depending on the luck of the draw, you would think you had done either quite well or quite badly. Notice, however, that you wouldn't really have any standard for judging your performance—maybe getting 4 correct would be considered a great performance.

At the same time you were given your score, however, you would also be given your "partner's

score," although both the "partners" and their "scores" would also be computerized fictions. (Subjects were told they would be communicating with their partners via computer terminals but would not be allowed to see each other.) If you were assigned a score of 14, you would be told your partner had a score of 6; if you were assigned 6, you would be told your partner had 14.

This procedure meant that you would enter the teamwork phase of the experiment believing either (1) you had done better than your partner or (2) you had done worse than your partner. This information constituted part of the "standard" you would be operating under in the experiment. In addition, half of each group was told that a score between 12 and 20 meant the subject *definitely* had pattern recognition ability; the other subjects were told that a score of 14 wasn't really high enough to prove anything definite. Thus, you would emerge from this with one of the following beliefs:

1. You are *definitely better* at pattern recognition than your partner.
2. You are *possibly better* than your partner.
3. You are *possibly worse* than your partner.
4. You are *definitely worse* than your partner.

The control group for this experiment was told nothing about their own abilities or those of their partners. In other words, they had no expectations.

The final step in the experiment was to set the "teams" to work. As before, you and your partner would be given an initial pattern, followed by a comparison pair to choose from. When you entered your choice in this round, however, you would be told what your partner had answered; then you would be asked to choose again. In your final choice, you could either stick with your original choice or switch. The "partner's" choice was, of course, created by the computer, and as you can guess, there were often a disagreements in the teams: 16 out of 20 times, in fact.

The dependent variable in this experiment was the extent to which subjects would switch their choices to match those of their partners. The researchers hypothesized that the *definitely better* group would switch least often, followed by the *probably better* group, followed by the *control group*, followed by the *probably worse* group, followed by the *definitely worse* group, who would switch most often.

The number of times subjects in the five groups switched their answers follows. Realize that each had 16 opportunities to do so. These data indicate that each of the researchers' expectations was correct—with the exception of the comparison between the *possibly worse* and *definitely worse* groups. Although the latter group was in fact the more likely to switch, the difference was too small to be taken as a confirmation of the hypothesis. (Chapter 16 will discuss the statistical tests that let researchers make decisions like this.)

Group	Mean Number of Switches
Definitely better	5.05
Possibly better	6.23
Control group	7.95
Possibly worse	9.23
Definitely worse	9.28

In more-detailed analyses, it was found that the same basic pattern held for both men and women, though it was somewhat clearer for women than for men. Here are the actual data:

	Mean Number of Switches	
	Women	Men
Definitely better	4.50	5.66
Possibly better	6.34	6.10
Control group	7.68	8.34
Possibly worse	9.36	9.09
Definitely worse	10.00	8.70

Because specific research efforts like this one sometimes seem extremely focused in their scope, you might wonder about their relevance to anything. As part of a larger research effort, however, studies like this one add concrete pieces to our understanding of more-general social processes.

It's worth taking a minute to consider some of the life situations where "expectation states" might have very real and important consequences. I've mentioned the case of jury deliberations. How about all forms of prejudice and discrimination? Or, consider how expectation states figure into job interviews or meeting your heartthrob's parents. If you think about it, you'll undoubtedly see other situations where these laboratory concepts apply in real life.

Alternative Experimental Settings

Although we tend to equate the terms *experiment* and *laboratory experiment*, many important social science experiments occur outside controlled settings, as we've seen in our example of the Rosenthal-Jacobson study of the Pygmalion effect. Two other special circumstances deserve mention here: web-based experiments and "natural" experiments.

Web-Based Experiments

Increasingly, researchers are using the World Wide Web as a vehicle for conducting experiments. Because representative samples are not essential in most experiments, researchers can often use volunteers who respond to invitations online. One site you might visit to get a better idea of this form of experimentation is Online Social Psychology Studies. This website offers hot links to numerous professional and student research projects on such topics as "interpersonal relations," "beliefs and attitudes," and "personality and individual differences." In addition, the site offers some resources for conducting web experiments. (See the link on this book's website: http://www.cengage.com/sociology/babbie.)

"Natural" Experiments

Important social science experiments can occur in the course of normal social events, outside controlled settings. Sometimes nature designs and executes experiments that we can observe and analyze; sometimes social and political decision makers serve this natural function.

Imagine, for example, that a hurricane has struck a particular town. Some residents of the town suffer severe financial damages, and others escape relatively lightly. What, we might ask, are the behavioral consequences of suffering a natural disaster? Are those who suffer most more likely to take precautions against future disasters than those who suffer least are? To answer these questions, we might interview residents of the town some time after the hurricane. We might question them regarding their precautions before the hurricane and the ones they're currently taking, comparing the people who suffered greatly from the hurricane with those who suffered relatively little. In this fashion, we might take advantage of a natural experiment, which we could not have arranged even if we'd been perversely willing to do so.

A similar example comes from the annals of social research concerning World War II. After the war ended, social researchers undertook retrospective surveys of wartime morale among civilians in several German cities. Among other things, they wanted to determine the effect of mass bombing on the morale of civilians. They compared the reports of wartime morale of residents in heavily bombed cities with reports from cities that received relatively little bombing. (Bombing did not reduce morale.)

Because the researcher must take things pretty much as they occur, natural experiments raise many of the validity problems discussed earlier. Thus, when Stanislav Kasl, Rupert Chisolm, and Brenda Eskenazi (1981) chose to study the impact that the Three Mile Island (TMI) nuclear accident in Pennsylvania had on plant workers, they had to be especially careful in the study design:

> Disaster research is necessarily opportunistic, quasi-experimental, and after-the-fact. In the terminology of Campbell and Stanley's classical analysis of research designs, our study falls into the "static-group comparison" category, considered one of the weak research designs. However, the weaknesses are potential and

their actual presence depends on the unique circumstances of each study.

> *(1981: 474)*

The foundation of this study was a survey of the people who had been working at Three Mile Island on March 28, 1979, when the cooling system failed in the number 2 reactor and began melting the uranium core. The survey was conducted five to six months after the accident. Among other things, the survey questionnaire measured workers' attitudes toward working at nuclear power plants. If they had measured only the TMI workers' attitudes after the accident, the researchers would have had no idea whether attitudes had changed as a consequence of the accident. But they improved their study design by selecting another, nearby—seemingly comparable—nuclear power plant (abbreviated as PB) and surveyed workers there as a control group: hence their reference to a static-group comparison.

Even with an experimental and a control group, the authors were wary of potential problems in their design. In particular, their design was based on the idea that the two sets of workers were equivalent to each other, except for the single fact of the accident. The researchers could have assumed this if they had been able to assign workers to the two plants randomly, but of course that was not the case. Instead, they needed to compare characteristics of the two groups and infer whether or not they were equivalent. Ultimately, the researchers concluded that the two sets of workers were very much alike, and the plant the employees worked at was merely a function of where they lived.

Even granting that the two sets of workers were equivalent, the researchers faced another problem of comparability. They could not contact all the workers who had been employed at TMI at the time of the accident. The researchers discussed the problem as follows:

> One special attrition problem in this study was the possibility that some of the no-contact nonrespondents among the TMI subjects, but not PB subjects, had permanently left the area because of the accident. This biased attrition

would, most likely, attenuate the estimated extent of the impact. Using the evidence of disconnected or "not in service" telephone numbers, we estimate this bias to be negligible (1 percent).

(Kasl, Chisolm, and Eskenazi 1981: 475)

The TMI example points to both the special problems involved in natural experiments and the possibility for taking those problems into account. Social research generally requires ingenuity and insight; natural experiments call for a little more than the average.

Earlier in this chapter, we used a hypothetical example of studying whether an African American history film reduced prejudice. Sandra Ball-Rokeach, Joel Grube, and Milton Rokeach (1981) were able to address that topic in real life through a natural experiment. In 1977, the television dramatization of Alex Haley's *Roots,* a historical saga about African Americans, was presented by ABC on eight consecutive nights. It garnered the largest audiences in television history up to that time. Ball-Rokeach and her colleagues wanted to know whether *Roots* changed white Americans' attitudes toward African Americans. Their opportunity arose in 1979, when a sequel—*Roots: The Next Generation*—was televised. Although it would have been nice (from a researcher's point of view) to assign random samples of Americans either to watch or not to watch the show, that wasn't possible. Instead, the researchers selected four samples in Washington State and mailed questionnaires that measured attitudes toward African Americans. Following the last episode of the show, respondents were called and asked how many, if any, episodes they had watched. Subsequently, questionnaires were sent to respondents, remeasuring their attitudes toward African Americans.

By comparing attitudes before and after for both those who watched the show and those who didn't, the researchers reached several conclusions. For example, they found that people with already egalitarian attitudes were much more likely to watch the show than were those who were more prejudiced toward African Americans: a self-selection phenomenon. Comparing the before

and after attitudes of those who watched the show, moreover, suggested the show itself had little or no effect. Those who watched it were no more egalitarian afterward than they had been before.

This example anticipates the subject of Chapter 12, evaluation research, which can be seen as a special type of natural experiment. As you'll see, evaluation research involves taking the logic of experimentation into the field to observe and evaluate the effects of stimuli in real life. Because this is an increasingly important form of social research, an entire chapter is devoted to it.

Strengths and Weaknesses of the Experimental Method

Experiments are the primary tool for studying causal relationships. However, like all research methods, experiments have both strengths and weaknesses.

The chief advantage of a controlled experiment lies in the isolation of the experimental variable's impact over time. This is seen most clearly in terms of the basic experimental model. A group of experimental subjects are found, at the outset of the experiment, to have a certain characteristic; following the administration of an experimental stimulus, they are found to have a different characteristic. To the extent that subjects have experienced no other stimuli, we may conclude that the change of characteristics is attributable to the experimental stimulus.

Further, because individual experiments are often rather limited in scope, requiring relatively little time and money and relatively few subjects, we often can replicate a given experiment several times using several different groups of subjects. (This isn't always the case, of course, but it's usually easier to repeat experiments than, say, surveys.) As in all other forms of scientific research, replication of research findings strengthens our confidence in the validity and generalizability of those findings.

The greatest weakness of laboratory experiments lies in their artificiality. Social processes that occur in a laboratory setting might not necessarily occur in natural social settings. For example, an African American history film might genuinely reduce prejudice among a group of experimental subjects. This would not necessarily mean, however, that the same film shown in neighborhood movie theaters throughout the country would reduce prejudice among the general public. Artificiality is not as much of a problem, of course, for natural experiments as for those conducted in the laboratory.

In discussing several of the sources of internal and external invalidity mentioned by Campbell, Stanley, and Cook, we saw that we can create experimental designs that logically control such problems. This possibility points to one of the great advantages of experiments: They lend themselves to a logical rigor that is often much more difficult to achieve in other modes of observation.

Ethics and Experiments

As you've probably seen, researchers must consider many important ethical issues in conducting social science experiments. I'll mention only two here.

First, experiments almost always involve deception. In most cases, explaining the purpose of the experiment to subjects would probably cause them to behave differently—trying to look less prejudiced, for example. It's important, therefore, to determine (1) whether a particular deception is essential to the experiment and (2) whether the value of what may be learned from the experiment justifies the ethical violation.

Second, experiments are typically intrusive. Subjects often are placed in unusual situations and asked to undergo unusual experiences. Even when the subjects are not physically injured (don't do that, by the way), there is always the possibility that they will be psychologically damaged, as some of the previous examples in this chapter have illustrated. As with the matter of deception, you'll find yourself balancing the potential value of the research against the potential damage to subjects.

MAIN POINTS

Introduction

- In experiments, social researchers typically select a group of subjects, do something to them, and observe the effect of what was done.

Topics Appropriate for Experiments

- Experiments are an excellent vehicle for the controlled testing of causal processes.

The Classical Experiment

- The classical experiment tests the effect of an experimental stimulus (the independent variable) on a dependent variable through the pretesting and posttesting of experimental and control groups.

- It is generally less important that a group of experimental subjects be representative of some larger population than that experimental and control groups be similar to each other.

- A double-blind experiment guards against experimenter bias, because neither the experimenter nor the subject knows which subjects are in the control group(s) and which in the experimental group(s).

Selecting Subjects

- Probability sampling, randomization, and matching are all methods of achieving comparability in the experimental and control groups. Randomization is the generally preferred method. In some designs, it can be combined with matching.

Variations on Experimental Design

- Campbell and Stanley describe three forms of preexperiments: the one-shot case study, the one-group pretest-posttest design, and the static-group comparison. None of these designs features all the controls available in a true experiment.

- Campbell and Stanley list, among others, 12 sources of internal invalidity in experimental design. The classical experiment with random assignment of subjects guards against each of these problems.

- Experiments also face problems of external invalidity: Experimental findings may not reflect real life.

- The interaction of testing and stimulus is an example of external invalidity that the classical experiment does not guard against.
- The Solomon four-group design and other variations on the classical experiment can safeguard against external invalidity.
- Campbell and Stanley suggest that, given proper randomization in the assignment of subjects to the experimental and control groups, there is no need for pretesting in experiments.

An Illustration of Experimentation

- Experiments on "expectation states" demonstrate experimental designs and show how experiments can prove relevant to real-world concerns.

Alternative Experimental Settings

- More and more, researchers are using the Internet for conducting experiments.
- Natural experiments often occur in the course of social life in the real world, and social researchers can implement them in somewhat the same way they would design and conduct laboratory experiments.

Strengths and Weaknesses of the Experimental Method

- Like all research methods, experiments have strengths and weaknesses. Their primary weakness is artificiality: What happens in an experiment may not reflect what happens in the outside world. Their strengths include the isolation of the independent variable, which permits causal inferences; the relative ease of replication; and scientific rigor.

Ethics and Experiments

- Experiments typically involve deceiving subjects.
- By their intrusive nature, experiments open the possibility of inadvertently causing damages to subjects.

KEY TERMS

The following terms are defined in context in the chapter and at the bottom of the page where the term is introduced, as well as in the comprehensive glossary at the back of the book.

control group	matching
double-blind experiment	posttesting
experimental group	pretesting
external invalidity	randomization
internal invalidity	

PROPOSING SOCIAL RESEARCH: EXPERIMENTS

In the next series of exercises, we'll focus on specific data-collection techniques, beginning with experiments here. If you're doing these exercises as part of an assignment in the course, your instructor will tell you whether you should skip those chapters dealing with methods you won't use. If you're doing these exercises on your own, to improve your understanding of the topics in the book, you can temporarily modify your proposed data-collection method and explore how you would research your topic using the method at hand—in this case, experimentation.

In the proposal, you'll describe the experimental stimulus and how it will be administered, as well as detailing the experimental and control groups you'll use. You'll also describe the pretesting and posttesting that will be involved in your experiment. What will be the setting for your experiments: a laboratory or more natural circumstances?

It may be appropriate for you to conduct a double-blind experiment, in which case you should describe how you will accomplish it. You may also need to explore some of the internal and external problems of validity that might complicate your analysis of your results.

Finally, the experimental model is used to test specific hypotheses, so you should detail how you will accomplish that in terms of your study.

REVIEW QUESTIONS AND EXERCISES

1. In the library or on the web, locate a research report of an experiment. Identify the dependent variable and the stimulus.

2. Pick 6 of the 12 sources of internal invalidity discussed in this chapter and make up examples (not discussed in the chapter) to illustrate each.

3. Create a hypothetical experimental design that illustrates one of the problems of external invalidity.

4. Think of a recent natural disaster you've witnessed or read about. Frame a research question

that might be studied by treating that disaster as a natural experiment. In two or three paragraphs, outline how the study might be done.

5. In this chapter, we looked briefly at the problem of "placebo effects." On the web, find a study in which the placebo effect figured importantly. Write a brief report on the study, including the source of your information. (*Hint:* you might want to do a search on "placebo.")

SPSS EXERCISES

See the booklet that accompanies your text for exercises using SPSS (Statistical Package for the Social Sciences). There are exercises offered for each chapter, and you'll also find a detailed primer on using SPSS.

Online Study Resources

If your book came with an access code card, visit www.cengage.com/login to register. To purchase access, please visit www.ichapters.com.

1. Before you do your final review of the chapter, take the *CengageNOW* pretest to help identify the areas on which you should concentrate. You'll find information on this online tool, as well as instructions on how to access all of its great resources, in the front of the book.

2. As you review, take advantage of the *CengageNOW* personalized study plan, based on your quiz results. Use this study plan with its interactive exercises and other resources to master the material.

3. When you're finished with your review, take the posttest to confirm that you're ready to move on to the next chapter.

WEBSITE FOR *THE PRACTICE OF SOCIAL RESEARCH* 12TH EDITION

Go to your book's website at www.cengage.com/ sociology/babbie for tools to aid you in studying for your exams. You'll find *Tutorial Quizzes* with feedback, *Internet Exercises, Flash Cards, Glossaries,* and *Essay Quizzes,* as well as *InfoTrac College Edition* search terms, suggestions for additional reading, *Web Links,* and primers for using data-analysis software such as SPSS.

CHAPTER 9

Survey Research

CHAPTER OVERVIEW

Researchers have many methods for collecting data through surveys—from mail questionnaires to personal interviews to online surveys conducted over the Internet. Social researchers should know how to select an appropriate method and how to implement it effectively.

CengageNOW for Sociology

Use this online tool to help you make the grade on your next exam. After reading this chapter, go to "Online Study Resources" at the end of the chapter for instructions on how to benefit from *CengageNOW*.

Introduction

Surveys are a very old research technique. In the Old Testament, for example, we find the following:

> After the plague the Lord said to Moses and to Eleazar the son of Aaron, the priest, "Take a census of all the congregation of the people of Israel, from twenty years old and upward."
>
> *(Numbers 26: 1–2)*

Ancient Egyptian rulers conducted censuses to help them administer their domains. Jesus was born away from home because Joseph and Mary were journeying to Joseph's ancestral home for a Roman census.

A little-known survey was attempted among French workers in 1880. A German political sociologist mailed some 25,000 questionnaires to workers to determine the extent of their exploitation by employers. The rather lengthy questionnaire included items such as these:

> Does your employer or his representative resort to trickery in order to defraud you of a part of your earnings?
>
> If you are paid piece rates, is the quality of the article made a pretext for fraudulent deductions from your wages?

The survey researcher in this case was not George Gallup but Karl Marx ([1880] 1956: 208). Though 25,000 questionnaires were mailed out, there is no record of any being returned.

Today, survey research is a frequently used mode of observation in the social sciences. In a typical survey, the researcher selects a sample of respondents and administers a standardized questionnaire to them. Chapter 7 discussed sampling techniques in detail. This chapter discusses how to prepare a questionnaire and describes the various options for administering it so that respondents answer your questions adequately.

The chapter includes a short discussion of secondary analysis, the analysis of survey data collected by someone else. This use of survey results has become an important aspect of survey research in recent years, and it's especially useful for students and others with scarce research funds.

Let's begin by looking at the kinds of topics that researchers can appropriately study by using survey research.

Topics Appropriate for Survey Research

Surveys may be used for descriptive, explanatory, and exploratory purposes. They are chiefly used in studies that have individual people as the units of analysis. Although this method can be used for other units of analysis, such as groups or interactions, some individual persons must serve as **respondents** or informants. Thus, we could undertake a survey in which divorces were the unit of analysis, but we would need to administer the survey questionnaire to the participants in the divorces (or to some other respondents).

Survey research is probably the best method available to the social researcher who is interested in collecting original data for describing a population too large to observe directly. Careful probability sampling provides a group of respondents whose characteristics may be taken to reflect those of the larger population, and carefully constructed standardized questionnaires provide data in the same form from all respondents.

Surveys are also excellent vehicles for measuring attitudes and orientations in a large population. Public opinion polls—for example, Gallup, Harris, Roper, and Yankelovich—are well-known examples of this use. Indeed, polls have become so prevalent that at times the public seems unsure what to think of them. Pollsters are criticized by those who don't think (or want to believe) that polls are accurate (candidates who are "losing" in polls often tell voters not to trust the polls). But

---

respondent A person who provides data for analysis by responding to a survey questionnaire.

polls are also criticized for being too accurate—for example, when exit polls on election day are used to predict a winner before the actual voting is complete.

The general attitude toward public opinion research is further complicated by scientifically unsound "surveys" that nonetheless capture people's attention because of the topics they cover and/or their "findings." A good example is the "Hite Reports" on human sexuality. While enjoying considerable attention in the popular press, Shere Hite was roundly criticized by the research community for her data-collection methods. For example, a 1987 Hite report was based on questionnaires completed by women around the country—but which women? Hite reported that she distributed some 100,000 questionnaires through various organizations, and around 4,500 were returned.

Now 4,500 and 100,000 are large numbers in the context of survey sampling. However, given Hite's research methods, her 4,500 respondents didn't necessarily represent U.S. women any more than the *Literary Digest*'s enormous 1936 sample represented the U.S. electorate when their 2 million sample ballots indicated that Alf Landon would bury FDR in a landslide.

Sometimes, people use the pretense of survey research for quite different purposes. For example, you may have received a telephone call indicating you've been selected for a survey, only to find that the first question was "How would you like to make thousands of dollars a week right there in your own home?" Or you may have been told you could win a prize if you could name the president whose picture is on the penny. (Tell them it's Elvis.) Unfortunately, a few unscrupulous telemarketers try to prey on the general cooperation people have given to survey researchers.

By the same token, political parties and charitable organizations have begun conducting phony "surveys." Often under the guise of collecting public opinion about some issue, callers ultimately ask respondents for a monetary contribution.

Recent political campaigns have produced another form of bogus survey, the "push poll." Here's what the American Association for Public Opinion

Polling has said in condemning this practice (see also Figure 3-1):

> A "push poll" is a telemarketing technique in which telephone calls are used to canvass potential voters, feeding them false or misleading "information" about a candidate under the pretense of taking a poll to see how this "information" affects voter preferences. In fact, the intent is not to measure public opinion but to manipulate it—to "push" voters away from one candidate and toward the opposing candidate. Such polls defame selected candidates by spreading false or misleading information about them. The intent is to disseminate campaign propaganda under the guise of conducting a legitimate public opinion poll.
>
> *(Bednarz 1996)*

In short, the labels "survey" and "poll" are sometimes misused. Done properly, however, survey research can be a useful tool of social inquiry. Designing useful (and trustworthy) survey research begins with formulating good questions. Let's turn to that topic now.

Guidelines for Asking Questions

In social research, variables are often operationalized when researchers ask people questions as a way of getting data for analysis and interpretation. Sometimes the questions are asked by an interviewer; sometimes they are written down and given to respondents for completion. In other cases, several general guidelines can help researchers frame and ask questions that serve as excellent operationalizations of variables while avoiding pitfalls that can result in useless or even misleading information.

Surveys include the use of a **questionnaire**—an instrument specifically designed to elicit information that will be useful for analysis. Although some of the specific points to follow are more appropriate to structured questionnaires than to the more open-ended questionnaires used in qualitative, in-depth interviewing, the underlying logic is valuable whenever we ask people questions in order to gather data.

Choose Appropriate Question Forms

Let's begin with some of the options available to you in creating questionnaires. These options include using questions or statements and choosing **open-ended** or **closed-ended questions.**

Questions and Statements

Although the term *questionnaire* suggests a collection of questions, an examination of a typical questionnaire will probably reveal as many statements as questions. This is not without reason. Often, the researcher is interested in determining the extent to which respondents hold a particular attitude or perspective. If you can summarize the attitude in a fairly brief statement, you can present that statement and ask respondents whether they agree or disagree with it. As you may remember, Rensis Likert greatly formalized this procedure through the creation of the Likert scale, a format in which respondents are asked to strongly agree, agree, disagree, or strongly disagree, or perhaps strongly approve, approve, and so forth.

Both questions and statements can be used profitably. Using both in a given questionnaire gives you more flexibility in the design of items and can make the questionnaire more interesting as well.

Open-Ended and Closed-Ended Questions

In asking questions, researchers have two options. They can ask open-ended questions, in which case the respondent is asked to provide his or her own answers to the questions. For example, the respondent may be asked, "What do you feel is the most important issue facing the United States today?" and be provided with a space to write in the answer (or be asked to report it verbally to an interviewer). As we'll see in Chapter 10, in-depth, qualitative interviewing relies almost exclusively on open-ended questions. However, they are also used in survey research.

In the case of closed-ended questions, the respondent is asked to select an answer from among a list provided by the researcher. Closed-ended questions are very popular in survey research because they provide a greater uniformity of responses and are more easily processed than open-ended ones.

Open-ended responses must be coded before they can be processed for computer analysis, as we'll see in Chapter 14. This coding process often requires the researcher to interpret the meaning of responses, opening the possibility of misunderstanding and researcher bias. There is also a danger that some respondents will give answers that are essentially irrelevant to the researcher's intent. Closed-ended responses, on the other hand, can often be transferred directly into a computer format.

The chief shortcoming of closed-ended questions lies in the researcher's structuring of responses. When the relevant answers to a given question are relatively clear, there should be no problem. In other cases, however, the researcher's structuring of responses may overlook some important responses. In asking about "the most important issue facing the United States," for example, his or her checklist of issues might omit certain

---

questionnaire A document containing questions and other types of items designed to solicit information appropriate for analysis. Questionnaires are used primarily in survey research but also in experiments, field research, and other modes of observation.

open-ended questions Questions for which the respondent is asked to provide his or her own answers. In-depth, qualitative interviewing relies almost exclusively on open-ended questions.

closed-ended questions Survey questions in which the respondent is asked to select an answer from among a list provided by the researcher. Popular in survey research because they provide a greater uniformity of responses and are more easily processed than open-ended questions.

issues that respondents would have said were important.

The construction of closed-ended questions should be guided by two structural requirements. First, the response categories provided should be exhaustive: They should include all the possible responses that might be expected. Often, researchers ensure this by adding a category such as "Other (Please specify: _____)." Second, the answer categories must be mutually exclusive: The respondent should not feel compelled to select more than one. (In some cases, you may wish to solicit multiple answers, but these may create difficulties in data processing and analysis later on.) To ensure that your categories are mutually exclusive, carefully consider each combination of categories, asking yourself whether a person could reasonably choose more than one answer. In addition, it's useful to add an instruction to the question asking the respondent to select the one best answer, but this technique is not a satisfactory substitute for a carefully constructed set of responses.

Make Items Clear

It should go without saying that questionnaire items need to be clear and unambiguous, but the broad proliferation of unclear and ambiguous questions in surveys makes the point worth emphasizing. We can become so deeply involved in the topic under examination that opinions and perspectives are clear to us but not to our respondents—many of whom have paid little or no attention to the topic. Or, if we have only a superficial understanding of the topic, we may fail to specify the intent of a question sufficiently. The question "What do you think about the proposed peace plan?" may evoke in the respondent a counterquestion: "Which proposed peace plan?" Questionnaire items should be precise so that the respondent knows exactly what the researcher is asking. The possibilities for misunderstanding are endless, and no researcher is immune (Polivka and Rothgeb 1993).

One of the most established research projects in the United States is the Census Bureau's ongoing "Current Population Survey" or CPS, which measures, among other critical data, the nation's unemployment rate. A part of the measurement of employment patterns focuses on a respondent's activities during "last week," by which the Census Bureau means Sunday through Saturday. Studies undertaken to determine the accuracy of the survey found that more than half the respondents took "last week" to include only Monday through Friday. By the same token, whereas the Census Bureau defines "working full-time" as 35 or more hours a week, the same evaluation studies showed that some respondents used the more traditional definition of 40 hours per week. As a consequence, the wording of these questions in the CPS was modified in 1994 to specify the Census Bureau's definitions.

Similarly, the use of the term *Native American* to mean American Indian often produces an overrepresentation of that ethnic group in surveys. Clearly, many respondents understand the term to mean "born in the United States."

Avoid Double-Barreled Questions

Frequently, researchers ask respondents for a single answer to a question that actually has multiple parts. That seems to happen most often when the researcher has personally identified with a complex question. For example, you might ask respondents to agree or disagree with the statement "The United States should abandon its space program and spend the money on domestic programs." Although many people would unequivocally agree with the statement and others would unequivocally disagree, still others would be unable to answer. Some would want to abandon the space program and give the money back to the taxpayers. Others would want to continue the space program but also put more money into domestic programs. These latter respondents could neither agree nor disagree without misleading you.

As a general rule, whenever the word *and* appears in a question or questionnaire statement, check whether you're asking a double-barreled question. See "Double-Barreled and Beyond" for some imaginative variations on this theme.

Double-Barreled and Beyond

Even established, professional researchers have sometimes created double-barreled questions and worse. Consider this question, asked of U.S. citizens in April 1986, at a time when the country's relationship with Libya was at an especially low point. Some observers suggested that the United States might end up in a shooting war with the small North African nation. The Harris Poll sought to find out what U.S. public opinion was.

> If Libya now increases its terrorist acts against the U.S. and we keep inflicting more damage on Libya, then inevitably it will all end in the U.S. going to war and finally invading that country which would be wrong.

Respondents were given the opportunity of answering "Agree," "Disagree," or "Not sure." Notice the elements contained in the complex statement:

1. Will Libya increase its terrorist acts against the U.S.?
2. Will the U.S. inflict more damage on Libya?
3. Will the U.S. inevitably or otherwise go to war against Libya?
4. Would the U.S. invade Libya?
5. Would that be right or wrong?

These several elements offer the possibility of numerous points of view—far more than the three alternatives offered to the survey respondents. Even if we were to assume hypothetically that Libya would "increase its terrorist attacks" and the United States would "keep inflicting more damage" in return, you might have any one of at least seven distinct expectations about the outcome:

	U.S. will not go to war	War is probable but not inevitable	War is inevitable
U.S. will not invade Libya	1	2	3
U.S. will invade Libya but it would be wrong		4	5
U.S. will invade Libya and it would be right		6	7

The examination of prognoses about the Libyan situation is not the only example of double-barreled questions sneaking into public opinion research. Here are some questions the Harris Poll asked in an attempt to gauge U.S. public opinion about then Soviet General Secretary Gorbachev:

> He looks like the kind of Russian leader who will recognize that both the Soviets and the Americans can destroy each other with nuclear missiles so it is better to come to verifiable arms control agreements.
>
> He seems to be more modern, enlightened, and attractive, which is a good sign for the peace of the world.
>
> Even though he looks much more modern and attractive, it would be a mistake to think he will be much different from other Russian leaders.

How many elements can you identify in each of the questions? How many possible opinions could people have in each case? What does a simple "agree" or "disagree" really mean in such cases?

Sources: Reported in *World Opinion Update,* October 1985 and May 1986, respectively.

Respondents Must Be Competent to Answer

In asking respondents to provide information, you should continually ask yourself whether they can do so reliably. In a study of child rearing, you might ask respondents to report the age at which they first talked back to their parents. Quite aside from the problem of defining *talking back to parents,* it's doubtful that most respondents would remember with any degree of accuracy.

As another example, student government leaders occasionally ask their constituents to indicate how students' fees ought to be spent. Typically, respondents are asked to indicate the percentage of available funds that should be devoted to a long list of activities. Without a fairly good knowledge of the nature of those activities and the costs involved in them, the respondents cannot provide meaningful answers. Administrative costs, for example, will receive little support although they may be essential to the program as a whole.

One group of researchers examining the driving experience of teenagers insisted on asking an open-ended question concerning the number of miles driven since receiving a license. Although consultants argued that few drivers would be able to estimate such information with any accuracy, the question was asked nonetheless. In response, some teenagers reported driving hundreds of thousands of miles.

Respondents Must Be Willing to Answer

Often, we would like to learn things from people that they are unwilling to share with us. For example, Yanjie Bian indicates that it has often been difficult to get candid answers from people in China.

> [Here] people are generally careful about what they say on nonprivate occasions in order to survive under authoritarianism. During the Cultural Revolution between 1966 and 1976, for example, because of the radical political agenda and political intensity throughout the country, it was almost impossible to use survey techniques to collect valid and reliable data inside China about the Chinese people's life experiences, characteristics, and attitudes towards the Communist regime.
>
> *(1994: 19–20)*

Sometimes, U.S. respondents say they're undecided when, in fact, they have an opinion but think they're in a minority. Under that condition, they may be reluctant to tell a stranger (the interviewer) what that opinion is. Given this problem, the Gallup Organization, for example, has used a "secret ballot" format, which simulates actual election conditions, in that the "voter" enjoys complete anonymity. In an analysis of the Gallup Poll election data from 1944 to 1988, Andrew Smith and G. F. Bishop (1992) have found that this technique substantially reduced the percentage of respondents who said they were undecided about how they would vote.

This problem is not limited to survey research, however. Richard Mitchell (1991: 100) faced a similar problem in his field research among U.S. survivalists:

> Survivalists, for example, are ambivalent about concealing their identities and inclinations. They realize that secrecy protects them from the ridicule of a disbelieving majority, but enforced separatism diminishes opportunities for recruitment and information exchange. . . .
>
> "Secretive" survivalists eschew telephones, launder their mail through letter exchanges, use nicknames and aliases, and carefully conceal their addresses from strangers. Yet once I was invited to group meetings, I found them cooperative respondents.

Questions Should Be Relevant

Similarly, questions asked in a questionnaire should be relevant to most respondents. When attitudes are requested on a topic that few respondents have thought about or really care about, the results are not likely to be useful. Of course, because the respondents may express attitudes even though they've never given any thought to the issue, you run the risk of being misled.

This point is illustrated occasionally when researchers ask for responses relating to fictitious people and issues. In one political poll I conducted, I asked respondents whether they were familiar with each of 15 political figures in the community. As a methodological exercise, I made up a name: Tom Sakumoto. In response, 9 percent of the respondents said they were familiar with him. Of those respondents familiar with him, about half reported seeing him on television and reading about him in the newspapers.

When you obtain responses to fictitious issues, you can disregard those responses. But when the issue is real, you may have no way of telling which responses genuinely reflect attitudes and which reflect meaningless answers to an irrelevant question.

Ideally, we would like respondents to simply report that they don't know, have no opinion, or are undecided in those instances where that is the case. Unfortunately, however, they often make up answers.

Short Items Are Best

In the interests of being unambiguous and precise and of pointing to the relevance of an issue, researchers tend to create long and complicated items. That should be avoided. Respondents are often unwilling to study an item in order to understand it. The respondent should be able to read an item quickly, understand its intent, and select or provide an answer without difficulty. In general, assume that respondents will read items quickly and give quick answers. Accordingly, provide clear, short items that will not be misinterpreted under those conditions.

Avoid Negative Items

The appearance of a negation in a questionnaire item paves the way for easy misinterpretation. Asked to agree or disagree with the statement "The United States should not recognize Cuba," a sizable portion of the respondents will read over the word *not* and answer on that basis. Thus, some will agree with the statement when they're in favor of recognition, and others will agree when they oppose it. And you may never know which are which.

Similar considerations apply to other "negative" words. In a study of support for civil liberties, for example, respondents were asked whether they felt "the following kinds of people should be *prohibited* from teaching in public schools" and were presented with a list including such items as a Communist, a Ku Klux Klansman, and so forth. The response categories "yes" and "no" were given beside each entry. A comparison of the responses to this item with other items reflecting support for civil liberties strongly suggested that many respondents gave the answer "yes" to indicate willingness for such a person to teach, rather than to indicate

that such a person should be prohibited from teaching. (A later study in the series using the answer categories "permit" and "prohibit" produced much clearer results.)

In 1993 a national survey commissioned by the American Jewish Committee produced shocking results: One American in five believed that the Nazi Holocaust—in which six million Jews were reportedly killed—never happened; further, one in three Americans expressed some doubt that it had occurred. This research finding suggested that the Holocaust Revisionist movement in America was powerfully influencing public opinion ("1 in 5 Polled Voices Doubt on Holocaust" 1993).

In the aftermath of this shocking news, researchers reexamined the actual question that had been asked: "Does it seem possible or does it seem impossible to you that the Nazi extermination of the Jews never happened?" On reflection, it seemed clear that the complex, double-negative question could have confused some respondents.

A new survey was commissioned and asked, "Does it seem possible to you that the Nazi extermination of the Jews never happened, or do you feel certain that it happened?" In the follow-up survey, only 1 percent of the respondents believed the Holocaust never happened, and another 8 percent said they weren't sure ("Poll on Doubt of Holocaust Is Corrected" 1994).

Avoid Biased Items and Terms

Recall from our discussion of conceptualization and operationalization in Chapter 5 that there are no ultimately true meanings for any of the concepts we typically study in social science. *Prejudice* has no ultimately correct definition; whether a given person is prejudiced depends on our definition of that term. The same general principle applies to the responses we get from people completing a questionnaire.

The meaning of someone's response to a question depends in large part on its wording. This is true of every question and answer. Some questions seem to encourage particular responses more than other questions do. In the context of questionnaires, **bias**

---

bias That quality of a measurement device that tends to result in a misrepresentation of what is being measured in a particular direction. For example, the questionnaire item "Don't you agree that the president is doing a good job?" would be biased in that it would generally encourage more favorable responses.

refers to any property of questions that encourages respondents to answer in a particular way.

Most researchers recognize the likely effect of a question that begins, "Don't you agree with the President of the United States that . . ." No reputable researcher would use such an item. Unhappily, the biasing effect of items and terms is far subtler than this example suggests.

The mere identification of an attitude or position with a prestigious person or agency can bias responses. The item "Do you agree or disagree with the recent Supreme Court decision that . . ." would have a similar effect. Such wording may not produce consensus or even a majority in support of the position identified with the prestigious person or agency, but it will likely increase the level of support over what would have been obtained without such identification.

Sometimes the impact of different forms of question wording is relatively subtle. For example, when Kenneth Rasinski (1989) analyzed the results of several General Social Survey studies of attitudes toward government spending, he found that the way programs were identified had an impact on the amount of public support they received. Here are some comparisons:

More Support	Less Support
"Assistance to the poor"	"Welfare"
"Halting rising crime rate"	"Law enforcement"
"Dealing with drug addiction"	"Drug rehabilitation"
"Solving problems of big cities"	"Assistance to big cities"
"Improving conditions of blacks"	"Assistance to blacks"
"Protecting social security"	"Social security"

In 1986, for example, 62.8 percent of the respondents said too little money was being spent on "assistance to the poor," whereas in a matched survey that year, only 23.1 percent said we were spending too little on "welfare."

In this context, be wary of what researchers call the *social desirability* of questions and answers. Whenever we ask people for information, they answer through a filter of what will make them look good. This is especially true if they're interviewed face-to-face. Thus, for example, during the 2008 Democratic primary, many voters who might have been reluctant to vote for an African American (Barack Obama) or a woman (Hillary Clinton) might have also been reluctant to admit their racial or gender prejudice to a survey interviewer. (Some, to be sure, were not reluctant to say how they felt.)

The best way to guard against this problem is to imagine how you would feel giving each of the answers you intend to offer to respondents. If you would feel embarrassed, perverted, inhumane, stupid, irresponsible, or otherwise socially disadvantaged by any particular response, give serious thought to how willing others will be to give those answers.

The biasing effect of particular wording is often difficult to anticipate. For example, in both surveys and experiments, researchers sometimes ask respondents to consider hypothetical situations and say how they think they would behave. Those situations often involve other people, however, and the names used can affect responses. For instance, researchers have long known that male names for the hypothetical people can produce different responses than female names do. Research by Joseph Kasof (1993) points to the importance of what the specific names are: whether they generally evoke positive or negative images in terms of attractiveness, age, intelligence, and so forth. Kasof's review of past research suggests there has been a tendency to use more positively valued names for men than for women.

The Center for Disease Control (Choi and Pak 2005) has provided an excellent analysis of various ways in which your choice of terms can bias and otherwise confuse responses to questionnaires. Among other things, they warn against using ambiguous, technical, uncommon, or vague words. Their thorough analysis provides many concrete illustrations.

As in all other research, carefully examine the purpose of your inquiry and construct items that will be most useful to it. You should never be misled into thinking there are ultimately "right" and "wrong" ways of asking the questions. When in doubt about the best question to ask, moreover, remember that you should ask more than one.

These, then, are some general guidelines for writing questions to elicit data for analysis and interpretation. Next we look at how to construct questionnaires.

Questionnaire Construction

Questionnaires are used in connection with many modes of observation in social research. Although structured questionnaires are essential to and most directly associated with survey research, they are also widely used in experiments, field research, and other data-collection activities. For this reason, questionnaire construction can be an important practical skill for researchers. As we discuss the established techniques for constructing questionnaires, let's begin with some issues of questionnaire format.

General Questionnaire Format

The format of a questionnaire is just as important as the nature and wording of the questions asked. An improperly laid out questionnaire can lead respondents to miss questions, confuse them about the nature of the data desired, and even lead them to throw the questionnaire away.

As a general rule, a questionnaire should be spread out and uncluttered. If a self-administered questionnaire is being designed, inexperienced researchers tend to fear that their questionnaire will look too long; as a result, they squeeze several questions onto a single line, abbreviate questions, and try to use as few pages as possible. These efforts are ill-advised and even dangerous. Putting more than one question on a line will cause some respondents to miss the second question altogether. Some respondents will misinterpret abbreviated questions. More generally, respondents who find they have spent considerable time on the first page of what seemed like a short questionnaire will be more demoralized than respondents who quickly complete the first several pages of what initially seemed like a rather long form. Moreover, the latter will have made fewer errors and will not

have been forced to reread confusing, abbreviated questions. Nor will they have been forced to write a long answer in a tiny space.

Similar problems can arise for interviewers in a face-to-face or telephone interview. Like respondents to a self-administered questionnaire, interviewers may miss questions, lose their place, and generally become frustrated and flustered. Interview questionnaires need to be laid out in a way that supports the interviewer's work, including special instructions and guidelines that go beyond what respondents to a self-administered questionnaire would need.

The desirability of spreading out questions in the questionnaire cannot be overemphasized. Squeezed-together questionnaires are disastrous, whether completed by the respondents themselves or administered by trained interviewers. The processing of such questionnaires is another nightmare; I'll have more to say about that in Chapter 14.

Formats for Respondents

In one of the most common types of questionnaire items, the respondent is expected to check one response from a series. For this purpose my experience has been that boxes adequately spaced apart are the best format. Word processing makes the use of boxes a practical technique these days; setting boxes in type can be accomplished easily and neatly. You can approximate boxes by using brackets: [ ], but if you're creating a questionnaire on a computer, you should take the few extra minutes to use genuine boxes that will give your questionnaire a more professional look. Here are some easy examples:

□ ○ ❑

Rather than providing boxes to be checked, you might print a code number beside each response and ask the respondent to circle the appropriate number (see Figure 9-1). This method has the added advantage of specifying the code number to be entered later in the processing stage (see Chapter 14). If numbers are to be circled, however, you should provide clear and prominent instructions to the respondent, because many will be

Did you happen to vote in the last presidential election?

1. Yes
2. No
3. Don't know

Have you ever felt you were the victim of sexual discrimination?

1. Yes
2. No
3. Don't know

FIGURE 9-1

Circling the Answer

23. Have you ever smoked marijuana?

☐ Yes
☐ No

If yes: About how many times have you smoked marijuana?

☐ Once
☐ 2 to 5 times
☐ 6 to 10 times
☐ 11 to 20 times
☐ More than 20 times

FIGURE 9-2

Contingency Question Format. Contingency questions offer a structure for exploring subject areas logically in some depth.

tempted to cross out the appropriate number, which makes data processing more difficult. (Note that the technique can be used more safely when interviewers administer the questionnaires, because the interviewers themselves record the responses.)

Contingency Questions

Quite often in questionnaires, certain questions will be relevant to some of the respondents and irrelevant to others. In a study of birth control methods, for instance, you would probably not want to ask men if they take birth control pills.

This sort of situation often arises when researchers wish to ask a series of questions about a certain topic. You may want to ask whether your respondents belong to a particular organization and, if so, how often they attend meetings, whether they have held office in the organization, and so forth. Or, you might want to ask whether respondents have heard anything about a certain political issue and then learn the attitudes of those who have heard of it.

Each subsequent question in series such as these is called a **contingency question:** Whether it is to be asked and answered is contingent on responses to the first question in the series. The proper use of contingency questions can facilitate the respondents' task in completing the questionnaire, because they are not faced with trying to answer questions irrelevant to them.

There are several formats for contingency questions. The one shown in Figure 9-2 is probably the clearest and most effective. Note two key elements in this format. First, the contingency question is isolated from the other questions by being set off to the side and enclosed in a box. Second, an arrow connects the contingency question to the answer on which it is contingent. In the illustration, only those respondents answering yes are expected to answer the contingency question. The rest of the respondents should simply skip it.

Note that the questions shown in Figure 9-2 could have been dealt with in a single question. The question might have read, "How many times, if any, have you smoked marijuana?" The response categories, then, might have read: "Never," "Once," "2 to 5 times," and so forth. This single question would apply to all respondents, and each would find an appropriate answer category. Such a question, however, might put some pressure on

contingency question A survey question intended for only some respondents, determined by their responses to some other question. For example, all respondents might be asked whether they belong to the Cosa Nostra, and only those who said yes would be asked how often they go to company meetings and picnics. The latter would be a contingency question.

FIGURE 9-3

Contingency Table. Sometimes it will be appropriate for certain kinds of respondents to skip over inapplicable questions. To avoid confusion, you should be sure to provide clear instructions to that end.

respondents to report having smoked marijuana, because the main question asks how many times they have smoked it, even though it allows for those *exceptional* cases *who have never smoked marijuana even once.* (The emphases used in the previous sentence give a fair indication of how respondents might read the question.) The contingency question format illustrated in Figure 9-2 should reduce the subtle pressure on respondents to report having smoked marijuana.

Used properly, even rather complex sets of contingency questions can be constructed without confusing the respondent. Figure 9-3 illustrates a more complicated example.

Sometimes a set of contingency questions is long enough to extend over several pages. Suppose you're studying political activities of college students, and you wish to ask a large number of questions of those students who have voted in a national, state, or local election. You could separate out the relevant respondents with an initial question such as "Have you ever voted in a national, state, or local election?" but it would be confusing to place the contingency questions in a box stretching over several pages. It would make more sense to enter instructions, in parentheses after each answer, telling respondents to answer or skip the

FIGURE 9-4

Instructions to Skip

contingency questions. Figure 9-4 provides an illustration of this method.

In addition to these instructions, it's worthwhile to place an instruction at the top of each page containing only the contingency questions. For example, you might say, "This page is only for respondents who have voted in a national, state, or local election." Clear instructions such as these spare respondents the frustration of reading and puzzling over questions irrelevant to them and increase the likelihood of responses from those for whom the questions are relevant.

Matrix Questions

Quite often, you'll want to ask several questions that have the same set of answer categories. This is typically the case whenever the Likert response categories are used. In such cases, it is often possible to construct a matrix of items and answers as illustrated in Figure 9-5.

This format offers several advantages over other formats. First, it uses space efficiently. Second, respondents will probably find it faster to complete a set of questions presented in this fashion than in other ways. In addition, this format may increase the comparability of responses given to different questions for the respondent as well as for the researcher. Because respondents can quickly review their answers to earlier items in the set, they might choose between, say, "strongly agree" and "agree" on a given statement by comparing the strength of their agreement with their earlier responses in the set.

There are some dangers inherent in using this format, however. Its advantages may encourage you to structure an item so that the responses fit

17. Beside each of the statements presented below, please indicate whether you Strongly Agree (**SA**), Agree (**A**), Disagree (**D**), Strongly Disagree (**SD**), or are Undecided (**U**).

	SA	A	D	SD	U
a. What this country needs is more law and order.........	☐	☐	☐	☐	☐
b. The police should be disarmed in America..........	☐	☐	☐	☐	☐
c. During riots, looters should be shot on sight.........	☐	☐	☐	☐	☐
etc.					

FIGURE 9-5

Matrix Question Format. Matrix questions offer an efficient format for presenting a set of closed-ended questionnaire items that have the same response categories.

into the matrix format when a different, more idiosyncratic set of responses might be more appropriate. Also, the matrix question format can foster a response-set among some respondents: They may develop a pattern of, say, agreeing with all the statements. This would be especially likely if the set of statements began with several that indicated a particular orientation (for example, a liberal political perspective) with only a few later ones representing the opposite orientation. Respondents might assume that all the statements represented the same orientation and, reading quickly, misread some of them, thereby giving the wrong answers. This problem can be reduced somewhat by alternating statements representing different orientations and by making all statements short and clear.

Ordering Items in a Questionnaire

The order in which questionnaire items are presented can also affect responses. First, the appearance of one question can affect the answers given to later ones. For example, if several questions have been asked about the dangers of terrorism to the United States and then a question asks respondents to volunteer (open-endedly) what they believe to represent dangers to the United States, terrorism will receive more citations than would otherwise be the case. In this situation, it's preferable to ask the open-ended question first.

Similarly, if respondents are asked to assess their overall religiosity ("How important is your religion to you in general?"), their responses to later questions concerning specific aspects of religiosity will be aimed at consistency with the prior assessment. The converse is true as well. If respondents are first asked specific questions about different aspects of their religiosity, their subsequent overall assessment will reflect the earlier answers. The order of responses within a question can also make a difference (Bishop and Smith 2001).

The impact of item order is not uniform. When J. Edwin Benton and John Daly (1991) conducted a local government survey, they found that the less-educated respondents were more influenced by the order of questionnaire items than those with more education were.

Some researchers attempt to overcome this effect by randomizing the order of items. This effort is usually futile. In the first place, a randomized set of items will probably strike respondents as chaotic and worthless. The random order also makes it more difficult for respondents to answer, because they must continually switch their attention from one topic to another. Finally, even a randomized ordering of items will have the effect discussed previously—except that you'll have no control over the effect.

The safest solution is sensitivity to the problem. Although you cannot avoid the effect of item order, try to estimate what that effect will be so that you can interpret results meaningfully. If the order of items seems especially important in a given study, you might construct more than one version of the questionnaire with different orderings of the items. You will then be able to determine the effects by comparing responses to the various versions. At the very least, you should pretest your questionnaire in the different forms. (We'll discuss pretesting in a moment.)

The desired ordering of items differs between interviews and self-administered questionnaires. In the latter, it's usually best to begin the questionnaire with the most interesting set of items. The potential respondents who glance casually over the first few items should want to answer them. Perhaps the items will ask for attitudes they're aching to express. At the same time, however, the initial items should not be threatening. (It might be a bad idea to begin with items about sexual behavior or drug use.) Requests for duller, demographic data (age, gender, and the like) should generally be placed at the end of a self-administered questionnaire. Placing these items at the beginning, as many inexperienced researchers are tempted to do, gives the questionnaire the initial appearance of a routine form, and the person receiving it may not be motivated to complete it.

Just the opposite is generally true for interview surveys. When the potential respondent's door first opens, the interviewer must gain rapport quickly. After a short introduction to the study, the interviewer can best begin by enumerating the members of the household, getting demographic data about each. Such items are easily answered and generally nonthreatening. Once the initial rapport has been established, the interviewer can then move into the area of attitudes and more sensitive matters. An interview that began with the question "Do you believe in witchcraft?" would probably end rather quickly.

Questionnaire Instructions

Every questionnaire, whether it is to be completed by respondents or administered by interviewers, should contain clear instructions and introductory comments where appropriate.

It's useful to begin every self-administered questionnaire with basic instructions for completing it. Although many people these days have experience with forms and questionnaires, begin by telling them exactly what you want: that they are to indicate their answers to certain questions by placing a check mark or an X in the box beside the appropriate answer or by writing in their answer when asked to do so. If many open-ended questions are used, respondents should be given some guidelines about whether brief or lengthy answers are expected. If you wish to encourage your respondents to elaborate on their responses to closed-ended questions, that should be noted.

If a questionnaire has subsections—political attitudes, religious attitudes, background data—introduce each with a short statement concerning its content and purpose. For example, "In this section, we would like to know what people consider to be the most important community problems." Demographic items at the end of a self-administered questionnaire might be introduced thus: "Finally, we would like to know just a little about you so we can see how different types of people feel about the issues we have been examining."

Short introductions such as these help the respondent make sense of the questionnaire. They make the questionnaire seem less chaotic, especially when it taps a variety of data. And they help put the respondent in the proper frame of mind for answering the questions.

Some questions may require special instructions to facilitate proper answering. This is especially true if a given question varies from the general instructions pertaining to the whole questionnaire. Some specific examples will illustrate this situation.

Despite attempts to provide mutually exclusive answers in closed-ended questions, often more than one answer will apply for respondents. If you want a single answer, you should make this perfectly clear in the question. An example would be "From the list below, please check the primary reason for your decision to attend college." Often the main question can be followed by a parenthetical note: "Please check the one best answer." If, on the other hand, you want the respondent to check as many answers as apply, you should make this clear.

When the respondent is supposed to rank-order a set of answer categories, the instructions should indicate this, and a different type of answer format should be used (for example, blanks instead of boxes). These instructions should indicate how many answers are to be ranked (for example: all; only the first and second; only the first and last; the most important and least important). These instructions should also spell out the order of ranking (for example: "Place a 1 beside the most important item, a 2 beside the next most important, and so forth"). Rank-ordering of responses is often difficult for respondents, however, because they may have to read and reread the list several times, so this technique should be used only in those situations where no other method will produce the desired result.

In multiple-part matrix questions, giving special instructions is useful unless the same format is used throughout the questionnaire. Sometimes respondents will be expected to check one answer in each column of the matrix; in other questionnaires they'll be expected to check one answer in each row. Whenever the questionnaire contains both formats, it's useful to add an instruction clarifying which is expected in each case.

Pretesting the Questionnaire

No matter how carefully researchers design a data-collection instrument such as a questionnaire, there is always the possibility—indeed the certainty—of error. They will always make some mistake: an ambiguous question, one that people cannot answer, or some other violation of the rules just discussed.

The surest protection against such errors is to pretest the questionnaire in full or in part. Give the questionnaire to the ten people in your bowling league, for example. It's not usually essential that the pretest subjects comprise a representative sample, although you should use people for whom the questionnaire is at least relevant.

By and large, it's better to ask people to complete the questionnaire than to read through it looking for errors. All too often, a question seems to make sense on a first reading, but it proves to be impossible to answer.

Stanley Presser and Johnny Blair (1994) describe several different pretesting strategies and report on the effectiveness of each. They also provide data on the cost of the various methods. Paul Beatty and Gordon Willis (2007) offer a useful review of "cognitive interviewing." In this technique, the pretest includes gathering respondents' comments about the questionnaire itself, so that the researchers can see which questions are communicating effectively and collecting the information sought.

There are many more tips and guidelines for questionnaire construction, but covering them all would take a book in itself. For now I'll complete this discussion with an illustration of a real questionnaire, showing how some of these comments find substance in practice.

Before turning to the illustration, however, I want to mention a critical aspect of questionnaire design: precoding. Because the information collected by questionnaires is typically transformed into some type of computer format, it's usually appropriate to include data-processing instructions on the questionnaire itself. These instructions indicate where specific pieces of information will be stored in the machine-readable data files. Notice that the following illustration has been precoded with the mysterious numbers that appear near questions and answer categories.

A Composite Illustration

Figure 9-6 is part of a questionnaire used by the University of Chicago's National Opinion Research Center in its General Social Survey. The questionnaire dealt with people's attitudes toward the government and was designed to be self-administered, though most of the GSS is conducted in face-to-face interviews.

Self-Administered Questionnaires

So far we've discussed how to formulate questions and how to design effective questionnaires. As important as these tasks are, the labor will be wasted

(Text continues on p. 270.)

10. Here are some things the government might do for the economy. Circle one number for each action to show whether you are in favor of it or against it.

> 1. **Strongly in favor of**
> 2. **In favor of**
> 3. **Neither in favor of nor against**
> 4. **Against**
> 5. **Strongly against**

PLEASE CIRCLE A NUMBER

a. Control of wages by legislation	1	2	3	4	5	28/
b. Control of prices by legislation	1	2	3	4	5	29/
c. Cuts in government spending	1	2	3	4	5	30/
d. Government financing of projects to create new jobs	1	2	3	4	5	31/
e. Less government regulation of business	1	2	3	4	5	32/
f. Support for industry to develop new products and technology	1	2	3	4	5	33/
g. Supporting declining industries to protect jobs	1	2	3	4	5	34/
h. Reducing the work week to create more jobs	1	2	3	4	5	35/

11. Listed below are various areas of government spending. Please indicate whether you would like to see more or less government spending in each area. Remember that if you say "much more," it might require a tax increase to pay for it.

> 1. **Spend much more**
> 2. **Spend more**
> 3. **Spend the same as now**
> 4. **Spend less**
> 5. **Spend much less**
> 8. **Can't choose**

PLEASE CIRCLE A NUMBER

a. The environment	1	2	3	4	5	8	36/
b. Health	1	2	3	4	5	8	37/
c. The police and law enforcement	1	2	3	4	5	8	38/
d. Education	1	2	3	4	5	8	39/
e. The military and defense	1	2	3	4	5	8	40/
f. Retirement benefits	1	2	3	4	5	8	41/
g. Unemployment benefits	1	2	3	4	5	8	42/
h. Culture and the arts	1	2	3	4	5	8	43/

12. If the government *had* to choose between keeping down inflation or keeping down unemployment, to which do you think it should give highest priority?

Keeping down inflation	1	44/
Keeping down unemployment	2	
Can't choose	8	

13. Do you think that labor unions in this country have too much power or too little power?

Far too much power	1	45/
Too much power	2	
About the right amount of power	3	
Too little power	4	
Far too little power	5	
Can't choose	8	

FIGURE 9-6

A Sample Questionnaire. This questionnaire excerpt is from the General Social Survey, a major source of data for analysis by social researchers around the world.

14. How about business and industry, do they have too much power or too little power?

Far too much power 1   46/
Too much power 2
About the right amount of power 3
Too little power 4
Far too little power 5
Can't choose 8

15. And what about the federal government, does it have to much power or too little power?

Far too much power 1   47/
Too much power 2
About the right amount of power 3
Too little power 4
Far too little power 5
Can't choose 8

16. In general, how good would you say labor unions are for the country as a whole?

Excellent 1   48/
Very good 2
Fairly good 3
Not very good 4
Not good at all 5
Can't choose 8

17. What do you think the government's role in each of these industries should be?

> 1. **Own it**
> 2. **Control prices and profits but not own it**
> 3. **Neither own it nor control its prices and profits**
> 8. **Can't choose**

PLEASE CIRCLE A NUMBER

a. Electric power	1	2	3	8	49/
b. The steel industry	1	2	3	8	50/
c. Banking and insurance	1	2	3	8	51/

18. On the whole, do you think it should or should not be the government's responsibility to . . .

> 1. **Definitely should be**
> 2. **Probably should be**
> 3. **Probably should not be**
> 4. **Definitely should not be**
> 8. **Can't choose**

PLEASE CIRCLE A NUMBER

a. Provide a job for everyone who wants one	1	2	3	4	8	52/
b. Keep prices under control	1	2	3	4	8	53/
c. Provide health care for the sick	1	2	3	4	8	54/
d. Provide a decent standard of living for the old	1	2	3	4	8	55/

FIGURE 9-6

(Continued)

unless the questionnaire produces useful data—which means that respondents actually complete the questionnaire. We turn now to the major methods for getting responses to questionnaires.

I've referred several times in this chapter to interviews and self-administered questionnaires. Actually, there are three main methods of administering survey questionnaires to a sample of respondents: self-administered questionnaires, in which respondents are asked to complete the questionnaire themselves; surveys administered by interviewers in face-to-face encounters; and surveys conducted by telephone. This section and the next two discuss each of these methods in turn. A fourth section addresses online surveys, a new technique growing in popularity.

The most common form of self-administered questionnaire is the mail survey. However, there are several other techniques that are often used as well. At times, it may be appropriate to administer a questionnaire to a group of respondents gathered at the same place at the same time. For example, a survey of students taking introductory psychology might be conducted during class. High school students might be surveyed during homeroom period.

Some recent experimentation has been conducted with regard to the home delivery of questionnaires. A research worker delivers the questionnaire to the home of sample respondents and explains the study. Then the questionnaire is left for the respondent to complete, and the researcher picks it up later.

Home delivery and the mail can also be used in combination. Questionnaires are mailed to families, and then research workers visit homes to pick up the questionnaires and check them for completeness. Just the opposite technique is to have questionnaires hand-delivered by research workers with a request that the respondents mail the completed questionnaires to the research office.

On the whole, when a research worker either delivers the questionnaire, picks it up, or both, the completion rate seems higher than it is for straightforward mail surveys. Additional experimentation with this technique is likely to point to other ways to improve completion rates while reducing costs. The remainder of this section, however, is devoted specifically to the mail survey, which is still the typical form of self-administered questionnaire.

Mail Distribution and Return

The basic method for collecting data through the mail has been to send a questionnaire accompanied by a letter of explanation and a self-addressed, stamped envelope for returning the questionnaire. The respondent is expected to complete the questionnaire, put it in the envelope, and return it. If, by any chance, you've received such a questionnaire and failed to return it, it would be valuable to recall the reasons you had for not returning it and keep them in mind any time you plan to send questionnaires to others.

A common reason for not returning questionnaires is that it's too much trouble. To overcome this problem, researchers have developed several ways to make returning them easier. For instance, a self-mailing questionnaire requires no return envelope: When the questionnaire is folded a particular way, the return address appears on the outside. The respondent therefore doesn't have to worry about losing the envelope.

More-elaborate designs are available also. The university student questionnaire to be described later in this chapter was bound in a booklet with a special, two-panel back cover. Once the questionnaire was completed, the respondent needed only to fold out the extra panel, wrap it around the booklet, and seal the whole thing with the adhesive strip running along the edge of the panel. The foldout panel contained my return address and postage. When I repeated the study a couple of years later, I improved on the design. Both the front and back covers had foldout panels: one for sending the questionnaire out and the other for getting it back—thus avoiding the use of envelopes altogether.

The point here is that anything you can do to make the job of completing and returning the questionnaire easier will improve your study. Imagine receiving a questionnaire that made no provisions for its return to the researcher. Suppose you had to (1) find an envelope, (2) write the address on it, (3) figure out how much postage it required, and (4) put the stamps on it. How likely is it that you would return the questionnaire?

A few brief comments on postal options are in order. You have options for mailing questionnaires out and for getting them returned. On outgoing mail, your choices are essentially between first-class postage and bulk rate. First class is more certain, but bulk rate is far cheaper. (Check your local post office for rates and procedures.) On return mail, your choice is between postage stamps and business-reply permits. Here, the cost differential is more complicated. If you use stamps, you pay for them whether people return their questionnaires or not. With the business-reply permit, you pay for only those that are used, but you pay an additional surcharge of about a nickel. This means that stamps are cheaper if a lot of questionnaires are returned, but business-reply permits are cheaper if fewer are returned (and you won't know in advance how many will be returned).

There are many other considerations involved in choosing among the several postal options. Some researchers, for example, feel that using postage stamps communicates more "humanness" and sincerity than using bulk rate and business-reply permits does. Others worry that respondents will steam off the stamps and use them for some purpose other than returning the questionnaires. Because both bulk rate and business-reply permits require establishing accounts at the post office, you'll probably find stamps much easier for small surveys.

Monitoring Returns

The mailing of questionnaires sets up a new research question that may prove valuable to a study. Researchers shouldn't sit back idly as questionnaires are returned; instead, they should undertake a careful recording of the varying rates of return among respondents.

An invaluable tool in this activity is a return rate graph. The day on which questionnaires were mailed is labeled Day 1 on the graph, and every day thereafter the number of returned questionnaires is logged on the graph. It's usually best to compile two graphs. One shows the number returned each day—rising over time, then dropping. The second

reports the cumulative number or percentage. In part, this activity provides the researchers with gratification, as they get to draw a picture of their successful data collection. More important, however, it serves as their guide to how the data collection is going. If follow-up mailings are planned, the graph provides a clue about when such mailings should be launched. (The dates of subsequent mailings should be noted on the graph.)

As completed questionnaires are returned, each should be opened, scanned, and assigned an identification (ID) number. These numbers should be assigned serially as the questionnaires are returned, even if other identification numbers have already been assigned. Two examples should illustrate the important advantages of this procedure.

Let's assume you're studying attitudes toward a political figure. In the middle of the data collection, the media break the story that the politician is having extramarital affairs. By knowing the date of that public disclosure and the dates when questionnaires were received, you'll be in a position to determine the effects of the disclosure. (Recall from Chapter 8 the discussion of history in connection with experiments.)

In a less sensational way, serialized ID numbers can be valuable in estimating non-response biases in the survey. Barring more-direct tests of bias, you may wish to assume that those who failed to answer the questionnaire will be more like respondents who delayed answering than like those who answered right away. An analysis of questionnaires received at different points in the data collection might then be used for estimates of sampling bias. For example, if the grade point averages (GPAs) reported by student respondents decrease steadily through the data collection, with those replying right away having higher GPAs and those replying later having lower GPAs, you might tentatively conclude that those who failed to answer at all have lower GPAs yet. Although it would not be advisable to make statistical estimates of bias in this fashion, you could take advantage of approximate estimates based on the patterns you've observed.

If respondents have been identified for purposes of follow-up mailing, then preparations for those mailings should be made as the questionnaires are returned. The case study later in this section discusses this process in greater detail.

Follow-up Mailings

Follow-up mailings may be administered in several ways. In the simplest, nonrespondents are simply sent a letter of additional encouragement to participate. A better method, however, is to send a new copy of the survey questionnaire with the follow-up letter. If potential respondents have not returned their questionnaires after two or three weeks, the questionnaires have probably been lost or misplaced. Receiving a follow-up letter might encourage them to look for the original questionnaire, but if they can't find it easily, the letter may go for naught.

The methodological literature strongly suggests that follow-up mailings provide an effective method for increasing return rates in mail surveys. In general, the longer a potential respondent delays replying, the less likely he or she is to do so at all. Properly timed follow-up mailings, then, provide additional stimuli to respond.

The effects of follow-up mailings will be seen in the response rate curves recorded during data collection. The initial mailings will be followed by a rise and subsequent subsiding of returns; the follow-up mailings will spur a resurgence of returns; and more follow-ups will do the same. In practice, three mailings (an original and two follow-ups) seem the most efficient.

The timing of follow-up mailings is also important. Here the methodological literature offers less-precise guides, but I've found that two or three weeks is a reasonable space between mailings. (This period might be increased by a few days if the mailing time—out and in—is more than two or three days.)

If the individuals in the survey sample are not identified on the questionnaires, it may not be possible to remail only to nonrespondents. In such a case, send your follow-up mailing to all members of the sample, thanking those who may have already participated and encouraging those who have not to do so. (The case study reported later describes yet another method you can use in an anonymous mail survey.)

Response Rates

A question that new survey researchers frequently ask concerns the percentage return rate, or the response rate, that should be achieved in a survey. The body of inferential statistics used in connection with survey analysis assumes that *all* members of the initial sample complete the survey. Because this almost never happens, non-response bias becomes a concern, with the researcher testing (and hoping) for the possibility that the respondents look essentially like a random sample of the initial sample, and thus a somewhat smaller random sample of the total population.

Nevertheless, overall **response rate** is one guide to the representativeness of the sample respondents. If a high response rate is achieved, there is less chance of significant non-response bias than with a low rate. Conversely, a low response rate is a danger signal, because the nonrespondents are likely to differ from the respondents in ways other than just their willingness to participate in the survey. Richard Bolstein (1991), for example, found that those who did not respond to a preelection political poll were less likely to vote that those who did participate. Estimating the turnout rate from just the survey respondents, then, would have overestimated the number who would show up at the polls. Ironically, of course, since the non-respondents were unlikely to vote, the preferences of the survey participants might offer a good estimate of the election results.

As you can imagine, one of the more persistent discussions among survey researchers concerns

response rate The number of people participating in a survey divided by the number selected in the sample, in the form of a percentage. This is also called the *completion rate* or, in self-administered surveys, the *return rate:* the percentage of questionnaires sent out that are returned.

ways of increasing response rates. You'll recall that this was a chief concern in the earlier discussion of options for mailing out and receiving questionnaires. Survey researchers have developed many ingenious techniques addressing this problem. Some have experimented with novel formats. Others have tried paying respondents to participate. The problem with paying, of course, is that it's expensive to make meaningfully high payment to hundreds or thousands of respondents, but some imaginative alternatives have been used. Some researchers have said, "We want to get your two-cents' worth on some issues, and we're willing to pay"—enclosing two pennies. Another enclosed a quarter, suggesting that the respondent make some little child happy. Still others have enclosed paper money. Similarly, Michael Davern and his colleagues (2003) found that financial incentives also increased completion rates in face-to-face interview surveys (discussed in the next section).

Don Dillman (2007) has spent decades painstakingly assessing the various techniques that survey researchers have used to increase return rates on mail surveys, and he evaluates the impact of each. More important, Dillman stresses the necessity of paying attention to all aspects of the study—what he calls the "Tailored Design Method"—rather than one or two special gimmicks.

Having said all this, there is no absolutely acceptable level of response to a mail survey, except for 100 percent. While it is possible to achieve response rates of 70 percent or more, most mail surveys probably fall below that level. Thus, it's important to test for non-response bias wherever possible.

A Case Study

The steps involved in the administration of a mail survey are many and can best be appreciated in a walk-through of an actual study. Accordingly, this section concludes with a detailed description of how the student survey we discussed in Chapter 7 as an illustration of systematic sampling was administered. This study did not represent the theoretical ideal for such studies, but in that regard it serves our present purposes all the better. The

study was conducted by the students in my graduate seminar in survey research methods.

As you may recall, 1,100 students were selected from the university registration database through a stratified, systematic sampling procedure. For each student selected, six self-adhesive mailing labels were printed by the computer.

By the time we were ready to distribute the questionnaires, it became apparent that our meager research funds wouldn't cover several mailings to the entire sample of 1,100 students (questionnaire printing costs were higher than anticipated). As a result, we chose a systematic two-thirds sample of the mailing labels, yielding a subsample of 733 students.

Earlier, we had decided to keep the survey anonymous in the hope of encouraging more-candid responses to some sensitive questions. (Later surveys of the same issues among the same population indicated this anonymity was unnecessary.) Thus, the questionnaires would carry no identification of students on them. At the same time, we hoped to reduce the follow-up mailing costs by mailing only to nonrespondents.

To achieve both of these aims, a special postcard method was devised. Each student was mailed a questionnaire that carried no identifying marks, plus a postcard addressed to the research office—with one of the student's mailing labels affixed to the reverse side of the card. The introductory letter asked the student to complete and return the questionnaire—assuring anonymity—and to return the postcard simultaneously. Receiving the postcard would tell us—without indicating which questionnaire it was—that the student had returned his or her questionnaire. This procedure would then facilitate follow-up mailings.

The 32-page questionnaire was printed in booklet form. The three-panel cover described earlier in this chapter permitted the questionnaire to be returned without an additional envelope.

A letter introducing the study and its purposes was printed on the front cover of the booklet. It explained why the study was being conducted (to learn how students feel about a variety of issues), how students had been selected for the study, the importance of each student's responding, and the mechanics of returning the questionnaire.

Students were assured that their responses to the survey were anonymous, and the postcard method was explained. A statement followed about the auspices under which the study was being conducted, and a telephone number was provided for those who might want more information about the study. (Five students called for information.)

By printing the introductory letter on the questionnaire, we avoided the necessity of enclosing a separate letter in the outgoing envelope, thereby simplifying the task of assembling mailing pieces.

The materials for the initial mailing were assembled as follows. (1) One mailing label for each student was stuck on a postcard. (2) Another label was stuck on an outgoing manila envelope. (3) One postcard and one questionnaire were placed in each envelope—with a glance to ensure that the name on the postcard and on the envelope were the same in each case.

The distribution of the survey questionnaires had been set up for a bulk rate mailing. Once the questionnaires had been stuffed into envelopes, they were grouped by zip code, tied in bundles, and delivered to the post office.

Shortly after the initial mailing, questionnaires and postcards began arriving at the research office. Questionnaires were opened, scanned, and assigned identification numbers as described earlier in this chapter. For every postcard received, a search was made for that student's remaining labels, and they were destroyed.

After two or three weeks, the remaining mailing labels were used to organize a follow-up mailing. This time a special, separate letter of appeal was included in the mailing piece. The new letter indicated that many students had returned their questionnaires already, and it was very important for all others to do so as well.

The follow-up mailing stimulated a resurgence of returns, as expected, and the same logging procedures continued. The returned postcards told us which additional mailing labels to destroy. Unfortunately, time and financial pressures made a third mailing impossible, despite initial plans to do so, but the two mailings resulted in an overall return rate of 62 percent.

This illustration should give you a fairly good sense of what's involved in the execution of mailed self-administered questionnaires. Let's turn now to the second principal method of conducting surveys, in-person interviews.

Interview Surveys

The **interview** is an alternative method of collecting survey data. Rather than asking respondents to read questionnaires and enter their own answers, researchers send interviewers to ask the questions orally and record respondents' answers. Interviewing is typically done in a face-to-face encounter, but telephone interviewing, discussed in the next section, follows most of the same guidelines.

Most interview surveys require more than one interviewer, although you might undertake a small-scale interview survey yourself. Portions of this section will discuss methods for training and supervising a staff of interviewers assisting you with a survey.

This section deals specifically with survey interviewing. Chapter 10 discusses the less-structured, in-depth interviews often conducted in qualitative field research.

The Role of the Survey Interviewer

There are several advantages to having a questionnaire administered by an interviewer rather than a respondent. To begin with, interview surveys typically attain higher response rates than mail surveys do. A properly designed and executed interview survey ought to achieve a completion rate of at least 80 to 85 percent. (Federally funded

---

interview A data-collection encounter in which one person (an interviewer) asks questions of another (a respondent). Interviews may be conducted face-to-face or by telephone.

surveys often require one of these response rates.) Respondents seem more reluctant to turn down an interviewer standing on their doorstep than to throw away a mail questionnaire.

The presence of an interviewer also generally decreases the number of "don't knows" and "no answers." If minimizing such responses is important to the study, the interviewer can be instructed to probe for answers ("If you had to pick one of the answers, which do you think would come closest to your feelings?").

Further, if a respondent clearly misunderstands the intent of a question or indicates that he or she does not understand, the interviewer can clarify matters, thereby obtaining relevant responses. (As we'll discuss shortly, such clarifications must be strictly controlled through formal specifications.)

Finally, the interviewer can observe respondents as well as ask questions. For example, the interviewer can note the respondent's race if this is considered too delicate a question to ask. Similar observations can be made regarding the quality of the dwelling, the presence of various possessions, the respondent's ability to speak English, the respondent's general reactions to the study, and so forth. In one survey of students, respondents were given a short, self-administered questionnaire to complete—concerning sexual attitudes and behavior—during the course of the interview. While respondents completed the questionnaire, the interviewer made detailed notes regarding their dress and grooming.

This procedure raises an ethical issue. Some researchers have objected that such practices violate the spirit of the agreement by which the respondent has allowed the interview. Although ethical issues seldom are clear-cut in social research, it's important to be sensitive to them, as we saw in Chapter 3.

Survey research is of necessity based on an unrealistic stimulus-response theory of cognition and behavior. Researchers must assume that a questionnaire item will mean the same thing to every respondent, and every given response must mean the same when given by different respondents. Although this is an impossible goal, survey questions are drafted to approximate the ideal as closely as possible.

The interviewer must also fit into this ideal situation. The interviewer's presence should affect neither a respondent's perception of a question nor the answer given. In other words, the interviewer should be a neutral medium through which questions and answers are transmitted.

As such, different interviewers should obtain exactly the same responses from a given respondent. (Recall our earlier discussions of reliability.) This neutrality has a special importance in area samples. To save time and money, a given interviewer is typically assigned to complete all the interviews in a particular geographic area—a city block or a group of nearby blocks. If the interviewer does anything to affect the responses obtained, the bias thus interjected might be interpreted as a characteristic of that area.

Let's suppose that a survey is being done to determine attitudes toward low-cost housing in order to help in the selection of a site for a new government-sponsored development. An interviewer assigned to a given neighborhood might—through word or gesture—communicate his or her own distaste for low-cost housing developments. Respondents might therefore tend to give responses in general agreement with the interviewer's own position. The results of the survey would indicate that the neighborhood in question strongly resists construction of the development in its area when in fact their apparent resistance simply reflects the interviewer's attitudes.

General Guidelines for Survey Interviewing

The manner in which interviews ought to be conducted will vary somewhat by survey population and survey content. Nevertheless, some general guidelines apply to most interviewing situations.

Appearance and Demeanor

As a rule, interviewers should dress in a fashion similar to that of the people they'll be interviewing. A richly dressed interviewer will probably have difficulty getting good cooperation and responses from poorer respondents; a poorly dressed interviewer will have similar difficulties with richer respondents. To the extent that the interviewer's dress and grooming differ from those of the respondents, it should be in the direction of cleanliness and neatness in modest apparel. If cleanliness is not next to godliness, it appears at least to be next to neutrality. Although middle-class neatness and cleanliness may not be accepted by all sectors of U.S. society, they remain the primary norm and are the most likely to be acceptable to the largest number of respondents.

Dress and grooming are typically regarded as signs of a person's attitudes and orientations. Torn jeans, green hair, and razor blade earrings may communicate—correctly or incorrectly—that the interviewer is politically radical, sexually permissive, favorable to drug use, and so forth. Any of these impressions could bias responses or affect the willingness of people to be interviewed.

In demeanor, interviewers should be pleasant if nothing else. Because they'll be prying into a respondent's personal life and attitudes, they must communicate a genuine interest in getting to know the respondent, without appearing to spy. They must be relaxed and friendly, without being too casual or clinging. Good interviewers also have the ability to determine very quickly the kind of person the respondent will feel most comfortable with, the kind of person the respondent would most enjoy talking to. Clearly, the interview will be more successful in this case. Further, because respondents are asked to volunteer a portion of their time and to divulge personal information, they deserve the most enjoyable experience the researcher and interviewer can provide.

Familiarity with the Questionnaire

If an interviewer is unfamiliar with the questionnaire, the study suffers and the respondent faces an unfair burden. The interview is likely to take more time than necessary and be unpleasant. Moreover, the interviewer cannot acquire familiarity by skimming through the questionnaire two or three times. He or she must study it carefully, question by question, and must practice reading it aloud.

Ultimately, the interviewer must be able to read the questionnaire items to respondents without error, without stumbling over words and phrases. A good model is the actor reading lines in a play or movie. The lines must be read as though they constituted a natural conversation, but that conversation must follow exactly the language set down in the questionnaire.

By the same token, the interviewer must be familiar with the specifications prepared in conjunction with the questionnaire. Inevitably some questions will not exactly fit a given respondent's situation, and the interviewer must determine how the question should be interpreted in that situation. The specifications provided to the interviewer should give adequate guidance in such cases, but the interviewer must know the organization and contents of the specifications well enough to refer to them efficiently. It would be better for the interviewer to leave a given question unanswered than to spend five minutes searching through the specifications for clarification or trying to interpret the relevant instructions.

Following Question Wording Exactly

The first part of this chapter discussed the significance of question wording for the responses obtained. A slight change in the wording of a given question may lead a respondent to answer "yes" rather than "no." It follows that interviewers must be instructed to follow the wording of questions exactly. Otherwise all the effort that the developers have put into carefully phrasing the questionnaire items to obtain the information they need and to ensure that respondents interpret items precisely as intended will be wasted.

While I hope the logic of this injunction is clear, it is not necessarily a closed discussion. For example, Giampietro Gobo (2006) argues that we might consider giving interviewers more latitude, suggesting that respondents sometimes make errors that may be apparent to the interviewer on the spot. Allowing the interviewer to intervene, as he notes, does increase the possibility that the interviewer will impact the data collected.

Recording Responses Exactly

Whenever the questionnaire contains open-ended questions (ones soliciting the respondent's own answers), the interviewer must record those answers exactly as given. No attempt should be made to summarize, paraphrase, or correct bad grammar.

This exactness is especially important because the interviewer will not know how the responses are to be coded. Indeed, the researchers themselves may not know the coding until they've read a hundred or so responses. For example, the questionnaire might ask respondents how they feel about the traffic situation in their community. One respondent might answer that there are too many cars on the roads and that something should be done to limit their numbers. Another might say that more roads are needed. If the interviewer recorded these two responses with the same summary—"congested traffic"—the researchers would not be able to take advantage of the important differences in the original responses.

Sometimes, verbal responses are too inarticulate or ambiguous to permit interpretation. However, the interviewer may be able to understand the intent of the response through the respondent's gestures or tone. In such a situation, the interviewer should still record the exact verbal response but also add marginal comments giving both the interpretation and the reasons for arriving at it.

More generally, researchers can use any marginal comments explaining aspects of the response not conveyed in the verbal recording, such as the respondent's apparent anger, embarrassment, uncertainty in answering, and so forth. In each case, however, the exact verbal response should also be recorded.

Probing for Responses

Sometimes respondents in an interview will give an inappropriate or incomplete answer. In such cases, a **probe,** or request for an elaboration, can be useful. For example, a closed-ended question may present an attitudinal statement and ask the respondent to strongly agree, agree somewhat, disagree somewhat, or strongly disagree. The respondent, however, may reply: "I think that's true." The interviewer should follow this reply with "Would you say you strongly agree or agree somewhat?" If necessary, interviewers can explain that they must check one or the other of the categories provided. If the respondent adamantly refuses to choose, the interviewer should write in the exact response given by the respondent.

Probes are more frequently required in eliciting responses to open-ended than closed-ended questions. For example, in response to a question about traffic conditions, the respondent might simply reply, "Pretty bad." The interviewer could obtain an elaboration on this response through a variety of probes. Sometimes the best probe is silence; if the interviewer sits quietly with pencil poised, the respondent will probably fill the pause with additional comments. (This technique is used effectively by newspaper reporters.) Appropriate verbal probes might be "How is that?" or "In what ways?" Perhaps the most generally useful probe is "Anything else?"

Often, interviewers need to probe for answers that will be sufficiently informative for analytical purposes. In every case, however, such probes must be completely neutral; they must not in any way affect the nature of the subsequent response. Whenever you anticipate that a given question may require probing for appropriate responses, you should provide one or more useful probes next to the question in the questionnaire. This practice has two important advantages. First, you'll have more time to devise the best, most neutral probes. Second, all interviewers will use the same probes whenever they're needed. Thus, even if the probe isn't perfectly neutral, all respondents will be presented with the same stimulus. This is the same logical guideline discussed for question wording. Although a question should not be loaded or

probe A technique employed in interviewing to solicit a more complete answer to a question. It is a nondirective phrase or question used to encourage a respondent to elaborate on an answer. Examples include "Anything more?" and "How is that?"

biased, it's essential that every respondent be presented with the same question, even if it's biased.

Coordination and Control

Most interview surveys require the assistance of several interviewers. In large-scale surveys, interviewers are hired and paid for their work. Student researchers might find themselves recruiting friends to help them interview. Whenever more than one interviewer is involved in a survey, their efforts must be carefully controlled. This control has two aspects: training interviewers and supervising them after they begin work.

The interviewers' training session should begin with a description of what the study is all about. Even though the interviewers may be involved only in the data-collection phase of the project, it will be useful to them to understand what will be done with the interviews they conduct and what purpose will be served. Morale and motivation are usually lower when interviewers don't know what's going on.

The training on how to interview should begin with a discussion of general guidelines and procedures, such as those discussed earlier in this section. Then the whole group should go through the questionnaire together—question by question. Don't simply ask if anyone has any questions about the first page of the questionnaire. Read the first question aloud, explain the purpose of the question, and then entertain any questions or comments the interviewers may have. Once all their questions and comments have been handled, go on to the next question in the questionnaire.

It's always a good idea to prepare specifications to accompany an interview questionnaire. *Specifications* are explanatory and clarifying comments about handling difficult or confusing situations that may occur with regard to particular questions in the questionnaire. When drafting the questionnaire, try to think of all the problem cases that might arise—the bizarre circumstances that might make a question difficult to answer. The survey specifications should provide detailed guidelines on how to handle such situations. For example, even as simple a matter as age might

present problems. Suppose a respondent says he or she will be 25 next week. The interviewer might not be sure whether to take the respondent's current age or the nearest one. The specifications for that question should explain what should be done. (Probably, you would specify that the age as of last birthday should be recorded in all cases.)

If you've prepared a set of specifications, review them with the interviewers when you go over the individual questions in the questionnaire. Make sure your interviewers fully understand the specifications and the reasons for them as well as the questions themselves.

This portion of the interviewer training is likely to generate many troublesome questions from your interviewers. They'll ask, "What should I do if . . . ?" In such cases, avoid giving a quick, offhand answer. If you have specifications, show how the solution to the problem could be determined from the specifications. If you do not have specifications, show how the preferred handling of the situation fits within the general logic of the question and the purpose of the study. Giving unexplained answers to such questions will only confuse the interviewers and cause them to take their work less seriously. If you don't know the answer to such a question when it's asked, admit it and ask for some time to decide on the best answer. Then think out the situation carefully and be sure to give all the interviewers your answer, explaining your reasons.

Once you've gone through the whole questionnaire, conduct one or two demonstration interviews in front of everyone. Preferably, you should interview someone other than one of the interviewers. Realize that your interview will be a model for those you're training, so make it good. It would be best, moreover, if the demonstration interview were done as realistically as possible. Don't pause during the demonstration to point out how you've handled a complicated situation: Handle it, and then explain later. It's irrelevant if the person you're interviewing gives real answers or takes on some hypothetical identity for the purpose, as long as the answers are consistent.

After the demonstration interviews, pair off your interviewers and have them practice on each other. When they've completed the questionnaire, have them reverse roles and do it again. Interviewing is the best training for interviewing. As your interviewers practice on each other, wander around, listening in on the practice so you'll know how well they're doing. Once the practice is completed, the whole group should discuss their experiences and ask any other questions they may have.

The final stage of the training for interviewers should involve some "real" interviews. Have them conduct some interviews under the actual conditions that will pertain to the final survey. You may want to assign them people to interview, or perhaps they may be allowed to pick people themselves. Don't have them practice on people you've selected in your sample, however. After each interviewer has completed three to five interviews, have him or her check back with you. Look over the completed questionnaires for any evidence of misunderstanding. Again, answer any questions that the interviewers have. Once you're convinced that a given interviewer knows what to do, assign some actual interviews, using the sample you've selected for the study.

It's essential to continue supervising the work of interviewers over the course of the study. You should check in with them after they conduct no more than 20 or 30 interviews. You might assign 20 interviews, have the interviewer bring back those questionnaires when they're completed, look them over, and assign another 20 or so. Although this may seem overly cautious, you must continually protect yourself against misunderstandings that may not be evident early in the study. Moreover, Kristen Olson and Andy Peytchev (2007) have discovered that interviewers' behavior continues to change over the course of a survey project. For example, interviewers speed through the interview more quickly and are more likely to judge respondents as uninterested in it.

If you're the only interviewer in your study, these comments may not seem relevant. However, it would be wise, for example, to prepare specifications for potentially troublesome questions in your questionnaire. Otherwise, you run the risk of making ad hoc decisions, during the course of the study, that you'll later regret or forget. Also, the

emphasis on practice applies equally to the one-person project and to the complex funded survey with a large interviewing staff.

Telephone Surveys

For years telephone surveys had a rather bad reputation among professional researchers. By definition, telephone surveys are limited to people who have telephones. Years ago, this method produced a substantial social-class bias by excluding poor people from the surveys. This was vividly demonstrated by the *Literary Digest* fiasco of 1936. Recall that, even though voters were contacted by mail, the sample was partially selected from telephone subscribers, who were hardly typical in a nation just recovering from the Great Depression. By 2003, however, the U.S. Bureau of the Census (2006: 737, Table 1117) estimated that 95.5 percent of all housing units had telephones, so the earlier form of class bias has substantially diminished.

A related sampling problem involves unlisted numbers. A survey sample selected from the pages of a local telephone directory would omit all those people—typically richer—who requested that their numbers not be published. This potential bias has been erased through a technique that has advanced telephone sampling substantially: **random-digit dialing (RDD).**

Imagine that you were to select a set of seven-digit telephone numbers at random. Even those whose numbers were unlisted would have the same chance of selection as those who were in the directory would. However, if you were to start dialing randomly selected numbers, a high proportion of those would turn out to be "not in service," government offices, commercial enterprises, and so forth. Fortunately, you can obtain ranges of numbers that are (mostly) active residential numbers. Selecting a set of those numbers at random

random-digit dialing (RDD) A sampling technique in which random numbers are selected from within the ranges of numbers assigned to active telephones.

will provide a representative sample of residential households. As a consequence, random-digit dialing has become a standard procedure in telephone surveys.

The growth in popularity of cell phones has become a new source of concern for survey researchers, however, since cell phone numbers are typically not included in phone surveys. Those who use cell phones exclusively, moreover, tend to be younger. This, of course, can affect survey outcomes. For example, younger voters in 2004 were more likely to vote for John Kerry than older voters were. In 2008 they were more likely than the average voter to support Barack Obama. Further, in a study of this matter, Scott Keeter and his colleagues (2008) found a distinct bias by age and the variables closely related to it (such as *marital status*) distinguishing those who were reachable only by cell phone and those reachable by landline:

> One of the most striking differences between cell-only respondents and people reached on a landline telephone is their age. Nearly half of the cell-only respondents (46%) are under age 30 compared to only 12% in the landline sample. Related to their younger age, only 26% of cell-only respondents are married, compared with 57% percent of those in the landline sample. Similarly, about half of cell-only respondents have never been married (51%), compared with only 16% in the landline sample.
>
> *(Keeter et al. 2008)*

At the 2008 meetings of the American Association for Public Opinion Research (AAPOR), several research papers examined the implications of cell phone popularity. Overall, most of the researchers found that, for most purposes, ignoring those with only cell phones does not seriously bias survey results, because these customers represent a relatively small portion of all telephone customers. However, virtually all of the researchers concluded by saying that this situation was likely to change in the years ahead. The role of cell phones is clearly a development that social researchers will continue to examine and deal with.

Telephone surveys offer many advantages that underlie the popularity of this method. Probably the greatest advantages are money and time, in that order. In a face-to-face, household interview, you may drive several miles to a respondent's home, find no one there, return to the research office, and drive back the next day—possibly finding no one there again. It's cheaper and quicker to let your fingers make the trips.

Interviewing by telephone, you can dress any way you please without affecting the answers respondents give. And sometimes respondents will be more honest in giving socially disapproved answers if they don't have to look you in the eye. Similarly, it may be possible to probe into more-sensitive areas, though this isn't necessarily the case. People are, to some extent, more suspicious when they can't see the person asking them questions—perhaps a consequence of "surveys" aimed at selling magazine subscriptions and time-share condominiums.

Interviewers can communicate a lot about themselves over the phone, however, even though they can't be seen. For example, researchers worry about the impact of an interviewer's name (particularly if ethnicity is relevant to the study) and debate the ethics of having all interviewers use bland "stage names" such as Smith or Jones. (Female interviewers sometimes ask permission to do this, to avoid subsequent harassment from men they interview.)

Telephone surveys can allow greater control over data collection if several interviewers are engaged in the project. If all the interviewers are calling from the research office, they can get clarification from the person in charge whenever problems occur, as they inevitably do. Alone in the boondocks, an interviewer may have to wing it between weekly visits with the interviewing supervisor.

Finally, another important factor involved in the growing use of telephone surveys has to do with personal safety. Don Dillman (1978: 4) describes the situation this way:

> Interviewers must be able to operate comfortably in a climate in which strangers are viewed with distrust and must successfully counter respondents' objections to being interviewed. Increasingly, interviewers must be willing to

work at night to contact residents in many households. In some cases, this necessitates providing protection for interviewers working in areas of a city in which a definite threat to the safety of individuals exists.

Concerns for safety, thus, work in two ways to hamper face-to-face interviews. Potential respondents may refuse to be interviewed, fearing the stranger-interviewer. And the interviewers themselves may incur some risks. All this is made even worse by the possibility of the researchers being sued for huge sums if anything goes wrong.

Telephone interviewing presents its own problems, however. As I've already mentioned, the method is hampered by the proliferation of bogus "surveys" that are actually sales campaigns disguised as research. If you have any questions about any such call you receive, by the way, ask the interviewer directly whether you've been selected for a survey only or if a sales "opportunity" is involved. It's also a good idea, if you have any doubts, to get the interviewer's name, phone number, and company. Hang up if the caller refuses to provide any of these.

For the researcher, the ease with which people can hang up is another shortcoming of telephone surveys. Once you've been let inside someone's home for an interview, the respondent is unlikely to order you out of the house in midinterview. It's much easier to terminate a telephone interview abruptly, saying something like, "Whoops! Someone's at the door. I gotta go." or "Omigod! The pigs are eating my Volvo!" (That sort of thing is much harder to fake when the interviewer is sitting in your living room.)

Another potential problem for telephone interviewing is the prevalence of answering machines or voicemail. A study conducted by Walker Research (1988) found that half of the owners of answering machines acknowledged using their machines to "screen" calls at least some of the time. Research by Peter Tuckel and Barry Feinberg (1991), however, showed that answering machines had not yet had a significant effect on the ability of telephone researchers to contact prospective respondents. Nevertheless, the researchers concluded that as answering machines continue to proliferate, "the sociodemographic characteristics of owners will change." This fact made it likely that "different behavior patterns associated with the utilization of the answering machine" could emerge (1991: 216).

More-recent research has shown that several factors, including answering machines, have reduced response rates in telephone surveys. Peter Tuckel and Harry O'Neill (2002) and others have examined the impact of such factors as Caller ID, answering machines, telemarketing, and phone lines being tied up by faxes and Internet access. All these constitute difficulties modern survey researchers must deal with.

Computer-Assisted Telephone Interviewing (CATI)

In Chapter 14, we'll see some of the ways computers have influenced the conduct of social research—particularly data processing and analysis. Computers are also changing the nature of telephone interviewing. One innovation is **computer-assisted telephone interviewing (CATI).** This method is increasingly used by academic, government, and commercial survey researchers. Though there are variations in practice, here's what CATI can look like.

Imagine an interviewer wearing a telephone headset, sitting in front of a computer terminal and its video screen. The central computer selects a telephone number at random and dials it. (Recall that random-digit dialing avoids the problem of unlisted telephone numbers.) On the video screen is an introduction ("Hello, my name is . . .") and the first question to be asked ("Could you tell me how many people live at this address?").

When the respondent answers the phone, the interviewer says hello, introduces the study, and asks the first question displayed on the screen.

---

computer-assisted telephone interviewing (CATI) A data-collection technique in which a telephone-survey questionnaire is stored in a computer, permitting the interviewer to read the questions from the monitor and enter the answers on the computer keyboard.

Voice Capture

James E. Dannemiller

SMS Research, Honolulu

The development of various CATI techniques has been a boon to survey and marketing research, though mostly it has supported the collection, coding, and analysis of "data as usual." The Voice Capture technique developed by Survey Systems, however, offers quite unusual possibilities, which we are only beginning to explore.

In the course of a CATI-based telephone interview, the interviewer can trigger the computer to begin digitally recording the conversation with the respondent. Having determined that the respondent has recently changed his or her favorite TV news show, for example, the interviewer can ask, "Why did you change?" and begin recording the verbatim response. (Early in the interview, the interviewer has asked permission to record parts of the interview.)

Later on, coders can play back the responses and code them—much as they would do with the interviewer's typescript of the responses. This offers an easier and more accurate way of accomplishing a conventional task. But that's a tame use of the new capability.

It's also possible to incorporate such oral data as parts of a cross-tabulation during analysis. We may create a table of gender by age by reasons for switching TV news shows. Thus, we can hear, in turn, the responses of the young men, young women, middle-aged men, and so forth. In one such study we found the younger and older men tending to watch one TV news show, while the middle-aged men watched something else. Listening to the responses of the middle-aged men, one after another, we heard a common comment: "Well, now that I'm older . . ." This kind of aside might have been lost in the notes hastily typed by interviewers, but such comments stood out dramatically in the oral data. The middle-aged men seemed to be telling us they felt "maturity" required them to watch a particular show, while more years under their belts let them drift back to what they liked in the first place.

These kinds of data are especially compelling to clients, particularly in customer satisfaction studies. Rather than summarize what we feel a client's customers like and don't like, we can let the respondents speak directly to the client in their own words. It's like a focus group on demand. Going one step further, we have found that letting line employees (bank tellers, for example) listen to the responses has more impact than having their supervisors tell them what they are doing right or wrong.

As exciting as these experiences are, I have the strong feeling that we have scarcely begun to tap into the possibilities for such unconventional forms of data.

When the respondent answers the question, the interviewer types that answer into the computer terminal—either the verbatim response to an open-ended question or the code category for the appropriate answer to a closed-ended question. The answer is immediately stored in the computer. The second question appears on the video screen, is asked, and the answer is entered into the computer. Thus, the interview continues.

In addition to the obvious advantages in terms of data collection, CATI automatically prepares the data for analysis; in fact, the researcher can begin analyzing the data before the interviewing is complete, thereby gaining an advanced view of how the analysis will turn out. Sill another innovation that computer technology makes possible is described in "Voice Capture."

It's also possible to set up questionnaires in personal data assistants (PDAs) for use by an interviewer or for direct data entry by respondents. Some of these systems include the possibility of voice capture, as described in the box.

Response Rates in Interview Surveys

Earlier in this chapter we looked at the issue of response rates in mail surveys, and this is an equally important issue for interview surveys. In Chapter 7, when we discussed formulas for calculating sampling error to determine the accuracy of survey estimates, the implicit assumption was that everyone selected in a sample would participate—which is almost never the case. Lacking perfection, researchers must maximize participation by those selected. Although interview surveys tend to produce higher response rates than mail surveys do, interview success has recently declined.

By analyzing response-rate trends in the University of Michigan's Survey of Consumer Attitudes, Richard Curtin, Stanley Presser, and Eleanor Singer (2005) have sketched a pattern of general decline over recent years. Between 1979 and 1996, the response rate in this telephone survey dropped from 72 to 60 percent, representing an average annual decline of three-quarters of a percent. Since 1996, the rate of decline has doubled. The increased non-responses reflected both refusals and those they were unable to contact.

By contrast, the General Social Survey, using personal interviews, experienced response rates between 73.5 and 82.4 percent in the years from 1975 to 1998. In the 2000 and 2002 surveys, however, the GSS completion rate was 70 percent. Their decline came primarily from refusals rather than being unable to contact respondents, because household interviews produce higher rates of contact than telephone surveys do.

In recent years, both household and telephone surveys have experienced a decline in response rates. A special issue of the *Public Opinion Quarterly* (2006) was devoted entirely to analyzing the many dimensions of the decline in response rates in household surveys. As the analyses show, lower response rates do not necessarily produce inaccurate estimates of the population being studied, but the variations on this issue defy a simple summary.

Many researchers believe that the widespread growth of telemarketing has been a big part of the problems experienced by legitimate telephone surveys, and there are hopes that the state and national "do not call" lists may ease that problem. Further, we saw that other factors such as answering machines and voicemail also contribute to these problems (Tuckel and O'Neill 2002). Response rate is likely to remain an issue of high concern in survey research.

As a consumer of social research, you should be wary of "surveys" whose apparent purpose is to raise money for the sponsor. This practice has already invaded the realm of "fax surveys,"

evidenced by a fax entitled "Should Hand Guns Be Outlawed?" Two fax numbers were provided for expressing either a "Yes" or "No" opinion. The smaller print noted, "Calls to these numbers cost $2.95 per minute, a small price for greater democracy. Calls take approx. 1 or 2 minutes." You can imagine where the $2.95 went.

Online Surveys

An increasingly popular method of survey research involves the use of the Internet and the World Wide Web—two of the most far-reaching developments of the late twentieth century. Some researchers feel that the Internet can be used to conduct meaningful survey research, and this technique has been getting especially popular in marketing research, for example. Some online surveys are conducted completely via e-mail; others are conducted via websites. Commonly, potential respondents will receive an e-mail asking them to go to a web link where the survey resides.

Mick Couper (2001: 464) provides an excellent overview of the issues concerning the present and prospective state of online surveys.

> The rapid development of surveys on the World Wide Web (WWW) is leading some to argue that soon Internet (and, in particular, Web) surveys will replace traditional methods of survey data collection. Others are urging caution or even voicing skepticism about the future role Web surveys will play. Clearly, we stand at the threshold of a new era for survey research, but how this will play out is not yet clear.

As we've seen, one immediate objection that many social researchers make to online surveys concerns representativeness: Will the people who can be surveyed online be representative of meaningful populations, such as all U.S. adults, all voters, and so on? This is the criticism raised with regard to surveys via fax or by telephone interviewers.

Camilo Wilson (1999), founder of Cogix (see the link on this book's website), points out that some populations are ideally suited to online surveys: specifically, those who visit a particular website. For example, Wilson indicates that market research for online companies should be conducted online, and his firm has developed software called ViewsFlash for precisely that purpose. Although website surveys could easily collect data from all who visit a particular site, Wilson suggests that survey-sampling techniques can provide sufficient consumer data without irritating thousands or millions of potential customers.

But how about general population surveys? How about political polling? These are probably the main issues raised regarding online surveys today. Not everyone of interest can be reached via Internet nor feels comfortable using it for participation in surveys. Moreover, people who are less available to online surveys do not represent a random segment of the overall population. The poor and the elderly, for example, are likely to be underrepresented in online surveys. At the same time, as more and more people gain access to the Internet, this problem will decline. (An early criticism of telephone surveys was that not everyone had a phone.)

In one solution to this problem, the National Opinion Research Center, which conducts the periodic General Social Survey (GSS), used probability sampling methods to create a representative sample of potential respondents (T. Smith 2001). Each person in the sample was provided with WebTV access to the Internet, with an agreement that they would participate in polls from time to time. While these online respondents were demographically representative, there were differences in their responses on survey issues that will require further study. For example, the online respondents were more likely to choose extreme responses (such as "strongly agree") than those surveyed in face-to-face interviews were.

Commercial research firms, such as Harris Interactive and Knowledge Networks (see the links on this book's website) report they have developed large-scale panels of online respondents from whom they are able to select samples that are representative of whatever populations are of interest for study. Because their specific methods are proprietary, assessing their methodological strengths and weaknesses is difficult. However,

Harris Interactive has demonstrated success in predicting election results.

Many of the cautions urged in relation to online surveys today are similar to those urged in relation to telephone surveys in the first edition of this book, in 1975. Mick Couper (2001: 466) makes a similar observation:

> Several years ago, I predicted that the rapid spread of electronic data collection methods such as the Internet would produce a bifurcation in the survey industry between high-quality surveys based on probability samples and using traditional data collection methods, on the one hand, and surveys focused more on low cost and rapid turnaround than on representativeness and accuracy on the other. In hindsight, I was wrong, and I underestimated the impact of the Web on the survey industry. It has become much more of a fragmentation than a bifurcation (in terms of Web surveys at least), with vendors trying to find or create a niche for their particular approach or product. No longer is it just "quick and dirty" in one corner and "expensive but high quality" in the other; rather, there is a wide array of approaches representing varying levels of quality and cost.

In the meantime, researchers are amassing a body of experience with this new technique, yielding lessons for increasing success. For example, Survey Sampling, Inc., suggests the following dos and don'ts for conducting online surveys:

> **Do** use consistent wording between the invitation and the survey. Don't use terms such as "unique ID number" in the invitation, then ask respondents to type their "password" when they get to the survey. Changing terminology can be confusing.

> **Do** use plain, simple language.

> **Don't** force the respondent to scroll down the screen for the URL for the study location.

> **Do** offer to share selected results from the study with everyone who completes the survey. Respondents will often welcome information as

a reward for taking the study, especially when they are young adults and teens.

Do plan the time of day and day of week to mail, depending on the subject of the study and type of respondent. Send the invitation late afternoon, evening, or weekend, when respondents are most likely to be reading mail at home, especially if the study requests respondents to check an item in the kitchen or other area in the home. If a parent-child questionnaire is planned, send the invitation late afternoon when children are home, not early in the day, when respondents can't complete the study because children are at school.

Do be aware of technical limitations. For example, WebTV users currently cannot access surveys using Java. If respondents' systems need to be Java-enabled or require access to streaming video, alert panelists at the beginning of the study, not midway through.

Do test incentives, rewards, and prize drawings to determine the optimal offer for best response. Longer surveys usually require larger incentives.

Do limit studies to 15 minutes or less.*

Over the years, members of industrialized nations have become familiar with the format and process of self-administered questionnaires, but the web presents a new challenge for many. Leah Christian, Don Dillman, and Jolene Smyth (2007) provide a wealth of guidance on the formatting of web surveys. Their aim is, as their article title suggests, "helping respondents get it right the first time."

The web is already seeing extensive use as a marketplace for surveys and other research techniques. As only a few illustrative examples, see the following links on this book's website: http://www.cengage.com/sociology/babbie:

The Gallup Organization

SMS Research

The Survey/Marketing Research e-Store

Zogby International

Online surveys appear to have response rates approximately comparable to mail surveys, according to a large-scale study of Michigan State University students (Kaplowitz, Hadlock, and Levine 2004), especially when the online survey is accompanied by a postcard reminder encouraging respondents to participate. While producing a comparable response rate, the online survey costs substantially less than a conventional mail survey. The cost of paper, printing, and postage alone can constitute a large expense.

In another study of ways to improve response rates in online surveys, Stephen Porter and Michael Whitcomb (2003) found that some of the techniques effective in mail surveys, such as personalizing the appeal or varying the apparent status of the researcher, had little or no impact in the new medium. At the same time, specifying that the respondents had been specially selected for the survey and setting a deadline for participation did increase response rates. The years ahead will see many experiments aimed at improving the effectiveness of online surveys. See "How to Do It: Conducting an Online Survey" for more.

During the 2004 election campaign, the role and nature of polls—in-person, telephone, and online—drew considerable public attention. Mark Blumenthal (2005) reviewed the public discussion and considered the implications for the future of polling. Clearly, this discussion will not end any time soon.

Comparison of the Different Survey Methods

Now that we've seen several ways to collect survey data, let's take a moment to compare them directly.

Self-administered questionnaires are generally cheaper and quicker than face-to-face interview surveys. These considerations are likely to be

Source: http://www.worldopinion.com/the_frame/frame4.html. Reprinted with permission.

How to Do It: Conducting an Online Survey

If you're interested in conducting an online survey, you can experiment with a limited version of an online program called Survey Monkey, at no charge. To get started, go to the Survey Monkey link on this book's website (http://www.cengage.com/sociology/babbie) and click "Create Survey."

The program is quite user-friendly with regard to designing questionnaire items. To reach your intended respondents, you enter their e-mail addresses, and they then receive an e-mail invitation to visit the survey web page and participate. The free beginner package will also provide you with a basic analysis of the survey results.

important for an unfunded student wishing to undertake a survey for a term paper or thesis. Moreover, if you use the self-administered mail format, it costs no more to conduct a national survey than a local one of the same sample size. In contrast, a national interview survey (either face-to-face or by telephone) would cost far more than a local one. Also, mail surveys typically require a small staff: You could conduct a reasonable mail survey by yourself, although you shouldn't underestimate the work involved. Further, respondents are sometimes reluctant to report controversial or deviant attitudes or behaviors in interviews but are willing to respond to an anonymous self-administered questionnaire.

Interview surveys also offer many advantages. For example, they generally produce fewer incomplete questionnaires. Although respondents may skip questions in a self-administered questionnaire, interviewers are trained not to do so. In CATI surveys, the computer offers a further check on this. Interview surveys, moreover, have typically achieved higher completion rates than self-administered questionnaires have.

Although self-administered questionnaires may be more effective for sensitive issues, interview surveys are definitely more effective for complicated ones. Prime examples include the enumeration of household members and the determination of whether a given address corresponds to more than one housing unit. Although the concept of housing unit has been refined and standardized by the Census Bureau and interviewers can be trained to deal with the concept, it's extremely difficult to communicate this idea in a self-administered questionnaire. This advantage of interview surveys pertains generally to all complicated contingency questions.

With interviews, you can conduct a survey based on a sample of addresses or phone numbers rather than on names. An interviewer can arrive at an assigned address or call the assigned number, introduce the survey, and even—following instructions—choose the appropriate person at that address to respond to the survey. In contrast, self-administered questionnaires addressed to "occupant" receive a notoriously low response.

Finally, as we've seen, interviewers questioning respondents face-to-face can make important observations aside from responses to questions asked in the interview. In a household interview, they may note the characteristics of the neighborhood, the dwelling unit, and so forth. They can also note characteristics of the respondents or the quality of their interaction with the respondents—whether the respondent had difficulty communicating, was hostile, seemed to be lying, and so on.

The chief advantages of telephone surveys over those conducted face-to-face center primarily on time and money. Telephone interviews are much cheaper and can be mounted and executed quickly. Also, interviewers are safer when interviewing people living in high-crime areas. Moreover, the impact of the interviewers on responses is somewhat lessened when the respondents can't see them. As only one indicator of the popularity of telephone interviewing, when Johnny Blair and his colleagues (1995) compiled a bibliography on sample designs for telephone interviews, they listed over 200 items.

Online surveys have many of the strengths and weaknesses of mail surveys. Once the available software has been further developed, they will likely be substantially cheaper. An important weakness, however, lies in the difficulty of assuring that respondents to an online survey will be representative of some more general population.

Clearly, each survey method has its place in social research. Ultimately, you must balance the advantages and disadvantages of the different methods in relation to your research needs and your resources.

Strengths and Weaknesses of Survey Research

Regardless of the specific method used, surveys—like other modes of observation in social research—have special strengths and weaknesses. You should keep these in mind when determining whether a survey is appropriate for your research goals.

Surveys are particularly useful in describing the characteristics of a large population. A carefully selected probability sample in combination with a standardized questionnaire offers the possibility of making refined descriptive assertions about a student body, a city, a nation, or any other large population. Surveys determine unemployment rates, voting intentions, and so forth with uncanny accuracy. Although the examination of official documents—such as marriage, birth, or death records—can provide equal accuracy for a few topics, no other method of observation can provide this general capability.

Surveys—especially self-administered ones—make large samples feasible. Surveys of 2,000 respondents are not unusual. A large number of cases is very important for both descriptive and explanatory analyses, especially wherever several variables are to be analyzed simultaneously.

In one sense, surveys are flexible. Many questions can be asked on a given topic, giving you considerable flexibility in your analyses. Whereas an experimental design may require you to commit yourself in advance to a particular operational definition of a concept, surveys let you develop operational definitions from actual observations.

Finally, standardized questionnaires have an important strength in regard to measurement generally. Earlier chapters have discussed the ambiguous nature of most concepts: They have no ultimately real meanings. One person's religiosity is quite different from another's. Although you must be able to define concepts in those ways most relevant to your research goals, you may not find it easy to apply the same definitions uniformly to all subjects. The survey researcher is bound to this requirement by having to ask exactly the same questions of all subjects and having to impute the same intent to all respondents giving a particular response.

Survey research also has several weaknesses. First, the requirement of standardization often seems to result in the fitting of round pegs into square holes. Standardized questionnaire items often represent the least common denominator in assessing people's attitudes, orientations, circumstances, and experiences. By designing questions that will be at least minimally appropriate to all respondents, you may miss what is most appropriate to many respondents. In this sense, surveys often appear superficial in their coverage of complex topics. Although this problem can be partly offset by sophisticated analyses, it is inherent in survey research.

Similarly, survey research can seldom deal with the context of social life. Although questionnaires can provide information in this area, the survey researcher rarely develops the feel for the total life situation in which respondents are thinking and acting that, say, the participant observer can (see Chapter 10).

In many ways, surveys are inflexible. Studies involving direct observation can be modified as field conditions warrant, but surveys typically require that an initial study design remain unchanged throughout. As a field researcher, for example, you can become aware of an important new variable operating in the phenomenon you're studying and begin making careful observations of it. The survey researcher would probably be unaware of the new variable's importance and could do nothing about it in any event.

Finally, surveys are subject to the artificiality mentioned earlier in connection with experiments. Finding out that a person gives conservative answers in a questionnaire does not necessarily mean the person is conservative; finding out that a person gives prejudiced answers in a questionnaire does not necessarily mean the person is prejudiced. This shortcoming is especially salient in the realm of action. Surveys cannot measure social action; they can only collect self-reports of recalled past action or of prospective or hypothetical action.

The problem of artificiality has two aspects. First, the topic of study may not be amenable to measurement through questionnaires. Second, the act of studying that topic—an attitude, for example—may affect it. A survey respondent may have given no thought to whether the governor should be impeached until asked for his or her opinion by an interviewer. He or she may, at that point, form an opinion on the matter.

Survey research is generally weak on validity and strong on reliability. In comparison with field research, for example, the artificiality of the survey format puts a strain on validity. As an illustration, people's opinions on issues seldom take the form of strongly agreeing, agreeing, disagreeing, or strongly disagreeing with a specific statement. Their survey responses in such cases must be regarded as approximate indicators of what the researchers had in mind when they framed the questions. This comment, however, needs to be held in the context of earlier discussions of the ambiguity of validity itself. To say something is a valid or an invalid measure assumes the existence of a "real" definition of what's being measured, and many scholars now reject that assumption.

Reliability is a clearer matter. By presenting all subjects with a standardized stimulus, survey research goes a long way toward eliminating unreliability in observations made by the researcher. Moreover, careful wording of the questions can also significantly reduce the subject's own unreliability.

secondary analysis A form of research in which the data collected and processed by one researcher are reanalyzed—often for a different purpose—by another. This is especially appropriate in the case of survey data. Data archives are repositories or libraries for the storage and distribution of data for secondary analysis.

As with all methods of observation, a full awareness of the inherent or probable weaknesses of survey research can partially resolve them in some cases. Ultimately, though, researchers are on the safest ground when they can employ several research methods in studying a given topic.

Secondary Analysis

As a mode of observation, survey research involves the following steps: (1) questionnaire construction, (2) sample selection, and (3) data collection, through either interviewing or self-administered questionnaires. As you've gathered, surveys are usually major undertakings. It's not unusual for a large-scale survey to take several months or even more than a year to progress from conceptualization to data in hand. (Smaller-scale surveys can, of course, be done more quickly.) Through a method called secondary analysis, however, researchers can pursue their particular social research interests—analyzing survey data from, say, a national sample of 2,000 respondents—while avoiding the enormous expenditure of time and money such a survey entails.

Secondary analysis is a form of research in which the data collected and processed by one researcher are reanalyzed—often for a different purpose—by another. Beginning in the 1960s, survey researchers became aware of the potential value that lay in archiving survey data for analysis by scholars who had nothing to do with the survey design and data collection. Even when one researcher had conducted a survey and analyzed the data, those same data could be further analyzed by others who had slightly different interests. Thus, if you were interested in the relationship between political views and attitudes toward gender equality, you could examine that research question through the analysis of any data set that happened to contain questions relating to those two variables.

The initial data archives were very much like book libraries, with a couple of differences. First, instead of books, the data archives contained data sets: first as punched cards, then as magnetic tapes. Today they're typically contained on computer disks, CD-ROMs, or online servers. Second, whereas you're expected to return books to a conventional library, you can keep the data obtained from a data archive.

The best-known current example of a resource for secondary analysis is the General Social Survey (GSS). The National Opinion Research Center (NORC) at the University of Chicago conducts this major national survey, currently every other year, to collect data on a large number of social science variables. These surveys are conducted precisely for the purpose of making data available to scholars at little or no cost and are supported by a combination of private and government funding. Recall that the GSS was created by James A. Davis in 1972; it is currently directed by Davis, Tom W. Smith, and Peter V. Marsden. Their considerable ongoing efforts make an unusual contribution to social science research and to education in social science. You can learn more about the GSS at the link on this book's website: http://www.cengage.com/sociology/babbie.

Numerous other resources are available for identifying and acquiring survey data for secondary analysis. The Roper Center for Public Opinion Research at the University of Connecticut is one excellent resource. The center also publishes the journal *Public Perspective,* on public opinion polling. Polling the Nations is an online repository for thousands of polls conducted in the United States and 70 other nations. A paid subscription allows users to obtain specific data results from studies they specify, rather than obtaining whole studies. Although the cost of the subscription may be too steep for the average student, you might check to see if your school's library has subscribed. (See the links for the Roper Center and Polling the Nations on this book's website.)

Because secondary analysis has typically involved obtaining a data set and undertaking an extensive analysis, I would like you to consider another approach as well. Often you can do limited analyses by investing just a little time. Let's say you're writing a term paper about the impact of religion in contemporary American life. You want to comment on the role of the Roman Catholic church in the debate over abortion. Although you

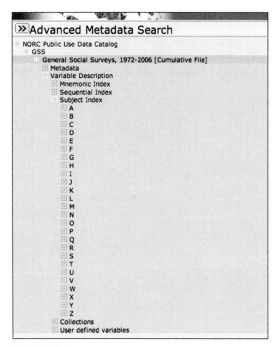

FIGURE 9-7

Subject Index for the General Social Survey

might get away with an offhand, unsubstantiated assertion, imagine how much more powerful your paper would be with the following addition.

1. Access the General Social Survey data set at the National Opinion Research Center, at the link on this book's website.

2. Follow the instructions by clicking successively, in the left column,

   a. The plus sign beside "GSS"

   b. The plus sign beside "General Social Surveys, 1972–2006"

   c. The plus sign beside "Variable Description"

   d. The plus sign beside "Subject Index"

Your screen should now contain the excerpt shown in Figure 9-7.

In the header to the right of this excerpt on your screen, you'll see three tabs: DESCRIPTION, TABULATION, AND ANALYSIS. You'll see that DESCRIPTION has been selected. To begin creating

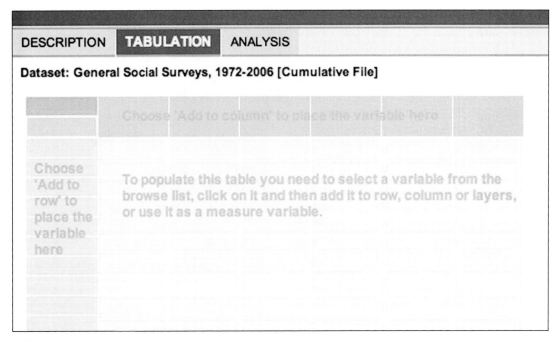

FIGURE 9-8

Instructions for Creating a Table

the analysis for your term paper, click the TABULA-TION tab. A portion of your screen will now look like Figure 9-8.

Let's select "Religious Preference" as the column variable in our table. You'll have to explore a bit, beginning with the plus sign beside the letter R in the Subject Index. Once you've opened that letter, open the entry for "Religion." Then open "R's Religious Preference." Finally, open "Respondent (current)." I know these are a lot of steps, but it demonstrates the wealth of data available for analysis. Once you get more familiar with the data set and the program, you'll move around deftly.

At this point, you'll find several options under "Respondent (current)." We're interested in the first of these: RS RELIGIOUS PREFERENCE. Notice that instead of a plus sign, this notation is preceded by a right-pointing arrow. Click on the arrow, and you should see the display shown in Figure 9-9.

Click on the phrase "Add to column." Your table is halfway complete. To add attitudes toward abortion, repeat the process just outlined but select the following:

A

 Abortion

 In case of . . .

 any reason

This will show the GSS questionnaire items that asked respondents whether a woman should have the right to a legal abortion for any reason. Opening the arrow next to ABORTION IF WOMAN WANTS FOR ANY REASON should give you the display shown in Figure 9-10.

Click on the phrase "add to row" and you'll be presented with a two-variable table, a portion of which is shown in Figure 9-11. (I've limited the display to those religious preferences with more than a few responses.) In reviewing these analytic results, you should realize that they represent the cumulation of surveys done between 1972 and 2006. If you want to limit your analysis to a single year, instructions on the website will tell you how to do that.

The results of this analysis may surprise you. Whereas Catholics are less supportive of abortion

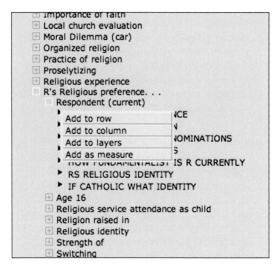

FIGURE 9-9

Selecting RS RELIGIOUS PREFERENCE as the Column Variable

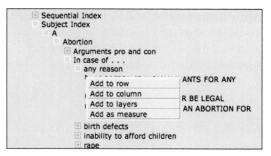

FIGURE 9-10

Selecting ABORTION IF WOMAN WANTS FOR ANY REASON as the Row Variable

RS RELIGIOUS PREFERENCE	PROTESTANT	CATHOLIC	JEWISH	NONE
ABORTION IF WOMAN WANTS FOR ANY REASON				
YES	36.0	35.4	77.8	66.3
NO	64.0	64.6	22.2	33.7
Total	100.0	100.0	100.0	100.0
N=	17,032	6,833	544	2,687

FIGURE 9-11

The Relationship between Religion and Attitudes toward Abortion

(35.4 percent) than Jews (77.8 percent) and those with no religion (66.3 percent) are, they do not differ substantially from Protestants (36.0 percent).

Imagine a term paper that says, "Whereas the Roman Catholic Church has taken a strong, official position on abortion, many Catholics do not necessarily agree, as shown in Table . . ." Moreover, this might be just the beginning of an analysis that looks a bit more deeply into the matter, as described in Chapter 14, on quantitative analysis.

The key advantage of secondary analysis is that it's cheaper and faster than doing original surveys, and, depending on who did the original survey, you may benefit from the work of topflight professionals. The ease of secondary analysis has also enhanced the possibility of meta-analysis, in which a researcher brings together a body of past research on a particular topic. To gain confidence in your understanding of the relationship between religion and abortion, for example, you could go beyond the GSS to analyze similar data collected in dozens or even hundreds of other studies.

There are disadvantages inherent in secondary analysis, however. The key problem involves the recurrent question of validity. When one researcher collects data for one particular purpose, you have no assurance that those data will be appropriate for your research interests. Typically, you'll find that the original researcher asked a question that "comes close" to measuring what you're interested in, but you'll wish the question had been asked just a little differently—or that another, related question had also been asked. For example, you may want to study how religious various people are and the survey data available to you only asked about attendance at worship services. Your quandary, then, is whether the question that was asked provides a valid measure of the variable you want to analyze. Nevertheless, secondary analysis can be immensely useful. Moreover, it illustrates once again the range of possibilities available in finding the answers to questions about social life. Although no single method unlocks all puzzles, there is no limit to the ways you can find out about things. And when you zero in on an issue from several independent directions, you gain that much more expertise.

I've discussed secondary analysis in this chapter on survey research because it's the type of analysis most associated with the technique. However, there is no reason that the reanalysis of social research data needs to be limited to those collected in surveys. Nigel Fielding (2004), for example, has examined the possibilities for the archiving and reanalysis of qualitative data as well.

Ethics and Survey Research

Survey research almost always involves a request that people provide us with information about themselves that is not readily available. Sometimes, we ask for information (about attitudes and behaviors, for example) that would be embarrassing to the respondents if that information became publicly known. In some cases, such revelations could result in the loss of a job or a marriage. Hence, maintaining the norm of confidentiality, mentioned earlier in the book, is particularly important in survey research.

Another ethical concern relates to the possibility of psychological injury to respondents. Even if the information they provide is kept confidential, simply forcing them to think about some matters can be upsetting. Imagine asking people for their attitudes toward suicide when one of them has recently experienced the suicide of a family member or close friend. Or asking people to report on their attitudes about different racial groups, which may cause them to reflect on whether they might be racists or at least appear as such to the interviewers. The possibilities for harming survey respondents are endless. While this fact should not prevent you from doing surveys, it should increase your considered efforts to avoid the problem wherever possible.

MAIN POINTS

Introduction

- Survey research, a popular social research method, is the administration of questionnaires to a sample of respondents selected from some population.

Topics Appropriate for Survey Research

- Survey research is especially appropriate for making descriptive studies of large populations; survey data may be used for explanatory purposes as well.

- Questionnaires provide a method of collecting data by (1) asking people questions or (2) asking them to agree or disagree with statements representing different points of view. Questions may be open-ended (respondents supply their own answers) or closed-ended (they select from a list of provided answers).

Guidelines for Asking Questions

- Items in a questionnaire should follow several guidelines: (1) The form of the items should be appropriate to the project; (2) the items must be clear and precise; (3) the items should ask only about one thing (that is, double-barreled questions should be avoided); (4) respondents must be competent to answer the item; (5) respondents must be willing to answer the item; (6) questions should be relevant to the respondent; (7) items should ordinarily be short; (8) negative terms should be avoided so as not to confuse respondents; (9) the items should be worded to avoid biasing responses.

Questionnaire Construction

- The format of a questionnaire can influence the quality of data collected.
- A clear format for contingency questions is necessary to ensure that the respondents answer all the questions intended for them.
- The matrix question is an efficient format for presenting several items sharing the same response categories.
- The order of items in a questionnaire can influence the responses given.
- Clear instructions are important for getting appropriate responses in a questionnaire.
- Questionnaires should be pretested before being administered to the study sample.
- Questionnaires are usually administered in one of three main ways: through self-administered questionnaires, face-to-face interviews, or telephone surveys. Researchers are exploring online surveys as well.

Self-Administered Questionnaires

- It's generally advisable to plan follow-up mailings in the case of self-administered questionnaires, sending new questionnaires to those respondents who fail to respond to the initial appeal. Properly monitoring questionnaire returns will provide a good guide to when a follow-up mailing is appropriate.

Interview Surveys

- Interviewers must be neutral in appearance and actions; their presence in the data-collection process must have no effect on the responses given to questionnaire items.
- Interviewers must be carefully trained to be familiar with the questionnaire, to follow the question wording and question order exactly, and to record responses exactly as they are given.
- Interviewers can use probes to elicit an elaboration on an incomplete or ambiguous response. Probes should be neutral. Ideally, all interviewers should use the same probes.

Telephone Surveys

- Telephone surveys can be cheaper and more efficient than face-to-face interviews, and they can permit greater control over data collection.
- Random-digit dialing (RDD) is a useful technique for eliminating potential bias in selecting numbers.
- The development of computer-assisted telephone interviewing (CATI) is especially promising.

Online Surveys

- New technologies offer additional opportunities for social researchers, surveys over the Internet. This method, however, must be used with caution because respondents may not be representative of the intended population.

Comparison of the Different Survey Methods

- The advantages of a self-administered questionnaire over an interview survey are economy, speed, lack of interviewer bias, and the possibility of anonymity and privacy to encourage candid responses on sensitive issues.
- The advantages of an interview survey over a self-administered questionnaire are fewer incomplete questionnaires and fewer misunderstood questions, generally higher completion rates, and greater flexibility in terms of sampling and special observations.
- The principal advantages of telephone surveys over face-to-face interviews are the savings in cost and time. Telephone interviewers have more safety than in-person interviewers do, and they may have a smaller effect on the interview itself.
- Online surveys have many of the strengths and weaknesses of mail surveys. Although they're cheaper to conduct, ensuring that the respondents represent a more general population can be difficult.

Strengths and Weaknesses of Survey Research

- Survey research in general offers advantages in terms of economy, the amount of data that can be collected, and the chance to sample a large population. The standardization of the data collected represents another special strength of survey research.
- Survey research has several weaknesses: It is somewhat artificial, potentially superficial, and relatively inflexible. Using surveys to gain a full sense of social processes in their natural settings is difficult. In general, survey research is comparatively weak on validity and strong on reliability.

Secondary Analysis

- Secondary analysis provides social researchers with an important option for "collecting" data cheaply and easily but at a potential cost in validity.

Ethics and Survey Research

- Surveys often ask for private information, and researchers must keep such information confidential.
- Because asking questions can cause psychological discomfort or harm to respondents, the researcher should minimize this risk.

KEY TERMS

The following terms are defined in context in the chapter and at the bottom of the page where the term is introduced, as well as in the comprehensive glossary at the back of the book.

bias	probe
closed-ended questions	questionnaire
computer-assisted telephone interviewing (CATI)	random-digit dialing (RDD)
contingency question	respondent
interview	response rate
open-ended questions	secondary analysis

PROPOSING SOCIAL RESEARCH: SURVEY RESEARCH

If you're planning a survey, you'll have already described the sampling you'll employ, and your discussion of measurement will have presented at least portions of your questionnaire. At this point you need to describe the type of survey you'll conduct: self-administered, telephone, face-to-face, or Internet. Whichever you plan, there will be numerous logistical details to spell out in the proposal. How will you deal with non-respondents, for example? Will you have follow-up mailing in a self-administered questionnaire, follow-up calls in a telephone survey, and so forth? Will you have a target completion rate?

In the case of interview surveys, you should say something about the way you'll select and train the interviewers. You should also say something about the time frame within which the survey will be conducted.

REVIEW QUESTIONS AND EXERCISES

1. For each of the following open-ended questions, construct a closed-ended question that could be used in a questionnaire.
   a. What was your family's total income last year?
   b. How do you feel about the space shuttle program?
   c. How important is religion in your life?
   d. What was your main reason for attending college?
   e. What do you feel is the biggest problem facing your community?

2. Construct a set of contingency questions for use in a self-administered questionnaire that would solicit the following information:
   a. Is the respondent employed?
   b. If unemployed, is the respondent looking for work?
   c. If the unemployed respondent is not looking for work, is he or she retired, a student, or a homemaker?
   d. If the respondent is looking for work, how long has he or she been looking?

3. Find a questionnaire printed in a magazine or newspaper (for a reader survey, for example).

Consider at least five of the questions in it and critique each one.

4. Look at your appearance right now. Identify aspects of your appearance that might create a problem if you were interviewing a general cross section of the public.

5. Locate a survey being conducted on the web. Briefly describe the survey and discuss its strengths and weaknesses.

SPSS EXERCISES

See the booklet that accompanies your text for exercises using SPSS (Statistical Package for the Social Sciences). There are exercises offered for each chapter, and you'll also find a detailed primer on using SPSS.

Online Study Resources

If your book came with an access code card, visit www.cengage.com/login to register. To purchase access, please visit www.ichapters.com.

1. Before you do your final review of the chapter, take the *CengageNOW* pretest to help identify the areas on which you should concentrate. You'll find information on this online tool, as well as instructions on how to access all of its great resources, in the front of the book.

2. As you review, take advantage of the *Cengage-NOW* personalized study plan, based on your quiz results. Use this study plan with its interactive exercises and other resources to master the material.

3. When you're finished with your review, take the posttest to confirm that you're ready to move on to the next chapter.

WEBSITE FOR *THE PRACTICE OF SOCIAL RESEARCH* 12TH EDITION

Go to your book's website at www.cengage.com/sociology/babbie for tools to aid you in studying for your exams. You'll find *Tutorial Quizzes* with feedback, *Internet Exercises*, *Flash Cards*, *Glossaries*, and *Essay Quizzes*, as well as *InfoTrac College Edition* search terms, suggestions for additional reading, *Web Links*, and primers for using data-analysis software such as SPSS.

Qualitative Field Research

CengageNOW for Sociology

Use this online tool to help you make the grade on your next exam. After reading this chapter, go to "Online Study Resources" at the end of the chapter for instructions on how to benefit from *CengageNOW*.

Introduction

Several chapters ago, I suggested that you've been doing social research all your life. This idea should become even clearer as we turn to what probably seems like the most obvious method of making observations: qualitative field research. In a sense, we do field research whenever we observe or participate in social behavior and try to understand it, whether in a college classroom, in a doctor's waiting room, or on an airplane. Whenever we report our observations to others, we're reporting our field research efforts.

Such research is at once very old and very new in social science, stretching at least from the nineteenth-century studies of preliterate societies, through firsthand examinations of urban community life in the "Chicago School" of the 1930s and 1940s, to contemporary observations of chat-room interactions on the web. Many of the techniques discussed in this chapter have been used by social researchers for centuries. Within the social sciences, anthropologists are especially associated with this method and have contributed to its development as a scientific technique. Moreover, something similar to this method is employed by many people who might not, strictly speaking, be regarded as social science researchers. Newspaper reporters are one example; welfare department case workers are another.

Although these are "natural" activities, they are also skills to be learned and honed. This chapter discusses these skills in some detail, examining some of the major paradigms of field research and describing some of the specific techniques that make scientific field research more useful than the casual observation we all engage in.

I use the term *qualitative field research* to distinguish this type of observational method from methods designed to produce data appropriate for quantitative (statistical) analysis. Thus, surveys provide data from which to calculate the percentage unemployed in a population, mean incomes, and so forth. Field research more typically yields qualitative data: observations not easily reduced to

numbers. Thus, for example, a field researcher may note the "paternalistic demeanor" of leaders at a political rally or the "defensive evasions" of a public official at a public hearing without trying to express either the paternalism or the defensiveness as a numerical quantity or degree. Although field research can be used to collect quantitative data—for example, noting the number of interactions of various specified types within a field setting—typically, field research is qualitative.

Field observation also differs from some other models of observation in that it's not just a data-collecting activity. Frequently, perhaps typically, it's a theory-generating activity as well. As a field researcher, you'll seldom approach your task with precisely defined hypotheses to be tested. More typically, you'll attempt to make sense out of an ongoing process that cannot be predicted in advance—making initial observations, developing tentative general conclusions that suggest particular types of further observations, making those observations and thereby revising your conclusions, and so forth. In short, the alternation of induction and deduction discussed in Part 1 of this book is perhaps nowhere more evident and essential than in good field research. For expository purposes, however, this chapter focuses primarily on some of the theoretical foundations of field research and on techniques of data collection. Chapter 13 discusses how to analyze qualitative data.

Topics Appropriate for Field Research

One of the key strengths of field research is how comprehensive a perspective it can give researchers. By going directly to the social phenomenon under study and observing it as completely as possible, researchers can develop a deeper and fuller understanding of it. As such, this mode of observation is especially, though not exclusively, appropriate to research topics and social studies that appear to defy simple quantification. Field researchers may recognize several nuances of attitude and behavior that might escape researchers using other methods.

Field research is especially appropriate for the study of those attitudes and behaviors best understood within their natural setting, as opposed to the somewhat artificial settings of experiments and surveys. For example, field research provides a superior method for studying the dynamics of religious conversion at a revival meeting, just as a statistical analysis of membership rolls would be a better way of discovering whether men or women were more likely to convert.

Finally, field research is well suited to the study of social processes over time. Thus, the field researcher might be in a position to examine the rumblings and final explosion of a riot as events actually occur rather than afterward in a reconstruction of the events.

Other good places to apply field research methods include campus demonstrations, courtroom proceedings, labor negotiations, public hearings, or similar events taking place within a relatively limited area and time. Several such observations must be combined in a more comprehensive examination over time and space.

In *Analyzing Social Settings* (2006:123–132), John Lofland and his colleagues discuss several elements of social life appropriate to field research:

1. *Practices:* Various kinds of behavior, such as talking or reading a book

2. *Episodes:* A variety of events such as divorce, crime, and illness

3. *Encounters:* Two or more people meeting and interacting

4. *Roles and social types:* The analysis of the positions people occupy and the behavior associated with those positions: occupations, family roles, ethnic groups

5. *Social and personal relationships:* Behavior appropriate to pairs or sets of roles: mother–son relationships, friendships, and the like

6. *Groups and cliques:* Small groups, such as friendship cliques, athletic teams, and work groups

7. *Organizations:* Formal organizations, such as hospitals or schools

8. *Settlements and habitats:* Small-scale "societies" such as villages, ghettos, and neighborhoods, as opposed to large societies such as nations, which are difficult to study

9. *Social worlds:* Ambiguous social entities with vague boundaries and populations, such as "the sports world" and "Wall Street"

10. *Subcultures and lifestyles:* How large numbers of people adjust to life in groups such as a "ruling class" or an "urban underclass"

In all these social settings, field research can reveal things that would not otherwise be apparent. Here's a concrete example.

One issue I'm particularly interested in (Babbie 1985) is the nature of responsibility for public matters: Who's responsible for making the things that we share work? Who's responsible for keeping public spaces—parks, malls, buildings, and so on—clean? Who's responsible for seeing that broken street signs get fixed? Or, if a strong wind knocks over garbage cans and rolls them around the street, who's responsible for getting them out of the street?

On the surface, the answer to these questions is pretty clear. We have formal and informal agreements in our society that assign responsibility for these activities. Government custodians are responsible for keeping public places clean. Transportation department employees are responsible for the street signs, and perhaps the police are responsible for the garbage cans rolling around on a windy day. And when these responsibilities are not fulfilled, we tend to look for someone to blame.

What fascinates me is the extent to which the assignment of responsibility for public things to specific individuals not only relieves others of the responsibility but actually prohibits them from taking responsibility. It's my notion that it has become unacceptable for someone like you or me to take personal responsibility for public matters that haven't been assigned to us.

Let me illustrate what I mean. If you were walking through a public park and you threw down a bunch of trash, you'd discover that your action was unacceptable to those around you. People would glare at you, grumble to each other; perhaps someone would say something to you about it. Whatever the form, you'd be subjected to definite, negative sanctions for littering. Now here's the irony. If you were walking through that same park, came across a bunch of trash that someone else had dropped, and cleaned it up, it's likely that your action would also be unacceptable to those around you. You'd probably face negative sanctions for cleaning it up.

When I first began discussing this pattern with students, most felt the notion was absurd. Although we would be negatively sanctioned for littering, cleaning up a public place would obviously bring positive sanctions: People would be pleased with us for doing it. Certainly, all my students said they would be pleased if someone cleaned up a public place. It seemed likely that everyone else would be pleased, too, if we asked them how they would react to someone's cleaning up litter in a public place or otherwise taking personal responsibility for fixing some social problem.

To settle the issue, I suggested that my students start fixing the public problems they came across in the course of their everyday activities. As they did so, I asked them to note the answers to two questions:

1. How did they feel while they were fixing a public problem they had not been assigned responsibility for?

2. How did others around them react?

My students picked up litter, fixed street signs, put knocked-over traffic cones back in place, cleaned and decorated communal lounges in their dorms, trimmed trees that blocked visibility at intersections, repaired public playground equipment, cleaned public restrooms, and took care of a hundred other public problems that weren't "their responsibility."

Most reported feeling very uncomfortable doing whatever they did. They felt foolish, goody-goody, conspicuous, and all the other feelings that keep us from performing these activities normally. In almost every case, their personal feelings of discomfort were increased by the reactions of those around them. One student was removing a damaged and long-unused newspaper box from the bus stop, where it had been a problem for months, when the police arrived, having been summoned

by a neighbor. Another student decided to clean out a clogged storm drain on his street and found himself being yelled at by a neighbor who insisted that the mess should be left for the street cleaners. Everyone who picked up litter was sneered at, laughed at, and generally put down. One young man was picking up litter scattered around a trash can when a passerby sneered, "Clumsy!" It became clear to us that there are only three acceptable explanations for picking up litter in a public place:

1. You did it and got caught—somebody forced you to clean up your mess.

2. You did it and felt guilty.

3. You're stealing litter.

In the normal course of things, it's simply not acceptable for people to take responsibility for public things.

Clearly, we could not have discovered the nature and strength of agreements about taking personal responsibility for public things except through field research. Social norms suggest that taking responsibility is a good thing, sometimes referred to as good citizenship. Asking people what they thought about taking responsibility would have produced a solid consensus that it was good. Only going out into life, doing it, and watching what happened gave us an accurate picture.

As an interesting footnote to this story, my students and I found that whenever people could get past their initial reactions and discover that the students were simply taking responsibility for fixing things for the sake of having them work, the passersby tended to assist. Although there are some very strong agreements making it "unsafe" to take responsibility for public things, the willingness of one person to rise above those agreements seemed to make it safe for others to do so, and they did.

Field research is not to be confused with journalism. Social scientists and journalists may use similar techniques, but they have quite a different relationship to data. For instance, individual interviewing is a common technique in journalism and sociology; nevertheless, sociologists are not simply concerned with reporting about a subject's attitude, belief, or experience. A sociologist's goal is to treat an interview as data that need to be analyzed to understand social life more generally.

Two important aspects of qualitative research need to be stressed. First, a wide range of studies fall under the umbrella "qualitative field research." As we'll see in this chapter, various epistemologies within different paradigms have quite different approaches to basic questions such as "What are data?" "How should we collect data?" and "How should we analyze data?" Second, we should remember that the questions we want to answer in our research determine the types of methods we need to use. A question such as "How do women construct their everyday lives in order to perform their roles as mothers, partners, and breadwinners?" could be addressed by in-depth interviews and direct observations. The assessment of advertising campaigns might profit from focus group discussions. In most cases, we'll find that researchers have alternate methods to choose from.

In summary, then, field research offers the advantage of probing social life in its natural habitat. Although some things can be studied adequately through questionnaires or in the laboratory, others cannot. And direct observation in the field lets researchers observe subtle communications and other events that might not be anticipated or measured otherwise.

Special Considerations in Qualitative Field Research

There are specific things to take into account in every research method, and qualitative field research is no exception. When you use field research methods, you're confronted with decisions about the role you'll play as an observer and your relations with the people you're observing. Let's examine some of the issues involved in these decisions.

The Various Roles of the Observer

In field research, observers can play any of several roles, including participating in what they want to observe (this was the situation of the students who

fixed public things). In this chapter, I've used the term *field research* rather than the frequently used term *participant observation,* because field researchers need not always participate in what they're studying, though they usually do study it directly at the scene of the action. As Catherine Marshall and Gretchen Rossman point out:

> The researcher may plan a role that entails varying degrees of "participantness"—that is, the degree of actual participation in daily life. At one extreme is the full participant, who goes about ordinary life in a role or set of roles constructed in the setting. At the other extreme is the complete observer, who engages not at all in social interaction and may even shun involvement in the world being studied. And, of course, all possible complementary mixes along the continuum are available to the researcher.
>
> *(1995: 60)*

The complete participant, in this sense, may be a genuine participant in what he or she is studying (for example, a participant in a campus demonstration) or may pretend to be a genuine participant. In any event, whenever you act as the complete participant, you must let people see you only as a participant, not as a researcher. For instance, if you're using this technique to study a group made up of uneducated and inarticulate people, it would not be appropriate for you to talk and act like a university professor or student.

This type of research introduces an ethical issue, one on which social researchers themselves are divided. Is it ethical to deceive the people you're studying in the hope that they will confide in you as they will not confide in an identified researcher? Do the potential benefits to be gained from the research offset such considerations? Although many professional associations have addressed this issue, the norms to be followed remain somewhat ambiguous when applied to specific situations.

Related to this ethical consideration is a scientific one. No researcher deceives his or her subjects solely for the purpose of deception. Rather, it's done in the belief that the data will be more valid and reliable, that the subjects will be more natural and honest if they do not know the researcher is doing a research project. If the people being studied know they're being studied, they might modify their behavior in a variety of ways. This is known as the problem of **reactivity.**

First, they might expel the researcher. Second, they might modify their speech and behavior to appear more "respectable" than would otherwise be the case. Third, the social process itself might be radically changed. Students making plans to burn down the university administration building, for example, might give up the plan altogether once they learn that one of their group is a social scientist conducting a research project.

On the other side of the coin, if you're a complete participant, you may affect what you're studying. Suppose, for example, that you're asked for your ideas about what the group should do next. No matter what you say, you will affect the process in some fashion. If the group follows your suggestion, your influence on the process is obvious. If the group decides not to follow your suggestion, the process whereby the suggestion is rejected may affect what happens next. Finally, if you indicate that you just don't know what should be done next, you may be adding to a general feeling of uncertainty and indecisiveness in the group.

Ultimately, anything the participant-observer does or does not do will have some effect on what's being observed; it's simply inevitable. More seriously, there is no complete protection against this effect, though sensitivity to the issue may provide a partial protection. (This influence, called the Hawthorne effect, was discussed more fully in Chapter 8.)

Because of these several considerations, ethical and scientific, the field researcher frequently chooses a different role from that of complete participant. You could participate fully with the group under study but make it clear that you were also undertaking research. As a member of the volleyball team, for example, you might use your position to launch a study in the sociology of sports, letting

your teammates know what you're doing. There are dangers in this role also, however. The people being studied may shift much of their attention to the research project rather than focusing on the natural social process, making the process being observed no longer typical. Or, conversely, you yourself may come to identify too much with the interests and viewpoints of the participants. You may begin to "go native" and lose much of your scientific detachment.

At the other extreme, the complete observer studies a social process without becoming a part of it in any way. Quite possibly, because of the researcher's unobtrusiveness, the subjects of study might not realize they're being studied. Sitting at a bus stop to observe jaywalking at a nearby intersection is one example. Although the complete observer is less likely to affect what's being studied and less likely to "go native" than the complete participant, she or he is also less likely to develop a full appreciation of what's being studied. Observations may be more sketchy and transitory.

Fred Davis (1973) characterizes the extreme roles that observers might play as "the Martian" and "the Convert." The latter involves delving more and more deeply into the phenomenon under study, running the risk of "going native." We'll examine this risk further in the next section.

To appreciate the "Martian" approach, imagine that you were sent to observe some newfound life on Mars. Probably you would feel yourself inescapably separate from the Martians. Some social scientists adopt this degree of separation when observing cultures or social classes different from their own.

Marshall and Rossman (1995: 60–61) also note that the researcher can vary the amount of time spent in the setting being observed: You can be a full-time presence on the scene or just show up now and then. Moreover, you can focus your attention on a limited aspect of the social setting or seek to observe all of it—framing an appropriate role to match your aims.

When Jeffrey Kidder set out to study the culture of bike messengers in New York City, he found it appropriate to identify his research role to some of those he observed but not others (2005: 349):

reactivity The problem that the subjects of social research may react to the fact of being studied, thus altering their behavior from what it would have been normally.

While I did have an academic motivation in working as a messenger, it should be made clear that my participation within the messenger world was neither forced nor faked. To the contrary, my lifelong interest in bicycles and alternative transportation melded seamlessly with the messenger lifestyle.

During the course of my fieldwork, most of the messengers with whom I came in contact were unaware of my research; this was a matter of necessity. In New York City, a messenger crosses paths with hundreds of messengers a day. The numerous individuals that helped form my understandings of messenger style could not all be approached to sign consent forms. Messengers with whom I had reoccurring contact were informed of my sociological interest.

Different situations ultimately require different roles for the researcher. Unfortunately, there are no clear guidelines for making this choice—you must rely on your understanding of the situation and your own good judgment. In making your decision, however, you must be guided by both methodological and ethical considerations. Because these often conflict, your decision will frequently be difficult, and you may find sometimes that your role limits your study.

Relations to Subjects

Having introduced the different roles field researchers might play in connection with their observations, we now focus more specifically on how researchers may relate to the subjects of their study and to the subjects' points of view.

We've already noted the possibility of pretending to occupy social statuses we don't really occupy. Consider now how you would think and feel in such a situation.

Suppose you've decided to study a religious cult that has enrolled many people in your neighborhood. You might study the group by joining it or pretending to join it. Take a moment to ask yourself what the difference is between "really" joining and "pretending" to join. The main difference is whether or not you actually take on the beliefs, attitudes, and other points of view shared by the "real" members. If the cult members believe that Jesus will come next Thursday night to destroy the world and save the members of the cult, do you believe it or do you simply pretend to believe it?

Traditionally, social scientists have tended to emphasize the importance of "objectivity" in such matters. In this example, that injunction would be to avoid getting swept up in the beliefs of the group. Without denying the advantages associated with such objectivity, social scientists today also recognize the benefits gained by immersing themselves in the points of view they're studying, what John Lofland and his colleagues (2006: 70) refer to as "selective competence" or "insider knowledge, skill, or understanding." Ultimately, you won't be able to fully understand the thoughts and actions of the cult members unless you can adopt their points of view as true—at least temporarily. To fully appreciate the phenomenon you've set out to study, you need to believe that Jesus is coming Thursday night. In some settings, this can also help you gain rapport with your subjects (see the discussion on rapport later in this chapter).

Adopting an alien point of view is an uncomfortable prospect for most people. It can be hard enough merely to learn about views that seem strange to you; you may sometimes find it hard just to tolerate certain views. But to take them on as your own can be ten times worse. Robert Bellah (1970, 1974) has offered the term *symbolic realism* to indicate the need for social researchers to treat the beliefs they study as worthy of respect rather than as objects of ridicule. The difficulty of adopting other's views led William Shaffir and Robert Stebbins (1991: 1) to conclude that "fieldwork must certainly rank with the more disagreeable activities that humanity has fashioned for itself."

There is, of course, a danger in adopting the points of view of the people you're studying. When you abandon your objectivity in favor of adopting such views, you lose the possibility of seeing and understanding the phenomenon within frames of reference unavailable to your subjects. On the one hand, accepting the belief that the world will end Thursday night allows you to appreciate aspects of that belief available only to believers; stepping outside that view, however, makes it possible for you to consider some reasons why people might adopt such a view. You may discover that some did so as a consequence of personal trauma (such as unemployment or divorce), whereas others were brought into the fold through their participation in particular social networks (for example, their whole bowling team joined the cult). Notice that the cult members might disagree with those "objective" explanations, and you might not come up with them to the extent that you had operated legitimately within the group's views.

Anthropologists sometimes use the term *emic perspective* in reference to taking on the point of view of those being studied. In contrast, the *etic perspective* maintains a distance from the native point of view in the interest of achieving more objectivity.

The apparent dilemma here is that both of these postures offer important advantages but also seem mutually exclusive. In fact, it's possible to assume both postures. Sometimes you can simply shift viewpoints at will. When appropriate, you can fully assume the beliefs of the cult; later, you can step outside those beliefs (more accurately, you can step inside the viewpoints associated with social science). As you become more adept at this kind of research, you may come to hold contradictory viewpoints simultaneously, rather than switching back and forth.

During my study of trance channeling—in which people allow spirits to occupy their bodies and speak through them—I found I could participate fully in channeling sessions without becoming alienated from conventional social science. Rather than "believing" in the reality of channeling, I found it possible to suspend beliefs in that realm: neither believing it to be genuine (like most of the other participants) nor disbelieving it (like most scientists). Put differently, I was open to either possibility. Notice how this differs from our normal need to "know" whether such things are legitimate or not.

Social researchers often refer to the concerns just discussed as a matter of *reflexivity*, in the sense of things acting on themselves. Thus, your own characteristics can affect what you see and how you interpret it. The issue is broader than that, however, and applies to the subjects as well as to the researcher. Imagine yourself interviewing a homeless person (1) on the street, (2) in a homeless shelter, or (3) in a social welfare office. The research setting could affect the person's responses. In other words, you might get different results depending on where you conducted the interview. Moreover, you might act differently as a researcher in those different settings. If you reflect on this issue, you'll be able to identify other aspects of the research encounter that complicate the task of "simply observing what's so."

The problem we've just been discussing could be seen as psychological, occurring mostly inside the researchers' or subjects' heads. There is a corresponding problem at a social level, however. When you become deeply involved in the lives of the people you're studying, you're likely to be moved by their personal problems and crises. Imagine, for example, that one of the cult members becomes ill and needs a ride to the hospital. Should you provide transportation? Sure. Suppose someone wants to borrow money to buy a stereo. Should you loan it? Probably not. Suppose they need the money for food?

There are no black-and-white rules for resolving situations such as these, but you should realize that you'll need to deal with them regardless of whether or not you reveal that you're a researcher. Such problems do not tend to arise in other types of research—surveys and experiments, for example—but they are part and parcel of field research.

Caroline Knowles (2006) raises a somewhat different issue with regard to the researcher's relationship to subjects in the field. In her interview study of British expatriates living in Hong Kong, she noticed that some were particularly difficult for her to deal with. When she found herself writing research notes explaining why the project would not profit from her interviewing them further, she forced herself to look more deeply into the interactional dynamics in question—with an emphasis on her side of the relationships. She examined *why* certain informants made her uncomfortable and then pressed through the discomfort to continue interviewing. She found that factors such as the attitudes they expressed, their rude interaction styles, and the nature of the relationship she was establishing with them contributed to her reaction. In the end, she gained a much deeper understanding of her subjects than would have been possible if she had limited herself to those who were cooperative and nice.

This discussion of the field researcher's relationships to subjects flies in the face of the usual view of "scientific objectivity." Before concluding this section, let's take the issue one step further.

In the conventional view of science, differences of power and status separate the researcher from the subjects of research. When we discussed experimental designs in Chapter 8, for example, it was obvious who was in charge: the experimenter, who organized things and told the subjects what to do. Often the experimenter was the only person who even knew what the research was really about. Something similar might be said about survey research. The person running the survey designs the questions, decides who will be selected for questioning, and analyzes the data collected.

Sociologists often look at these sorts of relationships as power or status relationships. In experimental and survey designs, the researcher clearly has more power and a higher status than the people being studied do. The researchers have a special knowledge that the subjects don't enjoy. They're not so crude as to say they're superior to their subjects, but there is a sense in which that's implicitly assumed. (Notice that there is a similar, implicit assumption about the writers and readers of textbooks.)

In field research, such assumptions can be problematic. When the early European anthropologists set out to study what were originally called "primitive" societies, there was no doubt that the anthropologists knew best. Whereas the natives "believed" in witchcraft, for example, the anthropologists "knew" it wasn't really true. Whereas the natives said some of their rituals would appease the gods, the anthropologists explained that the "real" functions of these rituals were the creation of social identity, the establishment of group solidarity, and so on.

The more social researchers have gone into the field to study their fellow humans face-to-face, however, the more they have become conscious of these implicit assumptions about researcher superiority, and the more they have considered alternatives. As we turn now to the various paradigms of field research, we'll see some of the ways in which that ongoing concern has worked itself out. See "Keeping Humanity in Focus" for an example of field research on status.

Some Qualitative Field Research Paradigms

Although I've described field research as simply going where the action is and observing it, there are actually many different approaches to this research method. This section examines several field research paradigms: naturalism, ethnomethodology, grounded theory, case studies and the extended case method, institutional ethnography, and participatory action research. Although this survey won't exhaust the variations on the method, it should give you a broad appreciation of the possibilities.

It's important to recognize that there are no specific methods attached to these paradigms. You could do ethnomethodology or institutional ethnography by analyzing court hearings or conducting group interviews, for example. The important distinctions of this section are epistemological, having to do with what data mean, regardless of how they were collected.

Naturalism

Naturalism is an old tradition in qualitative research. The earliest field researchers operated on the positivist assumption that social reality was "out

---

naturalism An approach to field research based on the assumption that an objective social reality exists and can be observed and reported accurately.

Keeping Humanity In Focus

What could seem like a clearer status relationship than between a guest in a luxury hotel and the room service and other staff who serve that guest's needs? In fact, Rachel Sherman has found a far more complex process than you might imagine. She is particularly interested in how service workers balance their relationships with management and their relationships with guests. Unlike manufacturing workers, the hotel service staff must deal with both supervisors and consumers, even when the demands of the two conflict. In part, she discovered that service workers in hotels often receive more discretion regarding how to serve guests' needs than we might expect. This has a positive impact on the worker's sense of self as well as providing a good experience for guests.

Sherman's observations and conclusions came from months spent as a service worker in two luxury hotels. She made her research identity known to management and was able to move around through many of the different service jobs: making reservations, delivering room-service meals, parking cars, carrying bags, housekeeping, and many other tasks that the guests in luxury hotels expect. Her immersion in the research allowed her access to data she would not have found out otherwise.

CLASS ACTS

SERVICE AND INEQUALITY
IN LUXURY HOTELS

RACHEL SHERMAN

Class Acts: Service and Inequality in Luxury Hotels by Rachel Sherman. © 2007 Rachel Sherman. Published by the University of California Press. Photograph © Karen Beard/Getty Images

---

there," ready to be naturally observed and reported by the researcher as it "really is" (Gubrium and Holstein 1997). This tradition started in the 1930s and 1940s at the University of Chicago's sociology department, whose faculty and students fanned out across the city to observe and understand local neighborhoods and communities. The researchers of that era and their research approach are now often referred to as the Chicago School.

One of the earliest and best-known studies that illustrates this research tradition is William Foote Whyte's ethnography of Cornerville, an Italian American neighborhood, in his book *Street Corner Society* (1943). An **ethnography** is a study that focuses on detailed and accurate description rather than explanation. Like other naturalists, Whyte

believed that in order to learn fully about social life on the streets, he needed to become more of an insider. He made contact with "Doc," his key informant, who appeared to be one of the street-gang leaders. Doc let Whyte enter his world, and Whyte got to participate in the activities of the people of Cornerville. His study offered something that surveys could not: a richly detailed picture of life among the Italian immigrants of Cornerville.

An important feature of Whyte's study is that he reported the reality of the people of Cornerville on their terms. The naturalist approach is based on telling "their" stories the way they "really are," not the way the ethnographer understands "them." The narratives collected by Whyte are taken at face value as the social "truth" of the Cornerville residents.

Forty-five years later, David Snow and Leon Anderson (1987) conducted exploratory field research into the lives of homeless people in Austin, Texas. Their main task was to understand how the

---

ethnography A report on social life that focuses on detailed and accurate description rather than explanation.

homeless construct and negotiate their identity while knowing that the society they live in attaches a stigma to homelessness. Snow and Anderson believed that, to achieve this goal, the collection of data had to arise naturally. Like Whyte in *Street Corner Society*, they found some key informants whom they followed in their everyday journeys, such as at their day-labor pickup sites or under bridges. Snow and Anderson chose to memorize the conversations they participated in or the "talks" that homeless people had with each other. At the end of the day, the two researchers debriefed and wrote detailed field notes about all the "talks" they encountered. They also taped in-depth interviews with their key informants.

Snow and Anderson reported "hanging out" with homeless people over the course of 12 months for a total of 405 hours in 24 different settings. Out of these rich data, they identified three related patterns in homeless people's conversations. First, the homeless showed an attempt to "distance" themselves from other homeless people, from the low-status job they currently had, or from the Salvation Army they depended on. Second, they "embraced" their street-life identity—their group membership or a certain belief about why they are homeless. Third, they told "fictive stories" that always contrasted with their everyday life. For example, they would often say that they were making much more money than they really were, or even that they were "going to be rich."

Richard Mitchell (2002) offers another, timely illustration of the power of ethnographic reporting. Recent U.S. history has raised the specter of violence from secretive survivalist groups, dramatized by the 1992 siege at Ruby Ridge, Idaho, which left the wife and son of the white supremacist Randy Weaver dead; the 1993 shootout with David Koresh and his Branch Davidians in Waco, Texas; and Timothy McVeigh's 1995 bombing, which left 168 dead under the rubble of the nine-story Murrah Federal Building in Oklahoma City.

Mitchell describes a variety of survivalist individuals and groups, seeking to understand their reasoning, their plans, and the threat they may pose for the rest of us. He found the survivalists disillusioned with and uncertain about the future of U.S. society, but most were more interested in creating alternative lives and cultures for themselves than in blowing anyone up. That's not to suggest none of the survivalists poses a threat, but Mitchell's examination moves beyond the McVeighs, Koreshes, and Weavers to draw a broader picture of the whole phenomenon.

While ethnographers seek to discover and understand the patterns of living among those they are studying, Mitchell Duneier (1999) has warned against what he calls the "ethnographic fallacy." This refers to an overgeneralization and oversimplification of the patterns observed. Despite the existence of patterns within groups, there is also diversity, and you need to be wary of broad assertions suggesting that "the poor," "the French," or "cheerleaders" act or think in certain ways as though all members of the group do so.

Whereas this chapter aims at introducing you to some of the different approaches available to you in qualitative field research, please realize that this discussion of ethnography merely sketches some of the many avenues social researchers have established. If you're interested in this general approach, you might want to explore the idea of *virtual ethnography*, which uses ethnographic techniques for inquiry into cyberspace. Or, in a different direction, *autoethnography* intentionally assumes a personal stance, breaking with the general proscription against the researcher getting involved at that level. Lest autoethnography seem a simple and/or trivial undertaking, you might look at Sarah Wall's 2008 article, "Easier Said than Done: Writing an Autoethnography."

You can learn more about these variants on ethnography by searching the web or your campus library. A later section of this chapter will examine *institutional ethnography*, which links individuals and organizations.

In Chapter 9, we saw how the Internet is affecting survey research. Eric Anderson used the Internet to launch a qualitative, in-depth interviewing study of male cheerleaders:

> Twelve collegiate male cheerleaders were contacted for interviews by using the member profile search on America Online which provides a

search engine for accessing the stated interests of AOL's 33 million subscribers. After communicating with these cheerleaders through instant messaging, I asked them for in-depth, taped telephone interviews.

(2005: 340)

Anderson then used snowball sampling to increase the number of cheerleaders to study. This is just another example of the wide variety of venues for ethnographic study.

Ethnomethodology

Ethnomethodology, which I introduced as a research paradigm in Chapter 2, is a unique approach to qualitative field research. It has its roots in the philosophical tradition of phenomenology, which can explain why ethnomethodologists are skeptical about the way people report their experience of reality (Gubrium and Holstein 1997). Alfred Schutz (1967, 1970), who introduced phenomenology, argued that reality was socially constructed rather than being "out there" for us to observe. People describe their world not "as it is" but "as they make sense of it." Thus, phenomenologists would argue that Whyte's street-corner men were describing their gang life as it made sense to them. Their reports, however, would not tell us how and why it made sense to them. For this reason, researchers cannot rely on their subjects' stories to depict social realities accurately.

Whereas traditional ethnographers believe in immersing themselves in a particular culture and reporting their informants' stories as if they represented reality, phenomenologists see a need to "make sense" out of the informants' perceptions of the world. Following in this tradition, some field researchers have felt the need to devise techniques that reveal how people make sense of their every-

day world. As we saw in Chapter 2, the sociologist Harold Garfinkel suggested that researchers *break the rules* so that people's taken-for-granted expectations would become apparent. This is the technique that Garfinkel called ethnomethodology.

Garfinkel became known for engaging his students to perform a series of what he called "breaching experiments" designed to break away from the ordinary (Heritage 1984). For instance, Garfinkel (1967) asked his students to do a "conversation clarification experiment." Students were told to engage in an ordinary conversation with an acquaintance or a friend and to ask for clarification about any of this person's statements. Through this technique, they uncovered elements of conversation that are normally taken for granted. Here are two examples of what Garfinkel's students reported (1967: 42):

Case 1

The subject was telling the experimenter, a member of the subject's car pool, about having had a flat tire while going to work the previous day.

I had a flat tire.

(E) What do you mean, you had a flat tire? She appeared momentarily stunned. Then she answered in a hostile way: "What do you mean, 'What do you mean?' A flat tire is a flat tire. That is what I meant. Nothing special. What a crazy question."

Case 6

The victim waved his hand cheerily.

(S) How are you?

(E) How I am in regard of what? My health, my finances, my school work, my peace of mind, my . . . ?

(S) (Red in the face and suddenly out of control.) Look I was just trying to be polite. Frankly, I don't give a damn how you are.

By setting aside or "bracketing" their expectations from these everyday conversations, the experimenters made visible the subtleties of mundane interactions. For example, although "How are you?" has many possible meanings, none of us has any trouble knowing what it means in casual

---

ethnomethodology An approach to the study of social life that focuses on the discovery of implicit, usually unspoken assumptions and agreements; this method often involves the intentional breaking of agreements as a way of revealing their existence.

interactions, as the unsuspecting subject revealed in his final comment.

Ethnomethodologists, then, are not simply interested in subjects' perceptions of the world. In these cases, we could imagine that the subjects may have thought that the experimenters were rude, stupid, or arrogant. The conversation itself, not the informants, is the object of ethnomethodological studies. In general, ethnomethodology focuses on the "underlying patterns" of interactions that regulate our everyday lives.

Ethnomethodologists believe that researchers who use a naturalistic analysis "[lose] the ability to analyze the commonsense world and its culture if [they use] analytical tools and insights that are themselves part of the world or culture being studied" (Gubrium and Holstein 1997: 43). D. L. Wieder provides an excellent example of how different a naturalistic approach is from an ethnomethodological approach (Gubrium and Holstein 1997). In his study *Language and Social Reality: The Case of Telling the Convict Code* (1988), Wieder started to approach convicts in a halfway house in a traditional ethnographic style: He was going to become an insider by befriending the inmates and by conducting participant observations. He took careful notes and recorded interactions among inmates and between inmates and staff. His first concern was to describe the life of the convicts of the halfway house the way it "really was" for them. Wieder's observations allowed him to report on a "convict code" that he thought was the source of the deviant behavior expressed by the inmates toward the staff. This code, which consisted of a series of rules such as "Don't kiss ass," "Don't snitch," and "Don't trust the staff," was followed by the inmates who interfered with the staff members' attempts to help them make the transition between prison and the community.

It became obvious to Wieder that the code was more than an explanation for the convicts' deviant behavior; it was a "method of moral persuasion and justification" (Wieder 1988: 175). At this point he changed his naturalistic approach to an ethnomethodological one. Whereas naturalistic field researchers aim to understand social life as the participants understand it, ethnomethodologists

are more intent on identifying the methods through which understanding occurs. In the case of the convict code, Wieder came to see that convicts used the code to make sense of their own interactions with other convicts and with the staff. The ethnography of the halfway house thus shifted to an ethnography of the code. For instance, the convicts would say, "You know I won't snitch," referring to the code as a way to justify their refusal to answer Wieder's question (p. 168). According to Wieder, the code "operated as a device for stopping or changing the topic of conversation" (p. 175). Even the staff would refer to the code to justify their reluctance to help the convicts. Although the code was something that constrained behavior, it also functioned as a tool for the control of interactions.

Grounded Theory

Grounded theory originated from the collaboration of Barney Glaser and Anselm Strauss, sociologists who brought together two main traditions of research: positivism and interactionism. Essentially, **grounded theory** is the attempt to derive theories from an analysis of the patterns, themes, and common categories discovered in observational data. The first major presentation of this method can be found in Glaser and Strauss's book, *The Discovery of Grounded Theory* (1967). Grounded theory can be described as an approach that attempts to combine a naturalist approach with a positivist concern for a "systematic set of procedures" in doing qualitative research.

Anselm Strauss and Juliet Corbin (1998: 43–46) have suggested that grounded theory allows the researcher to be scientific and creative at the same time, as long as the researcher follows these guidelines:

---

grounded theory An inductive approach to the study of social life that attempts to generate a theory from the constant comparing of unfolding observations. This is very different from hypothesis testing, in which theory is used to generate hypotheses to be tested through observations.

Anselm L. Strauss, a pioneer qualitative researcher, was a principal founder of the Grounded Theory Method.

- *Think comparatively:* The authors suggest that it is essential to compare numerous incidents as a way of avoiding the biases that can arise from interpretations of initial observations.

- *Obtain multiple viewpoints:* In part this refers to the different points of view of participants in the events under study, but Strauss and Corbin suggest that different observational techniques may also provide a variety of viewpoints.

- *Periodically step back:* As data accumulate, you'll begin to frame interpretations about what is going on, and it's important to keep checking your data against those interpretations. As Strauss and Corbin (1998: 45) say, "The data themselves do not lie."

- *Maintain an attitude of skepticism:* As you begin to interpret the data, you should regard all those interpretations as provisional, using new observations to test those interpretations, not just confirm them.

- *Follow the research procedures:* Grounded theory allows for flexibility in data collection as theories evolve, but Strauss and Corbin (1998: 46) stress that three techniques are essential: "making comparisons, asking questions, and sampling."

Grounded theory emphasizes research procedures. In particular, systematic coding is important for achieving validity and reliability in the data analysis. Because of this somewhat positivistic view of data, grounded theorists are quite open to the use of qualitative studies in conjunction with quantitative ones. Here are two examples of the implementation of this approach.

Studying Academic Change

Clifton Conrad's (1978) study of academic change in universities is an early example of the grounded theory approach. Conrad hoped to uncover the major sources of changes in academic curricula and at the same time understand the process of change. Using the grounded theory idea of theoretical sampling—whereby groups or institutions are selected on the basis of their theoretical relevance—Conrad chose four universities for the purpose of his study. In two, the main vehicle of change was the formal curriculum committee; in the other two, the vehicle of change was an ad hoc group.

Conrad explained, step by step, the advantage of using the grounded theory approach in building his theory of academic change. He described the process of systematically coding data in order to create categories that must "emerge" from the data and then assessing the fitness of these categories with each other. Going continuously from data to theory and theory to data allowed him to reassess the validity of his initial conclusions about academic change.

For instance, it first seemed that academic change was mainly caused by an administrator who was pushing for it. By reexamining the data and looking for more-plausible explanations, Conrad found the pressure of interest groups a more convincing source of change. The emergence of these interest groups actually allowed the administrator to become an agent of change.

Assessing how data from each of the two types of universities fit with the other helped refine the theory building. Conrad concluded that changes in university curricula are based on the following process: Conflict and interest groups emerge because of internal and external social structural forces;

they push for administrative intervention and recommendation to make changes in the current academic program; these changes are then made by the most powerful decision-making body.

Shopping Romania

Much has been written about large-scale changes caused by the shift from socialism to capitalism in the former USSR and its Eastern European allies. Patrick Jobes and his colleagues (1997) wanted to learn about the transition on a smaller scale among average Romanians. They focused on the task of shopping.

Noting that shopping is normally thought of as a routine, relatively rational activity, the researchers suggested that it could become a social problem in a radically changing economy. They used the Grounded Theory Method to examine Romanian shopping as a social problem, looking for the ways in which ordinary people solved the problem.

Their first task was to learn something about how Romanians perceived and understood the task of shopping. The researchers—participants in a social problems class—began by interviewing 40 shoppers and asking whether they had experienced problems in connection with shopping and what actions they had taken to cope with those problems.

Once the initial interviews were completed, the researchers reviewed their data, looking for categories of responses—the shoppers' most common problems and solutions. One of the most common problems was a lack of money. This led to the researchers' first working hypothesis: The "socio-economic position of shoppers would be associated with how they perceived problems and sought solutions" (1997: 133). This and other hypotheses helped the researchers focus their attention on more-specific variables in subsequent interviewing.

As they continued, they also sought to interview other types of shoppers. When they interviewed students, for example, they discovered that different types of shoppers were concerned with different kinds of goods, which in turn affected the problems faced and the solutions tried.

As the researchers developed additional hypotheses in response to the continued interviewing,

they also began to develop a more or less standardized set of questions to ask shoppers. Initially, all the questions were open-ended, but they eventually developed closed-ended items as well.

This study illustrates the key, inductive principles of grounded theory: Data are collected in the absence of hypotheses. The initial data are used to determine the key variables as perceived by those being studied, and hypotheses about relationships among the variables are similarly derived from the data collected. Continuing data collection yields refined understanding and, in turn, sharpens the focus of data collection itself.

Case Studies and the Extended Case Method

Social researchers often speak of case studies. A **case study** focuses attention on a single instance of some social phenomenon, such as a village, a family, or a juvenile gang. As Charles Ragin and Howard Becker (1992) point out, there is little consensus on what may constitute a "case," and the term is used broadly. The case being studied, for example, might be a period of time rather than a particular group of people. The limitation of attention to a particular instance of something is the essential characteristic of the case study.

The chief purpose of case studies may be descriptive, as when an anthropologist describes the culture of a preliterate tribe. Or the in-depth study of a particular case can yield explanatory insights, as when the community researchers Robert and Helen Lynd (1929, 1937) and W. Lloyd Warner (1949) sought to understand the structure and process of social stratification in small-town USA.

Case study researchers may seek only an idiographic understanding of the particular case under examination, or—as we've seen with grounded theory—case studies can form the basis for the development of more-general, nomothetic theories.

---

case study The in-depth examination of a single instance of some social phenomenon, such as a village, a family, or a juvenile gang.

Michael Burawoy and his colleagues (1991) have suggested a somewhat different relationship between case studies and theory. For them, the **extended case method** has the purpose of discovering flaws in, and then modifying, existing social theories. This approach differs importantly from some of the others already discussed.

Whereas the grounded theorists seek to enter the field with no preconceptions about what they'll find, Burawoy suggests just the opposite: to try "to lay out as coherently as possible what we expect to find in our site *before* entry" (Burawoy et al. 1991: 9). Burawoy sees the extended case method as a way to rebuild or improve theory instead of approving or rejecting it. Thus, he looks for all the ways in which observations conflict with existing theories and what he calls "theoretical gaps and silences" (1991: 10). This orientation to field research implies that knowing the literature beforehand is actually a must for Burawoy and his colleagues, whereas grounded theorists would worry that knowing what others have concluded might bias their observations and theories.

To illustrate the extended case method, I'll present two examples of studies by Burawoy's students.

Teacher–Student Negotiations

Leslie Hurst (1991) set out to study the patterns of interaction between teachers and students of a junior high school. She went into the field armed with existing contradictory theories about the "official" functions of the school. Some theories suggested that the purpose of schools is to promote social mobility, whereas others suggested that schools mainly reproduce the status quo in the form of a stratified division of labor. The official roles assigned to teachers and students could be interpreted in terms of either view.

Hurst was struck, however, by the contrast between these theories and the types of interactions she observed in the classroom. In her own experiences as a student, teachers had total rights over the minds, bodies, and souls of their pupils. She observed something quite different at a school in a lower-middle-class neighborhood in Berkeley, California—Emerald Junior High School, where she volunteered as a tutor. She had access to the classroom of Mr. Henry (an eighth-grade English teacher) as well as other teachers' classrooms, the lunchroom, and English Department meetings. She wrote field notes based on the negotiations between students and teachers. She explained the nature of the student–teacher negotiations she witnessed by focusing on the separation of functions among the school, the teacher, and the family.

In Hurst's observation, the school fulfilled the function of controlling its students' "bodies"—for example, by regulating their general movements and activities within the school. The students' "minds" were to be shaped by the teacher, whereas students' families were held responsible for their "souls"; that is, families were expected to socialize students regarding personal values, attitudes, sense of property, and sense of decorum. When students don't come to school with these values in hand, the teacher, according to Hurst, "must first negotiate with the students some compromise on how the students will conduct themselves and on what will be considered classroom decorum" (1991: 185).

Hurst explained that the constant bargaining between teachers and students is an expression of the separation between "the body," which is the school's concern, and "the soul" as family domain. The teachers, who had limited sanctioning power to control their students' minds in the classroom, were using forms of negotiations with students so that they could "control . . . the student's body and sense of property" (1991: 185), or as Hurst defines it, "babysit" the student's body and soul.

Hurst says she differs from the traditional sociological perspectives as follows:

> I do not approach schools with a futuristic eye. I do not see the school in terms of training, socializing, or slotting people into future hierarchies. To approach schools in this manner is to miss the negotiated, chaotic aspects of the classroom and educational experience.

extended case method A technique developed by Michael Burawoy in which case study observations are used to discover flaws in and to improve existing social theories.

A futurist perspective tends to impose an order and purpose on the school experience, missing its day-to-day reality.

(1991: 186)

In summary, what emerges from Hurst's study is an attempt to improve the traditional sociological understanding of education by adding the idea that classroom, school, and family have separate functions, which in turn can explain the emergence of "negotiated order" in the classroom.

The Fight against AIDS

Katherine Fox (1991) set out to study an agency whose goal was to fight the AIDS epidemic by bringing condoms and bleach for cleaning needles to intravenous drug users. It's a good example of finding the limitations of well-used models of theoretical explanation in the realm of understanding deviance—specifically, the "treatment model" that predicted that drug users would come to the clinic and ask for treatment. Fox's interactions with outreach workers—most of whom were part of the community of drug addicts or former prostitutes—contradicted that model.

To begin, it was necessary to understand the drug users' subculture and use that knowledge to devise more-realistic policies and programs. The target users had to be convinced, for example, that the program workers could be trusted, that they were really interested only in providing bleach and condoms. The target users needed to be sure they were not going to be arrested.

Fox's field research didn't stop with an examination of the drug users. She also studied the agency workers, discovering that the outreach program meant different things to the research directors and the outreach workers. Some of the volunteers who were actually providing the bleach and condoms were frustrated about the minor changes they felt they could make. Many thought the program was just a bandage on the AIDS and drug-abuse problems. Some resented having to take field notes. Directors, on the other hand, needed reports and field notes so that they could validate their research in the eyes of the federal and state agencies that financed the project. Fox's study

showed how the AIDS research project developed the bureaucratic inertia typical of established organizations: Its goal became that of sustaining itself.

Both of these studies illustrate how the extended case method can operate. The researcher enters the field with full knowledge of existing theories but aims to uncover contradictions that require the modification of those theories.

One criticism of the case study method is the limited generalizability of what may be observed in a single instance of some phenomenon. This risk is reduced, however, when more than one case is studied in depth: the *comparative* case study method. You can find examples of this in the discussion of comparative and historical methods in Chapter 11 of this book.

Institutional Ethnography

Institutional ethnography is an approach originally developed by Dorothy Smith (1978) to better understand women's everyday experiences by discovering the power relations that shape those experiences. Today this methodology has been extended to the ideologies that shape the experiences of any oppressed subjects.

Smith and other sociologists believe that if researchers ask women or other members of subordinated groups about "how things work," they can discover the institutional practices that shape their realities (M. L. Campbell 1998; D. Smith 1978). The goal of such inquiry is to uncover forms of oppression that more-traditional types of research often overlook.

Dorothy Smith's methodology is similar to ethnomethodology in the sense that the subjects themselves are not the focus of the inquiry. The institutional ethnographer starts with the personal

---

institutional ethnography A research technique in which the personal experiences of individuals are used to reveal power relationships and other characteristics of the institutions within which they operate.

Dorothy Smith, a pioneering social researcher and founder of institutional ethnography.

experiences of individuals but proceeds to uncover the institutional power relations that structure and govern those experiences. In this process, the researcher can reveal aspects of society that would have been missed by an inquiry that began with the official purposes of institutions.

This approach links the "microlevel" of everyday personal experiences with the "macrolevel" of institutions. As M. L. Campbell puts it,

> Institutional ethnography, like other forms of ethnography, relies on interviewing, observations, and documents as data. Institutional ethnography departs from other ethnographic approaches by treating those data not as the topic or object of interest, but as "entry" into the social relations of the setting. The idea is to tap into people's expertise.
>
> *(1998: 57)*

Here are two examples of this approach.

Mothering, Schooling, and Child Development

Our first example of institutional ethnography is a study by Alison Griffith (1995), who collected data with Dorothy Smith on the relationship among mothering, schooling, and children's development. Griffith started by interviewing mothers from three cities of southern Ontario about their everyday work of creating a relationship between their families and the school. This was the starting point for other interviews with parents, teachers, school administrators, social workers, school psychologists, and central office administrators.

In her findings, Griffith explained how the discourse about mothering had shifted its focus over time from a mother–child interaction to "child-centered" recommendations. She saw a distinct similarity in the discourse used by schools, the media (magazines and television programs), the state, and child-development professionals.

Teachers and child-development professionals saw the role of mothers in terms of a necessary collaboration between mothers and schools for the child to succeed not only in school but also in life. Because of unequal resources, all mothers do not participate in this discourse of "good" child development the same way. Griffith found that working-class mothers were perceived as weaker than middle-class mothers in the "stimulation" effort of schooling. Griffith argues that this child-development discourse, embedded in the school institution, perpetuates the reproduction of class by making middle-class ideals for family–school relations the norm for everyone.

Compulsory Heterosexuality

The second illustration of institutional ethnography is taken from Didi Khayatt's (1995) study of the institutionalization of compulsory heterosexuality in schools and its effects on lesbian students. In 1990, Khayatt began her research by interviewing 12 Toronto lesbians, 15 to 24 years of age. Beginning with the young women's viewpoint, she expanded her inquiry to other students, teachers, guidance counselors, and administrators.

Khayatt found that the school's administrative practices generated a *compulsory heterosexuality,* which produced a sense of marginality and vulnerability among lesbian students. For example, the school didn't punish harassment and name-calling directed at gay students. The issue of homosexuality was excluded from the curriculum lest it appear to students as an alternative to heterosexuality.

In both of the studies I've described, the inquiry began with the women's standpoint—mothers and lesbian students. However, instead of emphasizing the subjects' viewpoints, both analyses focused on the power relations that shaped these women's experiences and reality.

Participatory Action Research

Our final field research paradigm takes us further along in our earlier discussion of the status and power relationships linking researchers to the subjects of their research. Within the **participatory action research (PAR)** paradigm, the researcher's function is to serve as a resource to those being studied—typically, disadvantaged groups—as an opportunity for them to act effectively in their own interest. The disadvantaged subjects define their problems, define the remedies desired, and take the lead in designing the research that will help them realize their aims.

This approach began in Third World research development, but it spread quickly to Europe and North America (Gaventa 1991). It comes from a vivid critique of classical social science research. According to the PAR paradigm, traditional research is perceived as an "elitist model" (Whyte, Greenwood, and Lazes 1991) that reduces the "subjects" of research to "objects" of research. According to many advocates of the PAR perspective, the distinction between the researcher and the researched should disappear. They argue that the subjects who will be affected by research should also be responsible for its design.

Implicit in this approach is the belief that research functions not only as a means of knowledge production but also as a "tool for the education and development of consciousness as well as mobilization for action" (Gaventa 1991: 121–22). Advocates of participatory action research equate access to information with power and argue that this power has been kept in the hands of the dominant class, sex, ethnicity, or nation. Once people see themselves as researchers, they automatically regain power over knowledge.

Participatory action research poses a special challenge to researchers. On the one hand, participants in the social situation ideally become empowered to frame research relevant to their needs, as they define those needs. At the same time, the researcher brings special skills and insights that nonresearchers lack. So who should be in charge? Andrew Sense (2006: 1) suggests that this decision may have to be made in the moment: "Do I take the 'passenger' position on the bus or do I take the 'driver' seat and be a little more provocative to energise the session[?] My view at this moment is to judge it on the day."

Examples of this approach include research on community power structures, corporate research, and "right-to-know" movements (Whyte, Greenwood, and Lazes 1991). Here are two examples of corporate research that used a PAR approach.

The Xerox Corporation

A participatory action research project took place at the Xerox corporation at the instigation of leaders of both management and the union. Management's goal was to lower costs so that the company could thrive in an increasingly competitive market. The union suggested a somewhat broader scope: improving the quality of working life while lowering manufacturing costs and increasing productivity.

Company managers began by focusing attention on shop-level problems; they were less concerned with labor contracts or problematic managerial policies. At the time, management had a plan to start an "outsourcing" program that would lay off 180 workers, and the union had

---

participatory action research (PAR) An approach to social research in which the people being studied are given control over the purpose and procedures of the research; intended as a counter to the implicit view that researchers are superior to those they study.

begun mobilizing to oppose the plan. Peter Lazes, a consultant hired by Xerox, spent the first month convincing management and the union to create a "cost study team" (CST) that included workers in the wire harness department.

Eight full-time workers were assigned to the CST for six months. Their task was to study the possibilities of making changes that would save the company $3.2 million and keep the 180 jobs. The team had access to all financial information and was authorized to call on anyone within the company. This strategy allowed workers to make suggestions outside the realm usually available to them. According to Whyte and his colleagues, "reshaping the box enabled the CST to call upon management to explain and justify all staff services" (1991: 27). Because of the changes suggested by the CST and implemented by management, the company saved the targeted $3.2 million.

Management was so pleased by this result that it expanded the wire harness CST project to three other departments that were threatened by competition. Once again, management was happy about the money saved by the teams of workers.

The Xerox case study is an interesting example of participatory action research because it shows how the production of knowledge does not always have to be an elitist enterprise. The "experts" do not necessarily have to be the professionals. According to Whyte and his colleagues, "At Xerox, participatory action research created and guided a powerful process of organizational learning—a process whereby leaders of labor and management learned from each other and from the consultant/facilitator, while he learned from them" (1991: 30).

PAR and Welfare Policy

Participatory action research often involves poor people, as they are typically less able than other groups to influence the policies and actions that affect their own lives. Bernita Quoss, Margaret Cooney, and Terri Longhurst (2000) report a research project involving welfare policy in Wyoming. University students, many of them welfare recipients, undertook research and lobbying efforts aimed at getting Wyoming to accept postsecondary education as "work" under the state's new welfare regulations.

This project began against the backdrop of the 1996 Personal Responsibility and Work Opportunity Act (PRWORA), which

> eliminated education waivers that had been available under the previous welfare law, the 1988 Family Support Act (FSA). These waivers had permitted eligible participants in the cash assistance AFDC program to attend college as an alternative to work training requirements. Empirical studies of welfare participants who received these waivers have provided evidence that education, in general, is the most effective way to stay out of poverty and achieve self-sufficiency.
>
> (Quoss, Cooney, and Longhurst 2000: 47)

The students began by establishing an organization called Empower and by making presentations on campus to enlist broad student and faculty support. They compiled existing research relevant to the issue and established relationships with members of the state legislature. By the time the 1997 legislative session opened, the students were actively engaged in the process of modifying state welfare laws to offset the shift in federal policy.

The students prepared and distributed fact sheets and other research reports that would be relevant to the legislators' deliberations. They attended committee meetings and lobbied legislators on a one-to-one basis. When erroneous or misleading data were introduced into the discussions, the student-researchers were on hand to point out the errors and offer corrections.

Ultimately, they succeeded. Welfare recipients in Wyoming were allowed to pursue postsecondary education as an effective route out of poverty.

Some researchers speak of **emancipatory research,** which Ardha Danieli and Carol Woodhams (2005: 284) define as "first and foremost a

---

emancipatory research Research conducted for the purpose of benefiting disadvantaged groups.

process of producing knowledge which will be of benefit to oppressed people; a political outcome." Both qualitative and quantitative methods can be used to pursue this goal, but it goes well beyond simply learning what's so, even as seen from the subjects' point of view. The authors focus on the study of disability, and they note similarities in the development of emancipatory research and early feminist research.

Blocking a Demolition

In another example of researchers being directly involved in what they study, John Lofland (2003) detailed the demolition of a historic building in Davis, California, and community attempts to block the demolition. One thing that makes the book especially unusual is its reliance on photographs and facsimile news articles and government documents as raw data for the analysis (and for the reader): what Lofland refers to as "documentary sociology."

As Lofland explains, he was involved in the issue first as an active participant, joining with other community members in the attempt to block demolition of the Hotel Aggie (also known as the "Terminal Building" and "Terminal Hotel"). Built in 1924 in a town of around a thousand inhabitants, the hotel fell victim to population growth and urban development. Lofland says his role as researcher began on September 18, 2000, as the demolition of the building began.

> Before that, I was only and simply an involved citizen. Along with many other people, I was attempting to preserve the Terminal Building in some manner. This also explains why there are so few photographs in this book taken by me before that date, but many after that date. I had then begun seriously to document what was going on with a camera and field notes.
>
> Therefore, questions of "informed consent" (now so often raised regarding research) were not pertinent before September 18. After that day, it was my practice to indicate to everyone I encountered that I was "writing a book" about the building.
>
> (Lofland 2003: 20)

Recall the discussion of informed consent in Chapter 3, a method of protecting research subjects. In this case, as Lofland notes elsewhere, explicit consent was not necessarily needed here, because the behavior being studied was public. Still, his instincts as a social researcher were to ensure that he treat subjects appropriately.

One of Lofland's purposes was to study this failed attempt to secure "historic preservation" status for a building, thus providing useful information to activists in the future. This indicates that there can be many different forms of participatory action research.

At the same time, this is a valuable study for a study of research methods, because Lofland, as the author of research methods textbooks, is particularly sensitive to the methodological aspects of the study.

> The deptth and intensity of my involvement is a two-edged sword. On the one edge, my involvement provided me with a view closer than that of some other people. I was one type of "insider." This means I could gather data of certain sorts that were not available to the less involved.
>
> On the other edge, my partisanship clearly poses the threat of bias. I have always been aware of this, and I have tried my best to correct for it. But, in the end, I cannot be the final judge. Each reader will have to form her or his own assessment. I can hope, however, that the "digital documentary" evidence I mention above helps the study tell itself, so to speak. It makes the reader less dependent on me than is the case with some other methods of representing what happened.
>
> (Lofland 2003: 20)

As you can see, the seemingly simple process of observing social action as it occurs has subtle though important variations. As we saw in Chapter 2, all our thoughts occur within and are shaped by paradigms, whether we're conscious of it or not. Qualitative field researchers have been unusually deliberate in framing a variety of paradigms to enrich the observation of social life.

The impact of researcher paradigms on the conduct of research is nowhere more explicitly recognized than in the case of *kaupapa Maori research,* a form of participatory action research developed within the indigenous Maori community of New Zealand. As Shayne Walker, Anaru Eketone, and Anita Gibbs (2006) report, an adherence to Maori culture shapes not only the purposes of such research but also its processes and practices. In a study of foster care, for example, the purpose of the study was established by those most directly concerned. The method of collecting data conformed to Maori practices, including public gatherings. The implications derived from the analysis of data were tailored to Maori ways of doing things.

Conducting Qualitative Field Research

So far in this chapter we've examined the kinds of topics appropriate to qualitative field research, special considerations in doing this kind of research, and a sampling of paradigms that direct different types of research efforts. Along the way we've seen some examples that illustrate field research in action. To round out the picture, we turn now to specific ideas and techniques for conducting field research, beginning with how researchers prepare for work in the field.

Preparing for the Field

Suppose for the moment that you've decided to undertake field research on a campus political organization. Let's assume further that you're not a member of that group, that you do not know a great deal about it, and that you'll identify yourself to the participants as a researcher. This section will use this example and others to discuss some of the ways you might prepare yourself before undertaking direct observations.

As is true of all research methods, you would be well advised to begin with a search of the relevant literature, filling in your knowledge of the subject

and learning what others have said about it. (Library research is discussed at length in Appendix A.)

In the next phase of your research, you might wish to discuss the student political group with others who have already studied it or with anyone else likely to be familiar with it. In particular, you might find it useful to discuss the group with one or more informants (discussed in Chapter 7). Perhaps you have a friend who is a member, or you can meet someone who is. This aspect of your preparation is likely to be more effective if your relationship with the informant extends beyond your research role. In dealing with members of the group as informants, you should take care that your initial discussions do not compromise or limit later aspects of your research. Keep in mind that the impression you make on the informant, the role you establish for yourself, may carry over into your later effort. For example, creating the initial impression that you may be an undercover FBI agent is unlikely to facilitate later observations of the group.

You should also be wary about the information you get from informants. Although they may have more direct, personal knowledge of the subject under study than you do, what they "know" is probably a mixture of fact and point of view. Members of the political group in our example (as well as members of opposing political groups) would be unlikely to provide completely unbiased information. Before making your first contact with the student group, then, you should already be quite familiar with it, and you should understand its general philosophical context.

There are many ways to establish your initial contact with the people you plan to study. How you do it will depend, in part, on the role you intend to play. Especially if you decide to take on the role of complete participant, you must find a way to develop an identity with the people to be studied. If you wish to study dishwashers in a restaurant, the most direct method would be to get a job as a dishwasher. In the case of the student political group, you might simply join the group.

Many of the social processes appropriate to field research are open enough to make your contact with the people to be studied rather simple and straightforward. If you wish to observe a

How to Do It: Establishing Rapport

In qualitative field research, it's almost always vital that you be able to establish rapport with those you're observing, especially if your observations include in-depth interviews and interactions. **Rapport** might be defined as an open and trusting relationship. But how do you do that?

Let's assume that you'll be identifying yourself as a researcher. You'll need to explain your research purpose in a nonthreatening way. Say that you are there to learn about them and understand them, not to judge them or cause them any problems. This will work best if you actually have a genuine interest in understanding the people you're observing and can communicate that interest to them. This gives them a sense of self-worth, which will increase their willingness to open up to you. Pretending to be interested is not the same as really being interested. In fact, if you aren't interested in learning what things look like from the point of view of those you're observing, you might consider another activity and not waste their time and your own.

It follows that you'll function best as an attentive listener rather than as a talker. You should not remain mute, of course, but you should talk primarily (1) to elicit more information from the other person or (2) to answer questions they may have about you and your research. While you don't have to agree with any points of view expressed by your subjects, you should never argue with them nor try to change their minds. Keep reminding yourself that your genuine purpose is to understand their world and how it makes sense to them—whether it works for you or not. A little humility may help with this. You'll be able to hear and understand people better if you don't start out feeling superior to them.

Be relaxed and appropriate to the setting. Some people are more formal or informal than others, and you'll do well to take on their general style or at least find a way to relax with whatever style is most comfortable for them. If you can get them to relax and enjoy the interaction, you'll have achieved the rapport you need. And you'll probably enjoy the interaction yourself.

---

mass demonstration, just be there. If you wish to observe patterns in jaywalking, hang around busy streets.

Whenever you wish to make more-formal contact with the people, identifying yourself as a researcher, you must establish a rapport with them. You might contact a participant with whom you feel comfortable and gain that person's assistance. In studying a formal group, you might approach the groups' leaders, or you might find that one of your informants can introduce you. (See "How to Do It: Establishing Rapport" for more on this.)

Although you'll probably have many options in making your initial contact with the group, realize that your choice can influence your subsequent observations. Suppose, for example, that you're studying a university and begin with high-level administrators. This choice is likely to have a couple of important consequences. First, your initial impressions of the university will be shaped to some extent by the administrators' views, which will differ quite a bit from those of students or faculty. This initial impression may influence the way

you observe and interpret events subsequently— especially if you're unaware of the influence.

Second, if the administrators approve of your research project and encourage students and faculty to cooperate with you, the latter groups will probably look on you as somehow aligned with the administration, which can affect what they say to you. For example, faculty members might be reluctant to tell you about plans to organize through the teamsters' union.

In making direct, formal contact with the people you want to study, you'll be required to give them some explanation of the purpose of your study. Here again, you face an ethical dilemma. Telling them the complete purpose of your research might eliminate their cooperation altogether or significantly affect their behavior. On the other hand, giving only what you believe would be an

---

rapport An open and trusting relationship; especially important in qualitative research between researchers and the people they're observing.

acceptable explanation may involve outright deception. Your decisions in this and other matters will probably be largely determined by the purpose of your study, the nature of what you're studying, the observations you wish to use, and similar factors, but you must also take ethical considerations into account.

Previous field research offers no fixed rule—methodological or ethical—to follow in this regard. Your appearance as a researcher, regardless of your stated purpose, may result in a warm welcome from people who are flattered that a scientist finds them important enough to study. Or, it may result in your being totally ostracized or worse. It probably wouldn't be a good idea, for example, to burst into a meeting of an organized crime syndicate and announce that you're writing a term paper on organized crime. See "Pencils and Photos in the Hands of Research Subjects" for a look at some inventive data-collection techniques.

Qualitative Interviewing

In part, field research is a matter of going where the action is and simply watching and listening. As the baseball legend Yogi Berra said, "You can see a lot just by observing"—provided that you're paying attention. At the same time, as I've already indicated, field research can involve more-active inquiry. Sometimes it's appropriate to ask people questions and record their answers. Your on-the-spot observations of a full-blown riot will lack something if you don't know why people are rioting. Ask somebody.

When Cecilia Menjívar (2000) wanted to learn about the experiences of Salvadoran immigrants in San Francisco, she felt in-depth interviews would be a useful technique, along with personal observations. Before she was done, she had discovered a much more complex system of social processes and

structures than we would have imagined. Although it was important for new immigrants to have a support structure of family members already in the United States, Menjívar found that her interviewees were often reluctant to call on relatives for help, for several reasons. On the one hand, they might jeopardize those family members who were here illegally and living in poverty. At the same time, asking for help would put them in debt to those helping them out. Menjívar also discovered that Salvadoran gender norms put women immigrants in an especially difficult situation, because they were largely prohibited from seeking the help of men they weren't related to, lest they seem to obligate themselves sexually. These are the kinds of discoveries that can emerge from open-ended, in-depth interviewing.

We've already discussed interviewing in Chapter 9, and much of what was said there applies to qualitative field interviewing. The interviewing you'll do in connection with field observation, however, is different enough to demand a separate treatment. In surveys, questionnaires are rigidly structured; however, less-structured interviews are more appropriate to field research. Herbert and Riene Rubin (1995: 43) describe the distinction as follows: "Qualitative interviewing design is flexible, iterative, and continuous, rather than prepared in advance and locked in stone." They elaborate in this way:

> Design in qualitative interviewing is iterative. That means that each time you repeat the basic process of gathering information, analyzing it, winnowing it, and testing it, you come closer to a clear and convincing model of the phenomenon you are studying. . . .
>
> The continuous nature of qualitative interviewing means that the questioning is redesigned throughout the project.
>
> *(1995: 46–47)*

Unlike a survey, a **qualitative interview** is an interaction between an interviewer and a respondent in which the interviewer has a general plan of inquiry, including the topics to be covered, but not a set of questions that must be asked with particular words and in a particular order. At the

qualitative interview Contrasted with survey interviewing, the qualitative interview is based on a set of topics to be discussed in depth rather than based on the use of standardized questions.

Pencils and Photos in the Hands of Research Subjects

How would you go about studying the life conditions of Peruvian Indians living in the Amazon rain forest? With minimal telecommunications infrastructure and a slow ferry-based postal service in the vast region, a mail or telephone survey wouldn't be the best approach. It might occur to you to conduct in-depth interviews in which you would work from an outline of topics to be covered. Arvind Singhal and Elizabeth Rattine-Flaherty (2006) opted for a very different approach, which put the subjects of study more in control of the research and allowed for important but unexpected discoveries. They derived their inspiration from the work of the renowned Brazilian educator, Paulo Freire, who once set out to measure exploitation among street children. Instead of interviewing them, he gave them cameras and asked them to bring back photographs of exploitation. As Singhal and Rattine-Flaherty report:

> One child took a photo of a nail on a wall. It made no sense to adults, but other children were in strong agreement. The ensuing discussions showed that many young boys of that neighborhood worked in the shoe-shine business. Their clients were mainly in the city, not in the barrio where they lived. As their shoe-shine boxes were too heavy for them to carry, these boys rented a nail on a wall (usually in a shop), where they could hang their boxes for the night. To them, that nail on the wall represented "exploitation." The "nail on the wall" photograph spurred widespread discussions in the Peruvian barrio about other forms of institutionalized exploitation, including ways to overcome them.
>
> (2006: 314)

Singhal and Rattine-Flaherty's research involved gauging the quality of life in the Peruvian Amazon and assessing the impact of programs launched by a Peruvian NGO, Minga Peru. To view society through the eyes of children, the researchers set up drawing sessions with colored pencils. In the spirit of reciprocity, one of the authors sketched pictures of snowmen and jack-o'-lanterns that were a part of her growing up in the Midwest. In addition to depicting life in their villages and their close relationship with the natural environment, the children's sketches often featured examples of social change being brought about by the NGO's developmental programs.

> These include sketches of chicken coops, fish farms, and agroforestry projects. These enterprises, all launched by Minga Peru, began in the Peruvian Amazon only in the past few years. For

children to sketch these 'new' initiatives in their pictures on their own, without prompts, is noteworthy.

> (2006: 322)

The photographs taken by the adult women were equally revealing. Several drew attention to the patriarchal social structure. As the authors report:

> Several photographs depicted the subservient position of the Amazonian women relative to men, a situation that Minga Peru seeks to address. For instance, Adela's picture shows a middle-aged Amazonian woman and her husband sitting on their porch and having a conversation. The woman, sporting a forlorn expression, sits with her legs crossed while her husband stares directly into the camera, squatting with his arms and feet spread in an open position. Especially noticeable is the physical distance of about 10 feet that separates the woman and the man. When Adela was asked why she took the picture and why were the man and woman sitting so far apart, she noted: "The woman is sitting at one side of the house and he is on the other and this was not anything unusual." Upon probing, we learned that Amazonian men determine how close the couple sits. If they are sitting closer, and if the man has his arm around his partner, it is his decision to do so. This authority also applies to initiation of sex: The man determines if and when sex will happen.
>
> (2006: 323–24)

This research not only illustrates some unusual data-collection techniques, it also represents the spirit of participatory action research, discussed earlier in this chapter. With a very different setting and purpose, Pat O'Connor (2006) asked Irish adolescents to write essays about themselves and about Ireland, including drawings, poems, and songs, looking for evidence of the impact of globalization in Ireland. Both studies demonstrate that qualitative field research can involve a lot more than just observing and interviewing.

Sources: Arvind Singhal and Elizabeth Rattine-Flaherty, "Pencils and Photos as Tools of Communicative Research and Praxis: Analyzing Minga Peru's Quest for Social Justice in the Amazon," *International Communication Gazette* 68, no. 4 (2006): 313–30; Pat O'Connor, "Globalization, Individualization and Gender in Adolescents' Texts," *International Journal of Social Research Methodology* 9, no. 4 (October 2006): 261–77.

same time, the qualitative interviewer, like the survey interviewer, must be fully familiar with the questions to be asked. This allows the interview to proceed smoothly and naturally.

A qualitative interview is essentially a conversation in which the interviewer establishes a general direction for the conversation and pursues specific topics raised by the respondent. Ideally, the respondent does most of the talking. If you're talking more than 5 percent of the time, that's probably too much.

Steinar Kvale (1996: 3–5) offers two metaphors for interviewing: the interviewer as a "miner" or as a "traveler." The first model assumes that the subject possesses specific information and that the interviewer's job is to dig it out. By contrast, in the second model, the interviewer

> wanders through the landscape and enters into conversations with the people encountered. The traveler explores the many domains of the country, as unknown territory or with maps, roaming freely around the territory. . . . The interviewer wanders along with the local inhabitants, asks questions that lead the subjects to tell their own stories of their lived world.

Asking questions and noting answers is a natural human process, and it seems simple enough to add it to your bag of tricks as a field researcher. Be a little cautious, however. Wording questions is a tricky business. All too often, the way we ask questions subtly biases the answers we get. Sometimes we put our respondent under pressure to look good. Sometimes we put the question in a particular context that omits altogether the most relevant answers.

Suppose, for example, that you want to find out why a group of students is rioting and pillaging on campus. You might be tempted to focus your questioning on how students feel about the dean's recent ruling that requires students always to carry *The Practice of Social Research* with them on campus. (Makes sense to me.) Although you may collect a great deal of information about students' attitudes toward the infamous ruling, they may be rioting for some other reason. Perhaps most are simply joining

in for the excitement. Properly done, field research interviewing enables you to find out.

Although you may set out to conduct interviews with a reasonably clear idea of what you want to ask, one of the special strengths of field research is its flexibility. In particular, the answers evoked by your initial questions should shape your subsequent ones. It doesn't work merely to ask preestablished questions and record the answers. Instead, you need to ask a question, listen carefully to the answer, interpret its meaning for your general inquiry, and then frame another question either to dig into the earlier answer or to redirect the person's attention to an area more relevant to your inquiry. In short, you need to be able to listen, think, and talk almost at the same time.

The discussion of probes in Chapter 9 provides a useful guide to getting answers in more depth without biasing later answers. More generally, field interviewers need the skills involved in being a good listener. Be more interested than interesting. Learn to say things like "How is that?" "In what ways?" "How do you mean that?" "What would be an example of that?" Learn to look and listen expectantly, and let the person you're interviewing fill in the silence.

At the same time, you can't afford to be a totally passive receiver. You'll go into your interviews with some general (or specific) questions you want answered and some topics you want addressed. At times you'll need the skill of subtly directing the flow of conversation.

There's something we can learn in this regard from the martial arts. The aikido master never resists an opponent's blow but instead accepts it, joins with it, and then subtly redirects it in a more appropriate direction. Field interviewing requires an analogous skill. Instead of trying to halt your respondent's line of discussion, learn to take what he or she has just said and branch that comment back in the direction appropriate to your purposes. Most people love to talk to anyone who's really interested. Stopping their line of conversation tells them that you are not interested; asking them to elaborate in a particular direction tells them that you are.

Consider this hypothetical example in which you're interested in why college students chose their majors.

YOU: What are you majoring in?

RESP: Engineering.

YOU: I see. How did you come to choose engineering?

RESP: I have an uncle who was voted the best engineer in Arizona in 1981.

YOU: Gee, that's great.

RESP: Yeah. He was the engineer in charge of developing the new civic center in Tucson. It was written up in most of the engineering journals.

YOU: I see. Did you talk to him about your becoming an engineer?

RESP: Yeah. He said that he got into engineering by accident. He needed a job when he graduated from high school, so he went to work as a laborer on a construction job. He spent eight years working his way up from the bottom, until he decided to go to college and come back nearer the top.

YOU: So is your main interest civil engineering, like your uncle, or are you more interested in some other branch of engineering?

RESP: Actually, I'm leaning more toward electrical engineering—computers, in particular. I started messing around with microcomputers when I was in high school, and my long-term plan is . . .

Notice how the interview begins by wandering off into a story about the respondent's uncle. The first attempt to focus things back on the student's own choice of major ("Did you talk to your uncle . . . ?") fails. The second attempt ("So is your main interest . . . ?") succeeds. Now the student is providing the kind of information you're looking for. It's important for field researchers to develop the ability to "control" conversations in this fashion. At the same time, of course, you need to be on the alert for "distractions" that point to unexpectedly important aspects of your research interest.

Herbert and Riene Rubin offer several ways to control a "guided conversation," including the following:

> If you can limit the number of main topics, it is easier to maintain a conversational flow from one topic to another. Transitions should be smooth and logical. "We have been talking about mothers, now let's talk about fathers," sounds abrupt. A smoother transition might be, "You mentioned your mother did not care how you performed in school—was your father more involved?" The more abrupt the transition, the more it sounds like the interviewer has an agenda that he or she wants to get through, rather than wanting to hear what the interviewee has to say.
>
> *(1995: 123)*

Because field research interviewing is so much like normal conversation, researchers must keep reminding themselves that they are not having a normal conversation. In normal conversations, each of us wants to come across as an interesting, worthwhile person. If you watch yourself the next time you chat with someone you don't know too well, you'll probably find that much of your attention is spent on thinking up interesting things to say—contributions to the conversation that will make a good impression. Often, we don't really hear each other, because we're too busy thinking of what we'll say next. As an interviewer, the desire to appear interesting is counterproductive. The interviewer needs to make the other person seem interesting, by being interested—and by listening more than talking. (Do this in ordinary conversations, and people will actually regard you as a great conversationalist.)

John Lofland and his colleagues (2006: 69–70) suggest that researchers should adopt the role of the "socially acceptable incompetent" when interviewing. That is, offer yourself as someone who does not understand the situation you find yourself in and must be helped to grasp even the most basic and obvious aspects of that situation: "A naturalistic investigator, almost by definition, is one who does not understand. She or he is 'ignorant' and needs to be 'taught.' This role of watcher and asker of questions is the quintessential *student* role" (Lofland et al. 2006: 69).

Interviewing needs to be an integral part of the entire field research process. Later, I'll stress the need to review your observational notes every night—making sense out of what you've observed, getting a clearer feel for the situation you're studying, and finding out what you should pay more attention to in further observations. In the same fashion, you'll need to review your notes on interviews, recording especially effective questions and detecting all those questions you should have asked but didn't. Start asking such questions the next time you interview. If you've recorded the interviews, replay them as a useful preparation for future interviews.

Steinar Kvale (1996: 88) details seven stages in the complete interviewing process:

1. *Thematizing:* Clarifying the purpose of the interviews and the concepts to be explored

2. *Designing:* Laying out the process through which you'll accomplish your purpose, including a consideration of the ethical dimension

3. *Interviewing:* Doing the actual interviews

4. *Transcribing:* Creating a written text of the interviews

5. *Analyzing:* Determining the meaning of gathered materials in relation to the purpose of the study

6. *Verifying:* Checking the reliability and validity of the materials

7. *Reporting:* Telling others what you've learned

As with all other aspects of field research, interviewing improves with practice. Fortunately, it's something you can practice any time you want. Practice on your friends.

---

focus group A group of subjects interviewed together, prompting a discussion. The technique is frequently used by market researchers, who ask a group of consumers to evaluate a product or discuss a type of commodity, for example.

Focus Groups

Although our discussions of field research so far have focused on studying people in the process of living their lives, researchers sometimes bring people into the laboratory for qualitative interviewing and observation. The focus group method, which is also called group interviewing, is essentially a qualitative method. It is based on structured, semistructured, or unstructured interviews. It allows the researcher/interviewer to question several individuals systematically and simultaneously. This data-collection technique is frequently used in political and market research but is used for other purposes as well. In *Silent Racism,* for example, Barbara Trepagnier (2006) used focus groups to examine the persistence of racism among "well-meaning white people."

In a hypothetical market-research example, imagine that you're thinking about introducing a new product. Let's suppose that you've invented a new computer that not only does word processing, spreadsheets, data analysis, and the like but also contains a fax machine, AM/FM/TV tuner, CD and DVD player/recorder, microwave oven, denture cleaner, and coffeemaker. To highlight its computing and coffee-making features, you're thinking of calling it "The Compulator." You figure the new computer will sell for about $28,000, and you want to know whether people are likely to buy it. Your prospects might be well served by focus groups.

In a **focus group,** typically 5 to 15 people are brought together in a private, comfortable environment to engage in a guided discussion of some topic—in this case, the acceptability and salability of the Compulator. The subjects are selected on the basis of relevance to the topic under study. Given the likely cost of the Compulator, your focus group participants would probably be limited to upper-income groups, for example. Other, similar considerations might figure into the selection.

Participants in focus groups are not likely to be chosen through rigorous probability-sampling methods. This means that the participants do not statistically represent any meaningful population.

However, the purpose of the study is to explore rather than to describe or explain in any definitive sense. Nevertheless, typically more than one focus group is convened in a given study because of the serious danger that a single group of 7 to 12 people will be too atypical to offer any generalizable insights.

William Gamson (1992) used focus groups to examine how U.S. citizens frame their views of political issues. Having picked four issues—affirmative action, nuclear power, troubled industries, and the Arab-Israeli conflict—Gamson undertook a content analysis of press coverage to get an idea of the media context within which we think and talk about politics. Then the focus groups were convened for a firsthand observation of the process of people discussing issues with their friends.

Richard Krueger points to five advantages of focus groups:

1. The technique is a socially oriented research method capturing real-life data in a social environment.
2. It has flexibility.
3. It has high face validity.
4. It has speedy results.
5. It is low in cost.

(1988: 47)

In addition to these advantages, group dynamics frequently bring out aspects of the topic that would not have been anticipated by the researcher and would not have emerged from interviews with individuals. In a side conversation, for example, a couple of the participants might start joking about the results of leaving out one letter from a product's name. This realization might save the manufacturer great embarrassment later on.

Krueger also notes some disadvantages of the focus group method, however:

1. Focus groups afford the researcher less control than individual interviews.
2. Data are difficult to analyze.
3. Moderators require special skills.
4. Difference between groups can be troublesome.
5. Groups are difficult to assemble.
6. The discussion must be conducted in a conducive environment.

(1988: 48)

As we've seen, the group interview presents several advantages, but it also has its challenges. In a focus group interview, much more than in any other type of interview, the interviewer has to develop the skills of a moderator. Controlling the dynamic within the group is a major challenge. Letting one interviewee dominate the focus group interview reduces the likelihood that the other subjects will express themselves. This can generate the problem of group conformity or groupthink, which is the tendency for people in a group to conform with the opinions and decisions of the most outspoken members of the group. This danger is compounded by the possibility that only one or two people sometimes dominate the conversation. Interviewers need to be aware of this phenomenon and try to get everyone to participate fully on all the issues brought in the interview. Adding to the challenge, of course, is that the interviewer must resist overdirecting the interview and the interviewees, thus bringing her or his own views into play.

Although focus group research differs from other forms of qualitative field research, it further illustrates the possibilities for doing social research face-to-face with those we wish to understand. In addition, David Morgan (1993) suggests that focus groups are an excellent device for generating questionnaire items for a subsequent survey.

Because they center on a particular topic and take relatively little time, focus groups are typically regarded as an "in-depth" research technique. However, Carolina Överlien, Karin Aronsson, and Margareta Hydén (2005) have used the technique successfully for extended discussions of sexuality, among Swedish teenagers in a youth detention home.

Like other social research techniques, focus groups are adapting to new communication modalities. George Silverman (2005), for example, offers a discussion of telephone and online focus groups.

Recording Observations

The greatest advantage of the field research method is the presence of an observing, thinking researcher on the scene of the action. Even tape recorders and cameras cannot capture all the relevant aspects of social processes, although both of those devices can be quite useful to the field researcher. Consequently, in both direct observation and interviewing, it's vital to make full and accurate notes of what goes on. If possible, take notes on your observations as you observe. When that's not feasible, write down your notes as soon as possible afterward.

In your notes, include both your empirical observations and your interpretations of them. In other words, record what you "know" has happened and what you "think" has happened. Be sure to identify these different kinds of notes for what they are. For example, you might note that Person X spoke out in opposition to a proposal made by a group leader (an observation), that you think this represents an attempt by Person X to take over leadership of the group (an interpretation), and that you think you heard the leader comment to that effect in response to the opposition (a tentative observation).

Of course, you can't observe everything; nor can you record everything you do observe. Just as your observations will represent a sample of all possible observations, your notes will represent a sample of your observations. The idea, of course, is to record the most pertinent ones.

"Interview Transcript Annotated with Researcher Memos" provides an extract from an in-depth interview with a woman film director, given by Sandrine Zerbib. Notice that the illustration contains a portion of an in-depth interview, along with some of Zerbib's memos, written during her review of the interview later on. Chapter 13 will present extensive, computerized analyses from this study on women film directors.

Some of your most important observations can be anticipated before you begin the study; others will become apparent as your observations progress. Sometimes you can make note taking easier by preparing standardized recording forms in advance. In a study of jaywalking, for example, you might anticipate the characteristics of pedestrians that are most likely to be useful for analysis—age, gender, social class, ethnicity, and so forth—and prepare a form in which observations of these variables can be recorded easily. Alternatively, you might develop a symbolic shorthand in advance to speed up recording. For studying audience participation at a mass meeting, you might want to construct a numbered grid representing the different sections of the meeting room; then you could record the location of participants easily, quickly, and accurately.

None of this advance preparation should limit your recording of unanticipated events and aspects of the situation. Quite the contrary, speedy handling of anticipated observations can give you more freedom to observe the unanticipated.

You're already familiar with the process of taking notes, just as you already have at least informal experience with field research in general. Like good field research, however, good note taking requires careful and deliberate attention and involves specific skills. Some guidelines follow. (You can learn even more from Lofland et al. 2006: 110–17.)

First, don't trust your memory any more than you have to—it's untrustworthy. To illustrate this point, try this experiment. Recall the last three or four movies you saw that you really liked. Now, name five of the actors or actresses. Who had the longest hair? Who was the most likely to start conversations? Who was the most likely to make suggestions that others followed? Now, if you didn't have any trouble answering any of those questions, how sure are you of your answers? Would you be willing to bet a hundred dollars that a panel of impartial judges would observe what you recall?

Even if you pride yourself on having a photographic memory, it's a good idea to take notes either during the observation or as soon afterward as possible. If you take notes during observation, do it unobtrusively, because people are likely to behave differently if they see you taking down everything they say or do.

Second, it's usually a good idea to take notes in stages. In the first stage, you may need to take sketchy notes (words and phrases) in order to

Interview Transcript Annotated with Researcher Memos

Thursday August 26, 12:00–1:00

R: What is challenging for women directors on a daily experience, on a daily life?

J: Surviving.

R: OK. Could you develop a little bit on that? [I need to work on my interview schedule so that my interviewee answers with more elaboration without having to probe.]

J: Yeah, I mean it's all about trying to get, you know, in, trying to get the job, and try, you know, to do a great job so that you are invited back to the next thing. And particularly since they are so many, you know, difficulties in women directing. It makes it twice as hard to gain into this position where you do an incredible job, because … you can't just do an average job, you have to [347] do this job that just knocks your socks off all the time, and sometimes you don't get the opportunity to do that, because either you don't have a good producer or you have so many pressures that you can't see straight or your script is lousy, and you have to make a silk purse out of sow's hair. You know, you have a lot of extra strikes against you than the average guy who has similar problems, because you are a woman and they look at it, and women are more visible than men … in unique positions.

[It seems that Joy is talking about the particularities of the film industry. There are not that many opportunities and in order to keep working, she needs to build a certain reputation. It is only by continuing to direct that she can maintain or improve her reputation. She thinks that it is even harder for women but does not explain it.]

R: Hum … what about on the set did you experience, did it feel … did people make it clear that you were a woman, and you felt treated differently? [I am trying to get her to speak about more specific and more personal experiences without leading her answer]

J: Yeah, oh yeah, I mean … a lot of women have commiserated about, you know when you have to walk on the set for the first time,

they're all used to working like a well-oiled machine and they say, "Oh, here is the woman, something different" and sometimes they can be horrible, they can resist your directing and they can, they can sabotage you, by taking a long time to light, or to move sets, or to do something … and during that time you're wasting time, and that goes on a report, and the report goes to the front [368] office, and, you know, and so on and so on and so on and so forth. And people upstairs don't know what the circumstances are, and they are not about to fire a cinematographer that is on their show for ever and ever … nor do they want to know that this guy is a real bastard, and making your life a horror. They don't want to know that, so therefore, they go off, because she's a woman let's not hire any more women, since he has problems with women. You know, so, there is that aspect.

[I need to review the literature on institutional discrimination. It seems that the challenges that Joy is facing are not a matter of a particular individual. She is in a double bind situation where whether she complains or not, she will not be treated equal to men. Time seems to be one quantifiable measurement of how well she does her job and, as observed in other professions, the fact that she is a woman is perceived as a handicap. Review literature on women in high management position. I need to keep asking about the dynamics between my interviewees and the crewmembers on the set. The cinematographer has the highest status on the set under the director. Explore other interviews about reasons for conflict between them.]

[Methods (note to myself for the next interviews): try to avoid phone interviews unless specific request from the interviewee. It is difficult to assess how the interviewee feels with the questions. Need body language because I become more nervous about the interview process.]

Note: A number in brackets represents a word that was inaudible from the interview. It is the number that appeared on the transcribing machine, with each interview starting at count 0. The numbers help the researcher locate a passage quickly when he or she reviews the interview.

keep abreast of what's happening. Then go off by yourself and rewrite your notes in more detail. If you do this soon after the events you've observed, the sketchy notes should allow you to recall most of the details. The longer you delay, the less likely you'll be able to recall things accurately and fully.

In his study of bike messengers in New York City, mentioned earlier, Jeffrey Kidder reports on this process (2005: 349):

I obtained the vast majority of data for this article through informal interviews. I unobtrusively took notes throughout the day and at social events. Upon returning home, these data were compiled into my field notes. During the workday and during races, parties, and other social gatherings, casual conversations provided the truest glimpses into messenger beliefs, ideologies, and opinions. To this end, I avoided formal interviews and instead allowed my questions to be answered by normal talk within the social world.

I know this method sounds logical, but it takes self-discipline to put it into practice. Careful observation and note taking can be tiring, especially if it involves excitement or tension and if it extends over a long period. If you've just spent eight hours observing and making notes on how people have been coping with a disastrous flood, your first desire afterward will likely be to get some sleep, dry clothes, or a drink. You may need to take some inspiration from newspaper reporters who undergo the same sorts of hardships then write their stories to meet their deadlines.

Third, you'll inevitably wonder how much you should record. Is it really worth the effort to write out all the details you can recall right after the observational session? The general guideline is yes. Generally, in field research you can't be really sure of what's important and what's unimportant until you've had a chance to review and analyze a great volume of information, so you should record even things that don't seem important at the outset. They may turn out to be significant after all. Also, the act of recording the details of something "unimportant" may jog your memory on something that is important.

Realize that most of your field notes will not be reflected in your final report on the project. Put more harshly, most of your notes will be "wasted." But take heart: Even the richest gold ore yields only about 30 grams of gold per metric ton, meaning that 99.997 percent of the ore is wasted. Yet, that 30 grams of gold can be hammered out to cover an area 18 feet square—the equivalent of about 685 book pages. So take a ton of notes, and plan to select and use only the gold.

Like other aspects of field research (and all research for that matter), proficiency comes with practice. The nice thing about field research is

you can begin practicing now and can continue practicing in almost any situation. You don't have to be engaged in an organized research project to practice observation and recording. You might start by volunteering to take the minutes at committee meetings, for example. Or just pick a sunny day on campus, find a shady spot, and try observing and recording some specific characteristics of the people who pass by. You can do the same thing at a shopping mall or on a busy street corner. Remember that observing and recording are professional skills and, like all worthwhile skills, they improve with practice.

Strengths and Weaknesses of Qualitative Field Research

Like all research methods, qualitative field research has distinctive strengths and weaknesses. As I've already indicated, field research is especially effective for studying subtle nuances in attitudes and behaviors and for examining social processes over time. As such, the chief strength of this method lies in the depth of understanding it permits. Whereas other research methods may be challenged as "superficial," this charge is seldom lodged against field research.

Flexibility is another advantage of field research. As discussed earlier, you can modify your field research design at any time. Moreover, you're always prepared to engage in field research, whenever the occasion should arise, whereas you could not as easily initiate a survey or an experiment.

Field research can be relatively inexpensive as well. Other social science research methods may require costly equipment or an expensive research staff, but field research typically can be undertaken by one researcher with a notebook and a pencil. This is not to say that field research is never expensive. The nature of the research project may require a large number of trained observers, for example. Expensive recording equipment may be needed. Or you may wish to undertake participant observation of interactions in pricey Paris nightclubs.

Field research has several weaknesses as well. First, being qualitative rather than quantitative, it's not an appropriate means for arriving at statistical descriptions of a large population. Observing casual political discussions in Laundromats, for example, would not yield trustworthy estimates of the future voting behavior of the total electorate. Nevertheless, the study could provide important insights into how political attitudes are formed.

To assess field research further, let's focus on the issues of validity and reliability. Recall that validity and reliability are both qualities of measurements. Validity concerns whether measurements actually measure what they're supposed to rather than something else. Reliability, on the other hand, is a matter of dependability: If you made the same measurement again and again, would you get the same result? Let's see how field research stacks up in these respects.

Validity

Field research seems to provide measures with greater validity than do survey and experimental measurements, which are often criticized as superficial and not really valid. Let's review a couple of field research examples to see why this is so.

"Being there" is a powerful technique for gaining insights into the nature of human affairs in all their rich complexity. Listen, for example, to what this nurse reports about the impediments to patients' coping with cancer:

> Common fears that may impede the coping process for the person with cancer can include the following:
> —Fear of death—for the patient, and the implications his or her death will have for significant others.
> —Fear of incapacitation—because cancer can be a chronic disease with acute episodes that may result in periodic stressful periods, the variability of the person's ability to cope and constantly adjust may require a dependency upon others for activities of daily living and may consequently become a burden.
> —Fear of alienation—from significant others and health care givers, thereby creating helplessness and hopelessness.

> —Fear of contagion—that cancer is transmissible and/or inherited.
> —Fear of losing one's dignity—losing control of all bodily functions and being totally vulnerable.
>
> *(Garant 1980: 2167)*

Observations and conceptualizations such as these are valuable in their own right. In addition, they can provide the basis for further research— both qualitative and quantitative.

Now listen to what Joseph Howell has to say about "toughness" as a fundamental ingredient of life on Clay Street, a white, working-class neighborhood in Washington, D.C.:

> Most of the people on Clay Street saw themselves as fighters in both the figurative and literal sense. They considered themselves strong, independent people who would not let themselves be pushed around. For Bobbi, being a fighter meant battling the welfare department and cussing out social workers and doctors upon occasion. It meant spiking Barry's beer with sleeping pills and bashing him over the head with a broom. For Barry it meant telling off his boss and refusing to hang the door, an act that led to his being fired. It meant going through the ritual of a duel with Al. It meant pushing Bubba around and at times getting rough with Bobbi.
> June and Sam had less to fight about, though if pressed they both hinted that they, too, would fight. Being a fighter led Ted into near conflict with Peg's brothers, Les into conflict with Lonnie, Arlene into conflict with Phyllis at the bowling alley, etc.
>
> *(1973: 292)*

Even without having heard the episodes Howell refers to in this passage, you have the distinct impression that Clay Street is a tough place to live in. That "toughness" comes through far more powerfully through these field observations than it would in a set of statistics on the median number of fistfights occurring during a specified period.

These examples point to the superior validity of field research, as compared with surveys and experiments. The kinds of comprehensive measurements available to the field researcher tap a depth of meaning in concepts such as common fears of cancer patients and "toughness" (or concepts such as liberal and conservative) that are generally unavailable to surveys and experiments. Instead of specifying concepts, field researchers commonly give detailed illustrations.

Reliability

Field research, however, can pose problems of reliability. Suppose you were to characterize your best friend's political orientations according to everything you know about him or her. Your assessment of your friend's politics would appear to have considerable validity; certainly it's unlikely to be superficial. We couldn't be sure, however, that another observer would characterize your friend's politics the same way you did, even with the same amount of observation.

Although they are in-depth, field research measurements are also often very personal. How I judge your friend's political orientation depends greatly on my own, just as your judgment depends on your political orientation. Conceivably, then, you could describe your friend as middle-of-the-road, although I might feel that I've been observing a fire-breathing radical.

As I suggested earlier, researchers who use qualitative techniques are conscious of this issue and take pains to address it. Individual researchers often sort out their own biases and points of view, and the communal nature of science means that their colleagues will help them in that regard. Nevertheless, it's prudent to be wary of purely descriptive measurements in field research—your own, or someone else's. If a researcher reports that the members of a club are fairly conservative, such a judgment is unavoidably linked to the researcher's own politics. You can be more trusting of comparative evaluations: identifying who is more conservative than who, for example. Even if you and I had different political orientations, we would probably agree pretty much in ranking the relative conservatism of the members of a group.

As we've seen, field research is a potentially powerful tool for social scientists, one that provides a useful balance to the strengths and weaknesses of experiments and surveys. The remaining chapters of Part 3 present additional modes of observation available to social researchers.

Ethics and Qualitative Field Research

As I've noted repeatedly, all forms of social research raise ethical issues. By bringing researchers into direct and often intimate contact with their subjects, field research raises ethical concerns in a particularly dramatic way. Here are some of the issues mentioned by John and Lyn Lofland (1995: 63):

- Is it ethical to talk to people when they do not know you will be recording their words?
- Is it ethical to get information for your own purposes from people you hate?
- Is it ethical to see a severe need for help and not respond to it directly?
- Is it ethical to be in a setting or situation but not commit yourself wholeheartedly to it?
- Is it ethical to develop a calculated stance toward other humans, that is, to be strategic in your relations?
- Is it ethical to take sides or to avoid taking sides in a factionalized situation?
- Is it ethical to "pay" people with trade-offs for access to their lives and minds?
- Is it ethical to "use" people as allies or informants in order to gain entree to other people or to elusive understandings?

Planning and conducting field research in a responsible way requires attending to these and other ethical concerns.

▪ MAIN POINTS

Introduction

- Field research involves the direct observation of social phenomena in their natural settings. Typically, field research is qualitative rather than quantitative.

- In field research, observation, data processing, and analysis are interwoven, cyclical processes.

Topics Appropriate for Field Research

- Field research is especially appropriate for topics and processes that are not easily quantifiable, that are best studied in natural settings, or that change over time. Among these topics are practices, episodes, encounters, roles, relationships, groups, organizations, settlements, social worlds, and lifestyles or subcultures.

Special Considerations in Qualitative Field Research

- Among the special considerations involved in field research are the various possible roles of the observer and the researcher's relationships with subjects. As a field researcher, you must decide whether to observe as an outsider or as a participant, whether or not to identify yourself as a researcher, and how to negotiate your relationships with subjects.

Some Qualitative Field Research Paradigms

- Field research can be guided by any one of several paradigms, such as naturalism, ethnomethodology, grounded theory, case studies and the extended case method, institutional ethnography, and participatory action research.

Conducting Qualitative Field Research

- Preparing for the field involves doing background research, determining how to make contact with subjects, and resolving issues of what your relationship to your subjects will be.
- Field researchers often conduct in-depth interviews that are much less structured than those conducted in survey research. Qualitative interviewing is more of a guided conversation than a search for specific information. Effective interviewing involves skills of active listening and the ability to direct conversations unobtrusively.
- To create a focus group, researchers bring subjects together and observe their interactions as they explore a specific topic.
- Whenever possible, field observations should be recorded as they are made; otherwise, they should be recorded as soon afterward as possible.
- Among the advantages of field research are the depth of understanding it can provide, its flexibility, and (usually) its inexpensiveness.

Strengths and Weaknesses of Qualitative Field Research

- Compared with surveys and experiments, field research measurements generally have more validity but less reliability. Also, field research is generally not appropriate for arriving at statistical descriptions of large populations.

Ethics and Qualitative Field Research

- Conducting field research responsibly involves confronting several ethical issues that arise from the researcher's direct contact with subjects.

KEY TERMS

The following terms are defined in context in the chapter and at the bottom of the page where the term is introduced, as well as in the comprehensive glossary at the back of the book.

case study	naturalism
emancipatory research	participatory
ethnography	action research (PAR)
ethnomethodology	qualitative
extended case method	interview
focus group	rapport
grounded theory	reactivity
institutional ethnography	

PROPOSING SOCIAL RESEARCH: QUALITATIVE FIELD RESEARCH

This chapter has laid out a large number of different possibilites for conducting field research. If you're doing field research, you should indicate the kind of study you plan to do. Will you be the sole observer in the study? If not, how will you select and train the other observers?

Will you be a participant in the events you are observing and, if so, will you identify yourself as a researcher to those you are observing? You might say something about how these choices may affect what you observe, as well as discussing the ethical issues involved.

In earlier exercises, you dealt with the variables you'll examine and the ways you'll select informants and/or people to observe, as well as the times and places for your observations. As this chapter has demonstrated, there are other logistical issues to be worked out. It may be appropriate to describe your note-taking plans if that's likely to be difficult (for example, if you're a participant not identified as a researcher).

If you'll be conducting in-depth interviews, you should include an outline of the topics to be covered in those interviews. Are there topics or questions that must be addressed in each interview and others that will be pursued only if appropriate?

Compared with experiments and surveys, field research allows more flexibility as to the timing of the research. Depending on how things go, you may find yourself concluding earlier or later than you had planned. Nevertheless, you should say something in the proposal regarding the schedule you are planning.

REVIEW QUESTIONS AND EXERCISES

1. Think of some group or activity you participate in or are very familiar with. In two or three paragraphs, describe how an outsider might effectively go about studying that group or activity. What should he or she read, what contacts should be made, and so on?

2. Choose any two of the paradigms discussed in this chapter. Then describe how your hypothetical study from Exercise 1 might be conducted if you followed each. Compare and contrast the way these paradigms might work in the context of your study.

3. To explore the strengths and weaknesses of experiments, surveys, and field research, choose a general research area (such as prejudice, political orientation, education) and write brief descriptions of studies in that area that could be conducted using each of these three methods. In each case, explain why the chosen method is the most appropriate for the study you describe.

4. Return to the example you devised in response to Exercise 1 and list five ethical issues that you can imagine having to confront if you were to undertake your study.

5. Using InfoTrac College Edition, find a research report using the Grounded Theory Method. Summarize the study design and main findings.

SPSS EXERCISES

See the booklet that accompanies your text for exercises using SPSS (Statistical Package for the Social Sciences). There are exercises offered for each chapter, and you'll also find a detailed primer on using SPSS.

Online Study Resources

If your book came with an access code card, visit www.cengage.com/login to register. To purchase access, please visit www.ichapters.com.

1. Before you do your final review of the chapter, take the *CengageNOW* pretest to help identify the areas on which you should concentrate. You'll find information on this online tool, as well as instructions on how to access all of its great resources, in the front of the book.

2. As you review, take advantage of the *CengageNOW* personalized study plan, based on your quiz results. Use this study plan with its interactive exercises and other resources to master the material.

3. When you're finished with your review, take the posttest to confirm that you're ready to move on to the next chapter.

WEBSITE FOR *THE PRACTICE OF SOCIAL RESEARCH* 12TH EDITION

Go to your book's website at www.cengage.com/sociology/babbie for tools to aid you in studying for your exams. You'll find *Tutorial Quizzes* with feedback, *Internet Exercises, Flash Cards, Glossaries,* and *Essay Quizzes,* as well as *InfoTrac College Edition* search terms, suggestions for additional reading, *Web Links,* and primers for using data-analysis software such as SPSS.

CHAPTER ELEVEN

Unobtrusive Research

CHAPTER OVERVIEW

This chapter presents overviews of three unobtrusive research methods: content analysis, the analysis of existing statistics, and comparative and historical research. Each of these methods allows researchers to study social life from afar, without influencing it in the process.

CengageNOW for Sociology

Use this online tool to help you make the grade on your next exam. After reading this chapter, go to "Online Study Resources" at the end of the chapter for instructions on how to benefit from *CengageNOW*.

Introduction

With the exception of the complete observer in field research, each of the modes of observation discussed so far requires the researcher to intrude to some degree on whatever he or she is studying. This is most obvious in the case of experiments, followed closely by survey research. Even the field researcher, as we've seen, can change things in the process of studying them.

At least one previous example in this book, however, was totally exempt from that danger. Durkheim's analysis of suicide did nothing to affect suicides one way or the other (see Chapter 5). His study is an example of **unobtrusive research,** or methods of studying social behavior without affecting it. As you'll see, unobtrusive measures can be qualitative or quantitative.

This chapter examines three types of unobtrusive research methods: content analysis, analysis of existing statistics, and comparative and historical research. In content analysis, researchers examine a class of social artifacts that usually are written documents such as newspaper editorials. Next, the Durkheim study is an example of the analysis of existing statistics. As you'll see, there are great masses of data all around you, awaiting your use in the understanding of social life. Finally, comparative and historical research, a form of research with a venerable history in the social sciences, is currently enjoying a resurgence of popularity. Like field research, comparative and historical research is a usually qualitative method, one in which the main resources for observation and analysis are historical records. The method's name includes the word *comparative* because social scientists—in contrast to historians who may simply describe a particular set of events— seek to discover common patterns that recur in different times and places.

To set the stage for our examination of these three research methods, I want to draw your attention to an excellent book that should sharpen your senses about the potential for unobtrusive measures in general. It is, among other things, the book from which I take the term *unobtrusive measures.*

In 1966, Eugene Webb and three colleagues published an ingenious little book on social research (revised in 2000) that has become a classic. It focuses on the idea of unobtrusive or nonreactive research. Webb and his colleagues have played freely with the task of learning about human behavior by observing what people inadvertently leave behind them. Do you want to know what exhibits are the most popular at a museum? You could conduct a poll, but people might tell you what they thought you wanted to hear or what might make them look intellectual and serious. You could stand by different exhibits and count the viewers that came by, but people might come over to see what you were doing. Webb and his colleagues suggest that you check the wear and tear on the floor in front of various exhibits. Those that have the most-worn tiles are probably the most popular. Want to know which exhibits are popular with little kids? Look for mucus on the glass cases. To get a sense of the most popular radio stations, you could arrange with an auto mechanic to check the radio's tuner-button settings for cars brought in for repair.

The possibilities are limitless. Like a detective investigating a crime, the social researcher looks for clues. If you stop to notice, you'll find that clues of social behavior are all around you. In a sense, everything you see represents the answer to some important social science question—all you have to do is think of the question.

Although problems of validity and reliability crop up in unobtrusive measures, a little ingenuity can either handle them or put them in perspective. I encourage you to look at Webb's book. It's enjoyable reading, and it can be a source of stimulation and insight for social inquiry through data that already exist. For now, let's turn our attention to

---

unobtrusive research Methods of studying social behavior without affecting it. Such methods can be qualitative or quantitative.

three unobtrusive methods often employed by social scientists, beginning with content analysis.

Content Analysis

As I mentioned in the chapter introduction, **content analysis** is the study of recorded human communications. Among the forms suitable for study are books, magazines, web pages, poems, newspapers, songs, paintings, speeches, letters, e-mail messages, bulletin board postings on the Internet, laws, and constitutions, as well as any components or collections thereof. Shulamit Reinharz points out that feminist researchers have used content analysis to study "children's books, fairy tales, billboards, feminist nonfiction and fiction books, children's art work, fashion, fat-letter postcards, Girl Scout Handbooks, works of fine art, newspaper rhetoric, clinical records, research publications, introductory sociology textbooks, and citations, to mention only a few" (1992: 146–47). In another example, when William Mirola set out to discover the role of religion in the movements to establish the eight-hour working day in America, his data were taken "from Chicago's labor, religious, and secular presses, from pamphlets, and from speeches given by eight-hour proponents from three representative factions within the movement" (2003: 273).

Topics Appropriate for Content Analysis

Content analysis is particularly well suited to the study of communications and to answering the classic question of communications research: "Who says what, to whom, why, how, and with what effect?" Are popular French novels more concerned with love than novels in the United States are? Was the popular British music of the 1960s more politically cynical than the popular German music during that period? Do political candidates who primarily address "bread and butter" issues get elected more often than those who address issues of high principle? Each of these questions addresses a social science research topic: The first might address

national character, the second political orientations, and the third political process. Although you might study such topics by observing individual people, content analysis provides another approach.

An early example of content analysis is the work of Ida B. Wells. In 1891, Wells, whose parents had been slaves, wanted to test the widely held assumption that African American men were being lynched in the South primarily for raping white women. As a research method, she examined newspaper articles on the 728 lynchings reported during the previous ten years. In only a third of the cases were the lynching victims even accused of rape, much less proved guilty. Primarily, they were charged with being insolent, not staying in "their place" (cited in Reinharz 1992: 146).

More recently, the best-selling book *Megatrends 2000* (Naisbitt and Aburdene 1990) used content analysis to determine the major trends in modern U.S. life. The authors regularly monitored thousands of local newspapers a month in order to discover local and regional trends for publication in a series of quarterly reports. Their book examines some of the trends they observed in the nation at large. In a follow-up book (Aburdene 2005), this kind of analysis pointed to such trends as "The Power of Spirituality" and "The Rise of Conscious Capitalism."

Some topics are more appropriately addressed by content analysis than by any other method of inquiry. Suppose that you're interested in violence on television. Maybe you suspect that the manufacturers of men's products are more likely to sponsor violent TV shows than other kinds of sponsors are. Content analysis would be the best way of finding out.

Briefly, here's what you'd do. First, you'd develop operational definitions of the two key variables in your inquiry: *men's products* and *violence*. The section on coding, later in this chapter, will discuss some of the ways you could do that. Ultimately, you'd need a plan that would allow you

---

content analysis The study of recorded human communications, such as books, websites, paintings, and laws.

to watch TV, classify sponsors, and rate the degree of violence on particular shows.

Next, you'd have to decide what to watch. Probably you'd decide (1) what stations to watch, (2) for what period, and (3) at what hours. Then, you'd stock up on beer and potato chips and start watching, classifying, and recording. Once you'd completed your observations, you'd be able to analyze the data you collected and determine whether men's product manufacturers sponsored more blood and gore than other sponsors did.

Gabriel Rossman (2002) had a somewhat different question regarding the mass media. Public concern over the concentration of media in fewer and fewer corporate hands has grown, so Rossman decided to ask the following question: If a newspaper is owned by the same conglomerate that owns a movie production company, can you trust that newspaper's movie reviews of its parent company's productions?

You can't, according to Rossman's findings. Because many newspapers rate movies somewhat quantitatively (for example, three stars out of four), he could perform a simple quantitative analysis. For each movie review, he asked two main questions: (1) Was the movie produced by the same company that owned the newspaper? and (2) What rating did the film receive? He found that, indeed, movies produced by the parent company received higher ratings than other movies did. Further, the ratings given to movies by newspapers with the same parent company were higher than the ratings those movies received from other newspapers. This discrepancy, moreover, was strongest in the case of big-budget movies in which the parent company had invested heavily.

As a mode of observation, content analysis requires a thoughtful handling of the "what" that is being communicated. The analysis of data collected in this mode, as in others, addresses the "why" and "with what effect."

Sampling in Content Analysis

In the study of communications, as in the study of people, you often can't observe directly all you would like to explore. In your study of TV violence

and sponsorship, for example, I'd advise against attempting to watch everything that's broadcast. It wouldn't be possible, and your brain would probably short-circuit before you came close to discovering that for yourself. Usually, it's appropriate to sample. Let's begin by revisiting the idea of units of analysis. We'll then review some of the sampling techniques that might be applied to such units in content analysis.

Units of Analysis

As I discussed in Chapter 4, determining appropriate units of analysis—the individual units that we make descriptive and explanatory statements about—can be a complicated task. For example, if we wish to compute average family income, the individual family is the unit of analysis. But we'll have to ask individual members of families how much money they make. Thus, individuals will be the units of observation, even though the individual family remains the unit of analysis. Similarly, we may wish to compare crime rates of different cities in terms of their size, geographic region, racial composition, and other differences. Even though the characteristics of these cities are partly a function of the behaviors and characteristics of their individual residents, the cities would ultimately be the units of analysis.

The complexity of this issue is often more apparent in content analysis than in other research methods, especially when the units of observation differ from the units of analysis. A few examples should clarify this distinction.

Let's suppose we want to find out whether criminal law or civil law makes the most distinctions between men and women. In this instance, individual laws would be both the units of observation and the units of analysis. We might select a sample of a state's criminal and civil laws and then categorize each law by whether or not it makes a distinction between men and women. In this fashion, we could determine whether criminal or civil law distinguishes by gender the most.

Somewhat differently, we might wish to determine whether states that enact laws distinguishing between different racial groups are also more likely than other states to enact laws distinguishing between men and women. Although the examination of this question would also involve the coding of individual acts of legislation, the unit of analysis in this case is the individual state, not the law.

Or, changing topics radically, let's suppose we're interested in representationalism in painting. If we wish to compare the relative popularity of representational and nonrepresentational paintings, the individual paintings will be our units of analysis. If, on the other hand, we wish to discover whether representationalism in painting is more characteristic of wealthy or impoverished painters, of educated or uneducated painters, of capitalist or socialist painters, the individual painters will be our units of analysis.

It's essential that this issue be clear, because sample selection depends largely on what the unit of analysis is. If individual writers are the units of analysis, the sample design should select all or a sample of the writers appropriate to the research question. If books are the units of analysis, we should select a sample of books, regardless of their authors. Bruce Berg (1989: 112–13) points out that even if you plan to analyze some body of textual materials, the units of analysis might be words, themes, characters, paragraphs, items (such as a book or letter), concepts, semantics, or combinations of these. Figure 11-1 illustrates some of those possibilities.

I'm not suggesting that sampling should be based solely on the units of analysis. Indeed, we may often subsample—select samples of subcategories—for each individual unit of analysis. Thus, if writers are the units of analysis, we might (1) select a sample of writers from the total population of writers, (2) select a sample of books written by each writer selected, and (3) select portions of each selected book for observation and coding.

Finally, let's look at a trickier example: the study of TV violence and sponsors. What's the unit of analysis for the research question "Are the manufacturers of men's products more likely to sponsor violent shows than other sponsors are?" Is it the TV show? The sponsor? The instance of violence?

In the simplest study design, it would be none of these. Though you might structure your inquiry in various ways, the most straightforward design

FIGURE 11-1

A Few Possible Units of Analysis for Content Analysis

would be based on the commercial as the unit of analysis. You would use two kinds of observational units: the commercial and the program (the show that gets squeezed in between commercials). You would want to observe both units. You would classify commercials by whether they advertised men's products and the programs by their violence. The program classifications would be transferred to the commercials occurring near them. Figure 11-2 provides an example of the kind of record you might keep.

Notice that in the research design illustrated in Figure 11-2, all the commercials occurring in the same program break are grouped and get the same scores. Also, the number of violent instances recorded as following one commercial break is the same as the number preceding the next break. This

Commercial Break	Sponsor	Men's Product?			Number of Instances of Violence	
		Yes	No	?	Before the Commercial Break	After the Commercial Break
1st	Grunt Aftershave	✓			6	4
	Brute Jock Straps	✓			6	4
2nd	Bald-No-More Lotion	✓			4	3
	Grunt Aftershave	✓			3	0
3rd	Snowflake Toothpaste		✓		3	0
	Godliness Cleanser		✓		3	0
4th	Big Thumb Hammers			✓	0	1
5th	Snowflake Toothpaste		✓		1	0
	Big Thumb Hammers			✓	1	0
6th	Buttercup Bras		✓		0	0

FIGURE 11-2

Example of Recording Sheet for TV Violence

simple design allows us to classify each commercial by its sponsorship and the degree of violence associated with it. Thus, for example, the first Grunt Aftershave commercial is coded as being a men's product and as having 10 instances of violence associated with it. The Buttercup Bra commercial is coded as not being a men's product and as having no violent instances associated with it.

In the illustration, we have four men's product commercials with an average of 7.5 violent instances each. The four commercials classified as definitely not men's products have an average of 1.75, and the two that might or might not be considered men's products have an average of 1 violent instance each. If this pattern of differences persisted across a much larger number of observations, we'd probably conclude that manufacturers of men's products are more likely to sponsor TV violence than other sponsors are.

The point of this illustration is to demonstrate how units of analysis figure into the data collection and analysis. You need to be clear about your unit of analysis before planning your sampling strategy, but in this case you can't simply sample commercials. Unless you have access to the stations' broadcasting logs, you won't know when the commercials are going to occur. Moreover, you need to observe the programming as well as the commercials. As a result, you must set up a sampling design that will include everything you need in order to observe enough.

In designing the sample, you'd need to establish the universe to be sampled from. In this case, which TV stations will you observe? What will be the period of the study—the number of days? And during which hours of each day will you observe? Then, how many commercials do you want to observe and code for analysis? Watch television for a while and find out how many commercials occur each hour; then you can figure out how many hours of observation you'll need (and can stand).

Now you're ready to design the sample selection. As a practical matter, you wouldn't have to sample among the different stations if you had assistants—each of you could watch a different channel during the same period. But let's suppose you're working alone. Your final sampling frame, from which a sample will be selected and watched, might look something like this:

Jan. 7, Channel 2, 7–9 P.M.

Jan. 7, Channel 4, 7–9 P.M.

Jan. 7, Channel 9, 7–9 P.M.

Jan. 7, Channel 2, 9–11 P.M.

Jan. 7, Channel 4, 9–11 P.M.

Jan. 7, Channel 9, 9–11 P.M.

Jan. 8, Channel 2, 7–9 P.M.

Jan. 8, Channel 4, 7–9 P.M.

Jan. 8, Channel 9, 7–9 P.M.

Jan. 8, Channel 2, 9–11 P.M.

Jan. 8, Channel 4, 9–11 P.M.

Jan. 8, Channel 9, 9–11 P.M.

Jan. 9, Channel 2, 7–9 P.M.

Jan. 9, Channel 4, 7–9 P.M.

etc.

Notice that I've made several decisions for you in the illustration. First, I've assumed that channels 2, 4, and 9 are the ones appropriate to your study. I've assumed that you found the 7–11 P.M. prime-time hours to be the most relevant and that two-hour periods will do the job. I picked January 7 out of the hat for a starting date. In practice, of course, all these decisions should be based on your careful consideration of what would be appropriate to your particular study.

Once you have become clear about your units of analysis and the observations best suited to those units and have created a sampling frame like the one I've illustrated, sampling is simple and straightforward. The alternative procedures available to you are the same ones described in Chapter 7: random, systematic, stratified, and so on.

Sampling Techniques

As we've seen, in the content analysis of written prose, sampling may occur at any or all of several levels, including the contexts relevant to the works. Other forms of communication may also be sampled at any of the conceptual levels appropriate to them.

In content analysis, we could employ any of the conventional sampling techniques discussed in Chapter 7. We might select a random or systematic sample of French and U.S. novelists, of laws passed in the state of Mississippi, or of Shakespearean soliloquies. We might select (with a random start) every 23rd paragraph in Tolstoy's War and Peace. Or we might number all of the songs recorded by the Beatles and select a random sample of 25.

Stratified sampling is also appropriate for content analysis. To analyze the editorial policies of U.S. newspapers, for example, we might first group all newspapers by the region of the country or size of the community in which they are published, frequency of publication, or average circulation. We might then select a stratified random or systematic sample of newspapers for analysis. Having done so, we might select a sample of editorials from each selected newspaper, perhaps stratified chronologically.

Cluster sampling is equally appropriate to content analysis. Indeed, if individual editorials are our units of analysis, then the selection of newspapers at the first stage of sampling would be a cluster sample. In an analysis of political speeches, we might begin by selecting a sample of politicians; each politician would represent a cluster of political speeches. The TV commercial study described previously is another example of cluster sampling.

It should be repeated that sampling need not end when we reach the unit of analysis. If novels are the unit of analysis in a study, we might select a sample of novelists, a subsample of novels written by each selected author, and a subsample of paragraphs within each novel. We would then analyze the content of the paragraphs for the purpose of describing the novels themselves. (If you haven't realized this yet, researchers speak of samples within samples as "subsamples.")

Let's turn now to the coding or classification of the material being observed. Part 4 discusses the manipulation of such classifications to draw descriptive and explanatory conclusions.

Coding in Content Analysis

Content analysis is essentially a coding operation. **Coding** is the process of transforming raw data into a standardized form. In content analysis, communications—oral, written, or other—are coded or classified according to some conceptual framework. Thus, for example, newspaper editorials may be coded as liberal or conservative. Radio broadcasts may be coded as propagandistic or not, novels as romantic or not, paintings as representational or not, and political speeches as containing character assassinations or not. Recall that because terms such as these are subject to many interpretations, the researcher must specify definitions clearly.

Coding in content analysis involves the logic of conceptualization and operationalization, which I discussed in Chapter 5. As in other research methods, you must refine your conceptual framework and develop specific methods for observing in relation to that framework.

Manifest and Latent Content

In the earlier discussions of field research, we found that the researcher faces a fundamental choice between depth and specificity of understanding. Often, this represents a choice between validity and reliability, respectively. Typically, field researchers opt for depth, preferring to base their judgments on a broad range of observations and informa-

---

coding The process whereby raw data are transformed into standardized form suitable for machine processing and analysis.

manifest content In connection with content analysis, the concrete terms contained in a communication, as distinguished from *latent content*.

latent content In connection with content analysis, the underlying meaning of communications, as distinguished from their *manifest content*.

---

tion, even at the risk that another observer might reach a different judgment of the same situation. Survey research—through the use of standardized questionnaires—represents the other extreme: total specificity, even though the specific measures of variables may not be adequately valid reflections of those variables. The content analyst has some choice in this matter, however.

Coding the **manifest content**—the visible, surface content—of a communication is analogous to using a standardized questionnaire. To determine, for example, how erotic certain novels are, you might simply count the number of times the word *love* appears in each novel or the average number of appearances per page. Or, you might use a list of words, such as *love, kiss, hug,* and *caress,* each of which might serve as an indicator of the erotic nature of the novel. This method would have the advantage of ease and reliability in coding and of letting the reader of the research report know precisely how eroticism was measured. It would have a disadvantage, on the other hand, in terms of validity. Surely the phrase *erotic novel* conveys a richer and deeper meaning than the number of times the word *love* is used.

Alternatively, you could code the **latent content** of the communication: its underlying meaning. In the present example, you might read an entire novel or a sample of paragraphs or pages and make an overall assessment of how erotic the novel was. Although your total assessment might very well be influenced by the appearance of words such as *love* and *kiss,* it would not depend fully on their frequency.

Clearly, this second method seems better designed for tapping the underlying meaning of communications, but its advantage comes at a cost to reliability and specificity. Especially if more than one person is coding the novel, somewhat different definitions or standards may be employed. A passage that one coder regards as erotic may not seem erotic to another. Even if you do all of the coding yourself, there is no guarantee that your definitions and standards will remain constant throughout the enterprise. Moreover, the reader of your research report will likely be uncertain about the definitions you've employed.

Wherever possible, the best solution to this dilemma is to use both methods. For example, Carol Auster was interested in changes in the socialization of young women in Girl Scouts. To explore this, she undertook a content analysis of the Girl Scout manuals as revised over time. In particular, Auster was interested in the view that women should be limited to homemaking. Her analysis of the manifest content suggested a change: "I found that while 23% of the badges in 1913 centered on home life, this was true of only 13% of the badges in 1963 and 7% of the badges in 1980" (1985: 361).

An analysis of the latent content also pointed to an emancipation of Girl Scouts, similar to that occurring in U.S. society at large. The change of uniform was one indicator: "The shift from skirts to pants may reflect an acknowledgement of the more physically active role of women as well as the variety of physical images available to modern women" (Auster 1985: 362). Supporting evidence was found in the appearance of badges such as "Science Sleuth," "Aerospace," and "Ms. Fix-It."

Conceptualization and the Creation of Code Categories

For all research methods, conceptualization and operationalization typically involve the interaction of theoretical concerns and empirical observations. If, for example, you believe some newspaper editorials to be liberal and others to be conservative, ask yourself why you think so. Read some editorials, asking yourself which ones are liberal and which ones are conservative. Was the political orientation of a particular editorial most clearly indicated by its manifest content or by its tone? Was your decision based on the use of certain terms (for example, *leftist, fascist*, and so on) or on the support or opposition given to a particular issue or political personality?

Both inductive and deductive methods should be used in this activity. If you're testing theoretical propositions, your theories should suggest empirical indicators of concepts. If you begin with specific empirical observations, you should attempt to derive general principles relating to them and then apply those principles to the other empirical observations.

Bruce Berg (1989: 111) places code development in the context of grounded theory and likens it to solving a puzzle:

> Coding and other fundamental procedures associated with grounded theory development are certainly hard work and must be taken seriously, but just as many people enjoy finishing a complicated jigsaw puzzle, many researchers find great satisfaction in coding and analysis. As researchers . . . begin to see the puzzle pieces come together to form a more complete picture, the process can be downright thrilling.

Throughout this activity, remember that the operational definition of any variable is composed of the attributes included in it. Such attributes, moreover, should be mutually exclusive and exhaustive. A newspaper editorial, for example, should not be described as both liberal and conservative, though you should probably allow for some to be middle-of-the-road. It may be sufficient for your purposes to code novels as erotic or nonerotic, but you may also want to consider that some could be anti-erotic. Paintings might be classified as representational or not, if that satisfied your research purpose, or you might wish to classify them as impressionistic, abstract, allegorical, and so forth.

Realize further that different levels of measurement can be used in content analysis. You might, for example, use the nominal categories of liberal and conservative for characterizing newspaper editorials, or you might wish to use a more refined ordinal ranking, ranging from extremely liberal to extremely conservative. Bear in mind, however, that the level of measurement implicit in your coding methods—nominal, ordinal, interval, or ratio—does not necessarily reflect the nature of your variables. If the word *love* appeared 100 times in Novel A and 50 times in Novel B, you would be justified in saying that the word *love* appeared twice as often in Novel A, but not that Novel A was twice as erotic as Novel B. Similarly, agreeing with twice as many anti-Semitic statements in a questionnaire as someone else does not necessarily make one twice as anti-Semitic as that other person.

Counting and Record Keeping

If you plan to evaluate your content analysis data quantitatively, your coding operation must be amenable to data processing. This means, first, that the end product of your coding must be numerical. If you're counting the frequency of certain words, phrases, or other manifest content, the coding is necessarily numerical. But even if you're coding latent content on the basis of overall judgments, it will be necessary to represent your coding decision numerically: 1 = very liberal, 2 = moderately liberal, 3 = moderately conservative, and so on.

Second, your record keeping must clearly distinguish between units of analysis and units of observation, especially if these two are different. The initial coding, of course, must relate to the units of observation. If novelists are the units of analysis, for example, and you wish to characterize them through a content analysis of their novels, your primary records will represent novels as the units of observation. You may then combine your scoring of individual novels to characterize each novelist, the unit of analysis.

Third, while you're counting, it will normally be important to record the base from which the counting is done. It would probably be useless to know the number of realistic paintings produced by a given painter without knowing the number he or she has painted all together; the painter would be regarded as realistic if a high percentage of paintings were of that genre. Similarly, it would tell us little that the word *love* appeared 87 times in a novel if we did not know about how many words there were in the entire novel. The issue of observational base is most easily resolved if every observation is coded in terms of one of the attributes making up a variable. Rather than simply counting the number of liberal editorials in a given collection, for example, code each editorial by its political orientation, even if it must be coded "no apparent orientation."

Let's suppose we want to describe and explain the editorial policies of different newspapers. Figure 11-3 presents part of a tally sheet that might result from the coding of newspaper editorials. Note that newspapers are the units of analysis. Each newspaper has been assigned an identification number to facilitate mechanized processing. The second column has a space for the number of editorials coded for each newspaper. This will be an important piece of information, because we want to be able to say, for example, "Of all the editorials, 22 percent were pro–United Nations," not just "There were eight pro–United Nations editorials."

One column in Figure 11-3 is for assigning a subjective overall assessment of each newspaper's editorial policies. (Such assignments might later be compared with the several objective measures.) Other columns provide space for recording numbers of editorials reflecting specific editorial positions. In a real content analysis, there would be spaces for recording other editorial positions plus noneditorial information about each newspaper, such as the region in which it is published, its circulation, and so forth.

As in other forms of social research, there are now computer programs available to assist in content analysis. Two you might want to explore are Yoshikoder and T-LAB. There are many more programs available as well; the psychologist Matthias Romppel has provided an excellent online review of them, as well as other content analysis resources. (For more, see the links on this book's website: http://www.cengage.com/sociology/babbie.)

Qualitative Data Analysis

Not all content analysis results in counting. Sometimes a qualitative assessment of the materials is most appropriate, as in Carol Auster's examination of changes in Girl Scout uniforms and handbook language.

Bruce Berg (1989: 123–25) discusses "negative case testing" as a technique for qualitative hypothesis testing. First, in the grounded theory tradition, you begin with an examination of the data, which may yield a general hypothesis. Let's say that you're examining the leadership of a new community association by reviewing the minutes of meetings to see who made motions that were subsequently passed. Your initial examination of the data sug-

Newspaper ID	Number of editorials evaluated	SUBJECTIVE EVALUATION 1. Very liberal 2. Moderately liberal 3. Middle-of-road 4. Moderately conservative 5. Very conservative	Number of "isolationist" editorials	Number of "pro–United Nations" editorials	Number of "anti–United Nations" editorials
001	37	2	0	8	0
002	26	5	10	0	6
003	44	4	2	1	2
004	22	3	1	2	3
005	30	1	0	6	0

FIGURE 11-3

Sample Tally Sheet (Partial)

gests that the wealthier members are the most likely to assume this leadership role.

The second stage in the analysis is to search your data to find all the cases that contradict the initial hypothesis. In this instance, you would look for poorer members who made successful motions and wealthy members who never did. Third, you must review each of the disconfirming cases and either (1) give up the hypothesis or (2) see how it needs to be fine-tuned.

Let's say that in your analysis of disconfirming cases, you notice that each of the unwealthy leaders has a graduate degree, whereas each of the wealthy nonleaders has very little formal education. You may revise your hypothesis to consider both education and wealth as routes to leadership in the association. Perhaps you'll discover some threshold for leadership (a white-collar job, a level of income, and a college degree) beyond which those with the most money, education, or both are the most active leaders.

This process is an example of what Barney Glaser and Anselm Strauss (1967) called *analytic*

induction. It is inductive in that it begins primarily with observations, and it is analytic because it goes beyond description to find patterns and relationships among variables.

There are, of course, dangers in this form of analysis, as in all others. The chief risk is misclassifying observations so as to support an emerging hypothesis. For example, you may erroneously conclude that a nonleader didn't graduate from college or you may decide that the job of factory foreman is "close enough" to being white-collar.

Berg (1989: 124) offers techniques for avoiding these errors:

1. If there are sufficient cases, select some at random from each category in order to avoid merely picking those that best support the hypothesis.

2. Give at least three examples in support of every assertion you make about the data.

3. Have your analytic interpretations carefully reviewed by others uninvolved in the research project to see whether they agree.

4. Report whatever inconsistencies you do discover—any cases that simply do not fit your hypotheses. Realize that few social patterns are 100 percent consistent, so you may have discovered something important even if it doesn't apply to absolutely all of social life. However, you should be honest with your readers in that regard.

There are computer programs now available for qualitative content analysis. For example, you can try out MAXQDA online. Also, T-LAB, mentioned earlier, provides for some interesting qualitative analyses, such as mapping word associations in a political speech. Matthias Romppel has provided an excellent online review of qualitative content analysis programs (see the links on this book's website: http://www.cengage.com/sociology/babbie). Some of the programs appropriate for content analysis are discussed in Chapter 13 in connection with other kinds of qualitative data analysis.

Illustrations of Content Analysis

Several studies have indicated that women are stereotyped on television. R. Stephen Craig (1992) took this line of inquiry one step further to examine the portrayal of both men and women during different periods of television programming.

To study gender stereotyping in television commercials, Craig selected a sample of 2,209 network commercials during several periods between January 6 and 14, 1990.

> The weekday day part (in this sample, Monday–Friday, 2–4 P.M.) consisted exclusively of soap operas and was chosen for its high percentage of women viewers. The weekend day part (two consecutive Saturday and Sunday afternoons during sports telecasts) was selected for its high percentage of men viewers. Evening "prime time" (Monday–Friday, 9–11 P.M.) was chosen as a basis for comparison with past studies and the other day parts.
>
> *(1992: 199)*

Each of the commercials was coded in several ways. "Characters" were coded as

All male adults

All female adults

All adults, mixed gender

Male adults with children or teens (no women)

Female adults with children or teens (no men)

Mixture of ages and genderst

In addition, Craig's coders noted which character was on the screen longest during the commercial—the "primary visual character"—as well as the roles played by the characters (such as spouse, celebrity, parent), the type of product advertised (such as body product, alcohol), the setting (such as kitchen, school, business), and the voice-over narrator.

Table 11-1 indicates the differences in the times when men and women appeared in commercials. Women appeared most during the daytime (with its soap operas), men predominated during the weekend (with its sports programming), and men and women were equally represented during evening prime time.

Craig found other differences in the ways men and women were portrayed.

> Further analysis indicated that male primary characters were proportionally more likely than females to be portrayed as celebrities and professionals in every day part, while women were proportionately more likely to be portrayed as interviewer/demonstrators, parent/ spouses, or sex object/models in every day part. . . . Women were proportionately more likely to appear as sex object/models during the weekend than during the day.
>
> *(1992: 204)*

The research also showed that different products were advertised during different time periods. As you might imagine, almost all the daytime commercials dealt with body, food, or home products. These products accounted for only one in three on the weekends. Instead, weekend commercials stressed automotive products (29 percent), business products or services (27 percent), or alcohol (10 percent). There were virtually no alcohol ads during evenings and daytime.

TABLE 11-1

Percent of Adult Primary Visual Characters by Sex Appearing in Commercials in Three Day Parts

	Weekend	Daytime	Evening
Adult male	40	52	80
Adult female	60	48	20

Source: R. Stephen Craig, "The Effect of Television Day Part on Gender Portrayals in Television Commercials: A Content Analysis," *Sex Roles* 26, nos. 5/6 (1992): 204.

As you might suspect, women were most likely to be portrayed in home settings, men most likely to be shown away from home. Other findings dealt with the different roles played by men and women.

> The women who appeared in weekend ads were almost never portrayed without men and seldom as the commercial's primary character. They were generally seen in roles subservient to men (e.g., hotel receptionist, secretary, or stewardess), or as sex objects or models in which their only function seemed to be to lend an aspect of eroticism to the ad.
>
> *(Craig 1992: 208)*

Although some of Craig's findings may seem unsurprising, remember that "common knowledge" does not always correspond with reality. It's always worthwhile to check out widely held assumptions. And even when we think we know about a given situation, it's often useful to know specific details such as those provided by a content analysis like this one.

In another content analysis that drew on popular culture for content, Charis Kubrin (2005) chose a primarily qualitative approach. Kubrin was interested in the themes put forth in rap music, particularly in gangsta rap, and the relationship of those themes to neighborhood culture and the "street code."

> In response to societal and neighborhood conditions, black youth in disadvantaged communities have created a substitute social order governed by their own code—a street code— and rituals of authenticity. . . . This social order reflects the subcultural locus of interests that emerges from pervasive race and class

inequality and the social isolation of poor black communities.

> *(2005: 439)*

She began her study by identifying all the platinum rap albums released between 1992 and 2000: 130 albums containing a total of 1,922 songs. She then drew a simple random sample of one-third of the songs (632) and set about the task of listening to each. She did this twice with each song.

> First, I listened to a song in its entirety while reading the printed lyrics to determine what the song was about. Second, I listened to the song again and coded each line to determine whether the street code elements described earlier were present: (1) respect, (2) willingness to fight or use violence, (3) material wealth, (4) violent retaliation, (5) objectification of women, and (6) nihilism.
>
> *(2005: 443)*

Kubrin was particularly interested in the theme of nihilism—the rejection of traditional moral principles and a fundamental skepticism about the meaning of life. She was interested in how that theme was portrayed in gangsta rap and how it fit into the street code.

Though she began with a sample of 632 songs, she found that no new themes appeared to be showing up after about 350 songs had been analyzed. To be safe, she coded another 50 songs and found no new themes, completing her coding process at that point.

Kubrin notes that rap music is typically regarded as antisocial and resistant to organized society, but her in-depth analysis of lyrics suggests something different:

> Rap music does not exist in a cultural vacuum; rather it expresses the cultural crossing, mixing, and engagement of black youth culture with the values, attitudes and concerns of the white majority. Many of the violent (and patriarchical, materialistic, sexist, etc.) ways of thinking that are glorified in gangsta rap are a reflection of the prevailing values created and sustained in the larger society.
>
> *(2005: 454)*

She traces the implications of this for understanding street life as well as for the likely success of various crime-control strategies.

Strengths and Weaknesses of Content Analysis

Probably the greatest advantage of content analysis is its economy in terms of both time and money. A college student might undertake a content analysis, whereas undertaking a survey, for example, might not be feasible. There is no requirement for a large research staff; no special equipment is needed. As long as you have access to the material to be coded, you can undertake content analysis.

Content analysis also has the advantage of allowing the correction of errors. If you discover you've botched up a survey or an experiment, you may be forced to repeat the whole research project with all its attendant costs in time and money. If you botch up your field research, it may be impossible to redo the project; the event under study may no longer exist. In content analysis, it's usually easier to repeat a portion of the study than it is in other research methods. You might be required, moreover, to recode only a portion of your data rather than all of it.

A third advantage of content analysis is that it permits the study of processes occurring over a long time. You might focus on the imagery of African Americans conveyed in U.S. novels written between 1850 and 1860, for example, or you might examine how such imagery has changed from 1850 to the present.

Finally, content analysis has the advantage of all unobtrusive measures, namely, that the content analyst seldom has any effect on the subject being studied. Because the novels have already been written, the paintings already painted, the speeches already presented, content analyses can have no effect on them.

Content analysis has disadvantages as well. For one thing, it's limited to the examination of recorded communications. Such communications may be oral, written, or graphic, but they must be recorded in some fashion to permit analysis.

As we've seen, content analysis has both advantages and disadvantages in terms of validity and reliability. Problems of validity are likely unless you happen to be studying communication processes per se.

On the other side of the ledger, the concreteness of materials studied in content analysis strengthens the likelihood of reliability. You can always code your data and then recode the original documents from scratch. And you can repeat the process as many times as you want. In field research, by contrast, there's no way to return to the original events that were observed, recorded, and categorized.

Let's move from content analysis now and turn to a related research method: the analysis of existing data. Although numbers rather than communications are analyzed in this case, I think you'll see the similarity to content analysis.

Analyzing Existing Statistics

Frequently you can or must undertake social science inquiry through the use of official or quasi-official statistics. This differs from secondary analysis, in which you obtain a copy of someone else's data and undertake your own statistical analysis. In this section, we're going to look at ways of using the data analyses that others have already done.

This method is particularly significant because existing statistics should always be considered as at least a supplemental source of data. If you were planning a survey of political attitudes, for example, you would do well to examine and present your findings within a context of voting patterns, rates of voter turnout, or similar statistics relevant to your research interest. Or, if you were doing evaluation research on an experimental morale-building program on an assembly line, then statistics on absenteeism, sick leave, and so on would probably be interesting and revealing in connection with the data from your own research. Existing statistics, then, can often provide a historical or conceptual context within which to locate your original research.

Existing statistics can also provide the main data for a social science inquiry. An excellent example is the classic study mentioned at the beginning of this chapter, Emile Durkheim's *Suicide* ([1897] 1951). Let's take a closer look at Durkheim's work before considering some of the special problems this method presents.

Durkheim's Study of Suicide

Why do people kill themselves? Undoubtedly every suicide case has a unique history and explanation, yet all such cases could no doubt be grouped according to certain common causes: financial failure, trouble in love, disgrace, and other kinds of personal problems. The French sociologist Emile Durkheim had a slightly different question in mind when he addressed the matter of suicide, however. He wanted to discover the environmental conditions that encouraged or discouraged it, especially social conditions.

The more Durkheim examined the available records, the more patterns of differences became apparent to him. One of the first things to attract his attention was the relative stability of suicide rates. Looking at several countries, he found suicide rates to be about the same year after year. He also discovered that a disproportionate number of suicides occurred in summer, leading him to hypothesize that temperature might have something to do with suicide. If this were the case, suicide rates should be higher in the southern European countries than in the temperate ones. However, Durkheim discovered that the highest rates were found in countries in the central latitudes, so temperature couldn't be the answer.

He explored the role of age (35 was the most common suicide age), gender (men outnumbered women around four to one), and numerous other factors. Eventually, a general pattern emerged from different sources.

In terms of the stability of suicide rates over time, for instance, Durkheim found that the pattern was not totally stable. There were spurts in the rates during times of political turmoil, which occurred in several European countries around 1848.

This observation led him to hypothesize that suicide might have something to do with "breaches in social equilibrium." Put differently, social stability and integration seemed to be a protection against suicide.

This general hypothesis was substantiated and specified through Durkheim's analysis of a different set of data. The different countries of Europe had radically different suicide rates. The rate in Saxony, for example, was about ten times that of Italy, and the relative ranking of various countries persisted over time. As Durkheim considered other differences among the various countries, he eventually noticed a striking pattern: Predominantly Protestant countries had consistently higher suicide rates than Catholic ones did. The predominantly Protestant countries had 190 suicides per million population; mixed Protestant-Catholic countries, 96; and predominantly Catholic countries, 58 (Durkheim [1897] 1951: 152).

Although suicide rates thus seemed to be related to religion, Durkheim reasoned that some other factor, such as level of economic and cultural development, might explain the observed differences among countries. If religion had a genuine effect on suicide, then the religious difference would have to be found *within* given countries as well. To test this idea, Durkheim first noted that the German state of Bavaria had both the most Catholics and the lowest suicide rates in that country, whereas heavily Protestant Prussia had a much higher suicide rate. Not content to stop there, however, Durkheim examined the provinces composing each of those states.

Table 11-2 shows what he found. As you can see, in both Bavaria and Prussia, provinces with the highest proportion of Protestants also had the highest suicide rates. Increasingly, Durkheim became confident that religion played a significant role in the matter of suicide.

Returning eventually to a more general theoretical level, Durkheim combined the religious findings with the earlier observation about increased suicide rates during times of political turmoil. As we've seen, Durkheim suggested that many suicides are a product of *anomie*, that is, "normlessness," or a general sense of social

TABLE 11-2

Suicide Rates in Various German Provinces, Arranged in Terms of Religious Affiliation

Religious Character of Province	Suicides per Million Inhabitants
Bavarian Provinces (1867–1875)*	
Less than 50% Catholic	
Rhenish Palatinate	167
Central Franconia	207
Upper Franconia	204
Average	**192**
50% to 90% Catholic	
Lower Franconia	157
Swabia	118
Average	**135**
More than 90% Catholic	
Upper Palatinate	64
Upper Bavaria	114
Lower Bavaria	19
Average	**75**
Prussian Provinces (1883–1890)	
More than 90% Protestant	
Saxony	309.4
Schleswig	312.9
Pomerania	171.5
Average	**264.6**
68% to 89% Protestant	
Hanover	212.3
Hesse	200.3
Brandenburg and Berlin	296.3
East Prussia	171.3
Average	**220.0**
40% to 50% Protestant	
West Prussia	123.9
Silesia	260.2
Westphalia	107.5
Average	**163.6**
28% to 32% Protestant	
Posen	96.4
Rhineland	100.3
Hohenzollern	90.1
Average	**95.6**

*Note: The population below 15 years has been omitted.

Source: Adapted from Emile Durkheim, *Suicide* (Glencoe, IL: Free Press, [1897] 1951), 153.

instability and disintegration. During times of political strife, people may feel that the old ways of society are collapsing. They become demoralized and depressed, and suicide is one answer to the severe discomfort. Seen from the other direction, social integration and solidarity—reflected in personal feelings of being part of a coherent, enduring social whole—offer protection against depression and suicide. That was where the religious difference fit in. Catholicism, as a far more structured and integrated religious system, gave people a greater sense of coherence and stability than did the more loosely structured Protestantism.

From these theories, Durkheim created the concept of *anomic suicide.* More importantly, as you know, he added the concept of *anomie* to the lexicon of the social sciences.

This account of Durkheim's classic study is greatly simplified, of course. Anyone studying social research would profit from studying the original. For our purposes, Durkheim's approach provides a good illustration of the possibilities for research contained in the masses of data regularly gathered and reported by government agencies and other organizations.

The Consequences of Globalization

The notion of "globalization" has become increasingly controversial in the United States and around the world, with reactions ranging from scholarly debates to violent confrontations in the streets. One point of view sees the spread of U.S.-style capitalism to developing countries as economic salvation for those countries. A very different point of view sees globalization as essentially neocolonial exploitation, in which multinational conglomerates exploit the resources and people of poor countries. And, of course, there are numerous variations on these contradictory views.

Jeffrey Kentor (2001) wanted to bring data to bear on the question of how globalization affects the developing countries that host the process. To that end, he used data available from the World Bank's "World Development Indicators." (You can learn more about these data at the link on this book's website.) Noting past variations in the way globalization was measured, Kentor used the amount of foreign investment in a country's economy as a percentage of that country's whole economy. He reasoned that dependence on foreign investments was more important that the amount of the investment.

In his analysis of 88 countries with a per capita gross domestic product (the total goods and services produced in a country) of less that $10,000, Kentor found that dependence on foreign investment tended to increase income inequality among the citizens of a country. The greater the degree of dependence, the greater the income inequality. Kentor reasoned that globalization produced well-paid elites who, by working with the foreign corporations, maintained a status well above that of the average citizen. But because the profits derived from the foreign investments tended to be returned to the investors' countries rather than enriching the poor countries, the great majority of the population in the latter reaped little or no economic benefit.

Income inequality, in turn, was found to increase birth rates and, hence, population growth, in a process too complex to summarize here. Population growth, of course, brings a wide range of problems to countries already too poor to provide for the basic needs of their people.

This research example, along with our brief look at Durkheim's studies, should broaden your understanding of the kinds of social phenomena that we can study through data already collected and compiled by others.

Units of Analysis

The unit of analysis involved in the analysis of existing statistics is often not the individual. Durkheim, for example, was required to work with political-geographic units: countries, regions, states, and cities. The same situation would probably appear if you were to undertake a study of crime rates, accident rates, or disease. By their nature, most existing statistics are aggregated: They describe groups.

The aggregate nature of existing statistics can present a problem, though not an insurmountable one. As we saw, for example, Durkheim wanted to determine whether Protestants or Catholics were more likely to commit suicide. The difficulty was that none of the records available to him indicated the religion of those people who committed suicide. Ultimately, then, it was not possible for him to say whether Protestants committed suicide more often than Catholics did, though he inferred as much. Because Protestant countries, regions, and states had higher suicide rates than did Catholic countries, regions, and states, he drew the obvious conclusion.

There's danger in drawing this kind of conclusion, however. It's always possible that patterns of behavior at a group level do not reflect corresponding patterns on an individual level. Such errors are due to an ecological fallacy, which was discussed in Chapter 4. In the case of Durkheim's study, it was altogether possible, for example, that it was Catholics who committed suicide in the predominantly Protestant areas. Perhaps Catholics in predominantly Protestant areas were so badly persecuted that they were led into despair and suicide. In that case it would be possible for Protestant countries to have high suicide rates without any Protestants committing suicide.

Durkheim avoided the danger of the ecological fallacy in two ways. First, his general conclusions were based as much on rigorous theoretical deductions as on the empirical facts. The correspondence between theory and fact made a counterexplanation, such as the one I just made up, less likely. Second, by extensively retesting his conclusions in a variety of ways, Durkheim further strengthened the likelihood that they were correct. Suicide rates were higher in Protestant countries than in Catholic ones; higher in Protestant regions of Catholic countries than in Catholic regions of Protestant countries; and so forth. The replication of findings added to the weight of evidence in support of his conclusions.

Problems of Validity

Whenever we base research on an analysis of data that already exist, we're obviously limited to what exists. Often, the existing data do not cover exactly what we're interested in, and our measurements may not be altogether valid representations of the variables and concepts we want to make conclusions about.

Two characteristics of science are used to handle the problem of validity in analysis of existing statistics: *logical reasoning* and *replication*. Durkheim's strategy provides an example of logical reasoning. Although he could not determine the religion of people who committed suicide, he reasoned that most of the suicides in a predominantly Protestant region would be Protestants.

Replication can be a general solution to problems of validity in social research. Recall the earlier discussion of the interchangeability of indicators (Chapter 5). Crying in sad movies isn't necessarily a valid measure of compassion; nor is putting little birds back in their nests nor giving money to charity. None of these things, taken alone, would prove that one group (women, say) was more compassionate than another (men). But if women appeared more compassionate than men by all these measures, that would create a weight of evidence in support of the conclusion. In the analysis of existing statistics, a little ingenuity and reasoning can usually turn up several independent tests of a given hypothesis. If all the tests seem to confirm the hypothesis, then the weight of evidence supports the validity of the measure.

Problems of Reliability

The analysis of existing statistics depends heavily on the quality of the statistics themselves: Do they accurately report what they claim to report? This can be a substantial problem sometimes, because the weighty tables of government statistics, for example, are sometimes grossly inaccurate.

Consider research into crime. Because a great deal of this research depends on official crime statistics, this body of data has come under critical evaluation. The results have not been too encouraging. As an illustration, suppose you were interested in tracing long-term trends in marijuana use in the United States. Official statistics on the numbers of people arrested for selling or possessing marijuana would seem to be a reasonable measure of use, right? Not necessarily.

To begin, you face a hefty problem of validity. Before the passage of the Marihuana Tax Act in 1937, "grass" was legal in the United States, so arrest records would not give you a valid measure of use. But even if you limited your inquiry to the times after 1937, you would still have problems of reliability that stem from the nature of law enforcement and crime recording.

Law enforcement, for example, is subject to various pressures. A public outcry against marijuana, led perhaps by a vocal citizens' group, often results in a police crackdown on drug trafficking—especially during an election or budget year. A sensational story in the press can have a similar effect. In addition, the volume of other business facing the police can affect marijuana arrests.

In tracing the pattern of drug arrests in Chicago between 1942 and 1970, Lois DeFleur (1975) demonstrates that the official records present a far less accurate history of drug use than of police practices and political pressure on police. On a different level of analysis, Donald Black (1970) and others have analyzed the factors influencing whether an offender is actually arrested by police or let off with a warning. Ultimately, official crime statistics are influenced by whether specific offenders are well or poorly dressed, whether they are polite or abusive to police officers, and so forth. When we consider unreported crimes, sometimes estimated to be as much as ten times the number of crimes known to police, the reliability of crime statistics gets even shakier.

These comments concern crime statistics at a local level. Often it's useful to analyze national crime statistics, such as those reported in the FBI's annual *Uniform Crime Reports*. Additional problems are introduced at the national level. For example, different local jurisdictions define crimes differently. Also, participation in the FBI program is voluntary, so the data are incomplete.

Finally, the process of record keeping affects the data available to researchers. Whenever a law-enforcement unit improves its record-keeping system—computerizes it, for example—the apparent crime rates increase dramatically. This can happen even if the number of crimes committed, reported, and investigated does not increase.

Is America #1?

On September 19, 1999, ABC-TV broadcast a special show, hosted by John Stossel, to examine where the United States stood in the ranking of the world's societies. As the show unfolded, it became clear that the United States was doing okay—arguably #1—and that the key to our success was primarily our laissez-faire capitalist system. To make the latter point more strongly, Stossel pointed to other success stories that also owed their success to laissez-faire capitalism.

According to Stossel, Hong Kong stood out among the world's nations as the leader of free-market economics. As evidence of Hong Kong's success, Stossel reported that it had "the only government in the world that makes a surplus, a big surplus." What do you think about that conclusion? Is it convincing to you?

Here's what the media watchdog, Fairness and Accuracy in Reporting (FAIR), had to say about Stossel's assertion:

> As anyone who pays attention to Washington politics knows, the U.S. government has been running a federal budget surplus for more than a year; it amounted to $70 billion last year. Other countries with budget surpluses last year included the United Kingdom, Canada, Australia, Denmark, Finland, Iceland, Ireland, New Zealand, Norway and Sweden.

Stossel went on to contrast Hong Kong (a capitalist success story) with the alternative to a free economy:"stagnation, and often poverty. Consider China, now mired in Third World poverty. They were once the leader of the world." Again, FAIR suggests a different assessment:

> Actually, China's economy is anything but "stagnant." As the Treasury Department's Lawrence Summers said in a speech last year,"China has been the fastest growing economy in history since [economic] reform began in 1980." While China has adopted some aspects of market economics, a large proportion of its business firms are still owned by the government.

In the media and elsewhere, you'll often find assertions of fact that appear to be based on statistical analyses. However, it's usually a good idea to check the facts.

Source: Fairness and Accuracy in Reporting,"Action Alert: ABC News Gives up on Accuracy?" September 28, 1999 (http://www.fair.org/activism/stossel-america .html).

Researchers' first protection against the problems of reliability in the analysis of existing statistics is knowing that the problem may exist. Investigating the nature of the data collection and tabulation may enable you to assess the nature and degree of unreliability so that you can judge its potential impact on your research interest. If you also use logical reasoning and replication, you can usually cope with the problem. "Is America #1?" provides an example of what you might discover by carefully examining the use of existing statistics.

Sources of Existing Statistics

It would take a whole book just to list the sources of data available for analysis. In this section, I want to mention a few sources and point you in the direction of finding others relevant to your research interest. (See the links on this book's website for more on these sources.)

Undoubtedly, the single most valuable book you can buy is the annual *Statistical Abstract of the United States*, published by the United States Department of Commerce. Unquestionably the best source of data about the United States, it includes statistics on the individual states and (less extensively) cities, as well as on the nation as a whole. Where else can you find the number of work stoppages in the country year by year, the residential property taxes of major cities, the number of water-pollution discharges reported around the country, the number of business proprietorships in the nation, and hundreds of other such handy bits of information? To make things even better, Hoover's Business Press offers the same book in soft cover for less cost. This commercial version, entitled *The American Almanac*, shouldn't be confused with other almanacs that are less reliable and less useful for social science research. Better yet, you can buy the *Statistical Abstract* on a CD-ROM, making the search for and transfer of data quite easy. Best of all, you can download the *Statistical Abstract* from the web for free (your tax dollars at work for you).

Federal agencies—the Departments of Labor, Agriculture, Transportation, and so forth—publish numerous data series. To find out what's available, go to your library, find the government documents section, and spend a few hours browsing through the shelves. You can also visit the U.S. Government Printing Office website and look around.

As you can see, the web serves as a great resource for finding existing statistics. Here are just a few organizations you can access online, through the links on this book's website:

- Bureau of the Census
- Bureau of Labor Statistics
- Bureau of Transportation Statistics
- Centers for Disease Control and Prevention
- Central Intelligence Agency
- Department of Education
- Federal Bureau of Investigation
- The World Bank

If you find none of these interesting, you should turn to the vast listing of data sources provided, by topic, at the University of Michigan's website "Statistical Resources on the Web."

World statistics are available through the United Nations. Its *Demographic Yearbook* presents annual vital statistics (births, deaths, and other data relevant to population) for the individual nations of the world. Other publications report a variety of other kinds of data. Again, a trip to your library, along with a web search, is the best introduction to what's available.

The amount of data provided by nongovernment agencies is as staggering as the amount your taxes buy. Chambers of commerce often publish data reports on businesses, as do private consumer groups. Ralph Nader has information on automobile safety, and Common Cause covers politics and government. The Gallup Organization publishes reference volumes on public opinion as tapped by Gallup Polls since 1935.

Organizations such as the Population Reference Bureau publish a variety of demographic data, U.S. and international, that a secondary analyst could use. Their *World Population Data Sheet* and *Population Bulletin* are resources heavily used by social scientists. Social indicator data can be found in the journal *SINET: A Quarterly Review of Social Reports and Research on Social Indicators, Social Trends, and the Quality of Life.*

The sources I've listed represent only a tiny fraction of the thousands that are available. With so much data already collected, the lack of funds to support expensive data collection is no reason for not doing good and useful social research. Moreover, this research method need not be limited to tables of numbers. There are graphic resources available as well, such as the *Social Explorer* (see the link on this book's website). A wide range of data about the United States can be represented on a map of congressional districts or census tracts. You can examine aspects of population, religion, economy, and many other variables. For example, you can easily find the geographic concentrations of unmarried partners: male/female, male/male, and female/female.

You can do similar kinds of map-based examinations through the Census Bureau by clicking on "Maps" at their website (see the link at this book's site). Once you've displayed a variable such as multiracial marriages state-by-state, you can click on a particular state and get a detailed graph of the racial marriages in that state.

The availability of existing statistics also makes creating some fairly sophisticated measures possible. "Suffering around the World" (pp. 352–353) describes an analysis published by the Population Crisis Committee based on the kinds of data available in government practice.

Let's move now from an inherently quantitative method to one that is typically qualitative: comparative and historical research.

Comparative and Historical Research

Comparative and historical research differs substantially from the methods discussed so far, though it overlaps somewhat with field research, content analysis, and the analysis of existing statistics. It involves the use of historical methods by sociologists, political scientists, and other social scientists to examine societies (or other social units) over time and in comparison with one another.

The discussion of longitudinal research designs in Chapter 4 notwithstanding, our examination of research methods so far has focused primarily on studies anchored in one point in time and in one locale, whether a small group or a nation. Although accurately portraying the main thrust of contemporary social science research, this focus conceals the fact that social scientists are also interested in tracing the development of social forms over time and comparing those developmental processes across cultures. James Mahoney and Dietrich Rueschemeyer (2003: 4) suggest that current comparative and historical researchers "focus on a wide range of topics, but they are united by a commitment to providing historically grounded explanations of large-scale and substantively important outcomes." Thus, you find comparative and historical studies dealing with the topics social class, capitalism, religion, revolution, and the like.

After describing some major instances of comparative and historical research, past and present, this section discusses some of the key elements of this method.

Examples of Comparative and Historical Research

August Comte, who coined the term *sociologie,* saw that new discipline as the final stage in a historical development of ideas. With his broadest brush, he painted an evolutionary picture that took humans from a reliance on religion to metaphysics to science. With a finer brush, he portrayed science as evolving from the development of biology and the other natural sciences to the development of psychology and, finally, to the development of scientific sociology.

A great many later social scientists have also turned their attention to broad historical processes. Several have examined the historical progression of social forms from the simple to the complex, from rural-agrarian to urban-industrial societies. The

U.S. anthropologist Lewis Morgan, for example, saw a progression from "savagery" to "barbarism" to "civilization" (1870). Robert Redfield, another anthropologist, wrote more recently of a shift from "folk society" to "urban society" (1941). Emile Durkheim saw social evolution largely as a process of ever-greater division of labor ([1893] 1964). In a more specific analysis, Karl Marx examined economic systems progressing historically from primitive to feudal to capitalistic forms ([1867] 1967). All history, he wrote in this context, was a history of class struggle—the "haves" struggling to maintain their advantages and the "have-nots" struggling for a better lot in life. Looking beyond capitalism, Marx saw the development of socialism and finally communism.

Not all historical studies in the social sciences have had this evolutionary flavor, however. Some social science readings of the historical record, in fact, point to grand cycles rather than to linear progressions. No scholar better represents this view than Pitirim A. Sorokin. A participant in the Russian Revolution of 1917, Sorokin served as secretary to Prime Minister Kerensky. Both Kerensky and Sorokin fell from favor, however, and Sorokin began his second career—as a U.S. sociologist.

Whereas Comte read history as a progression from religion to science, Sorokin (1937–1940) suggested that societies alternate cyclically between two points of view, which he called "ideational" and "sensate." Sorokin's sensate point of view defined reality in terms of sense experiences. The ideational, by contrast, placed a greater emphasis on spiritual and religious factors. Sorokin's reading of the historical record further indicated that the passage between the ideational and sensate was through a third point of view, which he called the "idealistic." This third view combined elements of the sensate and ideational in an integrated, rational view of the world.

These examples indicate some of the topics comparative and historical researchers have

---

comparative and historical research The examination of societies (or other social units) over time and in comparison with one another.

Suffering around the World

In 1992, the Population Crisis Committee, a nonprofit organization committed to combating the population explosion, undertook to analyze the relative degree of suffering in nations around the world. Every country with a population of one million or more was evaluated in terms of the following ten indicators—with a score of 10 on any indicator representing the highest level of adversity:

Life expectancy

Daily per capita calorie supply

Percentage of the population with access to clean drinking water

Proportion of infant immunization

Rate of secondary school enrollment

Gross national product

Inflation

Number of telephones per 1,000 people

Political freedom

Civil rights

Here's how the world's nations ranked in terms of these indicators. Remember, high scores are signs of overall suffering.

Extreme Human Suffering

93—Mozambique

92—Somalia

89—Afghanistan, Haiti, Sudan

88—Zaire

87—Laos

86—Guinea, Angola

85—Ethiopia, Uganda

84—Cambodia, Sierra Leone

82—Chad, Guinea-Bissau

81—Ghana, Burma

79—Malawi

77—Cameroon, Mauritania

76—Rwanda, Vietnam, Liberia

75—Burundi, Kenya, Madagascar, Yemen

High Human Suffering

74—Ivory Coast

73—Bhutan, Burkina Faso, Central African Republic

71—Tanzania, Togo

70—Lesotho, Mali, Niger, Nigeria

69—Guatemala, Nepal

68—Bangladesh, Bolivia, Zambia

67—Pakistan

66—Nicaragua, Papua-New Guinea, Senegal, Swaziland, Zimbabwe

65—Iraq

64—Gambia, Congo, El Salvador, Indonesia, Syria

---

examined. To get a better sense of what comparative and historical research entails, let's look at a few examples in somewhat more detail.

Weber and the Role of Ideas

In his analysis of economic history, Karl Marx put forward a view of economic determinism. That is, he postulated that economic factors determined the nature of all other aspects of society. For example, Marx's analysis showed that a function of European churches was to justify and support the capitalist status quo—religion was a tool of the powerful in maintaining their dominance over the powerless. "Religion is the sigh of the oppressed creature," Marx wrote in a famous passage, "the sentiment of

a heartless world, and the soul of soulless conditions. It is the opium of the people" (Bottomore and Rubel [1843] 1956: 27).

Max Weber, a German sociologist, disagreed. Without denying that economic factors could and did affect other aspects of society, Weber argued that economic determinism did not explain everything. Indeed, Weber said, economic forms could come from noneconomic ideas. In his research in the sociology of religion, Weber examined the extent to which religious institutions were the source of social behavior rather than mere reflections of economic conditions. His most noted statement of this side of the issue is found in *The Protestant Ethic and the Spirit of Capitalism* ([1905] 1958). Here's a brief overview of Weber's thesis.

63—Comores, India, Paraguay, Peru

62—Benin, Honduras

61—Lebanon, China, Guyana, South Africa

59—Egypt, Morocco

58—Ecuador, Sri Lanka

57—Botswana

56—Iran

55—Suriname

54—Algeria, Thailand

53—Dominican Republic, Mexico, Tunisia, Turkey

51—Libya, Colombia, Venezuela

50—Brazil, Oman, Philippines

Moderate Human Suffering

49—Solomon Islands

47—Albania

45—Vanuatu

44—Jamaica, Romania, Saudi Arabia, Seychelles, Yugoslavia (former)

43—Mongolia

41—Jordan

40—Malaysia, Mauritius

39—Argentina

38—Cuba, Panama

37—Chile, Uruguay, North Korea

34—Costa Rica, South Korea, United Arab Emirates

33—Poland

32—Bulgaria, Hungary, Qatar

31—Soviet Union (former)

29—Bahrain, Hong Kong, Trinidad and Tobago

28—Kuwait, Singapore

25—Czechoslovakia, Portugal, Taiwan

Minimal Human Suffering

21—Israel

19—Greece

16—United Kingdom

12—Italy

11—Barbados, Ireland, Spain, Sweden

8—Finland, New Zealand

7—France, Iceland, Japan, Luxembourg

6—Austria, Germany

5—United States

4—Australia, Norway

3—Canada, Switzerland

2—Belgium, Netherlands

1—Denmark

John Calvin (1509–1564), a French theologian, was an important figure in the Protestant reformation of Christianity. Calvin taught that the ultimate salvation or damnation of every individual had already been decided by God; this idea is called *predestination.* Calvin also suggested that God communicated his decisions to people by making them either successful or unsuccessful during their earthly existence. God gave each person an earthly "calling"—an occupation or profession—and manifested their success or failure through that medium. Ironically, this point of view led Calvin's followers to seek proof of their coming salvation by working hard, saving their money, and generally striving for economic success.

In Weber's analysis, Calvinism provided an important stimulus for the development of capitalism. Rather than "wasting" their money on worldly comforts, the Calvinists reinvested it in their economic enterprises, thus providing the capital necessary for the development of capitalism. In arriving at this interpretation of the origins of capitalism, Weber researched the official doctrines of the early Protestant churches, studied the preaching of Calvin and other church leaders, and examined other relevant historical documents.

In three other studies, Weber conducted detailed historical analyses of Judaism ([1934] 1952) and the religions of China ([1934] 1951) and India ([1934] 1958). Among other things, Weber wanted to know why capitalism had not developed in the ancient societies of China, India, and Israel. In none of the three religions did he find any teaching that would have supported the accumulation and reinvestment of capital—strengthening his conclusion about the role of Protestantism in that regard.

Japanese Religion and Capitalism

Weber's thesis regarding Protestantism and capitalism has become a classic in the social sciences. Not surprisingly, other scholars have attempted to test it in other historical situations. No analysis has been more interesting, however, than Robert Bellah's examination of the growth of capitalism in Japan during the late nineteenth and early twentieth centuries, *Tokugawa Religion* (1957).

As both an undergraduate and a graduate student, Bellah had developed interests in Weber and in Japanese society. Given these two interests, it was perhaps inevitable that he would, in 1951, first conceive his Ph.D. thesis topic as "nothing less than an 'Essay on the Economic Ethic of Japan' to be a companion to Weber's studies of China, India, and Judaism: *The Economic Ethic of the World Religions*" (recalled in Bellah 1967: 168). Originally, Bellah sketched his research design as follows:

> Problems would have to be specific and limited—no general history would be attempted—since time span is several centuries. Field work in Japan on the actual economic ethic practiced by persons in various situations, with, if possible, controlled matched samples from the U.S. (questionnaires, interviews, etc.).
>
> *(1967: 168)*

Bellah's original plan, then, called for surveys of contemporary Japanese and Americans. However, he did not receive the financial support necessary for the study as originally envisioned. So instead, he immersed himself in the historical records of Japanese religion, seeking the roots of the rise of capitalism in Japan.

In the course of several years' research, Bellah uncovered numerous leads. In a 1952 term paper on the subject, Bellah felt he had found the answer in the samurai code of Bushido and in the Confucianism practiced by the samurai class:

> Here I think we find a real development of this worldly asceticism, at least equaling anything found in Europe. Further, in this class the idea of duty in occupation involved achievement without traditionalistic limits, but to the limits of one's capacities, whether in the role of bureaucrat, doctor, teacher, scholar, or other role open to the Samurai.
>
> *(Quoted in Bellah 1967: 171)*

The samurai, however, made up only a portion of Japanese society. So Bellah kept looking at the religions among the Japanese generally. His understanding of the Japanese language was not yet very good, but he wanted to read religious texts in the original. Under these constraints and experiencing increased time pressure, Bellah decided to concentrate his attention on a single group: Shingaku, a religious movement among merchants in the eighteenth and nineteenth centuries. He found that Shingaku had two influences on the development of capitalism. It offered an attitude toward work similar to the Calvinist notion of a "calling," and it had the effect of making business a more acceptable calling for Japanese. Previously, commerce had held a very low standing in Japan.

In other aspects of his analysis, Bellah examined the religious and political roles of the Emperor and the economic impact of periodically appearing emperor cults. Ultimately, Bellah's research pointed to the variety of religious and philosophical factors that laid the groundwork for capitalism in Japan. It seems unlikely that he would have achieved anything approaching that depth of understanding if he had been able to pursue his original plan to interview matched samples of U.S. and Japanese citizens.

I've presented these two studies in some depth to demonstrate the way comparative and historical researchers dig down into the variables relevant to their analyses. Here are a few briefer examples to illustrate some of the topics interesting to comparative and historical scholars today.

- *The Rise of Christianity:* Rodney Stark (1997) lays out his research question in the book's subtitle: *How the Obscure, Marginal Jesus Movement Became the Dominant Religious Force in the Western*

World in a Few Centuries. For many people the answer to this puzzle is a matter of faith in the miraculous destiny of Christianity. Without debunking Christian faith, Stark looks for a scientific explanation, undertaking an analysis of existing historical records that sketch out the population growth of Christianity during its early centuries. He notes, among other things, that the early growth rate of Christianity, rather than being unaccountably rapid, was very similar to the contemporary growth of Mormonism. He then goes on to examine elements in early Christian practice that gave it growth advantages over the predominant paganism of the Roman Empire. For example, the early Christian churches were friendlier to women than paganism was, and much of the early growth occurred among women—who often converted their husbands later on. And in an era of deadly plagues, the early Christians were more willing to care for stricken friends and family members, which not only enhanced the survival of Christians but also made it a more attractive conversion prospect. At every turn in the analysis, Stark makes rough calculations of the demographic impact of cultural factors. This study is an illustration of how social research methods can shed light on nonscientific realms such as faith and religion.

- *Policing World Society:* Mathieu Deflem (2002) set out to learn how contemporary systems of international cooperation among police agencies came about. All of us have heard movie and TV references to Interpol. Deflem went back to the middle of the nineteenth century and traced its development through World War II. In part, his analysis examines the strains between the bureaucratic integration of police agencies in their home governments and the need for independence from those governments.

- *Organizing America:* Charles Perrow (2002) wanted to understand the roots of the uniquely American form of capitalism. Compared with European nations, the United States has shown less interest in providing for the needs of average citizens and has granted greater power to gigantic corporations. Perrow feels the die was pretty much cast by the end of the nineteenth century, resting primarily on Supreme Court decisions in favor of corporations and the experiences of the textile and railroad industries.

- *Diminished Democracy:* Theda Skocpol (2003) turns her attention to something that fascinated Alexis de Tocqueville in his 1840 *Democracy in America:* the grassroots commitment to democracy, which appeared in all aspects of American community life. It almost seemed as though democratic decision making was genetic in the new world, but what happened? Skocpol's analysis of contemporary U.S. culture suggests a "diminished democracy" that cannot be easily explained by the ideologies of either the right or the left.

These examples of comparative and historical research should give you some sense of the potential power of the method. Let's turn now to an examination of the sources and techniques used in this method.

Sources of Comparative and Historical Data

As we saw in the case of existing statistics, there is no end of data available for analysis in historical research. To begin, historians may have already reported on whatever it is you want to examine, and their analyses can give you an initial grounding in the subject, a jumping-off point for more in-depth research.

Most likely you'll ultimately want to go beyond others' conclusions and examine some "raw data" to draw your own conclusions. These data vary, of course, according to the topic under study. In Bellah's study of Tokugawa religion, raw data included the sermons of Shingaku teachers. When W. I. Thomas and Florian Znaniecki (1918) studied the adjustment process for Polish peasants coming to the United States early in this century, they examined letters written by the immigrants to their families in Poland. (They obtained the letters through newspaper advertisements.) Other researchers have analyzed old diaries. Such personal documents only scratch the surface, however. In discussing procedures for studying the history of family life, Ellen Rothman points to the following sources:

> In addition to personal sources, there are public records which are also revealing of family history. Newspapers are especially rich in evidence on the educational, legal, and recreational aspects of family life in the past as seen from a local point of view. Magazines reflect more general patterns of family life; students often find them interesting to explore for data on perceptions and expectations of mainstream family values. Magazines offer several different kinds of sources at once: visual materials (illustrations and advertisements), commentary (editorial and advice columns), and fiction. Popular periodicals are particularly rich in the last two. Advice on many questions of concern to families— from the proper way to discipline children to the economics of wallpaper—fills magazine columns from the early nineteenth century to the present. Stories that suggest common experiences or perceptions of family life appear with the same continuity.
>
> *(1981: 53)*

Organizations generally document themselves, so if you're studying the development of some organization—as Bellah studied Shingaku, for example—you should examine its official documents: charters, policy statements, speeches by leaders, and so on. Once, when I was studying the rise of a contemporary Japanese religious group—Sokagakkai—I discovered not only weekly newspapers and magazines published by the group but also a published collection of all the speeches given by the original leaders. With these sources, I could trace changes in recruitment patterns over time. At the outset, followers were enjoined to enroll all the world. Later, the emphasis shifted specifically to Japan. Once a sizable Japanese membership had been established, an emphasis on enrolling all the world returned (Babbie 1966).

Often, official government documents provide the data needed for analysis. To better appreciate the history of race relations in the United States, A. Leon Higginbotham, Jr. (1978) examined 200 years of laws and court cases involving race. Himself the first African American appointed a federal judge, Higginbotham found that, rather than protecting African Americans, the law embodied bigotry and oppression. In the earliest court cases, there was considerable ambiguity over whether African Americans were indentured servants or, in fact, slaves. Later court cases and laws clarified the matter—holding African Americans to be something less than human.

The sources of data for historical analysis are too extensive to cover even in outline here, though the examples we've looked at should suggest some ideas. Whatever resources you use, however, a couple of cautions are in order.

As we saw in the case of existing statistics, you can't trust the accuracy of records—official or unofficial, primary or secondary. Your protection lies in replication: In the case of historical research, that means corroboration. If several sources point to the same set of "facts," your confidence in them might reasonably increase.

At the same time, you need always be wary of bias in your data sources. If all your data on the development of a political movement are taken from the movement itself, you're unlikely to gain a well-rounded view of it. The diaries of well-to-do gentry of the Middle Ages may not give you an accurate view of life in general during those times. Where possible, obtain data from a variety of sources representing different points of view. Here's what Bellah said regarding his analysis of Shingaku:

> One could argue that there would be a bias in what was selected for notice by Western scholars. However, the fact that there was material from Western scholars with varied interests from a number of countries and over a period of nearly a century reduced the probability of bias.
>
> *(Bellah 1967: 179)*

The issues raised by Bellah are important ones. As Ron Aminzade and Barbara Laslett indicate in "Reading and Evaluating Documents," there is an art to knowing how to regard such documents and what to make of them.

Reading and Evaluating Documents

Ron Aminzade
and Barbara Laslett

University of Minnesota

The purpose of the following comments is to give you some sense of the kind of interpretive work historians do and the critical approach they take toward their sources. It should help you to appreciate some of the skills historians develop in their efforts to reconstruct the past from residues, to assess the evidentiary status of different types of documents, and to determine the range of permissible inferences and interpretations. Here are some of the questions historians ask about documents:

1. Who composed the documents? Why were they written? Why have they survived all these years? What methods were used to acquire the information contained in the documents?

2. What are some of the biases in the documents and how might you go about checking or correcting them? How inclusive or representative is the sample of individuals, events, and so on, contained in the document? What were the institutional constraints and the general organizational routines under which the document was prepared? To what extent does the document provide more of an index of institutional activity than of the phenomenon being studied? What is the time lapse between the observation of the events documented and the witnesses' documentation of them? How confidential or public was the document meant to be? What role did etiquette, convention, and custom play in the presentation of the material contained within the document? If you relied solely upon the evidence contained in these documents, how might your vision of the past be distorted? What other kinds of documents might you look at for evidence on the same issues?

3. What are the key categories and concepts used by the writer of the document to organize the information presented? What selectivities or silences result from these categories of thought?

4. What sorts of theoretical issues and debates do these documents cast light on? What kinds of historical and/or sociological questions do they help to answer? What sorts of valid inferences can one make from the information contained in these documents? What sorts of generalizations can one make on the basis of the information contained in these documents?

Incidentally, the critical review that Aminzade and Laslett urge for the reading of historical documents is useful in many areas of your life besides the pursuit of comparative and historical research. Consider applying some of their questions to presidential press conferences, advertising, or (gasp) college textbooks. None of these offers a direct view of reality; all have human authors and human subjects.

Analytic Techniques

The analysis of comparative and historical data is another large subject that I can't cover exhaustively here. Moreover, because comparative and historical research is usually a qualitative method, there are no easily listed steps to follow in the analysis of historical data. Nevertheless, a few comments are in order.

Max Weber used the German term *verstehen*— "understanding"—in reference to an essential quality of social research. He meant that the researcher must be able to take on, mentally, the circumstances, views, and feelings of those being studied, so that the researcher can interpret their actions appropriately. Certainly this concept applies to comparative and historical research. The researcher's imaginative understanding is what breathes life and meaning into the evidence being analyzed.

The comparative and historical researcher must find patterns among the voluminous details describing the subject matter of study. Often this takes the form of what Weber called *ideal types:* conceptual models composed of the essential characteristics of social phenomena. Thus, for example, Weber himself did considerable research on bureaucracy. Having observed numerous actual bureaucracies, Weber ([1925] 1946) detailed those qualities essential to bureaucracies in general: jurisdictional areas,

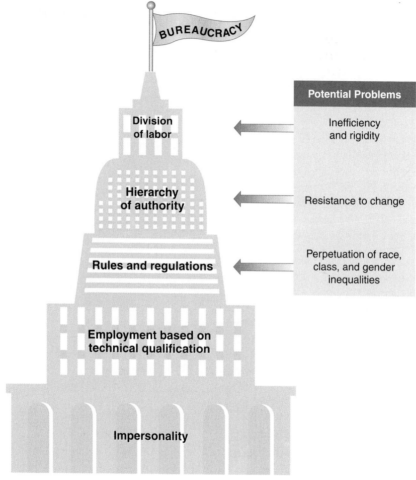

FIGURE 11-4

Some Positive and Negative Aspects of Bureaucracy

Source: Diana Kendall, *Sociology in Our Times*, 5th ed. (Belmont, CA: Wadsworth, ©2005). Used by permission.

hierarchically structured authority, written files, and so on. Weber did not merely list those characteristics common to all the actual bureaucracies he observed. Rather, to create a theoretical model of the "perfect" (ideal type) bureaucracy, he needed to understand fully the essentials of bureaucratic operation. Figure 11-4 offers a more recent, graphic portrayal of some positive and negative aspects of bureaucracy as a general social phenomenon.

Often, comparative and historical research is informed by a particular theoretical paradigm. Thus, Marxist scholars may undertake historical analyses of particular situations—such as the history of Latinos and Latinas in the United States—to determine whether they can be understood in terms of the Marxist version of conflict theory. Sometimes, comparative and historical researchers attempt to replicate prior studies in new situations—for example, Bellah's study of Tokugawa religion in the context of Weber's studies of religion and economics.

Although comparative and historical research is often regarded as a qualitative rather than quantitative technique, this is by no means necessary.

Historical analysts sometimes use time-series data to monitor changing conditions over time, such as data on population, crime rates, unemployment, infant mortality rates, and so forth. The analysis of such data sometimes requires sophistication, however. For example, Larry Isaac and Larry Griffin (1989) discuss the uses of a variation on regression techniques (see Chapter 16) in determining the meaningful breaking points in historical processes, as well as for specifying the periods within which certain relationships occur among variables. Criticizing the tendency to regard history as a steadily unfolding process, the authors focus their attention on the statistical relationship between unionization and the frequency of strikes, demonstrating that the relationship has shifted importantly over time.

Isaac and Griffin raise several important issues regarding the relationship among theory, research methods, and the "historical facts" they address. Their analysis, once again, warns against the naive assumption that history as documented necessarily coincides with what actually happened.

Ethics and Unobtrusive Measures

The use of unobtrusive measures avoids many of the ethical issues we've discussed in connection with other data-collection techniques, but if you reflect on the general principles we've discussed, I think you'll see that there are potential risks to guard against.

The general principle of confidentiality may be relevant in some projects, for example. Let's suppose you want to examine an immigrant subculture through a content analysis of letters written back to the old country, as was the case in the Thomas and Znaniecki (1918) study of Polish peasants, mentioned earlier in the chapter. To begin, you should obtain those letters legally and ethically (no getting a government agency to intercept the letters for you), and you need to protect the privacy of the letter writers and recipients.

As with all other research techniques, you're obliged to collect data, analyze them, and report your findings honestly, with the purpose of discovering what's so rather than attempting to support a favored hypothesis or personal agenda. While it may be easy to agree with such a principle, you're likely to find it somewhat more difficult to apply when you actually conduct research. Your ethical sensibilities will be more challenged by the vast gray areas than by those of black and white.

MAIN POINTS

Introduction

- Unobtrusive measures are ways of studying social behavior without affecting it in the process.

Content Analysis

- Content analysis is a social research method appropriate for studying human communications through social artifacts. Researchers can use it to study not only communication processes but other aspects of social behavior as well.
- Common units of analysis in content analysis include elements of communications—words, paragraphs, books, and so forth. Standard probability-sampling techniques are sometimes appropriate in content analysis.
- Content analysis involves coding—transforming raw data into categories based on some conceptual scheme. Coding may attend to both manifest and latent content. The determination of latent content requires judgments by the researcher.
- Both quantitative and qualitative techniques are appropriate for interpreting content analysis data.
- The advantages of content analysis include economy, safety, and the ability to study processes occurring over a long time. Its disadvantages are that it is limited to recorded communications and can raise issues of reliability and validity.

Analyzing Existing Statistics

- A variety of government and nongovernment agencies provide aggregate statistical data for studying aspects of social life.
- Problems of validity in the analysis of existing statistics can often be handled through logical reasoning and replication.
- Existing statistics often have problems of reliability, so they must be used with caution.

Comparative and Historical Research

- Social scientists use comparative and historical methods to discover patterns in the histories of different cultures.
- Although often regarded as a qualitative method, comparative and historical research can make use of quantitative techniques.

Ethics and Unobtrusive Measures

- Sometimes even unobtrusive measures can raise the possibility of violating subjects' privacy.
- The general principles of honest observation, analysis, and reporting apply to all research techniques.

The following terms are defined in context in the chapter and at the bottom of the page where the term is introduced, as well as in the comprehensive glossary at the back of the book.

coding	latent content
comparative and historical research	manifest content
content analysis	unobtrusive research

PROPOSING SOCIAL RESEARCH: UNOBTRUSIVE RESEARCH

This chapter has provided an overview of three major types of unobtrusive research: content analysis, analyzing existing statistics, and comparative and historical research. While existing statistics represent, by their nature, a quantitative method, the other two can be done with a qualitative and/or quantitative approach. In this exercise, you need to identify which method and orientation you'll use. If you're doing these exercises in order to understand the topics of the book better, you could try your hand at each of these methods.

You need to describe the data you'll use and detail anything special about your access to those data. Whether you're studying newspaper editorials, infant mortality rates, or accounts of political revolutions, you'll likely face potential problems of validity and reliability. Unobtrusive methods involve the use of available data, which often offer approximations of the observations you might ideally like to make. For example, you may need to use drug-arrest rates as an approximation of drug-use rates. You should discuss how you'll deal with any such approximations.

REVIEW QUESTIONS AND EXERCISES

1. Outline a content analysis design to determine whether the Republican or the Democratic party is the more supportive of a basic constitutional right such as free speech, freedom of religion, or protection against self-incrimination. Be sure to specify units of analysis and sampling methods. Describe a coding scheme that you could use for the content analysis.

2. Identify an international news story involving a conflict between two nations or cultural groups, such as clashes between Israelis and Palestinians. Locate on the web a newspaper report of the event from within each of the countries or cultures involved. Note differences in the way the event is reported. Now, find a report of the event in a newspaper in a third, distant country. (For example, compare reports from the *Jerusalem Post,* the *Palestine Chronicle,* and the *New York Times.*) Does the third report seem to favor one of the two original reports? If so, would you conclude that the third report is biased toward one side or that one of the original reports was simply inaccurate? Explain how and why you reached that conclusion. (You might use *World Press Review* as an alternative source of data; they present contrasting articles on a given story. See the link on this book's website.)

3. Using the web, find out how many countries have a higher "expected life expectancy" than the United States does. (You might want to try the Population Reference Bureau at the link on this book's website.)

4. Max Weber undertook extensive studies of some of the world's major religions. Create an annotated bibliography of his works in this area.

5. On the web, locate the American Sociological Association's section called "Comparative and Historical Sociology" (check out the link on this book's website). Summarize an article in the section's newsletter.

SPSS EXERCISES

See the booklet that accompanies your text for exercises using SPSS (Statistical Package for the Social Sciences). There are exercises offered for each chapter, and you'll also find a detailed primer on using SPSS.

Online Study Resources

If your book came with an access code card, visit www.cengage.com/login to register. To purchase access, please visit www.ichapters.com.

1. Before you do your final review of the chapter, take the *CengageNOW* pretest to help identify the areas on which you should concentrate. You'll find information on this online tool, as well as instructions on how to access all of its great resources, in the front of the book.

2. As you review, take advantage of the *CengageNOW* personalized study plan, based on your quiz results. Use this study plan with its interactive exercises and other resources to master the material.

3. When you're finished with your review, take the posttest to confirm that you're ready to move on to the next chapter.

WEBSITE FOR *THE PRACTICE OF SOCIAL RESEARCH* 12TH EDITION

Go to your book's website at www.cengage.com/sociology/babbie for tools to aid you in studying for your exams. You'll find *Tutorial Quizzes* with feedback, *Internet Exercises, Flash Cards, Glossaries,* and *Essay Quizzes,* as well as *InfoTrac College Edition* search terms, suggestions for additional reading, *Web Links,* and primers for using data-analysis software such as SPSS.

CHAPTER TWELVE

Evaluation Research

Now you're going to see one of the most rapidly growing uses of social research: the evaluation of social interventions. You'll come away from this chapter able to judge whether social programs have succeeded or failed.

CengageNOW for Sociology

Use this online tool to help you make the grade on your next exam. After reading this chapter, go to "Online Study Resources" at the end of the chapter for instructions on how to benefit from *CengageNOW*.

Introduction

You may not be familiar with *Twende na Wakati* ("Let's Go with the Times"), but it's the most popular radio show in Tanzania. It's a soap opera. The main character, Mkwaju, is a truck driver with some pretty traditional ideas about gender roles and sex. By contrast, Fundi Mitindo, a tailor, and his wife, Mama Waridi, have more-modern ideas regarding the roles of men and women, particularly in relation to the issues of overpopulation and family planning.

Twende na Wakati was the creation of Population Communications International (PCI) and other organizations working in conjunction with the Tanzanian government in response to two problems facing that country: (1) a population growth rate over twice that of the rest of the world and (2) an AIDS epidemic particularly heavy along the international truck route, where more than a fourth of the truck drivers and over half the commercial sex workers were found to be HIV positive in 1991. The prevalence of contraceptive use was 11 percent (Rogers et al. 1996: 5–6).

The purpose of the soap opera was to bring about a change in knowledge, attitudes, and practices (KAP) relating to contraception and family planning. Rather than instituting a conventional educational campaign, PCI felt it would be more effective to illustrate the message through entertainment.

Between 1993 and 1995, 208 episodes of *Twende na Wakati* were aired, aiming at the 67 percent of Tanzanians who listen to the radio. Eighty-four percent of the radio listeners reported listening to the PCI soap opera, making it the most popular show in the country. Ninety percent of the show's listeners recognized Mkwaju, the sexist truck driver, and only 3 percent regarded him as a positive role model. Over two-thirds identified Mama Waridi, a businesswoman, and her tailor husband as positive role models.

Surveys conducted to measure the impact of the show indicated it had affected knowledge, attitudes, and behavior. For example, 49 percent of the married women who listened to the show said they now practiced family planning, compared with only 19 percent of the nonlisteners. There were other impacts:

> Some 72 percent of the listeners in 1994 said that they adopted an HIV/AIDS prevention behavior because of listening to "Twende na Wakati," and this percentage increased to 82 percent in our 1995 survey. Seventy-seven percent of these individuals adopted monogamy, 16 percent began using condoms, and 6 percent stopped sharing razors and/or needles.
>
> *(Rogers et al. 1996: 21)*

We can judge the effectiveness of the soap opera because of a particular form of social science. *Evaluation research* refers to a research purpose rather than a specific method. This purpose is to evaluate the impact of social interventions such as new teaching methods or innovations in parole. Many methods—surveys, experiments, and so on—can be used in evaluation research.

Evaluation research is appropriate whenever some social intervention occurs or is planned. A *social intervention* is an action taken within a social context for the purpose of producing some intended result. In its simplest sense, **evaluation research** is the process of determining whether a social intervention has produced the intended result. Peter Rossi, Mark Lipsey, and Howard Freeman (2002: 4) define it as follows:

> *Program evaluation is the use of social research procedures to systematically investigate the effectiveness of social intervention programs.* More specifically, evaluation researchers [evaluators] use social research methods to study, appraise, and help improve social programs in all their important aspects, including the diagnosis of the social

evaluation research Research undertaken for the purpose of determining the impact of some social intervention, such as a program aimed at solving a social problem.

364 ■ Chapter 12: Evaluation Research

problems they address, their conceptualization and design, their implementation and administration, their outcomes, and their efficiency.

Evaluation research is probably as old as social research itself. Whenever people have instituted a social reform for a specific purpose, they have paid attention to its actual consequences, even if they have not always done so in a conscious, deliberate, or sophisticated fashion. In recent years, however, the field of evaluation research has become an increasingly popular and active research specialty, as reflected in textbooks, courses, and projects. Moreover, the growth of evaluation research points to a more general trend in the social sciences. As a researcher, you'll likely be asked to conduct evaluations of your own.

In part, the growth of evaluation research reflects social researchers' increasing desire to make a difference in the world. At the same time, we can't discount the influence of (1) an increase in federal requirements that program evaluations must accompany the implementation of new programs and (2) the availability of research funds to fulfill those requirements. In any case, it seems clear that social researchers will be bringing their skills into the real world more than ever before.

This chapter looks at some of the key elements in this form of social research. After considering the kinds of topics commonly subjected to evaluation, we'll move through some of its main operational aspects: measurement, study design, and execution. As you'll see, formulating questions is as important as answering them. Because it occurs within real life, evaluation research has its own problems, some of which we'll examine. Logistical problems arise from evaluation research generally and from its specific, technical procedures. The use of research results also presents certain concerns. As you review reports of program evaluations, you should be especially sensitive to these issues.

Evaluation is a form of applied research—that is, it's intended to have some real-world effect. It will be useful, therefore, to consider whether and how it's actually applied. As you'll see, the obvious implications of an evaluation research project do not necessarily affect real life. They may become the focus of ideological, rather than scientific, debates. They may simply be denied out of hand, for political or other reasons. Perhaps most typically, they may simply be ignored and forgotten, left to collect dust in bookcases across the land.

The chapter continues with a look at a particular resource for large-scale evaluation—social indicators research. This type of research is also a rapidly growing specialty. Essentially it involves the creation of aggregated indicators of the "health" of society, similar to the economic indicators that give diagnoses and prognoses of economies. The chapter then concludes with a look at the special ethical concerns that arise in evaluation research.

Topics Appropriate for Evaluation Research

The topics appropriate for evaluation research are limitless. When the federal government abolished the selective service system (the draft), military researchers began paying special attention to the impact on enlistment. As individual states have liberalized their marijuana laws, researchers have sought to learn the consequences, both for marijuana use and for other forms of social behavior. Do no-fault divorce reforms increase the number of divorces, and do related social problems decrease or increase? Has no-fault automobile insurance really brought down insurance policy premiums? Agencies providing foreign aid also conduct evaluations to determine whether the desired effects were produced. Has the "No Child Left Behind" program improved the quality of education in America? Have "Just Say No" abstinence programs reduced rates of sexual activity and pregnancies among young people? These are the kinds of questions that evaluation research can address.

There are many variations in the intent of evaluation research. **Needs assessment studies** aim to determine the existence and extent of problems, typically among a segment of the population,

needs assessment studies Studies that aim to determine the existence and extent of problems, typically among a segment of the population, such as the elderly.

such as the elderly. **Cost-benefit studies** determine whether the results of a program justify its expense (both financial and other). **Monitoring studies** provide a steady flow of information about something of interest, such as crime rates or the outbreak of an epidemic. Sometimes the monitoring involves incremental interventions. Read this description of "adaptive management" by the Nature Conservancy, a public-interest group seeking to protect natural areas:

> First, partners assess assumptions and set management goals for the conservation area. Based on this assessment, the team takes action, then monitors the environment to see how it responds. After measuring results, partners refine their assumptions, goals and monitoring regimen to reflect what they've learned from past experiences. With refinements in place, the entire process begins again.
>
> *(2005: 3)*

Much of evaluation research is referred to as **program evaluation** or **outcome assessment:** the determination of whether a social intervention is producing the intended result. Here's an example.

Some years ago, a project evaluating the nation's drivers' education programs, conducted by the National Highway and Transportation Safety Administration (NHTSA), stirred up a controversy. Philip Hilts (1981: 4) reported on the study's findings:

> For years the auto insurance industry has given large insurance discounts for children who take drivers' education courses, because statistics show that they have fewer accidents.
>
> The preliminary results of a new major study, however, indicate that drivers' education does not prevent or reduce the incidence of traffic accidents at all.

Based on an analysis of 17,500 young people in DeKalb County, Georgia (including Atlanta), the preliminary findings indicated that students who took drivers' education had just as many accidents and traffic violations as those who didn't take it. The study also seemed to reveal some subtle aspects of driver training.

First, it suggested that the apparent impact of drivers' education was largely a matter of self-selection. The kind of students who took drivers' education were less likely to have accidents and traffic violations—with or without driver training. Students with high grades, for example, were more likely to sign up for driver training, and they were also less likely to have accidents.

More startling, however, was the suggestion that driver-training courses may have actually increased traffic accidents! The existence of drivers' education may have encouraged some students to get their licenses earlier than if there were no such courses. In a study of ten Connecticut towns that discontinued driver training, about three-fourths of those who probably would have been licensed through their classes delayed getting licenses until they were 18 or older (Hilts 1981: 4).

As you might imagine, these results were not well received by those most closely associated with driver training. This matter was complicated, moreover, by the fact that the NHTSA study was also evaluating a new, more intensive training program—and the preliminary results showed that the new program was effective.

Here's a very different example of evaluation research. Rudolf Andorka, a Hungarian sociologist, has been particularly interested in his country's shift to a market economy. Even before the dramatic events in Eastern Europe in 1989, Andorka and his colleagues had been monitoring the nation's "second economy"—jobs pursued outside the socialist economy. Their surveys followed the rise and fall of such jobs and examined their impact within Hungarian society. One conclusion was that "the second economy, which earlier probably tended to diminish income inequalities or at least improved the standard of living of the poorest part of the

---

cost-benefit studies Studies that determine whether the results of a program justify its expense (both financial and other).

monitoring studies Studies that provide a steady flow of information about something of interest, such as crime rates or the outbreak of an epidemic.

program evaluation/outcome assessment The determination of whether a social intervention is producing the intended result.

population, in the 1980s increasingly contributed to the growth of inequalities" (Andorka 1990: 111).

Whereas evaluation research is basically a matter of discovering whether social interventions make a difference, it is not surprising that it is sometimes coupled with the intentions of participatory action research, discussed in Chapter 10. Since PAR has been particularly strong among Australian researchers, it's not surprising to find Wayne Miller and June Lennie (2005) speaking of "empowerment evaluation" to characterize their assessment of a national school-breakfast program. They say that this approach is

> distinguished by its clearly articulated underlying principles that allow for the extensive participation of program management and staff, funders, community members and other stakeholders in all stages of the evaluation. This approach can build evaluation capacities, give voice to a diversity of people involved, and enable open and honest discussion about the strengths and weaknesses of key program activities. It also enables collaborative planning and identification of the documentation or evidence required to assess the goals and strategies that participants develop to improve key program activities. The ultimate aim is for evaluation to become a normal part of planning and managing programs, resulting in ongoing improvement and learning.
>
> *(2005: 18)*

As you can see, the questions appropriate to evaluation research are of great practical significance: Jobs, programs, and investments as well as beliefs and values are at stake. Let's now examine how these questions are answered—how evaluations are conducted.

Formulating the Problem: Issues of Measurement

Several years ago, I headed an institutional research office that conducted research directly relevant to the operation of the university. Often,

we were asked to evaluate new programs in the curriculum. The following description is fairly typical of the problem that arose in that context, and it points to one of the key barriers to good evaluation research.

Faculty members would appear at my office to say they'd been told by the university administration to arrange for an evaluation of the new program they had permission to try. This points to a common problem: Often the people whose programs are being evaluated aren't thrilled at the prospect. For them, an independent evaluation threatens the survival of the program and perhaps even their jobs.

The main problem I want to introduce, however, has to do with the purpose of the intervention to be evaluated. The question "What is the intended result of the new program?" often produced a vague response such as "Students will get an in-depth and genuine understanding of mathematics, instead of simply memorizing methods of calculations." Fabulous! And how could we measure that "in-depth and genuine understanding"? Often, I was told that the program aimed at producing something that could not be measured by conventional aptitude and achievement tests. No problem there; that's to be expected when we're innovating and being unconventional. What would be an unconventional measure of the intended result? Sometimes this discussion came down to an assertion that the effects of the program would be "unmeasurable."

There's the common rub in evaluation research: measuring the "unmeasurable." Evaluation research is a matter of finding out whether something is there or not there, whether something happened or didn't happen. To conduct evaluation research, we must be able to operationalize, observe, and recognize the presence or absence of what is under study.

Often, intended outcomes can be derived from published program documents. Thus, when Edward Howard and Darlene Norman (1981) evaluated the performance of the Vigo County Public Library in Indiana, they began with the statement of purpose previously adopted by the library's Board of Trustees.

To acquire by purchase or gift, and by recording and production, relevant and potentially useful information that is produced by, about, or for the citizens of the community;

To organize this information for efficient delivery and convenient access, furnish the equipment necessary for its use, and provide assistance in its utilization; and

To effect maximum use of this information toward making the community a better place in which to live through aiding the search for understanding by its citizens.

(1981: 306)

As the researchers said, "Everything that VCPL does can be tested against the Statement of Purpose." They then set about creating operational measures for each of the purposes.

Although "official" purposes of interventions are often the key to designing an evaluation, they may not always be sufficient. Anna-Marie Madison (1992: 38), for example, warns that programs designed to help disadvantaged minorities do not always reflect what the proposed recipients of the aid may need and desire:

The cultural biases inherent in how middle-class white researchers interpret the experiences of low-income minorities may lead to erroneous assumptions and faulty propositions concerning causal relationships, to invalid social theory, and consequently to invalid program theory. Descriptive theories derived from faulty premises, which have been legitimized in the literature as existing knowledge, may have negative consequences for program participants.

In setting up an evaluation, then, researchers must pay careful attention to issues of measurement. Let's take a closer look at the types of measurements that evaluation researchers must deal with.

Specifying Outcomes

As I've already suggested, a key variable for evaluation researchers to measure is the outcome, or what is called the *response variable*. If a social pro-

gram is intended to accomplish something, we must be able to measure that something. If we want to reduce prejudice, we need to be able to measure prejudice. If we want to increase marital harmony, we need to be able to measure that.

It's essential to achieve agreements on definitions in advance:

The most difficult situation arises when there is disagreement as to standards. For example, many parties may disagree as to what defines serious drug abuse—is it defined best as 15% or more of students using drugs weekly, 5% or more using hard drugs such as cocaine or PCP monthly, students beginning to use drugs as young as seventh grade, or some combination of the dimensions of rate of use, nature of use, and age of user? . . . Applied researchers should, to the degree possible, attempt to achieve consensus from research consumers in advance of the study (e.g., through advisory groups) or at least ensure that their studies are able to produce data relevant to the standards posited by all potentially interested parties.

(Hedrick, Bickman, and Rog 1993: 27)

In some cases you may find that the definitions of a problem and a sufficient solution are defined by law or by agency regulations; if so, you must be aware of such specifications and accommodate them. Moreover, whatever the agreed-on definitions, you must also achieve agreement on how the measurements will be made. Because there are different possible methods for estimating the percentage of students "using drugs weekly," for example, you'd have to be sure that all the parties involved understood and accepted the method(s) you've chosen.

In the case of the Tanzanian soap opera, there were several outcome measures. In part, the purpose of the program was to improve knowledge about both family planning and AIDS. Thus, for example, one show debunked the belief that the AIDS virus was spread by mosquitoes and could be avoided by the use of insect repellant. Studies of listeners showed a reduction in that belief (Rogers et al. 1996: 21).

PCI also wanted to change Tanzanian attitudes toward family size, gender roles, HIV/AIDS, and other related topics; the research indicated that the show had affected these as well. Finally, the program aimed at affecting behavior. We've already seen that radio listeners reported changing their behavior with regard to AIDS prevention. They reported a greater use of family planning as well. However, because there's always the possibility of a gap between what people say they do and what they actually do, the researchers sought independent data to confirm their conclusions.

Tanzania's national AIDS-control program had been offering condoms free of charge to citizens. In the areas covered by the soap opera, the number of condoms given out increased sixfold between 1992 and 1994. This far exceeded the increase of 1.4 times in the control area, where broadcasters did not carry the soap opera.

Measuring Experimental Contexts

Measuring the dependent variables that are directly involved in the experimental program is only a beginning. As Henry Riecken and Robert Boruch (1974: 120–21) point out, it's often appropriate and important to measure those aspects of the context of an experiment researchers think might affect the experiment. Though external to the experiment itself, some variables may affect it.

Suppose, for example, that you were conducting an evaluation of a program aimed at training unskilled people for employment. The primary outcome measure would be their success at gaining employment after completing the program. You would, of course, observe and calculate the subjects' employment rate, but you should also determine what has happened to the employment/ unemployment rates of society at large during the evaluation. A general slump in the job market should be taken into account in assessing what might otherwise seem a pretty low employment rate for subjects. Or, if all the experimental subjects get jobs following the program, you should

consider any general increase in available jobs. Combining complementary measures with proper control-group designs should allow you to pinpoint the effects of the program you're evaluating.

Specifying Interventions

Besides making measurements relevant to the outcomes of a program, researchers must measure the program intervention—the experimental stimulus. In part, this measurement will be handled by the assignment of subjects to experimental and control groups, if that's the research design. Assigning a person to the experimental group is the same as scoring that person yes on the stimulus, and assignment to the control group represents a score of no. In practice, however, it's seldom that simple.

Let's stick with the job-training example. Some people will participate in the program; others will not. But imagine for a moment what job-training programs are probably like. Some subjects will participate fully; others will miss a lot of sessions or fool around when they are present. So you may need measures of the extent or quality of participation in the program. If the program is effective, you should find that those who participated fully have higher employment rates than those who participated less do.

Other factors may further confound the administration of the experimental stimulus. Suppose we're evaluating a new form of psychotherapy designed to cure sexual impotence. Several therapists administer it to subjects composing an experimental group. We plan to compare the recovery rate of the experimental group with that of a control group, which receives some other therapy or none at all. It may be useful to include the names of the therapists treating specific subjects in the experimental group, because some may be more effective than others. If this turns out to be the case, we must find out why the treatment worked better for some therapists than for others. What we learn will further develop our understanding of the therapy itself.

Specifying the Population

In evaluating an intervention, it's important to define the population of possible subjects for whom the program is appropriate. Ideally, all or a sample of appropriate subjects will then be assigned to experimental and control groups as warranted by the study design. Defining the population, however, can itself involve specifying measurements. If we're evaluating a new form of psychotherapy, for example, it's probably appropriate for people with mental problems. But how will "mental problems" be defined and measured? The job-training program mentioned previously is probably intended for people who are having trouble finding work, but what counts as "having trouble"?

Beyond defining the relevant population, then, the researcher should make fairly precise measurements of the variables considered in the definition. For example, even though the randomization of subjects in the psychotherapy study would ensure an equal distribution of those with mild and those with severe mental problems into the experimental and control groups, we'd need to keep track of the relative severity of different subjects' problems in case the therapy turns out to be effective only for those with mild disorders. Similarly, we should measure such demographic variables as gender, age, race, and so forth in case the therapy works only for women, the elderly, or some other group.

New versus Existing Measures

In providing for the measurement of these different kinds of variables, the researcher must continually choose whether to create new measures or use ones already devised by others. If a study addresses something that's never been measured before, the choice is easy. If it addresses something that others have tried to measure, the researcher will need to evaluate the relative worth of various existing measurement devices in terms of her or his specific research situations and purpose. Recall that this is a general issue in social research that applies well beyond evaluation research. Let's briefly compare creating new measures and using existing ones.

Creating measurements specifically for a study can offer greater relevance and validity than using existing measures would. If the psychotherapy we're evaluating aims at a specific aspect of recovery, we can create measures that pinpoint that aspect. We might not be able to find any standardized psychological measures that hit that aspect right on the head. However, creating our own measure will cost us the advantages to be gained from using preexisting measures. Creating good measures takes time and energy, both of which could be saved by adopting an existing technique. Of greater scientific significance, measures that have been used frequently by other researchers carry a body of possible comparisons that might be important to our evaluation. If the experimental therapy raises scores by an average of ten points on a standardized test, we'll be in a position to compare that therapy with others that had been evaluated using the same measure. Finally, measures with a long history of use usually have known degrees of validity and reliability, but newly created measures will require pretesting or will be used with considerable uncertainty.

Operationalizing Success/Failure

Potentially one of the most taxing aspects of evaluation research is determining whether the program under review succeeded or failed. The purpose of a foreign language program may be to help students better learn the language, but how much better is enough? The purpose of a conjugal visit program at a prison may be to raise morale, but how high does morale need to be raised to justify the program?

As you may anticipate, clear-cut answers to questions like these almost never arrive. This dilemma has surely been the source of what is generally called *cost-benefit analysis.* How much does the program cost in relation to what it returns in benefits? If the benefits outweigh the cost, keep the program going. If the reverse, junk it. That's simple enough, and it seems to apply in straightforward economic situations: If it cost you $20 to produce something and you can sell it for only $18, there's no way you can make up the difference in volume.

Unfortunately, the situations faced by evaluation researchers are seldom amenable to straightforward economic accounting. The foreign language program may cost the school district $100 per student, and it may raise students' performances on tests by an average of 15 points. Because the test scores can't be converted into dollars, there's no obvious ground for weighing the costs and benefits.

Sometimes, as a practical matter, the criteria of success and failure can be handled through competition among programs. If a different foreign language program costs only $50 per student and produces an increase of 20 points in test scores, it will undoubtedly be considered more successful than the first program—assuming that test scores are seen as an appropriate measure of the purpose of both programs and the less expensive program has no unintended negative consequences.

Ultimately, the criteria of success and failure are often a matter of agreement. The people responsible for the program may commit themselves in advance to a particular outcome that will be regarded as an indication of success. If that's the case, all you need to do is make absolutely certain that the research design will measure the specified outcome. I mention this obvious requirement simply because researchers sometimes fail to meet it, and there's little or nothing more embarrassing than that.

In summary, researchers must take measurement quite seriously in evaluation research, carefully determining all the variables to be measured and getting appropriate measures for each. As I've implied, however, such decisions are typically not purely scientific ones. Evaluation researchers often must work out their measurement strategy with the people responsible for the program being evaluated. It usually doesn't make sense to determine whether a program achieves Outcome X when its purpose is to achieve Outcome Y. (Realize, however, that evaluation designs sometimes have the purpose of testing for unintended consequences.)

There is a political aspect to these choices, also. Because evaluation research often affects other people's professional interests—their pet program may be halted, or they may be fired or lose professional standing—the results of evaluation research are often argued about.

Let's turn now to some of the research designs commonly employed by evaluators.

Types of Evaluation Research Designs

As I noted at the start of this chapter, evaluation research is not itself a method, but rather one application of social research methods. As such, it can involve any of several research designs. Here we'll consider three main types of research design that are appropriate for evaluations: experimental designs, quasi-experimental designs, and qualitative evaluations.

Experimental Designs

Many of the experimental designs introduced in Chapter 8 can be used in evaluation research. By way of illustration, let's see how the classical experimental model might be applied to our evaluation of a new psychotherapy treatment for sexual impotence.

In designing our evaluation, we should begin by identifying a population of patients appropriate for the therapy. This identification might be made by researchers experimenting with the new therapy. Let's say we're dealing with a clinic that already has 100 patients being treated for sexual impotence. We might take that group and the clinic's definition of sexual impotence as a starting point, and we should maintain any existing assessments of the severity of the problem for each specific patient.

For purposes of evaluation research, however, we would need to develop a more specific measure of impotence. Maybe it would involve whether patients have sexual intercourse at all within a specified time, how often they have intercourse, or whether and how often they reach orgasm. Alternatively, the outcome measure might be based on the assessments of independent therapists not involved in the therapy who interview the patients later. In any event, we would need to agree on the measures to be used.

In the simplest design, we would assign the 100 patients randomly to experimental and control groups; the former would receive the new therapy, and the latter would be taken out of therapy altogether during the experiment. Because ethical practice would probably prevent withdrawing therapy altogether from the control group, however, it's more likely that the control group would continue to receive their conventional therapy.

Having assigned subjects to the experimental and control groups, we would need to agree on the length of the experiment. Perhaps the designers of the new therapy feel it ought to be effective within two months, and an agreement could be reached. The duration of the study doesn't need to be rigid, however. One purpose of the experiment and evaluation might be to determine how long it actually takes for the new therapy to be effective. Conceivably, then, an agreement could be struck to measure recovery rates weekly, say, and let the ultimate length of the experiment rest on a continual review of the results.

Let's suppose the new therapy involves showing pornographic movies to patients. We'd need to specify that stimulus. How often would patients see the movies, and how long would each session be? Would they see the movies in private or in groups? Should therapists be present? Perhaps we should observe the patients while the movies are being shown and include our observations among the measurements of the experimental stimulus. Do some patients watch the movies eagerly but others look away from the screen? We'd have to ask these kinds of questions and create specific measurements to address them.

Having thus designed the study, all we have to do is "roll 'em." The study is set in motion, the observations are made and recorded, and the mass of data is accumulated for analysis. Once the study has run its course, we can determine whether the new therapy had its intended—or perhaps some unintended—consequences. We can tell whether the movies were most effective for mild problems or severe ones, whether they worked for young subjects but not older ones, and so forth.

This simple illustration shows how the standard experimental designs presented in Chapter 8 can be used in evaluation research. Many, perhaps most, of the evaluations reported in the research literature don't look exactly like this illustration, however. Because it's nested in real life, evaluation research often calls for quasi-experimental designs. Let's see what this means.

Quasi-Experimental Designs

Quasi experiments are distinguished from "true" experiments primarily by the lack of random assignment of subjects to an experimental and a control group. In evaluation research, it's often impossible to achieve such an assignment of subjects. Rather than forgo evaluation altogether, researchers sometimes create designs that give some evaluation of the program in question. This section describes some of these designs.

Time-Series Designs

To illustrate the **time-series design**—which involves measurements taken over time—I'll begin by asking you to assess the meaning of some hypothetical data. Suppose I come to you with what I say is an effective technique for getting students to participate in classroom sessions of a course I'm teaching. To prove my assertion, I tell you that on Monday only four students asked questions or made a comment in class; on Wednesday I devoted the class time to an open discussion of a controversial issue raging on campus; and on Friday, when we returned to the subject matter of the course, eight students asked questions or made comments. In other words, I contend, the discussion of

---

quasi experiments   Nonrigorous inquiries somewhat resembling controlled experiments but lacking key elements such as pre- and posttesting and/or control groups.

time-series design   A research design that involves measurements made over some period, such as the study of traffic accident rates before and after lowering the speed limit.

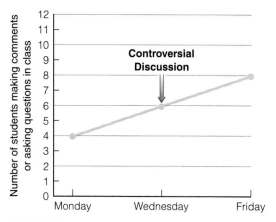

FIGURE 12-1

Two Observations of Class Participation: Before and After an Open Discussion

a controversial issue on Wednesday has doubled classroom participation. This simple set of data is presented graphically in Figure 12-1.

Have I persuaded you that the open discussion on Wednesday has had the consequence I claim for it? Probably you'd object that my data don't prove the case. Two observations (Monday and Friday) aren't really enough to prove anything. Ideally I should have had two classes, with students assigned randomly to each, held an open discussion in only one, and then compared the two on Friday. But I don't have two classes with random assignment of students. Instead, I've been keeping a record of class participation throughout the semester for the one class. This record allows you to conduct a time-series evaluation.

Figure 12-2 presents three possible patterns of class participation over time, both before and after the open discussion on Wednesday. Which of these patterns would give you some confidence that the discussion had the impact I contend it had?

nonequivalent control group A control group that is similar to the experimental group but is not created by the random assignment of subjects. This sort of control group differs significantly from the experimental group in terms of the dependent variable or variables related to it.

If the time-series results looked like the first pattern in Figure 12-2, you'd probably conclude that the process of greater class participation had begun on the Wednesday before the discussion and had continued, unaffected, after the day devoted to the discussion. The long-term data suggest that the trend would have occurred even without the discussion on Wednesday. The first pattern, then, contradicts my assertion that the special discussion increased class participation.

The second pattern contradicts my assertion by indicating that class participation has been bouncing up and down in a regular pattern throughout the semester. Sometimes it increases from one class to the next, and sometimes it decreases; the open discussion on that Wednesday simply came at a time when the level of participation was about to increase. More to the point, we note that class participation decreased again at the class following the alleged postdiscussion increase.

Only the third pattern in Figure 12-2 supports my contention that the open discussion mattered. As depicted there, the level of discussion before that Wednesday had been a steady four students per class. Not only did the level of participation double following the day of the discussion, but it continued to increase afterward. Although these data do not protect us against the possible influence of some extraneous factor (I might also have mentioned that participation would figure into students' grades), they do exclude the possibility that the increase results from a process of maturation (indicated in the first pattern) or from regular fluctuations (indicated in the second).

Nonequivalent Control Groups

The time-series design just described involves only an "experimental" group; it doesn't provide the value to be gained from having a control group. Sometimes, when researchers can't create experimental and control groups by random assignment from a common pool, they can find an existing "control" group that appears similar to the experimental group. Such a group is called a **nonequivalent control group.** If an innovative foreign language program is being tried in one class in a

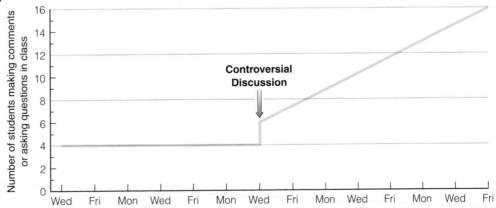

FIGURE 12-2

Three Patterns of Class Participation in a Longer Historical Period large high school, for example, you may be able to find another foreign language class in the same school that has a very similar student population: one that has about the same composition in terms of grade in school, gender, ethnicity, IQ, and so forth. The second class, then, could provide a point of comparison even though it is not formally part of the study. At the end of the semester, you could give both classes the same foreign language test and then compare performances.

Here's how two junior high schools were selected for purposes of evaluating a program aimed at discouraging tobacco, alcohol, and drug use:

> The pairing of the two schools and their assignment to "experimental" and "control" conditions was not random. The local Lung Association had identified the school where we delivered the program as one in which administrators were seeking a solution to admitted problems of smoking, alcohol, and drug abuse. The "control" school was chosen as a convenient and nearby demographic match where administrators were willing to allow our surveying and breath-testing procedures. The principal of that school considered the existing program of health education to be effective and believed that the onset of smoking was relatively uncommon among his students. The communities served by the two schools were very similar. The rate of parental smoking reported by the students was just above 40 percent in both schools.
>
> *(McAlister et al. 1980: 720)*

In the initial set of observations, the experimental and control groups reported virtually the same (low) frequency of smoking. Over the 21 months of the study, smoking increased in both groups, but it increased less in the experimental group than in the control group, suggesting that the program affected students' behavior.

---

multiple time-series designs The use of more than one set of data that were collected over time, as in accident rates over time in several states or cities, so that comparisons can be made.

Multiple Time-Series Designs

Sometimes the evaluation of processes occurring outside of "pure" experimental controls can be made easier by the use of more than one time-series analysis. **Multiple time-series designs** are an improved version of the nonequivalent control group design just described. Carol Weiss (1972: 69) presents a useful example:

> An interesting example of multiple time series was the evaluation of the Connecticut crackdown on highway speeding. Evaluators collected reports of traffic fatalities for several periods before and after the new program went into effect. They found that fatalities went down after the crackdown, but since the series had had an unstable up-and-down pattern for many years, it was not certain that the drop was due to the program. They then compared the statistics with time-series data from four neighboring states where there had been no changes in traffic enforcement. Those states registered no equivalent drop in fatalities. The comparison lent credence to the conclusion that the crackdown had had some effect.

Although this study design is not as good as one in which subjects are assigned randomly, it's nonetheless an improvement over assessing the experimental group's performance without any comparison. That's what makes these designs quasi experiments instead of just fooling around. The key in assessing this aspect of evaluation studies is comparability, as the following example illustrates.

Rural development, a growing concern in the poor countries of the world, has captured the attention and support of many rich countries. Through national foreign assistance programs and through international agencies such as the World Bank, the developed countries are in the process of sharing their technological knowledge and skills with the developing countries. Such programs have had mixed results, however. Often, modern techniques do not produce the intended results when applied in traditional societies.

Rajesh Tandon and L. Dave Brown (1981) undertook an experiment in which technological training would be accompanied by instruction in village organization. They felt it was important for poor farmers to learn how to organize and exert collective influence within their villages—getting needed action from government officials, for example. Only then would their new technological skills bear fruit.

Both intervention and evaluation were attached to an ongoing program in which 25 villages had been selected for technological training. Two poor farmers from each village had been trained in new agricultural technologies. Then they had been sent home to share their new knowledge with their village and to organize other farmers into "peer groups" who would assist in spreading that knowledge. Two years later, the authors randomly selected two of the 25 villages (subsequently called Group A and Group B) for special training and 11 other untrained groups as controls. A careful comparison of demographic characteristics showed the experimental and control groups to be strikingly similar, suggesting they were sufficiently comparable for the study.

The peer groups from the two experimental villages were brought together for special training in organization building. The participants were given some information about organizing and making demands on the government, and they were also given opportunities to act out dramas similar to the situations they faced at home. The training took three days.

The outcome variables considered by the evaluation all had to do with the extent to which members of the peer groups initiated group activities designed to improve their situation. Six types were studied. "Active initiative," for example, was defined as "active effort to influence persons or events affecting group members versus passive response or withdrawal" (Tandon and Brown 1981: 180). The data for evaluation came from the journals that the peer-group leaders had been keeping since their initial technological training. The researchers read through the journals and counted the number of initiatives taken by members of the peer groups. Two researchers coded the journals independently and compared their work to test the reliability of the coding process.

Figure 12-3 compares the number of active initiatives by members of the two experimental groups with those coming from the control groups. Similar results were found for the other outcome measures.

Notice two things about the graph. First, there is a dramatic difference in the number of initiatives by the two experimental groups as compared with the eleven controls. This would seem to confirm the effectiveness of the special training program. Second, notice that the number of initiatives also increased among the control groups. The researchers explain this latter pattern as a result of contagion. Because all the villages were near each other, the lessons learned by peer-group members in the experimental groups were communicated in part to members of the control villages.

This example illustrates the strengths of multiple time-series designs in situations where true experiments are inappropriate to the program being evaluated.

Qualitative Evaluations

Although I've laid out the steps involved in tightly structured, mostly quantitative evaluation research, evaluations can also be less structured and more qualitative. For example, Pauline Bart and Patricia O'Brien (1985) wanted to evaluate different ways to stop rape, so they undertook in-depth interviews with rape victims and with women who had successfully fended off rape attempts. As a general rule, they found that resistance (e.g., yelling, kicking, running away) was more likely to succeed than to make the situation worse, as women sometimes fear it will.

Sometimes even structured quantitative evaluations can yield unexpected qualitative results. Paul Steel is a social researcher specializing in the evaluation of programs aimed at pregnant drug users. One program he evaluated involved counseling by public-health nurses, who warned pregnant drug users that continued drug use would likely result in underweight babies whose skulls would be an average of 10 percent smaller than normal. In his in-depth interviews with program participants, however, he discovered that the program omitted

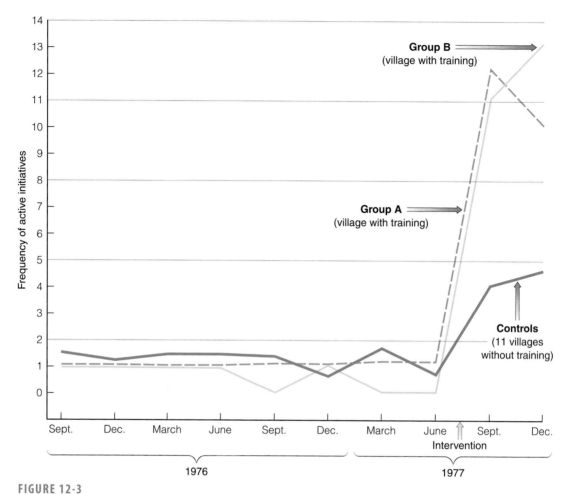

FIGURE 12-3

Active Initiatives over Time

Source: Rajesh Tandon and L. Dave Brown, "Organization-Building for Rural Development: An Experiment in India," *Journal of Applied Behavioral Science,* April–June 1981, p. 182.

one important piece of information: that under-sized babies were a bad thing. Many of the young women Steel interviewed thought that smaller babies would mean easier deliveries.

In another program, a local district attorney had instituted what would generally be regarded as a progressive, enlightened program. If a pregnant drug user were arrested, she could avoid prosecution if she would (1) agree to stop using drugs and (2) successfully complete a drug-rehabilitation program. Again, in-depth interviews suggested

that the program did not always operate on the ground the way it did in principle. Specifically, Steel discovered that whenever a young woman was arrested for drug use, her fellow inmates would advise her to get pregnant as soon as she was released on bail. That way, she would be able to avoid prosecution (personal communication, November 22, 1993).

The most effective evaluation research is one that combines qualitative and quantitative components. Making statistical comparisons is useful, and so is gaining an in-depth understanding of the processes producing the observed results—or preventing the expected results from appearing.

The evaluation of the Tanzanian soap opera, presented earlier in this chapter, employed several research techniques. I've already mentioned the listener surveys and data obtained from clinics. In addition, the researchers conducted numerous focus groups to probe more deeply into the impact the shows had on listeners. Also, content analyses were done on the soap opera episodes themselves and on the many letters received from listeners. Both quantitative and qualitative analyses were undertaken (Swalehe et al. 1995).

The soap opera research also offers an opportunity to see the impact of different cultures on the conduct of research. I had an opportunity to experience this firsthand when I consulted on the evaluation of soap operas being planned in Ethiopia. In contrast to the Western concern for confidentiality in social research, respondents selected for interviews in rural Ethiopian villages often took a special pride at being selected and wanted their answers broadly known in the community.

Or, sometimes, local researchers' desires to please the client got in the way of the evaluation. For example, some pilot episodes were tested in focus groups to determine whether listeners would recognize any of the social messages being communicated. The results were more encouraging than could have been expected. When I asked how the focus group subjects had been selected, the researcher described his introductory conversation: "We would like you to listen to some radio programs designed to encourage people to have small families, and we'd like you to tell us whether we've been successful." Not surprisingly, the small-family theme came through clearly to the focus group.

These experiences, along with earlier comments in previous sections, hint at the possibility of problems in the actual execution of evaluation research projects. Of course, all forms of research can run into problems, but evaluation research has a special propensity for it, as we shall now explore further.

The Social Context

This section looks at some of the logistical problems in evaluation research and presents some observations about using evaluation research results. The social context also raises special ethical issues; we'll discuss these at the end of the chapter.

Logistical Problems

In a military context, *logistics* refers to moving supplies around—making sure people have food, guns, and tent pegs when they need them. Here, I use it to refer to getting subjects to do what they're supposed to do, getting research instruments distributed and returned, and other seemingly simple tasks. These tasks are more challenging than you might guess!

Motivating Sailors

When Kent Crawford, Edmund Thomas, and Jeffrey Fink (1980) set out to find a way to motivate "low performers" in the U.S. Navy, they found out just how many problems can occur. The purpose of the research was to test a three-pronged program for motivating sailors who were chronically poor performers and often in trouble aboard ship. First, a workshop was to be held for supervisory personnel, training them in the effective leadership of low performers. Second, a few supervisors would be selected and trained as special counselors and role models—people the low performers could turn to for advice or just as sounding boards. Finally, the low performers themselves would participate in workshops aimed at training them to be more motivated and effective in their work and in their lives. The project was to be conducted aboard a particular ship, with a control group selected from sailors on four other ships.

To begin, the researchers reported that the supervisory personnel were not exactly thrilled with the program.

> Not surprisingly, there was considerable resistance on the part of some supervisors toward dealing with these issues. In fact, their reluctance to assume ownership of the

problem was reflected by "blaming" any of several factors that can contribute to their personnel problem. The recruiting system, recruit training, parents, and society at large were named as influencing low performance— factors that were well beyond the control of the supervisors.

(Crawford et al. 1980: 488)

Eventually, the reluctant supervisors came around and "this initial reluctance gave way to guarded optimism and later to enthusiasm" (1980: 489).

The low performers themselves were even more of a problem, however. The research design called for pre- and posttesting of attitudes and personalities, so that changes brought about by the program could be measured and evaluated.

Unfortunately, all of the LPs (Low Performers) were strongly opposed to taking these so-called personality tests and it was therefore concluded that the data collected under these circumstances would be of questionable validity. Ethical concerns also dictated that we not force "testing" on the LPs.

(Crawford et al. 1980: 490)

As a consequence, the researchers had to rely on interviews with the low performers and on the judgments of supervisors for their measures of attitude change. The subjects continued to present problems, however.

Initially, the ship's command ordered 15 low performers to participate in the experiment. Of the 15, however, one went into the hospital, another was assigned duties that prevented participation, and a third went "over the hill" (absent without leave). Thus, the experiment began with 12 subjects. But before it was completed, three more subjects completed their tour of duty and left the Navy, and another was thrown out for disciplinary reasons. The experiment concluded, then, with eight subjects. Although the evaluation pointed to positive results, the very small number of subjects warranted caution in any generalizations from the experiment.

The special, logistical problems of evaluation research grow out of the fact that it occurs within the context of real life. Although evaluation research is modeled after the experiment—which suggests that the researchers have control over what happens—it takes place within frequently uncontrollable daily life. Of course, the participant-observer in field research doesn't have control over what is observed either, but that method doesn't strive for control. Given the objectives of evaluation research, lack of control can create real dilemmas for the researcher.

Administrative Control

As suggested in the previous example, the logistical details of an evaluation project often fall to program administrators. Let's suppose you're evaluating the effects of a "conjugal visit" program on the morale of married prisoners. The program allows inmates periodic visits from their spouses during which they can have sexual relations. On the fourth day of the program, a male prisoner dresses up in his wife's clothes and escapes. Although you might be tempted to assume that his morale was greatly improved by escaping, that turn of events would complicate your study design in many ways. Perhaps the warden will terminate the program altogether, and where's your evaluation then? Or, if the warden is brave, he or she may review the files of all those prisoners you selected randomly for the experimental group and veto the "bad risks." There goes the comparability of your experimental and control groups. As an alternative, stricter security measures may be introduced to prevent further escapes, but the security measures may have a dampening effect on morale. So the experimental stimulus has changed in the middle of your research project. Some of the data will reflect the original stimulus; other data will reflect the modification. Although you'll probably be able to sort it all out, your carefully designed study has become a logistical snake pit.

Or suppose you've been engaged to evaluate the effect of race-relations lectures on prejudice in the army. You've carefully studied the soldiers available to you for study, and you've randomly assigned some to attend the lectures and others to stay away. The rosters have been circulated weeks in advance, and at the appointed day and hour, the lectures begin. Everything seems to be going smoothly until you begin processing the files: The names don't match. Checking around, you discover that military field exercises, KP duty, and a variety of emergencies required some of the experimental subjects to be elsewhere at the time of the lectures. That's bad enough, but then you learn that helpful commanding officers sent others to fill in for the missing soldiers. And whom do you suppose they picked to fill in? Soldiers who didn't have anything else to do or who couldn't be trusted to do anything important. You might learn this bit of information a week or so before the deadline for submitting your final report on the impact of the race-relations lectures.

These are some of the logistical problems confronting evaluation researchers. You need to be familiar with the problems to understand why some research procedures may not measure up to the design of the classical experiment. As you read reports of evaluation research, however, you'll find that—my earlier comments notwithstanding—it is possible to carry out controlled social research in conjunction with real-life experiments.

Use of Research Results

One more facts-of-life aspect of evaluation research concerns how evaluations are used. Because the purpose of evaluation research is to determine the success or failure of social interventions, you might think it reasonable that a program would automatically be continued or terminated based on the results of the research.

Reality isn't that simple and reasonable, however. Other factors intrude on the assessment of evaluation research results, sometimes blatantly and sometimes subtly. As president, Richard Nixon appointed a blue-ribbon national commission to study the consequences of pornography. After a diligent, multifaceted evaluation, the commission reported that pornography didn't appear to have any of the negative social consequences often attributed to it. Exposure to pornographic materials, for example, didn't increase the likelihood of sex crimes. You might have expected liberalized

legislation to follow from the research. Instead, the president said the commission was wrong.

Less-dramatic examples of the failure to follow the implications of evaluation research could be listed endlessly. Undoubtedly every evaluation researcher can point to studies he or she conducted—studies providing clear research results and obvious policy implications—that were ignored, as "The Impact of 'Three Strikes' Laws" illustrates.

There are three important reasons why the implications of the evaluation research results are not always put into practice. First, the implications may not always be presented in a way that the non-researchers can understand. Second, evaluation results sometimes contradict deeply held beliefs. That was certainly true in the case of the pornography commission. If "everybody knows" that pornography causes all manner of sexual deviance, then research results to the contrary will probably have little immediate impact. By the same token, people thought Copernicus was crazy when he said the earth revolved around the sun. Anybody could tell the earth was standing still. The third barrier to the use of evaluation results is vested interests. If I've devised a new rehabilitation program that I'm convinced will keep ex-convicts from returning to prison, and if people have taken to calling it "The Babbie Plan," how do you think I'm going to feel when your evaluation suggests the program doesn't work? I might apologize for misleading people, fold up my tent, and go into another line of work. More likely, I'd call your research worthless and begin intense lobbying with the appropriate authorities to have my program continue.

In the earlier example of the evaluation of drivers' education, Philip Hilts reported some of the reactions to the researchers' preliminary results:

> Ray Burneson, traffic safety specialist with the National Safety Council, criticized the study, saying that it was a product of a group (NHTSA) run by people who believe "that you can't do anything to train drivers. You can only improve medical facilities and build stronger cars for when the accidents happen. . . . This knocks the whole philosophy of education."
>
> *(1981: 4)*

The Impact of "Three Strikes" Laws

SACRAMENTO (AP)—The author of California's five-year-old "three strikes" law says it's prevented more than a million crimes and has saved $21.7 billion.

Secretary of State Bill Jones offered his interpretation of the "three strikes" results to a Doris Tate Crimes Victim Bureau conference on Friday in Sacramento.

(BayInsider, March 1, 1999)

The 1990s saw the passage of "three strikes" laws at the federal level and in numerous states. The intention was to reduce crime rates by locking up "career criminals." Under the 1994 California law, for example, having a past felony conviction would double your punishment when you were convicted of your second felony, and the third felony conviction would bring a mandatory sentence of 25 years to life. Over the years, only California has enforced such laws with any vigor.

Those who supported the passage of "three strikes" legislation, such as Bill Jones, quoted earlier, were quick to link the dramatic drop in crime rates during the 1990s to the new policy of getting tough with career criminals. While acknowledging that "three strikes" may not be the only cause of the drop in crime, Jones added, "If you can have a 51 percent reduction in the homicide rate in five years, I would guarantee you three strikes is a big part of the reason."

In spite of the politicians' guarantees, other observers have looked for additional evidence to support the impact of "three strikes" laws. Some critics of these laws, for example, have noted that crime rates have been dropping dramatically across the country, not only in California but in states that have no "three strikes" laws and in those where the courts have not enforced the "three strikes" laws that exist. In fact, crime rates have dropped in those California counties that have tended to ignore that state's law. Moreover, the drop in California crime rates began before the "three strikes" law went into effect.

In 1994, Peter Greenwood and his colleagues at the Rand Corporation estimated that implementation of the law would cost California's criminal justice system approximately $5.5 billion more per year, especially in prison costs as "career criminals" were sentenced to longer terms. Although the Rand group did not deny that the "three strikes" legislation would have some impact on crime—those serving long terms in prison can't commit crimes on the streets—a follow-up study (Greenwood, Rydell, and Model 1996) suggested it was an inefficient way of attacking crime. They estimated that a million dollars spent on "three strikes" would prevent 60 crimes, whereas the same amount spent on programs encouraging high school students to stay in school and graduate would prevent 258 crimes.

Criminologists have long recognized that most crimes are committed by young men. Focusing attention on older "career criminals" has little or no effect on the youthful offenders. In fact, "three strikes" sentences disproportionately fall on those approaching the end of their criminal careers by virtue of growing older.

In a more general critique, John Irwin and James Austin (1997) suggest that people in the United States tend to overuse prisons as a solution to crime, ignoring other, more effective, solutions. Often, imprisonment causes problems more serious than those it was intended to remedy.

As with many other social interventions, however, much of the support for "three strikes" laws in California and elsewhere stems mostly from public emotions about crime and the political implications of such emotions. Thus, evaluation research on these laws may eventually bring about changes, but its impact is likely to be much slower than you might logically expect.

Sources: Peter W. Greenwood, C. Peter Rydell, and Karyn Model, *Diverting Children from a Life of Crime: Measuring Costs and Benefits* (Santa Monica, CA: Rand Corporation, 1996); Peter W. Greenwood et al., *Three Strikes and You're Out: Estimated Benefits and Costs of California's New Mandatory-Sentencing Law* (Santa Monica, CA: Rand Corporation, 1994); John Irwin and James Austin, *It's About Time: America's Imprisonment Binge* (Belmont, CA: Wadsworth 1997); "State Saved $21.7 Billion with Five-Year-Old 'Three Strikes' Law," *BayInsider,* March 1, 1999.

By its nature, evaluation research takes place in the midst of real life, affecting it and being affected by it. Here's another example, well known to social researchers.

Rape Reform Legislation

For years, many social researchers and other observers have noted certain problems with the prosecution of rape cases. All too often, it is felt, the victim ends up suffering almost as much on the witness stand as in the rape itself. Frequently the defense lawyers portray her as having encouraged the sex act and being of shady moral character; other personal attacks are intended to deflect responsibility from the accused rapist.

Criticisms such as these have resulted in a variety of state-level legislation aimed at remedying the problems. Cassie Spohn and Julie Horney

(1990) were interested in tracking the impact of such legislation. The researchers summarize the ways in which new laws were intended to make a difference:

> The most changes are: (1) redefining rape and replacing the single crime of rape with a series of graded offenses defined by the presence or absence of aggravating conditions; (2) changing the consent standard by eliminating the requirement that the victim physically resist her attacker; (3) eliminating the requirement that the victim's testimony be corroborated; and (4) placing restrictions on the introduction of evidence of the victim's prior sexual conduct.
> *(1990: 2)*

It was generally expected that such legislation would encourage women to report being raped and would increase convictions when the cases were brought to court. To examine the latter expectation, the researchers focused on the period from 1970 to 1985 in Cook County, Illinois: "Our data file includes 4,628 rape cases, 405 deviate sexual assault cases, 745 aggravated criminal sexual assault cases, and 37 criminal sexual assault cases" (1990: 4). Table 12-1 shows some of what they discovered.

Spohn and Horney summarized these findings as follows:

> The only significant effects revealed by our analyses were increases in the average maximum prison sentences; there was an increase of almost 48 months for rape and of almost 36 months for sex offenses. Because plots of the data indicated an increase in the average sentence before the reform took effect, we modeled the series with the intervention moved back one year earlier than the actual reform date. The size of the effect was even larger and still significant, indicating that the effect should not be attributed to the legal reform.
> *(1990: 10)*

Notice in the table that there was virtually no change in the percentages of cases ending in conviction for rape or some other charge (e.g., assault). Hence the change in laws didn't have any effect on the likelihood of conviction. As the researchers

TABLE 12-1
Analysis of Rape Cases Before and After Legislation

	Rape	
	Before (N = 2,252)	*After* (N = 2,369)
Outcome of case		
Convicted of original charge	45.8%	45.4%
Convicted of another charge	20.6	19.4
Not convicted	33.6	35.1
Median prison sentence in months		
For those convicted of original charge	96.0	144.0
For those convicted of another charge	36.0	36.0

note, the one change that *is* evident—an increase in the length of sentences—cannot be attributed to the reform legislation itself.

In addition to the analysis of existing statistics, Spohn and Horney interviewed judges and lawyers to determine what they felt about the impact of the laws. Their responses were somewhat more encouraging.

> Judges, prosecutors and defense attorneys in Chicago stressed that rape cases are taken more seriously and rape victims treated more humanely as a result of the legal changes. These educative effects clearly are important and should please advocates of rape reform legislation.
> *(1990: 17)*

Thus, the study found other effects besides the qualitative results the researchers looked for. This study demonstrates the importance of following up on social interventions to determine whether, in what ways, and to what degree they accomplished their intended results.

Preventing Domestic Violence

In a somewhat similar study, researchers in Indianapolis focused their attention on the problem of wife battering, with a special concern for whether prosecuting the batterers can lead to subsequent violence. David Ford and Mary Jean Regoli (1992) set about studying the consequences of various options for prosecution allowed within the "Indianapolis Prosecution Experiment" (IPE).

Wife-battering cases can follow a variety of patterns, as Ford and Regoli summarize:

> After a violent attack on a woman, someone may or may not call the police to the scene. If the police are at the scene, they are expected to investigate for evidence to support probable cause for a warrantless arrest. If it exists, they may arrest at their discretion. Upon making such an on-scene arrest, officers fill out a probable cause affidavit and slate the suspect into court for an initial hearing. When the police are not called, or if they are called but do not arrest, a victim may initiate charges on her own by going to the prosecutor's office and swearing out a probable cause affidavit with her allegation against the man. Following a judge's approval, the alleged batterer may either be summoned to court or be arrested on a warrant and taken to court for his initial hearing.
>
> *(1992: 184)*

What if a wife brings charges against her husband and then reconsiders later on? Many courts have a policy of prohibiting such actions, in the belief that they are serving the interests of the victim by forcing the case to be pursued to completion. In the IPE, however, some victims are offered the possibility of dropping the charges if they so choose later in the process. In addition, the court offers several other options. Because wife battering is largely a function of sexism, stress, and an inability to deal with anger, some of the innovative possibilities in the IPE involve educational classes with anger-control counseling.

If the defendant admits his guilt and is willing to participate in an anger-control counseling program, the judge may postpone the trial for that purpose and can later dismiss the charges if the defendant successfully completes the program. Alternatively, the defendant may be tried and, if found guilty, be granted probation provided he participates in the anger-control program. Finally, the defendant can be tried and, if found guilty, be given a conventional punishment such as imprisonment.

Which of these possibilities most effectively prevents subsequent wife battering? That's the question Ford and Regoli addressed. Here are some of their findings.

First, their research shows that men who are brought to court for a hearing are less likely to continue beating their wives, no matter what the outcome of the hearing. Simply being brought into the criminal justice system has an impact.

Second, women who have the right to drop charges later on are less likely to be abused subsequently than those who do not have that right. In particular, the combined policies of arresting defendants by warrant and allowing victims to drop charges provides victims with greater security from subsequent violence than any of the other prosecution policies do.

However, giving victims the right to drop charges has a somewhat strange impact. Women who exercise that right are more likely to be abused later than those who insist on the prosecution proceeding to completion. The researchers interpret this as showing that future violence can be decreased when victims have a sense of control supported by a clear and consistent alliance with criminal justice agencies.

> A decisive system response to any violation of conditions for pretrial release, including of course new violence, should serve notice that the victim-system alliance is strong. It tells the defendant that the victim is serious in her resolve to end the violence and that the system is unwavering in its support of her interest in securing protection.
>
> *(Ford and Regoli 1992: 204)*

The effectiveness of anger-control counseling cannot be assessed simply. Policies aimed at getting defendants into anger-control counseling seem to be relatively ineffective in preventing new violence. The researchers noted, however, that the policy effects should not be confused with actual counseling outcomes. Some defendants scheduled for treatment never received it. Considerably more information on implementing counseling is needed for a proper evaluation.

Moreover, the researchers cautioned that their results point to general patterns, and that battered wives must choose courses of action appropriate to their particular situations and should not act blindly on the basis of the overall patterns. The research is probably more useful in what it says about ways of structuring the criminal justice system (giving victims the right to drop charges, for example) than in guiding the actions of individual victims.

Finally, the IPE offers an example of a common problem in evaluation research. Often, actual practices differ from what might be expected in principle. For example, the researchers considered the impact of different alternatives for bringing suspects into court: Specifically, the court can issue either a summons ordering the husband to appear in court or a warrant to have the husband arrested. The researchers were concerned that having the husband arrested might actually add to his anger over the situation. They were somewhat puzzled, therefore, to find no difference in the anger of husbands summoned or arrested.

The solution of the puzzle lay in the discrepancy between principle and practice:

> Although a warrant arrest should in principle be at least as punishing as on-scene arrest, in practice it may differ little from a summons. A man usually knows about a warrant for his arrest and often elects to turn himself in at his convenience, or he is contacted by the warrant service agency and invited to turn himself in. Thus, he may not experience the obvious punishment of, say, being arrested, handcuffed, and taken away from a workplace.
>
> *(Ford 1989: 9–10)*

In summary, many factors besides the scientific quality of evaluation research affect how its results are used. And, as we saw earlier, factors outside the evaluator's control can affect the quality of the study itself. But this "messiness" is balanced by the potential contributions that evaluation research can make toward the betterment of human life.

The Sabido Methodology

One of the clearest illustrations of the uses of evaluation research results can be found in the omnibus methodology developed by Miguel Sabido for the use of "Entertainment-Education" (E-E) projects to promote social programs. The example of *Twende na Wakati* at the outset of this chapter illustrated the methods initially developed by Sabido in the 1970s when he was Vice President for Research in Mexico's national broadcasting company Televisa. Sabido's first projects used television *novelas* to promote literacy and family planning. They were so successful that those methods have been used to promote a variety of social issues in the subsequent decades.

In part, the Sabido methodology concerns the nature of the radio or television dramas: particularly the kinds of characters portrayed. Some characters represent traditional points of view, some represent the modern views that the programming is designed to promote, and some represent a "transitional" point of view—they begin with traditional views but eventually shift to the modern views. Typically, when a transitional character signs up for literacy classes, thousands of audience members do the same shortly thereafter. When the transitional character begins using condoms for family planning or safe sex, family planning clinics are mobbed the next day by men wanting condoms.

The Sabido methodology extends beyond character definitions and plot structures. An E-E project begins with thorough research into the society where the change is being planned. A project in Ethiopia by the Population Media Center, for example, aimed to lower the birthrate, encourage safe-sex practices, and enhance the status of women. The production of radio serial dramas was preceded by extensive research into the existing situations regarding the project's aims. What was the birthrate? How did it differ in different regions of the country and among different ethnic groups? What were the attitudes toward family planning? In part these questions were answered through national surveys. At the same time, qualitative researchers went into the countryside to observe rural villages, talking with residents and sometimes recording the sounds of village life.

This formative research provided the writers with ideas about issues to be raised and how to raise them. For example, the research indicated that in some regions, abduction was still a common method of mate selection: A man would kidnap a young woman, sexually assaulting her and holding her prisoner until she would consent to be his wife. The formative research also revealed a widespread belief that condoms were infected with HIV, thus meaning that condom use increased the risk of AIDS rather than reducing it.

The initial research also provided a baseline for subsequent evaluations. By knowing public opinion toward family planning prior to the radio programs, researchers could determine how much these opinions had changed afterward. Prepro-gramming measures of the use of family planning centers could be compared with use levels afterward. Many of these evaluation efforts ran concurrently with the radio programming. For example, regular focus groups were used to monitor public reactions to each of the serial installments, examining whether people were reacting as intended.

The Sabido methodology provides an excellent illustration of how research methods can be used to construct and evaluate social action programs aimed at resolving social problems. To learn more about the Sabido methodology, see Barker and Sabido (2005), which can be downloaded from the link on this book's website: http://www.cengage.com/sociology/babbie.

Social Indicators Research

Let's now look at a type of research that combines what you've learned about evaluation research and about the analysis of existing data. A rapidly growing field in social research involves the development and monitoring of **social indicators,** aggregated statistics that reflect the social condition of a society or social subgroup. Researchers use social indicators to monitor aspects of social life in much

---

social indicators Measurements that reflect the quality or nature of social life, such as crime rates, infant mortality rates, number of physicians per 100,000 population, and so forth. Social indicators are often monitored to determine the nature of social change in a society.

---

the way that economists use indexes such as gross national product (GNP) per capita as an indicator of a nation's economic development.

Suppose we wanted to compare the relative health conditions in different societies. One strategy would be to compare their death rates (number of deaths per 1,000 population). Or, more specifically, we could look at infant mortality: the number of infants who die during their first year of life among every 1,000 births. Depending on the particular aspect of health conditions we were interested in, we could devise any number of other measures: physicians per capita, hospital beds per capita, days of hospitalization per capita, and so forth. Notice that intersocietal comparisons are facilitated by calculating per capita rates (dividing by the size of the population).

Before we go further, recall from Chapter 11 the problems involved in using existing statistics. In a word, they're often unreliable, reflecting their modes of collection, storage, and calculation. With this in mind, we'll look at some of the ways we can use social indicators for evaluation research on a large scale.

The Death Penalty and Deterrence

Does the death penalty deter capital crimes such as murder? This question is hotly debated every time a state considers eliminating or reinstating capital punishment and every time someone is executed. Those supporting capital punishment often argue that the threat of execution will deter potential murderers from killing people. Opponents of capital punishment often argue that it has no effect in that regard. Social indicators can help shed some light on the question.

If capital punishment actually deters people from committing murder, then we should expect to find murder rates lower in those states that have the death penalty than in those that do not. The relevant comparisons in this instance are not only possible, they've been compiled and published. Table 12-2 presents data compiled by William Bailey (1975) that directly contradict the view that

TABLE 12-2

Average Rate per 100,000 Population of First- and Second-Degree Murders for Capital-Punishment and Non-Capital-Punishment States, 1967 and 1968

	Non-Capital-Punishment States		Capital-Punishment States	
	1967	1968	1967	1968
First-degree murder	0.18	0.21	1.47	1.58
Second-degree murder	0.30	0.43	1.92	1.03
Total murders	0.48	0.64	1.38	1.59

Source: Adapted from William C. Bailey, "Murder and Capital Punishment," in William J. Chambliss, ed., *Criminal Law in Action*. Copyright © 1975 by John Wiley & Sons, Inc. Used by permission.

the death penalty deters murderers. In both 1967 and 1968, those states with capital punishment had dramatically *higher* murder rates than those without capital punishment did. Some people criticized the interpretation of Bailey's data, saying that most states had not used the death penalty in recent years, even when they had it on the books. That could explain why it didn't seem to work as a deterrent. Further analysis, however, contradicts this explanation. When Bailey compared those states that hadn't used the death penalty with those that had, he found no real difference in murder rates.

Another counterexplanation is possible, however. It could be the case that the interpretation given Bailey's data was *backward*. Maybe the existence of the death penalty as an option was a consequence of high murder rates: Those states with high rates instituted it; those with low rates didn't institute it or repealed it if they had it on the books. It could be the case, then, that instituting the death penalty would bring murder rates down, and repealing it would increase murders and still produce—in a broad aggregate—the data presented in Table 12-2. Not so, however. Analyses over time do not show an increase in murder rates when a state repeals the death penalty nor a decrease in murders when one is instituted.

Notice from the preceding discussion that it's possible to use social indicators data for comparison across groups either at one time or across some period of time. Often, doing both sheds the most light on the subject.

Though overall murder rates have increased substantially, by the way, the pattern observed by Bailey in 1967 and 1968 has persisted over time. In 1999, for example, the 38 death-penalty states had a combined murder rate of 5.86 per 100,000, compared with a combined murder rate of 3.84 among the 12 states that lack the death penalty (U.S. Bureau of the Census 2001: 22, 183).

At present, work on the use of social indicators is proceeding on two fronts. On the one hand, researchers are developing ever more-refined indicators—finding which indicators of a general variable are the most useful in monitoring social life. At the same time, research is being devoted to discovering the relationships among variables within whole societies.

As with many aspects of social research, the World Wide Web has become a valuable resource. To pursue the possibilities of social indicators, you might check out Sociometrics Corporation, for example (see the link on this book's website). Or simply search for "social indicators" using one of the web search engines.

Computer Simulation

One of the most exciting prospects for social indicators research lies in the area of computer simulation. As researchers begin compiling mathematical equations describing the relationships that link social variables to one another (for example, the relationship between growth in population and the number of automobiles), those equations can be stored and linked to one another in a computer. With a sufficient number of adequately accurate equations on tap, researchers one day will be able to test the implications of specific social changes by computer rather than in real life.

Suppose a state contemplated doubling the size of its tourism industry, for example. We could enter that proposal into a computer-simulation model and receive in seconds or minutes a description of all the direct and indirect consequences of the increase in tourism. We could know what new public facilities would be required, which public agencies such as police and fire departments would have to be increased and by how much, what the labor force would look like, what kind of training would be required to provide it, how much new income and tax revenue would be produced, and so forth, through all the intended and unintended consequences of the action. Depending on the results, the public planners might say, "Suppose we increased the industry only by half," and have a new printout of consequences immediately.

An excellent illustration of computer simulation linking social and physical variables is to be found in the research of Donella and Dennis Meadows and their colleagues at Dartmouth and the Massachusetts Institute of Technology (Meadows et al. 1972, 1992). They've taken as input data known and estimated reserves of various nonreplaceable natural resources (for example, oil, coal, iron), past patterns of population and economic growth, and the relationships between growth and use of resources. Using a complex computer-simulation model, they've been able to project, among other things, the probable number of years various resources will last in the face of alternative usage patterns in the future. Going beyond the initially gloomy projections, such models also make it possible to chart out less gloomy futures, specifying the actions required to achieve them. Clearly, the value of computer simulation is not limited to evaluation research, though it can serve an important function in that regard.

This potentiality points to the special value of evaluation research in general. Throughout human history, we've been tinkering with our social arrangements, seeking better results. Evaluation research provides a means for us to learn right away whether a particular tinkering really makes things better. Social indicators allow us to make that determination on a broad scale; coupling them with computer simulation opens up the possibility of knowing how much we would like a particular intervention, without having to experience its risks.

Ethics and Evaluation Research

As we have seen, evaluation research is by nature interwoven with real-world issues. Sometimes the social interventions being evaluated raise ethical issues. Evaluating the impact of busing school children to achieve educational integration will throw the researchers directly into the political, ideological, and ethical issues of busing itself. It's not possible to evaluate a sex-education program in elementary schools without becoming involved in the heated issues surrounding sex education itself, and the researcher will find remaining impartial difficult. The evaluation study design will *require* that some children receive sex education—in fact, you may very well be the one who decides which children do. (From a scientific standpoint, you *should* be in charge of selection.) This means that when parents become outraged that their child is being taught about sex, you'll be directly responsible.

Now let's look on the "bright" side. Maybe the experimental program is of great value to those participating in it. Let's say that the new industrial safety program being evaluated reduces injuries dramatically. What about the control-group members who were deprived of the program by the research design? The evaluators' actions could be an important part of the reason that a control-group subject suffered an injury.

Sometimes the name of evaluation research has actually served as a mask for unethical behavior. In Chapter 9 I discussed push polls, which pretend to evaluate the impact of various political campaign accusations but intend to spread malicious misinformation. That's not the worst example, however, as you'll recall from the discussion of the Tuskegee experiments, in Chapter 3.

Even in the most legitimate evaluation research, the researcher almost always faces pressure from the people affected by the evaluation. Often, as in the case of pharmaceutical testing, for example, those paying for the research may want a particular result. Evaluation researchers, therefore, often find themselves under pressure to produce a particular finding.

I'm sure there's no need to point out that researchers must not be swayed by personal desires or sponsors' demands in the design, execution, and analysis of results; manipulating research to produce a desired result is never acceptable. This is particularly important in the case of evaluation research, in that the real-world setting can create serious and far-reaching consequences for the people involved. Imagine a medical researcher slanting drug testing results to suggest a new drug is more effective than it is or covering up the negative side effects of the drug, so that the drug is given to patients who will not benefit from it or will actually be harmed by the "unknown" side effects. Or imagine that an evaluation of a prison rehabilitation program has been slanted to make the program seem more effective than it is. Limited resources might be diverted to support the ineffective program and possibly even harm the prisoners subjected to it.

My purpose in these comments has not been to cast a shadow on evaluation research. Rather, I want to bring home the real-life consequences of the evaluation researcher's actions. Ultimately, all social research has ethical components.

MAIN POINTS

Introduction

- Evaluation research is a form of applied research that studies the effects of social interventions.

Topics Appropriate for Evaluation Research

- Topics appropriate for evaluation research include needs assessment studies, cost-benefit studies, monitoring studies, and program evaluations/outcome assessments.
- Evaluation research is sometimes coupled with the intentions of participatory action research.

Formulating the Problem: Issues of Measurement

- A careful formulation of the problem, including relevant measurements and criteria of success or failure, is essential in evaluation research. In particular, evaluators must carefully specify outcomes, measure experimental contexts, specify the intervention being studied and the population targeted by the intervention, and decide whether to use existing measures or devise new ones.

Types of Evaluation Research Designs

- Evaluation researchers typically use experimental or quasi-experimental designs. Examples of quasi-experimental designs include time-series studies and the use of nonequivalent control groups.
- Evaluators can also use qualitative methods of data collection. Both quantitative and qualitative data analyses can be appropriate in evaluation research, sometimes in the same study.

The Social Context

- Evaluation research entails special logistical problems because it's embedded in the day-to-day events of real life.
- The implications of evaluation research won't necessarily be put into practice, especially if they conflict with official points of view.

Social Indicators Research

- Social indicators can provide an understanding of broad social processes.
- Computer-simulation models hold the promise of allowing researchers to study the possible results of social interventions without having to incur those results in real life.

Ethics and Evaluation Research

- Sometimes the social interventions being assessed in evaluation research themselves raise ethical issues.
- Evaluation research may entail added pressure to produce specific results, as desired by interested parties.
- Fraudulent research results in an evaluation study can have severer consequences than consequences produced by other types of research.

KEY TERMS

The following terms are defined in context in the chapter and at the bottom of the page where the term is introduced, as well as in the comprehensive glossary at the back of the book.

cost-benefit studies   monitoring studies
evaluation research   multiple time-series designs

needs assessment studies

nonequivalent control group

program evaluation/
outcome assessment

quasi experiments

social indicators

time-series design

5. The U.S. Bureau of Prisons engages in evaluation research regarding various aspects of prison operations. Locate one of their studies on the web and write a short summary of the study design and the findings. (See the link on this book's website.)

PROPOSING SOCIAL RESEARCH: EVALUATION RESEARCH

Evaluation research represents a research purpose rather than a particular method. In the proposal, you need to spell out the type of evaluation you're conducting and perhaps the implications of various possible outcomes.

In earlier assignments, you'll have spelled out the data-collection and measurement methods to be used in your study. If your study is designed to determine the success or failure of a program, you may also want to specify the research results that will be deemed a positive or negative assessment in that regard. This may not always be appropriate or possible, but it adds integrity to the evaluation process when it can be done.

REVIEW QUESTIONS AND EXERCISES

1. Suppose a community establishes an alcohol- and drug-free teen center as a way of reducing the use of alcohol and drugs by teenagers. Describe how you might go about evaluating the effectiveness of the center. Indicate whether your design would be experimental, quasi-experimental, or qualitative (or some combination of these).

2. Review the evaluation of the Navy low-performer program discussed in the chapter. Redesign the program and the evaluation to handle the problems that appeared in the actual study.

3. Discuss some of the potential political and ethical issues that might be involved in the study you described in Exercise 1.

4. Take a minute to think of the many ways your society has changed during your own lifetime. Specify three or four social indicators that could be used in monitoring the effects of at least one of those changes on the quality of life in your society.

SPSS EXERCISES

See the booklet that accompanies your text for exercises using SPSS (Statistical Package for the Social Sciences). There are exercises offered for each chapter, and you'll also find a detailed primer on using SPSS.

Online Study Resources

If your book came with an access code card, visit www.cengage.com/login to register. To purchase access, please visit www.ichapters.com.

1. Before you do your final review of the chapter, take the *CengageNOW* pretest to help identify the areas on which you should concentrate. You'll find information on this online tool, as well as instructions on how to access all of its great resources, in the front of the book.

2. As you review, take advantage of the *CengageNOW* personalized study plan, based on your quiz results. Use this study plan with its interactive exercises and other resources to master the material.

3. When you're finished with your review, take the posttest to confirm that you're ready to move on to the next chapter.

WEBSITE FOR *THE PRACTICE OF SOCIAL RESEARCH* 12TH EDITION

Go to your book's website at www.cengage.com/sociology/babbie for tools to aid you in studying for your exams. You'll find *Tutorial Quizzes* with feedback, *Internet Exercises*, *Flash Cards*, *Glossaries*, and *Essay Quizzes*, as well as *InfoTrac College Edition* search terms, suggestions for additional reading, *Web Links*, and primers for using data-analysis software such as SPSS.

In this part of the book, we'll discuss the analysis of social research data, and we'll examine the steps that separate observation from the final reporting of findings.

In Chapter 1, I made a fundamental distinction between qualitative and quantitative data. In the subsequent discussions, we've seen that many of the fundamental concerns in social research apply equally to both types of data. The analysis of qualitative and quantitative data, however, are quite different and will be discussed separately.

Before outlining the specifics of Part 4, I want to offer an observation about the ease or difficulty of producing high-quality data analyses, as represented in the following table, where "1" is the easiest to do and "4" is the hardest.

	Simplistic	*Sophisticated*
Qualitative	1	4
Quantitative	2	3

Analysis of Data: Quantitative and Qualitative

(1) In my work, I've seen that it's relatively easy to make some observations of social life and speculate about the meaning of what has been observed. Unfortunately, such speculation is unlikely to make much of a contribution to our understanding of social life.

(2) Doing even a simplistic quantitative data analysis is more difficult, because it requires at least some low-level statistical skills. All too often, however, we're confronted with statistical data analyses that don't really mean much. Terms such as *quantiphrenia* and *scientism* have sometimes been used in reference to attempts to mimic the physical sciences without any true meaning.

(3) Doing sophisticated, meaningful quantitative data analyses requires much thought and imagination. It does not necessarily require high-powered statistics, however, as much of the work of Paul Lazarsfeld and Sam Stouffer shows. What's needed instead is the willingness to search for, and the ability to recognize, meaningful patterns among variables. Although the many established techniques for quantitative data analysis are powerful tools to use in that pursuit, the really powerful discoveries are never produced by the rote administration of techniques.

(4) The most difficult task for social scientists lies in producing powerful analyses of qualitative data. This requires the same dedication and ability discussed in (3); however, qualitative analysis depends more on the individual insights of the researcher than on the tools available to support the analysis. Qualitative analysis remains today as much an art as a science.

I hope the chapters that make up this part of the book will give you some of the tools and sharpen the insights needed to produce sophisticated data analyses, whether qualitative or quantitative.

Chapter 13 examines qualitative data analysis. We'll begin by examining some of the theoretical groundings for this approach. Then we'll look at some conceptual procedures you should find useful in the search for meaning among qualitative data. I'll also demonstrate some of the computer programs that have been created specifically for qualitative data analysis. Near the conclusion of this chapter, we'll examine some attempts to create criteria for assessing the quality of qualitative research.

The first of several discussions on the logic of quantitative data analysis is presented in Chapter 14. We'll begin with an examination of methods of analyzing and presenting data related to a single variable. Then we'll turn to the relationship between two variables and learn how to construct and read simple percentage tables. The chapter ends with a preview of multivariate analysis and a discussion of sociological diagnostics, along with a look at ethical concerns.

Chapter 15 describes the elaboration model of data analysis developed by Paul Lazarsfeld at Columbia University. Here we'll look further into multivariate analysis. Chapter 15 also presents the logic of causal analysis through the use of percentage tables. We'll apply this same logic when we use other statistical techniques in Chapter 16. This logical model was developed for use

with quantitative data, but I think you'll see how appropriate it is for reasoning with qualitative data as well.

Chapter 16 provides an introduction to some of the more commonly used statistical methods in social science research, including an overview of some of the more advanced methods of multivariate analysis. Rather than merely showing how to compute statistics by these methods (computers can do that), I've attempted to place them in the context of earlier theoretical and logical discussions. Thus, you should come away from this chapter knowing when to use various statistical measures as well as how to compute them.

Finally, Chapter 17 addresses social research as literature: how to read it and how to write it. The materials of this chapter are essentially bookends for the research process: a review of the literature early in the project involves the skills of reading social research, and writing it comes into play in the communication of your results to others in the form of your research report. Again, we also look at the ethics involved in this endeavor.

CHAPTER THIRTEEN

Qualitative Data Analysis

CHAPTER OVERVIEW

Qualitative data analysis is the nonnumerical assessment of observations made through participant observation, content analysis, in-depth interviews, and other qualitative research techniques. Although qualitative analysis is as much an art as a science, it has its own logic and techniques, some of which are enhanced by special computer programs.

CengageNOW for Sociology

Use this online tool to help you make the grade on your next exam. After reading this chapter, go to "Online Study Resources" at the end of the chapter for instructions on how to benefit from *CengageNOW*.

Introduction

Later chapters in Part 4 of the book will deal with the *quantitative* analysis of social research data, sometimes called *statistical analysis*. Recent decades of social science research have tended to focus on quantitative data analysis techniques. This focus, however, sometimes conceals another approach to making sense of social observations: **qualitative analysis**—methods for examining social research data without converting them to a numerical format. This approach predates quantitative analysis. It remains a useful approach to data analysis and is enjoying a resurgence of interest among social scientists.

Learning qualitative analysis techniques requires a different approach than learning quantitative ones. Although statistical analyses may intimidate some students, the steps involved can sometimes be learned in a rote manner. That is, with practice, the rote exercise of quantitative skills can produce an ever more sophisticated understanding of the logic that lies behind those techniques.

It's much more difficult to teach qualitative analysis as a series of rote procedures. In this case, understanding must precede practice. In this chapter, we begin with the links between research and theory in qualitative analysis. Then we examine some procedures that have proved useful in pursuing the theoretical aims. After considering some simple manual techniques, we'll take some computer programs out for a spin.

Linking Theory and Analysis

As suggested in Chapter 10 and elsewhere in this book, qualitative research methods involve a continuing interplay between data collection and theory. As a result, I've already talked about qualitative data analysis in earlier discussions of field research and content analysis. In quantitative research, it's sometimes easy to get caught up in the logistics of data collection and in the statistical analysis of data, thereby losing sight of theory for a time. This is less likely in qualitative research, where data collection, analysis, and theory are more intimately intertwined.

In the discussions to follow, we'll use the image of theory offered by Anselm Strauss and Juliet Corbin (1994: 278) as consisting of *"plausible relationships proposed among concepts and sets of concepts."* They stress "plausible" to indicate that theories represent our best understanding of how life operates. The more our research confirms a particular set of relationships among particular concepts, however, the more confident we become that our understanding corresponds to social reality.

Whereas qualitative research is sometimes undertaken for purely descriptive purposes—such as the anthropologist's ethnography detailing ways of life in a previously unknown tribe—the rest of this chapter focuses primarily on the search for explanatory patterns. As we'll see, sometimes the patterns occur over time, and sometimes they take the form of causal relations among variables. Let's look at some of the ways qualitative researchers uncover such patterns.

Discovering Patterns

John Lofland and his colleagues (2006:149–65) suggest six different ways of looking for patterns in a particular research topic. Let's suppose you're interested in analyzing child abuse in a certain neighborhood. Here are some questions you might ask yourself to make sense out of your data:

1. *Frequencies:* How often does child abuse occur among families in the neighborhood under study? (Realize that there may be a difference between the frequency and what people are willing to tell you.)

---

qualitative analysis The nonnumerical examination and interpretation of observations, for the purpose of discovering underlying meanings and patterns of relationships. This is most typical of field research and historical research.

2. *Magnitudes:* What are the levels of abuse? How brutal are they?

3. *Structures:* What are the different types of abuse: physical, mental, sexual? Are they related in any particular manner?

4. *Processes:* Is there any order among the elements of structure? Do abusers begin with mental abuse and move on to physical and sexual abuse, or does the order of elements vary?

5. *Causes:* What are the causes of child abuse? Is it more common in particular social classes or among different religious or ethnic groups? Does it occur more often during good times or bad?

6. *Consequences:* How does child abuse affect the victims, in both the short and the long term? What changes does it cause in the abusers?

For the most part, in examining your data you'll look for patterns appearing across several observations that typically represent different cases under study, an approach called **cross-case analysis.** A. Michael Huberman and Matthew Miles (1994: 435f) offer two strategies for cross-case analysis: the variable-oriented and the case-oriented analysis. **Variable-oriented analysis** is similar to a model we've already discussed from time to time in this book. If we were trying to predict the decision to attend college, Huberman and Miles suggest, we might consider variables such as "gender, socio-economic status, parental expectations, school performance, peer support, and decision to attend college" (1994: 435). Thus, we would determine whether men or women were more likely to attend college. The focus of our analysis would be on inter-relations among variables, and the people observed would be primarily the carriers of those variables.

Variable-oriented analysis may remind you of the discussion in Chapter 1 that introduced the idea of nomothetic explanation. The aim here is to achieve a partial, overall explanation using relatively few variables. The political pollster who attempts to explain voting intentions on the basis of two or three key variables is using this approach. There is no pretense that the researcher can predict every individual's behavior nor even explain any one person's motivations in full. Sometimes,

though, it's useful to have even a partial explanation of overall orientations and actions.

You may also recall Chapter 1's introduction of idiographic explanation, wherein we attempt to understand a particular case fully. In the voting example, we would attempt to learn everything we could about all the factors that came into play in determining one person's decision on how to vote. This orientation lies at the base of what Huberman and Miles call a **case-oriented analysis.**

> In a case-oriented analysis, we would look more closely into a particular case, say, Case 005, who is female, middle-class, has parents with high expectations, and so on. These are, however, "thin" measures. To do a genuine case analysis, we need to look at a full history of Case 005; Nynke van der Molen, whose mother trained as a social worker but is bitter over the fact that she never worked outside the home, and whose father wants Nynke to work in the family florist shop. Chronology is also important: two years ago, Nynke's closest friend decided to go to college, just before Nynke began work in a stable and just before Nynke's mother showed her a scrapbook from social work school. Nynke then decided to enroll in veterinary studies.
>
> *(1994: 436)*

This abbreviated commentary should give you some idea of the detail involved in this type of analysis. Of course, an entire analysis would be more extensive and pursue issues in greater depth. This full, idiographic examination, however, tells us nothing about people in general. It offers nothing in the way of a theory about why people choose to attend college.

cross-case analysis An analysis that involves an examination of more than one case; this can be either a variable-oriented or case-oriented analysis.

variable-oriented analysis An analysis that describes and/or explains a particular variable.

case-oriented analysis An analysis that aims to understand a particular case or several cases by looking closely at the details of each.

Even so, in addition to understanding one person in great depth, the researcher sees the critical elements of the subject's experiences as instances of more-general social concepts or variables. For example, Nynke's mother's social work training can also be seen as "mother's education." Her friend's decision can be seen as "peer influence." More specifically, these could be seen as independent variables having an impact on the dependent variable of attending college.

Of course, one case does not a theory make—hence Huberman and Miles refer to cross-case analysis, in which the researcher turns to other subjects, looking into the full details of their lives as well but paying special attention to the variables that seemed important in the first case. How much and what kind of education did other subjects' mothers have? Is there any evidence of close friends attending college?

Some subsequent cases will closely parallel the first one in the apparent impact of particular variables. Other cases will bear no resemblance to the first. These latter cases may require the identification of other important variables, which may invite the researcher to explore why some cases seem to reflect one pattern while others reflect another.

Grounded Theory Method

The cross-case method just described should sound somewhat familiar. In the discussion of grounded theory in Chapter 10, we saw how qualitative researchers sometimes attempt to establish theories on a purely inductive basis. This approach begins with observations rather than hypotheses and seeks to discover patterns and develop theories

---

Grounded Theory Method (GTM) An inductive approach to research, introduced by Barney Glaser and Anselm Strauss, in which theories are generated solely from an examination of data rather than being derived deductively.

constant comparative method A component of the Grounded Theory Method in which observations are compared with one another and with the evolving inductive theory.

---

from the ground up, with no preconceptions, though some research may build and elaborate on earlier grounded theories.

Grounded theory was first developed by the sociologists Barney Glaser and Anselm Strauss (1967) in an attempt to come to grips with their clinical research in medical sociology. Since then, it has evolved as a method, with the cofounders taking it in slightly different directions. The following discussion will deal with the basic concepts and procedures of the **Grounded Theory Method (GTM).**

In addition to the fundamental, inductive tenet of building theory from data, GTM employs the **constant comparative method.** As Glaser and Strauss originally described this method, it involved four stages (1967: 105–13):

1. "Comparing incidents applicable to each category." As Glaser and Strauss researched the reactions of nurses to the possible death of patients in their care, the researchers found that the nurses were assessing the "social loss" attendant upon a patient's death. Once this concept arose in the analysis of one case, they looked for evidence of the same phenomenon in other cases. When they found the concept arising in the cases of several nurses, they compared the different incidents. This process is similar to conceptualization as described in Chapter 5—specifying the nature and dimensions of the many concepts arising from the data.

2. "Integrating categories and their properties." Here the researcher begins to note relationships among concepts. In the assessment of social loss, for example, Glaser and Strauss found that nurses took special notice of a patient's age, education, and family responsibilities. For these relationships to emerge, however, it was necessary for the researchers to have noticed all these concepts.

3. "Delimiting the theory." Eventually, as the patterns of relationships among concepts become clearer, the researcher can ignore some of the concepts that were initially noted but are evidently irrelevant to the inquiry. In addition

to the number of categories being reduced, the theory itself may become simpler. In the examination of social loss, for example, Glaser and Strauss found that the assessment processes could be generalized beyond nurses and dying patients: They seemed to apply to the ways all staff dealt with all patients (dying or not).

4. "Writing theory." Finally, the researcher must put his or her findings into words to be shared with others. As you may have already experienced for yourself, the act of communicating your understanding of something actually modifies and even improves your own grasp of the topic. In GTM, the writing stage is regarded as a part of the research process. A later section of this chapter (on memoing) elaborates on this point.

This brief overview should give you an idea of how grounded theory proceeds. The many techniques associated with GTM can be found both in print and on the web. One key publication is Anselm Strauss and Juliet Corbin's *Basics of Qualitative Research* (1998), which elaborates on and extends many of the concepts and techniques found in the original Glaser/Strauss volume. On the web, you can search for "grounded theory" to see a wealth of articles.

GTM is only one analytic approach to qualitative data. In the remainder of this section, we'll take a look at some other specialized techniques.

Semiotics

Semiotics is commonly defined as the "science of signs" and has to do with symbols and meanings. It's commonly associated with content analysis, which was discussed in Chapter 11, though it can be applied in a variety of research contexts.

Peter Manning and Betsy Cullum-Swan (1994: 466) offer some sense of the applicability of semiotics, as follows: "Although semiotics is based on language, language is but one of the many sign systems of varying degrees of unity, applicability, and complexity. Morse code, etiquette, mathematics, music, and even highway signs are examples of semiotic systems."

SIGN	MEANING
1. Poinsettia	a. Good luck
2. Horseshoe	b. First prize
3. Blue ribbon	c. Christmas
4. "Say cheese"	d. Acting
5. "Break a leg"	e. Smile for a picture

FIGURE 13-1

Matching Signs and their Meanings

There is no meaning inherent in any sign, however. Meanings reside in minds. So, a particular sign means something to a particular person. However, the agreements we have about the meanings associated with particular signs make semiotics a social science. As Manning and Cullum-Swan point out:

> For example, a lily is an expression linked conventionally to death, Easter, and resurrection as a content. Smoke is linked to cigarettes and to cancer, and Marilyn Monroe to sex. Each of these connections is social and arbitrary, so that many kinds of links exist between expression and content.
>
> *(1994: 466)*

To explore this contention, see if you can link the signs with their meanings in Figure 13-1. I'm confident enough that you know all the "correct" associations that there's no need for me to give the answers. (OK, you should have said 1c, 2a, 3b, 4e, 5d.) The point is this: What do any of these signs have to do with their "meanings"? Draft an e-mail message to a Martian social scientist explaining the logic at work here. (You might want to include some "emoticons" like :) —another example of semiotics.)

There is no doubt a story behind each of the linkages in Figure 13-1, and the meanings you and I "know" today have been socially constructed. Semiotic analysis involves a search for the

semiotics The study of signs and the meanings associated with them. This is commonly associated with content analysis.

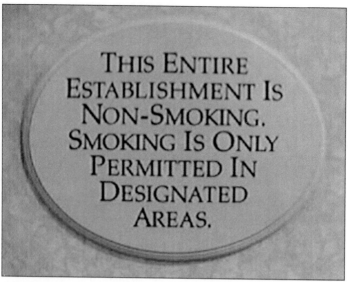

FIGURE 13-2
Mixed Signals?

meanings intentionally or unintentionally attached to signs.

Consider the sign shown in Figure 13-2, from a hotel lobby in Portland, Oregon. What's being communicated by the rather ambiguous sign? The first sentence seems to be saying that the hotel is up-to-date with the current move away from tobacco in the United States. Guests who want a smoke-free environment need look no farther: This is a healthy place to stay. At the same time, says the second sentence, the hotel would not like to be seen as inhospitable to smokers. There's room for everyone under this roof. No one need feel excluded. This sign is more easily understood within a marketing paradigm than one of logic.

The "signs" examined in semiotics, of course, are not limited to this kind of sign. Most are quite different, in fact. *Signs* are any things that are assigned special meanings. They can include logos, animals, people, and consumer products. Sometimes the symbolism is a bit subtle. A classic analysis can be found in Erving Goffman's *Gender Advertisements* (1979). Goffman focused on advertising pictures found in magazines and newspapers. The overt purpose of the ads, of course, was to sell specific products. But what else was communi-

cated, Goffman asked. What in particular did the ads say about men and women?

Analyzing pictures containing both men and women, Goffman was struck by the fact that men were almost always bigger and taller than the women accompanying them. (In many cases, in fact, the picture managed to convey the distinct impression that the women were merely accompanying the men.) Although the most obvious explanation is that men are, on average, heavier and taller than women, Goffman suggested the pattern had a different meaning: that size and placement implied *status*. Those larger and taller presumably had higher social standing—more power and authority (1979: 28). Goffman suggested that the ads communicated that men were more important than women.

In the spirit of Freud's comment that "sometimes a cigar is just a cigar" (he was a smoker), how would you decide whether the ads simply reflected the biological differences in the average sizes of men and women or whether they sent a message about social status? In part, Goffman's conclusion was based on an analysis of the exceptional cases: those in which the women appeared taller than the men. In these cases, the men were typically of a lower social status—the chef beside the society matron, for example. This confirmed Goffman's main point that size and height indicated social status.

The same conclusion was to be drawn from pictures with men of different heights. Those of higher status were taller, whether it was the gentleman speaking to a waiter or the boss guiding the work of his younger assistants. Where actual height was unclear, Goffman noted the placement of heads in the picture. The assistants were crouching down while the boss leaned over them. The servant's head was bowed so it was lower than that of the master.

The latent message conveyed by the ads, then, was that the higher a person's head appeared in the ad, the more important that person was. And in the great majority of ads containing men and women, the former were clearly portrayed as more important. The subliminal message in the ads, whether intended or not, was that men are more powerful than women and enjoy a higher status.

Goffman examined several differences besides physical size in the portrayal of men and women. As another example, men were typically portrayed in active roles, women in passive ones. The (male) doctor examined the child while the (female) nurse or mother looked on, often admiringly. A man guided a woman's tennis stroke (all the while keeping his head higher than hers). A man gripped the reins of his galloping horse, while a woman rode behind him with her arms wrapped around his waist. A woman held the football, while a man kicked it. A man took a photo, which contained only women.

Goffman suggested that such pictorial patterns subtly perpetuated a host of gender stereotypes. Even as people spoke publicly about gender equality, these advertising photos established a quiet backdrop of men and women in the "proper roles."

tiny of the way we converse with one another. In the examination of ethnomethodology in Chapter 10, you saw some examples of conversation analysis. Here we'll look a little more deeply into that technique.

David Silverman (1999), reviewing the work of other CA theorists and researchers, speaks of three fundamental assumptions. First, conversation is a socially structured activity. Like other social structures, it has established rules of behavior. For example, we're expected to take turns, with only one person speaking at a time. In telephone conversations, the person answering the call is expected to speak first (e.g., "Hello"). You can verify the existence of this rule, incidentally, by picking up the phone without speaking. You may recall that this is the sort of thing ethnomethodologists tend to do.

Second, Silverman points out that conversations must be understood contextually. The same utterance will have different meanings in different contexts. For example, notice how the meaning of "Same to you!" varies if preceded by "I don't like your looks" or by "Have a nice day."

Third, CA aims to understand the structure and meaning of conversation through excruciatingly accurate transcripts of conversations. Not only are the exact words recorded, but all the uhs, ers, bad grammar, and pauses are also noted. Pauses, in fact, are recorded to the nearest tenth of a second.

The practical uses of this type of analysis are many. Ann Marie Kinnell and Douglas Maynard (1996), for example, analyzed conversations between staff and clients at an HIV-testing clinic to examine how information about safe sex was communicated. Among other things, they found that the staff tended to provide standard information rather than try to speak directly to a client's specific circumstances. Moreover, they seemed reluctant to give direct advise about safe sex, settling for information alone.

Conversation Analysis

Ethnomethodology, as you'll recall, aims to uncover the implicit assumptions and structures in social life. **Conversation analysis (CA)** seeks to pursue that aim through an extremely close scru-

conversation analysis (CA) A meticulous analysis of the details of conversation, based on a complete transcript that includes pauses, hems, and also haws.

These discussions should give you some sense of the variety of qualitative analysis methods available to researchers. Now let's look at some of the data-processing and data-analysis techniques commonly used in qualitative research.

Qualitative Data Processing

Let me begin this section with a warning. The activity we're about to examine is as much art as science. At the very least, there are no cut-and-dried steps that guarantee success.

It's a lot like learning how to paint with watercolors or compose a symphony. Education in such activities is certainly possible, and university courses are offered in both. Each has its own conventions and techniques as well as tips you may find useful as you set out to create art or music. However, instruction can carry you only so far. The final product must come from you. Much the same can be said of qualitative data processing.

This section presents some ideas relating to the coding of qualitative data, writing memos, and mapping concepts graphically. Although far from a "how-to" manual, these ideas give a useful starting point for finding order in qualitative data.

Coding

Whether you've engaged in participant observation, in-depth interviewing, collecting biographical narratives, doing content analysis, or some other form of qualitative research, you'll now be in the possession of a growing mass of data—most typically in the form of textual materials. Now what do you do?

The key process in the analysis of qualitative social research data is *coding*—classifying or categorizing individual pieces of data—coupled with some kind of retrieval system (see Chapter 11). Together, these procedures allow you to retrieve materials you may later be interested in.

Let's say you're chronicling the growth of a social movement. You recall writing up some notes about the details of the movement's earliest

beginnings. Now you need that information. If all your notes have been catalogued by topic, retrieving those you need should be straightforward. As a simple format for coding and retrieval, you might have created a set of file folders labeled with various topics, such as "History." Data retrieval in this case means pulling out the "History" folder and rifling through the notes it contains until you find what you need.

As you'll see later in this chapter, there are now sophisticated computer programs that allow for a faster, more certain, and more precise retrieval process. Rather than looking through a "History" file, you can go directly to notes dealing with the "Earliest History" or the "Founding" of the movement.

Coding has another, even more important purpose. As discussed earlier, the aim of data analysis is the discovery of patterns among the data, patterns that point to theoretical understandings of social life. The coding and relating of concepts is key to this process and requires a more refined system than a set of manila folders. In this section, we'll assume that you'll be doing your coding manually. A later section of the chapter will illustrate the use of computer programs for qualitative data analysis.

Coding Units

As you may recall from the earlier discussion of content analysis, for statistical analysis it's important to identify a standardized unit of analysis prior to coding. If you were comparing American and French novels, for example, you might evaluate and code sentences, paragraphs, chapters, or whole books. It would be important, however, to code the same units for each novel analyzed. This uniformity is necessary in a quantitative analysis, as it allows us to report something like "23 percent of the paragraphs contained metaphors." This is only possible if we've coded the same unit—paragraphs—in each of the novels.

Coding data for a qualitative analysis, however, is quite different. The *concept* is the organizing principle for qualitative coding. Here the units of text appropriate for coding will vary within a given document. Thus, in a study of organizations, "Size" might require only a few words per coding unit, whereas "Mission" might take a few pages. Or, a lengthy description of a heated stockholders meeting might be coded as "Internal Dissent."

Realize also that a given code category may be applied to textual materials of quite different lengths. For example, some references to the organization's mission may be brief, others lengthy. Whereas standardization is a key principle in quantitative analysis, this is not the case in qualitative analysis.

Coding as a Physical Act

Before continuing with the logic of coding, let's take a moment to see what it actually looks like. Lofland and his colleagues offer this description of manual filing:

> Prior to the widespread availability of personal computers beginning in the late 1980s, coding frequently took the specific physical form of filing. The researcher established an expanding set of file folders with code names on the tabs and physically placed either the item of data itself or a note that referenced its location in another file folder. Before photocopying was easily available and cheap, some fieldworkers typed their fieldnotes with carbon paper, wrote codes in the margins of the copies of the notes, and cut them up with scissors. They then placed the resulting slips of paper in corresponding file folders.
>
> *(2006: 203)*

As Lofland and his colleagues point out, personal computers have greatly simplified this task. However, the image of slips of paper that contain text and are put in folders representing code categories is useful for understanding the process of coding. In the next section, when I suggest that we code a textual passage with a certain code, imagine that we have the passage typed on a slip of paper and that we place it in a file folder bearing the name of the code. Whenever we assign two codes to a passage, imagine placing duplicate copies of the passage in two different folders representing the two codes.

Creating Codes

So, what should your code categories be? Glaser and Strauss (1967: 101f) allow for the possibility of coding data for the purpose of testing hypotheses that have been generated by prior theory. In that case, then, the codes would be suggested by the theory, in the form of variables.

In this section, however, we're going to focus on the more common processes of open coding, axial coding, and selective coding. Strauss and Corbin (1998: 102) describe **open coding** as follows:

> To uncover, name, and develop concepts, we must open up the text and expose the thoughts, ideas, and meanings contained therein. Without this first analytic step, the rest of the analysis and the communication that follows could not occur. Broadly speaking, during open coding, data are broken down into discrete parts, closely examined, and compared for similarities and differences. Events, happenings, objects, and actions/interactions that are found to be conceptually similar in nature or related in meaning are grouped under more abstract concepts termed "categories."

Although the analysis of data will quickly advance to an iterative interplay of the three types of coding, open coding is the logical starting point. Beginning with some body of text (part of an interview, for example), you read and reread a passage, seeking to identify the key concepts contained within it. Any particular piece of data may be given several codes, reflecting as many concepts. For example, notice all the concepts contained in this comment by a student interviewee:

> I thought the professor should have given me at least partial credit for the homework I turned in.

Some obvious codes are "Professor," "Homework," and "Grading." The result of open coding is the

---

open coding The initial classification and labeling of concepts in qualitative data analysis. In open coding, the codes are suggested by the researchers' examination and questioning of the data.

identification of numerous concepts relevant to the subject under study. The open coding of more and more text will lengthen the list of codes.

Axial coding aims to identify the *core* concepts in the study. Although axial coding uses the results of open coding, more concepts can be identified through continued open coding after the axial coding has begun. Axial coding involves a regrouping of the data, in which the researcher uses the open-code categories and looks for more-analytic concepts. For example, the passage just given also carries the concept of "perceptions of fairness," which might appear frequently in the student interviews, thereby suggesting that it's an important element in understanding students' concerns. Another axial code reflected in the student comment might be "power relationships," because the professor is seen to exercise power over the student.

Selective coding seeks to identify *the* central code in the study: the one that the other codes all related to. Both of the axial codes just mentioned might be restructured as aspects of a more general concept: "professor–student relationships." Of course, in a real data analysis, decisions such as the ones we've been discussing would arise from masses of textual data, not from a single quotation. The basic notion of the Grounded Theory Method is that patterns of relationships can be teased out of an extensive, in-depth examination of a large body of observations.

Here's a concrete example to illustrate how you might engage in this form of analysis. Suppose you're interested in the religious bases for homophobia. You've interviewed some people opposed to homosexuality who cite a religious basis for their feelings. Specifically, they refer you

---

axial coding A reanalysis of the results of open coding in the Grounded Theory Method, aimed at identifying the important, general concepts.

selective coding In Grounded Method Theory, this analysis builds on the results of open coding and axial coding to identify the central concept that organizes the other concepts that have been identified in a body of textual materials.

---

to these passages in the Book of Leviticus (Revised Standard Version):

18:22   You shall not lie with a male as with a woman; it is an abomination.

20:13   If a man lies with a male as with a woman, both of them have committed an abomination; they shall be put to death, their blood is upon them.

Although the point of view expressed here seems unambiguous, you might decide to examine it in more depth. Perhaps a qualitative analysis of Leviticus can yield a fuller understanding of where these injunctions against homosexuality fit into the larger context of Judeo-Christian morality.

Let's start our analysis by examining the two passages just quoted. We might begin by coding each passage with the label "Homosexuality." This is clearly a key concept in our analysis. Whenever we focus on the issue of homosexuality in our analysis of Leviticus, we want to consider these two passages.

Because homosexuality is such a key concept, let's look more closely into what it means within the data under study. We first notice the way *homosexuality* is identified: a man lying with a man "as with a woman." Although we can imagine a lawyer seeking admission to heaven saying, "But here's my point; if we didn't actually lie down . . ." it seems safe to assume the passage refers to having sex, though what specific acts might or might not be included isn't clear.

Notice, however, that the injunctions appear to concern *male* homosexuality only; lesbianism is not mentioned. In our analysis, then, each of these passages might also be coded "Male Homosexuality." This illustrates two more aspects of coding: (1) Each unit can have more than one code and (2) hierarchical codes (one included within another) can be used. Now each passage has two codes assigned to it.

An even more general code might be introduced at this point: "Prohibited Behavior." This is important for two reasons. First, homosexuality is not inherently wrong, from an analytic standpoint. The purpose of the study is to examine the way it's made wrong by the religious texts in question. Second, our study of Leviticus may turn up other behaviors that are prohibited.

There are at least two more critical concepts in the passages: "Abomination" and "Put to Death." Notice that although these are clearly related to "Prohibited Behavior," they are hardly the same. Parking without putting money in the meter is prohibited, but few would call it an abomination and fewer still would demand the death penalty for that transgression. Let's assign these two new codes to our first two passages.

At this point, we want to branch out from the two key passages and examine the rest of Leviticus. We therefore examine and code each of the remaining chapters and verses. In our subsequent analyses, we'll use the codes we have already and add new ones as appropriate. When we do add new codes, it will be important to review the passages already coded to see whether the new codes apply to any of them.

Here are the passages we decide to code "Abomination." (I've boldfaced the abominations.)

7:18   If any of the flesh of the sacrifice of **his peace offering is eaten on the third day,** he who offers it shall not be accepted, neither shall it be credited to him; it shall be an abomination, and he who eats of it shall bear his iniquity.

7:21   And if any one **touches an unclean thing,** whether the uncleanness of man or an unclean beast or any unclean abomination, **and then eats of the flesh of the sacrifice** of the LORD's peace offerings, that person shall be cut off from his people.

11:10   But **anything in the seas or the rivers that has not fins and scales,** of the swarming creatures in the waters and of the living creatures that are in the waters, is an abomination to you.

11:11   They shall remain an abomination to you; **of their flesh you shall not eat, and their carcasses you shall have in abomination.**

11:12   **Everything in the waters that has not fins and scales** is an abomination to you.

11:13   And these you shall have in abomination among the birds, **they shall not be eaten,** they are an abomination: the **eagle,** the **vulture,** the **osprey,**

11:14   the **kite,** the **falcon** according to its kind,

11:15   every **raven** according to its kind,

11:16   the **ostrich,** the **nighthawk,** the **sea gull,** the **hawk** according to its kind,

11:17   the **owl,** the **cormorant,** the **ibis,**

11:18   the **water hen,** the **pelican,** the carrion vulture,

11:19   the **stork,** the **heron** according to its kind, the **hoopoe,** and the **bat.**

11:20   **All winged insects that go upon all fours** are an abomination to you.

11:41   **Every swarming thing** that swarms upon the earth is an abomination; it shall not be eaten.

11:42   Whatever goes on its belly, and whatever goes on all fours, or whatever has many feet, all the **swarming things** that swarm upon the earth, you shall not eat; for they are an abomination.

11:43   You shall not make yourselves abominable with any swarming thing that swarms; and you shall not defile yourselves with them, lest you become unclean.

18:22   You shall not **lie with a male as with a woman;** it is an abomination.

19:6   It shall be eaten the same day you offer it, or on the morrow; and anything left over until the third day shall be burned with fire.

19:7   **If it is eaten at all on the third day,** it is an abomination; it will not be accepted,

19:8   and every one who eats it shall bear his iniquity, because he has profaned a holy thing of the LORD; and that person shall be cut off from his people.

20:13   **If a man lies with a male as with a woman,** both of them have committed an abomination; they shall be put to death, their blood is upon them.

20:25　You shall therefore make a distinction between the clean beast and the unclean, and between the unclean bird and the clean; **you shall not make yourselves abominable by beast or by bird or by anything with which the ground teems,** which I have set apart for you to hold unclean.

Male homosexuality, then, isn't the only abomination identified in Leviticus. As you compare these passages, looking for similarities and differences, it will become apparent that most of the abominations have to do with dietary rules—specifically those potential foods deemed "unclean." Other abominations flow from the mishandling of ritual sacrifices. "Dietary Rules" and "Ritual Sacrifices" thus represent additional codes to be used in our analysis.

Earlier, I mentioned the death penalty as another concept to be explored in our analysis. When we take this avenue, we discover that many behaviors besides male homosexuality warrant the death penalty. Among them are these:

20:2　Giving your children to Molech (human sacrifice)

20:9　Cursing your father or mother

20:10　Adultery with your neighbor's wife

20:11　Adultery with your father's wife

20:12　Adultery with your daughter-in-law

20:14　Taking a wife and her mother also

20:15　Men having sex with animals (the animals are to be killed, also)

20:16　Women having sex with animals

20:27　Being a medium or wizard

24:16　Blaspheming the name of the Lord

24:17　Killing a man

---

memoing Writing memos that become part of the data for analysis in qualitative research such as grounded theory. Memos can describe and define concepts, deal with methodological issues, or offer initial theoretical formulations.

As you can see, the death penalty is broadly applied in Levicitus: everything from swearing to murder, including male homosexuality somewhere in between.

An extended analysis of prohibited behavior, short of abomination and death, also turns up a lengthy list. Among them are slander, vengeance, grudges, cursing the deaf, and putting stumbling blocks in front of blind people. In chapter 19, verse 19, Leviticus quotes God as ordering, "You shall not let your cattle breed with a different kind; you shall not sow your field with two kinds of seed; nor shall there come upon you a garment of cloth made of two kinds of stuff." Shortly thereafter, he adds, "You shall not eat any flesh with the blood in it. You shall not practice augury or witchcraft. You shall not round off the hair on your temples or mar the edges of your beard." Tattoos were prohibited, though Leviticus is silent on body piercing. References to all of these practices would be coded "Prohibited Acts" and perhaps given additional codes as well (recall "Dietary Rules").

I hope this brief glimpse into a possible analysis will give you some idea of the process by which codes are generated and applied. You should also have begun to see how such coding would allow you to better understand the messages being put forward in a text and to retrieve data appropriately as you need them.

Memoing

In the Grounded Theory Method, the coding process involves more than simply categorizing chunks of text. As you code data, you should also be using the technique of **memoing**—writing memos or notes to yourself and others involved in the project. Some of what you write during analysis may end up in your final report; much of it will at least stimulate what you write.

In GTM, these memos have a special significance. Strauss and Corbin (1998: 217) distinguish three kinds of memos: code notes, theoretical notes, and operational notes.

Code notes identify the code labels and their meanings. This is particularly important because, as in all social science research, most of the terms we use with technical meanings also have meanings in everyday language. It's essential, therefore, to write down a clear account of what you mean by the codes used in your analysis. In the Leviticus analysis, for example, you would want a code note regarding the meaning of "Abomination" and how you've used that code in your analysis of text.

Theoretical notes cover a variety of topics: reflections of the dimensions and deeper meanings of concepts, relationships among concepts, theoretical propositions, and so on. All of us have ruminated over the nature of something, trying to think it out, to make sense out of it. In qualitative data analysis, it's vital to write down these thoughts, even those you'll later discard as useless. They will vary greatly in length, though you should limit them to a single main thought so that you can sort and organize them later. In the Leviticus analysis, one theoretical note might discuss the way that most of the injunctions implicitly address the behavior of men, with women being mostly incidental.

Operational notes deal primarily with methodological issues. Some will draw attention to data-collection circumstances that may be relevant to understanding the data later on. Others will consist of notes directing future data collection.

Writing these memos occurs throughout the data-collection and analysis process. Thoughts demanding memos will come to you as you reread notes or transcripts, code chunks of text, or discuss the project with others. It's a good idea to get in the habit of writing out your memos as soon as possible after the thoughts come to you.

Notice that whereas we often think of writing as a linear process, starting at the beginning and moving through to the conclusion, memoing is very different. It might be characterized as a process of creating chaos and then finding order within it.

To explore this process further, refer to the works cited in this discussion and at the end of the chapter. You'll also find a good deal of information on the web. For example, see the link, on this book's website, for Barney Glaser's rules on memoing. Ultimately, the best education in this process

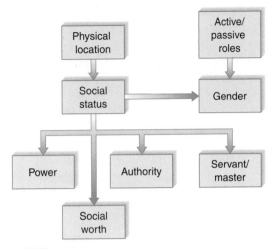

FIGURE 13-3

An Example of Concept Mapping

comes from practice. Even if you don't have a research project underway, you can practice now on class notes. Or start a journal and code it.

Concept Mapping

It should be clear by now that qualitative data analysts spend a lot of time committing thoughts to paper (or to a computer file), but this process is not limited to text alone. Often, we can think out relationships among concepts more clearly by putting the concepts in a graphic format, a process called **concept mapping.** Some researchers put all their major concepts on a single sheet of paper, whereas others spread their thoughts across several sheets of paper, blackboards, magnetic boards, computer pages, or other media. Figure 13-3 shows how we might think out some of the concepts of Goffman's examination of gender and advertising. (This image was created through the use of Inspiration, a concept-mapping computer program.)

concept mapping The graphic display of concepts and their interrelations, useful in the formulation of theory.

Incidentally, many of the topics discussed in this section have useful applications in quantitative as well as qualitative analyses. Certainly, concept mapping is appropriate in both types of analysis. The several types of memos would also be useful in both. And the discussion of coding readily applies to the coding of open-ended questionnaire responses for the purpose of quantification and statistical analysis. (We'll look at coding again in the next chapter, on quantifying data.)

Having noted the overlap of qualitative and quantitative techniques, it seems fitting now to address an instrument that is primarily associated with quantitative research but that is proving quite valuable for qualitative analysts as well—the personal computer.

Computer Programs for Qualitative Data

The advent of computers—mainframe and personal—has been a boon to quantitative research, allowing the rapid calculation of extremely complex statistics. The importance of the computer for qualitative research has been somewhat more slowly appreciated. Some qualitative researchers were quick to adapt the basic capacities of computers to nonnumerical tasks, but it took a bit longer for programmers to address the needs of qualitative research per se. Today, however, several powerful programs are available.

Let's start this section with a brief overview of some of the ways you can use basic computer tools in qualitative research. Perhaps only those who can recall hours spent with carbon paper and Wite-Out can fully appreciate the glory of computers in this regard. "Easier editing" and "easier duplication" simply do not capture the scope of the advance.

Moving beyond the basic recording and storage of data, simple word-processing programs can be used for some data analysis. The "find" or "search" command will take you to passages containing keywords. Or, going one step further, you can type code words alongside passages in your notes so that you can search for those keywords later.

Database and spreadsheet programs can also be used for processing and analyzing qualitative data. Figure 13-4 is a simple illustration of how some of the verses from Leviticus might be manipulated within a spreadsheet. The three columns to the left represent three of the concepts we've discussed. An "X" means that the passage to the right contains that concept. As shown, the passages are sorted in such a way as to gather all those dealing with punishment by death. Another simple "sort" command would gather all those dealing with sex, with homosexuality, or any of the other concepts coded.

QDA Programs

While the simple spreadsheet illustration just given touched on how computers might be used for analyzing qualitative social research data, there is now a long list of sophisticated computer programs available for this purpose. Where the analyst's problem used to be simply finding such a program, the problem now centers on choosing one. Here are a few commonly used qualitative data analysis (QDA) programs with online sites where you can learn more about them and, often, download demo copies (see all the links on this book's website: http://www.cengage.com/sociology/babbie).

AnSWR

Atlas.ti

Ethno

Ethnograph

HyperQual

HyperRESEARCH

HyperTRANSCRIBE

MAXQDA

NVivo 8 (NUD*IST)

QDA Miner

Qualrus

SPAD

TAMS

sex	homosex	death	Verse	Passage
X	X	X	20:13	If a man lies with a male as with a woman, both of them have committed an abomination; they shall be put to death, their blood is upon them.
X		X	20:12	If a man lies with his daughter-in-law, both of them shall be put to death; they have committed incest, their blood is upon them.
X		X	20:15	If a man lies with a beast, he shall be put to death; and you shall kill the beast.
		X	20:09	For every one who curses his father or his mother shall be put to death; he has cursed his father or his mother, his blood is upon him.
		X	20:02	Any man of the people of Israel, or of the strangers that sojourn in Israel, who gives any of his children to Molech shall be put to death.
X	X		18:22	You shall not lie with a male as with a woman; it is an abomination.

FIGURE 13-4

Using a Spreadsheet for Qualitative Analysis

T-LAB

Weft

There are also some powerful online resources to assist you in choosing the program best suited to your needs. Sociologists at the University of Surrey, England, have prepared an overview of these and other programs with descriptions and contact information. You can find this resource at the link on this book's website.

Another excellent resource is "Choosing a CAQDAS Software Package" by Ann Lewins and Christina Silver (2006), which can be found at the corresponding link on this book's website. This will familiarize you with some of the key features in QDA programs and will help you choose which one is best suited to your purposes. This website provides a brief description of each of the programs listed, along with contact links.

Let's turn now to a couple of illustrations of QDA programs at work. While all the available programs differ somewhat from one another, I think these illustrations will give you a general sense of how to use computers to analyze qualitative data. We'll begin with a brief examination of Leviticus, and then we'll examine a project focused

on understanding the experiences of women film directors.

Leviticus as Seen through NUD*IST

Let's take a closer look at how qualitative data analysis programs operate. We'll consider one of the programs just mentioned, NUD*IST (Nonnumeric Unstructured Data, Index Searching, and Theorizing). This popular program for teaching qualitative social research offers a fair representation of QDA programs. (Note that NVivo 8 has now superseded NUD*IST; however, some researchers still use the older variant of the software.)

Although it is possible to type directly into NUD*IST the text materials to be coded, usually materials already in existence—such as field notes or, in this case, the verses of Leviticus—are *imported* into the program. Menu-based commands do this easily, though the text must be in a *plaintext* format (that is, without word-processing or other formatting).

Figure 13-5 shows how the text is displayed within NUD*IST. For the illustrations in this section, I've used the Macintosh version of NUD*IST, but the Windows version is similar. The

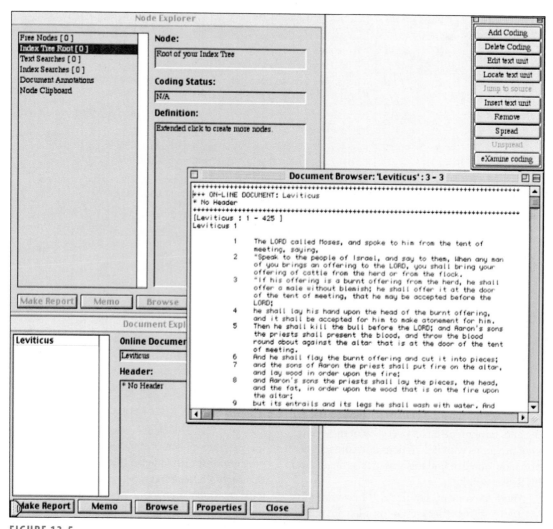

FIGURE 13-5

How Text Materials Are Displayed in NUD*IST

film director illustration uses the Windows version of NVivo, so you can note the difference and similarities in the two platforms.

To see the document, select its name in the "Document Explorer" window and click "Browse." The text window can be resized and moved around the screen to suit your taste.

Note the set of buttons in the upper right corner of the illustration. These allow you to select portions of the text for purposes of editing, coding, and other operations.

Now let's create a concept code: "homosex." This will stand for references to male homosexuality. Figure 13-6 shows what the creation of a concept code looks like.

As we create codes for our concepts, we can use them to code the text materials. Figure 13-7 illustrates how this is done. In the document browser, you can see that verse 20:13 has been selected (indicated by the box outline around this verse). Having done that, we click the button labeled "Add Coding" (not shown in this illustration).

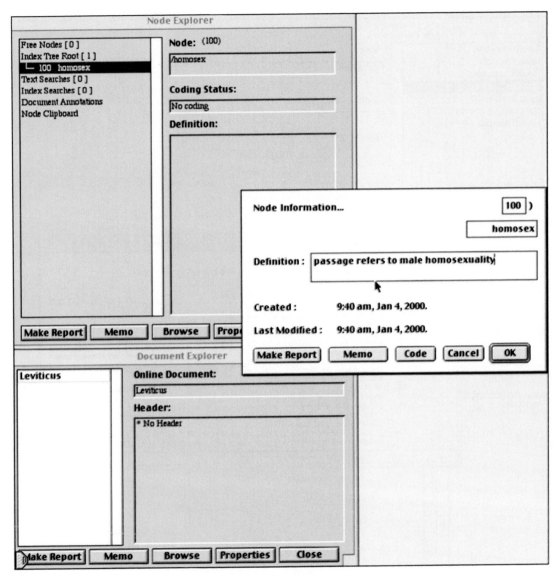

FIGURE 13-6
Creating the Code "homosex"

This prompts the computer to ask us to identify the appropriate code. The easiest way to respond is to click the "Browse" button, which presents you with a list of the current codes. In this example, I selected "homosex" and entered the code ID (100).

As text materials are coded, the program can then be used for purposes of analysis. As a simple example, we might want to pull together all the passages coded "homosex." This would allow us to see them all at once, looking for similarities and differences.

Figure 13-8 shows how NUD*IST would bring together the passages referring to male homosexuality. To do this, all you do is select the code name in the "Node Explorer" window and click the "Make Report" button.

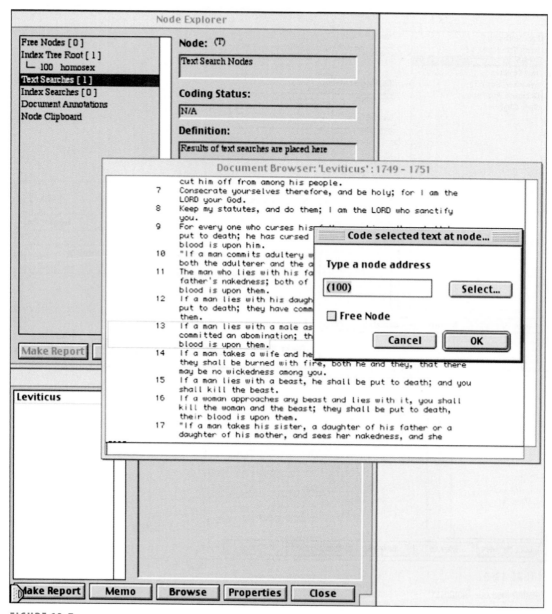

FIGURE 13-7

Coding a Text Passage

This simple example illustrates the possibilities opened up by a program designed specifically for qualitative data analysis. To get a little deeper into its use, let's shift to a different research example.

Sandrine Zerbib is a French sociologist interested in understanding the special difficulties faced by women breaking into the male-dominated world of film direction. To address this issue, she interviewed 30 women directors in depth. Having compiled hours of recorded interviews, she turned to NVivo as a vehicle for analysis. In the next section, she directly describes her experiences with the ongoing process of qualitative data analysis.

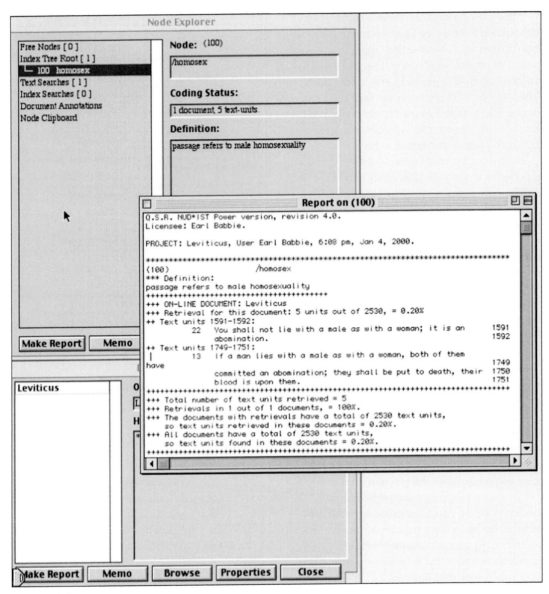

FIGURE 13-8

Reporting on "homosex"

Using NVivo to Understand Women Film Directors, by Sandrine Zerbib

Most software for qualitative analysis allows researchers to simultaneously analyze several interviews from different interviewers. However, I find it more efficient to start by importing only one interview into NVivo. Because you will have transcribed or at least read your interviews beforehand, you may be able to select the interview you think will be most fruitful. You should trust yourself, because you are becoming an expert in what you are currently studying and also because comparing and contrasting interviews should help you get a sense of how accurate your analysis is.

After having completed about 30 interviews with women filmmakers, I had a sense of what the main themes were, because they kept coming up in each interview. Nevertheless, I needed a tool for synthesizing those pages and pages of interviews. I chose to start with my interview with Joy. I had made a note to myself to use her interview as a starting point. An older film director, she seemed to have strong points she wanted to get across.

In Figure 13-9, my interview with Joy has been imported as a "text only" file. (Only part of the file is visible in the window.)

At this point you are ready to enjoy the coding process. You can simply highlight words, sentences, or sections and add nodes (i.e., codes) to it. The first step is to create "free nodes," that is, nodes independent of one another. How much text you should highlight per code is a decision you will have to make. However, keep in mind that you will have to use those quotes in the writing part of your research. You will need to be convincing. You also want to deconstruct the whole interview. Try to not leave anything out. It is easier to forgo using a quote because you have found a better one later than to have nothing to use because you were not consistent enough in your dissection of the interview.

When you create a node, you first want to use wide categories that would be more inclusive of other potential quotes. But you also want to be specific enough for your coding system to have validity. In Figure 13-10, for instance, I have created the free node "past" because my interviewee referred to the past as being extremely challenging for women who wanted to be film directors. There were very few women directors back then, many fewer than today. I decided to add a definition of this node so that I could remember why I used "past" as a node. I also anticipated having another free node called "today." Then I could move the "change" node to the index tree root and create "past" and "today" as subnodes under "change."

In Figure 13-11, I have highlighted a passage that deals with several things. Joy talks about the Directors Guild of America (DGA, or the directors' union) and more specifically about the efforts of its president. She also expresses her feelings toward

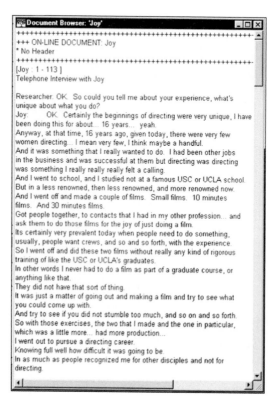

FIGURE 13-9

Text of Interview with Joy

gender inequalities. According to her, having talent is not enough in Hollywood if there is a bias against women. I decided to add two nodes to this quote, "DGA," which I needed to create, and "discrimination," which I had already created.

In Figure 13-12, I have attempted to transform some free nodes into index trees. The software is flexible enough for me to move nodes, rename them, or see what quotes are under each node. You can attach a different node to a quote you have wrongly coded. It is preferable to start with free nodes before you build a hierarchy of codes (or tree), because it takes time and patience to understand how categories are linked to one another. Coding other interviews should help you organize your coding system.

Figure 13-13 illustrates my decision to import two more interviews, Berta's and Queena's. I could browse all three interviews on the same screen.

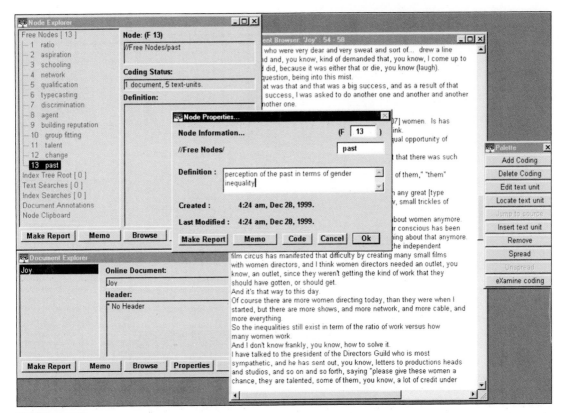

FIGURE 13-10

Creating the Code "past"

Because it was still early in the analysis process, I chose to analyze these two new interviews one by one. It was now starting to make sense; I was starting to see patterns. NVivo let me keep records of number of occurrences each node was attached to a quote, not only in Joy's interview but now also in Berta's and Queena's. With several nodes often attached to a single quote, the qualitative analysis allowed me to find out which nodes were more likely to overlap with one another.

One of my first observations was that the term *sabotage* was used fairly often by Joy and Queena. I decided to run a report that would synthesize all the quotes that I attached to the node "sabotage." Figure 13-14 shows the first page of the report created by NVivo. The program searched for all quotes under "sabotage," which is a subnode of "discrimination," for all online documents. It also provided the number assigned to each text unit, which allowed me to go back and see a quotation in the context of the whole document.

This procedure is only one of the many capabilities of this program. You may want to spend some time learning about this software before committing to it. What seems to be an efficient tool for me may not be for you. There is plenty qualitative research analysis software in the market; try to find out what works for you.

The Qualitative Analysis of Quantitative Data

Although it's important and appropriate to distinguish between qualitative and quantitative research, often to the point of discussing them

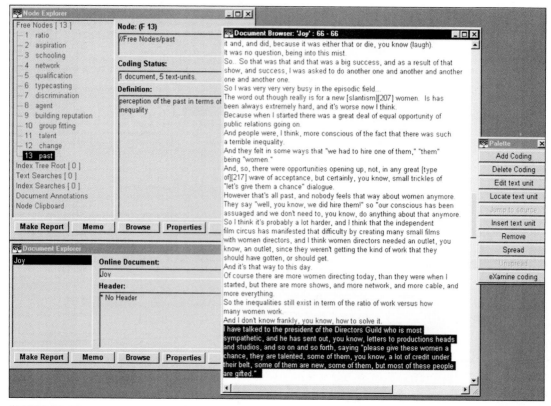

FIGURE 13-11

Coding a Passage in the Interview

separately, I don't want you to get the idea that they're incompatible or competing. Unless you can operate in both modes, you'll limit your potential as a social researcher.

In Chapter 14, I'll indicate some ways in which quantitative analyses can strengthen qualitative studies. Conversely, in this chapter I'll include an example of how quantitative data demand qualitative assessment.

Figure 13-15 presents FBI data on the hour and day of crimes committed in the United States (Maltz 1998: 401). These data are often presented in a tabular form, but notice how clearly the patterns of crime appear in this three-dimensional graph. The picture itself conveys the meaning of the statistical data. Summarizing it in the form of equations—while possibly useful for certain purposes—adds nothing to the clarity of the picture. Indeed, there hardly seems a need to describe the

pattern verbally. Here's a case where a picture is truly worth a thousand words.

Evaluating the Quality of Qualitative Research

As you've seen in earlier chapters, there are often clear guidelines for evaluating the quality of quantitative research. In the case of survey research, for example, we can note the size of the sample, the manner in which it was selected, and the completion rate achieved. The questionnaire items are standardized and open to scrutiny. And as you'll see in Chapters 14 and 16, researchers can use statistical tests to assess quantitative research findings.

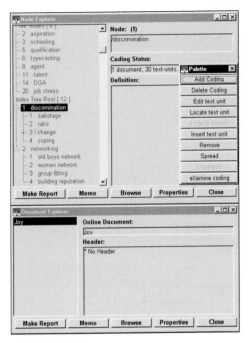

FIGURE 13-12

Creating an Index Tree

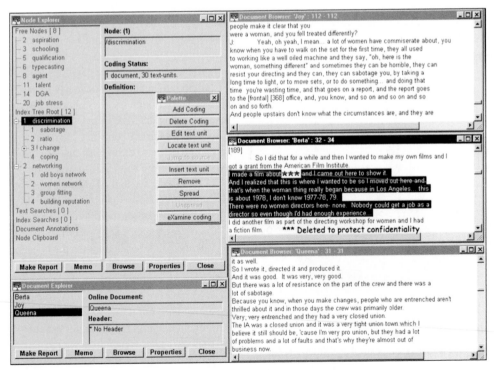

FIGURE 13-13

Adding Two More Cases to the Analysis

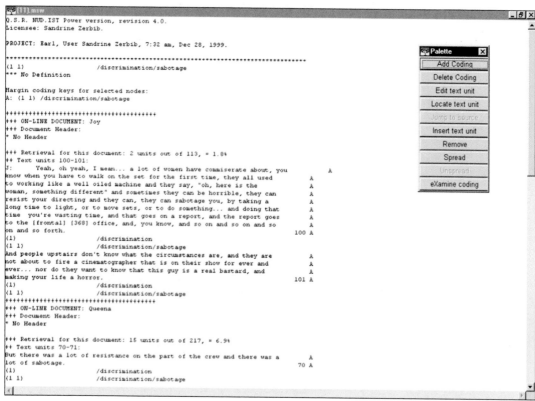

FIGURE 13-14

Analyzing the Node "sabotage"

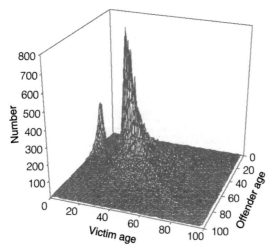

FIGURE 13-15

Number of One-on-One Homicides by Age of Victim and Age of Offender, Raw Data

Source: Michael D. Maltz, "Visualizing Homicide: A Research Note," *Journal of Quantitative Criminology* 15, no. 4 (1998): 401.

Judging the quality of qualitative research is more elusive, though no less important. Because there are many different forms of qualitative research, we'll examine some fairly general guidelines you can use to distinguish first-rate qualitative investigations from those not so well done.

In Chapter 5, we looked at two aspects of measurement quality: validity and reliability. That's a reasonable way to start our look at assessing qualitative research.

Validity, you'll recall, involves the question of whether you're measuring what you say you're measuring. Remember, most of the things social scientists measure are products of human thought and agreement, not things that exist independently of human judgment. Prejudice, for example, isn't real the way age or weight are. Nonetheless, we've all observed behaviors and orientations that we've gathered under the umbrella concept of "prejudice." To a degree, we mean the same general thing when we use the term, and we also have a lot of differences in that regard.

When you design a survey questionnaire to measure *prejudice,* it's important to assess the extent to which the questions asked and answers received actually reflect what we can agree to mean by the term. The same logic applies in qualitative research projects such as field observations or historical studies. If field researchers characterize a subject of observation as "prejudiced," you should examine their basis for saying that. Qualitative researchers, more than quantitative researchers, pay special attention to understanding life as the participants see it, so you may find the researchers in this case reporting that those who knew the subject in question also mentioned that he or she was prejudiced.

Some qualitative researchers prefer to use the term *credibility* in the place of validity in this context. This is done as a caution against the older, positivistic view that social concepts represent real phenomena that exist objectively and independently of human thought. Be warned, however, that some researchers use the term with other meanings that fall quite far from that of validity. Also, in this textbook, my use of the term *validity* explicitly denies objective reality for the concepts we use and study.

Reliability is also a reasonable criterion of quality with regard to qualitative research, though it needs to be applied appropriately. Recall this is a question of whether a measurement or observation technique would yield the same data if it were possible to measure or observe the same thing several times independently. In the case of categorizing raw data, such as data that in-depth interviews or even the open-ended answers to survey questions might produce, we can ask more than one person to undertake the coding or categorizing process independently and see if they all produce the same results. In most aspects of social research, however, the concept of reliability is more elusive, because (1) what we are observing may be constantly changing and/or (2) the act of measuring (for example, asking a question) may affect the person being studied. Still, the basic concept of reliability, which some qualitative researchers prefer to call *dependability,* is meaningful for qualitative research. Yvonna Lincoln and Egon Guba (1985), for example, proposed an "inquiry audit" for the purpose of assessing the consistency of both what was observed and the process by which it was observed.

Follow-up works by the same authors laid out several ways in which qualitative research could be assessed. Building on this foundation, several other researchers offered somewhat modified schemes for both assessing qualitative research and increasing its quality. A more recent effort, by Britain's National Centre for Social Research, sought to assist cabinet-level officials in assessing qualitative research projects that evaluated government programs. Although the study focused on the use of qualitative methods for purposes of evaluation research, the 18 questions that organized such assessments can be applied to most forms of qualitative research:

1. How credible are the findings?
2. How has knowledge or understanding been extended by the research?
3. How well does the evaluation address its original aims and purpose?
4. How well is the scope for drawing wider inference explained?
5. How clear is the basis of evaluative appraisal?
6. How defensible is the research design?
7. How well defended are the same design/target selection of cases/documents?
8. How well is the eventual sample composition and coverage described?
9. How well was the data collection carried out?
10. How well has the approach to, and formulation of, analysis been conveyed?
11. How well are the contexts of data sources retained and portrayed?
12. How well has diversity of perspective and content been explored?
13. How well has detail, depth and complexity (i.e., richness) of the data been conveyed?
14. How clear are the links between data, interpretation and conclusions—i.e., how well can the route to any conclusions be seen?
15. How clear and coherent is the reporting?
16. How clear are the assumptions/theoretical perspectives/values that have shaped the form and output of the evaluation?

17. What evidence is there of attention to ethical issues?

18. How adequately has the research process been documented?

(Spencer et al. 2003: 22–28)

The attempt to settle on criteria for evaluating qualitative social research is far from over. For example, some researchers are wary of the British effort just delineated: They express concern about the implications of a government body specifying research criteria and suggest that the list grows out of philosophical and political orientations that have not been made clear (J. Smith and Hodkinson 2005).

Ethics and Qualitative Data Analysis

At least two ethical issues cause special concern in the analysis and reporting of qualitative research. First, because it calls so directly on subjective judgments, researchers face an obvious risk of seeing what they are looking for or want to find. The risk is increased in the case of participatory action research or other projects involving an element of social justice. Researcher bias is hardly inevitable, however. Experienced qualitative analysts avoid this pitfall through a deliberate awareness of their own values and preferences as well as adherence to established techniques for data collection and analysis. And as an additional protection, the peer review inherent in the scientific research environment encourages colleagues to point out any failings in this regard.

Second, protecting subjects' privacy becomes a particularly important issue in qualitative research. The qualitative researcher will often analyze and report data collected from specific, identifiable individuals. Earlier, I indicated the importance of not revealing what we learn about subjects, though I mostly discussed it in the context of data collection. When writing up the results of your analyses, you will often have to actively conceal identities. Individuals, organizations, and communities are given pseudonyms to conceal their identities.

Sometimes, you may even need to suppress details that would let outsiders figure out who you are talking about. Thus, it may be appropriate to speak about interviewing "a church leader" rather than "the head deacon." And you may need to suppress or alter age, race, or gender references if any would give away a subject's identity. The key principle is to respect the privacy of those we study.

MAIN POINTS

Introduction

- Qualitative analysis is the nonnumerical examination and interpretation of observations.

Linking Theory and Analysis

- Qualitative analysis involves a continual interplay between theory and analysis. In analyzing qualitative data, we seek to discover patterns such as changes over time or possible causal links among variables.

- Examples of approaches to the discovery and explanation of such patterns are Grounded Theory Method (GTM), semiotics, and conversation analysis.

Qualitative Data Processing

- The processing of qualitative data is as much art as science. Three key tools for preparing data for analysis are coding, memoing, and concept mapping.

- In contrast to the standardized units used in coding for statistical analyses, the units to be coded in qualitative analyses may vary within a document. Although codes may be derived from the theory being explored, more often researchers use open coding, in which codes are suggested by the researchers' examination and questioning of the data.

- Memoing is appropriate at several stages of data processing to capture code meanings, theoretical ideas, preliminary conclusions, and other thoughts that will be useful during analysis.

- Concept mapping uses diagrams to explore relationships in the data graphically.

Computer Programs for Qualitative Data

- Many computer programs, such as NUD*IST, are designed specifically to assist researchers in the analysis of qualitative data. In addition, researchers can take advantage of the capabilities of common software tools such as word processors, database programs, and spreadsheets.

The Qualitative Analysis of Quantitative Data

- Although qualitative and quantitative methods of analysis may appear incompatible or in competition, research often demands that both kinds be used in the same project.

Evaluating the Quality of Qualitative Research

- Validity (credibility) and reliability (dependability) are reasonable criteria for evaluating qualitative research.

Ethics and Qualitative Data Analysis

- The subjective element in qualitative data analysis provides an added challenge to avoiding bias in the interpretation of data.
- Since the qualitative data analyst will know the identity of subjects, protecting their privacy requires special care.

KEY TERMS

The following terms are defined in context in the chapter and at the bottom of the page where the term is introduced, as well as in the comprehensive glossary at the back of the book.

axial coding	memoing
case-oriented analysis	open coding
concept mapping	qualitative analysis
constant comparative method	selective coding
conversation analysis (CA)	semiotics
cross-case analysis	variable-oriented analysis
Grounded Theory Method (GTM)	

PROPOSING SOCIAL RESEARCH: QUALITATIVE DATA ANALYSIS

In this chapter, we've seen some of the qualitative data analysis approaches that social researchers can use. Since you won't have analyzed your data when

you write this portion of the proposal, of course, you can't say anything about the conclusions you'll draw. However, you can describe your initial plans for the analysis. I say "initial" plans because you may change directions somewhat as the data accumulate and patterns begin to emerge. In some cases, your analysis will begin as observations are being made and/or other data being gathered, or you may plan to complete the data-collection phase before starting your data analysis.

This the place to indicate whether you plan to employ a particular method of analysis, such as grounded theory, semiotics, or conversational analysis. If you're planning to use one of the computer programs used for qualitative data analysis, mention that here as well.

REVIEW QUESTIONS AND EXERCISES

1. Review Goffman's examination of gender in advertising, then collect and analyze a set of advertising photos, from magazines or newspapers, that allow you to explore the relationship between gender and status.

2. Review the discussion of homosexuality in the Book of Leviticus and suggest ways that the examination might be structured as a cross-case analysis.

3. Imagine you were conducting a cross-case analysis of revolutionary documents such as the Declaration of Independence and the Declaration of the Rights of Man and of the Citizen (from the French Revolution). Identify the key concepts you might code in the following sentence:

   When in the Course of human events, it becomes necessary for one people to dissolve the political bands which have connected them with another, and to assume among the Powers of the earth, the separate and equal station to which the Laws of Nature and of Nature's God entitle them, a decent respect to the opinions of mankind requires that they should declare the causes which impel them to the separation.

4. Write one code note and one theoretical note for Exercise 3.

5. Using the library, InfoTrac College Edition, or the web, find a research report using conversation analysis. Summarize the main conclusions in your own words.

See the booklet that accompanies your text for exercises using SPSS (Statistical Package for the Social Sciences). There are exercises offered for each chapter, and you'll also find a detailed primer on using SPSS.

Online Study Resources

If your book came with an access code card, visit www.cengage.com/login to register. To purchase access, please visit www.ichapters.com.

1. Before you do your final review of the chapter, take the *CengageNOW* pretest to help identify the areas on which you should concentrate. You'll find information on this online tool, as well as instructions on how to access all of its great resources, in the front of the book.

2. As you review, take advantage of the *Cengage-eNOW* personalized study plan, based on your quiz results. Use this study plan with its interactive exercises and other resources to master the material.

3. When you're finished with your review, take the posttest to confirm that you're ready to move on to the next chapter.

WEBSITE FOR *THE PRACTICE OF SOCIAL RESEARCH* 12TH EDITION

Go to your book's website at www.cengage.com/sociology/babbie for tools to aid you in studying for your exams. You'll find *Tutorial Quizzes* with feedback, *Internet Exercises, Flash Cards, Glossaries,* and *Essay Quizzes,* as well as *InfoTrac College Edition* search terms, suggestions for additional reading, *Web Links,* and primers for using data-analysis software such as SPSS.

CHAPTER FOURTEEN

Quantitative Data Analysis

CHAPTER OVERVIEW

Often, social data are converted to numerical form for statistical analyses. In this chapter, we'll begin with the process of quantifying data, then turn to analysis. Quantitative analysis may be descriptive or explanatory; it may involve, one, two, or several variables. We begin our examination of how quantitative analyses are done with some simple but powerful ways of manipulating data in order to attain research conclusions.

CengageNOW for Sociology

Use this online tool to help you make the grade on your next exam. After reading this chapter, go to "Online Study Resources" at the end of the chapter for instructions on how to benefit from *CengageNOW*.

Introduction

In Chapter 13, we saw some of the logic and techniques by which social researchers analyze the qualitative data they've collected. This chapter examines quantitative analysis, or the techniques by which researchers convert data to numerical forms and subject them to statistical analyses.

To begin we'll look at *quantification*—the process of converting data to a numerical format. This involves converting social science data into a machine-readable form—a form that can be read and manipulated by computers and similar machines used in quantitative analysis.

The rest of the chapter will present the logic and some of the techniques of quantitative data analysis—starting with the simplest case, univariate analysis, which involves one variable, then discussing bivariate analysis, which involves two variables. We'll end with a brief introduction to multivariate analysis, or the examination of several variables simultaneously, such as *age, education,* and *prejudice.*

Before we can do any sort of analysis, we need to quantify our data. Let's turn now to the basic steps involved in converting data into machine-readable forms amenable to computer processing and analysis.

Quantification of Data

Today, **quantitative analysis** is almost always handled by computer programs such as SPSS and MicroCase. For those programs to work their magic, they must be able to read the data you've collected in your research. If you've conducted a survey, for example, some of your data are inherently numerical: age or income, for example.

---

quantitative analysis The numerical representation and manipulation of observations for the purpose of describing and explaining the phenomena that those observations reflect.

Whereas the writing and check marks on a questionnaire are qualitative in nature, a scribbled age is easily converted to quantitative data.

Other data are also easily quantified: transforming *male* and *female* into "1" and "2" is hardly rocket science. Researchers can also easily assign numerical representations to such variables as *religious affiliation, political party,* and *region of the country.*

Some data are more challenging, however. If a survey respondent tells you that he or she thinks the biggest problem facing Woodbury, Vermont, today is "the disintegrating ozone layer," the computer can't process that response numerically. You must translate by coding the responses. We've already discussed coding in connection with content analysis (Chapter 11) and again in connection with qualitative data analysis (Chapter 13). Now we look at coding specifically for quantitative analysis.

To conduct a quantitative analysis, researchers often must engage in a coding process after the data have been collected. For example, open-ended questionnaire items result in nonnumerical responses, which need to be coded before analysis. As with content analysis, the task is to reduce a wide variety of idiosyncratic items of information to a more limited set of attributes composing a variable. Suppose, for example, that a survey researcher asks respondents, "What is your occupation?" The responses to such a question will vary considerably. Although he or she can assign a separate numerical code to each reported occupation, this procedure will not facilitate analysis, which typically depends on several subjects having the same attribute.

The variable *occupation* has many preestablished coding schemes. One such scheme distinguishes professional and managerial occupations, clerical occupations, semiskilled occupations, and so forth. Another scheme distinguishes different sectors of the economy: manufacturing, health, education, commerce, and so forth. Still others combine both. Using an established coding scheme gives you the advantage of being able to compare your research results with those of other studies. (See, for instance, the Standard Occupational Classification [SOC] at the Bureau of Labor Statistics website,

via the link on this book's website: http://www
.cengage.com/sociology/babbie.)

The occupational coding scheme you choose
should be appropriate to the theoretical concepts
being examined in your study. For some studies,
coding all occupations as either white-collar or
blue-collar might be sufficient. For others, self-
employed and not self-employed might do. A peace
researcher might wish to know only whether the
occupation depended on the defense establishment
or not.

Although you should tailor the coding scheme
to meet particular requirements of the analysis,
you should also keep one general guideline in
mind. If the data are coded to maintain a great deal
of detail, code categories can always be combined
during an analysis that does not require such detail.
If the data are coded into relatively few, gross
categories, however, there's no way during analysis
to recreate the original detail. To keep your options
open, it's a good idea to code your data in greater
detail than you plan to use in the analysis.

Developing Code Categories

There are two basic approaches to the coding
process. First, you can begin with a relatively well-
developed coding scheme. You may choose to do
this because it serves your research purpose. Thus,
as suggested previously, the peace researcher might
code occupations in terms of their relationship to
the defense establishment. Or, you may want to
use an existing coding scheme because it allows
you to compare your findings with those of previous research.

Second, you can generate codes from your
data, as discussed in Chapter 13. Let's say we've
asked students in a self-administered campus
survey to state what they believe is the biggest
problem facing their college today. Here are a few
of the answers they might have written in.

   Tuition is too high
   Not enough parking spaces
   Faculty don't know what they are doing
   Advisors are never available

TABLE 14-1

Student Responses That Can Be Coded "Financial Concerns"

	Financial Concerns
Tuition is too high	X
Not enough parking spaces	
Faculty don't know what they are doing	
Advisors are never available	
Not enough classes offered	
Cockroaches in the dorms	
Too many requirements	
Cafeteria food is infected	
Books cost too much	X
Not enough financial aid	X

   Not enough classes offered
   Cockroaches in the dorms
   Too many requirements
   Cafeteria food is infected
   Books cost too much
   Not enough financial aid

Take a minute to review these responses and see
whether you can identify some categories represented. Realize that there is no right answer; you
could generate several coding schemes from these
answers.

Let's start with the first response: "Tuition is too
high." What general areas of concern does that response reflect? One obvious possibility is "Financial
Concerns." Are there other responses that would fit
into that category? Table 14-1 shows which of the
questionnaire responses could do just that.

In more general terms, the first answer can
also be seen as reflecting nonacademic concerns.
This categorization would be relevant if your research interest included the distinction between
academic and nonacademic concerns. If that were
the case, the responses might be coded as shown in
Table 14-2.

Notice that I didn't code the response "Books
cost too much" in Table 14-2, because this concern could be seen as representing both of the
categories. Books are part of the academic program,

TABLE 14-2

Student Concerns Coded as "Academic" and "Nonacademic"

	Academic	Nonacademic
Tuition is too high		X
Not enough parking spaces		X
Faculty don't know what they are doing	X	
Advisors are never available	X	
Not enough classes offered	X	
Cockroaches in the dorms		X
Too many requirements	X	
Cafeteria food is infected		X
Books cost too much		
Not enough financial aid		X

TABLE 14-3

Nonacademic Concerns Coded as "Administrative" or "Facilities"

	Academic	Administrative	Facilities
Tuition is too high		X	
Not enough parking spaces			X
Faculty don't know what they are doing	X		
Advisors are never available	X		
Not enough classes offered	X		
Cockroaches in the dorms			X
Too many requirements	X		
Cafeteria food is infected			X
Books cost too much	X		
Not enough financial aid		X	

but their cost is not. This signals the need to refine the coding scheme we're developing. Depending on our research purpose, we might be especially interested in identifying any problems that had an academic element; hence we'd code this one "Academic." Just as reasonably, however, we might be more interested in identifying nonacademic problems and would code the response accordingly. Or, as another alternative, we might create a separate category for responses that involved both academic and nonacademic matters.

As yet another alternative, we might want to separate nonacademic concerns into those involving administrative matters and those dealing with campus facilities. Table 14-3 shows how the first ten responses would be coded in that event.

As these few examples illustrate, there are many possible schemes for coding a set of data. Your choices should match your research purposes and reflect the logic that emerges from the data themselves. Often, you'll find yourself modifying the code categories as the coding process proceeds. Whenever you change the list of categories, however, you must review the data already coded to see whether changes are in order.

Like the set of attributes composing a variable, and like the response categories in a closed-ended questionnaire item, code categories should be both

exhaustive and mutually exclusive. Every piece of information being coded should fit into one and only one category. Problems arise whenever a given response appears to fit equally into more than one code category or whenever it fits into no category: Both signal a mismatch between your data and your coding scheme.

If you're fortunate enough to have assistance in the coding process, you'll need to train your coders—teaching them the definitions of code categories and showing them how to use those categories properly. To do so, explain the meaning of the code categories and give several examples of each. To make sure your coders fully understand what you have in mind, code several cases ahead of time. Then ask your coders to code the same cases without knowing how you coded them. Finally, compare your coders' work with your own. Any discrepancies will indicate an imperfect communication of your coding scheme to your coders. Even with perfect agreement between you and your coders, however, it's best to check the coding of at least a portion of the cases throughout the coding process.

If you're not fortunate enough to have assistance in coding, you should still obtain some verification of your own reliability as a coder.

POLVIEWS	ATTEND
We hear a lot of talk these days about liberals and conservatives. I'm going to show you a seven-point scale on which the political views that people might hold are arranged from extremely liberal—point 1—to extremely conservative—point 7. Where would you place yourself on this scale?	How often do you attend religious services?
1. Extremely liberal	0. Never
2. Liberal	1. Less than once a year
3. Slightly liberal	2. About once or twice a year
4. Moderate, middle of the road	3. Several times a year
5. Slightly conservative	4. About once a month
6. Conservative	5. 2–3 times a month
7. Extremely conservative	6. Nearly every week
8. Don't know	7. Every week
9. No answer	8. Several times a week
	9. Don't know, No answer

FIGURE 14-1

A Partial Codebook

Nobody's perfect, especially a researcher hot on the trail of a finding. Suppose that you're studying an emerging cult and that you have the impression that people who do not have a regular family will be the most likely to regard the new cult as a family substitute. The danger is that whenever you discover a subject who reports no family, you'll unconsciously try to find some evidence in the subject's comments that the cult is a substitute for family. If at all possible, then, get someone else to code some of your cases to see whether that person makes the same assignments you made.

Codebook Construction

The end product of the coding process is the conversion of data items into numerical codes. These codes represent attributes composing variables, which, in turn, are assigned locations within a data file. A **codebook** is a document that describes the locations of variables and lists the assignments of codes to the attributes composing those variables.

A codebook serves two essential functions. First, it's the primary guide used in the coding process. Second, it's your guide for locating variables and interpreting codes in your data file during analysis. If you decide to correlate two variables as a part of your analysis of your data, the codebook tells you where to find the variables and what the codes represent.

Figure 14-1 is a partial codebook created from two variables from the General Social Survey. Although there is no one right format for a codebook, this example presents some of the common elements.

Notice first that each variable is identified by an abbreviated variable name: *POLVIEWS, ATTEND*. We can determine the religious services attendance of respondents, for example, by referencing *ATTEND*. This example uses the format established by the General Social Survey, which has been carried over into SPSS. Other data sets and/or analysis programs might format variables differently. Some use

codebook The document used in data processing and analysis that tells the location of different data items in a data file. Typically, the codebook identifies the locations of data items and the meaning of the codes used to represent different attributes of variables.

numerical codes in place of abbreviated names, for example. You must, however, have some identifier that will allow you to locate and use the variable in question.

Next, every codebook should contain the full definition of the variable. In the case of a questionnaire, the definition consists of the exact wordings of the questions asked, because, as we've seen, the wording of questions strongly influences the answers returned. In the case of *POLVIEWS*, you know that respondents were handed a card containing the several political categories and asked to pick the one that best fit them.

The codebook also indicates the attributes composing each variable. In *POLVIEWS*, for example, respondents could characterize their political orientations as "Extremely liberal," "Liberal," "Slightly liberal," and so forth.

Finally, notice that each attribute also has a numerical label. Thus, in *POLVIEWS*, "Extremely liberal" is code category 1. These numerical codes are used in various manipulations of the data. For example, you might decide to combine categories 1 through 3 (all the "liberal" responses). It's easier to do this with code numbers than with lengthy names.

You can visit the GSS codebook online at the link on this book's website. Hold your cursor over the tab "BROWSE GSS VARIABLES" and select one of the browsing options. If you know the symbolic name (e.g., *POLVIEWS*), you can locate it in the "Mnemonic Index." Otherwise, you can browse the "Subject Index" to find all the different questions that have been asked regarding a particular topic.

Data Entry

In addition to transforming data into quantitative form, researchers interested in quantitative analysis also need to convert data into a machine-readable

format, so that computers can read and manipulate the data. There are many ways of accomplishing this step, depending on the original form of your data and also the computer program you choose for analyzing the data. I'll simply introduce you to the process here. If you find yourself undertaking this task, you should be able to tailor your work to the particular data source and program you're using.

If your data have been collected by questionnaire, you might do your coding on the questionnaire itself. Then, data-entry specialists (including yourself) could enter the data into, say, an SPSS data matrix or into an Excel spreadsheet to be imported later into SPSS.

Sometimes social researchers use optical scan sheets for data collection. These sheets can be fed into machines that convert the black marks into data, which can be imported into the analysis program. This procedure only works with subjects who are comfortable using such sheets, and it's usually limited to closed-ended questions.

Sometimes, data entry occurs in the process of data collection. In computer-assisted telephone interviewing, for example, the interviewer keys responses directly into the computer, where the data are compiled for analysis (see Chapter 9). Even more effortless, online surveys can be constructed so that the respondents enter their own answers directly into the accumulating database, without the need for an intervening interviewer or data-entry person.

Once data have been fully quantified and entered into the computer, researchers can begin quantitative analysis. Let's look at the three cases mentioned at the start of this chapter: univariate, bivariate, and multivariate analyses.

Univariate Analysis

The simplest form of quantitative analysis, **univariate analysis,** involves describing a case in terms of a single variable—specifically, the distribution of attributes that it comprises. For example, if *gender* were measured, we would look at how many of the subjects were men and how many were women.

---

univariate analysis The analysis of a single variable, for purposes of description. Frequency distributions, averages, and measures of dispersion would be examples of univariate analysis, as distinguished from bivariate and multivariate analysis.

TABLE 14-4

GSS Attendance at Religious Services, 2004

Attend		How Often R Attends Religious Services			
Value Label	Value	Frequency	Percent	Valid Percent	Cum Percent
NEVER	0	471	16.7	16.8	16.8
LT ONCE A YEAR	1	198	7.0	7.1	23.9
ONCE A YEAR	2	396	14.1	14.1	38.0
SEVRL TIMES A YR	3	371	13.2	13.2	51.3
ONCE A MONTH	4	191	6.8	6.8	58.1
2—3X A MONTH	5	255	9.1	9.1	67.2
NRLY EVERY WEEK	6	169	6.0	6.0	73.2
EVERY WEEK	7	508	18.1	18.1	91.4
MORE THN ONCE WK	8	242	8.6	8.6	100.0
DK,NA	9	11	0.4		
	Total	2,812	100.0	99.8	
Valid cases	2,812	Missing cases	11		

Source: General Social Survey, 2004, National Opinion Research Center.

Distributions

The most basic format for presenting univariate data is to report all individual cases, that is, to list the attribute for each case under study in terms of the variable in question. Let's take as an example the General Social Survey (GSS) data on attendance at religious services, *ATTEND*. Table 14-4 presents the results of an SPSS analysis of this variable.

Let's examine the table, piece by piece. First, if you look near the bottom of the table, you'll see that the sample being analyzed has a total of 2,812 cases. In the last row above the totals, you'll see that 11 of the 2,812 respondents either said they didn't know (DK) or gave no answer (NA) in response to this question. So our assessment of U.S. attendance at religious services in 2004 will be based on the 2,801 respondents who answered the question.

Go back to the top of the table now. You'll see that 471 people said they never went to religious services. This number in and of itself tells us noth-

ing about religious practices. It does not, in itself, gives us an idea of whether the "average American" attends religious services a little or a lot.

By analogy, suppose your best friend tells you that they drank a six-pack of beer. Is that a little beer or a lot? The answer, of course, depends on whether they consumed the beer in a month, a week, a day, or an hour. In the case of religious participation, similarly, we need some basis for assessing the number that represents the people who never attend religious services.

One way to assess the number is to calculate the percentage of all respondents who said they never go to religious services. If you were to divide 471 by the 2,801 who gave some answer, you would get 16.8 percent, which appears in the table as the "Valid Percent." Now we can say that 17 percent, or roughly one U.S. adult in six, reports never attending religious services.

This result is more meaningful, but does it suggest that people in the United States are generally nonreligious? A further look at Table 14-4 shows that the response category most often chosen was

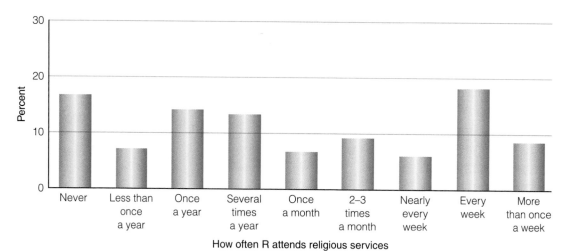

How often R attends religious services

FIGURE 14-2

Bar Chart of GSS ATTEND, 2004

"Every Week," with 18.1 percent of the respondents giving that answer. Add to that the 8.6 percent who report attending religious services more than once a week, and we find that over a fourth (26.7 percent) of U.S. adults say they attend religious services at least once a week. As you can see, each new comparison gives a more complete picture of the data.

A description of the number of times that the various attributes of a variable are observed in a sample is called a **frequency distribution.** Sometimes it's easiest to see a frequency distribution in a graph. Figure 14-2 was created by SPSS from the GSS data on *ATTEND.* The vertical scale on the left side of the graph indicates the percentage selecting each of the answers displayed along the horizontal

axis of the graph. Take a minute to notice how the percentages in Table 14-4 correspond to the heights of the bars in Figure 14-2.

Central Tendency

Beyond simply reporting the overall distribution of values, sometimes called the *marginal frequencies* or just the *marginals,* you may choose to present your data in the form of an **average** or measure of *central tendency.* You're already familiar with the concept of central tendency from the many kinds of averages you use in everyday life to express the "typical" value of a variable. For instance, in baseball a batting average of .300 says that a batter gets a hit three out of every ten opportunities—on average. Over the course of a season, a hitter might go through extended periods without getting any hits at all and go through other periods when he or she gets a bunch of hits all at once. Over time, though, the central tendency of the batter's performance can be expressed as getting three hits in every ten chances. Similarly, your grade point average expresses the "typical" value of all your grades taken together, even though some of them might be A's, others B's, and one or two might be C's (I know you never get anything lower than a C).

---

frequency distribution A description of the number of times the various attributes of a variable are observed in a sample. The report that 53 percent of a sample were men and 47 percent were women would be a simple example of a frequency distribution.

average An ambiguous term generally suggesting typical or normal—a central tendency. The mean, median, and mode are specific examples of mathematical averages.

Averages like these are more properly called the arithmetic **mean** (the result of dividing the sum of the values by the total number of cases). The mean is only one way to measure central tendency or "typical" values. Two other options are the **mode** (the most frequently occurring attribute) and the **median** (the middle attribute in the ranked distribution of observed attributes). Here's how the three averages would be calculated from a set of data.

Suppose you're conducting an experiment that involves teenagers as subjects. They range in age from 13 to 19, as indicated in the following table:

Age	Number
13	3
14	4
15	6
16	8
17	4
18	3
19	3

Now that you've seen the actual ages of the 31 subjects, how old would you say they are in general, or "on average"? Let's look at three different ways you might answer that question.

The easiest average to calculate is the mode, the most frequent value. As you can see, there were more 16-year-olds (eight of them) than any other age, so the modal age is 16, as indicated in Figure 14-3. Technically, the modal age is the category "16," which may include some people who are closer to 17 than 16 but who haven't yet reached that birthday.

Figure 14-3 also demonstrates the calculation of the mean. There are three steps: (1) multiply each age by the number of subjects who have that age, (2) total the results of all those multiplications, and (3) divide that total by the number of subjects.

In the case of age, a special adjustment is needed. As indicated in the discussion of the mode, those who call themselves "13" actually range from exactly 13 years old to just short of 14. It's reason-able to assume, moreover, that as a group the "13-year-olds" in the country are evenly distributed within that one-year span, making their average age 13.5 years. This is true for each of the age groups. Hence, it is appropriate to add 0.5 years to the final calculation, making the mean age 16.37, as indicated in Figure 14-3.

The third measure of central tendency, the median, represents the "middle" value: Half are above it, half below. If we had the precise ages of each subject (for example, 17 years and 124 days), we'd be able to arrange all 31 subjects in order by age, and the median for the whole group would be the age of the middle subject.

As you can see, however, we do not know precise ages; our data constitute "grouped data" in this regard. For example, three people who are not precisely the same age have been grouped in the category "13-year-olds."

Figure 14-3 illustrates the logic of calculating a median for grouped data. Because there are 31 subjects altogether, the "middle" subject would be subject number 16 if they were arranged by age—15 teenagers would be younger and 15 older. Look at the bottom portion of Figure 14-3, and you'll see that the middle person is one of the eight 16-year-olds. In the enlarged view of that group, we see that number 16 is the third from the left.

Because we do not know the precise ages of the subjects in this group, the statistical convention here is to assume they are evenly spread along the

mean An average computed by summing the values of several observations and dividing by the number of observations. If you now have a grade point average of 4.0 based on 10 courses, and you get an F in this course, your new grade point (mean) average will be 3.6.

mode An average representing the most frequently observed value or attribute. If a sample contains 1,000 Protestants, 275 Catholics, and 33 Jews, Protestant is the modal category.

median An average representing the value of the "middle" case in a rank-ordered set of observations. If the ages of five men are 16, 17, 20, 54, and 88, the median would be 20. (The mean would be 39.)

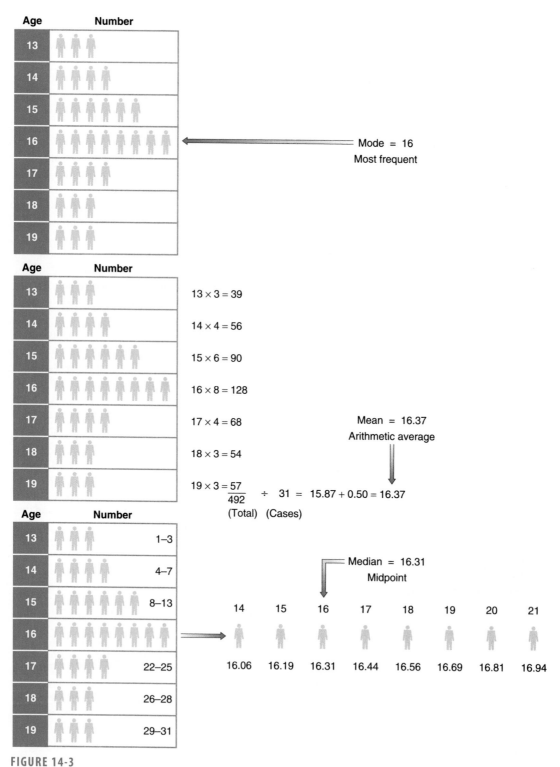

FIGURE 14-3

Three "Averages"

width of the group. In this instance, the possible ages of the subjects go from 16 years and no days to 16 years and 364 days. Strictly speaking, the range, then, is 364/365 days. As a practical matter, it's sufficient to call it one year.

If the eight subjects in this group were evenly spread from one limit to the other, they would be one-eighth of a year apart from each other—a 0.125-year interval. Look at the illustration and you'll see that if we place the first subject half the interval from the lower limit and add a full interval to the age of each successive subject, the final one is half an interval from the upper limit.

What we've done is calculate, hypothetically, the precise ages of the eight subjects—assuming their ages were spread out evenly. Having done this, we merely note the age of the middle subject—16.31—and that is the median age for the group.

Whenever the total number of subjects is an even number, of course, there is no middle case. To get the median, you merely calculate the mean of the two values on either side of the midpoint in the ranked data. Suppose, for example, that there was one more 19-year-old in our sample, giving us a total of 32 cases. The midpoint would then fall between subjects 16 and 17. The median would therefore be calculated as $(16.31 + 16.44) \div 2 = 16.38$.

As you can see in Figure 14-3, the three measures of central tendency produce three different values for our set of data, which is often (but not necessarily) the case. Which measure, then, best represents the "typical" value? More generally, which measure of central tendency should we prefer? The answer depends on the nature of your data and the purpose of your analysis. For example, whenever means are presented, you should be aware that they are susceptible to extreme values—a few very large or very small numbers. As only one example, the (mean) average person in Redmond, Washington, has a net worth in excess of a million dollars. If you were to visit Redmond, however, you would not find that the "average" resident lives up to your idea of a millionaire. The very high mean reflects the influence of one extreme case among Redmond's 40,000 residents—Bill Gates of Microsoft, who has a net worth (at the time this is being written) of tens of billions of dollars. Clearly, the median wealth would give you a more accurate picture of the residents of Redmond as a whole.

This example should illustrate the need to choose carefully among the various measures of central tendency. A course or textbook in statistics will give you a fuller understanding of the variety of situations in which each is appropriate.

Dispersion

Averages offer readers the advantage of reducing the raw data to the most manageable form: A single number (or attribute) can represent all the detailed data collected in regard to the variable. This advantage comes at a cost, of course, because the reader cannot reconstruct the original data from an average. Summaries of the dispersion of responses can somewhat alleviate this disadvantage.

Dispersion refers to the way values are distributed around some central value, such as an average. The simplest measure of dispersion is the range: the distance separating the highest from the lowest value. Thus, besides reporting that our subjects have a mean age of 15.87, we might also indicate that their ages range from 13 to 19.

A more sophisticated measure of dispersion is the **standard deviation.** This measure was briefly mentioned in Chapter 7 as the standard error of a sampling distribution. Essentially, the standard

---

dispersion The distribution of values around some central value, such as an average. The range is a simple example of a measure of dispersion. Thus, we may report that the mean age of a group is 37.9, and the range is from 12 to 89.

standard deviation A measure of dispersion around the mean, calculated so that approximately 68 percent of the cases will lie within plus or minus one standard deviation from the mean, 95 percent will lie within plus or minus two standard deviations, and 99.9 percent will lie within three standard deviations. Thus, for example, if the mean age in a group is 30 and the standard deviation is 10, then 68 percent have ages between 20 and 40. The smaller the standard deviation, the more tightly the values are clustered around the mean; if the standard deviation is high, the values are widely spread out.

deviation is an index of the amount of variability in a set of data. A higher standard deviation means that the data are more dispersed; a lower standard deviation means that they are more bunched together. Figure 14-4 illustrates the basic idea. Notice that the professional golfer not only has a lower mean score but is also more consistent—represented by the smaller standard deviation. The duffer, on the other hand, has a higher average and is also less consistent: sometimes doing much better, sometimes much worse.

There are many other measures of dispersion. In reporting intelligence test scores, for example, researchers might determine the *interquartile range*, the range of scores for the middle 50 percent of subjects. If the top one-fourth had scores ranging from 120 to 150, and if the bottom one-fourth had scores ranging from 60 to 90, the report might say that the interquartile range was from 90 to 120 (or 30 points) with a mean score of, let's say, 102.

Continuous and Discrete Variables

The preceding calculations are not appropriate for all variables. To understand this point, we must distinguish between two types of variables: continuous and discrete. A **continuous variable** (or ratio variable) increases steadily in tiny fractions. An example is *age*, which increases steadily with each increment of time. A **discrete variable** jumps

a. High standard deviation = spread-out values

Amateur Golfer's Scores

Number of games · Mean = 100

68% of values

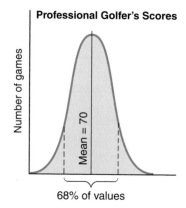

b. Low standard deviation = tightly clustered values

Professional Golfer's Scores

Number of games · Mean = 70

68% of values

FIGURE 14-4

High and Low Standard Deviations

---

continuous variable A variable whose attributes form a steady progression, such as *age* or *income*. Thus, the ages of a group of people might include 21, 22, 23, 24, and so forth and could even be broken down into fractions of years. Contrast this with discrete variables, such as *gender* or *religious affiliation*, whose attributes form discontinuous chunks.

discrete variable A variable whose attributes are separate from one another, or discontinuous, as in the case of *gender* or *religious affiliation*. Contrast this with continuous variables, in which one attribute shades off into the next. Thus, in *age* (a continuous variable), the attributes progress steadily from 21 to 22 to 23, and so forth, whereas there is no progression from male to female in the case of *gender*.

---

from category to category without intervening steps. Examples include *gender, military rank,* and *year in college* (you go from being a sophomore to a junior in one step).

In analyzing a discrete variable—a nominal or ordinal variable, for example—some of the techniques discussed previously do not apply. Strictly speaking, modes should be calculated for nominal data, medians for interval data, and means for ratio data, not for nominal data (see Chapter 5). If the variable in question is *gender*, for example, raw numbers (23 of the cross-dressing outlaw bikers in our sample are women) or percentages (7 percent are women) can be appropriate and useful analyses, but neither a median nor a mean would make any sense. Calculating the mode would be legitimate, though not very revealing,

TABLE 14-5

Marijuana Legalization by Age of Respondents, 2004

	Under 21	21–35	36–54	55 and older
Should be legalized	27%	40%	37%	24%
Should not be legalized	73	60	63	76
100% =	(34)	(238)	(338)	(265)

Source: General Social Survey, 2004, National Opinion Research Center.

because it would only tell us "most were men." However, the mode for data on religious affiliation might be more interesting, as in "most people in the United States are Protestant."

Detail versus Manageability

In presenting univariate and other data, you'll be constrained by two goals. On the one hand, you should attempt to provide your reader with the fullest degree of detail regarding those data. On the other hand, the data should be presented in a manageable form. As these two goals often directly counter each other, you'll find yourself continually seeking the best compromise between them. One useful solution is to report a given set of data in more than one form. In the case of age, for example, you might report the distribution of ungrouped ages *plus* the mean age and standard deviation.

As you can see from this introductory discussion of univariate analysis, this seemingly simple matter can be rather complex. In any event, the lessons of this section pave the way for a consideration of subgroup comparisons and bivariate analyses.

Subgroup Comparisons

Univariate analyses describe the units of analysis of a study and, if they are a sample drawn from some larger population, allow us to make descriptive inferences about the larger population. Bivariate and multivariate analyses are aimed primarily at expla-

TABLE 14-6

Marijuana Legalization by Political Orientation, 2004

	Should Legalize	Should Not Legalize	100% =
Extremely liberal	77%	23	(30)
Liberal	49%	51	(75)
Slightly liberal	35%	65	(92)
Moderate	33%	67	(326)
Slightly conservative	32%	68	(136)
Conservative	25%	75	(155)
Extremely conservative	16%	84	(37)

Source: General Social Survey, 2004, National Opinion Research Center.

nation. Before turning to explanation, however, we should consider the case of subgroup description.

Often it's appropriate to describe subsets of cases, subjects, or respondents. Here's a simple example from the General Social Survey. In 2004, respondents were asked, "Should marijuana be made legal?" In response, 33.4 percent said it should and 66.6 percent said it shouldn't. Table 14-5 presents the responses given to this question by respondents in different age categories.

Notice that the subgroup comparisons tell us how different groups in the population responded to this question. You can undoubtedly see a pattern in the results, though possibly not exactly what you expected; we'll return to that in a moment. First, let's see how another set of subgroups answered this question.

Table 14-6 presents different political subgroups' attitudes toward legalizing marijuana, based on whether respondents characterized themselves as conservative or liberal. Before looking at the table, you might try your hand at hypothesizing what the results are likely to be and why. Notice that I've changed the direction of percentaging this table, to make it easier to read. To compare the subgroups in this case, you would read down the columns, not across them.

Before examining the logic of causal analysis, let's consider another example of subgroup comparisons: one that will let us address some table-formatting issues.

TABLE 14-7

Attitudes toward the United Nations: "How is the UN doing in solving the problems it has had to face?"

	West Germany	Britain	France	Japan	United States
Very good job	2%	7%	2%	1%	5%
Good job	46	39	45	11	46
Poor job	21	28	22	43	27
Very poor job	6	9	3	5	13
Don't know	26	17	28	41	10

Source: "5-Nation Survey Finds Hope for U.N.," *New York Times,* June 26, 1985, p. 6.

TABLE 14-8

Collapsing Extreme Categories

	West Germany	Britain	France	Japan	United States
Good job or better	48%	46%	47%	12%	51%
Poor job or worse	27	37	25	48	40
Don't know	26	17	28	41	10

"Collapsing" Response Categories

"Textbook examples" of tables are often simpler than you'll typically find in published research reports or in your own analyses of data, so this section and the next one address two common problems and suggest solutions.

Let's begin by turning to Table 14-7, which reports data collected in a multinational poll conducted by the *New York Times,* CBS News, and the *Herald Tribune* in 1985, concerning attitudes about the United Nations. The question reported in Table 14-7 deals with general attitudes about the way the UN was handling its job.

Here's the question: How do people in the five nations reported in Table 14-7 compare in their support for the kind of job the UN was doing? As you review the table, you may find there are simply so many numbers that it's hard to see any meaningful pattern.

Part of the problem with Table 14-7 lies in the relatively small percentages of respondents selecting the two extreme response categories: the UN is doing a very good or a very poor job. Furthermore, although it might be tempting to read only the sec-

ond line of the table (those saying "good job"), that would be improper. Looking at only the second row, we would conclude that West Germany and the United States were the most positive (46 percent) about the UN's performance, followed closely by France (45 percent), with Britain (39 percent) less positive than any of those three and Japan (11 percent) the least positive of all.

This procedure is inappropriate in that it ignores all those respondents who gave the most positive answer of all: "very good job." In a situation like this, you should combine or "collapse" the two ends of the range of variation. In this instance, combine "very good" with "good" and "very poor" with "poor." If you were to do this in the analysis of your own data, it would be wise to add the raw frequencies together and recompute percentages for the combined categories, but in analyzing a published table such as this one, you can simply add the percentages as illustrated by the results shown in Table 14-8.

With the collapsed categories illustrated in Table 14-8, we can now rather easily read across the several national percentages of people who said the UN was doing at least a good job. Now the

TABLE 14-9

Omitting the "Don't Knows"

	West Germany	Britain	France	Japan	United States
Good job or better	65%	55%	65%	20%	57%
Poor job or worse	35%	45%	35%	81%	44%

United States appears the most positive; Germany, Britain, and France are only slightly less positive and are nearly indistinguishable from one another; and Japan stands alone in its quite low assessment of the UN's performance. Although the conclusions to be drawn now do not differ radically from what we might have concluded from simply reading the second line of Table 14-7, we should note that Britain now appears relatively more supportive.

Here's the risk I'd like to spare you. Suppose you had hastily read the second row of Table 14-7 and noted that the British had a somewhat lower assessment of the job the UN was doing than was true of people in the United States, West Germany, and France. You might feel obliged to think up an explanation for why that was so—possibly creating an ingenious psychohistorical theory about the painful decline of the once powerful and dignified British Empire. Then, once you had touted your "theory" about, someone else might point out that a proper reading of the data would show the British were actually not really less positive than the other three nations. This is not a hypothetical risk. Errors like these happen frequently, but they can be avoided by collapsing answer categories where appropriate.

Handling "Don't Knows"

Tables 14-7 and 14-8 illustrate another common problem in the analysis of survey data. It's usually a good idea to give people the option of saying "don't know" or "no opinion" when asking for their opinions on issues. But what do you do with those answers when you analyze the data?

Notice there is a good deal of variation in the national percentages saying "don't know" in this instance, ranging from only 10 percent in the United States to 41 percent in Japan. The presence of substantial percentages saying they don't know can confuse the results of tables like these. For example, was it simply because so many Japanese didn't express any opinion that they seemed so much less likely to say the UN was doing a good job?

Here's an easy way to recalculate percentages, with the "don't knows" excluded. Look at the first column of percentages in Table 14-8: West Germany's answers to the question about the UN's performance. Notice that 26 percent of the respondents said they didn't know. This means that those who said "good" or "bad" job—taken together—represent only 74 percent (100 minus 26) of the whole. If we divide the 48 percent saying "good job or better" by 0.74 (the proportion giving any opinion), we can say that 65 percent "of those with an opinion" said the UN was doing a good or very good job (48% ÷ 0.74 = 65%).

Table 14-9 presents the whole table with the "don't knows" excluded. Notice that these new data offer a somewhat different interpretation than the previous tables do. Specifically, it would now appear that France and West Germany were the most positive in their assessments of the UN, with the United States and Britain a bit lower. Although Japan still stands out as lowest in this regard, it has moved from 12 percent to 20 percent positive.

At this point, having seen three versions of the data, you may be asking yourself, Which is the right one? The answer depends on your purpose in analyzing and interpreting the data. For example, if it's not essential for you to distinguish "very good" from "good," it makes sense to combine them, because it's easier to read the table.

Whether to include or exclude the "don't knows" is harder to decide in the abstract. It may be a very important finding that such a large percentage of the Japanese had no opinion—if you wanted to find out whether people were familiar with the work of the UN, for example. On the other hand, if you wanted to know how people might vote on an issue, it might be more appropriate to exclude the "don't knows" on the assumption that they wouldn't vote or that ultimately they would be likely to divide their votes between the two sides of the issue.

In any event, the *truth* contained within your data is that a certain percentage said they didn't know and the remainder divided their opinions in whatever manner they did. Often, it's appropriate to report your data in both forms—with and without the "don't knows"—so your readers can draw their own conclusions.

Numerical Descriptions in Qualitative Research

Although this chapter deals primarily with quantitative research, the discussions also apply to qualitative studies. Numerical testing can often verify the findings of in-depth, qualitative studies. Thus, for example, when David Silverman wanted to compare the cancer treatments received by patients in private clinics with the cancer treatments in Britain's National Health Service, he primarily chose in-depth analyses of the interactions between doctors and patients:

> My method of analysis was largely qualitative and . . . I used extracts of what doctors and patients had said as well as offering a brief ethnography of the setting and of certain behavioural data. In addition, however, I constructed a coding form which enabled me to collate a number of crude measures of doctor and patient interactions.
>
> *(1993: 163)*

---

bivariate analysis The analysis of two variables simultaneously, for the purpose of determining the empirical relationship between them. The construction of a simple percentage table or the computation of a simple correlation coefficient are examples of bivariate analyses.

Not only did the numerical data fine-tune Silverman's impressions based on his qualitative observations, but his in-depth understanding of the situation allowed him to craft an ever more appropriate quantitative analysis. Listen to the interaction between qualitative and quantitative approaches in this lengthy discussion:

> My overall impression was that private consultations lasted considerably longer than those held in the NHS clinics. When examined, the data indeed did show that the former were almost twice as long as the latter (20 minutes as against 11 minutes) and that the difference was statistically highly significant. However, I recalled that, for special reasons, one of the NHS clinics had abnormally short consultations. I felt a fairer comparison of consultations in the two sectors should exclude this clinic and should only compare consultations taken by a single doctor in both sectors. This subsample of cases revealed that the difference in length between NHS and private consultations was now reduced to an average of under 3 minutes. This was still statistically significant, although the significance was reduced. Finally, however, if I compared only *new* patients seen by the same doctor, NHS patients got 4 minutes more on the average—34 minutes as against 30 minutes in the private clinic.
>
> *(1993: 163–64)*

This example further demonstrates the special power that can be gained from a combination of approaches in social research. The combination of qualitative and quantitative analyses can be especially potent.

Bivariate Analysis

In contrast to univariate analysis, subgroup comparisons involve two variables. In this respect subgroup comparisons constitute a kind of **bivariate analysis**—that is, the analysis of two variables simultaneously. However, as with univariate analysis, the purpose of subgroup comparisons is largely descriptive. Most bivariate analysis in social research adds another element: determining relationships between the variables themselves. Thus, univariate analysis and subgroup comparisons focus on describing the people (or other units of analysis) under study, whereas bivariate analysis focuses on the variables and their empirical relationships.

Table 14-10 could be regarded as an instance of subgroup comparison: It independently describes the religious services attendance of men and women, as reported in the 2004 General Social Survey. It shows—comparatively and descriptively—that the women under study attended church more often than the men did. However, the same table, seen as an explanatory bivariate analysis, tells a somewhat different story. It suggests that the variable *gender* has an effect on the variable *church attendance*. That is, we can view the behavior as a dependent variable that is partially determined by the independent variable, *gender.*

Explanatory bivariate analyses, then, involve the "variable language" introduced in Chapter 1. In a subtle shift of focus, we're no longer talking about men and women as different subgroups but about *gender* as a variable: one that has an influence on other variables. The theoretical interpretation of Table 14-10 might be taken from Charles Glock's Comfort Hypothesis as discussed in Chapter 2:

1. Women are still treated as second-class citizens in U.S. society.

2. People denied status gratification in the secular society may turn to religion as an alternative source of status.

3. Hence, women should be more religious than men.

The data presented in Table 14-10 confirm this reasoning. Thirty-one percent of the women attend religious services weekly, as compared with 22 percent of the men.

Using the logic of causal relationships among variables has an important implication for the construction and reading of percentage tables. One of the chief bugaboos for new-data analysts is deciding on the appropriate "direction of percentaging" for any given table. In Table 14-10, for example,

TABLE 14-10

Religious Attendance Reported by Men and Women in 2004

	Men	Women
Weekly	22%	31%
Less often	78	69
100% =	(1,276)	(1,525)

Note: Rounding to the nearest whole percentages may result in a total of 99% or 101 % in some cases. This is referred to as a "rounding error."

Source: General Social Survey, 2004, National Opinion Research Center.

I've divided the group of subjects into two subgroups—men and women—and then described the behavior of each subgroup. That is the correct method for constructing this table. Notice, however, that we could—however inappropriately—construct the table differently. We could first divide the subjects into different degrees of religious services attendance and then describe each of those subgroups in terms of the percentage of men and women in each. This method would make no sense in terms of explanation, however. Table 14-10 suggests that your gender will affect your frequency of religious services attendance. Had we used the other method of construction, the table would suggest that your religious services attendance affects whether you're a man or a woman—which makes no sense. Your behavior can't determine your gender.

A related problem complicates the lives of new-data analysts. How do you read a percentage table? There is a temptation to read Table 14-10 as follows: "Of the women, only 31 percent attended religious services weekly, and 69 percent said they attended less often; therefore, being a woman makes you less likely to attend religious services frequently." This is, of course, an incorrect reading of the table. Any conclusion that gender —as a variable—has an effect on religious service attendance must hinge on a comparison between men and women. Specifically, we compare the 31 percent with the 22 percent and note that women are more likely than men to attend religious services weekly. The comparison of subgroups, then, is essential in reading an explanatory bivariate table.

In constructing and presenting Table 14-10, I've used a convention called *percentage down*. This term means that you can add the percentages down each column to total 100 percent (with the possibility of a rounding error, as noted in the table). You read this form of table across a row. For the row labeled "weekly," what percentage of the men attend weekly? What percentage of the women attend weekly?

The direction of percentaging in tables is arbitrary, and some researchers prefer to percentage across. They would organize Table 14-10 so that "men" and "women" were shown on the left side of the table, identifying the two rows, and "weekly" and "less often" would appear at the top to identify the columns. The actual numbers in the table would be moved around accordingly, and each row of percentages would total approximately 100 percent. In that case, you would read the table down a column, still asking what percentage of men and women attended frequently. The logic and the conclusion would be the same in either case; only the form would differ.

In reading a table that someone else has constructed, therefore, you need to find out in which direction it has been percentaged. Usually this will be labeled or be clear from the logic of the variables being analyzed. As a last resort, however, you should add the percentages in each column and each row. If each of the columns totals 100 percent, the table has been percentaged down. If the rows total 100 percent each, it has been percentaged across. The rule, then, is as follows:

1. If the table is percentaged down, read across.
2. If the table is percentaged across, read down.

Percentaging a Table

Figure 14-5 reviews the logic by which we create percentage tables from two variables. I've used as variables *gender* and *attitudes toward equality for men and women*.

Here's another example. Suppose we're interested in learning something about newspaper editorial policies regarding the legalization of marijuana. We undertake a content analysis of editorials on this subject that have appeared during a given year in a sample of daily newspapers across the nation. Each editorial has been classified as favorable, neutral, or unfavorable toward the legalization of marijuana. Perhaps we wish to examine the relationship between editorial policies and the types of communities in which the newspapers are published, thinking that rural newspapers might be more conservative in this regard than urban ones. Thus, each newspaper (hence, each editorial) has been classified in terms of the population of the community in which it is published.

Table 14-11 presents some hypothetical data describing the editorial policies of rural and urban newspapers. Note that the unit of analysis in this example is the individual editorial. Table 14-11 tells us that there were 127 editorials about marijuana in our sample of newspapers published in communities with populations under 100,000. (Note that this cutting point is chosen for simplicity of illustration and does not mean that *rural* refers to a community of less than 100,000 in any absolute sense.) Of these, 11 percent (14 editorials divided by the base of 127) were favorable toward legalization of marijuana, 29 percent were neutral, and 60 percent were unfavorable. Of the 438 editorials that appeared in our sample of newspapers published in communities of more than 100,000 residents, 32 percent (140 editorials) were favorable toward legalizing marijuana, 40 percent were neutral, and 28 percent were unfavorable.

When we compare the editorial policies of rural and urban newspapers in our imaginary study, we find—as expected—that rural newspapers are less favorable toward the legalization of marijuana than urban newspapers are. We determine this by noting that a larger percentage (32 percent) of the urban editorials were favorable than the percentage of rural ones (11 percent). We might note as well that more rural than urban editorials were unfavorable (60 percent compared with 28 percent). Note that this table assumes that the size of a community might affect its newspapers' editorial policies on this issue, rather than that editorial policy might affect the size of communities.

a. Some men and women who either favor (+) gender equality or don't (−) favor it.

b. Separate the men from the women (the independent variable).

c. Within each gender group, separate those who favor equality from those who don't (the dependent variable).

d. Count the numbers in each cell of the table.

e. What percentage of the women favor equality?

f. What percentage of the men favor equality?

g. Conclusions.

While a majority of both men and women favored gender equality, women were more likely than men to do so.

Thus, gender appears to be one of the causes of attitudes toward sexual equality.

	Women	Men
Favor equality	80%	60%
Don't favor equality	20%	40%
Total	**100%**	**100%**

FIGURE 14-5

Percentaging a Table

TABLE 14-11

Hypothetical Data Regarding Newspaper Editorials on the Legalization of Marijuana

Editorial Policy toward Legalizing Marijuana	Community Size	
	Under 100,000	Over 100,000
Favorable	11%	32%
Neutral	29	40
Unfavorable	60	28
100% =	(127)	(438)

Constructing and Reading Bivariate Tables

Let's now review the steps involved in the construction of explanatory bivariate tables:

1. The cases are divided into groups according to the attributes of the independent variable.

2. Each of these subgroups is then described in terms of attributes of the dependent variable.

3. Finally, the table is read by comparing the independent variable subgroups with one another in terms of a given attribute of the dependent variable.

Following these steps, let's repeat the analysis of gender and attitude on sexual equality. For the reasons outlined previously, *gender* is the independent variable; *attitude toward sexual equality* constitutes the dependent variable. Thus, we proceed as follows:

1. The cases are divided into men and women.

2. Each gender subgrouping is described in terms of approval or disapproval of sexual equality.

3. Men and women are compared in terms of the percentages approving of sexual equality.

In the example of editorial policies regarding the legalization of marijuana, *size of community* is

contingency table A format for presenting the relationships among variables as percentage distributions.

the independent variable, and a *newspaper's editorial policy* the dependent variable. The table would be constructed as follows:

1. Divide the editorials into subgroups according to the sizes of the communities in which the newspapers are published.

2. Describe each subgroup of editorials in terms of the percentages favorable, neutral, or unfavorable toward the legalization of marijuana.

3. Compare the two subgroups in terms of the percentages favorable toward the legalization of marijuana.

Bivariate analyses typically have an explanatory causal purpose. These two hypothetical examples have hinted at the nature of causation as social scientists use it.

Tables such as the ones we've been examining are commonly called **contingency tables:** Values of the dependent variable are contingent on (depend on) values of the independent variable. Although contingency tables are common in social science, their format has never been standardized. As a result, you'll find a variety of formats in research literature. As long as a table is easy to read and interpret, there's probably no reason to strive for standardization. However, there are several guidelines that you should follow in the presentation of most tabular data.

1. A table should have a heading or a title that succinctly describes what is contained in the table.

2. The original content of the variables should be clearly presented—in the table itself if at all possible or in the text with a paraphrase in the table. This information is especially critical when a variable is derived from responses to an attitudinal question, because the meaning of the responses will depend largely on the wording of the question.

3. The attributes of each variable should be clearly indicated. Though complex categories will have to be abbreviated, their meaning should be clear in the table and, of course, the full description should be reported in the text.

4. When percentages are reported in the table, the base on which they are computed should be indicated. It's redundant to present all the raw numbers for each category, because these could be reconstructed from the percentages and the bases. Moreover, the presentation of both numbers and percentages often confuses a table and makes it more difficult to read.

5. If any cases are omitted from the table because of missing data ("no answer," for example), their numbers should be indicated in the table.

While I have introduced the logic of causal, bivariate analysis in terms of percentage tables, there are many other formats appropriate to this topic. Scatterplot graphs are one possibility, providing a visual display of the relationship between two variables. For an engaging example of this, you might check out the GapMinder software available on the web (see the link on this book's website). Using countries as the unit of analysis, you can examine the relationship between birthrate and infant mortality, for example. In fact, you can watch the relationship develop over time.

Introduction to Multivariate Analysis

The logic of **multivariate analysis,** or the analysis of more than two variables simultaneously, can be seen as an extension of bivariate analysis. Specifically, we can construct multivariate tables on the basis of a more complicated subgroup description by following essentially the same steps outlined for bivariate tables. Instead of one independent variable and one dependent variable, however, we'll have more than one independent variable. Instead of explaining the dependent variable on the basis of a single independent variable, we'll seek an explanation through the use of more than one independent variable.

Let's return to the example of religious services attendance. Suppose we believe that age would also affect such behavior (Glock's Comfort Hypothesis suggests that older people are more religious

TABLE 14-12

Multivariate Relationship: Religious Service Attendance, Gender, and Age

	"How often do you attend religious services?"			
	Under 40		*40 and Older*	
	Men	*Women*	*Men*	*Women*
About weekly*	22%	30%	33%	45%
Less often	78	70	67	55
100% =	(9,878)	(12,139)	(12,282)	(16,040)

*About weekly = "More than once a week," "Weekly," and "Nearly every week."
Source: General Social Survey, 1972–2006, National Opinion Research Center.

than younger people). As the first step in table construction, we would divide the total sample into subgroups based on the attributes of both independent variables simultaneously: younger men, older men, younger women, and older women. Then the several subgroups would be described in terms of the dependent variable, *religious services attendance,* and comparisons would be made. Table 14-12, from an analysis of the 1972–2006 General Social Survey data, is the result.

Table 14-12 has been percentaged down and therefore should be read across. The interpretation of this table warrants several conclusions:

1. Among both men and women, older people attend religious services more often than do younger people. Among women, 30 percent of those under 40 and 45 percent of those 40 and older attend religious services weekly. Among men, the respective figures are 22 and 33 percent.

2. Within each age group, women attend slightly more frequently than men. Among those respondents under 40, 30 percent of the women

multivariate analysis The analysis of the simultaneous relationships among several variables. Examining simultaneously the effects of *age, gender,* and *social class* on *religiosity* would be an example of multivariate analysis.

attend weekly, compared with 22 percent of the men. Among those 40 and over, 45 percent of the women and 33 percent of the men attend weekly.

3. As measured in the table, age appears to have a greater effect on attendance at religious services than does gender.

4. Age and gender have independent effects on religious service attendance. Within a given attribute of one independent variable, different attributes of the second still affect behaviors.

5. Similarly, the two independent variables have a cumulative effect on behaviors. Older women attend the most often (45 percent), and younger men attend the least often (22 percent).

Before I conclude this section, I want to note an alternative format for presenting such data. Several of the tables presented in this chapter are somewhat inefficient. When the dependent variable, *religious attendance,* is dichotomous (having exactly two attributes), knowing one attribute permits the reader to reconstruct the other easily. Thus, if we know that 30 percent of the women under 40 attend religious services weekly, then we know automatically that 70 percent attend less often. So reporting the percentages who attend less often is unnecessary.

On the basis of this recognition, Table 14-12 could be presented in the alternative format of Table 14-13. In Table 14-13, the percentages of people saying they attend religious services about weekly are reported in the cells representing the intersections of the two independent variables. The numbers presented in parentheses below each percentage represent the number of cases on which the percentages are based. Thus, for example, the reader knows there are 12,139 women under 40 years of age in the 1972–2006 surveys, and 30 percent of them attend religious services weekly. We can calculate from this that 3,642 of those 12,139 women said they attend religious services about weekly and that the other 8,497 younger women (or 70 percent) attend less frequently. This new table is easier to read than the former one, and it does not sacrifice any detail.

TABLE 14-13

A Simplification of Table 14-12

Percent Who Attend about Weekly

	Men	Women
Under 40	22	30
	(9,878)	(12,139)
40 and Older	33	45
	(12,282)	(16,040)

Source: General Social Survey, 1972–2006, National Opinion Research Center.

Sociological Diagnostics

The multivariate techniques we're now exploring can serve as powerful tools for diagnosing social problems. They can be used to replace opinions with facts and to settle ideological debates with data analysis.

For an example, let's return to the issue of gender and income. Many explanations have been advanced to account for the long-standing pattern of women in the labor force earning less than men. One explanation is that, because of traditional family patterns, women as a group have participated less in the labor force and many only begin working outside the home after completing certain child-rearing tasks. Thus, women as a group probably have less seniority at work than men do, and income increases with seniority. A 1984 study by the Census Bureau showed this reasoning to be partly true, as Table 14-14 shows.

Table 14-14 indicates, first of all, that job tenure does indeed affect income. Among both men and women, those with more years on the job earned more. This is seen by reading down the first two columns of the table.

The table also indicates that women earn less than men, regardless of job seniority. This can be seen by comparing average wages across the rows of the table, and the ratio of women-to-men wages is shown in the third column. Thus, years on the job is an important determinant of earnings, but seniority does not adequately explain the pattern of women earning less than men. In fact, we see that

TABLE 14-14

Gender, Job Tenure, and Income, 1984*

Years Working with Current Employer	Average Hourly Income		Women/Men Ratio
	Men	Women	
Less than 2 years	$8.46	$6.03	0.71
2–4 years	$9.38	$6.78	0.72
5–9 years	$10.42	$7.56	0.73
10 years or more	$12.38	$7.91	0.64

*Full-time workers 21–64 years of age

Source: U.S. Bureau of the Census, Current Population Reports, Series P-70, No. 10, Male-Female Differences in Work Experience, Occupation, and Earning, 1984 (Washington, DC: U.S. Government Printing Office, 1987), 4.

TABLE 14-15

Average Earnings of Year-Round, Full-Time Workers by Educational Attainment, 2003

	Men	Women	Ratio of Women/ Men Earnings
All workers	$53,039	$37,197	0.70
Less than 9th grade	23,972	20,979	0.88
9th–12th grades	29,100	21,426	0.74
HS graduates	38,331	27,956	0.73
Some college	46,332	31,655	0.68
Associate degree	48,683	36,528	0.75
Bachelors or more	81,007	53,215	0.66

Note: These data point to a persistent difference between the incomes of men and women, even when both groups have achieved the same levels of education.

Source: U.S. Bureau of the Census, Statistical Abstract of the United States (Washington, DC: U.S. Government Printing Office, 2006), Table 686, p. 467. You can also access this table online at the link on this book's website.

women with ten or more years on the job earn substantially less ($7.91/hour) than do men with less than two years ($8.46/hour).

Although years on the job does not fully explain the difference between men's and women's pay, there are other possible explanations: level of education, child care responsibilities, and so forth. The researchers who calculated Table 14-14 also examined some of the other variables that might reasonably explain the differences in pay without representing gender discrimination, including these:

- Number of years in the current occupation
- Total years of work experience (any occupation)
- Whether they have usually worked full time
- Marital status
- Size of city or town they live in
- Whether covered by a union contract
- Type of occupation
- Number of employees in the firm
- Whether private or public employer
- Whether they left previous job involuntarily
- Time spent between current and previous job
- Race
- Whether they have a disability
- Health status
- Age of children

- Whether they took an academic curriculum in high school
- Number of math, science, and foreign language classes in high school
- Whether they attended private or public high school
- Educational level achieved
- Percentage of women in the occupation
- College major

Each of the variables listed here might reasonably affect earnings and, if women and men differ in these regards, could help to account for male/female income differences. When all these variables were taken into account, the researchers could account for 60 percent of the discrepancy between the incomes of men and women. The remaining 40 percent, then, is a function of other "reasonable" variables and/or prejudice. This kind of conclusion can be reached only by examining the effects of several variables at the same time—that is, through multivariate analysis.

I hope this example shows how the logic implicit in day-to-day conversations can be represented and tested in a quantitative data analysis like this. Along those lines, you might be asking yourself, These data point to salary discrimination

Keeping Humanity in Focus

Transsexuals are individuals who choose to change their biological sex permanently though surgery and hormones. Clearly, such a radical change brings many adjustments and challenges that would make for interesting studies, but Kristen Schilt has taken an unusual tack.

While many kinds of research point to the disadvantaged status of women in the workplace, Schilt's research on transsexuals reveals the impact of gender on a personal level. In many of the cases, the subjects changed their sex while maintaining the same job in their employing

organization. Following their sex change, female-to-male transexuals were likely to enjoy pay raises and increased authority. In other studies, male-to-female transsexuals reported just the opposite experiences. Personal accounts such as these flesh out statistical studies that consistently show women earning less than men, even when they do the same work.

Source: Kristen Schilt,"Just One of the Guys? How Transmen Make Gender Visible in the Workplace," *Gender and Society* 20, no. 4 (2006): 465–90.

---

against women in 1984, but hasn't that been remedied? Not really, as indicated by more-recent data.

In 2003 the average full-time, year-round male worker earned $53,039. The average full-time, year-round female worker earned $37,197, or about 70 percent as much as her male counterpart (U.S. Bureau of the Census 2006: Table 686, p. 467). But does that difference represent sexual discrimination or does it reflect legitimate factors?

Some argue that education, for example, affects income and that in the past, women have gotten less education than men. We might start, therefore, by checking whether educational differences explain why women today earn less, on average, than men. Table 14-15 offers data to test this hypothesis.

As the table shows, at each level of comparable education, women earn substantially less than men do. Clearly, education does not explain the discrepancy.

This is the kind of analysis you are now equipped to undertake. See "Keeping Humanity in Focus" for more on gender discrimination in the workplace.

As another example of multivariate data analysis in real life, consider the common observation that minority group members are more likely to be denied bank loans than white applicants are. A counterexplanation might be that the minority applicants in question were more likely to have had a prior bankruptcy or that they had less collateral to guarantee the requested loan—both reasonable

bases for granting or denying loans. However, the kind of multivariate analysis we've just examined could easily resolve the disagreement.

Let's say we look only at those who have not had a prior bankruptcy and who have a certain level of collateral. Are whites and minorities equally likely to get the requested loan? We could conduct the same analysis in subgroups determined by level of collateral. If whites and minorities were equally likely to get their loans in each of the subgroups, we would need to conclude that there was no ethnic discrimination. If minorities were still less likely to get their loans, however, that would indicate that bankruptcy and collateral differences were not the explanation—strengthening the case that discrimination was at work.

All this should make it clear that social research can play a powerful role in serving the human community. It can help us determine the current state of affairs and can often point the way to where we want to go.

Welcome to the world of sociological diagnostics!

Ethics and Quantitative Data Analysis

In Chapter 13, I pointed out that the subjectivity present in qualitative data analysis increases the risk of biased analyses, which experienced researchers learn to avoid. Some people believe that *quantitative* analyses, however, are not susceptible to subjective biases. Unfortunately, this isn't exactly so. Even in the most mathematically explicit analysis, we can discover ample room for defining and measuring variables in ways that encourage one finding over another. Quantitative analysts need to guard against this. Sometimes, the careful specification of hypotheses in advance can offer protection, although this can also constitute a straitjacket hampering a full exploration of what data can tell us.

The quantitative analyst has an obligation to report formal hypotheses and less-formal expectations that didn't pan out. Let's suppose you think that a particular variable will prove a powerful cause of gender prejudice, but your data analysis contradicts that expectation. You should report the lack of correlation, since such information is useful to other researchers who will conduct research on this topic. While it would be more satisfying to discover what causes prejudice, it's very important to know what doesn't cause it.

The protection of subject privacy is as important in quantitative as in qualitative analysis. In the former case, however, it's often easier to collect and record data in ways that make subject identification more difficult. However, the first time public officials demand that you reveal the names of student-subjects who reported using illegal drugs in a survey, this issue will take on more salience. (Don't reveal the names, by the way. If necessary, burn the questionnaires—"accidentally.")

MAIN POINTS

Introduction

- Quantitative analysis involves the techniques by which researchers convert data to numerical forms and subject them to statistical analyses.

Quantification of Data

- Some data, such as age and income, are intrinsically numerical.
- Often, quantification involves coding into categories that are then given numerical representations.

- Researchers may use existing coding schemes, such as the Census Bureau's categorization of occupations, or develop their own coding categories. In either case, the coding scheme must be appropriate to the nature and objectives of the study.
- A codebook is the document that describes (1) the identifiers assigned to different variables and (2) the codes assigned to the attributes of those variables.

Univariate Analysis

- Univariate analysis is the analysis of a single variable. Because univariate analysis does not involve the relationships between two or more variables, its purpose is descriptive rather than explanatory.
- Several techniques allow researchers to summarize their original data to make them more manageable while maintaining as much of the original detail as possible. Frequency distributions, averages, grouped data, and measures of dispersion are all ways of summarizing data concerning a single variable.

Subgroup Comparisons

- Subgroup comparisons can be used to describe similarities and differences among subgroups with respect to some variable.

Bivariate Analysis

- Bivariate analysis focuses on relationships between variables rather than on comparisons of groups. Bivariate analysis explores the statistical association between the independent variable and the dependent variable. Its purpose is usually explanatory rather than merely descriptive.
- The results of bivariate analyses often are presented in the form of contingency tables, which are constructed to reveal the effects of the independent variable on the dependent variable.

Introduction to Multivariate Analysis

- Multivariate analysis is a method of analyzing the simultaneous relationships among several variables. It may also be used to understand the relationship between two variables more fully.
- The logic and techniques involved in quantitative research can also be valuable to qualitative researchers.

Sociological Diagnostics

- Sociological diagnostics is a quantitative analysis technique for determining the nature of social problems such as ethnic or gender discrimination.

Ethics and Quantitative Data Analysis

- Unbiased analysis and reporting is as much an ethical concern in quantitative analysis as in qualitative analysis.
- Subjects' privacy must be protected in quantitative data analysis and reporting.

KEY TERMS

The following terms are defined in context in the chapter and at the bottom of the page where the term is introduced, as well as in the comprehensive glossary at the back of the book.

average	mean
bivariate analysis	median
codebook	mode
contingency table	multivariate analysis
continuous variable	quantitative analysis
discrete variable	standard deviation
dispersion	univariate analysis
frequency distribution	

PROPOSING SOCIAL RESEARCH: QUANTITATIVE DATA ANALYSIS

See the exercise for Chapter 16.

REVIEW QUESTIONS AND EXERCISES

1. How might the various majors at your college be classified into categories? Create a coding system that would allow you to categorize them according to some meaningful variable. Then create a different coding system, using a different variable.

2. How many ways could you be described in numerical terms? What are some of your intrinsically numerical attributes? Could you express some of your qualitative attributes in quantitative terms?

3. How would you construct and interpret a contingency table from the following information: 150 Democrats favor raising the minimum wage, and 50 oppose it; 100 Republicans favor raising the minimum wage, and 300 oppose it?

4. Using the hypothetical data in the following table, how would you construct and interpret tables showing the following?

   a. The bivariate relationship between age and attitude toward abortion

   b. The bivariate relationship between political orientation and attitude toward abortion

   c. The multivariate relationship linking age, political orientation, and attitude toward abortion

Age	Political Orientation	Attitude toward Abortion	Frequency
Young	Liberal	Favor	90
Young	Liberal	Oppose	10
Young	Conservative	Favor	60
Young	Conservative	Oppose	40
Old	Liberal	Favor	60
Old	Liberal	Oppose	40
Old	Conservative	Favor	20
Old	Conservative	Oppose	80

SPSS EXERCISES

See the booklet that accompanies your text for exercises using SPSS (Statistical Package for the Social Sciences). There are exercises offered for each chapter, and you'll also find a detailed primer on using SPSS.

Online Study Resources

If your book came with an access code card, visit www.cengage.com/login to register. To purchase access, please visit www.ichapters.com.

1. Before you do your final review of the chapter, take the *CengageNOW* pretest to help identify the areas on which you should concentrate. You'll find information on this online tool, as well as

instructions on how to access all of its great resources, in the front of the book.

2. As you review, take advantage of the *Cengage-eNOW* personalized study plan, based on your quiz results. Use this study plan with its interactive exercises and other resources to master the material.

3. When you're finished with your review, take the posttest to confirm that you're ready to move on to the next chapter.

WEBSITE FOR *THE PRACTICE OF SOCIAL RESEARCH* 12TH EDITION

Go to your book's website at www.cengage.com/sociology/babbie for tools to aid you in studying for your exams. You'll find *Tutorial Quizzes* with feedback, *Internet Exercises, Flash Cards, Glossaries,* and *Essay Quizzes,* as well as *InfoTrac College Edition* search terms, suggestions for additional reading, *Web Links,* and primers for using data-analysis software such as SPSS.

CHAPTER FIFTEEN

The Elaboration Model

CHAPTER OVERVIEW

The elaboration model illustrates the fundamental logic of multivariate and causal analysis. Exploring applications of this logic in the form of simple percentage tables provides a foundation for making sense of more-complex analytic methods.

CengageNOW for Sociology

Use this online tool to help you make the grade on your next exam. After reading this chapter, go to "Online Study Resources" at the end of the chapter for instructions on how to benefit from *CengageNOW*.

Introduction

This chapter is devoted to a perspective on social science analysis that is referred to variously as the **elaboration model,** the interpretation method, the Lazarsfeld method, or the Columbia school. Its many names reflect the fact that it aims at elaborating on an empirical relationship among variables in order to interpret that relationship, in the manner developed by Paul Lazarsfeld while he was at Columbia University. As such, the elaboration model is one method for doing multivariate analysis.

Researchers use the logic of the elaboration model to understand the relationship between two variables through the simultaneous introduction of additional variables, though they may not always refer to the model by name. Though developed primarily through the medium of contingency tables, it can be used with other statistical techniques, as Chapter 16 will show.

I firmly believe that the elaboration model offers the clearest available picture of the logic of causal analysis in social research. Especially through the use of contingency tables, this method portrays the logical process of scientific analysis. Moreover, if you can comprehend fully the use of the elaboration model using contingency tables, you should greatly improve your ability to use and understand more-sophisticated statistical techniques, such as partial regressions and log-linear models, for example.

In a sense, this discussion of elaboration analysis is an extension of our earlier examination of spuriousness in Chapter 4. As you'll recall, one of the criteria of causal relations in social research is that the observed relationship between two variables not be an artifact caused by some other variable. In the case of the positive relationship between the number of fire trucks responding to a fire and the amount of damage done, for example, we saw that the size of the fire explained away the apparent relationship between trucks and damage. The bigger the fire, the more trucks responding to it; and the bigger the fire, the more damage

done. The logic used in that hypothetical example was the same as the logic of the elaboration model.

Using both hypothetical and real examples, we'll see that the testing of an observed relationship may result in a variety of discoveries and logical interpretations. Spuriousness is only one of the possibilities.

The accompanying box "Why Do Elaboration?" by one of the elaboration model's creators, Patricia Kendall, provides another powerful justification for using this model.

The Origins of the Elaboration Model

The historical origins of the elaboration model provide a good illustration of how scientific research works in practice. As I mentioned in Chapter 1, during World War II Samuel Stouffer organized and headed a special social research branch within the U.S. Army. Throughout the war, this group conducted a large number and variety of surveys among U.S. servicemen. Although the objectives of these studies varied somewhat, they generally focused on the factors affecting soldiers' combat effectiveness.

Several of the studies examined morale in the military. Because morale seemed to be related positively to combat effectiveness, improving morale would make the war effort more effective. Stouffer and his research staff sought to uncover some of the variables that affected morale. In part, the group sought to confirm empirically some

---

elaboration model A logical model for understanding the relationship between two variables by controlling for the effects of a third. Developed principally by Paul Lazarsfeld. The various outcomes of an elaboration analysis are replication, explanation, interpretation, and specification.

Why Do Elaboration?

Patricia L. Kendall

Department of Sociology, Queens College, CUNY

There are several aspects of a true controlled experiment. The most crucial are (1) creating experimental and control groups that are identical within limits of chance (this is done by assigning individuals to the two groups through processes of randomization: using tables of random numbers, flipping coins, etc.); (2) making sure that it is the experimenter who introduces the stimulus, not external events; and (3) waiting to see whether the stimulus has had its presumed effect.

We may have the hypothesis, for example, that attending Ivy League colleges leads to greater success professionally than attending other kinds of colleges and universities does. How would we study this through a true experiment? Suppose you said, "Take a group of people in their 40s, find out which ones went to Ivy League colleges, and see whether they are more successful than those who went to other kinds of colleges." If that is your answer, you are wrong.

A true experiment would require the investigator to select several classes of high school seniors, divide each class at random into experimental and control groups, send the experimental groups to Ivy League colleges (regardless of their financial circumstances or academic qualifications and regardless of the desire of the colleges to accept them) and the control group to other colleges and universities, wait 20 years or so until the two groups have reached professional maturity, and then measure the relative success of the two groups. Certainly a bizarre process.

Sociologists also investigate the hypothesis that coming from a broken home leads to juvenile delinquency. How would we go about studying this experimentally? If you followed the example above, you would see that studying this hypothesis through a true experiment would be totally impossible. Just think of what the experimenter would have to do!

The requirements of true experiments are so unrealistic in sociological research that we are forced to use other, and less ideal, methods in all but the most trivial situations. We can study experimentally whether students learn more from one type of lecture than another, or whether a film changes viewers' attitudes. But these are not always the sorts of questions in which we are truly interested.

We therefore resort to approximations—generally surveys—that have their own shortcomings. However, the elaboration model allows us to examine survey data, take account of their possible shortcomings, and draw rather sophisticated conclusions about important issues.

Source: From Patricia Kendall, "Why Do Elaboration?" from *Attending an Ivy League College and Success in Later Professional Life.*

commonly accepted propositions, including the following:

1. Promotions surely affect soldiers' morale, so soldiers serving in units with low promotion rates should have relatively low morale.

2. Given racial segregation and discrimination in the South, African American soldiers being trained in northern training camps should have higher morale than should those being trained in the South.

3. Soldiers with more education should be more likely to resent being drafted into the army as enlisted men than should those with less education.

Each of these propositions made sense logically, and common wisdom held each to be true. Stouffer decided to test each empirically. To his surprise, none of the propositions was confirmed.

We discussed the first proposition in Chapter 1. As you may recall, Stouffer found that soldiers serving in the Military Police (where promotions were the slowest in the army) had fewer complaints about the promotion system than did those serving in the Army Air Corps (where promotions were the fastest in the army). The other propositions fared just as badly. African American soldiers serving in northern training camps and those serving in southern training camps seemed to differ little if at all in their general morale. And less-educated soldiers were more likely to resent being drafted into the army than those with more education were.

Rather than trying to hide the findings or just running tests of statistical significance and publishing the results, Stouffer asked, "Why?" He found the answer to this question within the concepts of reference group and relative deprivation. Put simply, Stouffer suggested that soldiers did not evaluate their positions in life according to absolute, objective standards, but rather on the basis of their position relative to others around them. The people they compared themselves with were in their reference group, and they felt relative deprivation if they didn't compare favorably in that regard.

Following this logic, Stouffer found an answer to each of the anomalies in his empirical data. Regarding promotion, he suggested that soldiers judged the fairness of the promotion system based on their own experiences relative to others around them. In the Military Police, where promotions were few and slow, few soldiers knew of a less-qualified buddy who had been promoted faster than they had. In the Army Air Corps, however, the rapid promotion rate meant that many soldiers knew of less-qualified buddies who had been promoted faster than seemed appropriate. Thus, ironically, the MPs said the promotion system was generally fair, and the air corpsmen said it was not.

A similar analysis seemed to explain the case of the African American soldiers. Rather than comparing conditions in the North with those in the South, African American soldiers compared their own status with the status of the African American civilians around them. In the South, where discrimination was at its worst, they found that being a soldier insulated them somewhat from adverse cultural norms in the surrounding community. Whereas southern African American civilians were grossly discriminated against and denied self-esteem, good jobs, and so forth, African American soldiers had a slightly better status. In the North, however, many of the African American civilians they encountered held well-paying defense jobs. And with discrimination being less severe, being a soldier did not help one's status in the community.

Finally, the concepts of reference group and relative deprivation seemed to explain the anomaly of highly educated draftees accepting their induction more willingly than those with less education did. Stouffer reasoned as follows:

1. A person's friends, on the whole, have about the same educational status as that person does.

2. Draft-age men with less education are more likely to engage in semi-skilled production-line occupations and farming than more educated men.

3. During wartime, many production-line industries and farming are vital to the national interest; workers in those industries and farmers are exempted from the draft.

4. A man with little education is more likely to have friends in draft-exempt occupations than a man with more education.

5. When each compares himself with his friends, a less educated draftee is more likely to feel discriminated against than a draftee with more education.

(Stouffer et al. 1949–1950: 122–27)

Stouffer's explanations unlocked the mystery of the three anomalous findings. Because they were not part of a preplanned study design, however, he lacked empirical data for testing them. Nevertheless, Stouffer's logical exposition provided the basis for the later development of the elaboration model: understanding the relationship between two variables through the controlled introduction of other variables.

Paul Lazarsfeld and his associates at Columbia University formally developed the elaboration model in 1946. In a methodological review of Stouffer's army studies, Lazarsfeld and Patricia Kendall used the logic of the elaboration model to present hypothetical tables that would have proved Stouffer's contention regarding education and acceptance of induction had the empirical data been available (Kendall and Lazarsfeld 1950).

The central logic of the elaboration model begins with an observed relationship between two variables and the possibility that one variable may be causing the other. In the Stouffer example, the initial two variables were *educational level* and *acceptance of being drafted as fair*. Because the soldiers' educational levels were set before they were drafted (and thus having an opinion about being drafted) it would seem that *educational level* was the cause, or independent variable, and *acceptance of induction* was the effect, or dependent

TABLE 15-1

Summary of Stouffer's Data on Education and Acceptance of Induction

	High Ed.	Low Ed.
Should not have been deferred	88%	70%
Should have been deferred	12	30
	100	100
	(1,761)	(1,876)

Source: Tables 15-1, 15-2, 15-3, and 15-4 are modified with permission of Macmillan Publishing Co., Inc., from *Continuities in Social Research: Studies in the Scope and Method of "The American Soldier"* by Robert K. Merton and Paul F. Lazarsfeld (eds.). Copyright 1950 by The Free Press, a Corporation, renewed 1978 by Robert K. Merton.

TABLE 15-2

Hypothetical Relationship between Education and Deferment of Friends

Friends Deferred?	High Ed.	Low Ed.
Yes	19%	79%
No	81	21
	100	100
	(1,761)	(1,876)

TABLE 15-3

Hypothetical Relationship between Deferment of Friends and Acceptance of One's Own Induction

	Friends Deferred?	
	Yes	No
Should not have been deferred	63%	94%
Should have been deferred	37	6
	100	100
	(1,819)	(1,818)

variable. As we just saw, however, the observed relationship countered what the researchers had expected.

The elaboration model examines the impact of other variables on the relationship first observed. Sometimes this analysis reveals the mechanisms through which the causal relationship occurs. Other times an elaboration analysis disproves the existence of a causal relationship altogether.

In the present example, the additional variable was whether or not a soldier's friends were deferred or drafted. In Stouffer's speculative explanation, this variable showed how it was actually logical that soldiers with more education would be the more accepting of being drafted: because it was likely that their friends would have been drafted. Those with the least education were likely to have been in occupations that often brought deferments from the draft, leading those drafted to feel they had been treated unfairly.

Kendall and Lazarsfeld began with Stouffer's data showing the positive association between education and acceptance of induction (see Table 15-1). In this and the following tables, "should have been deferred" and "should not have been deferred" represent inductees' judgments of their own situation, with the latter group feeling it was fair for them to have been drafted.

Then, Kendall and Lazarsfeld created some hypothetical tables to represent what the analysis might have looked like had soldiers been asked

whether most of their friends had been drafted or deferred. In Table 15-2, 19 percent of those with high education hypothetically said their friends were deferred, compared with 79 percent of the soldiers with less education.

Notice that the numbers of soldiers with high and low education are the same as in Stouffer's real data. In later tables, you'll see that the numbers who accepted or resented being drafted remain true to the original data. Only the numbers saying that friends were or were not deferred were made up.

Stouffer's explanation next assumed that soldiers with friends who had been deferred would be more likely to resent their own induction than those who had no deferred friends would. Table 15-3 presents the hypothetical data that would have supported that assumption.

The hypothetical data in Tables 15-2 and 15-3 would confirm linkages that Stouffer had specified in his explanation. First, soldiers with low education were more likely to have friends who were deferred than soldiers with more education were. Second, having friends who were deferred made

TABLE 15-4

Hypothetical Data Relating Education to Acceptance of Induction through the Factor of Having Friends Who Were Deferred

	Friends Deferred		No Friends Deferred	
	High Ed.	*Low Ed.*	*High Ed.*	*Low Ed.*
Should not have been deferred	63%	63%	94%	95%
Should have been deferred	37	37	6	5
	100	100	100	100
100% =	(335)	(1,484)	(1,426)	(392)

a soldier more likely to think he should have been deferred. Stouffer had suggested that these two relationships would clarify the original relationship between education and acceptance of induction. Kendall and Lazarsfeld created a hypothetical table that would confirm Stouffer's explanation (see Table 15-4).

Recall that the original finding was that draftees with high education were more likely to accept their induction into the army as fair than those with less education were. In Table 15-4, however, we note that level of education has no effect on the acceptance of induction among those who report having friends deferred: 63 percent among both educational groups indicate that they accept their induction (that is, they say they should not have been deferred). Similarly, educational level has no significant effect on acceptance of induction among those who reported having no friends deferred: 94 and 95 percent say they should not have been deferred.

On the other hand, among those with high education the acceptance of induction is strongly related to whether or not friends were deferred: 63 percent versus 94 percent. And the same is true among those with less education. The hypothetical data in Table 15-4, then, would support Stouffer's contention that education affected acceptance of induction only through the medium of having friends deferred. Highly educated draftees were less likely to have friends deferred and, by virtue of that fact, were more likely to accept their own induction as fair. Those with less education were more likely to have friends deferred and, by virtue of that fact, were less likely to accept their own induction.

Recognize that neither Stouffer's explanation nor the hypothetical data denied the reality of the original relationship. As educational level increased, acceptance of one's own induction also increased. The nature of this empirical relationship, however, was interpreted through the introduction of a third variable. The variable, *deferment of friends,* did not deny the original relationship; it merely clarified the mechanism through which the original relationship occurred.

This, then, is the heart of the elaboration model and of multivariate analysis. Having observed an empirical relationship between two variables (such as *level of education* and *acceptance of induction*), we seek to understand the nature of that relationship through the effects produced by introducing other variables (such as *having friends who were deferred*). Mechanically, we accomplish this by first dividing our sample into subsets on the basis of the **test variable,** also called the *control variable.* In our example, having friends deferred or not is the test variable, and the sample is divided into those who have deferred friends and those who do not. The relationship between the original two variables (*acceptance of induction* and *level of education*) is then recomputed separately for each of the subsamples.

test variable A variable that is held constant in an attempt to clarify further the relationship between two other variables. Having discovered a relationship between *education* and *prejudice,* for example, we might hold *gender* constant by examining the relationship between education and prejudice among men only and then among women only. In this example, *gender* would be the test variable.

The tables produced in this manner are called the *partial tables,* and the relationships found in the partial tables are called the **partial relationships,** or *partials.* The partial relationships are then compared with the initial relationship discovered in the total sample, often referred to as the **zero-order relationship** to indicate that no test variables have been controlled for.

Although the elaboration was first demonstrated through the use of hypothetical data, it laid out a logical method for analyzing relationships among variables that have been actually measured. As we'll see, our first, hypothetical example describes only one possible outcome in the elaboration model. There are others.

The Elaboration Paradigm

This section presents guidelines for understanding an elaboration analysis. To begin, we must know whether the test variable is antecedent (prior in time) to the other two variables or whether it is intervening between them, because these positions suggest different logical relationships in the multivariate model. If the test variable is intervening, as in the case of *education, deferment of friends,* and *acceptance of induction,* then the analysis is based on the model shown in Figure 15-1. The logic of this multivariate relationship is that the independent variable (*educational level*) affects the intervening

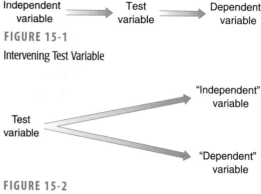

FIGURE 15-1

Intervening Test Variable

FIGURE 15-2

Antecedent Test Variable

test variable (*having friends deferred or not*), which in turn affects the dependent variable (*accepting induction*).

If the test variable is antecedent to both the independent and dependent variables, a different model must be used (see Figure 15-2). Here the test variable affects both the "independent" and "dependent" variables. Realize, of course, that the terms *independent variable* and *dependent variable* are, strictly speaking, used incorrectly in the diagram. In fact, we have one independent variable (the test variable) and two dependent variables. The incorrect terminology has been used only to provide continuity with the preceding example. Because of their individual relationships to the test variable, the "independent" and "dependent" variables are empirically related to each other, but there is no causal link between them. Their empirical relationship is merely a product of their coincidental relationships to the test variable. (Subsequent examples will further clarify this relationship.)

Table 15-5 provides a guide to understanding an elaboration analysis. The two columns in the table indicate whether the test variable is antecedent or intervening in the sense described previously. The left side of the table shows the nature of the partial relationships as compared with the original relationship between the independent and dependent variables. The body of the table gives the technical notations— replication, explanation, interpretation, and specification—assigned to each case. We'll discuss each in turn.

---

partial relationship In the elaboration model, this is the relationship between two variables when examined in a subset of cases defined by a third variable. Beginning with a zero-order relationship between *political party* and *attitudes toward abortion,* for example, we might want to see whether the relationship held true among both men and women (i.e., controlling for *gender*). The relationship found among men and the relationship found among women would be the partial relationships, sometimes simply called the partials.

zero-order relationship In the elaboration model, this is the original relationship between two variables, with no test variables controlled for.

TABLE 15-5

The Elaboration Paradigm

Partial Relationships Compared with Original	Test Variable	
	Antecedent	Intervening
Same Relationship	Replication	Replication
Less or none	Explanation	Interpretation
Split*	Specification	Specification

*One partial is the same or greater, and the other is less or none.

Replication

Whenever the partial relationships are essentially the same as the original relationship, the term **replication** is assigned to the result, regardless of whether the test variable is antecedent or intervening. This means that the original relationship has been replicated under test conditions. If, in our previous example, education still affected acceptance of induction both among those who had friends deferred and those who did not, then we would say the original relationship had been replicated. Note, however, that this finding would not confirm Stouffer's explanation of the original relationship. Having friends deferred or not would *not* be the mechanism through which education affected the acceptance of induction.

To see what a replication looks like, turn back to Tables 15-3 and 15-4. Imagine that our initial discovery was that having friends deferred strongly influenced how soldiers felt about being drafted, as shown in Table 15-3. Had we first discovered this relationship, we might have wanted to see whether it was equally true for soldiers of different educational backgrounds. To find out, we would have made *education* our control or test variable.

Table 15-4 contains the results of such an examination, though it is constructed somewhat differently from what we would have done had we used *education* as the test variable. Nevertheless, we see in the table that having friends deferred or not still influences attitudes toward being drafted among those soldiers with high education and those with low education. (Compare columns 1 and 3, then 2 and 4.) This result represents a rep-

lication of the relationship between having friends deferred and attitude toward being drafted.

Researchers frequently use the elaboration model rather routinely in the hope of replicating their findings among subsets of the sample. If we discovered a relationship between education and prejudice, for example, we might introduce such test variables as *age, region of the country, race, religion,* and so forth to test the stability of the original relationship. If the relationship were replicated among young and old, among people from different parts of the country, and so forth, we would have grounds for concluding that the original relationship was a genuine and general one.

Explanation

Explanation is the term used to describe a *spurious relationship:* an original relationship shown to be false through the introduction of a test variable. This requires two conditions: (1) The test variable must be antecedent to both the independent and dependent variables. (2) The partial relationships must be zero or significantly less than those found in the original. Several examples will illustrate this situation.

Let's look at an example we touched on in Chapter 4. There is an empirical relationship between the number of storks in different areas and the birthrates for those areas. The more storks in an area, the higher the birthrate. This empirical relationship might lead one to assume that the number of storks affects the birthrate. An antecedent test explains away this relationship, however. Rural areas have both more storks and higher birthrates

---

replication A technical term used in connection with the elaboration model, referring to the elaboration outcome in which the initially observed relationship between two variables persists when a control variable is held constant, thereby supporting the idea that the original relationship is genuine.

explanation An elaboration model outcome in which the original relationship between two variables is revealed to have been spurious, because the relationship disappears when an antecedent test variable is introduced.

I. Birthrates of Towns and Cities Having Few or Many Storks

Few Storks	Many Storks
L L LLL	H H
L LLL LH	H L H H H H H
L L L LL L L	H H H H

II. Birthrates of Towns and Cities Having Few or Many Storks, Controlling for Rural (Towns) and Urban (Cities)

	Few Storks	Many Storks
Rural	H	H H H H H H H H H H H
Urban	L L L L LL L L L L L L L L L L L L	L

H = Town or city with high birthrate
L = Town or city with low birthrate

FIGURE 15-3

The Facts of Life about Storks and Babies

than urban areas do. Within rural areas, there is no relationship between the number of storks and the birthrate; nor is there a relationship within urban areas.

Figure 15-3 illustrates how the rural/urban variable causes the apparent relationship between storks and birthrates. Part I of the figure shows the original relationship. Notice that all but one of the entries in the box for towns and cities with many storks have high birthrates and that all but one of those in the box for towns and cities with few storks have low birthrates. In percentage form, we say that 93 percent of the towns and cities with many storks also had high birthrates, contrasted with 7 percent of those with few storks. That's quite a large difference and represents a strong association between the two variables.

Part II of the figure separates the towns from the cities (the rural from urban areas) and examines storks and babies in each type of place separately. Now we can see that all the rural places have high birthrates, and all the urban places have low birth-

rates. Also notice that only one rural place had few storks and only one urban place had lots of storks.

Here's a similar example, also mentioned in Chapter 4 and at the beginning of this chapter. There is a positive relationship between the number of fire trucks responding to a fire and the amount of damage done. If more trucks respond, more damage is done. One might assume from this fact that the fire trucks themselves cause the damage. However, an antecedent test variable, the *size of the fire,* explains away the original relationship. Large fires do more damage than small ones do, and more fire trucks show up at large fires than at small ones. Looking only at large fires, we would see that the original relationship vanishes (or perhaps reverses itself); and the same would be true looking only at small fires.

Finally, let's take a real research example. Years ago, I found an empirical relationship between the region of the country in which medical school faculty members attended medical school and their attitudes toward Medicare (Babbie 1970). To simplify matters, only the East and the South will be examined. Of faculty members attending eastern medical schools, 78 percent said they approved of Medicare, compared with 59 percent of those attending southern medical schools. This finding made sense in view of the fact that the South seemed generally more resistant to such programs than the East did, and medical school training should presumably affect a doctor's medical attitudes. However, this relationship is explained away when we introduce an antecedent test variable: the region of the country in which the faculty member was raised. Of faculty members raised in the East, 89 percent attended medical school in the East and 11 percent in the South. Of those raised in the South, 53 percent attended medical school in the East and 47 percent in the South. Moreover, the areas in which faculty members were raised related to attitudes toward Medicare. Of those raised in the East, 84 percent approved of Medicare, as compared with 49 percent of those raised in the South.

Table 15-6 presents the three-variable relationship among (1) region in which raised, (2) region of medical school training, and (3) attitude toward Medicare. Faculty members raised in the East are quite likely to approve of Medicare, regardless of

TABLE 15-6

Region of Origin, Region of Medical School Training, and Attitude toward Medicare

		Percent Who Approve of Medicare	
		Region in Which Raised	
		East	South
Region of Medical School Training	East	84	50
	South	80	47

Source: Earl R. Babbie, *Science and Morality in Medicine* (Berkeley: University of California Press, 1970), 181.

where they attended medical school. Those raised in the South are relatively less likely to approve of Medicare, but, again, the region of their medical school training has little or no effect. These data indicate, therefore, that the original relationship between region of medical training and attitude toward Medicare was spurious; it was due only to the coincidental effect of region of origin on both region of medical training and attitude toward Medicare. When region of origin is held constant, as in Table 15-6, the original relationship disappears in the partials.

In "Attending an Ivy League College and Success in Later Professional Life," Patricia Kendall, one of the founders of the elaboration model, recalls a study in which the researcher suspected an explanation but found a replication. Though the data are no longer current, the topic is still of vital interest to students: To what extent does your professional success depend on attending the "right" school?

Interpretation

Interpretation is similar to explanation, except for the time placement of the test variable and the implications that follow from that difference. **Interpretation** represents the research outcome in which a test or control variable is discovered to be the mediating factor through which an independent variable has its effect on a dependent variable. The earlier example of education, friends deferred, and acceptance of induction is an excellent illustration of interpretation. In terms of the elaboration

model, the effect of education on acceptance of induction is not explained away; it is still a genuine relationship. In a real sense, educational differences cause differential acceptance of induction. The intervening variable, *deferment of friends,* merely helps to interpret the mechanism through which the relationship occurs. Thus, an interpretation does not deny the validity of the original causal relationship but simply clarifies the process through which that relationship functions.

Here's another example of interpretation. Researchers have observed that children from broken homes are more likely to become delinquent than those from intact homes are. This relationship may be interpreted, however, through the introduction of *supervision* as a test variable. Among children who are supervised, delinquency rates are not affected by whether or not their parents are divorced. The same is true among those who are not supervised. It is the relationship between broken homes and the lack of supervision that produced the original relationship.

Specification

Sometimes the elaboration model produces partial relationships that differ significantly from each other. For example, one partial relationship is the same as or stronger than the original two-variable relationship, and the second partial relationship is less than the original and may be reduced to zero. In the elaboration paradigm, this situation is referred to as **specification:** We have specified the

interpretation A technical term used in connection with the elaboration model. It represents the research outcome in which a control variable is discovered to be the mediating factor through which an independent variable has its effect on a dependent variable.

specification A technical term used in connection with the elaboration model, representing the elaboration outcome in which an initially observed relationship between two variables is replicated among some subgroups created by the control variable but not among others. In such a situation, you will have specified the conditions under which the original relationship exists: for example, among men but not among women.

Attending an Ivy League College and Success in Later Professional Life

Patricia L. Kendall

Department of Sociology, Queens College, CUNY

Probably the main danger for survey analysts is that a relationship they hope is causal will turn out to be spurious. That is, the original relationship between X and Y is explained by an antecedent test factor. More specifically, the partial relationships between X and Y reduce to 0 when that antecedent test factor is held constant.

This was a distinct possibility in a major finding from a study carried out several decades ago. One of my fellow graduate students at Columbia University, Patricia Salter West, based her dissertation on questionnaires obtained by *Time Magazine* from 10,000 of its male subscribers. Among many of the hypotheses developed by West was that male graduates of Ivy League schools (Brown, Columbia, Cornell, Dartmouth, Harvard, University of Pennsylvania, Princeton, and Yale) were more successful in their later professional careers, as defined by their annual earnings, than those who graduated from other colleges and universities.

The initial fourfold table (Table 1) supported West's expectation. Although I made up the figures, they conform closely to what West actually found in her study. Having attended an Ivy League school seems to lead to considerably greater professional success than does being a graduate of some other kind of college or university.

But wait a minute. Isn't this a relationship that typically could be spurious? Who can afford to send their sons to Ivy League schools? Wealthy families, of course.† And who can provide the business and professional connections that could help sons become successful in their careers? Again, wealthy or well-to-do families.

In other words, the socioeconomic status of the student's family may explain away the apparent causal relationship. In fact, some of West's findings suggest that this might indeed be the case.

TABLE 1*

Later Professional Success (Y)	College Attended (X)	
	Ivy League College	Other College or University
Successful (25%)	1,300 (65%)	2,000
Unsuccessful (75%)	1,700 (35%)	6,000
Total (100%)	3,000 (100%)	8,000

*I have had to invent relevant figures because the only published version of West's study contained no totals. See Ernest Havemann and Patricia Salter West, *They Went to College* (New York: Harcourt, Brace, 1952).

TABLE 2

Attendance at Ivy League Colleges According to Family Socioeconomic Status (SES)

College Attended (X)	Family SES (T)	
	High SES	Low SES
Ivy League colleges	1,500 (33%)	500 (9%)
Other colleges and universities	3,000 (67%)	5,000 (91%)
Total	4,500 (100%)	5,500 (100%)

According to Table 2, a third of those coming from families defined as wealthy, compared with 1 in 11 coming from less well-to-do backgrounds, attended Ivy League colleges. Thus there is a very high correlation between the two variables, X and T. (There is a similarly high

conditions under which the original relationship occurs.

Now recall the study, cited earlier in this book, of the sources of religious involvement (Glock, Ringer, and Babbie 1967: 92). It was discovered that among Episcopal church members, involvement decreased as social class increased. This finding is reported in Table 15-7, which examines mean levels of church involvement among women parishioners at different levels of social class.

Glock interpreted this finding in the context of others in the analysis and concluded that church involvement provides an alternative form of gratification for people who are denied gratification in the secular society. This conclusion explained why women were more religious than men, why old people were more religious than young people, and so forth. Glock reasoned that people of lower social class (measured by income and education) had fewer chances to gain self-esteem from the

TABLE 3

Partial Relationships between *X* and *Y* with *T* Held Constant

| | High Family SES (T) | | Low Family SES (T) | |
Later Success (Y)	Ivy League College (X)	Other College (X)	Ivy League College (X)	Other College (X)
Successful	1,000 (67%)	1,000 (33%)	300 (60%)	1,000 (20%)
Not successful	500 (33%)	2,000 (67%)	200 (40%)	4,000 (80%)
Total	1,500 (100%)	3,000 (100%)	500 (100%)	5,000 (100%)

correlation between family socioeconomic status [*T*] and later professional success [*Y*].)

The magnitude of these so-called marginal correlations suggest that West's hypothesis regarding the causal nature of having attended an Ivy League college might be incorrect; it suggests instead that the socioeconomic status of the students' families accounted for the original relationship she observed.

We are not done yet, however. The crucial question is what happens to the partial relationships once the test factor is controlled. These are shown in Table 3.

These partial relationships show that, even when family socioeconomic status is held constant, there is still a marked relationship between having attended an Ivy League college and success in later professional life. As a result, West's initial hypothesis received support from the analysis she carried out.

Despite this, West had in no way proved her hypothesis. There are almost always additional antecedent factors that might explain the

original relationship. Consider, for example, the intelligence of the students (as measured by IQ tests or SAT scores). Ivy League colleges pride themselves on the excellence of their student bodies. They may therefore be willing to award merit scholarships to students with exceptional qualifications but not enough money to pay tuition and board. Once admitted to these prestigious colleges, bright students may develop the skills—and connections—that will lead to later professional success. Since West had no data on the intelligence of the men she studied, she was unable to study whether the partial relationships disappeared once this test factor was introduced.

In sum, the elaboration paradigm permits the investigator to rule out certain possibilities and to gain support for others. It does not permit us to prove anything.

[†]Since she had no direct data on family socioeconomic status, West defined as wealthy or having high socioeconomic status those who supported their sons completely during all four years of college. She defined as less wealthy or having low socioeconomic status those whose sons worked their way through college, in part or totally.

secular society than people of higher social class did. To illustrate this idea, he noted that social class was strongly related to the likelihood that a woman had ever held an office in a secular organization (see Table 15-8).

Glock then reasoned that if social class were related to church involvement only by virtue of the fact that lower-class women would be denied opportunities for gratification in the secular society, the original relationship should not hold among women who were getting gratification. As a rough indicator of the receipt of gratification from the secular society, he used as a variable the holding of

secular office. In this test, social class should be unrelated to church involvement among those who had held such office.

Table 15-9 presents an example of a specification. Among women who have held office in secular organizations, there is essentially no relationship between social class and church involvement. In effect, the table specifies the conditions under which the original relationship holds: among those women lacking gratification in the secular society.

The term *specification* is used in the elaboration paradigm regardless of whether the test variable

TABLE 15-7

Social Class and Mean Church Involvement among Episcopal Women

	Social Class Levels				
	Low				High
	0	1	2	3	4
Mean involvement	0.63	0.58	0.49	0.48	0.45

Note: Mean scores rather than percentages have been used here.

Source: Tables 15-7, 15-8, and 15-9 are from Charles Y. Glock, Benjamin B. Ringer, and Earl R. Babbie, *To Comfort and to Challenge* (Berkeley: University of California Press, 1967). Used with permission of the Regents of the University of California.

TABLE 15-8

Social Class and the Holding of Office in Secular Organizations

	Social Class Levels				
	Low				High
	0	1	2	3	4
Percent who have held office in a secular organization	46	47	54	60	83

TABLE 15-9

Church Involvement by Social Class and Holding Secular Office

	Mean Church Involvement for Social Class Levels				
	Low				High
	0	1	2	3	4
Have held office	0.46	0.53	0.46	0.46	0.46
Have not held office	0.62	0.55	0.47	0.46	0.40

suppressor variable In the elaboration model, a test variable that prevents a genuine relationship from appearing at the zero-order level.

is antecedent or intervening. In either case, the meaning is the same. We have specified the particular conditions under which the original relationship holds.

Refinements to the Paradigm

The preceding sections have presented the primary logic of the elaboration model as developed by Lazarsfeld and his colleagues. Here we look at some logically possible variations, some of which can be found in a book by Morris Rosenberg (1968).

First, the basic paradigm assumes an initial relationship between two variables. It might be useful, however, in a more comprehensive model to differentiate between positive and negative relationships. Moreover, Rosenberg suggests using the elaboration model even with an original relationship of zero. He cites as an example a study of union membership and attitudes toward having Jews on the union staff (see Table 15-10). The initial analysis indicated that length of union membership did not relate to the attitude: Those who had belonged to the union less than four years were just as willing to accept Jews on the staff as were those who had belonged for more than four years. The *age* of union members, however, was found to *suppress* the relationship between length of union membership and attitude toward Jews. Overall, younger members were more favorable to Jews than older members were. At the same time, of course, younger members were not likely to have been in the union as long as the old members. Within specific age groups, however, those in the union longest were the most supportive of having Jews on the staff. Age, in this case, was a **suppressor variable,** concealing the relationship between length of membership and attitude toward Jews.

Second, the basic paradigm focuses on partials being the same as or weaker than the original relationship but does not provide guidelines for specifying what constitutes a significant difference between the original and the partials. When you use the elaboration model, you'll frequently find yourself making an arbitrary decision about

TABLE 15-10

Example of a Suppressor Variable

I: No Apparent Relationship between Attitudes toward Jews and Length of Time in the Union

	Length of Time in the Union	
	Less than four years	Four years or more
Percent who don't care if there are Jews on the union staff	49.2	50.5
	(126)	(256)

II: In Each Age Group, Length of Time in Union Increases Willingness to Have Jews on Union Staff

		Length of Time in the Union	
		Less than four years	Four years or more
Percent who don't care if there are Jews on the union staff, by age	29 years and under	56.4	62.7
		(78)	(51)
	30–49 years	37.1	48.3
		(35)	(116)
	50 years and older	38.4	56.1
		(13)	(89)

Source: Adapted from Morris Rosenberg, *The Logic of Survey Analysis* (New York: Basic Books, 1968), 88–89. Copyright © 1968 University of Minnesota Press.

whether a given partial is significantly weaker than the original. This, then, suggests another dimension that could be added to the paradigm.

Third, the limitation of the basic paradigm to partials that are the same as or weaker than the original neglects two other possibilities. A partial relationship might be stronger than the original. Or, on the other hand, a partial relationship might be the reverse of the original—for example, negative where the original was positive.

Rosenberg provides a hypothetical example of the latter possibility by first suggesting that a researcher might find that working-class respondents in his study are more supportive of the civil rights movement than middle-class respondents are (see Table 15-11). He further suggests that race might be a **distorter variable** in this instance, reversing the true relationship between class and attitudes. Presumably, African American respondents would be more supportive of the movement than whites would, but African Americans would also be overrepresented among working-class respondents and

underrepresented among the middle class. Middle-class African American respondents might be more supportive than working-class African Americans, however; and the same relationship might be found among whites. Holding race constant, then, the researcher would conclude that support for the civil rights movement was greater among the middle class than among the working class.

Here's another example of a distorter variable at work. When Michel de Seve set out to examine the starting salaries of men and women in the same organization, she was surprised to find the women were receiving higher starting salaries, on the average, than their male counterparts were. The distorter variable was *time of first hire*. Many of the women had been hired relatively recently, when salaries were higher overall than in the

---

distorter variable In the elaboration model, a test variable that reverses the direction of a zero-order relationship.

TABLE 15-11

Example of a Distorter Variable (Hypothetical)

I: Working-Class Subjects Appear More Liberal on Civil Rights than Middle-Class Subjects

Civil Rights Score	Middle Class	Working Class
High	37%	45%
Low	63	55
	100	100
100% =	(120)	(120)

II: Controlling for Race Shows the Middle Class to Be More Liberal than the Working Class

	Social Class			
	Blacks		Whites	
Civil Rights Score	Middle Class	Working Class	Middle Class	Working Class
High	70%	50%	30%	20%
Low	30	50	70	80
	100	100	100	100
100% =	(20)	(100)	(100)	(20)

Source: Morris Rosenberg, *The Logic of Survey Analysis* (New York: Basic Books, 1968), 94–95. Copyright © 1968 Basic Books.

earlier years when many of the men had been hired (reported in E. Cook 1995).

All these new dimensions further complicate the notion of specification. If one partial is the same as the original, and the other partial is even stronger, how should you react to that situation? You've specified one condition under which the original relationship holds up, but you've also specified another condition under which it holds even more clearly.

ex post facto hypothesis A hypothesis created after confirming data have already been collected. It is a meaningless construct because there is no way for it to be disconfirmed.

Finally, the basic paradigm focuses primarily on dichotomous test variables. In fact, the elaboration model is not so limited—either in theory or in use—but the basic paradigm becomes more complicated when the test variable divides the sample into three or more subsamples. And the paradigm becomes more complicated yet when more than one test variable is used simultaneously.

I'm not saying all this to fault the basic elaboration paradigm. To the contrary, I want to emphasize that the elaboration model is not a simple algorithm—a set of procedures through which to analyze research. Rather, it's primarily a logical device for assisting the researcher in understanding his or her data. A firm understanding of the elaboration model will make a sophisticated analysis easier. However, this model suggests neither which variables should be introduced as controls nor definitive conclusions about the nature of elaboration results. For all these things, you must look to your own ingenuity. Such ingenuity, moreover, will come only through extensive experience. By pointing to oversimplifications in the basic elaboration paradigm, I've sought to bring home the point that the model provides only a logical framework. You'll find sophisticated analyses far more complicated than the examples I've used to illustrate the basic paradigm.

At the same time, if you fully understand the basic model, you'll understand other techniques such as correlations, regressions, and factor analyses a lot more easily. Chapter 16 places such techniques as partial correlations and partial regressions in the context of the elaboration model.

Elaboration and Ex Post Facto Hypothesizing

Before we leave the discussion of the elaboration model, we should look at it in connection with a form of fallacious reasoning called **ex post facto hypothesizing.** Although the social science literature presents a host of references warning against it, inexperienced researchers can sometimes be confused about its implications.

"Ex post facto" means "after the fact." When you observe an empirical relationship between two variables and then simply suggest a reason for that relationship, that is sometimes called ex post facto hypothesizing. You've generated a hypothesis linking two variables after their relationship is already known. You'll recall, from an early discussion in this book, that all hypotheses must be subject to disconfirmation in order to be meaningful. Unless you can specify empirical findings that would *disprove* your hypothesis, it's not really a *hypothesis* as researchers use that term. You might reason, therefore, that once you've observed a relationship between two variables, any hypothesis regarding that relationship cannot be disproved.

This is a fair assessment if you're doing nothing more than dressing up your empirical observations with deceptive hypotheses after the fact. Having observed that women are more religious than men, you should not simply assert that women will be more religious than men because of some general dynamic of social behavior and then rest your case on the initial observation.

The unfortunate spin-off of the injunction against ex post facto hypothesizing is its inhibition of good, honest hypothesizing after the fact. Inexperienced researchers are often led to believe that they must make all their hypotheses before examining their data—even if that process means making a lot of poorly reasoned ones. Furthermore, they're led to ignore any empirically observed relationships that do not confirm some prior hypothesis.

Surely, few researchers would now wish that Samuel Stouffer had hushed up his anomalous findings regarding morale among soldiers in the army. Stouffer noted peculiar empirical observations and set about hypothesizing the reasons for those findings. And his reasoning has proved invaluable to researchers ever since. The key is that his "after the fact" hypotheses could themselves be tested.

There is another, more sophisticated point to be made here, however. Anyone can generate hypotheses to explain observed empirical relationships in a body of data, but the elaboration model provides the logical tools for *testing* those hypotheses within the same body of data. A good example of this testing may be found in the earlier discussion of social class and church involvement. Glock explained the original relationship in terms of social deprivation theory. If he had stopped at that point, his comments would have been interesting but hardly persuasive. He went beyond that point, however. He noted that if the hypothesis was correct, then the relationship between social class and church involvement should disappear among those women who were receiving gratification from the secular society—those who had held office in a secular organization. This hypothesis was then subjected to an empirical test. Had the new hypothesis not been confirmed by the data, he would have been forced to reconsider.

These additional comments should further illustrate the point that data analysis is a continuing process, demanding all the ingenuity and perseverance you can muster. The image of a researcher carefully laying out hypotheses and then testing them in a ritualistic fashion results only in ritualistic research.

In case you're concerned that the strength of ex post facto proofs seems to be less than that of the traditional kinds, let me repeat the earlier assertion that "scientific proof" is a contradiction in terms. Nothing is ever *proved* scientifically. Hypotheses, explanations, theories, or hunches can all escape a stream of attempts at disproof, but none can be proved in any absolute sense. The acceptance of a hypothesis, then, is really a function of the extent to which it has been tested and not disconfirmed. No hypothesis, therefore, should be considered sound on the basis of one test—whether the hypothesis was generated before or after the observation of empirical data. With this in mind, you should not deny yourself some of the most fruitful avenues available to you in data analysis. You should always try to reach an honest understanding of your data, develop meaningful theories for more general understanding, and not worry about the manner of reaching that understanding.

Introduction

- The elaboration model is a method of multivariate analysis appropriate for social research. It is primarily a logical model that can illustrate the basic logic of other multivariate methods.

The Origins of the Elaboration Model

- Paul Lazarsfeld and Patricia Kendall used the logic of the elaboration model to present hypothetical tables regarding Samuel Stouffer's work regarding education and acceptance of induction in the Army.

- A partial relationship (or "partial") is the observed relationship between two variables within a subgroup of cases based on some attribute of the test or control variable.

- A zero-order relationship is the observed relationship between two variables without a third variable being held constant or controlled.

The Elaboration Paradigm

- The basic steps in elaboration are as follows: (1) A relationship is observed to exist between two variables, (2) a third variable (the test variable) is held constant in the sense that the cases under study are subdivided according to the attributes of that third variable, (3) the original two-variable relationship is recomputed within each of the subgroups, and (4) the comparison of the original relationship with the relationships found within each subgroup (the partial relationships) provides a fuller understanding of the original relationship itself.

- The logical relationships of the variables differ depending on whether the test variable is antecedent to the other two variables or intervening between them.

- The outcome of an elaboration analysis may be replication (whereby a set of partial relationships is essentially the same as the corresponding zero-order relationship), explanation (whereby a set of partial relationships is reduced essentially to zero when an antecedent variable is held constant), interpretation (whereby a set of partial relationships is reduced essentially to zero when an intervening variable is held constant), or specification (whereby one partial relationship is reduced, ideally to zero, and the other remains about the same as the original relationship or is stronger).

- A suppressor variable conceals the relationship between two other variables; a distorter variable causes an apparent reversal in the relationship between two other variables (from negative to positive or vice versa).

Elaboration and Ex Post Facto Hypothesizing

- Ex post facto hypothesizing, or the development of hypotheses "predicting" relationships that have already been observed, is invalid in science, because disconfirming such hypotheses is impossible. Although nothing prevents us from suggesting reasons that observed relationships may be the way they are, we should not frame those reasons in the form of "hypotheses." More important, one observed relationship and possible reasons for it may suggest hypotheses about other relationships that have not been examined. The elaboration model is an excellent logical device for this kind of unfolding analysis of data.

The following terms are defined in context in the chapter and at the bottom of the page where the term is introduced, as well as in the comprehensive glossary at the back of the book.

distorter variable	replication
elaboration model	specification
ex post facto hypothesis	suppressor variable
explanation	test variable
interpretation	zero-order relationship
partial relationship	

See the exercise for Chapter 16.

1. Review the Stouffer-Kendall-Lazarsfeld example of education, friends deferred, and attitudes toward being drafted. Suppose they had begun with an association between friends deferred and attitudes toward being drafted, and then they had

controlled for education. What conclusion would they have reached?

2. In your own words describe the elaboration logic of (a) replication, (b) interpretation, (c) explanation, and (d) specification.

3. Review the box on Ivy League colleges and success in later professional life. In your own words, explain what Patricia Kendall means when she says, "Despite this [support from the analysis of partial relationships], West had in no way proved her hypothesis." What conclusions can one reasonably draw from West's study?

4. Construct hypothetical examples of suppressor and distorter variables.

5. Search the web for a research report on the discovery of a spurious relationship. Give the web address of the document and quote or paraphrase what was discovered.

SPSS EXERCISES

See the booklet that accompanies your text for exercises using SPSS (Statistical Package for the Social Sciences). There are exercises offered for each chapter, and you'll also find a detailed primer on using SPSS.

Online Study Resources

If your book came with an access code card, visit www.cengage.com/login to register. To purchase access, please visit www.ichapters.com.

1. Before you do your final review of the chapter, take the *CengageNOW* pretest to help identify the areas on which you should concentrate. You'll find information on this online tool, as well as instructions on how to access all of its great resources, in the front of the book.

2. As you review, take advantage of the *CengageNOW* personalized study plan, based on your quiz results. Use this study plan with its interactive exercises and other resources to master the material.

3. When you're finished with your review, take the posttest to confirm that you're ready to move on to the next chapter.

WEBSITE FOR *THE PRACTICE OF SOCIAL RESEARCH* 12TH EDITION

Go to your book's website at www.cengage.com/sociology/babbie for tools to aid you in studying for your exams. You'll find *Tutorial Quizzes* with feedback, *Internet Exercises*, *Flash Cards*, *Glossaries*, and *Essay Quizzes*, as well as *InfoTrac College Edition* search terms, suggestions for additional reading, *Web Links*, and primers for using data-analysis software such as SPSS.

CHAPTER SIXTEEN

Statistical Analyses

CHAPTER OVERVIEW

Statistics allow researchers to summarize data, measure associations between variables, and draw inferences from samples to populations. Getting acquainted with a few simple statistical procedures frequently used in social research is less painful (and less threatening to your social life) than you might believe.

CengageNOW for Sociology

Use this online tool to help you make the grade on your next exam. After reading this chapter, go to "Online Study Resources" at the end of the chapter for instructions on how to benefit from *CengageNOW*.

Introduction

It has been my experience over the years that many students are intimidated by statistics. Sometimes statistics makes them feel they're

- A few clowns short of a circus
- Dumber than a box of hair
- A few feathers short of a duck
- All foam, no beer
- Missing a few buttons on their remote control
- A few beans short of a burrito
- As screwed up as a football bat
- About as sharp as a bowling ball
- About four cents short of a nickel
- Not running on full thrusters*

Many people are intimidated by quantitative research because they feel uncomfortable with mathematics and statistics. And indeed, many research reports are filled with unspecified computations. The role of statistics in social research is often important, but it's equally important to see this role in its proper perspective.

Empirical research is first and foremost a logical rather than a mathematical operation. Mathematics is merely a convenient and efficient language for accomplishing the logical operations inherent in quantitative data analysis. *Statistics* is the applied branch of mathematics especially appropriate for a variety of research analyses. This textbook is not intended to teach you statistics or torture you with them. Rather, I want to sketch out a logical context within which you might learn and understand statistics.

We'll be looking at two types of statistics: descriptive and inferential. *Descriptive statistics* is a medium for describing data in manageable forms. *Inferential statistics,* on the other hand, assists researchers in drawing conclusions from their observations; typically, this involves drawing conclusions

about a population from the study of a sample drawn from it. After that discussion, I'll briefly introduce you to some of the analytic techniques you may come across in your reading of the social science literature.

Descriptive Statistics

As I've already suggested, **descriptive statistics** present quantitative descriptions in a manageable form. Sometimes we want to describe single variables, and sometimes we want to describe the associations that connect one variable with another. Let's look at some of the ways to do these things.

Data Reduction

Scientific research often involves collecting large masses of data. Suppose we surveyed 2,000 people, asking each of them 100 questions—not an unusually large study. We would then have a staggering 200,000 answers! No one could possibly read all those answers and reach any meaningful conclusion about them. Thus, much scientific analysis involves the reduction of data from unmanageable details to manageable summaries.

To begin our discussion, let's look briefly at the raw-data matrix created by a quantitative research project. Table 16-1 presents a partial data matrix. Notice that each row in the matrix represents a person (or other unit of analysis), each column represents a variable, and each cell represents the coded attribute or value a given person has on a given variable. The first column in Table 16-1 represents a person's gender. Let's say a "1"

---

descriptive statistics Statistical computations describing either the characteristics of a sample or the relationship among variables in a sample. Descriptive statistics merely summarize a set of sample observations, whereas inferential statistics move beyond the description of specific observations to make inferences about the larger population from which the sample observations were drawn.

---

*Thanks to the many contributors to humor lists on the Internet.

TABLE 16-1

Partial Raw-Data Matrix

	Gender	Age	Education	Income	Occupation	Political Affiliation	Political Orientation	Religious Affiliation	Importance of Religion
Person 1	1	3	2	4	1	2	3	0	4
Person 2	1	4	2	4	4	1	1	1	2
Person 3	2	2	5	5	2	2	4	2	3
Person 4	1	5	4	4	3	2	2	2	4
Person 5	2	3	7	8	6	1	1	5	1
Person 6	2	1	3	3	5	3	5	1	1

TABLE 16-2

Hypothetical Raw Data on Education and Prejudice

	Educational Level				
Prejudice	None	Grade School	High School	College	Graduate Degree
High	23	34	156	67	16
Medium	11	21	123	102	23
Low	6	12	95	164	77

represents male and a "2" represents female. This means that persons 1 and 2 are male, person 3 is female, and so forth.

In the case of age, person 1's "3" might mean 30–39 years old, person 2's "4" might mean 40–49. However age has been coded (see Chapter 14), the code numbers shown in Table 16-1 describe each of the people represented there.

Notice that the data have already been reduced somewhat by the time a data matrix like this one has been created. If age has been coded as suggested previously, the specific answer "33 years old" has already been assigned to the category "30–39." The people responding to our survey may have given us 60 or 70 different ages, but we've now reduced them to 6 or 7 categories.

Chapter 14 discussed some of the ways of further summarizing univariate data: averages such as the mode, median, and mean and measures of dispersion such as the range, the standard deviation, and so forth. It's also possible to summarize the associations among variables.

Measures of Association

The association between any two variables can also be represented by a data matrix, this time produced by the joint frequency distributions of the two variables. Table 16-2 presents such a matrix. It provides all the information needed to determine the nature and extent of the relationship between education and prejudice.

Notice, for example, that 23 people (1) have no education and (2) scored high on prejudice; 77 people (1) had graduate degrees and (2) scored low on prejudice.

Like the raw-data matrix in Table 16-1, this matrix provides more information than can easily be comprehended. A careful study of the table shows that as education increases from "None" to "Graduate Degree," there is a general tendency for prejudice to decrease, but no more than a general impression is possible. For a more precise summary of the data matrix, we need one of several types of descriptive statistics. Selecting the appropriate measure depends initially on the nature of the two variables.

We'll turn now to some of the options available for summarizing the association between two variables. Each of these measures of association is based on the same model—**proportionate reduction of error (PRE).**

To see how this model works, let's assume that I asked you to guess respondents' attributes on a given variable: for example, whether they answered yes or no to a given questionnaire item. To assist you, let's first assume you know the overall distribution of responses in the total sample—say, 60 percent said yes and 40 percent said no. You would make the fewest errors in this process if you always guessed the modal (most frequent) response: yes.

Second, let's assume you also know the empirical relationship between the first variable and some other variable: say, *gender*. Now, each time I ask you to guess whether a respondent said yes or no, I'll tell you whether the respondent is a man or a woman. If the two variables are related, you should make fewer errors the second time. It's possible, therefore, to compute the PRE by knowing the relationship between the two variables: the greater the relationship, the greater the reduction of error.

This basic PRE model is modified slightly to take account of different levels of measurement—nominal, ordinal, or interval. The following sections will consider each level of measurement and present one measure of association appropriate for each. Bear in mind that the three measures discussed are only an arbitrary selection from among many appropriate measures.

Nominal Variables

If the two variables consist of nominal data (for example, gender, religious affiliation, race), *lambda* (λ) would be one appropriate measure. (Lambda is a letter in the Greek alphabet corresponding to *l* in our alphabet. Greek letters are used for many concepts in statistics, which perhaps helps to account for the number of people who say of statistics, "It's all Greek to me.") Lambda is based on your ability to guess values on one of the variables: the

PRE achieved through knowledge of values on the other variable.

Imagine this situation. I tell you that a room contains 100 people and I would like you to guess the gender of each person, one at a time. If half are men and half women, you'll probably be right half the time and wrong half the time.

But suppose I tell you each person's occupation before you guess that person's gender. What gender would you guess if I said the person was a truck driver? You would probably be wise to guess "male"; although there are now plenty of women truck drivers, most are still men. If I said the next person was a nurse, you'd probably be wisest to guess "female," following the same logic. Although you would still make errors in guessing genders, you would clearly do better than you would if you didn't know their occupations. The extent to which you did better (the proportionate reduction of error) would be an indicator of the association that exists between gender and occupation.

Here's another simple hypothetical example that illustrates the logic and method of lambda. Table 16-3 presents hypothetical data relating gender to employment status. Overall, we note that 1,100 people are employed, and 900 are not employed. If you were to predict whether people were employed, and if you knew only the overall distribution on that variable, you would always predict "employed," because that would result in fewer errors than always predicting "not employed." Nevertheless, this strategy would result in 900 errors out of 2,000 predictions.

Let's suppose that you had access to the data in Table 16-3 and that you were told each person's gender before making your prediction of employment status. Your strategy would change

---

proportionate reduction of error (PRE) A logical model for assessing the strength of a relationship by asking how much knowing values on one variable would reduce our errors in guessing values on the other. For example, if we know how much education people have, we can improve our ability to estimate how much they earn, thus indicating there is a relationship between the two variables.

TABLE 16-3
Hypothetical Data Relating Gender to Employment Status

	Men	Women	Total
Employed	900	200	1,100
Unemployed	100	800	900
Total	1,000	1,000	2,000

in that case. For every man you would predict "employed," and for every woman you would predict "not employed." In this instance, you would make 300 errors—the 100 men who were not employed and the 200 employed women—or 600 fewer errors than you would make without knowing the person's gender.

Lambda, then, represents the reduction in errors as a proportion of the errors that would have been made on the basis of the overall distribution. In this hypothetical example, lambda would equal 0.67; that is, 600 fewer errors divided by the 900 total errors based on employment status alone. In this fashion, lambda measures the statistical association between gender and employment status.

If gender and employment status were statistically independent, we would find the same distribution of employment status for men and women. In this case, knowing each person's gender would not affect the number of errors made in predicting employment status, and the resulting lambda would be zero. If, on the other hand, all men were employed and none of the women were employed, by knowing gender you would avoid errors in predicting employment status. You would make 900 fewer errors (out of 900), so lambda would be 1.0—representing a perfect statistical association.

Lambda is only one of several measures of association appropriate for the analysis of two nominal variables. You could look at any statistics textbook for a discussion of other appropriate measures.

Ordinal Variables

If the variables being related are ordinal (for example, social class, religiosity, alienation), *gamma* (γ) is one appropriate measure of association. Like lambda, gamma is based on our ability to guess values on one variable by knowing values on another. However, whereas lambda is based on guessing exact values, gamma is based on guessing the ordinal arrangement of values. For any given pair of cases, we guess that their ordinal ranking on one variable will correspond (positively or negatively) to their ordinal ranking on the other.

Let's say we have a group of elementary students. It's reasonable to assume that there is a relationship between their ages and their heights. We can test this by comparing every pair of students: Sam and Mary, Sam and Fred, Mary and Fred, and so forth. Then we ignore all the pairs in which the students are the same age and/or the same height. We then classify each of the remaining pairs (those who differ in both age and height) into one of two categories: those in which the older child is also the taller ("same" pairs) and those in which the older child is the shorter ("opposite" pairs). So, if Sam is older and taller than Mary, the Sam-Mary pair is counted as a "same." If Sam is older but shorter than Mary, then that pair is an "opposite."

To determine whether age and height are related to each other, we compare the number of same and opposite pairs. If the same pairs outnumber the opposite pairs, we can conclude that there is a positive association between the two variables—as one increases, the other increases. If there are more opposites than sames, we can conclude that the relationship is negative. If there are about as many sames as opposites, we can conclude that age and height are not related to each another, that they're independent of each other.

Here's a social science example to illustrate the simple calculations involved in gamma. Let's say you suspect that religiosity is positively related to political conservatism, and if Person A is more religious than Person B, you guess that A is also more conservative than B. Gamma is the proportion of paired comparisons that fits this pattern.

Table 16-4 presents hypothetical data relating social class to prejudice. The general nature of the relationship between these two variables is that as social class increases, prejudice decreases. There is a negative association between social class and prejudice.

TABLE 16-4

Hypothetical Data Relating Social Class to Prejudice

Prejudice	Lower Class	Middle Class	Upper Class
Low	200	400	700
Medium	500	900	400
High	800	300	100

Gamma is computed from two quantities: (1) the number of pairs having the same ranking on the two variables and (2) the number of pairs having the opposite ranking on the two variables. The pairs having the same ranking are computed as follows. The frequency of each cell in the table is multiplied by the sum of all cells appearing below and to the right of it—with all these products being summed. In Table 16-4, the number of pairs with the same ranking would be 200(900 + 300 + 400 + 100) + 500(300 + 100) + 400(400 + 100) + 900(100), or 340,000 + 200,000 + 200,000 + 90,000 = 830,000.

The pairs having the opposite ranking on the two variables are computed as follows: The frequency of each cell in the table is multiplied by the sum of all cells appearing below and to the left of it—with all these products being summed. In Table 16-4, the numbers of pairs with opposite rankings would be 700(500 + 800 + 900 + 300) + 400(800 + 300) + 400(500 + 800) + 900(800), or 1,750,000 + 440,000 + 520,000 + 720,000 = 3,430,000. Gamma is computed from the numbers of same-ranked pairs and opposite-ranked pairs as follows:

$$gamma = \frac{same - opposite}{same + opposite}$$

In our example, gamma equals (830,000 − 3,430,000) divided by (830,000 + 3,430,000), or −0.61. The negative sign in this answer indicates the negative association suggested by the initial inspection of the table. Social class and prejudice, in this hypothetical example, are negatively associated with each other. The numerical figure for gamma indicates that 61 percent more of the pairs examined had the opposite ranking than the same ranking.

Note that whereas values of lambda vary from 0 to 1, values of gamma vary from −1 to +1, representing the direction as well as the magnitude of the association. Because nominal variables have no ordinal structure, it makes no sense to speak of the direction of the relationship. (A negative lambda would indicate that you made more errors in predicting values on one variable while knowing values on the second than you made in ignorance of the second, and that's not logically possible.)

Table 16-5 is an example of the use of gamma in social research. To study the extent to which widows sanctified their deceased husbands, Helena Lopata (1981) administered a questionnaire to a probability sample of 301 widows. In part, the questionnaire asked the respondents to characterize their deceased husbands in terms of the following semantic differentiation scale:

Characteristic

Positive Extreme								Negative Extreme
Good	1	2	3	4	5	6	7	Bad
Useful	1	2	3	4	5	6	7	Useless
Honest	1	2	3	4	5	6	7	Dishonest
Superior	1	2	3	4	5	6	7	Inferior
Kind	1	2	3	4	5	6	7	Cruel
Friendly	1	2	3	4	5	6	7	Unfriendly
Warm	1	2	3	4	5	6	7	Cold

Respondents were asked to describe their deceased spouses by circling a number for each pair of opposing characteristics. Notice that the series of numbers connecting each pair of characteristics is an ordinal measure.

Next, Lopata wanted to discover the extent to which the several measures were related to one another. Appropriately, she chose gamma as the measure of association. Table 16-5 shows how she presented the results of her investigation.

The format presented in Table 16-5 is called a *correlation matrix*. For each pair of measures, Lopata has calculated the gamma. Good and Useful, for example, are related to each other by a gamma

TABLE 16-5

Gamma Associations among the Semantic Differentiation Items of the Sanctification Scale

	Useful	Honest	Superior	Kind	Friendly	Warm
Good	0.79	0.88	0.80	0.90	0.79	0.83
Useful		0.84	0.71	0.77	0.68	0.72
Honest			0.83	0.89	0.79	0.82
Superior				0.78	0.60	0.73
Kind					0.88	0.90
Friendly						0.90

Source: Helena Znaniecki Lopata, "Widowhood and Husband Sanctification," *Journal of Marriage and the Family* (May 1981): 439–50. Copyright © 1981 NCFR.

equal to 0.79. The matrix is a convenient way of presenting the intercorrelations among several variables, and you'll find it frequently in the research literature. In this case, we see that all the variables are quite strongly related to one another, though some pairs are more strongly related than others.

Gamma is only one of several measures of association appropriate for ordinal variables. Again, any introductory statistics textbook will give you a more comprehensive treatment of this subject.

Interval or Ratio Variables

If interval or ratio variables (for example, *age, income, grade point average,* and so forth) are being associated, one appropriate measure of association is *Pearson's product-moment correlation* (r). The derivation and computation of this measure of association are complex enough to lie outside the scope of this book, so I'll make only a few general comments here.

Like both gamma and lambda, r is based on guessing the value of one variable by knowing another. For continuous interval or ratio variables, however, it's unlikely that you could predict the precise value of the variable. On the other hand, predicting only the ordinal arrangement of values on the two variables would not take advantage of the greater amount of information conveyed by an interval or ratio variable. In a sense, r reflects how closely you can guess the value of one variable through your knowledge of the value of another.

To understand the logic of r, consider the way you might hypothetically guess values that particular cases have on a given variable. With nominal variables, we've seen that you might always guess the modal value. But for interval or ratio data, you would minimize your errors by always guessing the mean value of the variable. Although this practice produces few if any perfect guesses, the extent of your errors will be minimized. Imagine the task of guessing peoples' incomes and how much better you would do if you knew how many years of education they had as well as the mean incomes for people with 0, 1, 2 (and so forth) years of education.

In the computation of lambda, we noted the number of errors produced by always guessing the modal value. In the case of r, errors are measured in terms of the sum of the squared differences between the actual value and the mean. This sum is called the *total variation*.

To understand this concept, we must expand the scope of our examination. Let's look at the logic of regression analysis and discuss correlation within that context.

Regression Analysis

The general formula for describing the association between two variables is $Y = f(X)$. This formula is read "Y is a function of X," meaning that values of Y can be explained in terms of variations in the values of X. Stated more strongly, we might say that X

causes Y, so the value of X determines the value of Y. **Regression analysis** is a method of determining the specific function relating Y to X. There are several forms of regression analysis, depending on the complexity of the relationships being studied. Let's begin with the simplest.

Linear Regression

The regression model can be seen most clearly in the case of a **linear regression analysis,** in which a perfect linear association between two variables exists or is approximated. Figure 16-1 is a scattergram presenting in graphic form the values of X and Y as produced by a hypothetical study. It shows that for the four cases in our study, the values of X and Y are identical in each instance. The case with a value of 1 on X also has a value of 1 on Y, and so forth. The relationship between the two variables in this instance is described by the equation $Y = X$; this is called the *regression equation.* Because all four points lie on a straight line, we could superimpose that line over the points; this is the *regression line.*

The linear regression model has important descriptive uses. The regression line offers a graphic picture of the association between X and Y, and the regression equation is an efficient form for summarizing that association. The regression model has inferential value as well. To the extent that the regression equation correctly describes the general association between the two variables, it may be used to predict other sets of values. If, for example, we know that a new case has a value of 3.5 on X, we can predict the value of 3.5 on Y as well.

In practice, of course, studies are seldom limited to four cases, and the associations between variables are seldom as clear as the one presented in Figure 16-1.

A somewhat more realistic example is presented in Figure 16-2, representing a hypothetical relationship between population and crime rate in small- to medium-size cities. Each dot in the scattergram is a city, and its placement reflects that city's population and its crime rate. As was the case in our previous example, the values of Y (crime rates) generally correspond to those of X (populations), and as values of X increase, so do values of Y. However, the association is not nearly as clear as it is in Figure 16-1.

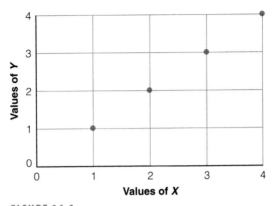

FIGURE 16-1

Simple Scattergram of Values of X and Y

In Figure 16-2 we can't superimpose a straight line that will pass through all the points in the scattergram. But we can draw an approximate line showing the best possible linear representation of the several points. I've drawn that line on the graph.

You may (or may not) recall from algebra that any straight line on a graph can be represented by an equation of the form $Y = a + bX$, where X and Y are values of the two variables. In this equation, a equals the value of Y when X is 0, and b represents the slope of the line. If we know the values of a and b, we can calculate an estimate of Y for every value of X.

We can now say more formally that regression analysis is a technique for establishing the regression equation representing the geometric line that comes closest to the distribution of points on a graph. The regression equation provides a mathematical *description* of the relationship between the variables, and it allows us to *infer* values of Y when we have values of X. Recalling Figure 16-2, we

regression analysis A method of data analysis in which the relationships among variables are represented in the form of an equation, called a regression equation.

linear regression analysis A form of statistical analysis that seeks the equation for the straight line that best describes the relationship between two ratio variables.

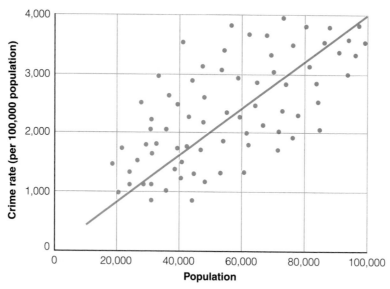

FIGURE 16-2

A Scattergram of the Values of Two Variables with Regression Line Added (Hypothetical)

could estimate crime rates of cities if we knew their populations.

To improve your guessing, you construct a *regression line*, stated in the form of a regression equation that permits the estimation of values on one variable from values on the other. The general format for this equation is $Y' = a + b(X)$, where a and b are computed values, X is a given value on one variable, and Y' is the estimated value on the other. The values of a and b are computed to minimize the differences between actual values of Y and the corresponding estimates (Y') based on the known value of X. The sum of squared differences between actual and estimated values of Y is called the *unexplained variation* because it represents errors that still exist even when estimates are based on known values of X.

The *explained variation* is the difference between the total variation and the unexplained variation. Dividing the explained variation by the total variation produces a measure of the *proportionate reduction of error* corresponding to the similar quantity in the computation of lambda. In the present case, this quantity is the *correlation squared*: r^2. Thus, if $r = 0.7$, then $r^2 = 0.49$, meaning that about half the

variation has been explained. In practice, we compute r rather than r^2, because the product-moment correlation can take either a positive or a negative sign, depending on the direction of the relationship between the two variables. (Computing r^2 and taking a square root would always produce a positive quantity.) You can consult any standard statistics textbook for the method of computing r, although I anticipate that most readers using this measure will have access to computer programs designed for this function.

Unfortunately—or perhaps fortunately—social life is so complex that the simple linear regression model often does not sufficiently represent the state of affairs. As we saw in Chapter 14, it's possible, using percentage tables, to analyze more than two variables. As the number of variables increases, such tables become increasingly complicated and hard to read. The regression model offers a useful alternative in such cases.

Multiple Regression

Very often, social researchers find that a given dependent variable is affected simultaneously by several independent variables. **Multiple regression**

analysis provides a means of analyzing such situations. This was the case when Beverly Yerg (1981) set about studying teacher effectiveness in physical education. She stated her expectations in the form of a multiple regression equation:

$$F = b_0 + b_1 I + b_2 X_1 + b_3 X_2 + b_4 X_3 + b_5 X_4 + e, \text{ where}$$

F = Final pupil-performance score
I = Initial pupil-performance score
X_1 = Composite of guiding and supporting practice
X_2 = Composite of teacher mastery of content
X_3 = Composite of providing specific, task-related feedback
X_4 = Composite of clear, concise task presentation
b = Regression weight
e = Residual

(Adapted from Yerg 1981: 42)

Notice that in place of the single X variable in a linear regression, there are several X's, and there are also several b's instead of just one. Also, Yerg has chosen to represent a as b_0 in this equation but with the same meaning as discussed previously. Finally, the equation ends with a residual factor (e), which represents the variance in Y that is not accounted for by the X variables analyzed.

Beginning with this equation, Yerg calculated the values of the several b's to show the relative contributions of the several independent variables in determining final student-performance scores. She also calculated the multiple-correlation coefficient as an indicator of the extent to which all six variables predict the final scores. This follows the same logic as the simple bivariate correlation discussed earlier, and it's traditionally reported as a capital R. In this case, $R = 0.877$, meaning that 77 percent of the variance ($0.877^2 = 0.77$) in final scores is explained by the six variables acting in concert.

Partial Regression

In exploring the elaboration model in Chapter 15, we paid special attention to the relationship between two variables when a third test variable was held constant. Thus, we might examine the effect of education on prejudice with age held constant, testing the independent effect of education. To do so, we would compute the tabular relationship between education and prejudice separately for each age group.

Partial regression analysis is based on this same logical model. The equation summarizing the relationship between variables is computed on the basis of the test variables remaining constant. As in the case of the elaboration model, the result may then be compared with the uncontrolled relationship between the two variables to clarify further the overall relationship.

Curvilinear Regression

Up to now, we've been discussing the association among variables as represented by a straight line. The regression model is even more general than our discussion thus far has implied.

You may already know that curvilinear functions, as well as linear ones, can be represented by equations. For example, the equation $X^2 + Y^2 = 25$ describes a circle with a radius of 5. Raising variables to powers greater than 1 has the effect of producing curves rather than straight lines. In the real world there is no reason to assume that the relationship among every set of variables will be linear. In some cases, then, **curvilinear regression analysis** can provide a better understanding of empirical relationships than any linear model can.

Recall, however, that a regression line serves two functions. It describes a set of empirical observations, and it provides a general model for

multiple regression analysis A form of statistical analysis that seeks the equation representing the impact of two or more independent variables on a single dependent variable.

partial regression analysis A form of regression analysis in which the effects of one or more variables are held constant, similar to the logic of the elaboration model.

curvilinear regression analysis A form of regression analysis that allows relationships among variables to be expressed with curved geometric lines instead of straight ones.

making inferences about the relationship between two variables in the general population that the observations represent. A very complex equation might produce an erratic line that would indeed pass through every individual point. In this sense, it would perfectly describe the empirical observations. There would be no guarantee, however, that such a line could adequately predict new observations or that it in any meaningful way represented the relationship between the two variables in general. Thus, it would have little or no inferential value.

Earlier in this book, we discussed the need for balancing detail and utility in data reduction. Ultimately, researchers attempt to provide the most faithful, yet also the simplest, representation of their data. This practice also applies to regression analysis. Data should be presented in the simplest fashion that best describes the actual data; as such, linear regressions are the ones most frequently used. Curvilinear regression analysis adds a new option to the researcher in this regard, but it does not solve the problems altogether. Nothing does that.

Cautions in Regression Analysis

The use of regression analysis for statistical inferences is based on the same assumptions made for correlational analysis: simple random sampling, the absence of nonsampling errors, and continuous interval data. Because social science research seldom completely satisfies these assumptions, you should use caution in assessing the results in regression analyses.

Also, regression lines—linear or curvilinear—can be useful for *interpolation* (estimating cases lying between those observed), but they are less trustworthy when used for *extrapolation* (estimating cases that lie beyond the range of observations). This limitation on extrapolations is important in two ways. First, you're likely to come across regression equations that seem to make illogical predictions. An equation linking population and

crimes, for example, might seem to suggest that small towns with, say, a population of 1,000 should produce 123 crimes a year. This failure in predictive ability does not disqualify the equation but dramatizes that its applicability is limited to a particular range of population sizes. Second, researchers sometimes overstep this limitation, drawing inferences that lie outside their range of observation, and you'd be right in criticizing them for that.

The preceding sections have introduced some of the techniques for measuring associations among variables at different levels of measurement. Matters become slightly more complex when the two variables represent different levels of measurement. Though we aren't going to pursue this issue in this textbook, "Measures of Association and Levels of Measurement," by Peter Nardi, may be a useful resource if you ever have to address such situations.

Inferential Statistics

Many, if not most, social science research projects involve the examination of data collected from a sample drawn from a larger population. A sample of people may be interviewed in a survey; a sample of divorce records may be coded and analyzed; a sample of newspapers may be examined through content analysis. Researchers seldom if ever study samples just to describe the samples per se; in most instances, their ultimate purpose is to make assertions about the larger population from which the sample has been selected. Frequently, then, you'll wish to interpret your univariate and multivariate sample findings as the basis for inferences about some population.

This section examines **inferential statistics**— the statistical measures used for making inferences from findings based on sample observations to a larger population. We'll begin with univariate data and move to multivariate.

Univariate Inferences

Chapter 14 dealt with methods of presenting univariate data. Each summary measure was intended as a method of describing the sample studied. Now

inferential statistics The body of statistical computations relevant to making inferences from findings based on sample observations to some larger population.

Measures of Association and Levels of Measurement

Peter Nardi

Pitzer College

Note that this table itself is set up with the dependent variables in the rows and the independent variable in the columns, as tables are commonly organized. Also, notice that the levels of measurement are themselves an ordinal scale.

If you want to use an interval/ratio-level variable in a crosstab, you must first recode it into an ordinal-level variable.

| | | Independent Variable | | |
		Nominal	*Ordinal*	*Interval/Ratio*
	Nominal	Crosstabs	Crosstabs	
		Chi square	Chi square	
		Lambda	Lambda	
	Ordinal	Crosstabs	Crosstabs	
		Chi square	Chi square	
		Lambda	Lambda	
			Gamma	
			Kendall's tau	
Dependent Variable			Sommers' d	
	Interval/Ratio	*Means*	*Means*	*Correlate*
		t-test	t-test	Pearson r
		ANOVA	ANOVA	Regression (R)

we'll use such measures to make broader assertions about a population. This section addresses two univariate measures: percentages and means.

If 50 percent of a sample of people say they had colds during the past year, 50 percent is also our best estimate of the proportion of colds in the total population from which the sample was drawn. (This estimate assumes a simple random sample, of course.) It's rather unlikely, however, that precisely 50 percent of the population had colds during the year. If a rigorous sampling design for random selection has been followed, however, we'll be able to estimate the expected range of error when the sample finding is applied to the population.

Chapter 7, on sampling theory, covered the procedures for making such estimates, so I'll only review them here. In the case of a percentage, the quantity

$$\sqrt{\frac{p \times q}{n}}$$

where p is a proportion, q equals $(1 - p)$, and n is the sample size, is called the *standard error*. As noted in Chapter 7, this quantity is very important in the estimation of sampling error. We may be 68 percent confident that the population figure falls within plus or minus one standard error of the sample figure; we may be 95 percent confident that it falls within plus or minus two standard errors; and we may be 99.9 percent confident that it falls within plus or minus three standard errors.

Any statement of sampling error, then, must contain two essential components: the *confidence level* (for example, 95 percent) and the *confidence interval* (for example, ±2.5 percent). If 50 percent of a sample of 1,600 people say they had colds during the year, we might say we're 95 percent confident that the population figure is between 47.5 percent and 52.5 percent.

In this example we've moved beyond simply describing the sample into the realm of making estimates (inferences) about the larger population. In doing so, we must take care in several ways.

First, the sample must be drawn from the population about which inferences are being made. A sample taken from a telephone directory cannot legitimately be the basis for statistical inferences about the population of a city, but only about the population of telephone subscribers with listed numbers.

Second, the inferential statistics assume several things. To begin with, they assume simple random sampling, which is virtually never the case in sample surveys. The statistics also assume sampling with replacement, which is almost never done— but this is probably not a serious problem. Although systematic sampling is used more frequently than random sampling, it, too, probably presents no serious problem if done correctly. Stratified sampling, because it improves representativeness, clearly presents no problem. Cluster sampling does present a problem, however, because the estimates of sampling error may be too small. Quite clearly, street-corner sampling does not warrant the use of inferential statistics. Finally, the calculation of standard error in sampling assumes a 100 percent

completion rate—that is, that everyone in the sample completed the survey. The seriousness of this problem increases as the completion rate decreases.

Third, inferential statistics are addressed to sampling error only, not **nonsampling error** such as coding errors or misunderstandings of questions by respondents. Thus, although we might state correctly that between 47.5 and 52.5 percent of the population (95 percent confidence) would *report* having colds during the previous year, we couldn't so confidently guess the percentage who had actually *had* them. Because nonsampling errors are probably larger than sampling errors in a respectable sample design, we need to be especially cautious in generalizing from our sample findings to the population.

Tests of Statistical Significance

There is no scientific answer to the question of whether a given association between two variables is significant, strong, important, interesting, or worth reporting. Perhaps the ultimate test of significance rests in your ability to persuade your audience (present and future) of the association's significance. At the same time, there is a body of inferential statistics to assist you in this regard called *parametric tests of significance*. As the name suggests, parametric statistics are those that make certain assumptions about the parameters describing the population from which the sample is selected. They allow us to determine the **statistical significance** of associations. "Statistical significance" does not imply "importance" or "significance" in any general sense. It refers simply to the likelihood that relationships observed in a sample could be attributed to sampling error alone. Researchers often distinguish between statistical significance and *substantive significance* in this regard, with the latter referring to whether the relationship between variables is big enough to make a meaningful difference. Whereas statistical significance can be calculated, substantive significance is always a judgment call.

Although **tests of statistical significance** are widely reported in social science literature, the logic underlying them is rather subtle and often misunderstood. Tests of significance are based on

---

nonsampling error Those imperfections of data quality that are a result of factors other than sampling error. Examples include misunderstandings of questions by respondents, erroneous recordings by interviewers and coders, and keypunch errors.

statistical significance A general term referring to the likelihood that relationships observed in a sample could be attributed to sampling error alone.

tests of statistical significance A class of statistical computations that indicate the likelihood that the relationship observed between variables in a sample can be attributed to sampling error only.

	Women	Men
Favor equality	50%	50%
Oppose equality	50%	50%
	100%	**100%**

Legend

♀ = Woman who favors equality

♂ = Man who favors equality

⚲ = Woman who opposes equality

⚲ = Man who opposes equality

FIGURE 16-3

A Hypothetical Population of Men and Women Who Either Favor or Oppose Sexual Equality

the same sampling logic discussed elsewhere in this book. To understand that logic, let's return for a moment to the concept of sampling error in regard to univariate data.

Recall that a sample statistic normally provides the best single estimate of the corresponding population parameter, but the statistic and the parameter seldom correspond precisely. Thus, we report the probability that the parameter falls within a certain range (confidence interval). The degree of uncertainty within that range is due to normal sampling error. The corollary of such a statement is, of course, that it is improbable that the parameter would fall outside the specified range *only* as a result of sampling error. Thus, if we estimate that a parameter (99.9 percent confidence) lies between 45 percent and 55 percent, we say by implication that it is extremely improbable that the parameter is actually, say, 90 percent if our *only* error of estimation is due to normal sampling. This is the basic logic behind tests of statistical significance.

The Logic of Statistical Significance

I think I can illustrate the logic of statistical significance best in a series of diagrams representing the selection of samples from a population. Here are the elements in the logic:

1. Assumptions regarding the independence of two variables in the population study

2. Assumptions regarding the representativeness of samples selected through conventional probability-sampling procedures

3. The observed joint distribution of sample elements in terms of the two variables

Figure 16-3 represents a hypothetical population of 256 people; half are women, half are men. The diagram also indicates how each person feels about seeing women as equal to men. In the diagram, those favoring equality have open circles, those opposing it have their circles filled in.

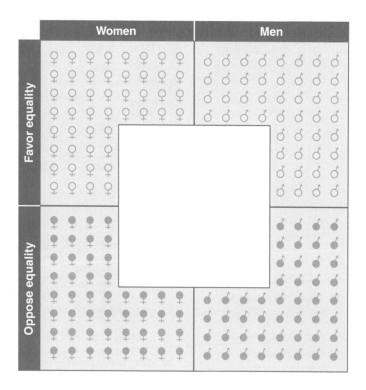

FIGURE 16-4

A Representative Sample

The question we'll be investigating is whether there is any relationship between gender and feelings about equality for men and women. More specifically, we'll see if women are more likely to favor equality than men are, because women would presumably benefit more from it. Take a moment to look at Figure 16-3 and see what the answer to this question is.

The illustration in the figure indicates no relationship between gender and attitudes about equality. Exactly half of each group favors equality and half opposes it. Recall the earlier discussion of proportionate reduction of error. In this instance, knowing a person's gender would not reduce the "errors" we'd make in guessing his or her attitude toward equality. The table in Figure 16-3 provides a tabular view of what you can observe in the graphic diagram.

Figure 16-4 represents the selection of a one-fourth sample from the hypothetical population. In terms of the graphic illustration, a "square" selection from the center of the population provides a representative sample. Notice that our sample contains 16 of each type of person: Half are men and half are women; half of each gender favors equality, and the other half opposes it.

The sample selected in Figure 16-4 would allow us to draw accurate conclusions about the relationship between gender and equality in the larger population. Following the sampling logic we saw in Chapter 7, we'd note there was no relationship between gender and equality in the sample; thus, we'd conclude there was similarly no relationship in the larger population—because we've presumably selected a sample in accord with the conventional rules of sampling.

Of course, real-life samples are seldom such perfect reflections of the populations from which they are drawn. It would not be unusual for us to have selected, say, one or two extra men who

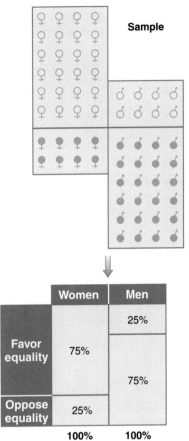

FIGURE 16-5

An Unrepresentative Sample

opposed equality and a couple of extra women who favored it—even if there was no relationship between the two variables in the population. Such minor variations are part and parcel of probability sampling, as we saw in Chapter 7.

Figure 16-5, however, represents a sample that falls far short of the mark in reflecting the larger population. Notice that it includes far too many supportive women and opposing men. As the table shows, three-fourths of the women in the sample support equality, but only one-fourth of the men do so. If we had selected this sample from a population in which the two variables were unrelated to each other, we'd be sorely misled by our sample.

As you'll recall, it's unlikely that a properly drawn probability sample would ever be as inaccu-

rate as the one shown in Figure 16-5. In fact, if we actually selected a sample that gave us the results this one does, we'd look for a different explanation. Figure 16-6 illustrates the more likely situation.

Notice that the sample selected in Figure 16-6 also shows a strong relationship between gender and equality. The reason is quite different this time. We've selected a perfectly representative sample, but we see that there is actually a strong relationship between the two variables in the population at large. In this latest figure, women are more likely to support equality than men are: That's the case in the population, and the sample reflects it.

In practice, of course, we never know what's so for the total population; that's why we select samples. So if we selected a sample and found the

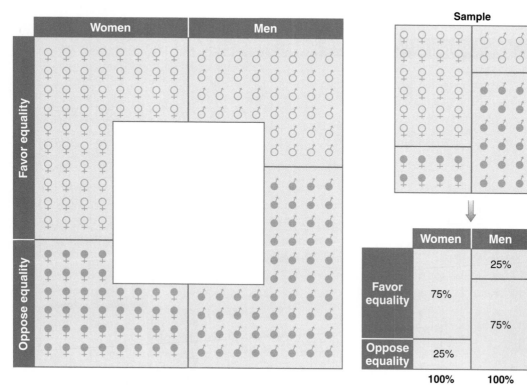

FIGURE 16-6

A Representative Sample from a Population in Which the Variables Are Related

strong relationship presented in Figures 16-5 and 16-6, we'd need to decide whether that finding accurately reflected the population or was simply a product of sampling error.

The fundamental logic of tests of statistical significance, then, is this: Faced with any discrepancy between the assumed independence of variables in a population and the observed distribution of sample elements, we may explain that discrepancy in either of two ways: (1) we may attribute it to an unrepresentative sample, or (2) we may reject the assumption of independence. The logic and statistics associated with probability sampling methods offer guidance about the varying probabilities of varying degrees of unrepresentativeness (expressed as sampling error). Most simply put, there is a *high* probability of a small degree of unrepresentativeness and a *low* probability of a large degree of unrepresentativeness.

The statistical significance of a relationship observed in a set of sample data, then, is always expressed in terms of probabilities. "Significant at the .05 level $(p \leq .05)$" simply means that the probability that a relationship as strong as the observed one can be attributed to sampling error alone is no more than 5 in 100. Put somewhat differently, if two variables are independent of each other in the population, and if 100 probability samples are selected from that population, no more than 5 of those samples should provide a relationship as strong as the one that has been observed.

There is, then, a corollary to confidence intervals in tests of significance, which represents the probability of the measured associations being due *only* to sampling error. This is called the **level of significance.** Like confidence intervals, levels of significance are derived from a logical model in which several samples are drawn from a given population. In the present case, we assume that there is no association between the variables in the population, and then we ask what proportion of the samples drawn from that population would produce associations at least as great as those measured in the empirical data. Three levels of significance are frequently used in research reports: .05, .01, and .001. These mean, respectively, that the chances of obtaining the measured association as a result of sampling error are 5/100, 1/100, and 1/1,000.

Researchers who use tests of significance normally follow one of two patterns. Some specify in advance the level of significance they'll regard as sufficient. If any measured association is statistically significant at that level, they'll regard it as representing a genuine association between the two variables. In other words, they're willing to discount the possibility of its resulting from sampling error only.

Other researchers prefer to report the specific level of significance for each association, disregarding the conventions of .05, .01, and .001. Rather than reporting that a given association is significant at the .05 level, they might report significance at the .023 level, indicating the chances of its having resulted from sampling error as 23 out of 1,000.

Chi Square

Chi square (χ^2) is a frequently used test of significance in social science. It's based on the *null hypothesis:* the assumption that there is no relationship between two variables in the total population (as you may recall from Chapter 2). Given the observed distribution of values on the two separate variables, we compute the conjoint distribution that would be expected if there were no relationship between the two variables. The result of this operation is a set of *expected frequencies* for all the cells in the contingency table. We then compare this expected distribution with the distribution of cases actually found in the sample data, and we determine the probability that the discovered discrepancy could have resulted from sampling error alone. An example will illustrate this procedure.

Let's assume we're interested in the possible relationship between church attendance and gender for the members of a particular church. To test this relationship, we select a sample of 100 church members at random. We find that our sample is made up of 40 men and 60 women and that 70 percent of our sample say they attended church during the preceding week, whereas the remaining 30 percent say they did not.

If there is no relationship between gender and church attendance, then 70 percent of the men in the sample should have attended church during the preceding week, and 30 percent should have stayed away. Moreover, women should have attended in the same proportion. Table 16-6 (part I) shows that, based on this model, 28 men and 42 women would have attended church, with 12 men and 18 women not attending.

Part II of Table 16-6 presents the observed attendance for the hypothetical sample of 100 church members. Note that 20 of the men report having attended church during the preceding week, and the remaining 20 say they did not. Among the women in the sample, 50 attended church and 10 did not. Comparing the expected and observed frequencies (parts I and II), we note that somewhat fewer men attended church than expected, whereas somewhat more women attended than expected.

Chi square is computed as follows. For each cell in the tables, the researcher (1) subtracts the expected frequency for that cell from the observed frequency, (2) squares this quantity, and (3) divides the squared difference by the expected frequency. This procedure is carried out for each cell in the tables; part III of Table 16-6 presents the cell-by-cell computations. The several results are then added together to find the value of chi square: 12.70 in the example.

---

level of significance In the context of tests of statistical significance, the degree of likelihood that an observed, empirical relationship could be attributable to sampling error. A relationship is significant at the .05 level if the likelihood of its being only a function of sampling error is no greater than 5 out of 100.

TABLE 16-6

A Hypothetical Illustration of Chi Square

I. Expected Cell Frequencies	Men	Women	Total
Attended church	28	42	70
Did not attend church	12	18	30
Total	40	60	100

II. Observed Cell Frequencies	Men	Women	Total
Attended church	20	50	70
Did not attend church	20	10	30
Total	40	60	100

III. (Observed − Expected)² ÷ Expected	Men	Women	
Attended church	2.29	1.52	$\chi^2 = 12.70$
Did not attend church	5.33	3.56	$p < .001$

This value is the overall discrepancy between the observed conjoint distribution in the sample and the distribution we would expect if the two variables were unrelated to each other. Of course, the mere discovery of a discrepancy does not prove that the two variables are related, because normal sampling error might produce discrepancies even when there is no relationship in the total population. The magnitude of the value of chi square, however, permits us to estimate the probability of that having happened.

To determine the statistical significance of the observed relationship, we must use a standard set of chi square values. This will require the computation of the *degrees of freedom*, which refer to the possibilities for variation within a statistical model. Suppose I challenge you to find three numbers whose mean is 11. There are infinite solutions to this problem: (11, 11, 11), (10, 11, 12), (−11, 11, 33), and so on. Now, suppose I require that one of the numbers be 7. There would still be an infinite number of possibilities for the other two numbers.

If I told you one number had to be 7 and another 10, however, there would be only one possible value for the third. If the average of three numbers is 11, their sum must be 33. If two of the numbers total 17, the third must be 16. In this situation, we say there are two degrees of free-

dom. Two of the numbers could have any values we choose, but once they are specified, the third number is determined.

More generally, whenever we're examining the mean of N values, we can see that the degrees of freedom equal $N - 1$. Thus, in the case of the mean of 23 values, we could make 22 of them anything we liked, but the 23rd would then be determined.

A similar logic applies to bivariate tables, such as those analyzed by chi square. Consider a table reporting the relationship between two dichotomous variables: *gender* (men/women) and *abortion attitude* (approve/disapprove). Notice that the table provides the marginal frequencies of both variables.

Abortion Attitude	Men	Women	Total
Approve			500
Disapprove			500
Total	500	500	1,000

Despite the conveniently round numbers in this hypothetical example, notice that there are numerous possibilities for the cell frequencies. For example, it could be the case that all 500 men approve and all 500 women disapprove, or it could be just the reverse. Or there could be 250 cases in each cell. Notice there are numerous other possibilities.

Now the question is, How many cells could we fill in pretty much as we choose before the remainder are determined by the marginal frequencies? The answer is only one. If we know that 300 men approved, for example, then 200 men would have had to disapprove, and the distribution would need to be just the opposite for the women.

In this instance, then, we say the table has one degree of freedom. Now, take a few minutes to construct a three-by-three table. Assume you know the marginal frequencies for each variable, and see if you can determine how many degrees of freedom it has.

For chi square, the degrees of freedom are computed as follows: the number of rows in the table of observed frequencies, minus 1, is multiplied by the number of columns, minus 1. This may be written as $(r - 1)(c - 1)$. For a three-by-three table, then, there are four degrees of freedom: $(3 - 1)(3 - 1) = (2)(2) = 4$.

In the example of gender and church attendance, we have two rows and two columns (discounting the totals), so there is one degree of freedom. Turning to a table of chi square values (see Appendix D), we find that for one degree of freedom and random sampling from a population in which there is no relationship between two variables, 10 percent of the time we should expect a chi square of at least 2.7. Thus, if we selected 100 samples from such a population, we should expect about 10 of those samples to produce chi squares equal to or greater than 2.7. Moreover, we should expect chi square values of at least 6.6 in only 1 percent of the samples and chi square values of 10.8 in only one tenth of a percent (.001) of the samples. The higher the chi square value, the less probable it is that the value could be attributed to sampling error alone.

In our example, the computed value of chi square is 12.70. If there were no relationship between gender and church attendance in the church-member population and a large number of samples had been selected and studied, then we would expect a chi square of this magnitude in fewer than 1/10 of 1 percent (.001) of those samples. Thus, the probability of obtaining a chi square of this magnitude is less than .001, if random sampling has been used and there is no relationship in the population. We report this finding by saying the relationship is statistically significant at the .001 level. Because it is so improbable that the observed relationship could have resulted from sampling error alone, we're likely to reject the null hypothesis and assume that there is a relationship between the two variables in the population of church members.

t-Test

Chi square is appropriate for testing the statistical association of relations found in typically nominal or ordinal tabular data, as in the example just discussed. Suppose your data represent a high level of measurement such as interval or ratio data. Let's say you want to know if men and women have significantly different weights. To determine this, you measure the weights of a sample of men and women and then calculate the mean average for each sex. Let's say the average weight for men is 170; for women, it's 135. That seems like a pretty substantial difference on the face of it. But what if your "sample" consists of two men and two women? Intuitively, you can see that even a difference of the observed magnitude could have resulted from your picking two big men and two small women, just by chance. We wouldn't want to conclude we had discovered something about men and women in general, simply based on four people who might not be typical.

The *t-test*, sometimes known as Student's *t*, is a commonly used measure for judging the statistical significance of differences in group means. The formula for calculating *t* involves some statistics we haven't discussed in this book, so let me give you a sense of the logic involved in this measure.

First, it makes sense that the value of *t* will increase with the size of the difference between the means.

The value of *t* will also increase with the size of the sample involved; hence, differences found in larger samples—as we saw in regard to chi square—are more likely to be judged statistically significant.

Finally, the value of *t* will be larger when variations of values within each group are smaller. In the case of sex and weight, the value of *t* will be greatest when

- The difference between the average weight of men and that of women is large.

- When we've examined a large sample.

- When most women's weights are clustered around their mean weight and most men's weights are clustered around the mean for men. In the extreme case, the heaviest woman would weigh less than the lightest man, though this is unlikely in any substantial sample.

Once you calculate a value for *t* in your data, you look that value up in a *t*-test table, found in any statistics textbook. This gives you the significance of that value, expressed as the probability that the observed difference might have been due to sampling error alone—the same logic used in the case of chi square.

Most measures of association can be tested for statistical significance in a similar manner. Standard tables of values permit us to determine whether a given association is statistically significant and at what level. Any standard statistics textbook provides instructions on the use of such tables.

Some Words of Caution

Tests of significance provide an objective yardstick that we can use to estimate the statistical significance of associations between variables. They help us rule out associations that may not represent genuine relationships in the population under study. However, the researcher who uses or reads reports of significance tests should remain wary of several dangers in their interpretation.

First, we've been discussing tests of statistical significance; there are no objective tests of *substantive* significance. Thus, we may be legitimately convinced that a given association is not due to sampling error, but we may be in the position of asserting without fear of contradiction that two variables are only slightly related to each other. Recall that sampling error is an inverse function

of sample size—the larger the sample, the smaller the expected error. Thus, a correlation of, say, .1 might very well be significant (at a given level) if discovered in a large sample, whereas the same correlation between the same two variables would not be significant if found in a smaller sample. This makes perfectly good sense given the basic logic of tests of significance: In the larger sample, there is less chance that the correlation could be simply the product of sampling error. In both samples, however, it might represent an essentially zero correlation.

The distinction between statistical and substantive significance is perhaps best illustrated by those cases where there is *absolute certainty* that observed differences cannot be a result of sampling error. This would be the case when we observe an entire population. Suppose we were able to learn the ages of every public official in the United States and of every public official in Russia. For argument's sake, let's assume further that the average age of U.S. officials was 45 years old compared with, say, 46 for the Russian officials. Because we would have the ages of all officials, there would be no question of sampling error. We would know with certainty that the Russian officials were older than their U.S. counterparts. At the same time, we would say that the difference was of no substantive significance. We would conclude, in fact, that they were essentially the same age.

Second, lest you be misled by this hypothetical example, realize that statistical significance should not be calculated on relationships observed in data collected from whole populations. Remember, tests of statistical significance measure the likelihood of relationships between variables being only a product of sampling error; if there's no sampling, there's no sampling error.

Third, tests of significance are based on the same sampling assumptions we used in computing confidence intervals. To the extent that these assumptions are not met by the actual sampling design, the tests of significance are not strictly legitimate.

We've examined statistical significance here in the form of chi square and *t*-tests, but social scientists commonly use several other measures as well.

Analysis of variance is one example you may run across in your studies.

As is the case for most matters covered in this book, I have a personal prejudice. In this instance, it's against tests of significance. I don't object to the statistical logic of those tests, because the logic is sound. Rather, I'm concerned that such tests seem to mislead more than they enlighten. Here are my principal reservations:

1. Tests of significance make sampling assumptions that are virtually never satisfied by actual sampling designs.

2. They depend on the absence of nonsampling errors, a questionable assumption in most actual empirical measurements.

3. In practice, they are too often applied to measures of association that have been computed in violation of the assumptions made by those measures (for example, product-moment correlations computed from ordinal data).

4. Statistical significance is too easily misinterpreted as "strength of association," or substantive significance.

These concerns are underscored by a study (Sterling, Rosenbaum, and Weinkam 1995) examining the publication policies of nine psychology and three medical journals. As the researchers discovered, the journals were quite unlikely to publish articles that did not report statistically significant correlations among variables. They quote the following from a rejection letter:

> Unfortunately, we are not able to publish this manuscript. The manuscript is very well written and the study was well documented. Unfortunately, the negative results translate into a minimal contribution to the field. We encourage you to continue your work in this area and we will be glad to consider additional manuscripts that you may prepare in the future.
>
> *(Sterling et al. 1995: 109)*

Let's suppose a researcher conducts a scientifically excellent study to determine whether *X* causes *Y*. The results indicate no statistically significant correlation. That's good to know. If we're interested in what causes cancer, war, or juvenile delinquency, it's good to know that a possible cause actually does *not* cause it. That knowledge would free researchers to look elsewhere for causes.

As we've seen, however, journals might very well reject such a study. Other researchers would likely continue testing whether *X* causes *Y*, not knowing that previous studies found no causal relationship. This would produce many wasted studies, none of which would see publication and draw a close to the analysis of *X* as a cause of *Y*.

From what you've learned about probabilities, however, you can understand that if enough studies are conducted, one will eventually measure a statistically significant correlation between *X* and *Y*. If there is absolutely no relationship between the two variables, we would expect a correlation significant at the .05 level five times out of a hundred, because that's what the .05 level of significance means. If a hundred studies were conducted, therefore, we could expect five to suggest a causal relationship where there was actually none—and those five studies would be published!

There are, then, serious problems inherent in too much reliance on tests of statistical significance. At the same time (perhaps paradoxically) I would suggest that tests of significance *can* be a valuable asset to the researcher—useful tools for understanding data. Although many of my comments suggest an extremely conservative approach to tests of significance—that you should use them only when all assumptions are met—my general perspective is just the reverse.

I encourage you to use any statistical technique—any measure of association or test of significance—if it will help you understand your data. If the computation of product-moment correlations among nominal variables and the testing of statistical significance in the context of uncontrolled sampling will meet this criterion, then I encourage such activities. I say this in the spirit of what Hanan Selvin, another pioneer in developing the elaboration model, referred to as "data-dredging techniques." Anything goes, if it leads ultimately to the understanding of data and of the social world under study.

The price of this radical freedom, however, is the giving up of strict, statistical interpretations. You will not be able to base the ultimate importance of your finding solely on a significant correlation at the .05 level. Whatever the avenue of discovery, empirical data must ultimately be presented in a legitimate manner, and their importance must be argued logically.

Other Multivariate Techniques

For the most part, this book has focused on rather rudimentary forms of data manipulation, such as the use of contingency tables and percentages. The elaboration model of analysis was presented in this form, as well as many of the examples of data analysis throughout the book.

This section of the chapter presents a cook's tour of several other multivariate techniques from the logical perspective of elaborating the relationships among social variables. This discussion is intended not to teach you how to use these techniques but rather to present sufficient information so that you can understand them if you run across them in a research report. The methods of analysis that we'll examine—path analysis, time-series analysis, factor analysis, analysis of variance, discriminant analysis, log-linear models, and Geographic Information Systems—are only a few of the many multivariate techniques used by social scientists.

Path Analysis

Path analysis is a causal model for understanding relationships between variables. Though based on regression analysis, it can provide a more useful graphic picture of relationships among several variables than other means can. Path analysis assumes that the values of one variable are caused by the values of another, so distinguishing independent and dependent variables is essential. This require-

---

path analysis A form of multivariate analysis in which the causal relationships among variables are presented in a graphic format.

---

ment is not unique to path analysis, of course, but path analysis provides a unique way of displaying explanatory results for interpretation.

Recall for a moment one of the ways I represented the elaboration model in Chapter 15 (Figure 15-1). Here's how we might diagram the logic of interpretation:

| Independent variable | $\rightarrow$ | Intervening variable | $\rightarrow$ | Dependent variable |

The logic of this presentation is that an independent variable has an impact on an intervening variable, which in turn has an impact on a dependent variable. The path analyst constructs similar patterns of relationships among variables, but the typical path diagram contains many more variables than shown in this diagram.

Besides diagramming a network of relationships among variables, path analysis also shows the strengths of those several relationships. The strengths of relationships are calculated from a regression analysis that produces numbers analogous to the partial relationships in the elaboration model. These *path coefficients,* as they're called, represent the strengths of the relationships between pairs of variables, with the effects of all other variables in the model held constant.

The analysis in Figure 16-7, for example, focuses on the religious causes of anti-Semitism among Christian church members. The variables in the diagram are, from left to right, (1) orthodoxy, or the extent to which the subjects accept conventional beliefs about God, Jesus, biblical miracles, and so forth; (2) particularism, the belief that one's religion is the "only true faith"; (3) acceptance of the view that the Jews crucified Jesus; (4) religious hostility toward contemporary Jews, such as believing that God is punishing them or that they will suffer damnation unless they convert to Christianity; and (5) secular anti-Semitism, such as believing that Jews cheat in business, are disloyal to their country, and so forth.

To start with, the researchers who conducted this analysis proposed that secular anti-Semitism was produced by moving through the five variables: Orthodoxy caused particularism, which caused the view of the historical Jews as crucifiers, which caused religious hostility toward contem-

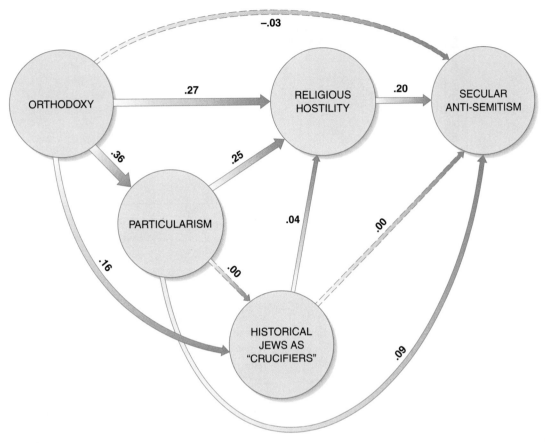

FIGURE 16-7

Diagramming the Religious Sources of Anti-Semitism

Source: Rodney Stark, Bruce D. Foster, Charles Y. Glock, and Harold E. Quinley, *Wayward Shepherds—Prejudice and the Protestant Clergy*. Copyright © 1971 by Anti-Defamation League of B'nai Brith. Reprinted by permission of Harper & Row, Publishers, Inc.

porary Jews, which resulted, finally, in secular anti-Semitism.

The path diagram tells a different story. The researchers found, for example, that belief in the historical role of Jews as the crucifiers of Jesus doesn't seem to matter in the process that generates anti-Semitism. And, although particularism is a part of one process resulting in secular anti-Semitism, the diagram also shows that anti-Semitism is created more directly by orthodoxy and religious hostility. Orthodoxy produces religious hostility even without particularism, and religious hostility generates secular hostility in any event.

One last comment on path analysis is in order. Although it's an excellent way of handling complex causal chains and networks of variables, path analysis itself does not tell the causal order of the variables. Nor was the path diagram in Figure 16-7 generated by computer. The researcher decided the structure of relationships among the variables and used computer analysis merely to calculate the path coefficients that applied to the structure.

Time-Series Analysis

The various forms of regression analysis are often used to examine time-series data, representing changes in one or more variables over time. As I'm sure you know, U.S. crime rates have generally

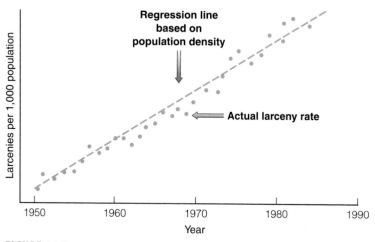

FIGURE 16-8

The Larceny Rates over Time in a Hypothetical City

increased over the years. A **time-series analysis** of crime rates could express the long-term trend in a regression format and provide a way of testing explanations for the trend—such as population growth or economic fluctuations—and could permit forecasting of future crime rates.

In a simple illustration, Figure 16-8 graphs the larceny rates of a hypothetical city over time. Each dot on the graph represents the number of larcenies reported to police during the year indicated.

Suppose we feel that larceny is partly a function of overpopulation. You might reason that crowding would lead to psychological stress and frustration, resulting in increased crimes of many sorts. Recalling the discussion of regression analysis, we could create a regression equation representing the relationship between larceny and population density—using the actual figures for each variable, with years as the units of analysis. Having created the best-fitting regression equation, we could then calculate a larceny rate for each year, based on that year's population density rate. For the sake of simplicity, let's assume that the city's population size (and hence density) has been

steadily increasing. This would lead us to predict a steadily increasing larceny rate as well. These regression estimates are represented by the dashed regression line in Figure 16-8.

Time-series relationships are often more complex than this simple illustration suggests. For one thing, there can be more than one causal variable. For example, we might find that unemployment rates also had a powerful impact on larceny. We might develop an equation to predict larceny on the basis of both of these causal variables. As a result, the predictions might not fall along a simple, straight line. Whereas population density was increasing steadily in the first model, unemployment rates rise and fall. As a consequence, our predictions of the larceny rate would similarly go up and down.

Pursuing the relationship between larceny and unemployment rates, we might reason that people do not begin stealing as soon as they become unemployed. Typically, they might first exhaust their savings, borrow from friends, and keep hoping for work. Larceny would be a last resort.

Time-lagged regression analysis could be used to address this more complex case. Thus, we might create a regression equation that predicted a given year's larceny rate based, in part, on the previous year's unemployment rate or perhaps on an aver-

---

time-series analysis An analysis of changes in a variable (such as *crime rates*) over time.

age of the two years' unemployment rates. The possibilities are endless.

If you think about it, a great many causal relationships are likely to involve a time lag. Historically, many of the world's poor countries have maintained their populations by matching high death rates with equally high birthrates. It has been observed repeatedly, moreover, that when a society's death rate is drastically reduced—through improved medical care, public sanitation, and improved agriculture, for example—that society's birthrate drops sometime later on, but with an intervening period of rapid population growth. Or, to take a very different example, a crackdown on speeding on a state's highways would likely reduce the average speed of cars. Again, however, the causal relationship would undoubtedly involve a time lag—days, weeks, or months, perhaps—as motorists began to realize the seriousness of the crackdown.

In all such cases, the regression equations generated might take many forms. In any event, the criterion for judging success or failure is the extent to which the researcher can account for the actual values observed for the dependent variable.

Factor Analysis

Factor analysis is a unique approach to multivariate analysis. Its statistical basis is complex enough and different enough from the foregoing discussions to suggest a general discussion here.

Factor analysis is a complex algebraic method used to discover patterns among the variations in values of several variables. This is done essentially through the generation of artificial dimensions (*factors*) that correlate highly with several of the real variables and that are independent of one another. A computer must be used to perform this complex operation.

Let's suppose that a data file contains several indicators of subjects' prejudice. Each item should provide some indication of prejudice, but none will give a perfect indication. All of these items, moreover, should be highly intercorrelated empirically. In a factor analysis of the data, the researcher would create an artificial dimension that would be highly correlated with each of the items measuring prejudice. Each subject would essentially receive a value on that artificial dimension, and the value assigned would be a good predictor of the observed attributes on each item.

Suppose now that the same study provided several indicators of subjects' mathematical ability. It's likely that the factor analysis would also generate an artificial dimension highly correlated with each of those items.

The output of a factor analysis program consists of columns representing the several factors (artificial dimensions) generated from the observed relations among variables plus the correlations between each variable and each factor—called the *factor loadings.*

In the preceding example, it's likely that one factor would more or less represent prejudice, and another would more or less represent mathematical ability. Data items measuring prejudice would have high loadings on (correlations with) the prejudice factor and low loadings on the mathematical ability factor. Data items measuring mathematical ability would have just the opposite pattern.

In practice, factor analysis does not proceed in this fashion. Rather, the variables are input to the program, and the program outputs a series of factors with appropriate factor loadings. The analyst must then determine the meaning of a given factor on the basis of those variables that load highly on it. The program's generation of factors, however, has no reference to the meaning of variables, only to their empirical associations. Two criteria are taken into account: (1) a factor must explain a relatively large portion of the variance found in the study variables, and (2) every factor must be more or less independent of every other factor.

Here's an example of the use of factor analysis. Many social researchers have studied the problem of delinquency. If you look deeply into the problem, however, you'll discover that there are

---

factor analysis A complex algebraic method for determining the general dimensions or factors that exist within a set of concrete observations.

TABLE 16-7

Factor Analysis: Delinquent Acts, Whites

Delinquent Act	Property Offenses Factor I	Incorrigibility Factor II	Drugs/Truancy Factor III	Fighting Factor IV
Broke street light, etc.	0.669	0.126	0.119	0.167
Broke windows	0.637	0.093	0.077	0.215
Broke down fences, clotheslines, etc.	0.621	0.186	0.186	0.186
Taken things worth $2 to $50	0.616	0.187	0.233	0.068
Let air out of tires	0.587	0.243	0.054	0.156
Taken things worth over $50	0.548	−0.017	0.276	0.034
Thrown eggs, garbage, etc.	0.526	0.339	−0.023	0.266
Taken things worth under $2	0.486	0.393	0.143	0.077
Taken things from desks, etc., at school	0.464	0.232	−0.002	0.027
Taken car without owner's permission	0.461	0.172	0.080	0.040
Put paint on something	0.451	0.237	0.071	0.250
Disobeyed parents	0.054	0.642	0.209	0.039
Marked on desk, wall, etc.	0.236	0.550	−0.061	0.021
Said mean things to get even	0.134	0.537	0.045	0.100
Disobeyed teacher, school official	0.240	0.497	0.223	0.195
Defied parents to their face	0.232	0.458	0.305	0.058
Made anonymous telephone calls	0.373	0.446	0.029	0.135
Smoked marijuana	0.054	0.064	0.755	−0.028
Used other drugs for kicks	0.137	0.016	0.669	0.004
Signed name to school excuse	0.246	0.249	0.395	0.189
Drank alcohol, parents absent	0.049	0.247	0.358	0.175
Skipped school	0.101	0.252	0.319	0.181
Beat up someone in a fight	0.309	0.088	0.181	0.843
Fought—hit or wrestled	0.242	0.266	0.070	0.602
Percent of variance	67.2	13.4	10.9	8.4

Source: Morris A. Forslund, *Patterns of Delinquency Involvement: An Empirical Typology,* paper presented at the Annual Meeting of the Western Association of Sociologists and Anthropologists, Lethbridge, Alberta, February 8, 1980. The table above is adapted from page 10.

many different types of delinquents. In a survey of high school students in a small Wyoming town, Morris Forslund (1980) set out to create a typology of delinquency. His questionnaire asked students to report whether they had committed a variety of delinquent acts. He then submitted their responses to factor analysis. The results are shown in Table 16-7.

As you can see in this table, the various delinquent acts are listed on the left. The numbers shown in the body of the table are the factor load-

ings on the four factors constructed in the analysis. You'll notice that after examining the dimensions, or factors, Forslund labeled them. I've bracketed the items on each factor that led to his choice of labels. Forslund summarized the results as follows:

For the total sample four fairly distinct patterns of delinquent acts are apparent. In order of variance explained, they have been labeled: 1) Property Offenses, including both vandalism and theft; 2) Incorrigibility; 3) Drugs/Truancy; and 4) Fighting. It is interesting, and perhaps surprising, to find both vandalism and theft appear together in the same factor. It would seem that those high school students who engage in property offenses tend to be involved in both vandalism and theft. It is also interesting to note that drugs, alcohol and truancy fall in the same factor.

(1980: 4)

Having determined this overall pattern, Forslund reran the factor analysis separately for boys and for girls. Essentially the same patterns emerged in both cases.

This example shows that factor analysis is an efficient method of discovering predominant patterns among a large number of variables. Instead of being forced to compare countless correlations—simple, partial, and multiple—to discover those patterns, researchers can use factor analysis for this task. Incidentally, this is a good example of a helpful use of computers.

Factor analysis also presents data in a form that the reader or researcher can interpret. For a given factor, the reader can easily discover the variables loading highly on it, thus noting clusters of variables. Or, the reader can easily discover which factors a given variable is or is not loaded highly on.

But factor analysis also has disadvantages. First, as noted previously, factors are generated with no regard to substantive meaning. Often researchers will find factors producing very high loadings for a group of substantively disparate variables. They might find, for example, that prejudice and religiosity have high positive loadings on a given factor, with education having an equally high negative loading. Surely the three variables are highly correlated, but what does the factor represent in the real world? All too often, inexperienced researchers will be led into naming such factors as "religio-prejudicial lack of education" or something similarly nonsensical.

Second, factor analysis is often criticized on basic philosophical grounds. Recall that to be useful, a hypothesis must be disprovable. If the researcher cannot specify the conditions under which the hypothesis would be disproved, the hypothesis is either a tautology or useless. In a sense, factor analysis suffers this defect. No matter what data are input, factor analysis produces a solution in the form of factors. Thus, if the researcher were asking, "Are there any patterns among these variables?" the answer always would be yes. This fact must also be taken into account in evaluating the results of factor analysis. The generation of factors by no means ensures meaning.

My personal view of factor analysis is the same as that for other complex modes of analysis. It can be an extremely useful tool for the social science researcher. Its use should be encouraged whenever such activity might assist researchers in understanding a body of data. As in all cases, however, such tools are only tools and never magical solutions.

Let me reiterate that the analytic techniques we've touched on are only a few of the many techniques commonly used by social scientists. As you pursue your studies, you may very well want to study this subject in more depth later.

Analysis of Variance

Analysis of variance (ANOVA) applies the logic of statistical significance, discussed earlier. Fundamentally, the cases under study are combined into groups representing an independent variable, and the extent to which the groups differ from one another is analyzed in terms of some dependent variable. The extent to which the groups differ is compared with the standard of random distribution: Could we expect to obtain such differences if we had assigned cases to the various groups through random selection?

We'll look briefly now at two common forms of ANOVA: one-way analysis of variance and two-way analysis of variance.

---

analysis of variance (ANOVA) Method of analysis in which cases under study are combined into groups representing an independent variable, and the extent to which the groups differ from one another is analyzed in terms of some dependent variable. Then, the extent to which the groups differ is compared with the standard of random distribution.

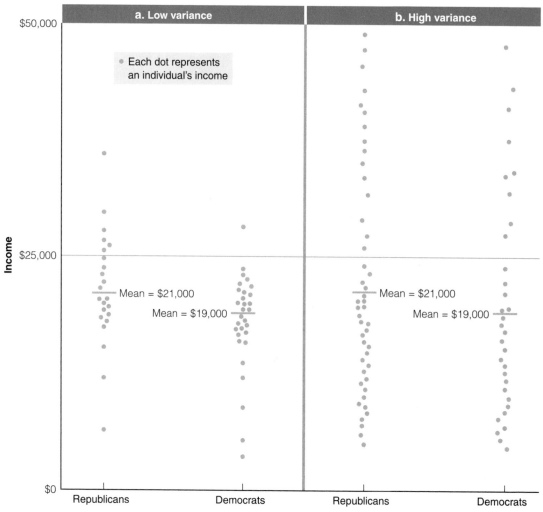

FIGURE 16-9

Two Distribution Patterns of the Incomes of Republicans and Democrats

One-Way Analysis of Variance

Suppose we want to compare income levels of Republicans and Democrats to see if Republicans are really richer. We select a sample of individuals for our study, and we ask them (1) which political party they identify with and (2) their total income for the last year. We calculate the mean or median incomes of each political group, finding that the Republicans in our sample have a mean income of, say, $21,000, compared with $19,000 for the Democrats. Clearly, our Republicans are richer than our Democrats, but is the difference "significant"?

Would we have been likely to get a $2,000 difference if we had created two groups by way of random selection?

ANOVA answers this question through the use of variance. Most simply put, the variance of a distribution (of incomes, for example) is a measurement of the extent to which a set of values are clustered close to the mean or range high and low away from it.

Figure 16-9 illustrates these two possibilities. Notice that in both distributions the Republicans have a mean income of $21,000 and the Democrats have $19,000. In part (a), most Republicans have incomes relatively close to the mean of $21,000, and most Democrats have incomes close to their party's mean of $19,000. Part (b), however, presents quite a different picture. Although the group means are the same as in part (a), both Republicans and Democrats have incomes ranging from very high to very low, with considerable overlap in the parties' distributions. In technical terms, there is a higher degree of variance in part (b) than in part (a). On the face of it, we'd conclude that part (a) of Figure 16-9 indicates a genuine difference in the incomes of Republicans and Democrats. With data like those presented in part (b), we wouldn't be so sure; in this case, there seems more likelihood that the normal variations produced by random sampling error could have produced means of $21,000 and $19,000.

In an actual ANOVA, statistical calculations rather than impressions are used to make this decision. The observed difference in means is expressed as standardized multiples and fractions of the observed variance. Because the variance in part (a) of Figure 16-9 is smaller than the variance in part (b), $2,000 would represent a larger difference in part (a) than in part (b). The resulting difference of means—standardized by the variance—would then be checked against a standard statistical table showing the theoretical distribution of such values, as in our earlier discussion of statistical significance. Ultimately, we'd conclude that the difference was significant at some level of significance. We might discover, for example, that sampling error would have produced a difference as large as the one observed only one time in a thousand. Thus, we would say that difference was "significant at the .001 level."

In the example just given, I've glossed over the actual calculations in favor of the basic logic of the procedure. In practice, such calculations are typically performed by computer programs.

This simplest case is often referred to as a *t*-test for the difference between two means. With more than two groups, the calculations become more complex, because more comparisons must be made. Basically, it's necessary to compare the differences separating group means with the variations found within each group. The end result of the analysis, as discussed in the simplest case, is expressed in terms of statistical significance—the likelihood of the observed differences resulting from sampling error in random selection.

Two-Way Analysis of Variance

One-way ANOVA represents a form of bivariate analysis (*political party* and *income* were the two variables in our example). As we've seen, however, social researchers often engage in multivariate analysis. Two-way ANOVA permits the simultaneous examination of more than two variables. Suppose, for example, that we suspect that the income differences between Republicans and Democrats are a function of education. Our hypothesis is that Republicans are better educated than Democrats and that educated people—regardless of party—earn more, on average, than people with less education do. A two-way ANOVA would sort out the effects of the two explanatory variables in a manner similar to that of the elaboration model discussed in Chapter 15 and following the same logic discussed in the case of partial correlations and regressions.

Discriminant Analysis

Discriminant analysis offers an interesting twist on several of the techniques we've already examined in this chapter. Its logic is similar to that of multiple regression, except that the dependent variable can be nominal; regression, you'll recall, requires interval variables. For an illustration, let's look at a simple example.

Figure 16-10 represents six writers. Three of the writers do their writing by hand (with a pencil), and three write on computers. (Yes, I know there are other alternatives, but I said this was going to be simple.) Our task is to account for the difference in writing method. Can we find a way of predicting whether a given writer uses a pencil or a computer?

---

discriminant analysis Method of analysis similar to multiple regression, except that the dependent variable can be nominal.

= Writers who write by hand

= Writers who use a computer

FIGURE 16-10

Six Writers: Three Who Write by Hand and Three Who Use Computers

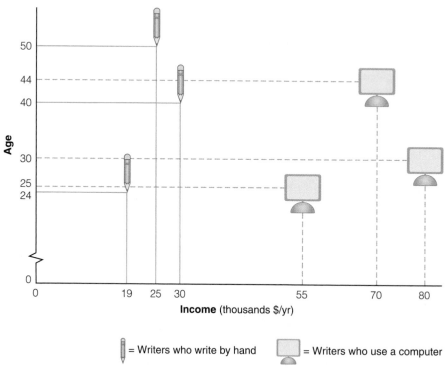

= Writers who write by hand   = Writers who use a computer

FIGURE 16-11

Plotting the Six Writers in Terms of Age and Income

Figure 16-11 explores two variables that we think will likely affect how the writers write. Age might make a difference, because the older writers might have grown accustomed to writing by hand and might have difficulty adapting to the new tech-nology, whereas the younger writers would have grown up with computers. Income could make a difference, because computers cost more than pen-cils. Figure 16-11 therefore plots each writer on the graph on the basis of his or her age and income.

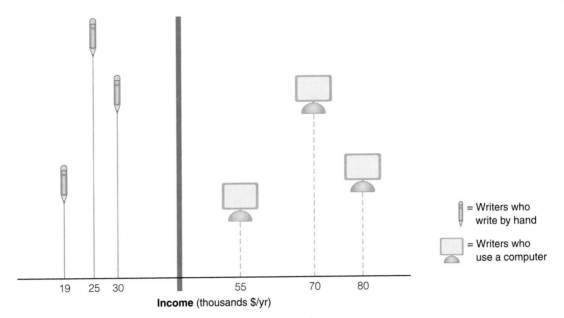

FIGURE 16-12

Income Alone Is Sufficient to Predict Writing Method

See if you can reach any conclusion from the graph about what might account for the difference in writing method.

Figure 16-12 further clarifies the conclusion you might have drawn. Income alone seems an adequate predictor, at least as far as these six writers are concerned. Writers earning $30,000 or less all use pencils, and those earning $30,000 or more all use computers.*

Life is seldom that simple, however, even in simplified illustrations. So let's muddy the water a bit. Figure 16-13 presents the six hypothetical writers in a somewhat more complicated configuration in terms of their ages and incomes. Notice that we cannot draw a line that would separate the pencils from the computers, using either age or income.

If you study Figure 16-13 a little more carefully, however, you'll discover that we can draw a line that separates the pencils from the computers. It's just not perpendicular to either axis of the graph. Figure 16-14 shows the line that achieves our aim. To take advantage of the line that separates the pencils from the computers, we need to find a way of predicting which writers fall on which side

of that line. Figure 16-15 illustrates how this is done.

By constructing a new line perpendicular to the dividing line, we can calculate where each writer would fall on the new, composite dimension. This calculation would take a form similar to the regression equations discussed earlier. The equation would look something like the following:

New dimension $= a + (b \times \text{Age}) + (c \times \text{Income})$

A discriminant analysis computer program would be able to take the values of age and income, examine their relationship to writing method, then generate an equation that would allow you to use the ages and incomes of additional writers to predict their writing methods.

*If you said, "Ah, but the relationship might go in the other direction—how you write determines how much you earn"—give yourself a pat on the back for an excellent insight, and then set it aside for purposes of this illustration. For now, let's assume that income causes writing method rather than the other way around.

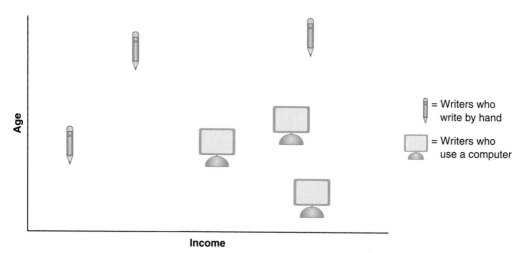

FIGURE 16-13

A Slightly More Complicated Pattern

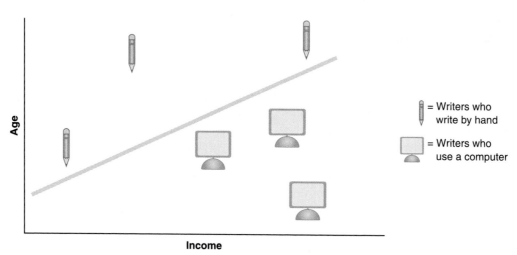

FIGURE 16-14

Separating the Pencils from the Computers

Log-Linear Models

Suppose we want to know whether political orientation is related to party affiliation. Are liberals, for example, more likely than conservatives to be Democrats? By dividing our sample into two groups—liberal and conservative—we can calculate the percentage of Democrats in each group.

If we find a higher percentage among the liberals, we conclude that political orientation and party affiliation are indeed related.

In this example, and in the tabular analyses of Chapters 14 and 15, all the dependent variables analyzed were dichotomous, that is, composed of two attributes. When the dependent variable is not dichotomous, however, matters become more

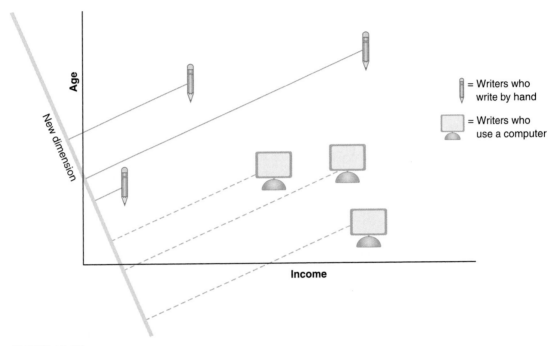

FIGURE 16-15

Plotting the Six Writers on the New Dimensions

complex. Suppose that besides Democrats and Republicans our sample includes Independents, Socialists, and Libertarians. It no longer makes sense to examine the percentage of liberals and conservatives who are Democrats, any more than it makes sense to look only at the percentages affiliated with any one of the other groups. Looking at each group independently would result in more tables than could be easily interpreted.

The complexity of this situation is increased if the explanatory variable is not dichotomous. Suppose we add moderates to the liberals and conservatives, or suppose we examine the interactive effects of other explanatory variables such as race and religion on the political equation. As you can imagine, the resulting percentage tables would become incredibly complicated.

Log-linear models offer a potential solution to this complexity. This technique, which involves rather elaborate logarithmic calculations, is based on specifying models that describe the interrelationships among variables and then comparing

expected and observed table-cell frequencies. (The logic here is similar to that for chi square, discussed earlier.) H. T. Reynolds describes the process:

> At the outset of log-linear analysis, as in most statistical procedures, the investigator proposes a model that he feels might fit the data. The model is a tentative statement about how a set of variables are interrelated. After choosing the model, he next estimates the frequencies expected in a sample of the given size if the model were true. He then compares these estimates, F, with the observed values.
>
> *(1977: 76–77)*

In specifying the models to be tested in a log-linear analysis, the researcher will consider direct

log-linear models Data-analysis technique based on specifying models that describe the interrelationships among variables and then comparing expected and observed table-cell frequencies.

relationships between the dependent variable and each independent variable, relationships between pairs of independent variables, and three-variable (and more, depending on the total number of variables) relationships similar to those already discussed in the elaboration model (Chapter 15). We'll consider a three-variable case taken from the preceding example.

We might suspect that a person's political party affiliation ("party") is a function of political orientation ("philosophy") and race. The components of this model, then, include (1) the direct effect of philosophy on party, (2) the direct effect of race on party, (3) the effect of race on philosophy, (4) the effect of race on the relationship between philosophy and party (as in the elaboration model), and (5) the effect of philosophy on the relationship between race and party. Though each of these components will have some explanatory power, log-linear analysis provides a means of identifying which are the most important and which can, as a practical matter, be ignored. Although the calculations involved in log-linear analysis are many and complex, computer programs can perform them all handily. If you find references in the research literature to *logit, probit,* or *multi-way frequency analysis* (MFA), those analyses are using this model.

Log-linear analysis has two main shortcomings. First, its logic makes certain mathematical assumptions that a particular set of data might not satisfy, but this issue is far too complex to be pursued here. Second, as with other summary techniques discussed, the results of log-linear analysis do not permit the immediate, intuitive grasp possible in simple comparisons of percentages or means. Because of this, log-linear methods would not be appropriate—even if statistically justified—in cases where the analysis can be managed through simple percentage tables. It's best reserved for complex situations in which tabular analyses are not powerful enough.

Geographic Information Systems (GIS) Analytic technique in which researchers map quantitative data that describe geographic units in a graphic display.

Geographic Information Systems (GIS)

Finally, let's examine a very different analytic technique: **Geographic Information Systems (GIS).** Much of the aggregated data of interest to social scientists describes geographic units: countries, states, counties, cities, census tracts, and the like. Whereas such data can and often are presented in statistical tables, the patterns they represent can often be grasped more readily in a graphic format. With this in mind, U.S. Census data are increasingly being made available in a mappable format.

Much of the analysis of the 2000 and 2004 presidential elections in the United States was couched in terms of red (Republican) and blue (Democratic) states, and I'm sure you've seen maps of the distribution of the two. Some researchers have pointed out that no state was completely red or blue, and they added purple for those fairly evenly divided in their support for the two major parties.

Other researchers have pointed out that counties are a more appropriate unit of analysis in this case, displaying the political diversity within a given state. As a general pattern, Republicans did better in rural counties, Democrats did better in the urban ones. Robert Vanderbei (2004) used GIS mapping to display 2004 presidential voting patterns in a way that reflected all these concerns. The unit of analysis in Figure 16-16 is the county. The relative tilt toward the Republicans or Democrats is indicated by shading, and the height of each county column reflects the number of voters per square mile: the higher the column, the more urban the county.

If you're interested in pursuing the possibilities of this analytic technique, you might try a web search for "GIS" or "Geographic Information Systems." By the time you read this paragraph, newer applications of the technique will have appeared. And you'll find that its use is hardly limited to the United States.

This completes our discussion of some of the analytic techniques commonly used by social scientists. I've merely brushed the surface of each, and there are many other techniques that I haven't

FIGURE 16-16

GIS Display of 2004 U.S. Presidential Election. White = Republican; blue = Democratic.

Source: Robert Vanderbei, "Election 2004 Results" (http://www.princeton.edu/~rvdb/JAVA/election2004/).

touched on at all. My purpose has been to give you a preview of some of the techniques you might want to study in more depth later on, as well as to familiarize you with them in case you run across them in reading the research reports of others.

MAIN POINTS

Introduction

- Statistics is the applied branch of mathematics especially appropriate for a variety of research analyses.

Descriptive Statistics

- Descriptive statistics are used to summarize data under study. Some descriptive statistics summarize the distribution of attributes on a single variable; others summarize the associations between variables.

- Descriptive statistics summarizing the relationships between variables are called measures of association.

- Many measures of association are based on a proportionate reduction of error (PRE) model. This model is based on a comparison of (1) the number of errors we would make in attempting to guess the attributes of a given variable for each of the cases under study—if we knew nothing but the distribution of attributes on that variable—and

(2) the number of errors we would make if we knew the joint distribution overall and were told for each case the attribute of one variable each time we were asked to guess the attribute of the other. These measures include *lambda* (λ), which is appropriate for the analysis of two nominal variables; *gamma* (γ), which is appropriate for the analysis of two ordinal variables; and *Pearson's product-moment correlation* (r), which is appropriate for the analysis of two interval or ratio variables.

- *Regression analysis* represents the relationships between variables in the form of equations, which can be used to predict the values of a dependent variable on the basis of values of one or more independent variables.

- Regression equations are computed on the basis of a regression line: that geometric line representing, with the least amount of discrepancy, the actual location of points in a scattergram.

- Types of regression analysis include linear regression analysis, multiple regression analysis, partial regression analysis, and curvilinear regression analysis.

Inferential Statistics

- Inferential statistics are used to estimate the generalizability of findings arrived at through the analysis of a sample to the larger population from which the sample has been selected. Some inferential statistics estimate the single-variable characteristics of the population; others—tests of statistical significance—estimate the relationships between variables in the population.

- Inferences about some characteristic of a population must indicate a confidence interval and a confidence level. Computations of confidence levels and intervals are based on probability theory and assume that conventional probability-sampling techniques have been employed in the study.

- Inferences about the generalizability, to a population, of the associations discovered between variables in a sample involve tests of statistical significance, which estimate the likelihood that an association as large as the observed one could result from normal sampling error if no such association exists between the variables in the larger population. Tests of statistical significance are also based on probability theory and assume that conventional probability-sampling techniques have been employed in the study.

- The level of significance of an observed association is reported in the form of the probability that the association could have been produced merely by sampling error. To say that an association is significant at the .05 level is to say that an association as large as the observed one could not be expected to result from sampling error more than 5 times out of 100.

- Social researchers tend to use a particular set of levels of significance in connection with tests of statistical significance: .05, .01, and .001. This is merely a convention, however.

- A frequently used test of statistical significance in tabular data is *chi square* (χ^2).

- The *t-test* is a frequently used test of statistical significance for comparing means.

- Statistical significance must not be confused with substantive significance, the latter meaning that an observed association is strong, important, meaningful, or worth writing home to your mother about.

- Tests of statistical significance, strictly speaking, make assumptions about data and methods that are almost never satisfied completely by real social research. Despite this, the tests can serve a useful function in the analysis and interpretation of data.

Other Multivariate Techniques

- *Path analysis* is a method of presenting graphically the networks of causal relationships among several variables. It illustrates the primary "paths" of variables through which independent variables cause dependent ones. Path coefficients are standardized regression coefficients that represent the partial relationships between variables.

- *Time-series analysis* is an analysis of changes in a variable (such as *crime rates*) over time.

- *Factor analysis*, feasible only with a computer, is an analytic method of discovering the general dimensions represented by a collection of actual variables. These general dimensions, or factors, are calculated hypothetical dimensions that are not perfectly represented by any of the empirical variables under study but are highly associated with groups of empirical variables. A factor loading indicates the degree of association between a given empirical variable and a given factor.

- *Analysis of variance* (ANOVA) is based on comparing variations between and within groups and determining whether between-group differences could reasonably have occurred in simple random sampling or whether they likely represent a genuine relationship between the variables involved.

- *Discriminant analysis* seeks to account for variation in some dependent variable by finding a hypothetical, composite dimension that separates categories of the dependent variable. It results in an equation that scores people on the basis of that hypothetical dimension and allows us to predict their values on the dependent variable.

- *Log-linear models* offer a method for analyzing complex relationships among several nominal variables having more than two attributes each.

- *Geographic Information Systems* (GIS) map quantitative data that describe geographic units for a graphic display.

KEY TERMS

The following terms are defined in context in the chapter and at the bottom of the page where the term is introduced, as well as in the comprehensive glossary at the back of the book.

analysis of variance (ANOVA)

curvilinear regression analysis

descriptive statistics

discriminant analysis

factor analysis

Geographic Information Systems (GIS)

inferential statistics

level of significance

linear regression analysis

log-linear models

multiple regression analysis

nonsampling error

partial regression analysis

path analysis

proportionate reduction of error (PRE)

regression analysis

statistical significance

tests of statistical significance

time-series analysis

PROPOSING SOCIAL RESEARCH: QUANTITATIVE DATA ANALYSIS

Chapters 14, 15, and 16 all discuss different aspects of quantitative data analysis. In this exercise, you should outline your plans for analysis.

In earlier exercises, you specified the variables to be analyzed, including precisely how you'll measure those variables. Now you need to present how you'll conduct your analysis. Here's where you should say whether you're planning a tabular analysis, multiple regression, factor analysis, or something else. It doesn't really matter which computer program you're using (e.g., SPSS, SAS) unless it's a specialized program or one that is not commonly used.

If you've derived precise hypotheses, you may want to specify levels of statistical significance that will determine the meaning of the outcomes. This is not always necessary, however.

REVIEW QUESTIONS AND EXERCISES

1. In your own words, explain the logic of proportionate reduction of error (PRE) measures of associations.

2. In your own words, explain the purpose of regression analyses.

3. In your own words, distinguish between measures of association and tests of statistical significance.

4. Find a study that reports the statistical significance of its findings and critique the clarity with which it is reported.

5. Use InfoTrac College Edition to locate a study that uses factor analysis and summarize the findings.

SPSS EXERCISES

See the booklet that accompanies your text for exercises using SPSS (Statistical Package for the Social Sciences). There are exercises offered for each chapter, and you'll also find a detailed primer on using SPSS.

Online Study Resources

If your book came with an access code card, visit www.cengage.com/login to register. To purchase access, please visit www.ichapters.com.

1. Before you do your final review of the chapter, take the *CengageNOW* pretest to help identify the areas on which you should concentrate. You'll find information on this online tool, as well as

instructions on how to access all of its great resources, in the front of the book.

2. As you review, take advantage of the *CengageNOW* personalized study plan, based on your quiz results. Use this study plan with its interactive exercises and other resources to master the material.

3. When you're finished with your review, take the posttest to confirm that you're ready to move on to the next chapter.

WEBSITE FOR *THE PRACTICE OF SOCIAL RESEARCH* 12TH EDITION

Go to your book's website at www.cengage.com/sociology/babbie for tools to aid you in studying for your exams. You'll find *Tutorial Quizzes* with feedback, *Internet Exercises, Flash Cards, Glossaries,* and *Essay Quizzes,* as well as *InfoTrac College Edition* search terms, suggestions for additional reading, *Web Links,* and primers for using data-analysis software such as SPSS.

CHAPTER SEVENTEEN

Reading and Writing Social Research

CHAPTER OVERVIEW

Social research is useless unless
communicated effectively to others.
Special skills are involved in reading
the research of others and writing
about your own.

CengageNOW for Sociology

Use this online tool to help you make the grade on your next exam. After
reading this chapter, go to "Online Study Resources" at the end of the
chapter for instructions on how to benefit from *CengageNOW*.

Introduction

Meaningful scientific research is inextricably wed to communication, but it's not always an easy or comfortable marriage. Scientists—social and other—are not necessarily good at communicating their methods and findings. Thus, reading and understanding the research of others is often difficult, and you may also struggle to write up your own research in ways that communicate your ideas effectively. This final chapter addresses these two problems.

We'll begin with reading social research, then we'll turn to writing it. Although my guidance on both topics will be helpful, you'll find that doing each well lies in practice. The more you read social science research, the easier it gets, and the same is true of writing it. The ethical dimension of doing so also becomes clearer, as the final section of this chapter discusses.

Reading Social Research

Before you read social research, you need to decide which studies to read. With the exception of some grounded theory methodologists, most social researchers begin the design of a research project with a review of the literature, as indicated in Chapter 4. Most original research is seen as an extension of what has previously been learned about a particular topic. A review of the literature is the way we learn what's already known and not known.

Organizing a Review of the Literature

In most cases, you should organize your search of the literature around the key concepts you wish to study; alternatively, you may want to study a certain population: Iraqi War veterans, computer hackers, Catholic priests, gay athletes, and so forth. In any case, you'll identify a set of terms that represent your core interests.

Your college or university library will probably have several search routines you can use at the library or online. Let's say you're interested in designing a study of attitudes toward capital punishment. If your library provides access to InfoTrac College Edition or a similar program, you might discover, as I just did, 8,735 newspaper references and 5,489 periodical references to capital punishment. In such situations, InfoTrac College Edition is indexed to allow narrowing the search, and I soon discovered 249 entries for "public opinion" on capital punishment. Some of the entries were bibliographic citations and some were full-text articles I could read online.

Another resource available to everyone is the Library of Congress, easily accessed online (see the link on this book's website: http://www.cengage .com/sociology/babbie). Clicking on "Basic Search" or "Guided Search" will open up a vast resource for you. When I specified the keyword as "capital punishment" and limited the search to English-language books published between 2000 and 2005, the site listed 3,674 entries, such as the following:

- Abolition of the death penalty : SAHRDC's submission to the National Commission for the Review of the Working of the Constitution.
- America's experiment with capital punishment : reflections on the past, present, and future of the ultimate penal sanction / [edited by] James R. Acker.
- Beyond repair? : America's death penalty / edited by Stephen P. Garvey.
- Capital punishment : a bibliography / C. Cliff, editor.
- Death penalty : influences and outcomes / edited by Austin Sarat.

Sometimes a simple web search is a useful way to begin. Use a search engine such as Google, Hot-Bot, or Yahoo to look for web resources on "capital punishment" or "death penalty." Be sure to use quotation marks to look for a phrase rather than two separate words. You might also add "public opinion" to the request to narrow the field of possible resources. In general, online searches tend to turn up huge numbers of entries, most of which will not help you much. You'll need some time to separate the wheat from the chaff. Later in this chapter, I'll give you more-detailed guidelines for searching the web.

No matter how you start the literature review process, you should always consider a technique akin to snowball sampling, discussed in Chapter 7. Once you identify a particularly useful book or article, note which publications its author cites. Some of these will likely be useful. In fact, you'll probably discover some citations that appear again and again, suggesting that they're core references within the subject matter area you're exploring. This last point is important, because the literature review is not about providing "window dressing" in the form of a few citations. Rather, it's about digging into the body of knowledge that previous researchers have generated—and taking advantage of that knowledge as you design your own inquiry.

Once you've identified some potential resources, you must read them to find anything of value to your project. Here are some guidelines for reading research publications.

Reading Journals versus Books

As you might have guessed, you don't read a social research report the way you'd read a novel. You can, of course, but it's not the most effective approach. Journal articles and books are laid out somewhat differently, so here are some initial guidelines for reading each.

Reading a Journal Article

In most journals, each article begins with an abstract. Read it first. It should tell you the purpose of the research, the methods used, and the major findings.

In a good detective or spy novel, the suspense builds throughout the book and is resolved in some kind of surprise ending. This is not the effect most scholarly writers are going for. Social research is purposely anticlimactic. Rather than stringing the reader along, dragging out the suspense over whether X causes Y, social researchers willingly give away the punch line in the abstract.

The **abstract** serves two major functions. First, it gives you a good idea as to whether you'll want to read the rest of the article. If you're reviewing the literature for a paper you're writing, the abstract tells you whether that particular article is relevant. Second, the abstract establishes a framework within which to read the rest of the article. It may raise questions in your mind regarding method or conclusions, thereby creating an agenda to pursue in your reading. (It's not a bad idea to jot those questions down, to be sure you get answers to them.)

After you've read the abstract, you might go directly to the summary and/or conclusions at the end of the article. That will give you a more detailed picture of what the article is all about. (You can also do this with detective and spy novels; it makes reading them a lot faster but maybe not as much fun.) Jot down any new questions or observations that occur to you.

Next, skim the article, noting the section headings and any tables or graphs. You don't need to study any of these things in your skimming, though it's okay to dally with anything that catches your attention. By the end of this step, you should start feeling familiar with the article. You should be pretty clear on the researcher's conclusions and have a general idea of the methods used in reaching them.

When you now carefully read the whole article, you'll have a good idea of where it's heading and how each section fits into the logic of the whole article. Keep taking notes. Mark any passages you think you might like to quote later on.

After carefully reading the article, it's a good idea to skim it quickly one more time. This way you get back in touch with the forest after having focused on the trees.

If you want to fully grasp what you've just read, find someone else to explain it to. If you're doing

---

abstract A summary of a research article. The abstract usually begins the article and states the purpose of the research, the methods used, and the major findings.

the reading in connection with a course, you should have no trouble finding someone willing to listen. If you can explain it coherently to someone who has no prior contact with the subject matter, you'll have an absolute lock on the material.

Reading a Book

The approach for articles can be adapted to reading a book-length report, sometimes also called a **research monograph.** These longer research reports cover the same basic terrain and roughly the same structure. Instead of an abstract, the preface and opening chapter of the book lay out the purpose, method, and main findings of the study. The preface tends to be written more informally and to be easier to understand than an abstract.

As with an article, it's useful to skim through the book, getting a sense of its organization; its use of tables, graphs, and other visuals; and so forth. You should come away from this step feeling somewhat familiar with the book. And as I suggested in connection with reading an article, you should take notes as you go along, writing down things you observe and questions that are raised.

As you settle in to read the book more carefully, you should repeat this same process with each chapter. Read the opening paragraphs to get a sense of what's to come and then skip to the concluding paragraphs for the summary. Skim the chapter to increase your familiarity with it, and then read more deliberately, taking notes as you go.

It's sometimes okay to skip portions of a scholarly book, unlike the way you were taught to read and appreciate literature. This all depends on your purpose in reading it in the first place. Perhaps there are only a few portions of the book that are relevant to your purposes. However, realize that if you're interested in the researcher's findings, you must pay some attention to the methods used (for example, who was studied, how, when?) in order

to judge the quality of the conclusions offered by the author. See "Keeping Humanity in Focus" to learn about a book-length report you might want to practice on.

Evaluating Research Reports

In this section, I've provided sets of questions you might ask in reading and evaluating a research report. I've organized these questions to parallel some of the preceding chapters in this book, to facilitate your getting more details on a topic if necessary. Although hardly exhaustive, I hope these sets of questions will help you grasp the meanings of research reports you read and alert you to potential problems in them.

Theoretical Orientations

- Is there a theoretical aspect to the study, or does it lack any reference to theory?
- Can you identify the researcher's chief paradigm or theoretical orientation? Authors cited in the report's review of the literature and elsewhere may offer a clue.
- On the other hand, is the author attempting to refute some paradigm or theory?
- Is a theory or hypothesis being tested?
- In what way has the theoretical orientation shaped the methodology used in the study, such as the data-collection technique and the choice of which data were collected and which ignored?
- Is the methodology used appropriate to the theoretical issues involved?

Research Design

- What was the purpose of the study: exploration, description, explanation, or a combination?
- Who conducted the research? Who paid for it, if anyone? What motivated the study? If the study's conclusions happen to correspond to the interests of the sponsor or researcher, this doesn't disqualify the conclusions, but you'll want to be especially wary.

---

research monograph A book-length research report, either published or unpublished. This is distinguished from a textbook, a book of essays, a novel, and so forth.

Keeping Humanity in Focus

Probably everyone has ideas about life in juvenile gangs, mostly as portrayed in movies and on TV. But Sudhir Venkatesh was interested in finding things out for himself. When Venkatesh was a graduate student, his professor suggested that he head off to the South Side of Chicago and interview people who live there (recall the Elijah Anderson book discussed in Chapter 5). Before long, the young sociologist found himself being challenged by a group of gang members who demanded to know what gang he belonged to. They were not immediately impressed by learning that he was a sociologist, but eventually the gang leader took an interest in him and began trying to answer the survey questions. Finally, he informed Venkatesh that the only way to understand life in the streets was to "hang out," which led to a six-year research project.

While we tend to think of urban gangs as a threat to social order, Venkatesh found that they could also be seen as the *source* of social order in many impoverished neighborhoods. The gang leader needed to be able to manage the members of his gang but also a broader constituency, including prostitutes and pimps, thieves, corrupt police, and others.

Cover of *Gang Leader of a Day: A Rogue Sociologist Takes to the Streets* by Sudhir Venkatesh. Penguin Group USA, Inc. (2008). Reproduced with permission. Jacket design by Darren Haggar. Cover photo © Melanie Dunea/Creative Photographers.

- What was the unit of analysis? Was it appropriate to the purpose of the study? Are the conclusions drawn from the research appropriate to the unit of analysis? For example, have the researchers studied cities and ended up with assertions about individuals?

- Is this a cross-sectional or a longitudinal study? Be especially wary of longitudinal assertions being made on the basis of cross-sectional observations.

- If longitudinal data have been collected, have comparable measurements been made at each point in time? In the case of survey data, have the same questions been asked each time? If the report compares, say, crime or poverty rates, are they defined the same way each time? (Definitions of poverty, for example, change frequently.)

- If a panel study has been conducted, how many people dropped out over the course of the study?

Measurement

- What are the names of the concepts under study?

- Has the researcher delineated different dimensions of the variables? Do the analysis and reporting maintain those distinctions?

- What indicators—either qualitative or quantitative—have been chosen as measures of those dimensions and concepts? Is each indicator a valid measure of what it's intended to measure? What else could the indicator be a measure of? Is it a reliable measure? Has the reliability been tested?

- What is the level of measurement of each variable: nominal, ordinal, interval, or ratio? Is it the appropriate level?

- Have composite measurements (indexes, scales, or typologies) been used? If so, are they appropriate to the purpose of the study? Have they been constructed correctly?

Sampling

- Was it appropriate to study a sample, or should all elements have been studied? Remember, it's not always feasible to select a random sample.

- If sampling was called for, were probability-sampling methods appropriate, or would a purposive, snowball, or quota sample have been appropriate? Has the appropriate sample design been used?

- What population does the researcher want to draw conclusions about?

- What is the researcher's purpose? If it's statistical description, then rigorous probability sampling methods are called for.

- If a probability sample has been selected, what sampling frame has been used? Does it appropriately represent the population that interests the researcher? What elements of the population have been omitted from the sampling frame, and what extraneous elements have been included?

- What specific sampling techniques have been employed: simple random sampling, systematic sampling, or cluster sampling? Has the researcher stratified the sampling frame prior to sampling? Have the stratification variables been chosen wisely? That is, are they relevant to the variables under study?

- How large a sample was selected? What percentage of the sample responded? Are there any likely differences between those who responded and those who didn't?

- Even assuming that the respondents are representative of those selected in the sample, what sampling error do you expect from a sample of this size?

- Has the researcher tested for representativeness: comparing the gender distribution of the population and of respondents, for example, or their ages, ethnicity, education, or income?

- Ultimately, do the studied individuals (or other units of analysis) represent the larger population from which they were chosen? That is, do conclusions drawn about the sample tell us anything about meaningful populations or about life in general?

- If probability sampling and statistical representation were not appropriate for the study—in a qualitative study, for example—have subjects and observations been selected in such a way as to provide a broad overview of the phenomenon being examined? Has the researcher paid special attention to deviant or disconfirming cases?

Experiments

- What is the primary dependent variable in the experiment? What effect is the experimenter trying to achieve, for example?

- What is the experimental stimulus?

- What other variables are relevant to the experiment? Have they been measured?

- How has each variable been defined and measured? What potential problems of validity and reliability do these definitions and measurements raise?

- Has a proper control group been used? Have subjects been assigned to the experimental and control groups through random selection or by matching? Has it been done properly? Has the researcher provided any evidence of the initial comparability of experimental and control-group subjects?

- Have there been pre- and posttest measurements of the dependent variable?

- What is the chance of a placebo (or "Hawthorne") effect in the experiment? Has any attention been given to the problem? Does the study employ a double-blind design, for example?

- Are there any problems of internal validity: history, maturation, testing, instrumentation, statistical regression, selection bias, experimental mortality, ambiguous causal time order, diffusion or imitation of treatments, compensation, compensatory rivalry, or demoralization?

- Are there issues of external validity? How has the experimenter ensured that the laboratory findings will apply to life in the real world?

Survey Research

- Does the study stand up to all the relevant questions regarding sampling?

- What questions were asked of respondents? What was the precise wording of the questions? Be wary of researcher reports that provide only paraphrases of the questions.

- If closed-ended questions were asked, were the answer categories provided appropriate, exhaustive, and mutually exclusive?

- If open-ended questions were asked, how have the answers been categorized? Has the researcher guarded against his or her own bias creeping in during the coding of open-ended responses?

- Are all the questions clear and unambiguous? Could they have been misinterpreted by respondents? If so, could the answers given mean something other than what the researcher has assumed?

- Were the respondents capable of answering the questions asked? If not, they may have answered anyway, but their answers might not mean anything.

- Are any of the questions double-barreled? Look for conjunctions (such as *and, or*). Are respondents being asked to agree or disagree with two ideas, when they might like to agree with one and disagree with the other?

- Do the questions contain negative terms? If so, respondents may have misunderstood them and answered inappropriately.

- Is there a danger of social desirability in any of the questions? Is any answer so right or so wrong that respondents may have answered on the basis of what people would think of them?

- How would you yourself answer each item? As a general rule, test all questionnaire items by asking yourself how you would answer. Any difficulty you might have in answering might also apply to others. Then, try to assume different points of view (for example, liberal and conservative, religious and irreligious) and ask how the questions might sound to someone with each point of view.

- Has the researcher conducted a secondary analysis of previously collected data? If so, determine the quality of the research that produced the data originally. Also, are the data available for analysis appropriate to the current purposes? Do the questions originally asked reflect adequately on the variables now being analyzed?

The National Council on Public Polls has created a list of 20 questions to ask about polls. You'll find these questions in Appendix G of this book.

Field Research

- What theoretical paradigm has informed the researcher's approach to the study?

- Has the research set out to test hypotheses or generate theory from the observations? Or is there no concern for theory in the study?

- What are the main variables in this study? How have they been defined and measured? Do you see any problems of validity?

- How about reliability? Would another researcher, observing the same events, classify things the same way?

- Is there any chance that the classification of observations has been influenced by the way those classifications will affect the research findings and/or the researcher's hypotheses?

- If descriptive conclusions have been drawn— for example, "the group's standards were quite conservative"—what are the implicit standards being used?

- How much can the study's findings be generalized to a broader sector of society? What claims has the researcher made in this regard? What is the basis for such claims?

- If people have been interviewed, how were they selected? Do they represent all appropriate types?

- How much did the researcher participate in the events under study? How might that participation have affected the events themselves?

- Did the researcher reveal his or her identity as a researcher? If so, what influence could that revelation have had on the behavior of those being observed?

- Does the research indicate any personal feelings—positive or negative—about those being observed? If so, what effect might these feelings have had on the observations that were made and the conclusions that were drawn from them?

- How has the researcher's own cultural identity or background affected the interpretation of what has been observed?

Content Analysis

- What are the key variables in the analysis? Are they appropriate to the research question being asked?

- What is the source and form of data being analyzed? Are they appropriate to the research questions being asked?

- Is the time frame of the data being analyzed appropriate to the research question?

- What is the unit of analysis?

- If a quantitative analysis has been conducted: (1) has an appropriate sample been selected from the data source and (2) have the appropriate statistical techniques been used?

- If a qualitative analysis has been conducted, (1) has an appropriate range of data been examined and (2) are the researcher's conclusions logically consistent with the data presented?

Analyzing Existing Statistics

- Who originally collected the data being reanalyzed? Were there any flaws in the data-collection methods? What was the original purpose of the data collection? Would that have affected the data that were collected?

- What was the unit of analysis of the data? Is it appropriate to the current research question and the conclusions being drawn? Is there a danger of the ecological fallacy?

- When were the data collected? Are they still appropriate to present concerns?

- What are the variables being analyzed in the present research? Were the definitions used by

the original researchers appropriate to present interests?

Comparative and Historical Research

- Is this a descriptive or an explanatory study? Does it involve cross-sectional comparisons or changes over time?

- What is the unit of analysis in this study (for instance, country, social movement)?

- What are the key variables under study? If it's an explanatory analysis, what causal relationships are examined?

- Does the study involve the use of other research techniques, such as existing statistics, content analysis, surveys, or field research? Use the guidelines elsewhere in this section to assess those aspects of the study.

- Is the range of data appropriate to the analysis—for example, the units being compared or the number of observations made for the purpose of characterizing units?

- If historical or other documents are used as a data source, who produced them and for what purposes? What biases might be embedded in them? Diaries kept by members of the gentry, for example, will not reflect the life of peasants of the same time and country.

Evaluation Research

- What is the social intervention being analyzed? How has it been measured? Are there any problems of validity or reliability?

- Have the appropriate people (or other units of analysis) been observed?

- How has "success" been defined? Where would the success be manifested—in individuals, in organizations, in crime rates? Has it been measured appropriately?

- Has the researcher judged the intervention a success or a failure? Is the judgment well founded?

- Who paid for the research, and who actually conducted it? Can you be confident of the researcher's objectivity? Did the sponsor interfere in any way?

Data Analysis

- Did the purpose and design of the study call for a qualitative or a quantitative analysis?

- How have nonstandardized data been coded? This question applies to both qualitative and quantitative analysis. To what extent were the codes (1) based on prior theory or (2) generated by the data?

- Has the researcher undertaken all relevant analyses? Have all appropriate variables been identified and examined? Could the correlation observed between two variables have been caused by a third, antecedent variable, making the observed relationship spurious?

- Does a particular research finding really matter? Is an observed difference between subgroups, for example, a large or meaningful one? Are there any implications for action?

- Has the researcher gone beyond the actual findings in drawing conclusions and implications?

- Are there logical flaws in the analysis and interpretation of data?

- Have the empirical observations of the study revealed new patterns of relationships, providing the bases for grounded theories of social life? Has the researcher looked for disconfirming cases that would challenge the new theories?

- Are the statistical techniques used in the analysis of data appropriate to the levels of measurement of the variables involved?

- If tests of statistical significance were used, have they been interpreted correctly? Has statistical significance been confused with substantive significance?

Reporting

- Has the researcher placed this particular project in the context of previous research on the topic? Does this research add to, modify, replicate, or contradict previous studies?

- In general, has the researcher reported the details of the study design and execution fully? Are there parts of the report that seem particularly vague or incomplete in the reporting of details?

- Has the researcher reported any flaws or shortcomings in the study design or execution? Are there any suggestions for improving research on the topic in the future?

I hope this section will prove useful to you in reading and understanding social research. The exercises at the end of this chapter will walk you through the reading of two journal articles: one qualitative and one quantitative. As I said earlier, you'll find that your proficiency in reading social research reports will mature with practice.

Before discussing how to go about creating social research reports for others to read, let's look at how to read and evaluate data from an increasingly popular source of information—the Internet.

Using the Internet Wisely

In the closing decade of the twentieth century, the World Wide Web developed into a profoundly valuable tool for social research. As it expands exponentially, the web is becoming the mind of humanity, the repository of human knowledge, opinions, and beliefs—carrying with it intellectual insights, misconceptions, and outright bigotry. Clearly, it will continue to evolve as an ever more powerful entity. As with gunpowder and television, there are no guarantees that it will always be used wisely. As I write this, a substantial number of faculty still prohibit their students from using web materials. I've opted to encourage use of the web rather than opposing it, but I'm mindful of the problems that make many of my colleagues more cautious.

In this section of the chapter, I address the major problems inherent in using the web and suggest ways to avoid them.

Some Useful Websites

The website associated with this book has up-to-date links to useful social research websites. I've placed the materials on the web instead of in an appendix, so they can be revised and updated before the next textbook revision. I want to mention a few key links here and, more importantly, offer advice on how to search the web.

The first website I'll mention is the one created to support this textbook and mentioned at the end of each chapter. You should consider it as an extension of the book. It can be found at http://www.cengage.com/sociology/babbie. As you've seen, in addition to tutoring you on this book and coaching you in your research methods course, the website also provides numerous links that will take you to other useful resources to aid you in both learning and doing social research.

For now, let me mention just a few generally useful websites that you might like to check out; see the links on this book's website.

- General Social Survey (GSS)
- U.S. Bureau of the Census
- USA Statistics in Brief
- Statistical Resources on the Web, University of Michigan
- Social Sciences Virtual Library
- Yahoo Social Sciences
- QUALPAGE: Resources for Qualitative Research
- Computer Assisted Qualitative Data Analysis Software, University of Surrey, England

Now, let's assume you need some information that you suspect is somewhere on the web, but you don't know where. Here are some ideas about becoming a web detective.

Searching the Web

I won't estimate the number of pages of information on the World Wide Web; its growth curve is so dramatic that any number I might give now would be embarrassingly low by the time you read this. Let's just say there are millions and millions of pages. Estimating the number of "facts" or pieces of data on the web would be impossible, but most of the factual questions you might have can be answered on the web. Finding them involves skill, however.

Let's say you want to know who was the thirteenth president of the United States. That's easily learned in several ways. The most straightforward would be to open one of the many **search engines** available to you; let's say you use Google. When I searched for "thirteenth president," my responses began with those shown in Figure 17-1. (Realize that if you replicate this procedure, you may get somewhat different responses, because the content of the web is continuously evolving.)

The first response in the list gives us the answer: Millard Fillmore. In this case, it's not even necessary to follow up on any of the web links given, unless we want to know something more about him. Notice that we've gotten the same answer from five different websites—each adding to our confidence that we found the right answer. Notice also that some of the answers reflect the ambiguity of our request in not specifying president of "what."

Here's a more elaborate example. Let's say you want to examine differences in the infant mortality rates of countries around the world. You may already know some websites that are likely to have that information, but let's assume you don't.

Let's go back to Google or another search engine. Search for "infant mortality rate." If you put your request inside quotation marks, as I did, the search engine will look for that exact phrase instead of reporting websites that happen to have all three words. Figure 17-2 presents the initial results I received.

Notice that several of the web links are probably more specific than we want—one deals only with Cuba, another gives data on the United States. Often an effective web search requires more than one attempt. In this case, I added the word *world* to the request: "world infant mortality rate."

Like many other search engines, Google interpreted this as a request to find websites that contain the word *world* plus the exact phrase *infant*

search engine A computer program designed to locate where specified terms appear on websites throughout the World Wide Web.

Millard Fillmore - **Thirteenth President** of the United States
Biographical fast facts about Millard Fillmore, the **thirteenth President** of the
United States.
americanhistory.about.com/library/fastfacts/blffpres 13.htm - 28k - Cached - Similar pages

Millard Fillmore - **Thirteenth President** of the United States
Learn all about Millard Fillmore, the **Thirteenth president** of the United States.
americanhistory.about.com/od/millardfillmore/ - 16k - Cached - Similar pages

Amazon.com: Millard Fillmore: **Thirteenth President** of the United ...
Amazon.com: Millard Fillmore: **Thirteenth President** of the United States (Encyclopedia
of Presidents): Books by Jane Clark Casey.
www.amazon.com/exec/obidos/tg/detail/-/051601353X?v=glance - 53k - Cached - Similar pages

President Ronald R. Thomas. **Thirteenth President** of the University ...
The University of Puget Sound is honored to welcome Ronald R. Thomas as its
thirteenth president, and only its fifth since 1913. ...
www2.ups.edu/inauguration/inaugurationPresident.html - 11k - Cached - Similar pages

Wilson, Woodrow
Wilson, [Thomas] Woodrow (1856-1924), **thirteenth president** of Princeton, was born
December 29, 1856, in Staunton, Virginia, the son of Joseph Ruggles Wilson ...
etc.princeton.edu/CampusWWW/Companion/wilson_woodrow.html - 13k - Cached - Similar pages

The New ASU Story: Leadership
John W. Schwada was ASU's **thirteenth president**, serving from 1971 to 1981.
He presided over one of the largest growth periods in university history, ...
www.asu.edu/lib/archives/asustory/pages/18lead.htm - 4k - Sep 14, 2005 - Cached - Similar pages

Millard Fillmore, **Thirteenth President** Of The United States
reference - author, title, language for ISBN051601353X Millard Fillmore, **Thirteenth
President** Of The United States.
my.linkbaton.com/isbn/051601353X - 3k - Cached - Similar pages

Thirteenth President of the United States Millard Fillmore - Books ...
journal articles on: **Thirteenth President** of the United States Millard Fillmore
... Take Millard Fillmore, who gave his...money, our **thirteenth president** ...
www.questia.com/search/Thirteenth-President-of-the-United-States-Millard-Fillmore - 41k - Cached - Similar pages

The Library of Congress Shop > Presidents, First Ladies > Millard ...
Millard Fillmore, **Thirteenth President** of the United States, Millard Fillmore,
Thirteenth President of the United States Click on image to enlarge ...
www.loc.gov/shop/index.php?action=cCatalog.showItem&cid=33&scid=229&iid=1019 - 12k - Sep 14, 2005 - Cached - Similar pages

FIGURE 17-1

Finding the Thirteenth President

©2005 Google. Downloaded September 15, 2005, 12:25 P.M.

mortality rate. Figure 17-3 presents the first set of results.

The first web link is to the *World Factbook,* by the CIA, which draws on data from a variety of sources. The second and third sources in this extract are commercial data sources, and Wikipedia is a free encyclopedia compiled by the web community.

The rapid growth of Wikipedia has been a source of conversation and concern among academics. No one questions how extensive or user-friendly it is, but some worry that entries are not always accurate and errors may go unnoticed. Rarely, true mischief has been perpetrated, with opposing political candidates maliciously altering each other's entries in the encyclopedia, for example. In one recent response to academic concerns, the history department at Middlebury College (2007), one of the nation's most highly rated liberal arts colleges, told students they could not cite Wikipedia as a source in term papers and exams. Lest this be seen as a condemnation of Wikipedia, however, Middlebury clarified:

Map & Graph: Countries by Health: **Infant mortality rate**
Our **infant mortality rate** is driven by our high accidental death rate and ... But our relatively high **infant mortality rate** relative to per capita income is ...
www.nationmaster.com/graph-T/hea_inf_mor_rat - 99k - Sep 13, 2005 -
Cached - Similar pages

GeographyIQ - World Atlas - Rankings - **Infant mortality rate** (All ...
Worldwide **Infant mortality rate** (All Ascending) ranking information.
www.geographyiq.com/ranking/ranking_Infant_Mortality_Rate_aall.htm - 94k -
Cached - Similar pages

United Nations Statistics Division - Millennium Indicators
Indicator. 14. **Infant mortality rate** (UNICEF-WHO) ... MDG, 1230, **Infant mortality rate** (0-1 year) per 1000 live births (UNICEF estimates) · View data ...
millenniumindicators.un.org/unsd/mi/mi_indicator_xrxx.asp?ind_code=14 - 16k -
Cached - Similar pages

CIA - The World Factbook -- Rank Order - **Infant mortality rate**
Infant mortality rate (deaths/1000 live births). Date of Information. 1. Angola, 187.49, 2005 est. 2. Afghanistan, 163.07, 2005 est. ...
www.cia.gov/cia/publications/factbook/rankorder/2091rank.html - 92k -
Cached - Similar pages

State Rankings--Statistical Abstract of the United States--Infant ...
rankings of states for **infant mortality rate**. ... **INFANT MORTALITY RATE** -- 2001. [When states share the same rank, the next lower rank is omitted. ...
www.census.gov/statab/ranks/rank17.html - 15k - Sep 14, 2005 - Cached - Similar pages

Human Development Reports
Infant mortality rate (per 1000 live births) The probability of dying between birth and exactly one year of age, expressed per 1000 live births. ...
www.undp.org/hdr2003/indicator/indic_289.html - 83k - Cached - Similar pages

Health, Cuba Reports Record Low **Infant Mortality Rate**: Cuba News ...
Health, Cuba Reports Record Low **Infant Mortality Rate**: Cuba News, Cuba Travel, cultural, business news.. Cuba Travel eXPlorer.
www.cubaxp.com/modules/news/article-447.html - 45k - Cached - Similar pages

Infant mortality rate - deaths per 1000 live births - Flags, Maps ...
Infant mortality rate - deaths per 1000 live births - Flags, Maps, Economy, Geography, Climate, Natural Resources, Current Issues, International Agreements, ...
www.photius.com/wfb/1999/rankings/infant_mortality_0.html - 52k - Cached - Similar pages

FIGURE 17-2

Search for "Infant Mortality Rate"

(c)2005 Google. Downloaded September 15, 2005, 12:30 P.M.

CIA - The **World** Factbook -- Rank Order - **Infant mortality rate**
Top banner The **World** Factbook Banner ... **Infant mortality rate** (deaths/1000 live births).
Date of Information. 1. Angola, 187. 49, 2005 est. ...
www.cia.gov/cia/publications/factbook/rankorder/2091rank.html - 92k -
Cached - Similar pages

CIA - The **World** Factbook -- United States
Buoyed by victories in **World** Wars I and II and the end of the Cold War in 1991, ... **Infant mortality rate**:. Definition · Field Listing · Rank Order ...
www.cia.gov/cia/publications/factbook/geos/us.html - 101k - Sep 13, 2005-
Cached - Similar pages

Global Geografia - **World**, Demographic statistics: **Infant Mortality** ...
www.globalgeografia.com - Website about geography.
www.globalgeografia.com/world/infant_mortality_rate.htm - 6k - Cached - Similar pages

GeographyIQ - **World** Atlas - Rankings - **Infant mortality rate** (All ...
Worldwide **Infant mortality rate** (All Ascending) ranking information.
www.geographyiq.com/ranking/ranking_Infant_Mortality_Rate_aall.htm - 94k -
Cached - Similar pages

GeographyIQ - **World** Atlas - Rankings - **Infant mortality rate** ...
Worldwide Infant mortality rate (Bottom 25) ranking information.
www.geographyiq.com/ranking/ranking_Infant_Mortality_Rate_bottom25.htm - 28k -
Cached - Similar pages

Infant mortality - Wikipedia, the free encyclopedia
World infant mortality rate declined from 198 in 1960 to 83 in 2001. However, IMR remained higher in LDCs. In 2001, the **Infant Mortality Rate** for Less ...
en.wikipedia.org/wiki/Infant_mortality_rate - 20k - Cached - Similar pages

List of countries by **infant mortality rate** (2005) - Wikipedia, the ...
This is a list of countries by **infant mortality rate**, based on The **World** Factbook, 2005 estimates.[1]. The **Infant mortality rate** (IMR) is reported as number ...
en.wikipedia.org/wiki/List_of_countries_by_infant_mortality_rate_(2005) - 35k -
Cached - Similar pages

FIGURE 17-3
Search for "World Infant Mortality Rate"
©2005 Google. Downloaded September 15, 2005, 12:33 P.M.

While the department did vote to restrict the use of the online encyclopedia as a source in course work, it did not suggest, as some reports had it, that students should be prevented from accessing Wikipedia or should not use it as a research tool. In fact, the department praised Wikipedia as "extraordinarily convenient and, for some general purposes, extremely useful."

Realize that Figure 17-3 only presents the first few websites returned by the Google search. Google reported that it had found about 1,630,000 websites that seemed to have the information we were seeking.

Conducting this search on your own and visiting the web links that result is a useful exercise. (See "How to Do It: Using Google Scholar" for

How to Do It: Using Google Scholar

In searching the web for research materials, you can narrow your focus with Google Scholar (see the link on this book's website). Let's say you're interested in studying "same-sex marriage" and want to know what research has already been done on that topic. Enter that phrase in the box and click the "Search" button. A regular Google search would have turned up many websites that used the words *same-sex marriage* but were not much use in a research literature review. Google Scholar,

though, will provide you with richer pickings, although you'll still need to judge the quality of the documents turned up.

You can also take advantage of the "Advanced Scholar Search" to specify a set of words, indicating that all must appear in an article—or just some of them. You can specify a particular author or journal, and you can indicate which scholarly field you are interested in, thus limiting the search to articles in that field.

more on searches.) You'll find that some of the sites are discussions of the topic rather than tables of data. Others present a limited set of data ("selected countries"). Thus, compiling a list of web links like this is a step along the way to obtaining relevant data, but it's not the final step.

Evaluating the Quality of Internet Materials

There are other tricks to conducting effective web searches, but you now know enough to begin learning through experience. You'll quickly learn that finding data on the web is relatively easy. Evaluating what you've found is a bit more difficult, however. I've already alluded to the matter of quality, but there's quite a bit to be said on the topic. In fact, many other people have said many other things about it. What do you suppose is your best source of such advice? If you said, "The web," you got it.

Open up a search engine and ask it to find websites having to do with "evaluating websites" or "evaluating web sites." Figure 17-4 gives you some idea of the extent of advice available to you.

As you can tell from the ".edu" in the addresses of most of these sites, this is a topic of concern for colleges and universities. Although each of the various sites takes a different approach to the topic, the guidance they offer has some elements in common. You would do well to study one or more of the sites in depth. In the meantime, here's

an overview of the most common questions and suggestions for evaluating the data presented on websites.

1. *Who/what is the author of the website?* The two biggest risks you face in getting information from the web are (1) bias and (2) sloppiness. The democratic beauty of the web stems from its accessibility to such a large proportion of the population and from the lack of censorship. These pluses also present dangers, however, in that just about anyone can put just about anything on the web. The first thing you should note, therefore, is who the author of the website is: either an organization or an individual. In some cases, you may find SourceWatch a useful tool to help you judge the trustworthiness of web sources (see the link on this book's website). Sometimes, you'll find that a "research team" is actually a public relations firm or that an individual "expert" always seems to report findings in support of a particular company or industry.

2. *Is the site advocating a particular point of view?* Many of the sites on the World Wide Web have been created to support a particular political, religious, nationalistic, or social point of view. This fact does not necessarily mean that the data they present are false, though that's sometimes the case. Beyond outright lying, however, you can be relatively sure that the website will only present data supporting its particular point of view. You can usually tell whether a

Evaluating Web Pages: Techniques to Apply & Questions to Ask
Includes checklist form (PDF) that can be used to analyze web sites and pages.
www.lib.berkeley.edu/TeachingLib/Guides/Internet/Evaluate.html - 46k -
Cached - Similar pages

Kathy Schrock's Guide for Educators - Critical Evaluation Surveys ...
... ...a great site which looks at the different types of pages; **Evaluating Web Sites** ...a rubric
and ... **Evaluating Web Sites**: What Makes a Web Site Good? ...
school.discovery.com/schrockguide/eval.html - 42k - Cached - Similar pages

Evaluating Web Sites
The User Context: The most important factor when **evaluating Web sites** is your search,
your needs. What are you using the Web for? Entertainment? ...
www.library.cornell.edu/olinuris/ref/research/webeval.html - 11k - Sep 13, 2005 -
Cached - Similar pages

Five criteria for evaluating Web pages
Evaluation of Web documents, How to interpret the basics. 1. Accuracy of Web Documents.
Who wrote the page and can you contact him or her? ...
www.library.cornell.edu/okuref/research/webcrit.html - 7k - Cached - Similar pages

Evaluating Web Sites
Lesley is a multi-site University with more than 150 locations throughout the continental
United States.
www.lesley.edu/library/guides/research/evaluating_web.html - 25k - Sep 13, 2005 -
Cached - Similar pages

Evaluation Criteria from "The Good, The Bad & The Ugly: or, Why ...
A easy to use guide for web evaluation. Lists evaluation criteria with links to actual pages that
illustrate each point. The Examples page can be used by ...
lib.nmsu.edu/instruction/evalcrit.html - 10k - Cached - Similar pages

Evaluating Web Sites for Educational Uses
This site contains a list of articles from librarians and other information specialists on Web
evaluations. In addition, a checklist for evaluating a Web ...
www.unc.edu/cit/guides/irg-49.html - 14k - Cached - Similar pages

Evaluating Web Sites for Accessibility
Goals for **evaluating Web sites** vary, and require different approaches to meet those goals:.
Preliminary review can:. identify general kinds of barries on a ...
www.w3.org/WAI/eval/Overview.html - 31k - Cached - Similar pages

FIGURE 17-4

Search for "Evaluating Web Sites"

website is reasonably objective or has an ax to grind, and you should be wary of those that go overboard to convince you of something.

3. *Does the website give accurate and complete references?* When data are presented, can you tell where they came from—how they were created? If the website is reporting data collected by someone else, does it give you enough information to locate the original researchers? Or, if the website authors themselves compiled the data, do the authors provide you with sufficiently detailed descriptions of their research methods? If they present the data without such clarifications, you should leave the information there and move on.

4. *Are the data up-to-date?* Another common problem on the web is that materials may be posted and forgotten. Hence, you may find data reporting crime rates, chronicles of peace negotiations, and so forth that are out-of-date. Be sure that the data you obtain are timely for your purposes.

5. *Are the data official?* It's often a good idea to find data at official, government research sites, such as the U.S. Bureau of the Census, the Bureau of Labor Statistics, the National Center for Health Statistics, and others. FedStats is a good launching point for finding data among some 100 federal research agencies. (See the links for all these agencies on this book's website.) As we saw in Chapter 11, data presented by official agencies are not necessarily "The Truth," but they are grounded in a commitment to objectivity and have checks and balances to support them in achieving that goal.

6. *Is it a university research site?* Like government research agencies, university research centers and institutes are usually safe resources, committed to conducting professional research and having checks and balances (such as peer review) to support their achieving that. Throughout

this book, I've mentioned the General Social Survey (GSS), conducted regularly by the National Opinion Research Center. You can trust the data presented there, using them with confidence in their legitimacy and knowing that your instructor will not question your use of that resource.

7. *Do the data seem consistent with data from other sites?* Verify (cross-check) data wherever possible. We've already seen that a web search is likely to turn up more than one possible source of data. Take the time to compare what they present. If several websites present essentially the same data, you can use any of those sources with confidence.

As with so many things, your effective use of the web will improve with practice. Moreover, the web itself will be evolving alongside your use of it.

Citing Internet Materials

If you use materials from the web, you must provide a bibliographic citation that allows your reader to locate the original materials—to see them in context. This also protects you from the serious problem of plagiarism, discussed a little later in this chapter.

There are many standardized formats for bibliographic citations, illustrated in the "How to Do It" box presented later on. Web materials, unfortunately, don't fit any of those familiar formats.

Many organizations, however, have risen to the challenge of web citations. If you don't believe me, go to your favorite search engine and look for "web citations." You'll find plenty of guidance.

Your instructor may prefer a specific format for web citations. However, here are the elements commonly suggested for inclusion:

● The **URL** or web address. For example, http://www.fedstats.gov/qf/states/50000.html provides demographic data for comparing Vermont with the United States as a whole. So if I tell you that Vermont grew 8.2 percent during the 1990s, you can go directly to the source of my data.

URL Web address, typically beginning with "http://"; stands for "uniform resource locator" or "universal resource locator."

- The date and time when the site was accessed. Many, like the one just cited, do not change, but many others do. It may be useful for the reader to know when you visited the site in question.

- If you're citing text materials, there may very well be an author and title, as well as publishing information. These should be cited the same way you would cite printed materials: for example, John Doe. 2003. "How I Learned to Love the Web." *Journal of Web Worship* 5 (3): 22–45.

- Sometimes, you'll use the web to read a published journal article, locating it with InfoTrac College Edition or another vehicle. Such materials may be presented in a print format, with page numbers. If so, cite the appropriate page number. Lacking that, you may be able to cite the section where the materials in question appeared. The goal in all this is to help your reader locate the original web materials you've used. Although you sometimes cannot give a precise location in an article posted to a website, most browsers allow users to search the site for a specified word or phrase and thus locate the materials being cited.

Writing Social Research

Unless research is properly communicated, all the efforts devoted to the various procedures discussed throughout this book will go for naught. This means, first and foremost, that good social reporting requires good English (or Spanish or whatever language you use). Whenever we ask the figures "to speak for themselves," they tend to remain mute. Whenever we use unduly complex terminology or construction, communication suffers.

My first advice to you is to read and reread (at approximately three-month intervals) an excellent small book by William Strunk, Jr., and E. B. White, *The Elements of Style* (2000; see also Birchfield 1998). If you do this faithfully, and if even 10 percent of the contents rub off, you stand a good chance of making yourself understood and your findings appreciated.

Next, you need to understand that scientific reporting has several functions. First, your report should communicate a body of specific data and ideas. You should provide those specifics clearly and with sufficient detail to permit an informed evaluation by others. Second, you should view your report as a contribution to the general body of scientific knowledge. While remaining appropriately humble, you should always regard your research report as an addition to what we know about social behavior. Finally, the report should stimulate and direct further inquiry.

Some Basic Considerations

Despite these general guidelines, different reports serve different purposes. A report appropriate for one purpose might be wholly inappropriate for another. This section deals with some of the basic considerations in this regard.

Audience

Before drafting your report, ask yourself who you hope will read it. Normally you should make a distinction between scientists and general readers. If the report is written for the former, you can make certain assumptions about their existing knowledge and therefore summarize certain points rather than explain them in detail. Similarly, you can use more technical language than would be appropriate for a general audience.

At the same time, remain aware that any science has its factions and cults. Terms, assumptions, and special techniques familiar to your immediate colleagues may only confuse other scientists. The sociologist of religion writing for a general sociology audience, for example, should explain previous findings in more detail than he or she would if addressing an audience of sociologists of religion.

Form and Length of Report

My comments here apply to both written and oral reports. Each form, however, affects the nature of the report.

It's useful to think about the variety of reports that might result from a research project. To begin, you may wish to prepare a short research note for publication in an academic or technical journal. Such reports are approximately one to five pages long (typed, double-spaced) and should be concise and direct. In a small amount of space, you can't present the state of the field in any detail, so your methodological notes must be abbreviated. Basically, you should tell the reader why you feel your findings justify a brief note, then tell what those findings are.

Often researchers must prepare reports for the sponsors of their research. These reports can vary greatly in length. In preparing such a report, you should bear in mind its audience—scientific or lay—and their reasons for sponsoring the project in the first place. It is both bad politics and bad manners to bore the sponsors with research findings that have no interest or value to them. At the same time, it may be useful to summarize how the research has advanced basic scientific knowledge (if it has).

Working papers are another form of research reporting. In a large and complex project especially, you'll find comments on your analysis and the interpretation of your data useful. A working paper constitutes a tentative presentation with an implicit request for comments. Working papers can also vary in length, and they may present all of the research findings of the project or only a portion of them. Because your professional reputation is not at stake in a working paper, feel free to present tentative interpretations that you can't altogether justify—identifying them as such and asking for evaluations.

Many research projects result in professional papers, or those delivered at professional meetings. Often, these serve the same purpose as working papers. You can present findings and ideas of possible interest to your colleagues and ask for their comments. Although the length of professional papers varies, depending on the organization of the meetings, it's best to say too little rather than too much. Although a working paper may ramble somewhat through tentative conclusions, conference participants should not be forced to sit through an oral unveiling of the same. Interested listeners can always ask for more details later, and uninterested ones can gratefully escape.

Probably the most popular research report is the article published in an academic journal. Again, lengths vary, and you should examine the lengths of articles previously published by the journal in question. As a rough guide, however, 25 typed, double-spaced pages is a good length. Student term papers should follow this model as well. As a general rule, a term paper that would make a good journal article also makes a good term paper. (You'll learn more on the structure of the journal article, on which the organization of a report is primarily based, a bit later on in this chapter.)

A book, of course, represents the most prestigious form of research report. It has the length and detail of a working paper but is more polished. Because publishing research findings as a book lends them greater substance and worth, you have a special obligation to your audience. Although some colleagues may provide comments, possibly leading you to revise your ideas, other readers may be led to accept your findings uncritically.

Aim of the Report

Earlier in this book, we considered the different purposes of social research projects. In preparing your report, keep these different purposes in mind.

Some reports focus primarily on the exploration of a topic. As such, their conclusions are tentative and incomplete. If you're writing this sort of report, clearly indicate to your audience the exploratory aim of the study and present the shortcomings of the particular project. An exploratory report points the way to more-refined research on the topic.

Most research reports have a descriptive element reflecting the descriptive purpose of the studies they document. In yours, carefully distinguish those descriptions that apply only to the sample and those that apply to the population. Give your audience some indication of the probable range of error in any inferential descriptions you make.

Many reports have an explanatory aim: pointing to causal relationships among variables. Depending on your probable audience, carefully delineate the rules of explanation that lie behind your computations and conclusions. Also, as in the case of description, give your readers some guide to the relative certainty of your conclusions.

If your intention is to test a hypothesis based in theory, you should make that hypothesis clear and succinct. Specify what will constitute acceptance or rejection of the hypothesis and how either of those reflects on the theoretical underpinnings.

Finally, some research reports propose action. For example, if you've studied prejudice, you may suggest in your report how prejudice can be reduced on the basis of your research findings. This suggestion may become a knotty problem for you, however, because your values and orientations may have interfered with your proposals. Although it's perfectly legitimate for such proposals to be motivated by personal values, you must ensure that the specific actions proposed are warranted by the data. Thus, you should be especially careful to spell out the logic by which you move from empirical data to proposed action.

Organization of the Report

Although the various forms and purposes of reports somewhat affect the way they are organized, knowing a general format for presenting research data can be helpful. The following comments apply most directly to journal articles, but with some modification they apply to most forms of research reports as well.

Purpose and Overview

It's always helpful if you begin with a brief statement of the purpose of the study and the main findings of the analysis. In a journal article, as we've seen, this overview sometimes takes the form of an abstract.

Some researchers find this difficult to do. For example, your analysis may have involved considerable detective work, with important findings revealing themselves only as a result of imaginative deduction and data manipulation. You may wish, therefore, to lead the reader through the same exciting process, chronicling the discovery process with a degree of suspense and surprise. To the extent that this form of reporting gives an accurate

picture of the research process, it has considerable instructional value. Nevertheless, many readers may not be interested in following your entire research account, and not knowing the purpose and general conclusions in advance may make it difficult for them to understand the significance of the study.

An old forensic dictum says, "Tell them what you're going to tell them; tell them; and tell them what you told them." You would do well to follow this dictum.

Review of the Literature

Next, you must indicate where your report fits into the context of the general body of scientific knowledge. After presenting the general purpose of your study, you should bring the reader up-to-date on the previous research in the area, pointing to general agreements and disagreements among the previous researchers. Your review of the literature should lay the groundwork for your own study, showing why your research may have value in the larger scheme of things.

In some cases, you may wish to challenge previously accepted ideas. Carefully review the studies that have led to the acceptance of those ideas, then indicate the factors that have not been previously considered or the logical fallacies present in the previous research.

When you're concerned with resolving a disagreement among previous researchers, you should summarize the research supporting one view, then summarize the research supporting the other, and then suggest the reasons for the disagreement.

Your review of the literature serves a bibliographic function for readers, by indexing the previous research on a given topic. This can be overdone, however, and you should avoid an opening paragraph that mentions every previous study in the field and goes on for three pages. The comprehensive bibliographic function can best be served by a bibliography at the end of the report, and the review of the literature should focus only on those studies that have direct relevance to the present one. See "How to Do It: Citing Bibliographic Sources" to learn how to do this in good form.

How to Do It: Citing Bibliographic Sources

It's important to cite the bibliographic sources that comprise your review of the literature and other readings that figure in your paper. It's just about as important to cite them properly. The good news is that proper citation isn't that hard to do. The bad news is that several formats are in common use. I'll illustrate a few of the most common formats here, but you should ask your instructor what version to use. I'll illustrate both a book and an article. Here is the pertinent information for each:

Book Information

Author: C. Wright Mills
Title: *The Power Elite*
City of publication: New York
Publisher: Oxford University Press
Year of publication: 1956

Article Information

Authors: Sharon Sassler and Anna Cunningham
Title: *How Cohabitors View Childbearing*
Journal name: *Sociological Perspectives*
Year of publication: 2008
Month/season of publication: Spring
Volume: 51
Number: 1
Pages: 3–28

With such "raw data" in hand, you can format them by following any of the following bibliographic styles.

ASA Style Guide (American Sociological Association)

Mills, C. Wright. 1956. *The Power Elite*. New York: Oxford University Press.
Sassler, Sharon and Anna Cunningham. 2008. "How Cohabitors View Childbearing." *Sociological Perspectives* 51:3–28.

MLA Style Guide (Modern Language Association)

Mills, C. Wright. *The Power Elite*. New York: Oxford University Press, 1956.
Sassler, Sharon, and Anna Cunningham. "How Cohabitors View Childbearing." *Sociological Perspectives* 51.1 (2008): 3–28.

APSA Style Guide (American Political Science Association)

Mills, C. Wright. 1956. *The Power Elite*. New York: Oxford University Press.
Sassler, Sharon, and Anna Cunningham. 2008. "How Cohabitors View Childbearing." *Sociological Perspectives* 51(Spring): 3–28.

APA Style Guide (American Psychological Association)

Mills, C. Wright. (1956). *The power elite*. New York: Oxford University Press.
Sassler, S., & Cunningham, A. (2008). How cohabitors view childbearing. *Sociological Perspectives, 51*(1), 3–28.

Avoiding Plagiarism

Whenever you're reporting on the work of others, you must be clear about who said what. That is, you must avoid **plagiarism:** the theft of another's words and/or ideas—whether intentional or accidental—through the presentation of those words and ideas as your own. Because this is a common and sometimes unclear problem for college students, especially in regard to the review of the literature, we'll consider the issue here. Realize, of course, that these concerns apply to everything you write.

Here are the ground rules regarding plagiarism:

- You cannot use another writer's exact words without using quotation marks and giving a complete citation that indicates the source of the quotation such that your reader could locate the quotation in its original context. As a general rule, taking a passage of eight or more words without citation is a violation of federal copyright laws.

- It's also not acceptable to edit or paraphrase another's words and present the revised version as your own work.

- Finally, it's not even acceptable to present another's ideas as your own—even if you use totally different words to express those ideas.

The following examples should clarify what is or is not acceptable in the use of another's work.

plagiarism Presenting someone else's words or thoughts as though they were your own, constituting intellectual theft.

The Original Work

Laws of Growth

Systems are like babies: once you get one, you have it. They don't go away. On the contrary, they display the most remarkable persistence. They not only persist; they grow. And as they grow, they encroach. The growth potential of systems was explored in a tentative, preliminary way by Parkinson, who concluded that administrative systems maintain an average growth of 5 to 6 percent per annum regardless of the work to be done. Parkinson was right so far as he goes, and we must give him full honors for initiating the serious study of this important topic. But what Parkinson failed to perceive, we now enunciate—the general systems analogue of Parkinson's Law.

The System Itself Tends to Grow at 5 to 6 Percent per Annum

Again, this Law is but the preliminary to the most general possible formulation, the Big-Bang Theorem of Systems Cosmology.

Systems Tend to Expand to Fill the Known Universe

(Gall 1975: 12–14)

Now let's look at some of the acceptable ways you might make use of Gall's work in a term paper.

- *Acceptable:* John Gall, in his work *Systematics,* draws a humorous parallel between systems and infants: "Systems are like babies: once you get one, you have it. They don't go away. On the contrary, they display the most remarkable persistence. They not only persist; they grow."*
- *Acceptable:* John Gall warns that systems are like babies. Create a system and it sticks around. Worse yet, Gall notes, systems keep growing larger and larger.†
- *Acceptable:* It has also been suggested that systems have a natural tendency to persist, even grow and encroach (Gall 1975: 12).

Note that the last format requires that you give a complete citation in your bibliography, as I do in this book. Complete footnotes or endnotes work as well. See the publication manuals of various organizations such as the APA or the ASA, as well as the *Chicago Manual of Style,* for appropriate citation formats.

Here now are some unacceptable uses of the same material, reflecting some common errors.

- *Unacceptable:* In this paper, I want to look at some of the characteristics of the social systems we create in our organizations. First, systems are like babies: Once you get one, you have it. They don't go away. On the contrary, they display the most remarkable persistence. They not only persist; they grow. [It's unacceptable to quote someone else's materials directly without using quotation marks and giving a full citation.]
- *Unacceptable:* In this paper, I want to look at some of the characteristics of the social systems we create in our organizations. First, systems are a lot like children: once you get one, it's yours. They don't go away; they persist. They not only persist, in fact: They grow. [It's unacceptable to edit another's work and present it as your own.]
- *Unacceptable:* In this paper, I want to look at some of the characteristics of the social systems we create in our organizations. One thing I've noticed is that once you create a system, it never seems to go away. Just the opposite, in fact: They have a tendency to grow. You might say systems are a lot like children in that respect. [It's unacceptable to paraphrase someone else's ideas and present them as your own.]

Each of the preceding unacceptable passages is an example of plagiarism and represents a serious offense. Admittedly, there are gray areas. Some ideas are more or less in the public domain, not belonging to any one person. Or you may reach an idea on your own that someone else has already

---

*John Gall, *Systematics: How Systems Work and Especially How They Fail* (New York: Quadrangle, 1975), 12.
†John Gall, *Systematics: How Systems Work and Especially How They Fail* (New York: Quadrangle, 1975), 12.

put in writing. If you have a question about a specific situation, discuss it with your instructor in advance.

I've discussed this topic in some detail because, although you must place your research in the context of what others have done and said, the improper use of their materials is a serious offense. Learning to avoid plagiarism is a part of your "coming of age" as a scholar.

Study Design and Execution

A research report containing interesting findings and conclusions will frustrate readers if they can't determine the methodological design and execution of the study. The worth of all scientific findings depends heavily on the manner in which the data were collected and analyzed.

In reporting the design and execution of a survey, for example, always include the following: the population, the sampling frame, the sampling method, the sample size, the data-collection method, the completion rate, and the methods of data processing and analysis. Comparable details should be given if other methods are used. The experienced researcher can report these details in a rather short space, without omitting anything required for the reader's evaluation of the study.

Analysis and Interpretation

Having set the study in the perspective of previous research and having described the design and execution of it, you should then present your data. This chapter momentarily will provide further guidelines in this regard. For now, a few general comments are in order.

The presentation of data, the manipulation of those data, and your interpretations should be integrated into a logical whole. It frustrates the reader to discover a collection of seemingly unrelated analyses and findings with a promise that all the loose ends will be tied together later in the report. Every step in the analysis should make sense at the time it's taken. You should present your rationale for a particular analysis, present the data relevant to it, interpret the results, and then indicate where that result leads next.

Summary and Conclusions

According to the forensic dictum mentioned earlier, summarizing the research report is essential. Avoid reviewing every specific finding, but review all the significant ones, pointing once more to their general significance.

The report should conclude with a statement of what you've discovered about your subject matter and where future research might be directed. Many journal articles end with the statement, "It is clear that much more research is needed." This conclusion is probably always true, but it has little value unless you can offer pertinent suggestions about the nature of that future research. You should review the particular shortcomings of your own study and suggest ways those shortcomings might be avoided.

Guidelines for Reporting Analyses

The presentation of data analyses should provide a maximum of detail without being cluttered. You can accomplish this best by continually examining your report to see whether it achieves the following aims.

If you're using quantitative data, present them so the reader can recompute them. In the case of percentage tables, for example, the reader should be able to collapse categories and recompute the percentages. Readers should receive sufficient information to permit them to compute percentages in the table in the direction opposite from that of your own presentation.

Describe all aspects of a quantitative analysis in sufficient detail to permit a secondary analyst to replicate the analysis from the same body of data. This means that he or she should be able to create the same indexes and scales, produce the same tables, arrive at the same regression equations, obtain the same factors and factor loadings, and so forth. This will seldom be done, of course, but if the report allows for it, the reader will be far better equipped to evaluate the report than if it does not.

Provide details. If you're doing a qualitative analysis, you must provide enough detail that your reader has a sense of having made the observations with you. Presenting only those data that support your interpretations is not sufficient; you must also share those data that conflict with the way you've made sense of things. Ultimately, you should provide enough information that the reader might reach a different conclusion than you did—though you can hope your interpretation will make the most sense. The reader, in fact, should be in position to replicate the entire study independently, whether it involves participant observation among heavy metal groupies, an experiment regarding jury deliberation, or any other study format. Recall that replicability is an essential norm of science. A single study does not prove a point; only a series of studies can begin to do so. And unless studies can be replicated, there can be no meaningful series of studies.

Integrate supporting materials. I've previously mentioned the importance of integrating data and interpretations in the report. Here is a more specific guideline for doing this. Tables, charts, and figures, if any, should be integrated into the text of the report—appearing near that portion of the text discussing them. Sometimes students describe their analyses in the body of the report and place all the tables in an appendix. This procedure greatly impedes the reader, however. As a general rule, it's best to (1) describe the purpose for presenting the table, (2) present it, and (3) review and interpret it.

Draw explicit conclusions. Although research is typically conducted for the purpose of drawing general conclusions, you should carefully note the specific basis for such conclusions. Otherwise you may lead your reader into accepting unwarranted conclusions.

Point to any qualifications or conditions that would help the reader evaluate your conclusions accurately. Typically, you know best the shortcomings and tentativeness of your conclusions, and you should give the reader the advantage of that knowledge. Failure to do so can misdirect future research and result in a waste of research funds.

As I said at the outset of this discussion, research reports should be written in the best possible literary style. Writing lucidly is easier for some people than for others, and it's always harder than writing poorly. I again refer you to the Strunk

and White book. Every researcher would do well to follow this procedure: Write. Read Strunk and White. Revise. Reread Strunk and White. Revise again. This will be a difficult and time-consuming endeavor, but so is science.

A perfectly designed, carefully executed, and brilliantly analyzed study will be altogether worthless unless you can communicate your findings to others. This section has attempted to provide some guidelines toward that end. The best guides are logic, clarity, and honesty. Ultimately, there is no substitute for practice.

Going Public

Though I have written this chapter with a particular concern for the research projects you may be called on to undertake in your research methods course, you should realize that graduate and even undergraduate students are increasingly presenting the results of their research as professional papers or published articles.

If you would like to explore these possibilities further, you may find state and regional associations more open to students than national associations are, although students may present papers to the American Sociological Association, for example. Some associations have special sessions and programs for student participants. You can learn more about these possibilities by visiting the associations' websites to learn of upcoming meetings and the topics for which papers are being solicited.

Typically, you'll submit your paper to someone who has agreed to organize a session with three to five papers on a particular topic. The organizer chooses which of the submissions will be accepted for presentation. Oral presentations at scholarly meetings are typically 15–20 minutes long, with the possibility of questions from the audience. Some presenters read a printed paper, whereas others speak from notes. Increasingly, presenters use computer slide shows, though such presentations are still in the minority.

To publish an article in a scholarly journal, you would do well to identify a journal that publishes articles on the topic of your research. Again, the journals published by state or regional associations may be more accessible to student authors. Each journal will contain instructions for submitting articles, including instructions for formatting your article. Typically, articles submitted to a journal are circulated among three or so anonymous reviewers, who make comments and recommendations to the journal's editor. This is referred to as the "peer review" process. Sometimes manuscripts are accepted pretty much as submitted, some are returned for revision and resubmission, and still others are rejected. The whole process from submission to a decision to publish or reject may take a few months, and there will be a further delay before the article is actually published.

To meet the costs of publication, a journal will sometimes require that authors pay a small fee on acceptance. Typically, authors receive extra copies of their article—called "reprints"—to give to friends and family and to satisfy requests from professional colleagues.

The Ethics of Reading and Writing Social Research

I've already commented on some ethical issues involved in writing research reports. However, there are also some ethical issues at play in terms of reading the research literature. In reviewing the literature, you might gravitate toward reports that support a point of view you may be fond of. Further, the power of the Internet for fast and expansive searches, while wonderful in most respects, can allow even more cherry-picking of supportive research literature. This places an ever-greater burden on researchers to exercise professional honesty in representing the history of research findings in a particular area.

Since this chapter concludes the main body of the book, I hope this final section makes clear that research ethics constitute a fundamental component of social science and not merely a nice thing to consider as long as it doesn't get in the way. Research ethics has not always been recognized in this fashion. When I first began writing textbooks there was some objection to including this topic.

It wasn't so much that researchers objected to the ethical treatment of subjects—it simply wasn't considered a proper topic for a book like this one. Attitudes have changed substantially over the years, however. I hope the discussions of ethics will help you as you practice social research, in whatever form, throughout your life.

This chapter, and indeed this book, has provided what I hope will be a springboard for you to engage in and enjoy the practice of social research. The next time you find yourself pondering the cause of prejudice, or observing a political rally, or just plain curious about the latest trends in television, I trust you'll have the tools to explore your world with a social scientific eye.

MAIN POINTS

Introduction
- Meaningful scientific research is inextricably wed to communication; knowing how to read and write such research requires practice.

Reading Social Research
- Social researchers can access many resources, including the library and the Internet, for organizing a review of the literature.
- Reading scholarly literature is different from reading other works, such as novels.
- In reading scholarly literature, you should begin by reading the abstract, skimming the piece, and reading the conclusion to get a good sense of what it's about.
- Readers of social science literature should form questions and take notes as they go along.
- The key elements to note in reading a research report include theoretical orientation, research design, measurement methods, sampling (if any), and other considerations specific to the several data-collection methods discussed in this book.

Using the Internet Wisely
- The Internet is a powerful tool for social researchers, but it also carries risks.
- Not everything you read on the web is necessarily true.

- Original sources of data are preferred over those that take data from elsewhere.
- In evaluating a web source, you should ask the following:

  Who or what is the author of the website?

  Is the site advocating a particular point of view?

  Does the site give accurate and complete references?

  Are the data up-to-date?

- Official agencies usually serve as a good source of data, although the data are subject to error.
- The reader of a report should verify (cross-check) data wherever possible.
- Web citations, like other bibliographic references, should be complete—allowing the reader to locate and review the materials cited.

Writing Social Research

- Good social research writing begins with good writing, which means, among other things, writing to communicate rather than to impress.
- Being mindful of your audience and purpose in writing the report is important.
- Avoiding plagiarism—that is, presenting someone else's words or thoughts as though they were your own—is essential. Whenever using someone else's exact words, you must be sure to use quotation marks or some other indication that you're quoting. In paraphrasing someone else's words or ideas, you must provide a full bibliographic citation of the source.
- The research report should include an account of the study design and execution.
- The analysis of a report should be clear at each step, and its conclusion should be specific but not overly detailed.
- To write good reports, researchers need to provide details, integrate supporting materials, and draw explicit conclusions.
- Increasingly, students are presenting papers at professional meetings and publishing articles in scholarly journals.

The Ethics of Reading and Writing Social Research

- A review of the literature should not be biased toward supporting a particular point of view.
- Research ethics is a fundamental component of social science, not a nice afterthought.

KEY TERMS

The following terms are defined in context in the chapter and at the bottom of the page where the term is introduced, as well as in the comprehensive glossary at the back of the book.

abstract   search engine

plagiarism   URL

research monograph

PROPOSING SOCIAL RESEARCH: PUTTING THE PROPOSAL TOGETHER

If you've been doing the Proposing Social Research exercises all through the book, you should have just about everything you need now to create the finished product. This chapter has given you some additional guidance on reviewing the literature—both printed and online—and on writing social research, so you can review what you've written already and tidy it up. Appendix A will provide further guidelines on the presentation of your bibliography.

Now you're ready to assemble the parts into a coherent whole. Here is the outline we discussed in Chapter 1.

> Introduction (Chapter 1)
>
> Review of the Literature (Chapters 2, 17; Appendix A)
>
> Specifying the Problem/Question/Topic (Chapters 5, 6, 12)
>
> Research Design (Chapter 4)
>> Data-Collection Method (Chapters 4, 8, 9, 10, 11)
>>
>> Selection of Subjects (Chapter 7)
>>
>> Ethical Issues (Chapter 3)
>
> Data Analysis (Chapters 13, 14, 15, 16)
>
> Bibliography (Chapter 17; Appendix A)

Perhaps you'll be able to present this proposal as evidence that you've mastered the materials of the textbook. Or, something similar to this could be used to propose a senior thesis or graduate dissertation. If you go on to a career in social research, you could use a proposal like this to obtain funding to support your research. If you're applying for funding, you should also include a project budget to indicate how much support you'll need and for what.

Whichever way you use this kind of document, I wish you every success.

REVIEW QUESTIONS AND EXERCISES

1. Analyze a quantitative research report: Stanley Lieberson, Susan Dumais, and Shyon Baumann, "The Instability of Androgynous Names: The Symbolic Maintenance of Gender Boundaries," *American Journal of Sociology* 105, no. 5 (March 2000): 1249 (can be accessed in print or online through InfoTrac College Edition, for example). Use the following questions as your guide:

   a. What are the theoretical underpinnings of the study?

   b. How are some of the key variables such as androgynous, racial, and gender segregation conceptualized and operationalized?

   c. What data is this research based on?

   d. Are there controlling variables?

   e. What is the unit of analysis?

   f. What type of analysis was done?

   g. What did the authors find?

   h. What are the strengths and weaknesses of this study?

2. Analyze a qualitative research report: Dingxin Zhao, "State-Society Relations and the Discourses and Activities of the 1989 Beijing Student Movement," *American Journal of Sociology* 105, no. 6 (May 2000): 1592 (can be accessed in print or online through InfoTrac College Edition, for example). Use the following questions as your guide:

   a. What is the author's main research question?

   b. What theoretical frameworks does he refer to, and which ones did he use?

   c. What methodology is the author using? What type of data collection did he choose? What is the unit of analysis?

   d. Does the author have a hypothesis? If so, what is it?

   e. How does the author conceptualize key terms such as *state*, *state-society*, and *traditionalism*? What new ideal types of states does he bring to the field?

   f. What are his findings?

   g. What is the significance of this study? Were you convinced by the author, or do you see weaknesses in the study?

SPSS EXERCISES

See the booklet that accompanies your text for exercises using SPSS (Statistical Package for the Social Sciences). There are exercises offered for each chapter, and you'll also find a detailed primer on using SPSS.

Online Study Resources

If your book came with an access code card, visit www.cengage.com/login to register. To purchase access, please visit www.ichapters.com.

1. Before you do your final review of the chapter, take the *CengageNOW* pretest to help identify the areas on which you should concentrate. You'll find information on this online tool, as well as instructions on how to access all of its great resources, in the front of the book.

2. As you review, take advantage of the *CengageNOW* personalized study plan, based on your quiz results. Use this study plan with its interactive exercises and other resources to master the material.

3. When you're finished with your review, take the posttest to confirm that you're ready to move on to the next chapter.

WEBSITE FOR *THE PRACTICE OF SOCIAL RESEARCH* 12TH EDITION

Go to your book's website at www.cengage.com/sociology/babbie for tools to aid you in studying for your exams. You'll find *Tutorial Quizzes* with feedback, *Internet Exercises, Flash Cards, Glossaries,* and *Essay Quizzes,* as well as *InfoTrac College Edition* search terms, suggestions for additional reading, *Web Links,* and primers for using data-analysis software such as SPSS.

Using the Library

Introduction

We live in a world filled with social science research reports. Our daily newspapers, magazines, professional journals, alumni bulletins, club newsletters—virtually everything we pick up to read may carry reports dealing with a particular topic. For formal explorations of a topic, of course, the best place to start is still a good college or university library. Today, there are two major approaches to finding library materials: the electronic route and the traditional paper system. Because I don't know what will be available to you, I'll address both methods, with special attention given to the electronic option.

Getting Help

When you want to find something in the library, your best friend is the reference librarian, who is specially trained to find things in the library. Some libraries have specialized reference librarians—for the social sciences, humanities, government documents, and so forth. Find the librarian who specializes in your field. Make an appointment. Tell the librarian what you're interested in. He or she will probably put you in touch with some of the many available reference sources.

Reference Sources

You've probably heard the expression "information explosion." Your library is ground zero. Fortunately, a large number of reference volumes exist to offer a guide to the information that's available.

Books in Print

This volume lists all the books currently in print in the United States—listed separately by author and by title. Out-of-print books can often be found in older editions of *Books in Print*.

Readers' Guide to Periodical Literature

This annual volume with monthly updates lists articles published in many journals and magazines. Because the entries are organized by subject matter, this is an excellent source for organizing your reading on a particular topic. Figure A-1 presents a sample page from the *Readers' Guide*.

In addition to these general reference volumes, you'll find a great variety of specialized references/databases, many of which are online. Here are just a few:

- *Sociological Abstracts*
- *Psychological Abstracts*
- *Social Science Index*
- *Social Science Citation Index*
- *Popular Guide to Government Publications*
- *New York Times Index*
- *Facts on File*
- *Editorial Research Reports*
- *Business Periodicals Index*
- *Monthly Catalog of Government Publications*

MUSIC—*cont.*

Study and teaching
See also
Guitar—Study and teaching

Themes, motives, etc.
See also
Automobiles in music

Theory
See also
Atonality

Japan
The Japanese and Western music. L. Futoransky. il *The Courier (Unesco)* 40:38+ D '87

MUSIC, AMERICAN
See also
Jazz music

MUSIC, ELECTRONIC
See also
Computers—Musical use
Musical instruments, Electronic

MUSIC AND STATE
Viewpoint [government subsidies of opera] J. L. Poole. Opera News 52:4 F 13 '88

Soviet Union
Gorbachev sets the beat for Soviet rock. il *U.S. News & World Report* 104:8-9 F 8 '88

MUSIC AND THE BLIND
Call him Doc [D. Watson] F. L. Schultz. il pors *Country Journal* 15:44-53 F '88

MUSIC AND THE HANDICAPPED
See also
Guitarists, Handicapped

MUSIC CORPORATION OF AMERICA *See* MCA Inc.

MUSIC CRITICS AND CRITICISM
See also
Opera reviews

MUSIC FESTIVALS

Austria
Bregenz. H. Koegler. il *Opera News* 52:38 F 13 '88

Germany (West)
Bayreuth. J. H. Sutcliffe. il *Opera News* 52:36 Ja 30 '88

Great Britain
Buxton. E. Forbes. *Opera News* 52:40-1 F 13 '88

Italy
Torre del Lago (Puccini Festival) M. Hamlet-Mets. *Opera News* 52:38-40 F 13 '88

Pennsylvania
Philadelphia [American Music Theater Festival] R. Baxter, *Opera News* 52:34 Ja 30 '88

MUSICAL COMEDIES, REVUES, ETC. *See* Musicals, revues, etc.

MUSICAL INSTRUMENTS, ELECTRONIC
It's alive with the sound of—well, just about everything (Synclavier synthesizer) L. Helm. il *Business Week* p75 F 8 '88

MUSICAL INSTRUMENTS INDUSTRY
See also
New England Digital Corporation

MUSICALS, REVUES, ETC.

Choreography
See Choreography

Reviews
Single works
Anything goes
    Dance Magazine il 62:52-7 Ja '88. J. Gruen
Cabaret
    Dance Magazine 62:73-4 Ja '88. H. M. Simpson
The chosen
    The Nation 246:176 F 6 '88. T. M Disch
Into the woods
    Dance Magazine 62:64 Ja '88. K. Grubb
Oil City Symphony
    The Nation 246:175-6 F 6 '88. T. M. Disch
The phantom of the opera
    Life il 11:88-92 F '88. M. Stasio
    Maclean's il 101:51 F 8 '88. L. Black
    New York il 21:89-90 F 8 '88. J. Simon
    The New Yorker 63:97-8 F 8 '88. M. Kramer
    Newsweek il por 111:68-70+ F 8 '88. J. Kroll
    Rolling Stone il p26 F 25 '88. D. Handelman
    Time il 131:83-4 F 8 '88. W. A. Henry

Stage setting and scenery
High-tech magic: follow that gondola [Phantom of the opera] J. Kroll. il *Newsweek* 111:70 F 8 '88

Writing
Changing the face of Broadway [A. Lloyd Webber] M. Stasio. il pors *Life* 11:88-92 F '88

MUSICIANS
See also

Drugs and musicians
Rock musicians

MUSKE, CAROL, 1945-
Skid [poem] *The New Yorker* 63:38 F 8 '88

MUSLIMS
See also
Islam

Afghanistan
Beyond the Afghan stalemate. L. Komisar. il *The New Leader* 71:5-6 Ja 11-25 '88

Middle East
The Islamic resurgence: a new phase? R. Wright. bibl f *Current History* 87:53-6+ F '88

MUTATION
See also
Transposons

MUTUAL FUNDS *See* Investment trusts

MUTUALISM (BIOLOGY) *See* Symbiosis

MUZIEKTHEATER (AMSTERDAM, NETHERLANDS)
See Opera houses

MYASTHENIA GRAVIS
Suzanne Rogers: "I looked at my face and thought, 'Who'd hire a freak?" A. W. Petrucelli. pors *Redbook* 170:104+ F '88

MYCOBACTERIAL DISEASES
See also
Tuberculosis

MYCOTOXINS *See* Toxins and antitoxins

N

N. W. AYER & SON, INC.
Ayer to the throne [Burger King ad campaign] B. Kanner. il *New York* 21:24+ F 29 '88

NADIS, STEVEN J.
Robot observatories. il *Omni (New York, N.Y.)* 10:24+ Ja '88

NAEP *See* National Assessment of Educational Progress

NAKAGAMI, KENJI, 1946-
about
Two contemporary writers. D. Palmé. *The Courier (Unesco)* 40:44 D '87

NAKED SHORT SELLING *See* Securities—Short selling

NANDINA
Nandina does the unexpected. il *Southern Living* 23:50 Ja '88

NAPLES (ITALY)

Music
See also
Opera—Italy

NARCOTIC ADDICTS *See* Drug abuse

NARCOTICS LAWS AND REGULATIONS
See also
Boats in narcotics regulation
Robots in narcotics regulation

Austria
A five-year penalty call [Czech hockey legend J. Bubla serving prison sentence for smuggling heroin] J. Holland. il por *Maclean's* 101:6 F 8 '88

Colombia
Battling the drug lords [Attorney General C. Hoyos murdered] E. Tolmie. il *Maclean's* 101:26 F 8 '88
Day of the assassins [Attorney General C. Hoyos murdered] M. S. Serrill. il por *Time* 131:42 F 8 '88
How cocaine rules the law in Colombia [assassination of Attorney General C. Hoyos] C. A. Robbins. il *U.S. News & World Report* 104:28-9 F 8 '88
Murderers of Medellín [assassination of Colombia's Attorney General C. Hoyos] F. Willey. il *Newsweek* 111:33 F 8 '88

NARCOTICS TRADE
See also
Boats in narcotics regulation
Narcotics laws and regulations
Robots in narcotics regulation
Teen drug dealers: uncovering the real story. W. White and K. Dickerson. il *'Teen* 32:36-9+ F '88

Panama
The dictator in the dock [M. A. Noriega] N. Cooper. il por *Newsweek* 111:33 F 22 '88
Drugs, money and death [cover story, special section] il pors map *Newsweek* 111:32-6+ F 15 '88
More bad news for Noriega. N. Cooper. il por *Newsweek* 111:37 F 8 '88
Noriega's money machine [aides testify before Senate subcommittee] M. S. Serrill. il *Time* 131:39-40 F 22 '88

FIGURE A-1

A Page from the *Readers' Guide to Periodical Literature*

Source: From H. W. Wilson, *Reader's Guide*, 1978, 3-4. © 1978 H. W. Wilson.

- *Public Affairs Information Service Bulletin*
- *Education Index*
- *Applied Science and Technology Index*
- *A Guide to Geographic Periodicals*
- *General Science Index*
- *Biological and Agricultural Index*
- *Nursing and Applied Health Index*
- *Nursing Studies Index*
- *Index to Little Magazines*
- *Popular Periodical Index*
- *Biography Index*
- *Congressional Quarterly Weekly Report*
- *Library Literature*
- *Bibliographic Index*

Using the Stacks

Serious research usually involves using the stacks, where most of the library's books are stored. This section provides information about finding books in the stacks.

The Electronic Catalog

Your library's catalog of holdings will be available electronically or through a printed card catalog (discussed later) or both. You can access an electronic catalog through a computer search, on either a library computer or your personal computer. Online catalog systems vary, but the following illustration from Chapman University's Leatherby Libraries will probably resemble what you'll find in our own library.

To start, let's look up a book I wrote: *The Sociological Spirit*. The library home page is shown in Figure A-2, and we begin our search by clicking on "Find Books . . ."

In this case, we're given several choices of libraries to search (Figure A-3). We'll choose the Leatherby Libraries, which are located on campus.

Selecting the Leatherby Libraries Catalog presents us with the screen shown in Figure A-4, which provides for several ways of searching: by author, by title, by subject, and so forth.

Clicking "AUTHOR" will present a screen (not shown) that asks for the author's name. Searching for the name "Babbie" in this system eventually presents a list of books with that author-name, including the one we're looking for, as shown in Figure A-5.

You'll notice two entries for the title we're looking for. These represent two editions of the same book. Let's click on the more recent edition, published in 1994. Figure A-6 is an electronic catalog "card" similar to the paper version illustrated in Figure A-7.

Notice the adjoining red bars marked "LOCATION," "CALL#," and "STATUS." Just below the bar, we learn that the book is on the second floor, in the social science collection. More specifically, the Library of Congress number (or call number)—HM51.B164 1994—will help us locate the book on the shelves, which we see is, in fact, available rather than checked out.

Here's a useful strategy to use when you're researching a topic. Once you've identified the call number for a particular book in your subject area, go to the stacks, find that book, and look over the other books on the shelves near it. Because the books are arranged by subject, this method will help you locate relevant books you didn't know about.

Alternatively, you may want to go directly to the stacks and look at books in your subject area. In most libraries, books are arranged by the Library of Congress numbers. (Some follow the Dewey decimal system.) The following is a selected list of Library of Congress categories.

CHAPMAN UNIVERSITY **LEATHERBY LIBRARIES**

Search this Site   Ask A Librarian   University Archives   Hours & Information   Suggestions   New Materials   Help

Quick Reference	**Library Notes**	Government Sources
Find Books...	Leatherby Libraries Events Calendar	Web Searching
Find Articles	Freshman Foundations Students	Instruction / Tutorials
Find by Subject	Fall 2008 Library Citation Workshops	Interlibrary Loan
University College Library Services	Find Reserve Items Course Reserves: Faculty and Student Information	Holocaust Education
	Renew Items Online	

LEATHERBY LIBRARIES
CHAPMAN UNIVERSITY

Help   Suggestions   Site Map   Search   Chapman University • Leatherby Libraries
One University Drive, Orange, CA 92866 • 714-532-7756

FIGURE A-2
Library Home Page

| Quick Reference |
| Find Books... |
| Find Articles |
| Find by Subject |
| Government Sources |
| Web Searching |
| Instruction/Tutorial |
| Interlibrary Loan |
| University College Library Services |

Find Books

Chapman University Library Catalogs:

Leatherby Libraries Catalog

This is the main Chapman University library catalog which allows you to locate books, videos, DVD's, CD's, periodicals, reserves, and a variety of other library related materials. You may also Renew Items Online.

Rinker Law Library Catalog

Electronic Book Collections

Blackwell Reference Online

With nearly 300 reference volumes, Blackwell Reference Online is the largest academic reference collection of its kind available online, and covers subject areas in Business, Cultural Studies, Economics, History, Language, Linguistics, Literature, Philosophy, Psychology, Religion and Sociology.

CQ Press Political Reference Suite

Provides online access to California Political Almanac, Political Handbook of the World, and Supreme Court Yearbook. Gift of the R.C. Hoiles Libertarian Collection.

Gale Virtual Reference

Gale Virtual Reference Library is a database of encyclopedias and specialized reference sources for multidisciplinary research.

NetLibrary eBook Catalog

Access our full-text online electronic book collection by clicking on the NetLibrary eBook Catalog link above and our new Online Reference Center!

Sage eReference

A collection of over 40 electronic reference sources covering some key subjects in the social sciences and humanities.

Springer Electronic Book Collection

The Leatherby Libraries provides access only to the two Springer Behavioral Sciences and Business and Economics e-book collections. Full-text access to other Springer e-books and journals is not currently available via this interface. For assistance on searching the Springer Electronic Book Collection, including how to limit to the accessible e-book collections, please call the Reference Desk at (714) 532-7714 or e-mail the Reference Librarian at libweb@chapman.edu

FIGURE A-3

A Choice of Libraries to Search

Leatherby Libraries Catalog

Search by:

- AUTHOR
- TITLE (Book, Journal, Video or CD)
- SUBJECT
- KEYWORD (SIMPLE)
- KEYWORD (ADVANCED)
- AUTHOR/TITLE
- CALL NUMBER
- STANDARD NUMBER, ISBN, ISSN

- Log In / Renew Checked-Out Items
- Course Reserves by Course Name
- Course Reserves by Instructor Name
- New Materials in the Leatherby Libraries
- Combined Search of Leatherby and Law Libraries

Back to Main Search Screen | Library Information | Find Articles | University Colleges | Rinker Law Library Catalog

| Help | Suggestions | Site Map | Search | **Chapman University • Leatherby Libraries**
One University Drive, Orange, CA 92866 • 714-532-7756 |

©2008 Chapman University.
Last updated: January 17, 2008. Comments to: libweb@chapman.edu

WebPAC PRO © Innovative Interfaces, Inc.

FIGURE A-4
Search Options

1
Save ☐
The practice of Social Research (11th ed.)
Babbie, E.
1st FL Circulation/Reserves Desk BOD 002

2
Save ☐
Adventures in social research : data analysis using SPSS for Windows / Earl Babbie, Fred Halley
Babbie, Earl R.
2nd FL Social Science Library Books HA32 .B283 1995
Thousand Oaks, Calif. : Pine Forge Press, c1995

3
Save ☐
Adventures in social research : data analysis using SPSS for Windows 95 / Earl Babbie, Fred Halley
Babbie, Earl R.
2nd FL Social Science Library Books HA32 .B284 1998
Thousand Oaks, Calif. : Pine Forge Press, c1998

4
Save ☐
The basics of social research / Earl Babbie.
Babbie, Earl R.
2nd FL Social Science Library Books H62 .B18 2005
Belmont, CA : Thomson/Wadsworth, c2005.

5
Save ☐
Observing ourselves : essays in social research / Earl Babbie
Babbie, Earl R.
2nd FL Social Science Library Books H62 .B2 1986 suppl
Belmont, Calif. : Wadsworth Pub. Co., 1985

6
Save ☐
The practice of social research / Earl Babbie
Babbie, Earl R.
2nd FL Social Science Library Books H62 .B2 1983
Belmont, Calif. : Wadsworth Pub. Co., c1983

7
Save ☐
The practice of social research / Earl Babbie
Babbie, Earl R.
2nd FL Social Science Library Books H62 .B2 1986
Belmont, Calif. : Wadsworth Pub. Co., c1986

8
Save ☐
The practice of social research / Earl Babbie
Babbie, Earl R.
2nd FL Social Science Library Books H62 .B2 1989
Belmont, Calif. : Wadsworth Pub. Co., c1989

9
Observing ourselves : essays in social research / Earl Babbie

FIGURE A-5

A Research Summary from *Sociological Abstracts*. Used by permission.

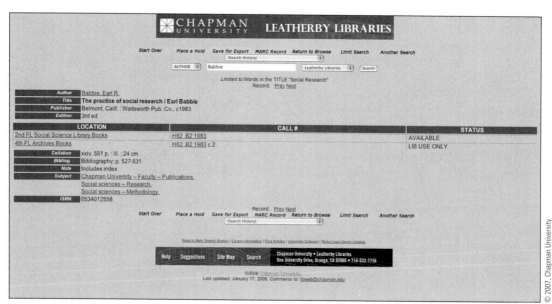

FIGURE A-6

Electronic Catalog "Card".

Library of Congress Classifications (partial)

A   GENERAL WORKS
B   PHILOSOPHY, PSYCHOLOGY, RELIGION
    B-BD   Philosophy
    BF   Psychology
    BL-BX   Religion
C   HISTORY-AUXILIARY SCIENCES
D   HISTORY (except America)
    DA-DR   Europe
    DS   Asia
    DT   Africa
E-F HISTORY (America)
    E   United States
    E51–99   Indians of North America
    E185   Negroes in the United States
    F101–1140 Canada
    F1201–3799 Latin America
G   GEOGRAPHY-ANTHROPOLOGY
    G-GF   Geography
    GC   Oceanology and oceanography
    GN   Anthropology
    GV   Sports, amusements, games
H   SOCIAL SCIENCES
    H62.B2   *The Practice of Social Research*
    HB-HJ   Economics and business
    HM-HX   Sociology
J   POLITICAL SCIENCE
    JK   United States
    JN   Europe
    JQ   Asia, Africa
    JX   International relations
K   LAW
L   EDUCATION
M   MUSIC
N   FINE ARTS
    NA   Architecture
    NB   Sculpture
    NC   Graphic arts
    ND   Painting
    NE   Engraving
    NK   Ceramics, textiles
P   LANGUAGE AND LITERATURE
    RE   English language
    PG   Slavic language

    PJ-PM   Oriental language
    PN   Drama, oratory, journalism
    PQ   Romance literature
    PR   English literature
    PS   American literature
    PT   Germanic literature
Q   SCIENCE
    QA   Mathematics
    QB   Astronomy
    QC   Physics
    QD   Chemistry
    QE   Geology
    QH-QR   Biology
R   MEDICINE
    RK   Dentistry
    RT   Nursing
S   AGRICULTURE—PLANT AND ANIMAL INDUSTRY
T   TECHNOLOGY
    TA-TL   Engineering
    TR   Photography
U   MILITARY SCIENCE
V   NAVAL SCIENCE
Z   BIBLIOGRAPHY AND LIBRARY SCIENCE

The Card Catalog

In the traditional paper system, the card catalog is the main reference system for finding out where books are stored. Each book is described on three separate 3-by-5 cards. The cards are then filed in three alphabetical sets. One set is arranged by author, another by title, and the third by subject matter.

If you want to find a particular book, you can look it up in either the author file or the title file. If you have only a general subject area of interest, you should thumb through the subject catalog. Figure A-7 presents a sample card in the card catalog. Notice the following elements:

1. Subject heading (always in capital letters)
2. Author's name (last name, first name)
3. Title of the book
4. Publisher
5. Date of publication

FIGURE A-7

Sample Subject Catalog Card

Source: Lilian L. Shapiro, *Teaching Yourself in Libraries* (New York: H.W. Wilson, 1978), 3–4. Used by permission.

6. Number of pages in the book plus other information (such as whether the book contains illustrations)

7. Call number needed to find a nonfiction book on the library shelves; fiction is generally found in alphabetical order by the author's last name

Searching the Periodical Literature

Earlier, we looked at some printed documents that help you search the articles published in academic journals and other periodicals. Electronic library systems make this process very powerful indeed.

Many college libraries now have access to the Education Resources Information Center (ERIC). (See the link on this book's website: http://www .cengage.com/sociology/babbie.) This computer-based system allows you to search through hundreds of major journals to find articles published in the subject area of your interest. As a rule, each library website should have a list of the databases that you can visit; they also list them by discipline, which may help you limit the number of titles related to a specific keyword. Make sure you narrow your search by limiting, for instance, the language or period of the publication. Once you identify the articles you're interested in, the computer will print out their abstracts.

Of particular value to social science researchers, the publications *Sociological Abstracts* and *Psychological Abstracts* present summaries of books and articles—often prepared by the original authors—so that you can locate a great many relevant references easily and effectively. As you find relevant references, you can track down the original works and see the full details. The summaries are available in both written and computerized forms.

Figure A-8 contains the abstract of an article obtained in a computer search of *Sociological Abstracts*. I began by asking for a list of articles dealing with sociology textbooks. After reviewing the list, I asked to see the abstracts of each of the listed articles. Here's an example of what I received seconds later: an article by the sociologist Graham

AU Kinloch-Graham-C.
TI   The Changing Definition and Content of Sociology in Introductory Textbooks, 1894–1981.
SO International Review of Modern Sociology. 1984, 14, 1, spring, 89–103.
DE Sociology-Education; (D810300). Textbooks; (D863400).
AB An analysis of 105 introductory sociology textbooks published between 1894 & 1981 reveals historical changes in definitions of the discipline & major topics in relation to professional factors & changing societal contexts. Predominant views of sociology in each decade are discussed, with the prevailing view being that of a "scientific study of social structure in order to decrease conflict & deviance, thereby increasing social control." Consistencies in this orientation over time, coupled with the textbooks' generally low sensitivity to social issues, are explored in terms of their authors' relative homogeneity in age & educational backgrounds. 1 Table, 23 References. Modified HA.

FIGURE A-8

A Research Summary from *Sociological Abstracts*. Used by permission.

C. Kinloch, published in the *International Review of Modern Sociology*.

In case the meaning of the abbreviations in Figure A-8 isn't immediately obvious, I should explain that AU is author; TI is title; SO is the source or location of the original publication; DE indicates classification codes under which the abstract is referenced; and AB is the abstract. The computerized availability of resources such as *Sociological Abstracts* provides a powerful research tool for modern social scientists. You'll have the option to download or print, with or without the abstract, any title you find through the library's browsers.

If a document is not available in the library itself or via the web, you always have the resource of interlibrary loans, which often are free. Libraries don't own every document or multimedia material (CD-ROMs, videocassettes, DVDs, films), but many have loan agreements that can serve your needs. You need to be aware of how much time it will actually take for you to receive the book or article from the time you made your request. In the case of a book that is located in another library close by, for example, it may be faster for you to get it directly yourself. The key to a good library search is to become well informed. So start networking with librarians, faculty, and peers!

ADDITIONAL READINGS

Bart, Pauline, and Linda Frankel. 1986. *The Student Sociologist's Handbook*. New York: Random House. A survival kit for doing sociological research. Contains a step-by-step guide for writing research papers; chapters on periodicals, abstract and indexing services, bibliographies, bibliographic aids, and other secondary sources; and a complete guide to government and nongovernment sources of data. Special section on sex roles and women's studies.

Li, Tze-chung. 2000. *Social Science Reference Sources: A Practical Guide*. Westport, CT: Greenwood Press. Lists and describes all types of reference materials, including databases and archives as well as published sources. Organized into two parts: social sciences in general and by discipline.

Richlin-Klonsky, Judith, and Ellen Strenski, eds. 1998. *A Guide to Writing Sociology Papers*. New York: St. Martin's Press. This is a great little book with good advice on doing research. It's particularly useful for those who are new to sociology or other social science disciplines and have to learn about the most rudimentary aspects of research.

GSS Household Enumeration Questionnaire

The National Opinion Research Center (NORC) at the University of Chicago is one of the primary social science research centers in the world. Its General Social Survey (GSS), moreover, has become a major source of social science data for researchers. You may have noticed that many of the examples in this book have been taken from the GSS. It is fitting, therefore, that we turn to the GSS for an example of an interview questionnaire. The following pages present the questionnaire used by interviewers in collecting basic demographic data about households selected in the sample. Questionnaires such as these are an important part of the scientific equipment used by social scientists. Just as a microscope is an appropriate instrument for observing cells and molecules and a telescope for observing distant planets and stars, a questionnaire such as this one is often the best instrument for observing the subject matter of social science.

GSS
AMERICA'S
SOCIAL
SURVEY

INTERVIEWER NAME

_____

INTERVIEWER ID#

| | | | | | | _ | |

5325 NORC
JANUARY 1990

HOUSEHOLD ENUMERATION FOLDER
COMPLETE SAMPLING REPORT

HEF

INTRODUCTION:

Hello, I'm (YOUR NAME) from the National Opinion Research Center at the University of Chicago (SHOW ID CARD).

(We recently sent you a letter explaining that) Your household has been selected to take part in this year's GSS: America's Social Survey.

We've been conducting this study all over the country for more than fifteen years, learning about how people feel about issues like schools, crime, government spending, and the military. This year several households in your community will be participating in this important research.

First, I'd like to make sure that I have your street address described correctly. Is it (READ FROM ASSIGNMENT LABEL OR BOX BELOW: STREET NUMBER AND NAME, APARTMENT NUMBER OR OTHER DESCRIPTION OF HU.) IF EXACTLY THE SAME, CHECK BOX:

☐(GO TO Q.1, P.2)

IF DIFFERENT IN ANY WAY, EXAMINE SEGMENT PRINTOUT AND RECONCILE. EXPLAIN THE DIFFERENCE HERE:

_____

_____

IF DIFFERENCE CAN'T BE RESOLVED, CALL YOUR FIELD MANAGER BEFORE CONTINUING.

INTERVIEWER: STICK ASSIGNMENT LABEL HERE
OR
IF THIS IS A CASE YOU'RE ADDING
TO YOUR ASSIGNMENT AS A MISSED HU,
COPY INFORMATION FROM ORIGINAL LABEL

SURVEY 5325   CASE#

PSU

SEG   PT 1   LINE   BLK

LOCALITY

HU ADDRESS

HU DESCRIPTION

A, B, OR C   X OR Y

(CASE #)   01-06/

(PSU)   07-09/

(SEG)   10-12/

(PT)   13/

(LINE #)   14-23/

(A, B, or C)   24/

(X or Y)   25/

(INTID)   26-31/

1. Only one member of your household will be eligible for this survey. In order to scientifically select that person, first I need to list the names of the people who usually live here.

2. Please tell me the names of the people who usually live in this household. Let's start with the head of the household. LIST ON LINES 01-10 BELOW.

3. Have we forgotten anyone: such as babies or small children; roomers; people who usually live here, but are away temporarily—on business trips, vacations, at school, temporarily in a hospital, and so on?

   Yes ☐ LIST ADDITIONAL PERSONS ON LINES 01-10 BELOW.
   No ☐ GO TO Q.4.

4. Are there any people currently staying here—visitors, friends or relatives—who do not usually live here?

   Yes ☐ LIST VISITORS ON LINES 11-14 BELOW.
   No ☐ GO TO Q.5.

AFTER QS. 1-4, ASK QS. 5-8 FOR EACH PERSON.

9. Are any of the people we have have listed staying somewhere else right now?
If NO, CHECK BOX AND SKIP TO Q. 12 ☐
IF YES, GO TO Q. 10 BELOW.

	5. What is (PERSON)'s relationship to (HEAD OF HOUSEHOLD)?	6. CODE SEX (ASK IF NOT OBVIOUS) M / F	7. How old was (HEAD/PERSON) on (his/her) last birthday?	8. IF 13 YRS OR OLDER ASK: Is (PERSON) now married, widowed, divorced, separated, or has (he/she) never been married? Ma Wi Di Se NM	10. Who is staying somewhere else right now? CHECK (√) LINE OF EACH PERSON WHO IS AWAY.	11. ASK FOR EACH PERSON CHECKED (√) IN Q. 10: Where is (PERSON) staying right now: Is (PERSON) staying at another household; Is (he/she) traveling; Is (he/she) in some institution or dormitory-like at college, or in a hospital or something; or what? CIRCLE CODE, FOLLOW INSTRUCTIONS Another Household	Travel-ing	Insti-tution	Other DK
First Name / Last name									
01	HEAD 49	50 1 2	51-52	53 1 2 3 4 5	54	1 CROSS OUT 55	2 LEAVE IN	3 CROSS OUT	4*
02	56	57 1 2	58-59	60 1 2 3 4 5	61	1 CROSS OUT 62	2 LEAVE IN	3 CROSS OUT	4*
03	63	64 1 2	65-66	67 1 2 3 4 5	68	1 CROSS OUT 69	2 LEAVE IN	3 CROSS OUT	4*
04	BEGIN DECK 03 07	08 1 2	09-10	11 1 2 3 4 5	12	1 CROSS OUT 13	2 LEAVE IN	3 CROSS OUT	4*

05		14	15 1 2	16–17	18 1 2 3 4 5	19	1 CROSS OUT	2 LEAVE IN	3 CROSS OUT	20 4*
06		21	22 1 2	23–24	25 1 2 3 4 5	26	1 CROSS OUT	2 LEAVE IN	3 CROSS OUT	27 4*
07		28	29 1 2	30–31	32 1 2 3 4 5	33	1 CROSS OUT	2 LEAVE IN	3 CROSS OUT	34 4*
08		35	36 1 2	37–38	39 1 2 3 4 5	40	1 CROSS OUT	2 LEAVE IN	3 CROSS OUT	41 4*
09		42	43 1 2	44–45	46 1 2 3 4 5	47	1 CROSS OUT	2 LEAVE IN	3 CROSS OUT	48 4*
10		49	50 1 2	51–52	53 1 2 3 4 5	54	1 CROSS OUT	2 LEAVE IN	3 CROSS OUT	55 4*
11		56	57 1 2	58–59	60 1 2 3 4 5	61	1 CROSS OUT	2 LEAVE IN	3 CROSS OUT	62 4*
12		63	64 1 2	65–66	67 1 2 3 4 5	68	1 CROSS OUT	2 LEAVE IN	3 CROSS OUT	69 4*
13	BEGIN DECK 04	07	08 1 2	09–10	11 1 2 3 4 5	12	1 CROSS OUT	2 LEAVE IN	3 CROSS OUT	13 4*
14		14	15 1 2	16–17	18 1 2 3 4 5	19	1 CROSS OUT	2 LEAVE IN	3 CROSS OUT	20 4*

VISITOR

IF MORE THAN 10 USUAL PERSONS AND/OR MORE THAN 4 VISITORS, USE A BLANK HEF FOR ADDITIONAL LISTING.

AFTER Q.8 FOR LAST PERSON 13 OR OLDER, ASK Q.9.

*PROBE FOR DETAILS AND CHECK INTERVIEWER MANUAL FOR HELP IF NECESSARY

SAMPLING TABLE

NUMBER OF ELIGIBLE PERSONS LISTED ABOVE	INTERVIEW PERSON ON LINE NUMBER
	STICK SAMPLING PART OF ASSIGNMENT LABEL HERE. (If this is a "missed HU" added to your assignment, copy sampling numbers from original case.)
TWO	
THREE	
FOUR	
FIVE	
SIX OR MORE	

STEP 3: USE SAMPLING TABLE TO THE RIGHT TO SELECT WHICH ELIGIBLE PERSON TO INTERVIEW.

STEP 4: CIRCLE SELECTED R'S LINE # IN SUMMARY BOX.

STEP 5: PRINT SELECTED R'S NAME HERE:

_____

STEP 6: ARRANGE TO INTERVIEW THIS PERSON. REQUEST PHONE AND MAILING INFORMATION (Qs.13 AND 14) TO HELP IN FOLLOW-UP.

12. Now I'm going to scientifically select the one person in this household chosen for this study. By interviewing only the person picked in this way, we can be sure the views we find do accurately represent the views of the country as a whole.

STEP 1: ELIGIBLE PEOPLE: 18 OR OVER
NAME NOT CROSSED OUT ON HOUSEHOLD ENUMERATION (P.2)

IF ONLY ONE ELIGIBLE PERSON, GO TO Q.13 AND MAKE APPOINTMENT TO INTERVIEW THAT PERSON.

IF MORE THAN ONE, CONTINUE WITH STEPS 2-6 BELOW.

STEP 2: LIST NAMES OF ELIGIBLE PEOPLE IN SUMMARY BOX, IN ORDER OF AGE.

SUMMARY BOX

	LINE #	AGE	NAMES OF ELIGIBLE PERSONS
OLDEST (Not necessarily the head)	1		
	2		
	3		
	4		
	5		
YOUNGEST	6		

13. If I have to talk with (SELECTED RESPONDENT), what phone numbers should I use?

Telephone number given: (   ) _____
            AREA CODE   NUMBER

A. Code location of phone:

    In household.......................... 1 22/
    In home of neighbor................... 2
    Other (SPECIFY) _____ 3

B. If no number given, code

    No phone.............................. 4
    Refused............................... 5

14. If I have to mail a note to (SELECTED RESPONDENT), what would be the best mailing address to use?

| # AND STREET/BOX #/RFD # |
| POST OFFICE, TOWN |
| STATE   ZIP CODE |

THANK RESPONDENT FOR THEIR TIME AND HELP.

TIME
HEF
ENDED: [   ]   AM
          PM

LENGTH
OF TIME
FOR HEF: [   ]   MIN

IMMEDIATELY AFTER LEAVING THE HOUSEHOLD, FILL OUT A-D BELOW AND COMPLETE THE SAMPLING REPORT.

A. Date HEF administered: [ ][ ] [ ]
                      MONTH   DAY   YEAR   23-25/

B. Name of HEF informant: _____

C. List line number of HEF informant from Household Enumeration (P.2, Q.4) .. [ ] 26-27/
   (If non-household informant, line # = 22)

D. Race of household (by observation)

    White............................ 1 28/
    Black/Negro...................... 2
    American Indian.................. 3
    Asiatic, Oriental............... 4
    Other, mixed, not able to observe. 5

E. List line number of selected respondent from Household Enumeration (P.2, Q.4) .. [ ] 29-30/

HEF-4 HEF-5
NON-INTERVIEW REPORT (NIR)

NIR

Name (if known): _____

Telephone number, (if available): _____

Please Circle Appropriate Code:

HEF not completed............................... 1 31/
HEF complete/interview incomplete............... 2

1. Why were you unable to complete HEF/Interview at this address?

NOT AN HU: .... (Ans. 32-33/
  Condemned .... (Ans. Q.2) 01
  Demolished .... (Ans. Q.2) 02
  Place of business .... (Ans. Q.2) 03
  No such address/no such HU .... (Ans. Q.2) 04
  Group quarters .... (Ans. Q.2) 05
  Vacation cabin .... (Ans. Q.2) 06
  Not usable as permanent residence (Ans. Q.2) 07
  Transient use (less than one month) (Ans. Q.2) 08
  Not an HU for other reason .... (Ans. Q.2) 09
  Still under construction .... (Ans. Q.2) 10
VACANT .... (Skip to Q.7) 11
REFUSED .... (Skip to Q.4) 12
BREAKOFF .... (Skip to Q.4) 13
NOT HOME AFTER 4 CALLS .... (Skip to Q.3) 14
RESPONDENT IS UNAVAILABLE FOR ENTIRE
  FIELD PERIOD .... (Skip to Q.5) 15
LANGUAGE PROBLEM (SPECIFY LANGUAGE SPOKEN)
  .... (SKIP TO Q.7) 16
TOO ILL (DESCRIBE SITUATION ON PAGE 7)
  .... (Skip to Q.7) 17
OTHER (DESCRIBE SITUATION ON PAGE 7)
  .... (Skip to Q.7) 18

2. IF NOT AN HU:
Describe the reason for this NIR fully, then go to Q.19.
34-35/

3. IF NOT AT HOME AFTER 4 CALLS:
Why do you think it has been so hard to find the occupants of this housing unit at home?
36-37/

A. Describe your efforts to obtain information about the occupants.
38-39/

4. IF REFUSED OR BREAKOFF:
Did the respondent give the refusal?
Yes............................. (Ans. A) ... 1 40/
No.............................. (Ans. B) ... 2
DK, HEF not complete........ (Skip to Q.6).. 8
A. IF YES: Why did the respondent (refuse/breakoff)?
(Report verbatim remarks and reasons.).. 41-42/
(SKIP TO Q. 6)
B. IF NO: Why were you unable to speak with the respondent?
43-44/
(SKIP TO Q.6)

5. IF TEMPORARILY UNAVAILABLE FOR THE ENTIRE PERIOD:
What is the reason for this status?
45-46/
A. When will R be available?
47-48/

6. Was a follow-up letter sent to the respondent?
Yes............................. (Ans. A) ... 1 49/
No.............................. (Ans. B) ... 2
DK 8
A. IF YES: Did you speak with R after receipt of letter?
Yes............................. (Ans. B) ... 1 50/
No.............................. 2
DK 8
B. IF NO: Why not?

ANSWER Q.7 UNLESS NOT AN HU
7. Were you ever able to talk with someone at this HU (not necessarily a resident)?

Yes......................... (GO TO Q.8) ... 1 51/
No...........................(Ans. A) ... 2

A. Why not?   52-53/

8. On any calls were there people in the HU who did not answer the door?

Yes, definitely 1 54/
Suspect so 2
No reason to think so 3

```
ANSWER Qs 9-17 FROM INFORMATION OBTAINED BY OBSERVATION
AND/OR CONTACTS WITH HM MEMBERS, NEIGHBORS AND OTHER
SOURCES.
```

9. What is the estimated income of R's family? (NOTE: IF MULTIPLE FAMILIES AND R NOT DETERMINED, ESTIMATE FOR PRIMARY FAMILY.)

Low................................... 1 55/
Medium................................ 2
High.................................. 3
Don't know/unable to observe 8

10. Were you able to complete the household listing?

Yes .... (Skip to Q.16) ... 1 56/
No............................(Ans. Qs. 11-20) ... 2

11. What is the race of the residents?

White........................ (definitely) ... 1 57/
White........................ (probably) ... 2
Black........................ (definitely) ... 3
Black........................ (probably) ... 4
Hispanic 5
Asiatic/Oriental 6
Other (DESCRIBE) 7
Could not determine 8

12. Estimate the number of adults living in the HU.

# of adults: 58-59/

13. Estimate the number of adult males living in the HU.

# of adults: 60-61/

14. Is there a married (or living as married) couple living in the HU?

Yes.. 1 62/
No... 2
Don't know................................... 3

15. What is your estimate of the age of the Household Head?

Under 30 .... 1 63/
30-64 2
65 or older 3
Don't know 8

16. Type of structure.   64-65

Trailer.. 01
Detached single family house 02
2 family house, 2 units side by side 03
2 family house, 2 units one above the other 04
Detached 3-4 family house 05
Rowhouse (3 or more units in an attached row) 06
Apartment house (5 or more units,
  3 stories or less) 07
Apartment house (5 or more units,
  4 stories or more), ... 08
Apartment in a partly commercial structure 09
Other (SPECIFY) _____   10

17. Compared to house/apartments in the neighborhood, would you say the house/apartment was ... 1 66/

Far above average 1
Above average 2
Average 3
Below average 4
Far below average 5

18. ADDITIONAL COMMENTS:

19. Interviewer Name:

20. Interviewer Number:

21. Enter Supervisor's name.

HEF-6
RECORD OF CALLS ON HOUSEHOLD

USE PENCIL

A. Day
M.1 Tu.2 W.3
Th.4 F.5 Sa.6
Su.7
B. Month
C. Date

D. Time
Pre3..1 3-6..2 Post 6..3
E. Type
Per..1 Tel..2 Mail..3
F. Outcome of Contact
If 35,37,45,47,82,92 Do G

G. Refusal Description

H. Results/Reason for Refusal. Give verbatim reasons and explain in all circumstances.

I. Purpose of Contact

J. Int'r Initials

A. ⊔⊔ 32/R	D. ⊔ 37/R	G. M.....F | Age ____	H.	I. 41/R J.
B. ⊔⊔⊔ 33-34/R	E. ⊔ 38/R	Other char:		HEF 1
C. ⊔⊔⊔ 35-36/R	F. ⊔⊔ 39-40/R	HEF Line # when complete ____		Quex 2
A. ⊔⊔ 42/R	D. ⊔ 47/R	G. M.....F | Age ____	H.	I. 51/R J.
B. ⊔⊔⊔ 43-44/R	E. ⊔ 48/R	Other char:		HEF 1
C. ⊔⊔⊔ 45-46/R	F. ⊔⊔ 49-50/R	HEF Line # when complete ____		Quex 2
A. ⊔⊔ 52/R	D. ⊔ 57/R	G. M.....F | Age ____	H.	I. 61/R J.
B. ⊔⊔⊔ 53-54/R	E. ⊔ 58/R	Other char:		HEF 1
C. ⊔⊔⊔ 55-56/R	F. ⊔⊔ 59-60/R	HEF Line # when complete ____		Quex 2
A. ⊔⊔ 62/R	D. ⊔ 67/R	G. M.....F | Age ____	H.	I. 71/R J.
B. ⊔⊔⊔ 63-64/R	E. ⊔ 68/R	Other char:		HEF 1
C. ⊔⊔⊔ 65-66/R	F. ⊔⊔ 69-70/R	HEF Line # when complete ____		Quex 2

BEGIN DECK 02

<table>
<tr><td>A.</td><td></td><td>07/R</td><td>D.</td><td></td><td>12/R</td><td>G. M.....F | Age _____</td><td>H.</td><td>I.</td><td>16/R</td><td>J.</td></tr>
<tr><td>B.</td><td>08-09/R</td><td></td><td>E.</td><td></td><td>13/R</td><td>Other char:</td><td></td><td>HEF ....1</td><td></td><td></td></tr>
<tr><td>C.</td><td>10-11/R</td><td></td><td>F.</td><td></td><td>14-15/R</td><td>HEF Line # when complete _____</td><td></td><td>Quex ....2</td><td></td><td></td></tr>
<tr><td>A.</td><td></td><td>17/R</td><td>D.</td><td></td><td>22/R</td><td>G. M.....F | Age _____</td><td>H.</td><td>I.</td><td>26/R</td><td>J.</td></tr>
<tr><td>B.</td><td>18-19/R</td><td></td><td>E.</td><td></td><td>23/R</td><td>Other char:</td><td></td><td>HEF ....1</td><td></td><td></td></tr>
<tr><td>C.</td><td>20-21/R</td><td></td><td>F.</td><td></td><td>24-25/R</td><td>HEF Line # when complete _____</td><td></td><td>Quex ....2</td><td></td><td></td></tr>
<tr><td>A.</td><td></td><td>27/R</td><td>D.</td><td></td><td>32/R</td><td>G. M.....F | Age _____</td><td>H.</td><td>I.</td><td>36/R</td><td>J.</td></tr>
<tr><td>B.</td><td>28-29/R</td><td></td><td>E.</td><td></td><td>33/R</td><td>Other char:</td><td></td><td>HEF ....1</td><td></td><td></td></tr>
<tr><td>C.</td><td>30-31/R</td><td></td><td>F.</td><td></td><td>34-35/R</td><td>HEF Line # when complete _____</td><td></td><td>Quex ....2</td><td></td><td></td></tr>
<tr><td>A.</td><td></td><td>37/R</td><td>D.</td><td></td><td>42/R</td><td>G. M.....F | Age _____</td><td>H.</td><td>I.</td><td>46/R</td><td>J.</td></tr>
<tr><td>B.</td><td>38-39/R</td><td></td><td>E.</td><td></td><td>43/R</td><td>Other char:</td><td></td><td>HEF ....1</td><td></td><td></td></tr>
<tr><td>C.</td><td>40-41/R</td><td></td><td>F.</td><td></td><td>44-45/R</td><td>HEF Line # when complete _____</td><td></td><td>Quex ....2</td><td></td><td></td></tr>
</table>

47-48

STATUS CODES FOR OUTCOME (COLUMN F)

TEMPORARY   FINAL (Requires supervisor's approval)

HEF
31 No action
32 No one home/No answer/Busy
35 Temporary refusal/breakoff
36 Appointment
37 Broken appointment
38 HH not accessible
39 Other

QUEX
43 R not home/unavailable
45 Temp refusal/breakoff
46 Appointment
47 Broken appointment
49 Other

HEF
10 No HU*
11 Vacant*
12 HH speaks no English*
82 Refusal/breakoff*
84 Not home/unavailable*
85 HH not accessible*
87 Other*

QUEX
13 R does not speak English*
92 Refusal/breakoff*
94 R absent entire field period*
97 Other*
60 Ballot Quex only
65 Ballot Quex & some, not all documents
67 Ballot Quex & all documents

EVERYONE COMPLETE SAMPLING REPORT   USE CONTINUATION SHEET IF NECESSARY   *FILL OUT NIR

Random Numbers

10480	15011	01536	02011	81647	91646	69179	14194	62590	36207	20969	99570	91291	90700
22368	46573	25595	85393	30995	89198	27982	53402	93965	34095	52666	19174	39615	99505
24130	48360	22527	97265	76393	64809	15179	24830	49340	32081	30680	19655	63348	58629
42167	93093	06243	61680	07856	16376	39440	53537	71341	57004	00849	74917	97758	16379
37570	39975	81837	16656	06121	91782	60468	81305	49684	60672	14110	06927	01263	54613
77921	06907	11008	42751	27756	53498	18602	70659	90655	15053	21916	81825	44394	42880
99562	72905	56420	69994	98872	31016	71194	18738	44013	48840	63213	21069	10634	12952
96301	91977	05463	07972	18876	20922	94595	56869	69014	60045	18425	84903	42508	32307
89579	14342	63661	10281	17453	18103	57740	84378	25331	12566	58678	44947	05585	56941
85475	36857	53342	53988	53060	59533	38867	62300	08158	17983	16439	11458	18593	64952
28918	69578	88231	33276	70997	79936	56865	05859	90106	31595	01547	85590	91610	78188
63553	40961	48235	03427	49626	69445	18663	72695	52180	20847	12234	90511	33703	90322
09429	93969	52636	92737	88974	33488	36320	17617	30015	08272	84115	27156	30613	74952
10365	61129	87529	85689	48237	52267	67689	93394	01511	26358	85104	20285	29975	89868
07119	97336	71048	08178	77233	13916	47564	81056	97735	85977	29372	74461	28551	90707
51085	12765	51821	51259	77452	16308	60756	92144	49442	53900	70960	63990	75601	40719
02368	21382	52404	60268	89368	19885	55322	44819	01188	65255	64835	44919	05944	55157
01011	54092	33362	94904	31273	04146	18594	29852	71585	85030	51132	01915	92747	64951
52162	53916	46369	58586	23216	14513	83149	98736	23495	64350	94738	17752	35156	35749
07056	97628	33787	09998	42698	06691	76988	13602	51851	46104	88916	19509	25625	58104
48663	91245	85828	14346	09172	30168	90229	04734	59193	22178	30421	61666	99904	32812
54164	58492	22421	74103	47070	25306	76468	26384	58151	06646	21524	15227	96909	44592
32639	32363	05597	24200	13363	38005	94342	28728	35806	06912	17012	64161	18296	22851
29334	27001	87637	87308	58731	00256	45834	15398	46557	41135	10367	07684	36188	18510
02488	33062	28834	07351	19731	92420	60952	61280	50001	67658	32586	86679	50720	94953
81525	72295	04839	96423	24878	82651	66566	14778	76797	14780	13300	87074	79666	95725
29676	20591	68086	26432	46901	20849	89768	81536	86645	12659	92259	57102	80428	25280
00742	57392	39064	66432	84673	40027	32832	61362	98947	96067	64760	64584	96096	98253
05366	04213	25669	26422	44407	44048	37937	63904	45766	66134	75470	66520	34693	90449
91921	26418	64117	94305	26766	25940	39972	22209	71500	64568	91402	42416	07844	69618
00582	04711	87917	77341	42206	35126	74087	99547	81817	42607	43808	76655	62028	76630
00725	69884	62797	56170	86324	88072	76222	36086	84637	93161	76038	65855	77919	88006
69011	65795	95876	55293	18988	27354	26575	08625	40801	59920	29841	80150	12777	48501
25976	57948	29888	88604	67917	48708	18912	82271	65424	69774	33611	54262	85963	03547
09763	83473	73577	12908	30883	18317	28290	35797	05998	41688	34952	37888	38917	88050
91567	42595	27958	30134	04024	86385	29880	99730	55536	84855	29080	09250	79656	73211
17955	56349	90999	49127	20044	59931	06115	20542	18059	02008	73708	83517	36103	42791
46503	18584	18845	49618	02304	51038	20655	58727	28168	15475	56942	53389	20562	87338
92157	89634	94824	78171	84610	82834	09922	25417	44137	48413	25555	21246	35509	20468
14577	62765	35605	81263	39667	47358	56873	56307	61607	49518	89656	20103	77490	18062
98427	07523	33362	64270	01638	92477	66969	98420	04880	45585	46565	04102	46880	45709
34914	63976	88720	82765	34476	17032	87589	40836	32427	70002	70663	88863	77775	69348
70060	28277	39475	46473	23219	53416	94970	25832	69975	94884	19661	72828	00102	66794
53976	54914	06990	67245	68350	82948	11398	42878	80287	88267	47363	46634	06541	97809
76072	29515	40980	07391	58745	25774	22987	80059	39911	96189	41151	14222	60697	59583
90725	52210	83974	29992	65831	38857	50490	83765	55657	14361	31720	57375	56228	41546
64364	67412	33339	31926	14883	24413	59744	92351	97473	89286	35931	04110	23726	51900
08962	00358	31662	25388	61642	34072	81249	35648	56891	69352	48373	45578	78547	81788
95012	68379	93526	70765	10592	04542	76463	54328	02349	17247	28865	14777	62730	92277
15664	10493	20492	38391	91132	21999	59516	81652	27195	48223	46751	22923	32261	85653
16408	81899	04153	53381	79401	21438	83035	92350	36693	31238	59649	91754	72772	02338
18629	81953	05520	91962	04739	13092	97662	24822	94730	06496	35090	04822	86774	98289
73115	35101	47498	87637	99016	71060	88824	71013	18735	20286	23153	72924	35165	43040
57491	16703	23167	49323	45021	33132	12544	41035	80780	45393	44812	12515	98931	91202
30405	83946	23792	14422	15059	45799	22716	19792	09983	74353	68668	30429	70735	25499

16631	35006	85900	98275	32388	52390	16815	69298	82732	38480	73817	32523	41961	44437
96773	20206	42559	78985	05300	22164	24369	54224	35083	19687	11052	91491	60383	19746
38935	64202	14349	82674	66523	44133	00697	35552	35970	19124	63318	29686	03387	59846
31624	76384	17403	53363	44167	64486	64758	75366	76554	31601	12614	33072	60332	92325
78919	19474	23632	27889	47914	02584	37680	20801	72152	39339	34806	08930	85001	87820
03931	33309	57047	74211	63445	17361	62825	39908	05607	91284	68833	25570	38818	46920
74426	33278	43972	10119	89917	15665	52872	73823	73144	88662	88970	74492	51805	99378
09066	00903	20795	95452	92648	45454	09552	88815	16553	51125	79375	97596	16296	66092
42238	12426	87025	14267	20979	04508	64535	31355	86064	29472	47689	05974	52468	16834
16153	08002	26504	41744	81959	65642	74240	56302	00033	67107	77510	70625	28725	34191
21457	40742	29820	96783	29400	21840	15035	34537	33310	06116	95240	15957	16572	06004
21581	57802	02050	89728	17937	37621	47075	42080	97403	48626	68995	43805	33386	21597
55612	78095	83197	33732	05810	24813	86902	60397	16489	03264	88525	42786	05269	92532
44657	66999	99324	51281	84463	60563	79312	93454	68876	25471	93911	25650	12682	73572
91340	84979	46949	81973	37949	61023	43997	15263	80644	43942	89203	71795	99533	50501
91227	21199	31935	27022	84067	05462	35216	14486	29891	68607	41867	14951	91696	85065
50001	38140	66321	19924	72163	09538	12151	06878	91903	18749	34405	56087	82790	70925
65390	05224	72958	28609	81406	39147	25549	48542	42627	45233	57202	94617	23772	07896
27504	96131	83944	41575	10573	08619	64482	73923	36152	05184	94142	25299	84387	34925
37169	94851	39117	89632	00959	16487	65536	49071	39782	17095	02330	74301	00275	48280
11508	70225	51111	38351	19444	66499	71945	05422	13442	78675	84081	66938	93654	59894
37449	30362	06694	54690	04052	53115	62757	95348	78662	11163	81651	50245	34971	52924
46515	70331	85922	38329	57015	15765	97161	17869	45349	61796	66345	81073	49106	79860
30986	81223	42416	58353	21532	30502	32305	86482	05174	07901	54339	58861	74818	46942
63798	64995	46583	09785	44160	78128	83991	42865	92520	83531	80377	35909	81250	54238
82486	84846	99254	67632	43218	50076	21361	64816	51202	88124	41870	52689	51275	83556
21885	32906	92431	09060	64297	51674	64126	62570	26123	05155	59194	52799	28225	85762
60336	98782	07408	53458	13564	59089	26445	29789	85205	41001	12535	12133	14645	23541
43937	46891	24010	25560	86355	33941	25786	54990	71899	15475	95434	98227	21824	19585
97656	63175	89303	16275	07100	92063	21942	18611	47348	20203	18534	03862	78095	50136
03299	01221	05418	38982	55758	92237	26759	86367	21216	98442	08303	56613	91511	75928
79626	06486	03574	17668	07785	76020	79924	25651	83325	88428	85076	72811	22717	50585
85636	68335	47539	03129	65651	11977	02510	26113	99447	68645	34327	15152	55230	93448
18039	14367	61337	06177	12143	46609	32989	74014	64708	00533	35398	58408	13261	47908
08362	15656	60627	36478	65648	16764	53412	09013	07832	41574	17639	82163	60859	75567
79556	29068	04142	16268	15387	12856	66227	38358	22478	73373	88732	09443	82558	05250
92608	82674	27072	32534	17075	27698	98204	63863	11951	34648	88022	56148	34925	57031
23982	25835	40055	67006	12293	02753	14827	23235	35071	99704	37543	11601	35503	85171
09915	96306	05908	97901	28395	14186	00821	80703	70426	75647	76310	88717	37890	40129
59037	33300	26695	62247	69927	76123	50842	43834	86654	70959	79725	93872	28117	19233
42488	78077	69882	61657	34136	79180	97526	43092	04098	73571	80799	76536	71255	64239
46764	86273	63003	93017	31204	36692	40202	35275	57306	55543	53203	18098	47625	88684
03237	45430	55417	63282	90816	17349	88298	90183	36600	78406	06216	95787	42579	90730
86591	81482	52667	61582	14972	90053	89534	76036	49199	43716	97548	04379	46370	28672
38534	01715	94964	87288	65680	43772	39560	12918	86537	62738	19636	51132	25739	56947

Abridged from *Handbook of Tables for Probability and Statistics,* 2nd ed., edited by William H. Beyer (Cleveland, OH The Chemical Rubber Company, 1968). Used by permission of The Chemical Rubber Company.

Distribution of Chi Square

			Probability				
df	.99	.98	.95	.90	.80	.70	.50
1	.0³157	.0³628	.00393	.0158	.0642	.148	.455
2	.0201	.0404	.103	.211	.446	.713	1.386
3	.115	.185	.352	.584	1.005	1.424	2.366
4	.297	.429	.711	1.064	1.649	2.195	3.357
5	.554	.752	1.145	1.610	2.343	3.000	4.351
6	.872	1.134	1.635	2.204	3.070	3.828	5.348
7	1.239	1.564	2.167	2.833	3.822	4.671	6.346
8	1.646	2.032	2.733	3.490	4.594	5.528	7.344
9	2.088	2.532	3.325	4.168	5.380	6.393	8.343
10	2.558	3.059	3.940	4.865	6.179	7.267	9.342
11	3.053	3.609	4.575	5.578	6.989	8.148	10.341
12	3.571	4.178	5.226	6.304	7.807	9.034	11.340
13	4.107	4.765	5.892	7.042	8.634	9.926	12.340
14	4.660	5.368	6.571	7.790	9.467	10.821	13.339
15	5.229	5.985	7.261	8.547	10.307	11.721	14.339
16	5.812	6.614	7.962	9.312	11.152	12.624	15.338
17	6.408	7.255	8.672	10.085	12.002	13.531	16.338
18	7.015	7.906	9.390	10.865	12.857	14.440	17.338
19	7.633	8.567	10.117	11.651	13.716	15.352	18.338
20	8.260	9.237	10.851	12.443	14.578	16.266	19.337
21	8.897	9.915	11.591	13.240	15.445	17.182	20.337
22	9.542	10.600	12.338	14.041	16.314	18.101	21.337
23	10.196	11.293	13.091	14.848	17.187	19.021	22.337
24	10.856	11.992	13.848	15.659	18.062	19.943	23.337
25	11.524	12.697	14.611	16.473	18.940	20.867	24.337
26	12.198	13.409	15.379	17.292	19.820	21.792	25.336
27	12.879	14.125	16.151	18.114	20.703	22.719	26.336
28	13.565	14.847	16.928	18.939	21.588	23.647	27.336
29	14.256	15.574	17.708	19.768	22.475	24.577	28.336
30	14.953	16.306	18.493	20.599	23.364	25.508	29.336

continued

For larger values of df, the expression $\sqrt{2\chi^2} - \sqrt{2df - 1}$ may be used as a normal deviate with unit variance, remembering that the probability of χ^2 corresponds with that of a single tail of the normal curve.

Source: I am grateful to the Literary Executor of the late Sir Ronald A. Fisher, F.R.S., to Dr. Frank Yates, F.R.S., and to Longman Group Ltd., London, for permission to reprint Table IV from their book *Statistical Tables for Biological, Agricultural, and Medical Research* (6th Edition, 1974).

Probability

df	.30	.20	.10	.05	.02	.01	.001
1	1.074	1.642	2.706	3.841	5.412	6.635	10.827
2	2.408	3.219	4.605	5.991	7.824	9.210	13.815
3	3.665	4.642	6.251	7.815	9.837	11.341	16.268
4	4.878	5.989	7.779	9.488	11.668	13.277	18.465
5	6.064	7.289	9.236	11.070	13.388	15.086	20.517
6	7.231	8.558	10.645	12.592	15.033	16.812	22.457
7	8.383	9.803	12.017	14.067	16.622	18.475	24.322
8	9.524	11.030	13.362	15.507	18.168	20.090	29.125
9	10.656	12.242	14.684	16.919	19.679	21.666	27.877
10	11.781	13.442	15.987	18.307	21.161	23.209	29.588
11	12.899	14.631	17.275	19.675	22.618	24.725	31.264
12	14.011	15.812	18.549	21.026	24.054	26.217	32.909
13	15.119	16.985	19.812	22.362	25.472	27.688	34.528
14	16.222	18.151	21.064	23.685	26.873	29.141	36.123
15	17.322	19.311	22.307	24.996	28.259	30.578	37.697
16	18.841	20.465	23.542	26.296	29.633	32.000	39.252
17	15.511	21.615	24.769	27.587	30.995	33.409	40.790
18	20.601	22.760	25.989	28.869	32.346	34.805	42.312
19	21.689	23.900	27.204	30.144	33.687	36.191	43.820
20	22.775	25.038	28.412	31.410	35.020	37.566	45.315
21	23.858	26.171	29.615	32.671	36.343	38.932	46.797
22	24.939	27.301	30.813	33.924	37.659	40.289	48.268
23	26.018	28.429	32.007	35.172	38.968	41.638	49.728
24	27.096	29.553	33.196	36.415	40.270	42.980	51.179
25	28.172	30.675	34.382	37.652	41.566	44.314	52.620
26	29.246	31.795	35.563	38.885	42.856	45.642	54.052
27	30.319	32.912	36.741	40.113	44.140	46.963	55.476
28	31.391	34.027	37.916	41.337	45.419	48.278	56.893
29	32.461	35.139	39.087	42.557	46.693	49.588	58.302
30	35.530	36.250	40.256	43.773	47.962	50.892	59.703

Normal Curve Areas

z	.00	.01	.02	.03	.04	.05	.06	.07	.08	.09
0.0	.0000	.0040	.0080	.0120	.0160	.0199	.0239	.0279	.0319	.0359
0.1	.0398	.0438	.0478	.0517	.0557	.0596	.0636	.0675	.0714	.0753
0.2	.0793	.0832	.0871	.0910	.0948	.0987	.1026	.1064	.1103	.1141
0.3	.1179	.1217	.1255	.1293	.1331	.1368	.1406	.1443	.1480	.1517
0.4	.1554	.1591	.1628	.1664	.1700	.1736	.1772	.1808	.1844	.1879
0.5	.1915	.1950	.1985	.2019	.2054	.2088	.2123	.2157	.2190	.2224
0.6	.2257	.2291	.2324	.2357	.2389	.2422	.2454	.2486	.2517	.2549
0.7	.2580	.2611	.2642	.2673	.2704	.2734	.2764	.2794	.2823	.2852
0.8	.2881	.2910	.2939	.2967	.2995	.3023	.3051	.3078	.3106	.3133
0.9	.3159	.3186	.3212	.3238	.3264	.3289	.3315	.3340	.3365	.3389
1.0	.3413	.3438	.3461	.3485	.3508	.3531	.3554	.3577	.3599	.3621
1.1	.3643	.3665	.3686	.3708	.3729	.3749	.3770	.3790	.3810	.3830
1.2	.3849	.3869	.3888	.3907	.3925	.3944	.3962	.3980	.3997	.4015
1.3	.4032	.4049	.4066	.4082	.4099	.4115	.4131	.4147	.4162	.4177
1.4	.4192	.4207	.4222	.4236	.4251	.4265	.4279	.4292	.4306	.4319
1.5	.4332	.4345	.4357	.4370	.4382	.4394	.4406	.4418	.4429	.4441
1.6	.4452	.4463	.4474	.4484	.4495	.4505	.4515	.4525	.4535	.4545
1.7	.4554	.4564	.4573	.4582	.4591	.4599	.4608	.4616	.4625	.4633
1.8	.4641	.4649	.4656	.4664	.4671	.4678	.4686	.4693	.4699	.4706
1.9	.4713	.4719	.4726	.4732	.4738	.4744	.4750	.4756	.4761	.4767
2.0	.4772	.4778	.4783	.4788	.4793	.4798	.4803	.4808	.4812	.4817
2.1	.4821	.4826	.4830	.4834	.4838	.4842	.4846	.4850	.4854	.4857
2.2	.4861	.4864	.4868	.4871	.4875	.4878	.4881	.4884	.4887	.4890
2.3	.4893	.4896	.4898	.4901	.4904	.4906	.4909	.4911	.4913	.4916
2.4	.4918	.4920	.4922	.4925	.4927	.4929	.4931	.4932	.4934	.4936
2.5	.4938	.4940	.4941	.4943	.4945	.4946	.4948	.4949	.4951	.4952
2.6	.4953	.4955	.4956	.4957	.4959	.4960	.4961	.4962	.4963	.4964
2.7	.4965	.4966	.4967	.4968	.4969	.4970	.4971	.4972	.4973	.4974
2.8	.4974	.4975	.4976	.4977	.4977	.4978	.4979	.4979	.4980	.4981
2.9	.4981	.4982	.4982	.4983	.4984	.4984	.4985	.4985	.4986	.4986
3.0	.4987	.4987	.4987	.4988	.4988	.4989	.4989	.4989	.4990	.4990

Abridged from Table I of *Statistical Tables and Formulas*, by A. Hald (New York: John Wiley & Sons, Inc., 1952). Used by permission of John Wiley & Sons, Inc.

Estimated Sampling Error

How to use this table: Find the intersection between the sample size and the approximate percentage distribution of the binomial in the sample. The number appearing at this intersection represents the estimated sampling error, at the 95 percent confidence level, expressed in percentage points (plus or minus).

Example: In the sample of 400 respondents, 60 percent answer yes and 40 percent answer no.

The sampling error is estimated at plus or minus 4.9 percentage points. The confidence interval, then, is between 55.1 percent and 64.9 percent. We would estimate (95 percent confidence) that the proportion of the total population who would say yes is somewhere within that interval.

Sample Size	Binomial Percentage Distribution				
	50/50	60/40	70/30	80/20	90/10
100	10	9.8	9.2	8	6
200	7.1	6.9	6.5	5.7	4.2
300	5.8	5.7	5.3	4.6	3.5
400	5	4.9	4.6	4	3
500	4.5	4.4	4.1	3.6	2.7
600	4.1	4	3.7	3.3	2.4
700	3.8	3.7	3.5	3	2.3
800	3.5	3.5	3.2	2.8	2.1
900	3.3	3.3	3.1	2.7	2
1000	3.2	3.1	2.9	2.5	1.9
1100	3	3	2.8	2.4	1.8
1200	2.9	2.8	2.6	2.3	1.7
1300	2.8	2.7	2.5	2.2	1.7
1400	2.7	2.6	2.4	2.1	1.6
1500	2.6	2.5	2.4	2.1	1.5
1600	2.5	2.4	2.3	2	1.5
1700	2.4	2.4	2.2	1.9	1.5
1800	2.4	2.3	2.2	1.9	1.4
1900	2.3	2.2	2.1	1.8	1.4
2000	2.2	2.2	2	1.8	1.3

Twenty Questions a Journalist Should Ask about Poll Results*

Sheldon R. Gawiser
G. Evans Witt

For journalists and for pollsters, questions are the most frequently used tool for gathering information. Here are 20 questions for the journalist to ask the pollster before reporting poll results. This publication is designed to help the working journalist do a thorough, professional job covering polls; it is not a primer on how to conduct a public opinion survey.

This work is about polls that are "scientific." A number of the questions will help you decide whether or not a poll is a "scientific" one worthy of coverage—or an unscientific survey that may be entertaining, but meaningless.

Of the scientific polls, the Gallup Poll is probably the best known and one of the oldest. There are many other excellent surveys conducted by reputable firms, as well.

The unscientific surveys are less well-known, if quite widespread. There are 900-number call-in polls, man-on-the-street surveys, shopping mall polls, and even the classic toilet tissue poll featuring pictures of the candidates on each sheet.

The major distinguishing difference between scientific and unscientific polls is who picks the respondents for the survey. In a scientific poll, the pollster identifies and seeks out the people to be interviewed. In an unscientific poll, the respondents usually "volunteer" their opinions, selecting themselves for the poll.

*From Sheldon Gawiser and G. Evans Witt, © National Council on Public Polls. Reprinted by permission.

The results of the well-conducted scientific poll can provide a reliable guide to the opinions of many people in addition to those interviewed—even the opinions of all Americans. The results of an unscientific poll tell you nothing more than simply what those respondents say.

With these 20 questions in hand, the journalist can seek the facts to decide just how to handle every poll that comes across the news desk each day.

1. Who did the poll? What polling firm, research house, political campaign, corporation, or other group conducted the poll? This is always the first question to ask.

If you don't know who did the poll, you can't get the answers to all the other questions listed here. And if the person providing poll results can't or won't tell you who did it, serious questions must be raised about the reliability and truthfulness of the results being presented.

In most cases, reputable polling firms will provide you with the information you need to evaluate the survey. And because reputation is important to a quality firm, a professionally conducted poll will avoid many errors.

2. Who paid for the poll and why was it done? You must know who paid for the survey, because that tells you—and your audience—who thought these topics were important enough to spend money finding out what people think. And that goes to the whole issue of why the poll was done.

Polls usually are not conducted for the good of the world. They are conducted for a reason—either to gain helpful information or to advance a particular cause.

It may be the news organization wants to develop a good story. It may be the politician wants to be re-elected. It may be that the corporation is trying to push sales of its new product. Or a special interest group may be trying to prove that its views are the views of the entire country.

All are legitimate reasons for doing a poll.

The important issue for you as a journalist is whether the motive for doing the poll creates such serious doubts about the validity of the poll results that the results should not be publicized.

An example of suspect polls are private polls conducted by a political campaign. These polls are conducted solely to help the candidate win—and for no other reason. The poll may have terrifically slanted questions or a strange sampling methodology, all with a tactical campaign purpose. For example, the campaign may be testing out new slogans or a new stance on a key issue or a new attack on the opponent.

But since accurately gauging the general public's sentiments is not the goal of the candidate's poll, the results should be reported with great care.

Likewise, reporting on a survey by a special interest group is tricky. For example, an environmental group trumpets a poll saying the American people support strong measures to protect the environment. That may be true, but the poll may have been conducted for a group with definite views. That may have swayed the question wording, the timing of the poll, the group interviewed, and the order of the questions. You should examine the poll to be certain that it accurately samples public opinion—and does not simply push a single viewpoint.

3. How many people were interviewed for the survey? This is another basic piece of information you should have. Because polls give approximate answers, the more people interviewed in a scientific poll, the smaller the error due to the size of the sample, all other things being equal.

However, a common trap to avoid is that "more is automatically better." It is absolutely true that the more people interviewed in a reliable survey, the smaller the sampling error—all other things being equal. But, other factors may be more important in judging the quality of a survey.

4. How were those people chosen? The key reason that some polls reflect public opinion accurately and other polls are unscientific junk is how the people were chosen to be interviewed. In scientific polls, the pollster uses a specific method for picking respondents. In unscientific polls, the person picks himself to participate.

The method pollsters use to pick interviewees relies on the bedrock of mathematical reality: when the chance of selecting each person in the target population is known, then and only then do the results of the sample survey reflect the entire population. This is called a random sample or a probability sample. This is the reason that interviews with 1000 American adults can accurately reflect the opinions of more than 185 million American adults.

Most scientific samples use special techniques to be economically feasible. For example, some sampling methods for telephone interviewing do not just pick randomly generated telephone numbers. Only telephone exchanges that are known to contain working residential numbers are selected—to reduce the number of wasted calls.

But even a random sample cannot be purely random in practice since some people don't have phones, refuse to answer, or aren't home.

5. What area: nation, state, or region—or what group: teachers, lawyers, Democratic voters, etc.—were these people chosen from? Although the results of probability samples can be projected to represent the larger population from which they were selected, the characteristics of the larger population must be specified; for example, you should know if a sample represents all people in the United States or just those in one state or one city. In another example, the case of telephone samples, the population is that of people living in households with telephones. For most purposes, telephone households may be similar to the general population. But, if you were reporting a poll on what it was like to be poor or homeless, this would not be the appropriate sample. Remember, the use of a scientific sampling technique does not mean that the correct population was interviewed.

It is absolutely critical to know from which group the interviewees were chosen.

For example, a survey of business people can reflect the opinions of business people—but not of all adults. Only if the interviewees were chosen from among all American adults can the poll reflect the opinions of all American adults.

Political polls are especially sensitive to this issue.

In pre-primary and pre-election polls, how the people are chosen as the base for poll results is critical. A poll of all adults, for example, is not very useful on a primary race where only 25 percent of the registered voters actually turn out. So look for polls based on registered voters, "likely voters," previous primary voters, and such. These distinctions are important and should be included in the story.

One of the most variable aspects of polling is trying to figure out who actually is going to vote.

6. Are the results based on the answers of all the people interviewed? One of the easiest ways to misrepresent the results of a poll is to report the answers of only a subgroup. For example, there is usually a substantial difference between the opinions of Democrats and Republicans on campaign-related matters. Reporting the opinions of only Democrats in a poll purported to be of all adults would substantially misrepresent the results.

Poll results based on Democrats must be identified as such and should be reported as representing only Democratic opinions.

Of course, reporting on just one subgroup can be exactly the right course. In polling on a Republican primary contest, it is the opinions of the Republicans who can vote in the primary that count—not those of Democrats who cannot vote in that GOP contest.

7. Who should have been interviewed and was not? You ought to know how many people refused to answer the survey or were never contacted.

The non-response rate is the percentage of people contacted who should have been interviewed, but were not. They may have refused attempts

to interview them. Or interviews may not have been attempted if people were not home when the interviewer called.

The results of a survey should be judged very differently if the 100 convention delegates interviewed were a random sample of 1000 delegates as compared to their being the only 100 out of the 1000 willing to participate. The same potential for distorted results occurs if some of the delegates who were in the sample were never actually contacted.

8. When was the poll done? Events have a dramatic impact on poll results. Your interpretation of a poll should depend on when it was conducted relative to key events. Even the freshest poll results can be overtaken by subsequent events. The President may have given a stirring speech to the nation, the stock market may have crashed, or an oil tanker may have sunk, spilling millions of gallons of crude on beautiful beaches.

Poll results that are several weeks or months old may be perfectly valid as history, but are not always newsworthy.

9. How were the interviews conducted? There are three main possibilities: in person at home, by telephone, or by mail.

Most surveys are now conducted by telephone, with the calls made from a central interviewing center. However, some surveys are still conducted by sending interviewers into people's homes to conduct the interviews.

Some surveys are conducted by mail. In scientific polls, the pollster picks the people to receive the mail questionnaires. The respondent fills out the questionnaire and returns it.

Mail surveys can be excellent sources of information, but it takes weeks to do a mail survey, meaning that the results cannot be as timely as a telephone survey. And mail surveys can be subject to other kinds of errors, particularly low response rates. In many mail surveys, more people fail to participate than do. This makes the results suspect.

Surveys done in shopping malls, in stores or restaurants or on the sidewalk may have their uses for their sponsors, but publishing the results in the media is not among them. These "man in the street" approaches may yield interesting human interest stories, but they should never be treated as if they represent a public opinion poll.

10. Is this a dial-in poll, a mail-in poll, or a subscriber coupon poll? If the poll you are looking at is a dial-in, mail-in, or coupon poll, don't report the results because the respondents are self-selected. These pseudo-polls have no validity. Remember, the purpose of a poll is to draw conclusions about the population, not about the sample. In these pseudo-polls there is no way to project the results to any larger group. Scientific polls usually show different results than pseudo-polls.

The 900-number dial-in polls may be fine for deciding whether or not Larry the Lobster should be cooked on *Saturday Night Live* or even for dedicated fans to express their opinions on who is the greatest quarterback in the National Football League, but they have only entertainment value. There is no way to tell who actually called in, how old they are, or how many times each person called.

Never be fooled by the number of responses. In some cases, a few people call in thousands of times. Even if 500,000 calls are tallied, no one has any real knowledge of what the results mean. If big numbers impress you, remember that the *Literary Digest*'s non-scientific sample of 12,000,000 people said Landon would beat Roosevelt.

The subscriber coupon polls are just as bad. In these cases, the magazine or newspaper includes a coupon to be mailed in with the answers to the questions. Again, there is no way to know who responded and how many times. These results are not projectable even to the subscribers of the publication that includes the coupon.

11. What is the sampling error for the poll results? Interviews with a scientific sample of 1000 adults can accurately reflect the opinions of more than 185 million American adults. That means interviews attempted with all 185 million adults—if such were possible—would give approximately the same results as a well-conducted survey.

But what happens if another carefully done poll of 1000 adults gives slightly different results from the first survey? Neither of the polls is "wrong." This range of results is called the error due to sampling, often called the margin of error.

This is not an "error" in the sense of making a mistake. Rather, it is a measure of the possible range of approximation in the results because a sample was used.

Pollsters express the size of the uncertainty caused by using a sample at a "confidence level." This means a sample is likely to be within so many points of the results one would have gotten if an interview was attempted with the entire target population. They usually say this with 95% confidence.

Thus, for example, a "3 percentage point margin of error" in a national poll means that if the attempt was made to interview every adult in the nation with the same questions in the same way at about the same time as the poll was taken, the poll's answers would fall within plus or minus 3 percentage points of the complete count result 95% of the time.

Please note that this does not address the issue of whether or not people cooperate with the survey, or if the questions are understood, or if any other methodological issue exists. The sampling error is only the portion of the potential error in a survey introduced by using a sample rather than the entire population. Sampling error tells us nothing about the refusals or those consistently unavailable for interview; it also tells us nothing about the biasing effects of a particular question wording or the bias a particular interviewer may inject into the interview situation.

Remember that the sampling error margin applies to each figure in the results—it is at least 3 percentage points plus or minus for each one. Thus, in a poll question matching two candidates for President, both figures are subject to sampling error.

This raises one of the thorniest problems in the presentation of poll results. For a horse-race poll, when is one candidate really ahead of the other?

Certainly, when the gap between the two candidates is more than twice the error margin— 6 percentage points in our example—you can say with confidence that the poll says Candidate A is leading Candidate B.

And just as certainly, if the gap between the two candidates is less than error margin then you should not say that one candidate is ahead of the other. Then, the race is "close"; the race is "roughly even"; or there is "little difference between the candidates." And bear in mind that when subgroup results are reported—women or blacks, or young people—the sampling error margin for those figures is greater than for results based on the survey as a whole.

12. What other kinds of mistakes can skew poll results? The margin of sampling error is just one source of inaccuracy in a poll and not necessarily the greatest source of error; we use it because it's the only one that can be quantified. Question phrasing and ordering are also a likely source of flaws. That's why you need to examine poll questions for bias.

You should always ask if the poll results have been "weighted." This process is usually used to account for unequal probabilities of selection and to correct demographics in the sample. However, you should be aware that a poll can also be unduly manipulated by weighting to produce some desired result.

And there are other possible sources of error. These include issues such as inadequate interviewer training and supervision, data processing errors, and other operational problems. Professional polling operations are less subject to these problems than volunteer-conducted polls, which are usually less trustworthy.

13. What questions were asked? You must find out the exact wording of the poll questions. Why? Because the very wording of questions can make major differences in the results.

Perhaps the best test of any poll question is your reaction to it. On the face of it, does the question seem fair and unbiased? Does it present a balanced set of choices? Would people you know be able to answer the question?

On sensitive questions—such as abortion—the complete wording of the question should probably be included in your story. But at the very least, you must have the exact wording as you are preparing the story.

It may well be worthwhile to compare the results of several different polls from different orga-

nizations on these sensitive questions. In that case, you should be careful to compare both the results and the exact wording of the questions.

14. In what order were the questions asked? Sometimes the very order of the questions can have an impact on the results. Often that impact is intentional; sometimes, it is not. The impact of order can often be subtle.

During troubled economic times, for example, if people are asked what they think of the economy before they are asked their opinion of the president, the presidential popularity rating will probably be lower than if you had reversed the questions. And in good economic times, the opposite is true.

In political polls, campaign consultants often ask a series of questions about various issue positions of the candidates—or various things that could be said about the candidates. After these questions are asked, the horse-race question is asked, usually for the second time in the poll. This second horse-race question is then examined to see if the questions about issues and positions swayed any opinions. This may be a good way to test issues. It is a poor way to test the candidates' true standings in the public's mind.

What is important here is whether the questions that went before the important question affect the results. If the poll asks questions about abortion just before a question about an abortion ballot measure, those previous questions could sway the results.

15. What other polls have been done on this topic? Do they say the same thing? If they are different, why are they different? Results of other polls—a candidate's opponent, public polls, media polls, or whatever—should be used to check and contrast poll results you have in hand.

If the polls differ, first check the timing of when the interviewing was done. The different poll results may demonstrate a swing in public opinion.

If the polls were done about the same time, and no other factor seems to explain the disagreement, go to each poll sponsor and ask for an explanation of the differences. Conflicting polls often make good stories.

16. So, the poll says the race is all over. What now? No matter how good the poll, no matter how wide the margin, no matter how big the sample, a pre-election poll does not show that one candidate has the race "locked up." Things change—often and dramatically in politics.

17. Was the poll part of a fund-raising effort? This is another example of a pseudo-poll. An organization sends out a survey form to a large list of people. The last question usually asks for a contribution from the respondent. He or she is expected to send money to support the organization or pay for tabulating the survey.

The people who respond to these types of surveys are likely to be those who agree with the organization's goals. Also, the questions are usually loaded and the results meaningless.

This technique is used by a wide variety of organizations from political parties and special interest groups to charitable organizations. Again,

if the poll in question is part of a fund-raising pitch, pitch it—in the waste basket.

18. So I've asked all the questions. The answers sound good. The poll is correct, right? Usually, yes. However, remember that the laws of chance alone say that the results of one poll out of 20 may be skewed away from the public's real views just because of sampling error.

19. With all these potential problems, should we ever report poll results? Yes. Because reputable polling organizations consistently do good work. In spite of the difficulties, the public opinion survey, correctly conducted, is still the best objective measure of the state of the views of the public.

20. Is this poll worth reporting? If the poll was conducted correctly, and you have been able to obtain the information outlined here, your news judgment and that of your editors should be applied to polls, as it is to every other element of a story.

Glossary

abstract   (1) A summary of a research article. The abstract usually begins the article and states the purpose of the research, the methods used, and the major findings. See Chapter 17. (2) An expensive painting you may not understand but may need to appreciate if you want to impress people at the art museum.

agreement reality   Those things we "know" as part of the culture we share with those around us. See Chapter 1.

analysis of variance (ANOVA)   Method of analysis in which cases under study are combined into groups representing an independent variable, and the extent to which the groups differ from one another is analyzed in terms of some dependent variable. Then, the extent to which the groups differ is compared with the standard of random distribution. See Chapter 16.

anonymity   Anonymity is achieved in a research project when neither the researchers nor the readers of the findings can identify a given response with a given respondent. See Chapter 3.

attributes   Characteristics of people or things. See *variables* and Chapter 1.

average   An ambiguous term generally suggesting typical or normal—a central tendency. The mean, median, and mode are specific examples of mathematical averages. See Chapter 14.

axial coding   A reanalysis of the results of *open coding* in the Grounded Theory Method, aimed at identifying the important, general concepts. See also *selective coding* and Chapter 13.

bias   (1) That quality of a measurement device that tends to result in a misrepresentation of what is being measured in a particular direction. For example, the questionnaire item "Don't you agree that the president is doing a good job?" would be biased in that it would generally encourage more favorable responses. See Chapter 9. (2) The thing inside you that makes other people or groups seem consistently better or worse than they really are. (3) What a nail looks like after you hit it crooked. (If you drink, don't drive.)

bivariate analysis   The analysis of two variables simultaneously, for the purpose of determining the empirical relationship between them. The construction of a simple percentage table or the computation of a simple correlation coefficient are examples of bivariate analyses. See Chapter 14 for more on this topic.

Bogardus social distance scale   (1) A measurement technique for determining the willingness of people to participate in social relations—of varying degrees of closeness—with other kinds of people. It is an especially efficient technique in that one can summarize several discrete answers without losing any of the original details of the data. See Chapter 6. (2) The distance you might be prepared to travel to see a rarely shown black-and-white movie of good ol' Humphrey.*

case study   The in-depth examination of a single instance of some social phenomenon, such as a village, a family, or a juvenile gang. See Chapter 10.

case-oriented analysis   (1) An analysis that aims to understand a particular case or several cases by looking closely at the details of each. See Chapter 13. (2) A private investigator's billing system.

closed-ended questions   Survey questions in which the respondent is asked to select an answer from among a list provided by the researcher. Popular in survey research because they provide

*Supplemental definitions marked with an asterisk have been provided courtesy of James Instone, University of New England, Armidale, NSW, Australia.

a greater uniformity of responses and are more easily processed than *open-ended questions*. See Chapter 9.

cluster sampling (1) A multistage sampling in which natural groups (clusters) are sampled initially, with the members of each selected group being subsampled afterward. For example, you might select a sample of U.S. colleges and universities from a directory, get lists of the students at all the selected schools, then draw samples of students from each. See Chapter 7. (2) Pawing around in a box of macadamia nut clusters to take all the big ones for yourself.

codebook (1) The document used in data processing and analysis that tells the location of different data items in a data file. Typically, the codebook identifies the locations of data items and the meaning of the codes used to represent different attributes of variables. See Chapter 14. (2) The document that cost you 38 box tops just to learn that Captain Marvelous wanted you to brush your teeth and always tell the truth. (3) The document that allows CIA agents to learn that Captain Marvelous wants them to brush their teeth.

coding (1) The process whereby raw data are transformed into standardized form suitable for machine processing and analysis. See Chapter 11. (2) A strong drug you may take when you hab a bad code.

cohort study A study in which some specific subpopulation, or cohort, is studied over time, although data may be collected from different members in each set of observations. For example, a study of the occupational history of the class of 1970 in which questionnaires were sent every five years would be a cohort study. See Chapter 4 for more on this topic (if you want more). See also *longitudinal study, panel study,* and *trend study*.

comparative and historical research The examination of societies (or other social units) over time and in comparison with one another. See Chapter 11.

completion rate See *response rate*.

computer-assisted telephone interviewing (CATI) A data-collection technique in which a telephone-survey questionnaire is stored in a computer, permitting the interviewer to read the questions from the monitor and enter the answers on the computer keyboard. See Chapter 9.

concept mapping (1) The graphic display of concepts and their interrelations, useful in the formulation of theory. See Chapter 13. (2) A masculine technique for finding locations by logic and will, without asking for directions.

conceptualization (1) The mental process whereby fuzzy and imprecise notions (concepts) are made more specific and precise. So you want to study prejudice. What do you mean by "prejudice"? Are there different kinds of prejudice? What are they? See Chapter 5, which is all about conceptualization and its pal, operationalization. (2) Sexual reproduction among intellectuals.

confidence interval (1) The range of values within which a population parameter is estimated to lie. A survey, for example, may show 40 percent of a sample favoring Candidate A (poor devil). Although the best estimate of the support existing among all voters would also be 40 percent, we would not expect it to be exactly that. We might, therefore, compute a confidence interval (such as from 35 to 45 percent) within which the actual percentage of the population probably lies. Note that we must specify a confidence level in connection with every confidence interval. See Chapters 7 and 16. (2) How close you dare to get to an alligator.

confidence level (1) The estimated probability that a population parameter lies within a given confidence interval. Thus, we might be 95 percent confident that between 35 and 45 percent of all voters favor Candidate A. See Chapters 7 and 16. (2) How sure you are that the ring you bought from a street vendor for $10 is really a three-carat diamond.

confidentiality A research project guarantees confidentiality when the researcher can identify a given person's responses but promises not to do so publicly. See Chapter 3.

conflict paradigm A paradigm that views human behavior as attempts to dominate others or avoid being dominated by others. See Chapter 2.

constant comparative method (1) A component of the Grounded Theory Method in which observations are compared with one another and with the evolving inductive theory. See Chapter 13. (2) A blind-dating technique.

construct validity The degree to which a measure relates to other variables as expected within a system of theoretical relationships. See Chapter 5.

content analysis The study of recorded human communications, such as books, websites, paintings, and laws. See Chapter 11.

content validity   The degree to which a measure covers the range of meanings included within a concept. See Chapter 5.

contingency question   A survey question intended for only some respondents, determined by their responses to some other question. For example, all respondents might be asked whether they belong to the Cosa Nostra, and only those who said yes would be asked how often they go to company meetings and picnics. The latter would be a contingency question. See Chapter 9.

contingency table   (1) A format for presenting the relationships among variables as percentage distributions. See Chapter 14. (2) The card table you keep around in case your guests bring their seven kids with them to dinner.

continuous variable   A variable whose attributes form a steady progression, such as *age* or *income*. Thus, the ages of a group of people might include 21, 22, 23, 24, and so forth and could even be broken down into fractions of years. Contrast this with *discrete variables,* such as *gender* or *religious affiliation,* whose attributes form discontinuous chunks. See Chapter 14.

control group   (1) In experimentation, a group of subjects to whom no experimental stimulus is administered and who should resemble the experimental group in all other respects. The comparison of the control group and the experimental group at the end of the experiment points to the effect of the experimental stimulus. See Chapter 8. (2) American Association of Managers.

control variable   See *test variable.*

conversation analysis (CA)   A meticulous analysis of the details of conversation, based on a complete transcript that includes pauses, hems, and also haws. See Chapter 13.

correlation   (1) An empirical relationship between two variables such that (a) changes in one are associated with changes in the other or (b) particular attributes of one variable are associated with particular attributes of the other. Thus, for example, we say that *education* and *income* are correlated in that higher levels of education are associated with higher levels of income. Correlation in and of itself does not constitute a causal relationship between the two variables, but it is one criterion of causality. See Chapter 4. (2) Someone you and your friend are both related to.

cost-benefit studies   Studies that determine whether the results of a program justify its expense (both financial and other). See Chapter 12.

criterion-related validity   The degree to which a measure relates to some external criterion. For example, the validity of College Board tests is shown in their ability to predict the college success of students. Also called *predictive validity.* See Chapter 5.

critical race theory   A paradigm grounded in race awareness and an intention to achieve racial justice. See Chapter 2.

critical realism   A paradigm that holds things are real insofar as they produce effects. See Chapter 2.

cross-case analysis   An analysis that involves an examination of more than one case; this can be either a variable-oriented or case-oriented analysis. See Chapter 13.

cross-sectional study   A study based on observations representing a single point in time. Contrasted with a *longitudinal study.* See Chapter 4.

curvilinear regression analysis   A form of regression analysis that allows relationships among variables to be expressed with curved geometric lines instead of straight ones. See Chapter 16.

debriefing   (1) Interviewing subjects to learn about their experience of participation in the project. Especially important if there's a possibility that they have been damaged by that participation. See Chapter 3. (2) Pulling someone's shorts down. Don't do that. It's not nice.

deduction   The logical model in which specific expectations of hypotheses are developed on the basis of general principles. Starting from the general principle that all deans are meanies, you might anticipate that this one won't let you change courses. This anticipation would be the result of deduction. See also *induction* and Chapters 1 and 2. (2) What the Internal Revenue Service said your good-for-nothing moocher of a brother-in-law technically isn't. (3) Of a duck.

dependent variable   (1) A variable assumed to depend on or be caused by another (called the independent variable). If you find that *income* is partly a function of *amount of formal education, income* is being treated as a dependent variable. See Chapter 1. (2) A wimpy variable.

descriptive statistics   Statistical computations describing either the characteristics of a sample or the relationship among variables in a sample.

Descriptive statistics merely summarize a set of sample observations, whereas inferential statistics move beyond the description of specific observations to make inferences about the larger population from which the sample observations were drawn. See Chapter 16.

dimension A specifiable aspect of a concept. "Religiosity," for example, might be specified in terms of a belief dimension, a ritual dimension, a devotional dimension, a knowledge dimension, and so forth. See Chapter 5.

discrete variable (1) A variable whose attributes are separate from one another, or discontinuous, as in the case of *gender* or *religious affiliation.* Contrast this with *continuous variables,* in which one attribute shades off into the next. Thus, in *age* (a continuous variable), the attributes progress steadily from 21 to 22 to 23, and so forth, whereas there is no progression from male to female in the case of *gender.* See Chapter 14. (2) A variable that doesn't undress in public.*

discriminant analysis Method of analysis similar to multiple regression, except that the dependent variable can be nominal. See Chapter 16.

dispersion The distribution of values around some central value, such as an average. The range is a simple example of a measure of dispersion. Thus, we may report that the mean age of a group is 37.9, and the range is from 12 to 89. See Chapter 14.

distorter variable In the elaboration model, a test variable that reverses the direction of a zero-order relationship. See Chapter 15.

double-blind experiment An experimental design in which neither the subjects nor the experimenters know which is the experimental group and which is the control. See Chapter 8.

ecological fallacy Erroneously drawing conclusions about individuals solely from the observation of groups. See Chapter 4.

elaboration model A logical model for understanding the relationship between two variables by controlling for the effects of a third. Developed principally by Paul Lazarsfeld. The various outcomes of an elaboration analysis are *replication, specification, explanation,* and *interpretation.* See Chapter 15.

element (1) That unit of which a population is composed and which is selected in a sample. Distinguished from *units of analysis,* which are used in data analysis. See Chapter 7. (2) What an elephant eats when it has bad breath.*

emancipatory research Research conducted for the purpose of benefiting disadvantaged groups. See Chapter 10.

epistemology The science of knowing; systems of knowledge. See Chapter 1.

EPSEM (equal probability of selection method) A sample design in which each member of a population has the same chance of being selected into the sample. See Chapter 7.

ethnography A report on social life that focuses on detailed and accurate description rather than explanation. See Chapter 10.

ethnomethodology An approach to the study of social life that focuses on the discovery of implicit, usually unspoken assumptions and agreements; this method often involves the intentional breaking of agreements as a way of revealing their existence. See Chapter 10.

evaluation research Research undertaken for the purpose of determining the impact of some social intervention, such as a program aimed at solving a social problem. See Chapter 12.

ex post facto hypothesis A hypothesis created after confirming data have already been collected. It is a meaningless construct because there is no way for it to be disconfirmed. See Chapter 15.

experimental group In experimentation, a group of subjects to whom an experimental stimulus is administered. Compare with *control group.* See Chapter 8.

explanation (1) An elaboration model outcome in which the original relationship between two variables is revealed to have been spurious, because the relationship disappears when an antecedent test variable is introduced. See Chapter 15. (2) "My little sister ate my homework."

extended case method A technique developed by Michael Burawoy in which case-study observations are used to discover flaws in and to improve existing social theories. See Chapter 10.

external invalidity Refers to the possibility that conclusions drawn from experimental results may not be generalizable to the "real" world. See Chapter 8 and also *internal invalidity.*

external validation The process of testing the validity of a measure, such as an index or scale, by examining its relationship to other, presumed indicators of the same variable. If the index really measures prejudice, for example, it should correlate with other indicators of prejudice. See Chapter 6.

face validity (1) That quality of an indicator that makes it seem a reasonable measure of some variable. That the frequency of attendance at religious services is some indication of a person's religiosity seems to make sense without a lot of explanation. It has face validity. See Chapter 5. (2) When your face looks like your driver's license photo (rare and perhaps unfortunate).

factor analysis A complex algebraic method for determining the general dimensions or factors that exist within a set of concrete observations. See Chapter 16.

feminist paradigms Paradigms that (a) view and understand society through the experiences of women and/or (b) examine the generally deprived status of women in society. See Chapter 2.

focus group A group of subjects interviewed together, prompting a discussion. The technique is frequently used by market researchers, who ask a group of consumers to evaluate a product or discuss a type of commodity, for example. See Chapter 10.

frequency distribution (1) A description of the number of times the various attributes of a variable are observed in a sample. The report that 53 percent of a sample were men and 47 percent were women would be a simple example of a frequency distribution. See Chapter 14. (2) A radio dial.

Geographic Information Systems (GIS) Analytic technique in which researchers map quantitative data that describe geographic units for a graphic display. See Chapter 16.

grounded theory (1) An inductive approach to the study of social life that attempts to generate a theory from the constant comparing of unfolding observations. This is very different from hypothesis testing, in which theory is used to generate hypotheses to be tested through observations. See Chapter 10. (2) A theory that is not allowed to fly.

Grounded Theory Method (GTM) An inductive approach to research, introduced by Barney Glaser and Anselm Strauss, in which theories are generated solely from an examination of data rather than being derived deductively. See Chapter 13.

Guttman scale (1) A type of composite measure used to summarize several discrete observations and to represent some more-general variable. See Chapter 6. (2) The device Louis Guttman weighs himself on.

hypothesis A specified testable expectation about empirical reality that follows from a more general proposition; more generally, an expectation about the nature of things derived from a theory. It is a statement of something that ought to be observed in the real world if the theory is correct. See *deduction* and Chapter 2.

idiographic An approach to explanation in which we seek to exhaust the idiosyncratic causes of a particular condition or event. Imagine trying to list all the reasons why you chose to attend your particular college. Given all those reasons, it's difficult to imagine your making any other choice. By contrast, see *nomothetic*. See also Chapter 1.

independent variable (1) A variable with values that are not problematic in an analysis but are taken as simply given. An independent variable is presumed to cause or determine a dependent variable. If we discover that religiosity is partly a function of gender—women are more religious than are men—*gender* is the independent variable and *religiosity* is the dependent variable. Note that any given variable might be treated as independent in one part of an analysis and dependent in another part of it. *Religiosity* might become an independent variable in the explanation of crime. See *dependent variable* and Chapter 1. (2) A variable that refuses to take advice.

index A type of composite measure that summarizes and rank-orders several specific observations and represents some more general dimension. Contrasted with *scale*. See Chapter 6.

indicator An observation that we choose to consider as a reflection of a variable we wish to study. Thus, for example, attending religious services might be considered an indicator of *religiosity*. See Chapter 5.

induction (1) The logical model in which general principles are developed from specific observations. Having noted that Jews and Catholics are more likely to vote Democratic than Protestants are, you might conclude that religious minorities in the United States are more affiliated with the Democratic party and then your task is to explain why. This would be an example of induction. See also *deduction* and Chapters 1 and 2. (2) The culinary art of stuffing ducks.

inferential statistics The body of statistical computations relevant to making inferences from findings based on sample observations to some larger population. See also *descriptive statistics* and Chapter 16. (Not to be confused with infernal statistics, a characterization sometimes invoked by frustrated statistics students.)

informant   Someone who is well versed in the social phenomenon that you wish to study and who is willing to tell you what he or she knows about it. If you were planning participant observation among the members of a religious sect, you would do well to make friends with someone who already knows about them—possibly a member of the sect—who could give you some background information about them. Not to be confused with a *respondent.* See Chapter 7.

informed consent   A norm in which subjects base their voluntary participation in research projects on a full understanding of the possible risks involved. See Chapter 3.

institutional ethnography   A research technique in which the personal experiences of individuals are used to reveal power relationships and other characteristics of the institutions within which they operate. See Chapter 10.

interest convergence   The thesis that majority group members will only support the interests of minorities when those actions also support the interests of the majority group. See Chapter 2.

internal invalidity   (1) Refers to the possibility that the conclusions drawn from experimental results may not accurately reflect what went on in the experiment itself. See Chapter 8 and also *external invalidity.* (2) What my grandad has and why he wears special "nappies."*

interpretation   A technical term used in connection with the elaboration model. It represents the research outcome in which a control variable is discovered to be the mediating factor through which an independent variable has its effect on a dependent variable. See Chapter 15.

interval measure   A level of measurement describing a variable whose attributes are rank-ordered and have equal distances between adjacent attributes. The Fahrenheit temperature scale is an example of this, because the distance between 17 and 18 is the same as that between 89 and 90. See also Chapter 5 and *nominal measure, ordinal measure,* and *ratio measure.*

interview   A data-collection encounter in which one person (an interviewer) asks questions of another (a *respondent*). Interviews may be conducted face-to-face or by telephone. See Chapter 9.

item analysis   An assessment of whether each of the items included in a composite measure makes an independent contribution or merely duplicates the contribution of other items in the measure. See Chapter 6.

latent content   (1) In connection with content analysis, the underlying meaning of communications, as distinguished from their *manifest content.* See Chapter 11. (2) What you need to make a latent.

level of significance   (1) In the context of tests of statistical significance, the degree of likelihood that an observed, empirical relationship could be attributable to sampling error. A relationship is significant at the .05 level if the likelihood of its being only a function of sampling error is no greater than 5 out of 100. See Chapter 16. (2) Height limits on outdoor advertising.

Likert scale   A type of composite measure developed by Rensis Likert, in an attempt to improve the levels of measurement in social research through the use of standardized response categories in survey questionnaires to determine the relative intensity of different items. Likert items are those using such response categories as strongly agree, agree, disagree, and strongly disagree. Such items may be used in the construction of true Likert scales as well as other types of composite measures. See Chapter 6.

linear regression analysis   A form of statistical analysis that seeks the equation for the straight line that best describes the relationship between two ratio variables. See Chapter 16.

log-linear models   Data-analysis technique based on specifying models that describe the interrelationships among variables and then comparing expected and observed table-cell frequencies. See Chapter 16.

longitudinal study   A study design involving the collection of data at different points in time, as contrasted with a *cross-sectional study.* See also Chapter 4 and *cohort study, panel study,* and *trend study.*

macrotheory   A theory aimed at understanding the "big picture" of institutions, whole societies, and the interactions among societies. Karl Marx's examination of the class struggle is an example of macrotheory. By contrast, see *microtheory.* See also Chapter 2.

manifest content   (1) In connection with content analysis, the concrete terms contained in a communication, as distinguished from *latent content.* See Chapter 11. (2) What you have after a manifest bursts.

matching In connection with experiments, the procedure whereby pairs of subjects are matched on the basis of their similarities on one or more variables, and one member of the pair is assigned to the *experimental group* and the other to the *control group.* See Chapter 8.

mean (1) An average computed by summing the values of several observations and dividing by the number of observations. If you now have a grade point average of 4.0 based on 10 courses, and you get an F in this course, your new grade point (mean) average will be 3.6. See Chapter 14. (2) The quality of the thoughts you might have if your instructor did that to you.

median (1) An average representing the value of the "middle" case in a rank-ordered set of observations. If the ages of five men are 16, 17, 20, 54, and 88, the median would be 20. (The mean would be 39.) See Chapter 14. (2) The dividing line between safe driving and exciting driving.

memoing Writing memos that become part of the data for analysis in qualitative research such as grounded theory. Memos can describe and define concepts, deal with methodological issues, or offer initial theoretical formulations. See Chapter 13.

methodology The science of finding out; procedures for scientific investigation. See Chapter 1.

microtheory A theory aimed at understanding social life at the intimate level of individuals and their interactions. Examining how the play behavior of girls differs from that of boys would be an example of microtheory. By contrast, see *macrotheory.* See also Chapter 2.

mode (1) An average representing the most frequently observed value or attribute. If a sample contains 1,000 Protestants, 275 Catholics, and 33 Jews, Protestant is the modal category. See Chapter 14 for more thrilling disclosures about averages. (2) Better than apple pie à la median.

monitoring studies Studies that provide a steady flow of information about something of interest, such as crime rates or the outbreak of an epidemic. See Chapter 12.

multiple regression analysis A form of statistical analysis that seeks the equation representing the impact of two or more independent variables on a single dependent variable. See Chapter 16.

multiple time-series designs The use of more than one set of data that were collected over time, as in accident rates over time in several states

or cities, so that comparisons can be made. See Chapter 12.

multivariate analysis The analysis of the simultaneous relationships among several variables. Examining simultaneously the effects of *age, gender,* and *social class* on *religiosity* would be an example of multivariate analysis. See Chapters 14, 15, and 16.

naturalism An approach to field research based on the assumption that an objective social reality exists and can be observed and reported accurately. See Chapter 10.

needs assessment studies Studies that aim to determine the existence and extent of problems, typically among a segment of the population, such as the elderly. See Chapter 12.

nominal measure A variable whose attributes have only the characteristics of exhaustiveness and mutual exclusiveness. In other words, a level of measurement describing a variable that has attributes that are merely different, as distinguished from ordinal, interval, or ratio measures. *Gender* is an example of a nominal measure. See Chapter 5.

nomothetic An approach to explanation in which we seek to identify a few causal factors that generally impact a class of conditions or events. Imagine the two or three key factors that determine which colleges students choose—proximity, reputation, and so forth. By contrast, see *idiographic.* See also Chapter 1.

nonequivalent control group A control group that is similar to the experimental group but is not created by the random assignment of subjects. This sort of control group differs significantly from the experimental group in terms of the dependent variable or variables related to it. See Chapter 12.

nonprobability sampling Any technique in which samples are selected in some way not suggested by probability theory. Examples include reliance on available subjects as well as *purposive* (judgmental), *quota,* and *snowball sampling.* See Chapter 7.

nonsampling error (1) Those imperfections of data quality that are a result of factors other than sampling error. Examples include misunderstandings of questions by respondents, erroneous recordings by interviewers and coders, and keypunch errors. See Chapter 16. (2) The mistake you made in deciding to interview everyone rather than selecting a sample.

null hypothesis (1) In connection with hypothesis testing and tests of statistical significance, that hypothesis that suggests there is no relationship among the variables under study. You may conclude that the variables are related after having statistically rejected the null hypothesis. See Chapter 2. (2) An expectation about nulls.

open coding   The initial classification and labeling of concepts in qualitative data analysis. In open coding, the codes are suggested by the researchers' examination and questioning of the data. See Chapter 13.

open-ended questions   Questions for which the respondent is asked to provide his or her own answers. In-depth, qualitative interviewing relies almost exclusively on open-ended questions. See Chapters 9 and 10.

operational definition   The concrete and specific definition of something in terms of the operations by which observations are to be categorized. The operational definition of "earning an A in this course" might be "correctly answering at least 90 percent of the final exam questions." See Chapter 2.

operationalization   (1) One step beyond conceptualization. Operationalization is the process of developing operational definitions, or specifying the exact operations involved in measuring a variable. See Chapters 2 and 5. (2) Surgery on intellectuals.

ordinal measure   A level of measurement describing a variable with attributes we can rank-order along some dimension. An example is *socioeconomic status* as composed of the attributes *high, medium, low.* See also Chapter 5 and *interval measure, nominal measure,* and *ratio measure.*

panel study   A type of longitudinal study, in which data are collected from the same set of people (the sample or panel) at several points in time. See Chapter 4 and *cohort, longitudinal,* and *trend study.*

paradigm   (1) A model or frame of reference through which to observe and understand. See Chapter 2. (2) (pl.) $0.20.

parameter   The summary description of a given variable in a population. See Chapter 7.

partial   See *partial relationship.*

partial regression analysis   (1) A form of regression analysis in which the effects of one or more variables are held constant, similar to the logic of the elaboration model. See Chapter 16. (2) A regression analysis you didn't have time to finish.

partial relationship   (1) In the elaboration model, this is the relationship between two variables when examined in a subset of cases defined by a third variable. Beginning with a zero-order relationship between *political party* and *attitudes toward abortion,* for example, we might want to see whether the relationship held true among both men and women (i.e., controlling for *gender*). The relationship found among men and the relationship found among women would be the partial relationships, sometimes simply called the *partials.* See Chapter 15. (2) Someone you would take to the opera but not to mud wrestling.

participatory action research (PAR)   An approach to social research in which the people being studied are given control over the purpose and procedures of the research; intended as a counter to the implicit view that researchers are superior to those they study. See Chapter 10.

path analysis   (1) A form of multivariate analysis in which the causal relationships among variables are presented in a graphic format. See Chapter 16. (2) Watching your step along a horse trail.

plagiarism   Presenting someone else's words or thoughts as though they were your own, constituting intellectual theft. See Chapter 17.

population   The theoretically specified aggregation of the elements in a study. See Chapter 7.

positivism   Introduced by August Comte, this philosophical system is grounded on the rational proof/disproof of scientific assertions; assumes a knowable, objective reality. See Chapter 2.

postmodernism   A paradigm that questions the assumptions of positivism and theories describing an "objective" reality. See Chapter 2

posttesting   (1) The remeasurement of a dependent variable among subjects after they've been exposed to an independent variable. See Chapter 8. (2) What my younger sister did when she was learning to drive.*

PPS (probability proportionate to size)   (1) A type of multistage cluster sample in which clusters are selected, not with equal probabilities (see *EPSEM*) but with probabilities proportionate to their sizes—as measured by the number of units to be subsampled. See Chapter 7. (2) The odds on who gets to go first: you or the 275-pound fullback.

predictive validity   See *criterion-related validity.*

pretesting   The measurement of a dependent variable among subjects. See Chapter 8.

probability sampling   The general term for samples selected in accord with probability theory, typically involving some random-selection mechanism. Specific types of probability sampling include *EPSEM, PPS, simple random sampling,* and *systematic sampling.* See Chapter 7.

probe A technique employed in interviewing to solicit a more complete answer to a question. It is a nondirective phrase or question used to encourage a respondent to elaborate on an answer. Examples include "Anything more?" and "How is that?" See Chapter 9.

program evaluation/outcome assessment The determination of whether a social intervention is producing the intended result. See Chapter 12.

proportionate reduction of error (PRE) A logical model for assessing the strength of a relationship by asking how much knowing values on one variable would reduce our errors in guessing values on the other. For example, if we know how much education people have, we can improve our ability to estimate how much they earn, thus indicating there is a relationship between the two variables. See Chapter 16.

purposive (judgmental) sampling A type of nonprobability sampling in which the units to be observed are selected on the basis of the researcher's judgment about which ones will be the most useful or representative. See Chapter 7.

qualitative analysis (1) The nonnumerical examination and interpretation of observations, for the purpose of discovering underlying meanings and patterns of relationships. This is most typical of field research and historical research. See Chapter 13. (2) A classy analysis.

qualitative interview Contrasted with survey interviewing, the qualitative interview is based on a set of topics to be discussed in depth rather than based on the use of standardized questions. See Chapter 10.

quantitative analysis (1) The numerical representation and manipulation of observations for the purpose of describing and explaining the phenomena that those observations reflect. See Chapter 14 especially, and also the remainder of Part 4. (2) A BIG analysis.

quasi experiments Nonrigorous inquiries somewhat resembling controlled experiments but lacking key elements such as pre- and posttesting and/or control groups. See Chapter 12.

questionnaire A document containing questions and other types of items designed to solicit information appropriate for analysis. Questionnaires are used primarily in survey research but also in experiments, field research, and other modes of observation. See Chapter 9.

quota sampling A type of nonprobability sampling in which units are selected into a sample on the basis of prespecified characteristics, so that the total sample will have the same distribution of characteristics assumed to exist in the population being studied. See Chapter 7.

random selection A sampling method in which each element has an equal chance of selection independent of any other event in the selection process. See Chapter 7.

random-digit dialing (RDD) A sampling technique in which random numbers are selected from within the ranges of numbers assigned to active telephones. See Chapter 9.

randomization A technique for assigning experimental subjects to experimental and control groups randomly. See Chapter 8.

rapport An open and trusting relationship; especially important in qualitative research between researchers and the people they're observing. See Chapter 10.

ratio measure A level of measurement describing a variable with attributes that have all the qualities of nominal, ordinal, and interval measures and in addition are based on a "true zero" point. *Age* is an example of a ratio measure. See also Chapter 5 and *nominal measure, interval measure,* and *ordinal measure.*

reactivity The problem that the subjects of social research may react to the fact of being studied, thus altering their behavior from what it would have been normally. See Chapter 10.

reductionism (1) A fault of some researchers: a strict limitation (reduction) of the kinds of concepts to be considered relevant to the phenomenon under study. See Chapter 4. (2) The cloning of ducks.

regression analysis (1) A method of data analysis in which the relationships among variables are represented in the form of an equation, called a regression equation. See Chapter 16 for a discussion of the different forms of regression analysis. (2) What seems to happen to your knowledge of social research methods just before an exam.

reliability (1) That quality of measurement method that suggests that the same data would have been collected each time in repeated observations of the same phenomenon. In the context of a survey, we would expect that the question "Did you attend religious services last week?" would have higher reliability than the question "About how many times have you attended religious services in your life?" This is not to be confused with *validity*. See Chapter 5. (2) Quality of repeatability in untruths.

replication (1) Repeating a research study to test and either confirm or question the findings of an earlier study. See Chapter 1. (2) A technical term used in connection with the elaboration model, referring to the elaboration outcome in which the initially observed relationship between two variables persists when a control variable is held constant, thereby supporting the idea that the original relationship is genuine. See Chapter 15.

representativeness (1) That quality of a sample of having the same distribution of characteristics as the population from which it was selected. By implication, descriptions and explanations derived from an analysis of the sample may be assumed to represent similar ones in the population. Representativeness is enhanced by probability sampling and provides for generalizability and the use of inferential statistics. See Chapter 7. (2) A noticeable quality in the presentation-of-self of some members of the U.S. Congress.

research monograph A book-length research report, either published or unpublished. This is distinguished from a textbook, a book of essays, a novel, and so forth. See Chapter 17.

respondent A person who provides data for analysis by responding to a survey questionnaire. See Chapter 9.

response rate The number of people participating in a survey divided by the number selected in the sample, in the form of a percentage. This is also called the *completion rate* or, in self-administered surveys, the *return rate:* the percentage of questionnaires sent out that are returned. See Chapter 9.

return rate See *response rate.*

sampling error The degree of error to be expected by virtue of studying a sample instead of everyone. For probability sampling, the maximum error depends on three factors: the sample size, the diversity of the population, and the confidence level. See Chapter 7.

sampling frame That list or quasi list of units composing a population from which a sample is selected. If the sample is to be representative of the population, it is essential that the sampling frame include all (or nearly all) members of the population. See Chapter 7.

sampling interval The standard distance between elements selected from a population for a sample. See Chapter 7.

sampling ratio The proportion of elements in the population that are selected to be in a sample. See Chapter 7.

sampling unit That element or set of elements considered for selection in some stage of sampling. See Chapter 7.

scale (1) A type of composite measure composed of several items that have a logical or empirical structure among them. Examples of scales include Bogardus social distance, Guttman, Likert, and Thurstone scales. Contrasted with *index*. See Chapter 6. (2) One of the less-appetizing parts of a fish.

search engine A computer program designed to locate where specified terms appear on websites throughout the World Wide Web. See Chapter 17.

secondary analysis (1) A form of research in which the data collected and processed by one researcher are reanalyzed—often for a different purpose—by another. This is especially appropriate in the case of survey data. Data archives are repositories or libraries for the storage and distribution of data for secondary analysis. See Chapter 9. (2) Estimating the weight and speed of an opposing team's linebackers.

selective coding In Grounded Method Theory, this analysis builds on the results of *open coding* and *axial coding* to identify the central concept that organizes the other concepts that have been identified in a body of textual materials. See also *axial coding* and Chapter 13.

semantic differential A questionnaire format in which the respondent is asked to rate something in terms of two, opposite adjectives (e.g., rate textbooks as "boring" or "exciting"), using qualifiers such as "very," "somewhat," "neither," "somewhat," and "very" to bridge the distance between the two opposites. See Chapter 6.

semiotics (1) The study of signs and the meanings associated with them. This is commonly associated with content analysis. See Chapter 13. (2) Antibiotics that only work half of the time.*

simple random sampling (SRS) (1) A type of probability sampling in which the units composing a population are assigned numbers. A set of random numbers is then generated, and the units having those numbers are included in the sample. Although probability theory and the calculations it provides assume this basic sampling method, it's seldom used, for practical reasons. An equivalent alternative is the systematic sample (with a random start). See Chapter 7. (2) A random sample with a low IQ.

snowball sampling (1) A nonprobability sampling method, often employed in field research, whereby each person interviewed may be asked to suggest additional people for interviewing. See Chapters 7 and 10. (2) Picking the icy ones to throw at your methods instructor.

social artifact Any product of social beings or their behavior. Can be a unit of analysis. See Chapter 4.

social indicators Measurements that reflect the quality or nature of social life, such as crime rates, infant mortality rates, number of physicians per 100,000 population, and so forth. Social indicators are often monitored to determine the nature of social change in a society. See Chapter 12.

sociobiology A paradigm based in the view that social behavior can be explained solely in terms of genetic characteristics and behavior. See Chapter 4.

specification (1) The process through which concepts are made more specific. See Chapter 5. (2) A technical term used in connection with the elaboration model, representing the elaboration outcome in which an initially observed relationship between two variables is replicated among some subgroups created by the control variable but not among others. In such a situation, you will have specified the conditions under which the original relationship exists: for example, among men but not among women. See Chapter 15.

spurious relationship (1) A coincidental statistical correlation between two variables, shown to be caused by some third variable. For example, there is a positive relationship between the number of fire trucks responding to a fire and the amount of damage done: the more trucks, the more damage. The third variable is the size of the fire. They send lots of fire trucks to a large fire and a lot of damage is done because of the size of the fire. For a little fire, they just send a little fire truck, and not much damage is done because it's a small fire. Sending more fire trucks does not cause more damage. For a given size of fire, in fact, sending more trucks would reduce the amount of damage. See Chapter 4. (2) You thought you were going steady but that @#*&@#&* thought you were "just friends."

standard deviation (1) A measure of dispersion around the mean, calculated so that approximately 68 percent of the cases will lie within plus or minus one standard deviation from the mean, 95 percent will lie within plus or minus two standard deviations, and 99.9 percent will lie within three standard deviations. Thus, for example, if the mean age in a group is 30 and the standard deviation is 10, then 68 percent have ages between 20 and 40. The smaller the standard deviation, the more tightly the values are clustered around the mean; if the standard deviation is high, the values are widely spread out. See Chapter 14. (2) Routine rule-breaking.

statistic The summary description of a variable in a sample, used to estimate a population parameter. See Chapter 7.

statistical significance (1) A general term referring to the likelihood that relationships observed in a sample could be attributed to sampling error alone. See *tests of statistical significance* and Chapter 16. (2) How important it would really be if you flunked your statistics exam. I mean, you could always be a poet.

stratification The grouping of the units composing a population into homogeneous groups (or strata) before sampling. This procedure, which may be used in conjunction with *simple random, systematic,* or *cluster sampling,* improves the representativeness of a sample, at least in terms of the stratification variables. See Chapter 7.

structural functionalism A paradigm that divides social phenomena into parts, each of which serves a function for the operation of the whole. See Chapter 2.

study population That aggregation of elements from which a sample is actually selected. See Chapter 7.

suppressor variable In the elaboration model, a test variable that prevents a genuine relationship from appearing at the zero-order level. See Chapter 15.

symbolic interactionism A paradigm that views human behavior as the creation of meaning through social interactions, with those meanings conditioning subsequent interactions. See Chapter 2.

systematic sampling (1) A type of probability sampling in which every *k*th unit in a list is selected for inclusion in the sample—for example, every 25th student in the college directory of students. You compute *k* by dividing the size of the population by the desired sample size; *k* is called the *sampling interval.* Within certain constraints, systematic sampling is a functional equivalent of simple random sampling and usually easier to do.

Typically, the first unit is selected at random. See Chapter 7. (2) Picking every third one whether it's icy or not. See *snowball sampling* (2).

test variable   A variable that is held constant in an attempt to clarify further the relationship between two other variables. Having discovered a relationship between *education* and *prejudice,* for example, we might hold *gender* constant by examining the relationship between education and prejudice among men only and then among women only. In this example, *gender* would be the test variable. See Chapter 15 to find out how important the proper use of test variables is in analysis.

tests of statistical significance   (1) A class of statistical computations that indicate the likelihood that the relationship observed between variables in a sample can be attributed to sampling error only. See *inferential statistics* and Chapter 16. (2) A determination of how important statistics have been in improving humankind's lot in life. (3) An examination that can radically affect your grade in this course and your GPA as well.

theory   A systematic explanation for the observations that relate to a particular aspect of life: juvenile delinquency, for example, or perhaps social stratification or political revolution. See Chapter 1.

Thurstone scale   A type of composite measure, constructed in accord with the weights assigned by "judges" to various indicators of some variables. See Chapter 6.

time-series analysis   An analysis of changes in a variable (such as *crime rates*) over time. See Chapter 16.

time-series design   A research design that involves measurements made over some period, such as the study of traffic accident rates before and after lowering the speed limit. See Chapter 12.

trend study   A type of longitudinal study in which a given characteristic of some population is monitored over time. An example would be the series of Gallup Polls showing the electorate's preferences for political candidates over the course of a campaign, even though different samples were interviewed at each point. See Chapter 4 and *cohort, longitudinal,* and *panel study.*

triangulation   Choking a triangle (submitted by Wendy Ogden, Mount Royal College, Calgary, Canada).

typology   (1) The classification (typically nominal) of observations in terms of their attributes on two or more variables. The classification of newspapers

as liberal-urban, liberal-rural, conservative-urban, or conservative-rural would be an example. See Chapter 6. (2) Apologizing for your neckwear.

units of analysis   The what or whom being studied. In social science research, the most typical units of analysis are individual people. See Chapter 4.

univariate analysis   The analysis of a single variable, for purposes of description. Frequency distributions, averages, and measures of dispersion would be examples of univariate analysis, as distinguished from *bivariate* and *multivariate analysis.* See Chapter 14.

unobtrusive research   Methods of studying social behavior without affecting it. Such methods can be qualitative or quantitative. See Chapter 11.

URL   (1) Web address, typically beginning with "http://"; stands for "uniform resource locator" or "universal resource locator." See Chapter 17. (2) Phonetic spelling of "Earl." (3) What my mum used to say to me when I sounded like I was getting a cold.*

validity   A term describing a measure that accurately reflects the concept it is intended to measure. For example, your IQ would seem a more valid measure of your intelligence than the number of hours you spend in the library would. Though the ultimate validity of a measure can never be proved, we may agree to its relative validity on the basis of *face validity, criterion-related validity, construct validity, content validity,* internal validation, and *external validation.* Validity must not be confused with *reliability.* See Chapters 5 and 6.

variable-oriented analysis   An analysis that describes and/or explains a particular variable. See Chapter 13.

variables   Logical sets of attributes. The variable *gender* is made of up of the attributes *male* and *female.* See Chapter 1.

weighting   Assigning different weights to cases that were selected into a sample with different probabilities of selection. In the simplest scenario, each case is given a weight equal to the inverse of its probability of selection. When all cases have the same chance of selection, no weighting is necessary. See Chapter 7.

zero-order relationship   (1) In the elaboration model, this is the original relationship between two variables, with no test variables controlled for. See Chapter 15. (2) A blind date that just didn't work out. Hang in there. You can always turn to social research methods.

Bibliography

Abdollahyan, Hamid, and Taghi Azadarmaki. 2000. *Sampling Design in a Survey Research: The Sampling Practice in Iran.* Paper presented at the meeting of the American Sociological Association, August 12–16, Washington, DC.

Abdulhadi, Rabab. 1998. "The Palestinian Women's Autonomous Movement: Emergence, Dynamics, and Challenges." *Gender and Society* 12 (6): 649–73.

Aburdene, Patricia. 2005. *Megatrends 2010.* Charlottesville, VA: Hampton Roads.

Anderson, Eric. 2005. "Orthodox and Inclusive Masculinity: Competing Masculinities among Heterosexual Men in a Feminized Terrain." *Sociological Perspectives* 48 (3): 337–55.

Anderson, Walt. 1990. *Reality Isn't What It Used to Be: Theatrical Politics, Ready-to-Wear Religion, Global Myths, Primitive Chic, and Other Wonders of the Postmodern World.* San Francisco: Harper & Row.

Andorka, Rudolf. 1990. "The Importance and the Role of the Second Economy for the Hungarian Economy and Society." *Quarterly Journal of Budapest University of Economic Sciences* 12 (2): 95–113.

Aneshensel, Carol S., Rosina Becerra, Eve Fielder, and Roberleigh Schuler. 1989. "Participation of Mexican American Female Adolescents in a Longitudinal Panel Survey." *Public Opinion Quarterly* 53:548–62.

Asch, Solomon. 1958. "Effects of Group Pressure upon the Modification and Distortion of Judgments." Pp. 174–83 in *Readings in Social Psychology,* 3rd ed., edited by Eleanor E. Maccoby et al. New York: Holt, Rinehart & Winston.

Asher, Ramona M., and Gary Alan Fine. 1991. "Fragile Ties: Sharing Research Relationships with Women Married to Alcoholics." Pp. 196–205 in *Experiencing Fieldwork: An Inside View of Qualitative Research,* edited by William B. Shaffir and Roberta A. Stebbins. Newbury Park, CA: Sage.

Auster, Carol J. 1985. "Manuals for Socialization: Examples from Girl Scout Handbooks 1913–1984." *Qualitative Sociology* 8(4): 359–67.

Babbie, Earl. 1966. "The Third Civilization." *Review of Religious Research,* Winter, pp. 101–21.

———. 1967. "A Religious Profile of Episcopal Churchwomen." *Pacific Churchman,* January, pp. 6–8, 12.

———. 1970. *Science and Morality in Medicine.* Berkeley: University of California Press.

———. 1982. *Social Research for Consumers.* Belmont, CA: Wadsworth.

———. 1985. *You Can Make a Difference.* New York: St. Martin's Press.

———. 2004. "Laud Humphreys and Research Ethics." *International Journal of Sociology and Social Policy* 24 (3/4/5): 12–18.

Bailey, William C. 1975. "Murder and Capital Punishment." In *Criminal Law in Action,* edited by William J. Chambliss. New York: Wiley.

Ball-Rokeach, Sandra J., Joel W. Grube, and Milton Rokeach. 1981. "Roots: The Next Generation—Who Watched and with What Effect." *Public Opinion Quarterly* 45:58–68.

Barker, Kriss, and Miguel Sabido, eds. 2005. *Soap Operas for Social Change to Prevent HIV/AIDS: A Training Guide for Journalists and Media Personnel.* Shelburne, VT: Population Media Center.

Bart, Pauline, and Patricia O'Brien. 1985. *Stopping Rape: Successful Survival Strategies.* New York: Pergamon.

Beatty, Paul C., and Gordon B. Willis. 2007. "Research Synthesis: The Practice of Cognitive Interviewing." *Public Opinion Quarterly* 71 (2): 287–311.

Becker, Penny Edgell. 1998. "Making Inclusive Communities: Congregations and the 'Problem' of Race.'" *Social Problems* 45 (4): 451–72.

Bednarz, Marlene. 1996. "Push Polls Statement." Report to the AAPORnet listserv, April 5. http://www.aapor.org/ethics/pushpoll.html.

Belenky, Mary Field, Blythe McVicker Clinchy, Nancy Rule Goldberger, and Jill Mattuck Tarule. 1986. *Women's Ways of Knowing: The Development of Self, Voice, and Mind.* New York: Basic Books.

Bell, Derrick A. 1980. "*Brown v. Board of Education* and the Interest-Convergence Dilemma." *Harvard Law Review* 93:518–33.

Bellah, Robert N. 1957. *Tokugawa Religion.* Glencoe, IL: Free Press.

———. 1970. "Christianity and Symbolic Realism." *Journal for the Scientific Study of Religion* 9:89–96.

———. 1974. "Comment on the Limits of Symbolic Realism." *Journal for the Scientific Study of Religion* 13:487–89.

Benton, J. Edwin, and John L. Daly. 1991. "A Question Order Effect in a Local Government Survey." *Public Opinion Quarterly* 55:640–42.

Berbrier, Mitch. 1998. "'Half the Battle': Cultural Resonance, Framing Processes, and Ethnic Affectations in Contemporary White Separatist Rhetoric." *Social Problems* 45 (4): 431–50.

Berg, Bruce L. 1989. *Qualitative Research Methods for the Social Sciences.* Boston: Allyn and Bacon.

Bian, Yanjie. 1994. *Work and Inequality in Urban China.* Albany: State University of New York Press.

Biddle, Stuart J. H., David Markland, David Gilbourne, Nikos L. D. Chatzisarantis, and Andrew C. Sparkes. 2001. "Research Methods in Sport and Exercise Psychology: Quantitative and Qualitative Issues." *Journal of Sports Sciences* 19 (10): 777.

Bielby, William T., and Denise Bielby. 1999. "Organizational Mediation of Project-Based Labor Markets: Talent Agencies and the Careers of Screenwriters." *American Sociological Review* 64:64–85.

Birchfield, R. W. 1998. *The New Fowler's Modern English Usage.* 3rd ed. New York: Oxford University Press.

Bishop, George, and Andrew Smith. 2001. "Response-Order Effects and the Early Gallup Split-Ballots." *Public Opinion Quarterly* 65:479–505.

Black, Donald. 1970. "Production of Crime Rates." *American Sociological Review* 35 (August): 733–48.

Blair, Johnny, Shanyang Zhao, Barbara Bickart, and Ralph Kuhn. 1995. *Sample Design for Household Telephone Surveys: A Bibliography 1949–1995.* College Park: Survey Research Center, University of Maryland.

Blaunstein, Albert, and Robert Zangrando, eds. 1970. *Civil Rights and the Black American.* New York: Washington Square Press.

Blumenthal, Mark M. 2005. "Toward an Open-Source Methodology: What We Can Learn from the Blogsphere." *Public Opinion Quarterly* 69 (5): 655–69. [Special issue].

Bolstein, Richard. 1991. "Comparison of the Likelihood to Vote among Preelection Poll Respondents and Nonrespondents." *Public Opinion Quarterly* 55:648–50.

Bottomore, T. B., and Maximilien Rubel, eds. [1843] 1956. *Karl Marx: Selected Writings in Sociology and Social Philosophy.* Translated by T. B. Bottomore. New York: McGraw-Hill.

Brown v. Board of Education, 347 U.S. 483 (1954).

Browne, Kath. 2005. "Snowball Sampling: Using Social Networks to Research Non-Heterosexual Women." *International Journal of Social Research Methodology* 8 (1): 47–60.

Burawoy, M., A. Burton, A. A. Ferguson, K. J. Fox, J. Gamson, N. Gartrell, L. Hurst, C. Kurzman, L. Salzinger, J. Schiffman, and S. Ui, eds. 1991. *Ethnography Unbound: Power and Resistance in the Modern Metropolis.* Berkeley: University of California Press.

Campbell, Donald, and Julian Stanley. 1963. *Experimental and Quasi-Experimental Designs for Research.* Chicago: Rand McNally.

Campbell, M. L. 1998. "Institutional Ethnography and Experience as Data." *Qualitative Sociology* 21 (1): 55–73.

Carpini, Michael X. Delli, and Scott Keeter. 1991. "Stability and Change in the U.S. Public's Knowledge of Politics." *Public Opinion Quarterly* 55:583–612.

Carr, C. Lynn. 1998. "Tomboy Resistance and Conformity: Agency in Social Psychological Gender Theory." *Gender and Society* 12 (5): 528–53.

Census Bureau. *See* U.S. Bureau of the Census.

Chirot, Daniel, and Jennifer Edwards. 2003. "Making Sense of the Senseless: Understanding Genocide." *Contexts* 2 (2): 12–19.

Choi, Bernard C. K., and Anita W. P. Pak. 2005. "A Catalog of Biases in Questionnaires." *Preventing Chronic Disease* 2 (1). http://www.cdc.gov/Pcd/issues/2005/jan/04_0050.htm.

Chossudovsky, Michel. 1997. *The Globalization of Poverty: Impacts of IMF and World Bank Reforms.* London: Zed Books.

Christian, Leah Melani, Don A. Dillman, and Jolene D. Smyth. 2007. "Helping Respondents Get It Right the First Time: The Influence of Words, Symbols, and Graphics in Web Surveys." *Public Opinion Quarterly* 71 (1): 113–25.

Clark, Roger, Rachel Lennon, and Leana Morris. 1993. "Of Caldecotts and Kings: Gendered Images in Recent American Children's Books by Black and Non-Black Illustrators." *Gender and Society* 7 (2): 227–45.

Coates, Rodney D. 2006. "Towards a Simple Typology of Racial Hegemony." *Societies without Borders* 1:69–91.

Coleman, James. 1966. *Equality of Educational Opportunity.* Washington, DC: U.S. Government Printing Office.

Collins, G. C., and Timothy B. Blodgett. 1981. "Sexual Harassment . . . Some See It . . . Some Won't." *Harvard Business Review,* March–April, pp. 76–95.

Comstock, Donald. 1980. "Dimensions of Influence in Organizations." *Pacific Sociological Review,* January, pp. 67–84.

Conrad, Clifton F. 1978. "A Grounded Theory of Academic Change." *Sociology of Education* 51:101–12.

Cook, Elizabeth. 1995. Communication to the METHODS listserv, April 25, from Michel de Seve (T120@music.ulaval.ca) to Cook (EC1645A@american.edu).

Cook, Thomas D., and Donald T. Campbell. 1979. *Quasi-Experimentation: Design and Analysis Issues for Field Settings.* Chicago: Rand McNally.

Cooper-Stephenson, Cynthia, and Athanasios Theologides. 1981. "Nutrition in Cancer: Physicians' Knowledge, Opinions, and Educational Needs." *Journal of the American Dietetic Association,* May, pp. 472–76.

Couper, Mick P. 2001. "Web Surveys: A Review of Issues and Approaches." *Public Opinion Quarterly* 64 (4): 464–94.

Craig, R. Stephen. 1992. "The Effect of Television Day Part on Gender Portrayals in Television Commercials: A Content Analysis." *Sex Roles* 26 (5/6): 197–211.

Crawford, Kent S., Edmund D. Thomas, and Jeffrey J. Fink. 1980. "Pygmalion at Sea: Improving the Work Effectiveness of Low Performers." *Journal of Applied Behavioral Science,* October–December, pp. 482–505.

Curtin, Richard, Stanley Presser, and Eleanor Singer. 2005. "Changes in Telephone Survey Nonresponse over the Past Quarter Century." *Public Opinion Quarterly* 69 (1): 87–98.

Danieli, Ardha, and Carol Woodhams. 2005. "Emancipatory Research Methodology and Disability: A Critique." *International Journal of Social Research* 8 (4): 281–96.

Davern, Michael, Todd H. Rockwood, Randy Sherrod, and Stephen Campbell. 2003. "Prepaid Monetary Incentives and Data Quality in Face-to-Face Interviews: Data from the 1996 Survey of Income and Program Participation Incentive Experiment." *Public Opinion Quarterly* 67:139–47.

Davis, Fred. 1973. "The Martian and the Convert: Ontological Polarities in Social Research." *Urban Life* 2 (3): 333–43.

Davis, James A. 1992. "Changeable Weather in a Cooling Climate atop the Liberal Plateau: Conversion and Replacement in Forty-two General Social Survey Items, 1972–1989." *Public Opinion Quarterly* 56:261–306.

De Coster, Stacy. 2005. "Depression and Law Violation: Gendered Responses to Gendered Stresses." *Sociological Perspectives* 48 (2): 155–87.

Deflem, Mathieu. 2002. *Policing World Society: Historical Foundations of International Police Cooperation.* New York: Oxford University Press.

DeFleur, Lois. 1975. "Biasing Influences on Drug Arrest Records: Implications for Deviance Research." *American Sociological Review* 40:88–103.

Delgado, Richard. 2002. "Explaining the Rise and Fall of African American Fortunes—Interest Convergence and Civil Rights Gains." *Harvard Civil Rights–Civil Liberties Law Review* 37:369–87.

Dillman, Don A. 1978. *Mail and Telephone Surveys: The Total Design Method.* New York: Wiley.

——— 2007. *Mail and Internet Surveys: The Tailored Design Method.* 2nd ed. Hoboken, NJ: Wiley.

Doyle, Sir Arthur Conan. [1891] 1892. "A Scandal in Bohemia." First published in *The Strand,* July 1891. Reprinted in *The Original Illustrated Sherlock Holmes,* pp. 11–25. Secaucus, NJ: Castle.

DuBois, W. E. B. 1903. *The Souls of Black Folk.* Chicago: McClurg. http://www.bartleby.com/114/.

Duneier, Mitchell. 1999. *Sidewalk.* New York: Farrar, Straus, and Giroux.

Durkheim, Emile. [1893] 1964. *The Division of Labor in Society.* Translated by George Simpson. New York: Free Press.

——— [1897] 1951. *Suicide.* New York: Free Press.

Eastman, Crystal. 1910. "Work-Accidents and Employers' Liability." *The Survey,* September 3, pp. 788–94.

Ellison, Christopher G., and Darren E. Sherkat. 1990. "Patterns of Religious Mobility among Black Americans." *Sociological Quarterly* 31 (4): 551–68.

Emerson, Robert M., Kerry O. Ferris, and Carol Brooks Gardner. 1998. "On Being Stalked." *Social Problems* 45 (3): 289–314.

Farquharson, Karen. 2005. "A Different Kind of Snowball: Identifying Key Policymakers." *International Journal of Social Research* 8 (4): 345–53.

Fausto-Sterling, Anne. 1992. "Why Do We Know So Little about Human Sex?" *Discover Archives,* June. http://cas.bellarmine.edu/tietjen/Human%20Nature%20S%201999/why_do_we_know_so_little_about_h.htm, accessed July 10, 2003.

Festinger, Leon, Henry W. Reicker, and Stanley Schachter. 1956. *When Prophecy Fails.* Minneapolis: University of Minnesota Press.

Fielding, Nigel. 2004. "Getting the Most from Archived Qualitative Data: Epistemological, Practical and Professional Obstacles." *International Journal of Social Research Methodology* 7 (1): 97–104.

Ford, David A. 1989. "Preventing and Provoking Wife Battery through Criminal Sanctioning: A Look at the Risks." September, unpublished manuscript.

Ford, David A., and Mary Jean Regoli. 1992. "The Preventive Impacts of Policies for Prosecuting Wife Batterers." Pp. 181–208 in *Domestic Violence: The Changing Criminal Justice Response,* edited by E. S. Buzawa and C. G. Buzawa. New York: Auburn.

Forslund, Morris A. 1980. *Patterns of Delinquency Involvement: An Empirical Typology.* Paper presented at the Annual Meeting of the Western Association of Sociologists and Anthropologists, Lethbridge, Alberta, February 8.

Foschi, Martha, G. Keith Warriner, and Stephen D. Hart. 1985. "Standards, Expectations, and Interpersonal Influence." *Social Psychology Quarterly* 48 (2): 108–17.

Fox, Katherine J. 1991. "The Politics of Prevention: Ethnographers Combat AIDS among Drug Users." Pp. 227–49 in *Ethnography Unbound: Power and Resistance in the Modern Metropolis,* edited by M. Burawoy et al. Berkeley: University of California Press.

Frankel, Mark S., and Sanyin Siang. 1999. *Ethical and Legal Aspects of Human Subjects Research on the Internet: A Report of a Workshop.* Washington, DC: American Association for the Advancement of Science, November. http://www.aaas.org/spp/dspp/sfrl/projects/intres/main.htm.

Gall, John. 1975. *Systemantics: How Systems Work and Especially How They Fail.* New York: Quadrangle.

Gamson, William A. 1992. *Talking Politics.* New York: Cambridge University Press.

Gans, Herbert. 2002. "More of Us Should Become Public Sociologists." In *Footnotes.* Washington, DC: American Sociological Association, July–August. http://www.asanet.org/footnotes/julyaugust02/fn10.html, accessed July 8, 2003.

Garant, Carol. 1980. "Stalls in the Therapeutic Process." *American Journal of Nursing,* December, pp. 2166–67.

Gard, Greta. 1993. *Ecofeminism: Women, Animals, Nature.* Philadelphia: Temple University Press.

Garfinkel, Harold. 1967. *Studies in Ethnomethodology.* Englewood Cliffs, NJ: Prentice-Hall.

Gaventa, J. 1991. "Towards a Knowledge Democracy: Viewpoints on Participatory Research in North America." Pp. 121–31 in *Action and Knowledge: Breaking the Monopoly with Participatory Action-Research,* edited by O. Fals-Borda and M. A. Rahman. New York: Apex Press.

Gee, Wilson. 1950. *Social Science Research Methods.* New York: Appleton-Century-Crofts.

Geertz, Clifford. 1973. *The Interpretation of Cultures.* New York: Basic Books.

Glaser, Barney G., and Anselm L. Strauss. 1967. *The Discovery of Grounded Theory: Strategies for Qualitative Research.* Chicago: Aldine.

Glock, Charles Y., Benjamin B. Ringer, and Earl R. Babbie. 1967. *To Comfort and to Challenge.* Berkeley: University of California Press.

Gobo, Giampietro. 2006. "Set Them Free: Improving Data Quality by Broadening the Interviewer's Tasks." *International Journal of Social Research Methodology* 9 (4): 279–301.

Goffman, Erving. 1961. *Asylums: Essays on the Social Situation of Mental Patients and Other Inmates.* Chicago: Aldine.

1963. *Stigma: Notes on the Management of a Spoiled Identity.* Englewood Cliffs, NJ: Prentice-Hall.

1974. *Frame Analysis.* Cambridge, MA: Harvard University Press.

1979. *Gender Advertisements.* New York: Harper & Row.

Gottlieb, Bruce. 1999. "Cooking the School Books: How *U.S. News* Cheats in Picking Its 'Best American Colleges.'" Slate, August 31. http://slate.msn.com/default.aspx?id=34027.

Graham, Laurie, and Richard Hogan. 1990. "Social Class and Tactics: Neighborhood Opposition to Group Homes." *Sociological Quarterly* 31 (4): 513–29.

Greenwood, Peter W., et al. 1994. *Three Strikes and You're Out: Estimated Benefits and Costs of California's New Mandatory-Sentencing Law.* Santa Monica, CA: Rand Corporation.

Greenwood, Peter W., C. Peter Rydell, and Karyn Model. 1996. *Diverting Children from a Life of Crime: Measuring Costs and Benefits.* Santa Monica, CA: Rand Corporation.

Griffith, Alison I. 1995. "Mothering, Schooling, and Children's Development." Pp. 108–21 in *Knowledge, Experience, and Ruling Relations: Studies in the Social Organization of Knowledge,* edited by M. Campbell and A. Manicom. Toronto: University of Toronto Press.

Gubrium, Jaber F., and James A. Holstein. 1997. *The New Language of Qualitative Method.* New York: Oxford University Press.

Hannon, Lance, and James Defronzo. 1998. "The Truly Disadvantaged, Public Assistance, and Crime." *Social Problems* 45 (3): 383–92.

Hartsock, N. 1983. "The Feminist Standpoint." Pp. 283–310 in *Discovering Reality,* edited by S. Harding and M. B. Hintikka. Boston: D. Riedel.

Hawking, Stephen. 2001. *The Universe in a Nutshell.* New York: Bantam.

Hedrick, Terry E., Leonard Bickman, and Debra J. Rog. 1993. *Applied Research Design: A Practical Guide.* Newbury Park, CA: Sage.

Hempel, Carl G. 1952. "Fundamentals of Concept Formation in Empirical Science." *International Encyclopedia of United Science II,* no. 7.

Heritage, J. 1984. *Garfinkel and Ethnomethodology.* Cambridge: Polity Press.

Heritage, Johen, and David Greatbatch. 1992. "On the Institutional Character of Institutional Talk." In *Talk at Work,* edited by P. Drew and J. Heritage. Cambridge: Cambridge University Press.

Higginbotham, A. Leon, Jr. 1978. *In the Matter of Color: Race and the American Legal Process.* New York: Oxford University Press.

Hill, Lewis. 2000. *Yankee Summer: The Way We Were: Growing Up in Rural Vermont in the 1930s.* Bloomington, IN: 1stBooks Library.

Hilts, Philip J. 1981. "Values of Driving Classes Disputed." *San Francisco Chronicle,* June 25, p. 4.

Hite, Shere. 1987. *The Hite Report: a Nationwide Study of Female Sexuality.* New York: Dell.

Hogan, Richard, and Carolyn C. Perrucci. 1998. "Producing and Reproducing Class and Status Differences: Racial and Gender Gaps in U.S. Employment and Retirement Income." *Social Problems* 45 (4): 528–49.

Howard, Edward N., and Darlene M. Norman. 1981. "Measuring Public Library Performance." *Library Journal,* February, pp. 305–8.

Howell, Joseph T. 1973. *Hard Living on Clay Street.* Garden City, NY: Doubleday Anchor.

Huberman, A. Michael, and Matthew B. Miles. 1994. "Data Management and Analysis Methods." Pp. 428–44 in *Handbook of Qualitative Research,* edited by Norman K. Denzin and Yvonna S. Lincoln. Thousand Oaks, CA: Sage.

Hughes, Michael. 1980. "The Fruits of Cultivation Analysis: A Reexamination of Some Effects of Television Watching." *Public Opinion Quarterly* 44:287–302.

Humphreys, Laud. 1970. *Tearoom Trade: Impersonal Sex in Public Places.* Chicago: Aldine.

Hurst, Leslie. 1991. "Mr. Henry Makes a Deal." Pp. 183–202 in *Ethnography Unbound: Power and Resistance in the Modern Metropolis,* edited by M. Burawoy et al. Berkeley: University of California Press.

Iannacchione, Vincent G., Jennifer M. Staab, and David T. Redden. 2003. "Evaluating the Use of Residential Mailing Addresses in a Metropolitan Household Survey." *Public Opinion Quarterly* 67:202–10.

Ibrahim, Saad Eddin. 2003. "Letter from Cairo." *Contexts* 2 (2): 68–72.

Irwin, John, and James Austin. 1997. *It's About Time: America's Imprisonment Binge.* Belmont, CA: Wadsworth.

Isaac, Larry W., and Larry J. Griffin. 1989. "Ahistoricism in Time-Series Analyses of Historical Process: Critique, Redirections, and Illustrations from U.S. Labor History." *American Sociological Review* 54:873–90.

Jackman, Mary R., and Mary Scheuer Senter. 1980. "Images of Social Groups: Categorical or Qualified?" *Public Opinion Quarterly* 44:340–61.

Jackson, Jonathan. 2005. "Validating New Measures of the Fear of Crime." *International Journal of Social Research* 8 (4): 297–315.

Jacobs, Bruce A., and Jody Miller. 1998. "Crack Dealing, Gender, and Arrest Avoidance." *Social Problems* 45 (4): 550–69.

Jasso, Guillermina. 1988. "Principles of Theoretical Analysis." *Sociological Theory* 6:1–20.

Jensen, Arthur. 1969. "How Much Can We Boost IQ and Scholastic Achievement?" *Harvard Educational Review* 39:273–74.

Jobes, Patrick C., Andra Aldea, Constantin Cernat, Ioana-Minerva Icolisan, Gabriel Iordache, Sabastian Lazeru, Catalin Stoica, Gheorghe Tibil, and Eugenia Edangiu. 1997. "Shopping as a Social Problem: A Grounded Theoretical Analysis of Experiences among Romanian Shoppers." *Journal of Applied Sociology* 14 (1): 124–46.

Johnson, Jeffrey C. 1990. *Selecting Ethnographic Informants.* Newbury Park, CA: Sage.

Johnston, Hank. 1980. "The Marketed Social Movement: A Case Study of the Rapid Growth of TM." *Pacific Sociological Review,* July, pp. 333–54.

Johnston, Hank, and David A. Snow. 1998. "Subcultures and the Emergence of the Estonian Nationalist Opposition 1945–1990." *Sociological Perspectives* 41 (3): 473–97.

Kaplan, Abraham. 1964. *The Conduct of Inquiry.* San Francisco: Chandler.

Kaplowitz, Michael D., Timothy D. Hadlock, and Ralph Levine. 2004. "A Comparison of Web and Mail Survey Response Rates." *Public Opinion Quarterly* 68 (1): 94–101.

Kasl, Stanislav V., Rupert F. Chisolm, and Brenda Eskenazi. 1981. "The Impact of the Accident at Three Mile Island on the Behavior and Well-Being of Nuclear Workers." *American Journal of Public Health,* May, pp. 472–95.

Kasof, Joseph. 1993. "Sex Bias in the Naming of Stimulus Persons." *Psychological Bulletin* 113 (1): 140–63.

Kebede, Alemseghed, and J. David Knottnerus. 1998. "Beyond the Pales of Babylon: The Ideational Components and Social Psychological Foundations of Rastafari." *Sociological Perspectives* 42 (3): 499–517.

Keeter, Scott, Michael Dimock, Leah Christian, and Courtney Kennedy. 2008. *The Impact of "Cell-Onlys" on Public Opinion Polls: Ways of Coping with a Growing Population Segment.* Pew Research Center Publications. http://pewresearch.org/pubs/714/the-impact-of-cell-onlys-on-public-opinion-polls, posted January 31.

Kendall, Patricia L., and Paul F. Lazarsfeld. 1950. "Problems of Survey Analysis." Pp. 133–96 in *Continuities in Social Research: Studies in the Scope and Method of "The American Soldier,"* edited by Robert K. Merton and Paul F. Lazarsfeld. New York: Free Press.

Kentor, Jeffrey. 2001. "The Long Term Effects of Globalization on Income Inequality, Population Growth, and Economic Development." *Social Problems* 48 (4): 435–55.

Khayatt, Didi. 1995. "Compulsory Heterosexuality: Schools and Lesbian Students." Pp. 149–63 in *Knowledge, Experience, and Ruling Relations: Studies in the Social Organization of Knowledge,* edited by M. Campbell and A. Manicom. Toronto: University of Toronto Press.

Kidder, Jeffrey L. 2005. "Style and Action: A Decoding of Bike Messenger Symbols." *Journal of Contemporary Ethnography* 34 (2): 344–67.

Kilburn, John C., Jr. 1998. "It's a Matter of Definition: Dilemmas of Service Delivery and Organizational Structure in a Growing Voluntary Organization." *Journal of Applied Sociology* 15 (1): 89–103.

Kinnell, Ann Marie, and Douglas W. Maynard. 1996. "The Delivery and Receipt of Safer Sex Advice in Pretest Counseling Sessions for HIV and AIDS." *Journal of Contemporary Ethnography* 24: 405–37.

Kinsey, Alfred C., et al. 1948. *Sexual Behavior in the Human Male.* Philadelphia: W. B. Saunders.

———. 1953. *Sexual Behavior in the Human Female.* Philadelphia: W. B. Saunders.

Kish, Leslie. 1965. *Survey Sampling.* New York: Wiley.

Knowles, Caroline. 2006. "Handling the Baggage in the Field Reflections on Research Relationships." *International Journal of Social Research Methodology* 9 (5): 393–404.

Krueger, Richard A. 1988. *Focus Groups.* Newbury Park, CA: Sage.

Kubrin, Charis E. 2005. "I See Death around the Corner: Nihilism in Rap Music." *Sociological Perspectives* 48 (4): 433–59.

Kubrin, Charis E., and Ronald Weitzer. 2003. "Retaliatory Homicide: Concentrated Disadvantage and Neighborhood Culture." *Social Problems* 50 (2): 157–80.

Kuhn, Thomas. 1970. *The Structure of Scientific Revolutions.* Chicago: University of Chicago Press.

Kvale, Steinar. 1996. *InterViews: An Introduction to Qualitative Research Interviewing.* Thousand Oaks, CA: Sage.

Laumann, Edward O., John H. Gagnon, Robert T. Michael, and Stuart Michaels. 1994. *The Social Organization of Sexuality.* Chicago: University of Chicago Press.

Lee, Motoko Y., Stephen G. Sapp, and Melvin C. Ray. 1996. "The Reverse Social Distance Scale." *Journal of Social Psychology* 136 (1): 17–24.

Lengermann, Patricia Madoo, and Jill Niebrugge-Brantley. 2002. "Back to the Future: Settlement Sociology, 1885–1930." *American Sociologist* 33 (3): 5–20.

Lever, Janet. 1986. "Sex Differences in the Complexity of Children's Play and Games." Pp. 74–89 in *Structure and Process,* edited by Richard J. Peterson and Charlotte A. Vaughan. Belmont, CA: Wadsworth.

Lewins, Ann, and Christina Silver. 2006. *Choosing a CAQDAS Software Package.* July, working paper, 5th ed., http://caqdas.soc.surrey.ac.uk/.

Libin, A., and J. Cohen-Mansfield. 2000. "Individual versus Group Approach for Studying and Intervening with Demented Elderly Persons: Methodological Perspectives." *Gerontologist,* October 15, p. 105.

Lincoln, Yvonna S., and Egon G. Guba. 1985. *Naturalistic Inquiry.* Beverly Hills, CA: Sage.

Link, Michael W., Michael P. Battaglia, Martin R. Frankel, Larry Osborn, and Ali H. Mokdad. 2008. "A Comparison of Address-Based Sampling (ABS) versus Random-Digit Dialing (RDD)." *Public Opinion Quarterly* 72 (1): 6–27.

Linton, Ralph. 1937. *The Study of Man.* New York: D. Appleton-Century.

Literary Digest. 1936a. "Landon, 1,293,669: Roosevelt, 972,897." October 31, pp. 5–6.

——— 1936b. "What Went Wrong with the Polls?" November 14, pp. 7–8.

Lofland, John. 2003. *Demolishing a Historic Hotel.* Davis, CA: Davis Research.

Lofland, John, and Lyn H. Lofland. 1995. *Analyzing Social Settings: A Guide to Qualitative Observation and Analysis.* 3rd. ed. Belmont, CA: Wadsworth.

Lofland, John, David Snow, Leon Anderson, and Lyn H. Lofland. 2006. *Analyzing Social Settings: A Guide to Qualitative Observation and Analysis.* 4th ed. Belmont, CA: Wadsworth.

Lopata, Helena Znaniecki. 1981. "Widowhood and Husband Sanctification." *Journal of Marriage and the Family,* May, pp. 439–50.

Lynd, Robert S., and Helen M. Lynd. 1929. *Middletown.* New York: Harcourt, Brace.

——— 1937. *Middletown in Transition.* New York: Harcourt, Brace.

Madison, Anna-Marie. 1992. "Primary Inclusion of Culturally Diverse Minority Program Participants in the Evaluation Process." *New Directions for Program Evaluation,* no. 53, pp. 35–43.

Mahoney, James, and Dietrich Rueschemeyer, eds. 2003. *Comparative Historical Analysis in the Social Sciences.* New York: Cambridge University Press.

Maltz, Michael D. 1998. "Visualizing Homocide: A Research Note. *Journal of Quantitative Criminology* 15 (4): 397–410.

Manning, Peter K., and Betsy Cullum-Swan. 1994. "Narrative, Content, and Semiotic Analysis." Pp. 463–77 in *Handbook of Qualitative Research,* edited by Norman K. Denzin and Yvonna S. Lincoln. Thousand Oaks, CA: Sage.

Marshall, Catherine, and Gretchen B. Rossman. 1995. *Designing Qualitative Research.* Thousand Oaks, CA: Sage.

Marx, Karl. [1867] 1967. *Capital.* New York: International Publishers.

——— [1880] 1956. *Revue Socialist,* July 5. Reprinted in *Karl Marx: Selected Writings in Sociology and Social Philosophy,* edited by T. B. Bottomore and Maximilien Rubel. New York: McGraw-Hill.

McAlister, Alfred, Cheryl Perry, Joel Killen, Lee Ann Slinkard, and Nathan Maccoby. 1980. "Pilot Study of Smoking, Alcohol, and Drug Abuse Prevention." *American Journal of Public Health,* July, pp. 719–21.

McGrane, Bernard. 1994. *The Un-TV and the 10 mph Car: Experiments in Personal Freedom and Everyday Life.* Fort Bragg, CA: The Small Press.

Meadows, Donella H., Dennis L. Meadows, and Jørgen Randers. 1992. *Beyond the Limits: Confronting Global Collapse, Envisioning a Sustainable Future.* Post Mills, VT: Chelsea Green.

Meadows, Donella, Dennis L. Meadows, Jørgen Randers, and William W. Behrens, III. 1972. *The Limits to Growth.* New York: Signet Books.

Menjívar, Cecilia. 2000. *Fragmented Ties: Salvadoran Immigrant Networks in America.* Berkeley: University of California Press.

Merton, Robert K. 1938. "Social Structure and Anomie." *American Sociological Review* 3:672–82.

——— 1957. *Social Theory and Social Structure.* New York: Free Press.

Middlebury College. 2007. http://www.middlebury.edu/about/newsevents/archive/2007/newsevents_633084484309809133.htm, accessed June 8, 2008.

Milgram, Stanley. 1963. "Behavioral Study of Obedience." *Journal of Abnormal Social Psychology* 67:371–78.

———. 1965. "Some Conditions of Obedience and Disobedience to Authority." *Human Relations* 18:57–76.

Miller, Wayne, and June Lennie. 2005. "Empowerment Evaluation: A Practical Method for Evaluating a National School Breakfast Program." *Evaluation Journal of Australiasia* 5 (2): 18–26. [new series]

Mirola, William A. 2003. "Asking for Bread, Receiving a Stone: The Rise and Fall of Religious Ideologies in Chicago's Eight-Hour Movement." *Social Problems* 50 (2): 273–93.

Mitchell, Richard G., Jr. 1991. "Secrecy and Disclosure in Field Work." Pp. 97–108 in *Experiencing Fieldwork: An Inside View of Qualitative Research,* edited by William B. Shaffir and Robert A. Stebbins. Newbury Park, CA: Sage.

———. 2002. *Dancing at Armageddon: Survivalism and Chaos in Modern Times.* Chicago: University of Chicago Press.

Morgan, David L., ed. 1993. *Successful Focus Groups: Advancing the State of the Art.* Newbury Park, CA: Sage.

Morgan, Lewis H. 1870. *Systems of Consanguinity and Affinity.* Washington, DC: Smithsonian Institution.

Moskowitz, Milt. 1981. "The Drugs That Doctors Order." *San Francisco Chronicle,* May 23, p. 33.

Moynihan, Daniel. 1965. *The Negro Family: The Case for National Action.* Washington, DC: U.S. Government Printing Office.

Myrdal, Gunnar. 1944. *An American Dilemma.* New York: Harper & Row.

Naisbitt, John, and Patricia Aburdene. 1990. *Megatrends 2000: Ten New Directions for the 1990's.* New York: Morrow.

Nature Conservancy. 2005. "The Nature of Science on the Sacramento River." *The Nature Conservancy,* Spring–Summer, p. 3. [newsletter]

Neuman, W. Lawrence. 1998. "Negotiated Meanings and State Transformation: The Trust Issue in the Progressive Era." *Social Problems* 45 (3): 315–35.

O'Connor, Pat. 2006. "Globalization, Individualization and Gender in Adolescents' Texts." *International Journal of Social Research Methodology* 9 (4): 261–77.

Olson, Kristen, and Andy Peytchev. 2007. "Effect of Interviewer Experience on Interview Pace and Interviewer Attitudes." *Public Opinion Quarterly* 71 (2): 273–86.

"1 in 5 Polled Voices Doubt on Holocaust." 1993. *New York Times,* April 20, p. A12.

Onwuegbuzie, Anthony J., and Nancy L. Leech. 2005. "On Becoming a Pragmatic Researcher: The Importance of Combining Quantitative and Qualitative Research Methodologies." *International Journal of Research Methodology* 8 (5): 375–87.

Överlien, Carolina, Karin Aronsson, and Margareta Hydén. 2005. "The Focus Group Interview as an In-depth Method? Young Women Talking about Sexuality." *International Journal of Social Research* 8 (4): 331–44.

Perinelli, Phillip J. 1986. "Nonsuspecting Public in TV Call-in Polls." *New York Times,* February 14, letter to the editor.

Perrow, Charles. 2002. *Organizing America: Wealth, Power, and the Origins of Corporate Capitalism.* Princeton, NJ: Princeton University Press.

Petersen, Larry R., and Judy L. Maynard. 1981. "Income, Equity, and Wives' Housekeeping Role Expectations." *Pacific Sociological Review,* January, pp. 87–105.

Picou, J. Steven. 1996a. "Compelled Disclosure of Scholarly Research: Some Comments on High Stakes Litigation." *Law and Contemporary Problems* 59 (3): 149–57.

———. 1996b. "Sociology and Compelled Disclosure: Protecting Respondent Confidentiality." *Sociological Spectrum* 16 (3): 207–38.

Picou, J. Steven, Duane A. Gill, and Maurie J. Cohen, eds. 1999. *The Exxon Valdez Disaster: Readings on a Modern Social Problem.* Dubuque, IA: Kendall-Hunt.

Plutzer, Eric, and Michael Berkman. 2005. "The Graying of America and Support for Funding the Nation's Support." *Public Opinion Quarterly* 69 (1): 66–86.

Polivka, Anne E., and Jennifer M. Rothgeb. 1993. "Redesigning the CPS Questionnaire." *Monthly Labor Review* 116 (9): 10–28.

"Poll on Doubt of Holocaust Is Corrected." 1993. *New York Times,* July 8, p. A7.

Population Communications International. 1996. *International Dateline* [February]. New York: Population Communications International.

Porter, Stephen R., and Michael E. Whitcomb. 2003. "The Impact of Contact Type on Web Survey Response Rates." *Public Opinion Quarterly* 67:579–88.

Powell, Elwin H. 1958. "Occupation, Status, and Suicide: Toward a Redefinition of Anomie." *American Sociological Review* 23:131–39.

Presser, Stanley, and Johnny Blair. 1994. "Survey Pretesting: Do Different Methods Produce Different Results?" Pp. 73–104 in *Sociological Methodology 1994,* edited by Peter Marsden. San Francisco, CA: Jossey-Bass.

Prewitt, Kenneth. 2003. "Partisan Politics in the 2000 U.S. Census." *Population Reference Bureau,* November 2003. http://www.prb.org/Template .cfm?Section=PRB&template=/Content/ ContentGroups/Articles/03/Partisan_Politics_in_ the_2000_U_S__Census.htm.

Public Opinion Quarterly. 2006. 70 (5). [Special issue: *Nonresponse Bias in Household Surveys*]

Quoss, Bernita, Margaret Cooney, and Terri Longhurst. 2000. "Academics and Advocates: Using Participatory Action Research to Influence Welfare Policy." *Journal of Consumer Affairs* 34 (1): 47.

Ragin, Charles C., and Howard S. Becker. 1992. *What Is a Case?: Exploring the Foundations of Social Inquiry.* Cambridge: Cambridge University Press.

Rasinski, Kenneth A. 1989. "The Effect of Question Wording on Public Support for Government Spending." *Public Opinion Quarterly* 53:388–94.

Redfield, Robert. 1941. *The Folk Culture of Yucatan.* Chicago: University of Chicago Press.

Reinharz, Shulamit. 1992. *Feminist Methods in Social Research.* New York: Oxford University Press.

Reynolds, H. T. 1977. *Analysis of Nominal Data.* Beverly Hills, CA: Sage.

Riecken, Henry W., and Robert F. Boruch. 1974. *Social Experimentation: A Method for Planning and Evaluating Social Intervention.* New York: Academic Press.

Roethlisberger, F. J., and W. J. Dickson. 1939. *Management and the Worker.* Cambridge, MA: Harvard University Press.

Rogers, Everett M., Peter W. Vaughan, Ramadhan M. A. Swalehe, Nagesh Rao, and Suruchi Sood. 1996. "Effects of an Entertainment-Education Radio Soap Opera on Family Planning and HIV/ AIDS Prevention Behavior in Tanzania." Report presented at a technical briefing on the Tanzania Entertainment-Education Project, Rockefeller Foundation, New York, March 27.

Rosenberg, Morris. 1968. *The Logic of Survey Analysis.* New York: Basic Books.

Rosenthal, Robert, and Lenore Jacobson. 1968. *Pygmalion in the Classroom.* New York: Holt, Rinehart & Winston.

Ross, Jeffrey Ian. 2004. "Taking Stock of Research Methods and Analysis on Oppositional Political Terrorism." *American Sociologist* 35 (2): 26–37.

Rossi, Peter H., Mark Lipsey, and Howard E. Freeman. 2002. *Evaluation: A Systematic Approach.* 6th ed. Thousand Oaks, CA: Sage.

Rossman, Gabriel. 2002. *The Qualitative Influence of Ownership on Media Content: The Case of Movie Reviews.* Paper presented to the American Sociological Association, Chicago. Reported in *Contexts* 2 (2): 7.

Rothman, Ellen K. 1981. "The Written Record." *Journal of Family History,* Spring, pp. 47–56.

Rubin, Herbert J., and Riene S. Rubin. 1995. *Qualitative Interviewing: The Art of Hearing Data.* Thousand Oaks, CA: Sage.

Rynbrandt, Linda J., and Mary Jo Deegan. 2002. "The Ecofeminist Pragmatism of Caroline Bartlett Crane, 1896–1935." *American Sociologist* 33 (3): 55–68.

Sacks, Jeffrey J., W. Mark Krushat, and Jeffrey Newman. 1980. "Reliability of the Health Hazard Appraisal." *American Journal of Public Health,* July, pp. 730–32.

Sanders, William B. 1994. *Gangbangs and Drive-bys: Grounded Culture and Juvenile Gang Violence.* New York: Aldine De Gruyter.

Saxena, Prem C., et al. 2004. "Nuptiality Transition and Marriage Squeeze in Lebanon—Consequences of Sixteen Years of Civil War." *Journal of Comparative Family Studies* 35 (2): 241–58. [Special issue on turbulence in the Middle East]

Scarce, Rik. 1990. *Ecowarriors: Understanding the Radical Environmental Movement.* Chicago: Noble Press.

1999. "Good Faith, Bad Ethics: When Scholars Go the Distance and Scholarly Associations Do Not." *Law and Social Inquiry* 24 (4): 967–76.

Schiflett, Kathy L., and Mary Zey. 1990. "Comparison of Characteristics of Private Product Producing Organizations and Public Service Organizations." *Sociological Quarterly* 31 (4): 569–83.

Schilt, Kristen. 2006. "Just One of the Guys? How Transmen Make Gender Visible in the Workplace." *Gender and Society* 20 (4): 465–90.

Schmitt, Frederika E., and Patricia Yancey Martin. 1999. "Unobtrusive Mobilization by an Institutionalized Rape Crisis Center: 'All We Do Comes from Victims.'" *Gender and Society* 13 (3): 364–84.

Schutz, Alfred. 1967. *The Phenomenology of the Social World.* Evanston, IL: Northwestern University Press.

1970. *On Phenomenology and Social Relations.* Chicago: University of Chicago Press.

Sense, Andrew J. 2006. "Driving the Bus from the Rear Passenger Seat: Control Dilemmas of Participative Action Research." *International Journal of Social Research Methodology* 9 (1): 1–13.

Shaffir, William B., and Robert A. Stebbins, eds. 1991. *Experiencing Fieldwork: An Inside View of Qualitative Research.* Newbury Park, CA: Sage.

Shea, Christopher. 2000. "Don't Talk to the Humans: The Crackdown on Social Science Research." *Lingua Franca* 10 (6).

Sherif, Muzafer. 1935. "A Study of Some Social Factors in Perception." *Archives of Psychology* 27:1–60.

Sherman, Rachel. 2005. *Class Acts: Service and Inequality in Luxury Hotels.* Berkeley: University of California Press.

Silverman, David. 1993. *Interpreting Qualitative Data: Methods for Analyzing Talk, Text, and Interaction.* Newbury Park, CA: Sage.

———. 1999. *Doing Qualitative Research: A Practical Handbook.* Thousand Oaks, CA: Sage.

Silverman, George. 2005. "Qualitative Research: Face-to-Face Focus Groups, Telephone Focus Groups, Online Focus Groups." http://www.mnav.com/ qualitative_research.htm, accessed June 1, 2005.

Singhal, Arvind, and Elizabeth Rattine-Flaherty. 2006. "Pencils and Photos as Tools of Communicative Research and Praxis: Analyzing Minga Peru's Quest for Social Justice in the Amazon." *International Communication Gazette* 68 (4): 313–30.

Skedsvold, Paula. 2002. "New Developments Concerning Public Use Data Files." *Footnotes* 30 (1): 3.

Skocpol, Theda. 2003. *Diminished Democracy: From Membership to Management in American Civic Life.* Norman: Oklahoma University Press.

Smith, Andrew E., and G. F. Bishop. 1992. *The Gallup Secret Ballot Experiments: 1944–1988.* Paper presented at the Annual Conference of the American Association for Public Opinion Research, St. Petersburg, FL, May.

Smith, Dorothy E. 1978. *The Everyday World as Problematic.* Boston: Northeastern University Press.

Smith, John K., and Phil Hodkinson. 2005. "Relativism, Criteria, and Politics." Pp. 915–32 in *Handbook of Qualitative Research*, 3rd ed., edited by Norman K. Denzin and Yvonna S. Lincoln. Thousand Oaks, CA: Sage.

Smith, Tom W. 2001. "Are Representative Internet Surveys Possible?" Paper presented at Achieving Data Quality in a Statistical Agency: A Methodological Perspective, Hull, Ontario, October.

Smith, Vicki. 1998. "The Fractured World of the Temporary Worker: Power, Participation, and Fragmentation in the Contemporary Workplace." *Social Problems* 45 (4): 411–30.

Snow, David A., and Leon Anderson. 1987. "Identity Work among the Homeless: The Verbal Construction and Avowal of Personal Identities." *American Journal of Sociology* 92:1336–71.

Sorokin, Pitirim A. 1937–1940. *Social and Cultural Dynamics.* 4 vols. Englewood Cliffs, NJ: Bedminster Press.

Spencer, Liz, Jane Ritchie, Jane Lewis, and Lucy Dillon. 2003. *Quality in Qualitative Evaluation: A Framework for Assessing Research Evidence.* London: National Centre for Social Research. http://www .cabinetoffice.gov.uk/strategy/~/media/assets/ www.cabinetoffice.gov.uk/strategy/qqe_rep% 20pdf.ashx.

Spohn, Cassie, and Julie Horney. 1990. "A Case of Unrealistic Expectations: The Impact of Rape Reform Legislation in Illinois." *Criminal Justice Policy Review* 4 (1): 1–18.

Srole, Leo. 1956. "Social Integration and Certain Corollaries: An Exploratory Study." *American Sociological Review* 21:709–16.

Stark, Rodney. 1997. *The Rise of Christianity: How the Obscure, Marginal Jesus Movement Became the Dominant Religious Force in the Western World in a Few Centuries.* San Francisco: HarperCollins.

"State Saved $21.7 Billion with Five-Year-Old 'Three Strikes' Law." 1999. *BayInsider*, March 1.

Stearns, Cindy A. 1999. "Breastfeeding and the Good Maternal Body." *Gender and Society* 13 (3): 308–26.

Sterling, T. D., W. L. Rosenbaum, and J. J. Weinkam. 1995. "Publication Decisions Revisited: The Effect of the Outcome of Statistical Tests on the Decision to Publish and Vice Versa." *American Statistician* 49 (1): 108–12.

Stouffer, Samuel, et al. 1949–1950. *The American Soldier.* 3 vols. Princeton, NJ: Princeton University Press.

Straight Dope. 2001. February 2. http://www .straightdope.com/columns/010202.html, accessed April 8, 2006.

Strang, David, and James N. Baron. 1990. "Categorical Imperatives: The Structure of Job Titles in California State Agencies." *American Sociological Review* 55:479–95.

Strauss, Anselm, and Juliet Corbin. 1994. "Grounded Theory Methodology: An Overview." Pp. 273–85 in *Handbook of Qualitative Research,* edited by Norman K. Denzin and Yvonna S. Lincoln. Thousand Oaks, CA: Sage.

———. 1998. *Basics of Qualitative Research: Techniques and Procedures for Developing Grounded Theory.* Thousand Oaks, CA: Sage.

Strunk, William, Jr., and E. B. White. 2000. *The Elements of Style.* 4th ed. New York: Macmillan.

Swalehe, Ramadhan, Everett M. Rogers, Mark J. Gilboard, Krista Alford, and Rima Montoya. 1995. "A Content Analysis of the Entertainment-Education Radio Soap Opera 'Twende na Wakati' (Let's Go with the Times) in Tanzania." Arusha, Tanzania: Population Family Life and Education Programme (POFLEP), Ministry of Community Development, Women Affairs, and Children, November 15.

Takeuchi, David. 1974. "Grass in Hawaii: A Structural Constraints Approach." M.A. thesis, University of Hawaii.

Tan, Alexis S. 1980. "Mass Media Use, Issue Knowledge and Political Involvement." *Public Opinion Quarterly* 44:241–48.

Tandon, Rajesh, and L. Dave Brown. 1981. "Organization-Building for Rural Development: An Experiment in India." *Journal of Applied Behavioral Science,* April–June, pp. 172–89.

Thomas, W. I., and Florian Znaniecki. 1918. *The Polish Peasant in Europe and America.* Chicago: University of Chicago Press.

Thompson, Paul. 2004. "Researching Family and Social Mobility with Two Eyes: Some Experiences of the Interaction between Qualitative and Quantitative Data." *International Journal of Social Research Methodology* 7 (3): 237–57.

Trepagnier, Barbara. 2006. *Silent Racism: How Well-Meaning White People Perpetuate the Racial Divide.* Boulder, CO: Paradigm.

Tuckel, Peter S., and Barry M. Feinberg. 1991. "The Answering Machine Poses Many Questions for Telephone Survey Researchers." *Public Opinion Quarterly* 55:200–217.

Tuckel, Peter, and Harry O'Neill. 2002. "The Vanishing Respondent in Telephone Surveys." *Journal of Advertising Research,* September–October, pp. 26–48.

Turk, Theresa Guminski. 1980. "Hospital Support: Urban Correlates of Allocation Based on Organizational Prestige." *Pacific Sociological Review,* July, pp. 315–32.

Turner, Jonathan H., ed. 1989. *Theory Building in Sociology: Assessing Theoretical Cumulation.* Newbury Park, CA: Sage.

Union of Concerned Scientists. 2005. "Political Interference in Science." October 23. http://www.ucsusa.org/scientific_integrity/interference/.

United Nations. 1995. "Human Development Report 1995, New York: United Nations Development Program." [Summarized in Population Communications International. 1996. *International Dateline,* February, pp. 1–4.]

U.S. Bureau of the Census. 2001. *Statistical Abstract of the United States.* Washington, DC: U.S. Government Printing Office.

———. 2006. *Statistical Abstract of the United States.* Washington, DC: U.S. Government Printing Office.

U.S. Department of Health and Human Services. 2002. http://grants.nih.gov/grants/policy/COC/background.htm?print=yes&.

U.S. News and World Report. 1999. "America's Best Colleges." August 30.

Vanderbei, Robert. 2004. "Election 2004 Results." http://www.princeton.edu/~rvdb/JAVA/election2004/.

Venkatesh, Sudhir. 2008. *Gang Leader for a Day: A Rogue Sociologist Takes to the Streets.* New York: Penguin.

Veroff, Joseph, Shirley Hatchett, and Elizabeth Douvan. 1992. "Consequences of Participating in a Longitudinal Study of Marriage." *Public Opinion Quarterly* 56:325–27.

Votaw, Carmen Delgado. 1979. "Women's Rights in the United States." United States Commission on Civil Rights, Inter-American Commission on Women. Washington, DC: Clearinghouse Publications.

Walker Research. 1988. *Industry Image Study.* 8th ed. Indianapolis, IN: Walker Research.

Walker, Shayne, Anaru Eketone, and Anita Gibbs. 2006. "An Exploration of Kaupapa Maori Research, Its Principles, Processes, and Applications." *International Journal of Social Research Methodology* 9 (4): 331–44.

Wall, Sarah. 2008. "Easier Said than Done: Writing an Autoethnography." *International Journal of Qualitative Methods* 7 (1): 38–3. http://ejournals.library.ualberta.ca/index.php/IJQM/article/view/1621/1144.

Warner, W. Lloyd. 1949. *Democracy in Jonesville.* New York: Harper.

Webb, Eugene J., Donald T. Campbell, Richard D. Schwartz, Lee Sechrest, and Janet Belew Grove.

2000. *Unobtrusive Measures.* Rev. ed. Thousand Oaks, CA: Sage.

Weber, Max. [1905] 1958. *The Protestant Ethic and the Spirit of Capitalism.* Translated by Talcott Parsons. New York: Scribner.

[1925] 1946. "Science as a Vocation." Pp. 129–56 in *From Max Weber: Essays in Sociology,* edited and translated by Hans Gerth and C. Wright Mills. New York: Oxford University Press.

[1934] 1951. *The Religion of China.* Translated by Hans H. Gerth. New York: Free Press.

[1934] 1952. *Ancient Judaism.* Translated by Hans H. Gerth and Don Martindale. New York: Free Press.

[1934] 1958. *The Religion of India.* Translated by Hans H. Gerth and Don Martindale. New York: Free Press.

Weiss, Carol. 1972. *Evaluation Research.* Englewood Cliffs, NJ: Prentice-Hall.

Weitzman, Lenore J., Deborah Eifler, Elizabeth Hokada, and Catherine Ross. 1972. "Sex-Role Socialization in Picture Books for Preschool Children." *American Journal of Sociology* 77:1125–50.

Wharton, Amy S., and James N. Baron. 1987. "So Happy Together? The Impact of Gender Segregation on Men at Work." *American Sociological Review* 52:574–87.

Whyte, W. F. 1943. *Street Corner Society.* Chicago: University of Chicago Press.

Whyte, W. F., D. J. Greenwood, and P. Lazes. 1991. "Participatory Action Research: Through Practice to Science in Social Research." Pp. 19–55 in *Participatory Action Research,* edited by W. F. Whyte. New York: Sage.

Wieder, D. L. 1988. *Language and Social Reality: The Case of Telling the Convict Code.* Landman, MD: University Press of America.

Wilson, Camilo. 1999. Private email, September 8.

Wilson, Edward O. 1975. *Sociobiology: The New Synthesis.* Cambridge, MA: Harvard University Press.

Yerg, Beverly J. 1981. "Reflections on the Use of the RTE Model in Physical Education." *Research Quarterly for Exercise and Sport,* March, pp. 38–47.

Yinger, J. Milton, et al. 1977. *Middle Start: An Experiment in the Educational Enrichment of Young Adolescents.* London: Cambridge University Press.

Index